THE QUANTUM PHASE OPERATOR
A Review

Series in Optics and Optoelectronics

Recent titles in the series

An Introduction to Biomedical Optics
R Splinter, B A Hooper

High-Speed Photonic Devices
Nadir Dagli

Lasers in the Preservation of Cultural Heritage: Principles and Applications
C Fotakis, D Anglos, V Zafiropulos, S Georgiou, V Tornari

Modeling Fluctuations in Scattered Waves
E Jakeman, K D Ridley

Fast Light, Slow Light and Left-Handed Light
P W Milonni

Diode Lasers
D Sands

Diffractional Optics of Millimetre Waves
I V Minin, O V Minin

Handbook of Electroluminescent Materials
D R Vij

Handbook of Moire Measurement
C A Walker

Next Generation Photovoltaics
A Martí, A Luque

Stimulated Brillouin Scattering
M J Damzen, V Vlad, A Mocofanescu, V Babin

Laser Induced Damage of Optical Materials
R M Wood

Optical Applications of Liquid Crystals
L Vicari

Optical Fibre Devices
J P Goure, I Verrier

Applications of Silicon-Germanium Heterostructure Devices
C K Maiti, G A Armstrong

Series in Optics and Optoelectronics

THE QUANTUM PHASE OPERATOR

A Review

Stephen M. Barnett
University of Strathclyde
Glasgow, Scotland

John A. Vaccaro
Griffith University
Brisbane, Australia

CRC Press
Taylor & Francis Group
Boca Raton London New York

CRC Press is an imprint of the
Taylor & Francis Group, an **informa** business
A TAYLOR & FRANCIS BOOK

CRC Press
Taylor & Francis Group
6000 Broken Sound Parkway NW, Suite 300
Boca Raton, FL 33487-2742

First issued in paperback 2019

ISBN-13: 978-1-58488-760-7 (hbk)
ISBN-13: 978-0-367-38914-7 (pbk)

Visit the Taylor & Francis Web site at
http://www.taylorandfrancis.com

and the CRC Press Web site at
http://www.crcpress.com

For David

Preface

It is most natural to define periodic phenomena in terms of a coordinate which tells us where we are positioned within the period. For example, the minute hand on a clock tells us how many minutes have elapsed past each hour. It does not tell us which hour we are experiencing but only where we are within an unspecified hour. Similarly the imaginary lines of longitude on the Earth's surface tell us only how far east or west we are relative to the Greenwich meridian. In physics we often describe periodic phenomena in terms of a phase or angular coordinate which takes a value in a specified 2π range. In simple harmonic motion, for example, the position and momentum of an oscillating body are most naturally expressed in terms of a common phase variable. Rotating systems are commonly described by an angular coordinate which measures the angular distance from some specified position.

Angle variables made an early entrance into quantum theory but pinning down the precise form that these should take has been a long process. Part of the reason for this was the widespread and long-held belief that such variables had no place in quantum theory and that there were no angle operators. In this volume we chart the development of phase and angle operators from their first appearance to the modern theory. We have not followed all the ideas that have been proposed in the long history of this problem. To do so would have undoubtedly been interesting but would have produced a far lengthier book. Our aim in selecting the papers was to present the developments of the ideas that led to the theory of the phase operator as it currently stands and to give as complete a picture as possible of the developments that have followed since then. To this end we have separated the papers into chapters, each one dealing with a different aspect of phase operators. Each of these chapters is preceded by a short introduction in which we have attempted to put the work in context.

Both editors wish to record their gratitude to David Pegg with whom they have enjoyed exploring this interesting topic and from whom they have learnt so much. We dedicate this book to him on the occasion of his 65th birthday and we wish him a long and happy retirement. We are also grateful to our many coworkers who have contributed so much to our understanding of phase and to the authors of the papers reproduced here for their kind permission to reproduce their work and for their generous encouragement. JAV also acknowledges support from the Australian Research Council and the State of Queensland.

Stephen M. Barnett, Glasgow

John A. Vaccaro, Brisbane

Contents

1

Precursors

Periodic systems are ubiquitous in physics and every field has its preferred models based on harmonic oscillators. The discussion of oscillators forms an important part of elementary courses on both classical and quantum mechanics. Their characteristic feature, namely periodic motion, is conveniently embodied in a periodic coordinate or *phase*. All physical properties of the oscillator are periodic in this phase and can be expressed in terms of periodic functions of it.

It comes as something of a surprise that the quantum description of the phase has been a problem for so long. In this book we present the theory and development of the phase operator $\hat{\phi}_\theta$ through the original papers. These have been collected into chapters, each of which begins with a brief introduction in which the principal results are reviewed and set into perspective. In this first chapter we examine the study of quantum phase prior to the discovery of the phase operator. The relevant period spans more than sixty years and in that time many ideas were suggested and avenues explored. Following these is an interesting intellectual exercise but here we concentrate on the main developments that led to the phase operator. For the interested reader we note that the tutorial review (Pegg and Barnett 1997a **Paper 2.6**) includes an extensive bibliography of papers on the subject, both old and new.

The problem of describing an angle operator dates back to earliest days of the new quantum theory. In 1926 it was suggested (London 1926) that periodic quantum systems could be described by a 2π-periodic coordinate w and a conjugate momentum J, the eigenstates of which could be expressed in the from

$$\Psi_n = \frac{1}{\sqrt{2\pi}} e^{\frac{i}{h} J w} = \frac{1}{\sqrt{2\pi}} e^{inw}, \qquad (1.1)$$

where n is the (integer) eigenvalue of J/h. London also introduced an operator E which in the basis of these eigenstates has the form

$$E = \begin{pmatrix} 0 & 0 & 0 & 0 & \cdots \\ 1 & 0 & 0 & 0 & \cdots \\ 0 & 1 & 0 & 0 & \cdots \\ 0 & 0 & 1 & 0 & \cdots \\ \vdots & \vdots & \vdots & \vdots & \end{pmatrix} \qquad (1.2)$$

and proposed that this be associated with the operator e^{iw}. The conjugate operator $E^\dagger$, which London denoted as E^{-1}, is similarly associated with e^{-iw}.

The angle operator makes an appearance in the earliest work on quantum electrodynamics. Dirac (1927 **Paper 1.1**) used it to derive the quantum form of the electromagnetic field amplitudes. His starting point was to use the commutator–Poisson bracket correspondence principle (Dirac 1925) to write a canonical commutation relation between the angle, which he denoted $\theta_r/\hbar$, and the excitation number N_r for a mode:

$$\theta_r N_r - N_r \theta_r = i\hbar. \tag{1.3}$$

(In his paper, Dirac actually used h to denote Planck's constant divided by 2π rather than the now familiar $\hbar$.) He used this to obtain the quantised field amplitudes

$$\begin{aligned} b_r &= (N_r + 1)^{1/2} e^{-i\theta_r/\hbar} = e^{-i\theta_r/\hbar} N_r^{1/2} \\ b_r^* &= N_r^{1/2} e^{i\theta_r/\hbar} = e^{i\theta_r/\hbar} (N_r + 1)^{1/2}. \end{aligned} \tag{1.4}$$

Note that the canonically-conjugate relationship between these is reflected in the fact that they do not commute and we recognise b_r and b_r^* as bosonic annihilation and creation operators with the associated commutation relation:

$$\begin{aligned} b_r b_r^* - b_r^* b_r &= (N_r + 1)^{1/2} e^{-i\theta_r/\hbar} \cdot e^{i\theta_r/\hbar} (N_r + 1)^{1/2} - N_r^{1/2} e^{i\theta_r/\hbar} \cdot e^{-i\theta_r/\hbar} N_r^{1/2} \\ &= (N_r + 1) - N_r = 1. \end{aligned} \tag{1.5}$$

We should note that Dirac's $e^{-i\theta_r/\hbar}$ plays the same role in his annihilation operator b_r as the operator $E^\dagger$ in London's paper: it is the bare lowering operator, which acts to lower the excitation number by unity. Changes of sign have occurred on a number of occasions in the development of the theory. For historical reasons we shall follow the opposite sign convention to that adopted by London and Dirac and we will also introduce carets to denote quantum operators. These lead us to rewrite Dirac's commutator (1.3) in the form

$$[\hat{\phi}, \hat{N}] = -i, \tag{1.6}$$

where $\hat{N}$ and $\hat{\phi}$ are the (excitation) number and phase operators for our electromagnetic field mode or harmonic oscillator.

The commutator (1.6) leads to problems if we take it seriously and calculate its matrix elements in the basis of eigenstates of $\hat{N}$ (Louisell 1963 **Paper 1.2**):

$$(n - n')\langle n'|\hat{\phi}|n\rangle = -i\langle n'|n\rangle = -i\delta_{nn'}. \tag{1.7}$$

This implies that the number state matrix elements of $\hat{\phi}$ are undefined. In particular the right hand side is zero unless $n = n'$ but this means that $\hat{\phi}$ is diagonal in the number state basis and so must commute with $\hat{N}$! When $n = n'$, moreover, the left hand side of (1.7) is zero. Clearly there is something seriously wrong with (1.6).

A straightforward application of Heisenberg's uncertainty principle to the commutation relation (1.6) leads to a number-phase uncertainty relation (Heitler 1954)

$$\Delta N \Delta \phi \geq \frac{1}{2}. \tag{1.8}$$

Contained within this uncertainty relation, however, is a further hint of the problems that were to come. The difficulty is that for states with rather small uncertainty in N the uncertainty in ϕ must become unbounded. It is difficult to reconcile this with the common sense notion that once the phase is truly random the phase probability takes the same value throughout the full 2π range of physically distinct values. Judge (1963 **Paper 1.3**) considered the analogous problem for rotation angle $\hat{\phi}$ and its conjugate angular momentum $\hat{L}_z$, which should satisfy a similar uncertainty relation to (1.6):

$$\Delta L_z \Delta \phi \geq \frac{\hbar}{2}. \tag{1.9}$$

A flat angle probability distribution, however, leads us to conclude that the angle uncertainty should be $\pi/\sqrt{3}$ rather than tending to infinity. The cause of the problem was suggested to lie in the possibility of including unphysical states which do not respect the required periodicity: $\psi(\phi + 2n\pi) = \psi(\phi)$ (Judge and Lewis 1963 **Paper 1.4**). This led to a modification of the commutation relation in the form

$$[\hat{L}_z, \hat{Y}(\phi)] = -i\hbar \left[1 - 2\pi \sum_{-\infty}^{\infty} \delta\{\phi - (2n+1)\pi\}\right], \tag{1.10}$$

where $\hat{Y}(\phi)$ is an operator in which the supposedly continuous values of ϕ are mapped onto a single 2π range.

The problems associated with phase and angle variables led Susskind and Glogower (1964 **Paper 1.5**) to examine the phase operator from the perspective of the well-behaved creation and annihilation operators. Like Dirac they wrote the creation operator as

$$\hat{a}^\dagger = \sqrt{\hat{N}}\widehat{e^{-i\phi}}, \tag{1.11}$$

the number-state matrix elements of which are

$$\langle n|\hat{a}^\dagger|m\rangle = \delta_{n,m+1}\sqrt{m+1} = \sqrt{n}\langle n|\widehat{e^{-i\phi}}|m\rangle. \tag{1.12}$$

The problem with this is that it does not determine matrix elements of $\widehat{e^{-i\phi}}$ for $n = 0$ but rather leaves us with

$$\widehat{e^{-i\phi}} = \begin{pmatrix} r_0 & r_1 & r_2 & r_3 & \cdots \\ 1 & 0 & 0 & 0 & \cdots \\ 0 & 1 & 0 & 0 & \cdots \\ 0 & 0 & 1 & 0 & \cdots \\ . & . & . & . & \cdots \end{pmatrix}, \tag{1.13}$$

where the r_i are unknown complex numbers. With no way to determine the required values of these they set them equal to zero. This makes $\widehat{e^{-i\phi}}$ the bare raising operator, equivalent to (1.2), and $\widehat{e^{i\phi}}$ the bare lowering operator:

$$\widehat{e^{-i\phi}} = \sum_{n=0}^{\infty} |n+1\rangle\langle n|$$
$$\widehat{e^{i\phi}} = \sum_{n=0}^{\infty} |n\rangle\langle n+1|. \tag{1.14}$$

We have placed the caret over the whole exponentials rather than the phase, ϕ. This is to draw attention to the fact that $\widehat{e^{-i\phi}}$ and $\widehat{e^{i\phi}}$ do not commute:

$$\widehat{e^{i\phi}}\widehat{e^{-i\phi}} = \sum_{n=0}^{\infty} |n\rangle\langle n| = \hat{1}$$
$$\widehat{e^{-i\phi}}\widehat{e^{i\phi}} = \sum_{n=0}^{\infty} |n+1\rangle\langle n+1| = \hat{1} - |0\rangle\langle 0| \tag{1.15}$$

and hence *cannot* be functions of a common phase operator $\hat{\phi}$. It seemed that the resolution of the problems with the commutator (1.6) is that there is, in fact, no phase operator and, by extension, that there is also no angle operator conjugate to the angular momentum. This view was reinforced in a number of articles and review articles in the following years (Carruthers and Nieto 1968, Lévy-Leblond 1976). In spite of this problem, many authors continued to use Susskind and Glogower's operators as representations of $e^{\pm i\phi}$ and in particular to write *non-commuting* operators for the sine and cosine of the phase:

$$\widehat{\cos\phi} = \frac{1}{2}\left(\widehat{e^{i\phi}} + \widehat{e^{-i\phi}}\right)$$
$$\widehat{\sin\phi} = \frac{1}{2i}\left(\widehat{e^{i\phi}} - \widehat{e^{-i\phi}}\right). \tag{1.16}$$

The original link between the phase observable and these operators was via the angle operator and with this gone it seems strange, with the benefit of hindsight, that these non-commuting operators were used to represent phase properties at all.

Susskind and Glogower also wrote down phase-like states in the form

$$|\theta\rangle = \sum_n e^{in\theta}|n\rangle, \tag{1.17}$$

but showed that these states are not the eigenstates of their $\widehat{\cos\phi}$ or $\widehat{\sin\phi}$ operators. Moreover these states are not even linearly independent. This seemed to add further credence to the idea that there is no Hermitian phase operator. Despite these difficulties, or perhaps because of them, some authors have favoured avoiding presenting a phase operator but rather to define just a phase probability density in the form (Shapiro *et al* 1989, Busch *et al* 1995)

$$P(\theta) = \frac{1}{2\pi}\mathrm{Tr}\,(\hat{\rho}|\theta\rangle\langle\theta|)\,, \tag{1.18}$$

where $\hat{\rho}$ is the density operator describing the state of the system. Loudon (1973 **Paper 1.6**), in the first edition of his celebrated textbook, sought to describe phase in terms of states that are better behaved than (1.17). To do this he introduced a limit into the definition of the phase states by writing

$$|\phi\rangle = \lim_{s\to\infty}(s+1)^{-1/2}\sum_{n=0}^{s}\exp(in\phi)|n\rangle. \tag{1.19}$$

Using this limit he was able to show that for this state the uncertainties in $\widehat{\cos\phi}$ or $\widehat{\sin\phi}$ tend to zero

$$\lim_{s\to\infty}\Delta\cos\phi = 0 = \lim_{s\to\infty}\Delta\sin\phi \tag{1.20}$$

and to compare and contrast the properties of these states with the more familiar number states. He presented a helpful, even if not strictly accurate, physical picture in terms of the evolution of the single-mode electric field for these states. For the number state the expectation value of the electric field is zero, but the associated energy is precisely $(n+\frac{1}{2})\hbar\omega$. Hence we can picture the number state as a field in which the amplitude of the electric field is reasonably well defined as

$$E_0 = \left(\frac{2\hbar\omega}{\epsilon_0 V}\right)^{1/2}\left(n+\frac{1}{2}\right)^{1/2}. \tag{1.21}$$

A pictorial representation of this is presented in Fig. 1.1. The phase state (1.19) has rather different properties. In particular the mean photon number for the mode diverges

$$\langle n|\hat{N}|n\rangle = \lim_{s\to\infty}(s+1)^{-1}\sum_{n=0}^{s} n = \lim_{s\to\infty}\frac{s}{2} \tag{1.22}$$

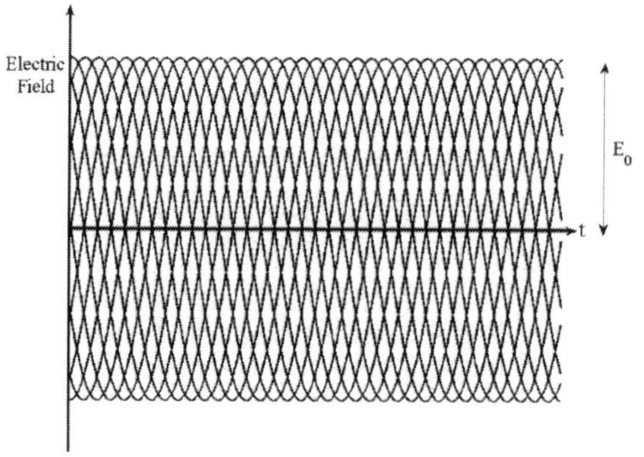

FIGURE 1.1
Pictorial representation of the variation of electric field with time at a fixed point in a cavity mode excited to the state $|n\rangle$. The sine waves should, more accurately, form a horizontal continuum. After Loudon (1973).

as does the spread in the photon number. The ratio of the uncertainty to the mean photon number is, however, finite

$$\frac{\Delta n}{\langle n|\hat{N}|n\rangle} = \frac{1}{\sqrt{3}}. \tag{1.23}$$

The expectation value of the single-mode electric field operator is

$$\begin{aligned}\langle\phi|\hat{E}|\phi\rangle &= -2\left(\frac{2\hbar\omega}{\epsilon_0 V}\right)^{1/2} \sin(\mathbf{k}.\mathbf{r} - \omega t + \phi) \lim_{s\to\infty}(s+1)^{-1}\sum_{n=0}^{s}(n+1)^{1/2} \\ &\approx -\frac{4}{3}\left(\frac{2\hbar\omega}{\epsilon_0 V}\right)^{1/2} \sin(\mathbf{k}.\mathbf{r} - \omega t + \phi) \lim_{s\to\infty}(s+1)^{1/2}.\end{aligned} \tag{1.24}$$

This varies periodically from a very large (divergent) positive quantity to an equally large negative one. The times at which the expectation value of the field crosses between these and takes the value zero are very well marked. Hence we can picture the field for a phase state as one where the amplitude is undefined but that the times at which the field changes sign are well determined. For this reason we can associate these states with a well-determined phase. A pictorial representation of the field for these states is present in Fig. 1.2.

We can try to understand the problem posed by the Susskind-Glogower operators (1.14) by recognising that they are not unitary operators: the product $\widehat{e^{i\phi}}\widehat{e^{-i\phi}}$ is the identity operator, as required by unitarity, but the product $\widehat{e^{-i\phi}}\widehat{e^{i\phi}}$ is not. One way to address this is to impose unitarity by expanding the space of states for the harmonic oscillator or field mode (Fain 1967, Newton 1980, Barnett and Pegg 1986 **Paper 1.7**, Stenholm 1993). These approaches differ in detail, but the general idea may be summarised as follows. The harmonic oscillator energy eigenstates are represented as $|n\rangle$ with n taking integer values greater than zero. To these we can add formally a set of states with *negative* values of n and within this enlarged state space we can introduce the unitary and commuting pair of operators

$$\begin{aligned}\widehat{e_u^{i\phi}} &= \sum_{n=-\infty}^{\infty} |n\rangle\langle n+1| \\ \widehat{e_u^{-i\phi}} &= \sum_{n=-\infty}^{\infty} |n+1\rangle\langle n|.\end{aligned} \tag{1.25}$$

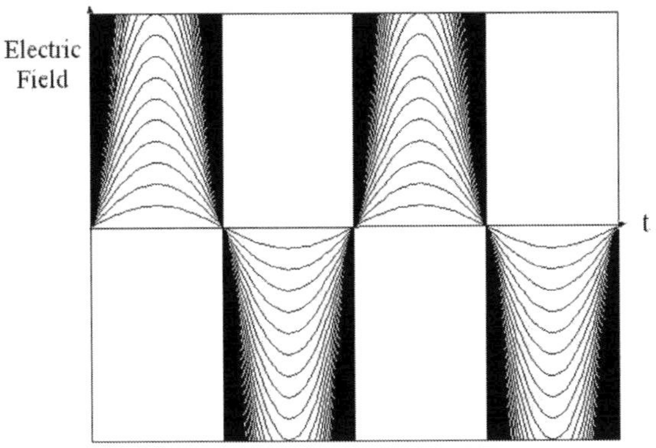

FIGURE 1.2
Pictorial representation of the variation of electric field with time at a fixed point in a cavity mode excite to state $|\phi\rangle$. The net effect of adding together the contributions shown is an electric field the expectation value of which changes abruptly between $+\infty$ and $-\infty$ every half cycle. After Loudon (1973).

The construction is completed by a mechanism that stops a physical system from accessing the states with negative values of n. One way to do this is to write the annihilation and creation operators in the form

$$\begin{aligned} \hat{a} &= \widehat{e_u^{i\phi}}\,|\hat{N}^{1/2}| \\ \hat{a}^{\dagger} &= |\hat{N}^{1/2}|\,\widehat{e_u^{-i\phi}}, \end{aligned} \tag{1.26}$$

where $|\hat{N}^{1/2}|$ is the Hermitian amplitude operator

$$|\hat{N}^{1/2}| = \sum_{n=-\infty}^{\infty} |n^{1/2}|\,|n\rangle\langle n|. \tag{1.27}$$

In this way the action of the annihilation operator on the physical ground state gives zero and denies access to the unphysical negative number states. The paper (Barnett and Pegg 1986 **Paper 1.7**) describes this approach in more detail and also proposes that it might be possible to define operators for the phase in terms of physical observables like the field quadratures. Furthermore, it provides a summary of the phase problem as it was at the beginning of 1986.

The Quantum Theory of the Emission and Absorption of Radiation.

By P. A. M. Dirac, St. John's College, Cambridge, and Institute for Theoretical Physics, Copenhagen.

(Communicated by N. Bohr, For. Mem. R.S.—Received February 2, 1927.)

§ 1. *Introduction and Summary.*

The new quantum theory, based on the assumption that the dynamical variables do not obey the commutative law of multiplication, has by now been developed sufficiently to form a fairly complete theory of dynamics. One can treat mathematically the problem of any dynamical system composed of a number of particles with instantaneous forces acting between them, provided it is describable by a Hamiltonian function, and one can interpret the mathematics physically by a quite definite general method. On the other hand, hardly anything has been done up to the present on quantum electrodynamics. The questions of the correct treatment of a system in which the forces are propagated with the velocity of light instead of instantaneously, of the production of an electromagnetic field by a moving electron, and of the reaction of this field on the electron have not yet been touched. In addition, there is a serious difficulty in making the theory satisfy all the requirements of the restricted principle of relativity, since a Hamiltonian function can no longer be used. This relativity question is, of course, connected with the previous ones, and it will be impossible to answer any one question completely without at the same time answering them all. However, it appears to be possible to build up a fairly satisfactory theory of the emission of radiation and of the reaction of the radiation field on the emitting system on the basis of a kinematics and dynamics which are not strictly relativistic. This is the main object of the present paper. The theory is non-relativistic only on account of the time being counted throughout as a c-number, instead of being treated symmetrically with the space co-ordinates. The relativity variation of mass with velocity is taken into account without difficulty.

The underlying ideas of the theory are very simple. Consider an atom interacting with a field of radiation, which we may suppose for definiteness to be confined in an enclosure so as to have only a discrete set of degrees of freedom. Resolving the radiation into its Fourier components, we can consider the energy and phase of each of the components to be dynamical variables describing the radiation field. Thus if E_r is the energy of a component labelled r and θ_r is the corresponding phase (defined as the time since the wave was in a standard phase), we can suppose each E_r and θ_r to form a pair of canonically conjugate variables. In the absence of any interaction between the field and the atom, the whole system of field plus atom will be describable by the Hamiltonian

$$H = \Sigma_r E_r + H_0 \tag{1}$$

equal to the total energy, H_0 being the Hamiltonian for the atom alone, since the variables E_r, θ_r obviously satisfy their canonical equations of motion

$$\dot{E}_r = -\frac{\partial H}{\partial \theta_r} = 0, \quad \dot{\theta}_r = \frac{\partial H}{\partial E_r} = 1.$$

When there is interaction between the field and the atom, it could be taken into account on the classical theory by the addition of an interaction term to the Hamiltonian (1), which would be a function of the variables of the atom and of the variables E_r, θ_r that describe the field. This interaction term would give the effect of the radiation on the atom, and also the reaction of the atom on the radiation field.

In order that an analogous method may be used on the quantum theory, it is necessary to assume that the variables E_r, θ_r are q-numbers satisfying the standard quantum conditions $\theta_r E_r - E_r \theta_r = ih$, etc., where h is $(2\pi)^{-1}$ times the usual Planck's constant, like the other dynamical variables of the problem. This assumption immediately gives light-quantum properties to the radiation.* For if ν_r is the frequency of the component r, $2\pi\nu_r\theta_r$ is an angle variable, so that its canonical conjugate $E_r/2\pi\nu_r$ can only assume a discrete set of values differing by multiples of h, which means that E_r can change only by integral multiples of the quantum $(2\pi h)\,\nu_r$. If we now add an interaction term (taken over from the clasical theory) to the Hamiltonian (1), the problem can be solved according to the rules of quantum mechanics, and we would expect to obtain the correct results for the action of the radiation and the atom on one another. It will be shown that we actually get the correct laws for the emission and absorption of radiation, and the correct values for Einstein's A's and B's. In the author's previous theory,† where the energies and phases of the components of radiation were c-numbers, only the B's could be obtained, and the reaction of the atom on the radiation could not be taken into account.

It will also be shown that the Hamiltonian which describes the interaction of the atom and the electromagnetic waves can be made identical with the Hamiltonian for the problem of the interaction of the atom with an assembly of particles moving with the velocity of light and satisfying the Einstein-Bose statistics, by a suitable choice of the interaction energy for the particles. The number of particles having any specified direction of motion and energy, which can be used as a dynamical variable in the Hamiltonian for the particles, is equal to the number of quanta of energy in the corresponding wave in the Hamiltonian for the waves. There is thus a complete harmony between the wave and light-quantum descriptions of the interaction. We shall actually build up the theory from the light-quantum point of view, and show that the Hamiltonian transforms naturally into a form which resembles that for the waves.

The mathematical development of the theory has been made possible by the author's general transformation theory of the quantum matrices.‡ Owing to the fact that we count the time as a c-number, we are allowed to use the notion of the value of any dynamical variable at any instant of time. This value is

* Similar assumptions have been used by Born and Jordan [' Z. f. Physik,' vol. 34, p. 886 (1925)] for the purpose of taking over the classical formula for the emission of radiation by a dipole into the quantum theory, and by Born, Heisenberg and Jordan [' Z. f. Physik,' vol. 35, p. 606 (1925)] for calculating the energy fluctuations in a field of black-body radiation.

† ' Roy. Soc. Proc.,' A, vol. 112, p. 661, § 5 (1926). This is quoted later by, *loc. cit.*, I.

‡ ' Roy. Soc. Proc.,' A, vol. 113, p. 621 (1927). This is quoted later by *loc. cit.*, II. An essentially equivalent theory has been obtained independently by Jordan [' Z. f. Physik,' vol. 40, p. 809 (1927)]. See also, F. London, ' Z. f. Physik,' vol. 40, p. 193 (1926).

a q-number, capable of being represented by a generalised " matrix " according to many different matrix schemes, some of which may have continuous ranges of rows and columns, and may require the matrix elements to involve certain kinds of infinities (of the type given by the δ functions*). A matrix scheme can be found in which any desired set of constants of integration of the dynamical system that commute are represented by diagonal matrices, or in which a set of variables that commute are represented by matrices that are diagonal at a specified time.† The values of the diagonal elements of a diagonal matrix representing any q-number are the characteristic values of that q-number. A Cartesian co-ordinate or momentum will in general have all characteristic values from $-\infty$ to $+\infty$, while an action variable has only a discrete set of characteristic values. (We shall make it a rule to use unprimed letters to denote the dynamical variables or q-numbers, and the same letters primed or multiply primed to denote their characteristic values. Transformation functions or eigenfunctions are functions of the characteristic values and not of the q-numbers themselves, so they should always be written in terms of primed variables.)

If $f(\xi, \eta)$ is any function of the canonical variables ξ_k, η_k, the matrix representing f at any time t in the matrix scheme in which the ξ_k at time t are diagonal matrices may be written down without any trouble, since the matrices representing the ξ_k and η_k themselves at time t are known, namely,

$$\left.\begin{aligned} \xi_k(\xi'\xi'') &= \xi_k'\,\delta(\xi'\xi''), \\ \eta_k(\xi'\xi'') &= -i\hbar\,\delta(\xi_1'-\xi_1'')\ldots\delta(\xi_{k-1}'-\xi_{k-1}'')\,\delta'(\xi_k'-\xi_k'')\,\delta(\xi_{k+1}'-\xi_{k+1}'')\ldots \end{aligned}\right\}. \tag{2}$$

Thus if the Hamiltonian H is given as a function of the ξ_k and η_k, we can at once write down the matrix $H(\xi'\ \xi'')$. We can then obtain the transformation function, (ξ'/α') say, which transforms to a matrix scheme (α) in which the Hamiltonian is a diagonal matrix, as (ξ'/α') must satisfy the integral equation

$$\int H(\xi'\xi'')\,d\xi''\,(\xi''/\alpha') = W(\alpha')\,.\,(\xi'/\alpha'), \tag{3}$$

of which the characteristic values $W(\alpha')$ are the energy levels. This equation is just Schrödinger's wave equation for the eigenfunctions (ξ'/α'), which becomes an ordinary differential equation when H is a simple algebraic function of the

* *Loc. cit.* II, § 2.

† One can have a matrix scheme in which a set of variables that commute are at all times represented by diagonal matrices if one will sacrifice the condition that the matrices must satisfy the equations of motion. The transformation function from such a scheme to one in which the equations of motion are satisfied will involve the time explicitly. See p. 628 in *loc. cit.*, II.

ξ_k and η_k on account of the special equations (2) for the matrices representing ξ_k and η_k. Equation (3) may be written in the more general form

$$\int \mathrm{H}\,(\xi'\xi'')\,d\xi''\,(\xi''/\alpha') = ih\,\partial\,(\xi'/\alpha')/\partial t, \tag{3'}$$

in which it can be applied to systems for which the Hamiltonian involves the time explicitly.

One may have a dynamical system specified by a Hamiltonian H which cannot be expressed as an algebraic function of any set of canonical variables, but which can all the same be represented by a matrix $\mathrm{H}(\xi'\xi'')$. Such a problem can still be solved by the present method, since one can still use equation (3) to obtain the energy levels and eigenfunctions. We shall find that the Hamiltonian which describes the interaction of a light-quantum and an atomic system is of this more general type, so that the interaction can be treated mathematically, although one cannot talk about an interaction potential energy in the usual sense.

It should be observed that there is a difference between a light-wave and the de Broglie or Schrödinger wave associated with the light-quanta. Firstly, the light-wave is always real, while the de Broglie wave associated with a light-quantum moving in a definite direction must be taken to involve an imaginary exponential. A more important difference is that their intensities are to be interpreted in different ways. The number of light-quanta per unit volume associated with a monochromatic light-wave equals the energy per unit volume of the wave divided by the energy $(2\pi h)\nu$ of a single light-quantum. On the other hand a monochromatic de Broglie wave of amplitude a (multiplied into the imaginary exponential factor) must be interpreted as representing a^2 light-quanta per unit volume for all frequencies. This is a special case of the general rule for interpreting the matrix analysis,* according to which, if (ξ'/α') or $\psi_{\alpha'}(\xi_k')$ is the eigenfunction in the variables ξ_k of the state α' of an atomic system (or simple particle), $|\psi_{\alpha'}(\xi_k')|^2$ is the probability of each ξ_k having the value ξ_k', [or $|\psi_{\alpha'}(\xi_k')|^2\,d\xi_1'\,d\xi_2'\ldots$ is the probability of each ξ_k lying between the values ξ_k' and $\xi_k' + d\xi_k'$, when the ξ_k have continuous ranges of characteristic values] on the assumption that all phases of the system are equally probable. The wave whose intensity is to be interpreted in the first of these two ways appears in the theory only when one is dealing with an assembly of the associated particles satisfying the Einstein-Bose statistics. There is thus no such wave associated with electrons.

* *Loc. cit.*, II, §§ 6, 7.

We shall now consider the transitions produced in an atomic system by an arbitrary perturbation. The method we shall adopt will be that previously given by the author,† which leads in a simple way to equations which determine the probability of the system being in any stationary state of the unperturbed system at any time.‡ This, of course, gives immediately the probable number of systems in that state at that time for an assembly of the systems that are independent of one another and are all perturbed in the same way. The object of the present section is to show that the equations for the rates of change of these probable numbers can be put in the Hamiltonian form in a simple manner, which will enable further developments in the theory to be made.

Let H_0 be the Hamiltonian for the unperturbed system and V the perturbing energy, which can be an arbitrary function of the dynamical variables and may or may not involve the time explicitly, so that the Hamiltonian for the perturbed system is $H = H_0 + V$. The eigenfunctions for the perturbed system must satisfy the wave equation

$$ih\, \partial\psi/\partial t = (H_0 + V)\, \psi,$$

where $(H_0 + V)$ is an operator. If $\psi = \Sigma_r a_r \psi_r$ is the solution of this equation that satisfies the proper initial conditions, where the ψ_r's are the eigenfunctions for the unperturbed system, each associated with one stationary state labelled by the suffix r, and the a_r's are functions of the time only, then $|a_r|^2$ is the probability of the system being in the state r at any time. The a_r's must be normalised initially, and will then always remain normalised. The theory will apply directly to an assembly of N similar independent systems if we multiply each of these a_r's by $N^{\frac{1}{2}}$ so as to make $\Sigma_r |a_r|^2 = N$. We shall now have that $|a_r|^2$ is the probable number of systems in the state r.

The equation that determines the rate of change of the a_r's is§

$$ih\dot{a}_r = \Sigma_s V_{rs} a_s, \tag{4}$$

where the V_{rs}'s are the elements of the matrix representing V. The conjugate imaginary equation is

$$-ih\dot{a}_r^* = \Sigma_s V_{rs}^* a_s^* = \Sigma_s a_s^* V_{sr}. \tag{4'}$$

† *Loc. cit.* I.

‡ The theory has recently been extended by Born [' Z. f. Physik,' vol. 40, p. 167 (1926)] so as to take into account the adiabatic changes in the stationary states that may be produced by the perturbation as well as the transitions. This extension is not used in the present paper.

§ *Loc. cit.*, I, equation (25).

If we regard a_r and $ih\, a_r^*$ as canonical conjugates, equations (4) and (4′) take the Hamiltonian form with the Hamiltonian function $F_1 = \Sigma_{rs} a_r^* V_{rs} a_s$, namely,

$$\frac{da_r}{dt} = \frac{1}{ih}\frac{\partial F_1}{\partial a_r^*}, \quad ih\frac{da_r^*}{dt} = -\frac{\partial F_1}{\partial a_r}.$$

We can transform to the canonical variables N_r, ϕ_r by the contact transformation

$$a_r = N_r^{\frac{1}{2}} e^{-i\phi_r/h}, \quad a_r^* = N_r^{\frac{1}{2}} e^{i\phi_r/h}.$$

This transformation makes the new variables N_r and ϕ_r real, N_r being equal to $a_r a_r^* = |a_r|^2$, the probable number of systems in the state r, and ϕ_r/h being the phase of the eigenfunction that represents them. The Hamiltonian F_1 now becomes

$$F_1 = \Sigma_{rs} V_{rs} N_r^{\frac{1}{2}} N_s^{\frac{1}{2}} e^{i(\phi_r - \phi_s)/h},$$

and the equations that determine the rate at which transitions occur have the canonical form

$$\dot{N}_r = -\frac{\partial F_1}{\partial \phi_r}, \quad \dot{\phi}_r = \frac{\partial F_1}{\partial N_r}.$$

A more convenient way of putting the transition equations in the Hamiltonian form may be obtained with the help of the quantities

$$b_r = a_r\, e^{-iW_r t/h}, \quad b_r^* = a_r^*\, e^{iW_r t/h},$$

W_r being the energy of the state r. We have $|b_r|^2$ equal to $|a_r|^2$, the probable number of systems in the state r. For b_r we find

$$\begin{aligned} ih\,\dot{b}_r &= W_r b_r + ih\dot{a}_r e^{-iW_r t/h} \\ &= W_r b_r + \Sigma_s V_{rs} b_s e^{i(W_s - W_r)t/h} \end{aligned}$$

with the help of (4). If we put $V_{rs} = v_{rs} e^{i(W_r - W_s)t/h}$, so that v_{rs} is a constant when V does not involve the time explicitly, this reduces to

$$\begin{aligned} ih\,\dot{b}_r &= W_r b_r + \Sigma_s v_{rs} b_s \\ &= \Sigma_s H_{rs} b_s, \end{aligned} \tag{5}$$

where $H_{rs} = W_r\,\delta_{rs} + v_{rs}$, which is a matrix element of the total Hamiltonian $H = H_0 + V$ with the time factor $e^{i(W_r - W_s)t/h}$ removed, so that H_{rs} is a constant when H does not involve the time explicitly. Equation (5) is of the same form as equation (4), and may be put in the Hamiltonian form in the same way.

It should be noticed that equation (5) is obtained directly if one writes down the Schrödinger equation in a set of variables that specify the stationary states of the unperturbed system. If these variables are ξ_h, and if $H(\xi'\xi'')$ denotes a matrix element of the total Hamiltonian H in the (ξ) scheme, this Schrödinger equation would be

$$ih\, \partial\psi(\xi')/\partial t = \Sigma_{\xi''}\, H(\xi'\xi'')\, \psi(\xi''), \tag{6}$$

like equation (3′). This differs from the previous equation (5) only in the notation, a single suffix r being there used to denote a stationary state instead of a set of numerical values ξ_k' for the variables ξ_k, and b_r being used instead of $\psi(\xi')$. Equation (6), and therefore also equation (5), can still be used when the Hamiltonian is of the more general type which cannot be expressed as an algebraic function of a set of canonial variables, but can still be represented by a matrix $H(\xi'\xi'')$ or H_{rs}.

We now take b_r and $ih\, b_r^*$ to be canonically conjugate variables instead of a_r and $ih\, a_r^*$. The equation (5) and its conjugate imaginary equation will now take the Hamiltonian form with the Hamiltonian function

$$F = \Sigma_{rs} b_r^* H_{rs} b_s. \tag{7}$$

Proceeding as before, we make the contact transformation

$$b_r = N_r^{\frac{1}{2}}\, e^{-i\theta_r/h}, \qquad b_r^* = N_r^{\frac{1}{2}}\, e^{i\theta_r/h}, \tag{8}$$

to the new canonical variables N_r, θ_r, where N_r is, as before, the probable number of systems in the state r, and θ_r is a new phase. The Hamiltonian F will now become

$$F = \Sigma_{rs} H_{rs} N_r^{\frac{1}{2}} N_s^{\frac{1}{2}} e^{i(\theta_r - \theta_s)/h},$$

and the equations for the rates of change of N_r and θ_r will take the canonical form

$$\dot{N}_r = -\frac{\partial F}{\partial \theta_r}, \qquad \dot{\theta}_r = \frac{\partial F}{\partial N_r}.$$

The Hamiltonian may be written

$$F = \Sigma_r W_r N_r + \Sigma_{rs} v_{rs} N_r^{\frac{1}{2}} N_s^{\frac{1}{2}} e^{i(\theta_r - \theta_s)/h}. \tag{9}$$

The first term $\Sigma_r W_r N_r$ is the total proper energy of the assembly, and the second may be regarded as the additional energy due to the perturbation. If the perturbation is zero, the phases θ_r would increase linearly with the time, while the previous phases ϕ_r would in this case be constants.

§ 3. *The Perturbation of an Assembly satisfying the Einstein-Bose Statistics.*

According to the preceding section we can describe the effect of a perturbation on an assembly of independent systems by means of canonical variables and Hamiltonian equations of motion. The development of the theory which naturally suggests itself is to make these canonical variables q-numbers satisfying the usual quantum conditions instead of c-numbers, so that their Hamiltonian equations of motion become true quantum equations. The Hamiltonian function will now provide a Schrödinger wave equation, which must be solved and interpreted in the usual manner. The interpretation will give not merely the probable number of systems in any state, but the probability of any given distribution of the systems among the various states, this probability being, in fact, equal to the square of the modulus of the normalised solution of the wave equation that satisfies the appropriate initial conditions. We could, of course, calculate directly from elementary considerations the probability of any given distribution when the systems are independent, as we know the probability of each system being in any particular state. We shall find that the probability calculated directly in this way does not agree with that obtained from the wave equation except in the special case when there is only one system in the assembly. In the general case it will be shown that the wave equation leads to the correct value for the probability of any given distribution when the systems obey the Einstein-Bose statistics instead of being independent.

We assume the variables b_r, ihb_r^* of § 2 to be canonical q-numbers satisfying the quantum conditions

$$b_r \, . \, ih\, b_r^* - ih\, b_r^* \, . \, b_r = ih$$

or

$$b_r b_r^* - b_r^* b_r = 1,$$

and

$$b_r b_s - b_s b_r = 0, \qquad b_r^* b_s^* - b_s^* b_r^* = 0,$$

$$b_r b_s^* - b_s^* b_r = 0 \qquad (s \neq r).$$

The transformation equations (8) must now be written in the quantum form

$$\left.\begin{aligned} b_r &= (\mathrm{N}_r + 1)^{\frac{1}{2}}\, e^{-i\theta_r/h} = e^{-i\theta_r/h}\, \mathrm{N}_r^{\frac{1}{2}} \\ b_r^* &= \mathrm{N}_r^{\frac{1}{2}} e^{i\theta_r/h} = e^{i\theta_r/h}\, (\mathrm{N}_r + 1)^{\frac{1}{2}}, \end{aligned}\right\} \tag{10}$$

in order that the N_r, θ_r may also be canonical variables. These equations show that the N_r can have only integral characteristic values not less than zero,† which provides us with a justification for the assumption that the variables are q-numbers in the way we have chosen. The numbers of systems in the different states are now ordinary quantum numbers.

† See § 8 of the author's paper 'Roy. Soc. Proc.,' A, vol. 111, p. 281 (1926). What are there called the c-number values that a q-number can take are here given the more precise name of the characteristic values of that q-number.

The Hamiltonian (7) now becomes

$$F = \Sigma_{rs} b_r^* H_{rs} b_s = \Sigma_{rs} N_r^{\frac{1}{2}} e^{i\theta_r/h} H_{rs} (N_s + 1)^{\frac{1}{2}} e^{-i\theta_s/h}$$
$$= \Sigma_{rs} H_{rs} N_r^{\frac{1}{2}} (N_s + 1 - \delta_{rs})^{\frac{1}{2}} e^{i(\theta_r - \theta_s)/h} \tag{11}$$

in which the H_{rs} are still c-numbers. We may write this F in the form corresponding to (9)

$$F = \Sigma_r W_r N_r + \Sigma_{rs} v_{rs} N_r^{\frac{1}{2}} (N_s + 1 - \delta_{rs})^{\frac{1}{2}} e^{i(\theta_r - \theta_s)/h} \tag{11'}$$

in which it is again composed of a proper energy term $\Sigma_r W_r N_r$ and an interaction energy term.

The wave equation written in terms of the variables N_r is†

$$ih \frac{\partial}{\partial t} \psi (N_1', N_2', N_s' \ldots) = F\psi (N_1', N_2', N_3' \ldots), \tag{12}$$

where F is an operator, each θ_r occurring in F being interpreted to mean $ih\, \partial/\partial N_r'$. If we apply the operator $e^{\pm i\theta_r/h}$ to any function $f(N_1', N_2', \ldots N_r', \ldots)$ of the variables $N_1', N_2', \ldots$ the result is

$$e^{\pm i\theta_r/h} f(N_1', N_2', \ldots N_r', \ldots) = e^{\mp \partial/\partial N_r'} f(N_1', N_2', \ldots N_r' \ldots)$$
$$= f(N_1', N_2', \ldots N_r' \mp 1, \ldots).$$

If we use this rule in equation (12) and use the expression (11) for F we obtain‡

$$ih \frac{\partial}{\partial t} \psi (N_1', N_2', N_3' \ldots)$$
$$= \Sigma_{rs} H_{rs} N_r'^{\frac{1}{2}} (N_s' + 1 - \delta_{rs})^{\frac{1}{2}} \psi (N_1', N_2' \ldots N_r' - 1, \ldots N_s' + 1, \ldots). \tag{13}$$

We see from the right-hand side of this equation that in the matrix representing F, the term in F involving $e^{i(\theta_r - \theta_s)/h}$ will contribute only to those matrix elements that refer to transitions in which N_r decreases by unity and N_s increases by unity, *i.e.*, to matrix elements of the type $F(N_1', N_2' \ldots N_r' \ldots N_s'; N_1', N_2' \ldots N_r' - 1 \ldots N_s' + 1 \ldots)$. If we find a solution $\psi (N_1', N_2' \ldots)$ of equation (13) that is normalised [*i.e.*, one for which $\Sigma_{N_1', N_2' \ldots} |\psi (N_1', N_2' \ldots)|^2 = 1$] and that satisfies the proper initial conditions, then $|\psi (N_1', N_2' \ldots)|^2$ will be the probability of that distribution in which N_1' systems are in state 1, N_2' in state 2, ... at any time.

Consider first the case when there is only one system in the assembly. The probability of its being in the state q is determined by the eigenfunction

† We are supposing for definiteness that the label r of the stationary states takes the values 1, 2, 3, ...

‡ When $s = r$, $\psi (N_1', N_2' \ldots N_r' - 1 \ldots N_s' + 1)$ is to be taken to mean $\psi (N_1' N_2' \ldots N_r' \ldots)$.

$\psi(N_1', N_2', \ldots)$ in which all the N′'s are put equal to zero except N_q', which is put equal to unity. This eigenfunction we shall denote by $\psi\{q\}$. When it is substituted in the left-hand side of (13), all the terms in the summation on the right-hand side vanish except those for which $r = q$, and we are left with

$$ih \frac{\partial}{\partial t} \psi\{q\} = \Sigma_r H_{qs} \psi\{s\},$$

which is the same equation as (5) with $\psi\{q\}$ playing the part of b_q. This establishes the fact that the present theory is equivalent to that of the preceding section when there is only one system in the assembly.

Now take the general case of an arbitrary number of systems in the assembly, and assume that they obey the Einstein-Bose statistical mechanics. This requires that, in the ordinary treatment of the problem, only those eigenfunctions that are symmetrical between all the systems must be taken into account, these eigenfunctions being by themselves sufficient to give a complete quantum solution of the problem.† We shall now obtain the equation for the rate of change of one of these symmetrical eigenfunctions, and show that it is identical with equation (13).

If we label each system with a number n, then the Hamiltonian for the assembly will be $H_A = \Sigma_n H(n)$, where $H(n)$ is the H of § 2 (equal to $H_0 + V$) expressed in terms of the variables of the nth system. A stationary state of the assembly is defined by the numbers $r_1, r_2 \ldots r_n \ldots$ which are the labels of the stationary states in which the separate systems lie. The Schrödinger equation for the assembly in a set of variables that specify the stationary states will be of the form (6) [with H_A instead of H], and we can write it in the notation of equation (5) thus :—

$$ih \dot{b}(r_1 r_2 \ldots) = \Sigma_{s_1, s_2 \ldots} H_A(r_1 r_2 \ldots ; s_1 s_2 \ldots) b(s_1 s_2 \ldots), \tag{14}$$

where $H_A(r_1 r_2 \ldots ; s_1 s_2 \ldots)$ is the general matrix element of H_A [with the time factor removed]. This matrix element vanishes when more than one s_n differs from the corresponding r_n; equals $H_{r_m s_m}$ when s_m differs from r_m and every other s_n equals r_n; and equals $\Sigma_n H_{r_n r_n}$ when every s_n equals r_n. Substituting these values in (14), we obtain

$$ih \dot{b}(r_1 r_2 \ldots) = \Sigma_m \Sigma_{s_m \neq r_m} H_{r_m s_m} b(r_1 r_2 \ldots r_{m-1} s_m r_{m+1} \ldots) + \Sigma_n H_{r_n r_n} b(r_1 r_2 \ldots). \tag{15}$$

We must now restrict $b(r_1 r_2 \ldots)$ to be a symmetrical function of the variables $r_1, r_2 \ldots$, in order to obtain the Einstein-Bose statistics. This is permissible since if $b(r_1 r_2 \ldots)$ is symmetrical at any time, then equation (15) shows that

† *Loc. cit.*, I, § 3.

$\dot{b}(r_1 r_2 \ldots)$ is also symmetrical at that time, so that $b\ (r_1 r_2 \ldots)$ will remain symmetrical.

Let N_r denote the number of systems in the state r. Then a stationary state of the assembly describable by a symmetrical eigenfunction may be specified by the numbers $N_1, N_2 \ldots N_r \ldots$ just as well as by the numbers $r_1, r_2 \ldots r_n \ldots$, and we shall be able to transform equation (15) to the variables $N_1, N_2 \ldots$. We cannot actually take the new eigenfunction $b\ (N_1, N_2 \ldots)$ equal to the previous one $b\ (r_1 r_2 \ldots)$, but must take one to be a numerical multiple of the other in order that each may be correctly normalised with respect to its respective variables. We must have, in fact,

$$\Sigma_{r_1, r_2 \ldots} |b(r_1 r_2 \ldots)|^2 = 1 = \Sigma_{N_1, N_2 \ldots} |b(N_1, N_2 \ldots)|^2,$$

and hence we must take $|b(N_1, N_2 \ldots)|^2$ equal to the sum of $|b(r_1 r_2 \ldots)|^2$ for all values of the numbers $r_1, r_2 \ldots$ such that there are N_1 of them equal to 1, N_2 equal to 2, etc. There are $N!/N_1!\,N_2!\ldots$ terms in this sum, where $N = \Sigma_r N_r$ is the total number of systems, and they are all equal, since $b(r_1 r_2 \ldots)$ is a symmetrical function of its variables $r_1, r_2 \ldots$. Hence we must have

$$b(N_1, N_2 \ldots) = (N!/N_1!\,N_2!\ldots)^{\frac{1}{2}}\, b(r_1 r_2 \ldots).$$

If we make this substitution in equation (15), the left-hand side will become $ih\,(N_1!\,N_2!\ldots/N!)^{\frac{1}{2}}\,\dot{b}\,(N_1, N_2 \ldots)$. The term $H_{r_m s_m} b\ (r_1 r_2 \ldots r_{m-1} s_m r_{m+1} \ldots)$ in the first summation on the right-hand side will become

$$[N_1!\,N_2!\ldots(N_r-1)!\ldots(N_s+1)!\ldots/N!]^{\frac{1}{2}}\,H_{rs} b\,(N_1, N_2 \ldots N_r-1 \ldots N_s+1 \ldots), \quad (16)$$

where we have written r for r_m and s for s_m. This term must be summed for all values of s except r, and must then be summed for r taking each of the values $r_1, r_2 \ldots$. Thus each term (16) gets repeated by the summation process until it occurs a total of N_r times, so that it contributes

$$\begin{aligned} N_r[N_1!\,N_2!\ldots(N_r-1)!\ldots(N_s+1)!\ldots/N!]^{\frac{1}{2}}\,H_{rs} b\,(N_1, N_2 \ldots N_r-1 \ldots N_s+1 \ldots) \\ = N_r^{\frac{1}{2}}(N_s+1)^{\frac{1}{2}}(N_1!\,N_2!\ldots/N!)^{\frac{1}{2}}\,H_{rs} b\,(N_1, N_2 \ldots N_r-1 \ldots N_s+1 \ldots) \end{aligned}$$

to the right-hand side of (15). Finally, the term $\Sigma_n H_{r_n r_n} b\,(r_1, r_2 \ldots)$ becomes

$$\Sigma_r N_r H_{rr} \,.\, b\,(r_1 r_2 \ldots) = \Sigma_r N_r H_{rr} \,.\, (N_1!\,N_2!\ldots/N!)^{\frac{1}{2}}\, b\,(N_1, N_2 \ldots).$$

Hence equation (15) becomes, with the removal of the factor $(N_1!\,N_2!\ldots/N!)^{\frac{1}{2}}$,

$$\begin{aligned} ih\dot{b}\,(N_1, N_2 \ldots) = \Sigma_r \Sigma_{s \neq r} N_r^{\frac{1}{2}}(N_s+1)^{\frac{1}{2}}\,H_{rs} b\,(N_1, N_2 \ldots N_r-1 \ldots N_s+1 \ldots) \\ + \Sigma_r N_r H_{rr} b\,(N_1, N_2 \ldots), \quad (17) \end{aligned}$$

which is identical with (13) [except for the fact that in (17) the primes have been omitted from the N's, which is permissible when we do not require to refer to the N's as q-numbers]. We have thus established that the Hamiltonian (11) describes the effect of a perturbation on an assembly satisfying the Einstein-Bose statistics.

§ 4. *The Reaction of the Assembly on the Perturbing System.*

Up to the present we have considered only perturbations that can be represented by a perturbing energy V added to the Hamiltonian of the perturbed system, V being a function only of the dynamical variables of that system and perhaps of the time. The theory may readily be extended to the case when the perturbation consists of interaction with a perturbing dynamical system, the reaction of the perturbed system on the perturbing system being taken into account. (The distinction between the perturbing system and the perturbed system is, of course, not real, but it will be kept up for convenience.)

We now consider a perturbing system, described, say, by the canonical variables J_k, ω_k, the J's being its first integrals when it is alone, interacting with an assembly of perturbed systems with no mutual interaction, that satisfy the Einstein-Bose statistics. The total Hamiltonian will be of the form

$$H_T = H_P(J) + \Sigma_n H(n),$$

where H_P is the Hamiltonian of the perturbing system (a function of the J's only) and $H(n)$ is equal to the proper energy $H_0(n)$ plus the perturbation energy $V(n)$ of the nth system of the assembly. $H(n)$ is a function only of the variables of the nth system of the assembly and of the J's and w's, and does not involve the time explicitly.

The Schrödinger equation corresponding to equation (14) is now

$$ih\dot{b}(J', r_1r_2\ldots) = \Sigma_{J''}\Sigma_{s_1,s_2}\ldots H_T(J', r_1r_2\ldots; J'', s_1s_2\ldots)\, b(J'', s_1s_2\ldots),$$

in which the eigenfunction b involves the additional variables J_k'. The matrix element $H_T(J', r_1r_2 \ldots; J'', s_1s_2 \ldots)$ is now always a constant. As before, it vanishes when more than one s_n differs from the corresponding r_n. When s_m differs from r_m and every other s_n equals r_n, it reduces to $H(J'r_m; J''s_m)$, which is the $(J'r_m; J''s_m)$ matrix element (with the time factor removed) of $H = H_0 + V$, the proper energy plus the perturbation energy of a single system of the assembly; while when every s_n equals r_n, it has the value $H_P(J')\,\delta_{J'J''} + \Sigma_n H(J'r_n; J''r_n)$. If, as before, we restrict the eigenfunctions to be symmetrical in the variables r_1, r_2 ..., we can again transform to the variables N_1, N_2 ..., which will lead, as before, to the result

$$i h \dot{b}(J', N_1', N_2' \ldots) = H_P(J')\, b(J', N'_1, N_2' \ldots)$$
$$+ \Sigma_{J''} \Sigma_{r,s} N_r'^{\frac{1}{2}} (N_s' + 1 - \delta_{rs})^{\frac{1}{2}} H(J'r; J''s)\, b(J'', N_1', N_2' \ldots N_r' - 1 \ldots N_s' + 1 \ldots) \quad (18)$$

This is the Schrödinger equation corresponding to the Hamiltonian function

$$F = H_P(J) + \Sigma_{r,s} H_{rs} N_r^{\frac{1}{2}} (N_s + 1 - \delta_{rs})^{\frac{1}{2}} e^{i(\theta_r - \theta_s)/h}, \quad (19)$$

in which H_{rs} is now a function of the J's and w's, being such that when represented by a matrix in the (J) scheme its (J′ J″) element is $H(J'r; J''s)$. (It should be noticed that H_{rs} still commutes with the N's and θ's.)

Thus the interaction of a perturbing system and an assembly satisfying the Einstein-Bose statistics can be described by a Hamiltonian of the form (19). We can put it in the form corresponding to (11′) by observing that the matrix element $H(J'r; J''s)$ is composed of the sum of two parts, a part that comes from the proper energy H_0, which equals W_r when $J_k'' = J_k'$ and $s = r$ and vanishes otherwise, and a part that comes from the interaction energy V, which may be denoted by $v(J'r; J''s)$. Thus we shall have

$$H_{rs} = W_r \delta_{rs} + v_{rs},$$

where v_{rs} is that function of the J's and w's which is represented by the matrix whose (J′ J″) element is $v(J'r; J''s)$, and so (19) becomes

$$F = H_P(J) + \Sigma_r W_r N_r + \Sigma_{r,s} v_{rs} N_r^{\frac{1}{2}} (N_s + 1 - \delta_{rs})^{\frac{1}{2}} e^{i(\theta_r - \theta_s)/h}. \quad (20)$$

The Hamiltonian is thus the sum of the proper energy of the perturbing system $H_P(J)$, the proper energy of the perturbed systems $\Sigma_r W_r N_r$ and the perturbation energy $\Sigma_{r,s} v_{rs} N_r^{\frac{1}{2}} (N_s + 1 - \delta_{rs})^{\frac{1}{2}} e^{i(\theta_r - \theta_s)/h}$.

§ 5. *Theory of Transitions in a System from One State to Others of the Same Energy.*

Before applying the results of the preceding sections to light-quanta, we shall consider the solution of the problem presented by a Hamiltonian of the type (19). The essential feature of the problem is that it refers to a dynamical system which can, under the influence of a perturbation energy which does not involve the time explicitly, make transitions from one state to others of the same energy. The problem of collisions between an atomic system and an electron, which has been treated by Born,* is a special case of this type. Born's method is to find a *periodic* solution of the wave equation which consists, in so far as it involves the co-ordinates of the colliding electron, of plane waves,

* Born, 'Z. f. Physik,' vol. 38, p. 803 (1926).

representing the incident electron, approaching the atomic system, which are scattered or diffracted in all directions. The square of the amplitude of the waves scattered in any direction with any frequency is then assumed by Born to be the probability of the electron being scattered in that direction with the corresponding energy.

This method does not appear to be capable of extension in any simple manner to the general problem of systems that make transitions from one state to others of the same energy. Also there is at present no very direct and certain way of interpreting a periodic solution of a wave equation to apply to a non-periodic physical phenomenon such as a collision. (The more definite method that will now be given shows that Born's assumption is not quite right, it being necessary to multiply the square of the amplitude by a certain factor.)

An alternative method of solving a collision problem is to find a *non-periodic* solution of the wave equation which consists initially simply of plane waves moving over the whole of space in the necessary direction with the necessary frequency to represent the incident electron. In course of time waves moving in other directions must appear in order that the wave equation may remain satisfied. The probability of the electron being scattered in any direction with any energy will then be determined by the rate of growth of the corresponding harmonic component of these waves. The way the mathematics is to be interpreted is by this method quite definite, being the same as that of the beginning of § 2.

We shall apply this method to the general problem of a system which makes transitions from one state to others of the same energy under the action of a perturbation. Let H_0 be the Hamiltonian of the unperturbed system and V the perturbing energy, which must not involve the time explicitly. If we take the case of a continuous range of stationary states, specified by the first integrals, α_k say, of the unperturbed motion, then, following the method of § 2, we obtain

$$ih\,\dot{a}(\alpha') = \int V(\alpha'\alpha'')\,d\alpha''\,.\,a(\alpha''), \tag{21}$$

corresponding to equation (4). The probability of the system being in a state for which each α_k lies between α_k' and $\alpha_k' + d\alpha_k'$ at any time is $|a(\alpha')|^2 d\alpha_1'\,.\,d\alpha_2' \ldots$ when $a(\alpha')$ is properly normalised and satisfies the proper initial conditions. If initially the system is in the state α^0, we must take the initial value of $a(\alpha')$ to be of the form $a^0\,.\,\delta(\alpha' - \alpha^0)$. We shall keep a^0 arbitrary, as it would be inconvenient to normalise $a(\alpha')$ in the present case. For a first approximation we may substitute for $a(\alpha'')$ in the right-hand side of (21) its initial value. This gives

$$ih\,\dot{a}(\alpha') = a^0 V(\alpha'\alpha^0) = \alpha^0 v(\alpha'\alpha^0)\, e^{i[W(\alpha')-W(\alpha^0)]t/h},$$

where $v(\alpha'\alpha^0)$ is a constant and $W(\alpha')$ is the energy of the state α'. Hence

$$ih\,a(\alpha') = a^0\,\delta(\alpha'-\alpha^0) + a^0 v(\alpha'\alpha^0)\frac{e^{i[W(\alpha')-W(\alpha^0)]t/h}-1}{i[W(\alpha')-W(\alpha^0)]/h}. \quad (22)$$

For values of the α_k' such that $W(\alpha')$ differs appreciably from $W(\alpha^0)$, $a(\alpha')$ is a periodic function of the time whose amplitude is small when the perturbing energy V is small, so that the eigenfunctions corresponding to these stationary states are not excited to any appreciable extent. On the other hand, for values of the α_k' such that $W(\alpha') = W(\alpha^0)$ and $\alpha_k' \neq \alpha_k^0$ for some k, $a(\alpha')$ increases uniformly with respect to the time, so that the probability of the system being in the state α' at any time increases proportionally with the square of the time. Physically, the probability of the system being in a state with exactly the same proper energy as the initial proper energy $W(\alpha^0)$ is of no importance, being infinitesimal. We are interested only in the integral of the probability through a small range of proper energy values about the initial proper energy, which, as we shall find, increases linearly with the time, in agreement with the ordinary probability laws.

We transform from the variables $\alpha_1, \alpha_2 \ldots \alpha_u$ to a set of variables that are arbitrary independent functions of the α's such that one of them is the proper energy W, say, the variables $W, \gamma_1, \gamma_2, \ldots \gamma_{u-1}$. The probability at any time of the system lying in a stationary state for which each γ_k lies between γ_k' and $\gamma_k' + d\gamma_k'$ is now (apart from the normalising factor) equal to

$$d\gamma_1' \cdot d\gamma_2' \ldots d\gamma_{u-1}' \int |a(\alpha')|^2 \frac{\partial(\alpha_1', \alpha_2' \ldots \alpha_u')}{\partial(W', \gamma_1' \ldots \gamma_{u-1}')}\, dW'. \quad (23)$$

For a time that is large compared with the periods of the system we shall find that practically the whole of the integral in (23) is contributed by values of W' very close to $W^0 = W(\alpha^0)$. Put

$$a(\alpha') = a(W', \gamma') \quad \text{and} \quad \partial(\alpha_1', \alpha_2' \ldots \alpha_u')/\partial(W', \gamma_1' \ldots \gamma_{u-1}') = J(W', \gamma').$$

Then for the integral in (23) we find, with the help of (22) (provided $\gamma_k' \neq \gamma_k^0$ for some k)

$$\int |a(W',\gamma')|^2 J(W',\gamma')\, dW'$$

$$= |a^0|^2 \int |v(W',\gamma'; W^0,\gamma^0)|^2 J(W',\gamma') \frac{[e^{i(W'-W^0)t/h}-1][e^{-i(W'-W^0)t/h}-1]}{(W'-W^0)^2} dW'$$

$$= 2|a^0|^2 \int |v(W',\gamma'; W^0,\gamma^0)|^2 J(W',\gamma')[1-\cos(W'-W^0)t/h]/(W'-W^0)^2 . dW'$$

$$= 2|a^0|^2 t/h . \int |v(W^0+hx/t,\gamma'; W^0,\gamma^0)|^2 J(W^0+hx/t,\gamma')(1-\cos x)/x^2 . dx,$$

if one makes the substitution $(W'-W^0)t/h = x$. For large values of t this reduces to

$$2\,|a^0|^2 t/h \,.\, |v(W^0, \gamma'\,;\; W^0, \gamma^0)|^2\, J(W^0, \gamma') \int_{-\infty}^{\infty} (1-\cos x)/x^2 \,.\, dx$$

$$= 2\pi\,|a^0|^2\, t/h \,.\, |v(W^0, \gamma'\,;\; W^0, \gamma^0)|^2\, J(W^0, \gamma').$$

The probability per unit time of a transition to a state for which each γ_k lies between γ_k' and $\gamma_k' + d\gamma_k'$ is thus (apart from the normalising factor)

$$2\pi\,|a^0|^2/h \,.\, |v(W^0, \gamma'\,;\; W^0, \gamma^0)|^2\, J(W^0, \gamma')\, d\gamma_1' \,.\, d\gamma_2' \ldots d\gamma_{u-1}', \tag{24}$$

which is proportional to the square of the matrix element associated with that transition of the perturbing energy.

To apply this result to a simple collision problem, we take the α's to be the components of momentum p_x, p_y, p_z of the colliding electron and the γ's to be θ and ϕ, the angles which determine its direction of motion. If, taking the relativity change of mass with velocity into account, we let P denote the resultant momentum, equal to $(p_x^2 + p_y^2 + p_z^2)^{\frac{1}{2}}$, and E the energy, equal to $(m^2c^4 + P^2c^2)^{\frac{1}{2}}$, of the electron, m being its rest-mass, we find for the Jacobian

$$J = \frac{\partial(p_x, p_y, p_z)}{\partial(E, \theta, \phi)} = \frac{EP}{c^2}\sin\theta.$$

Thus the $J(W^0, \gamma')$ of the expression (24) has the value

$$J(W^0, \gamma') = E'P' \sin\theta'/c^2, \tag{25}$$

where E′ and P′ refer to that value for the energy of the scattered electron which makes the total energy equal the initial energy W^0 (*i.e.*, to that value required by the conservation of energy).

We must now interpret the initial value of $a(\alpha')$, namely, $a^0\,\delta(\alpha'-\alpha^0)$, which we did not normalise. According to § 2 the wave function in terms of the variables α_k is $b(\alpha') = a(\alpha')\,e^{-iW't/h}$, so that its initial value is

$$a^0\,\delta(\alpha'-\alpha^0)\,e^{-iW't/h} = a^0\,\delta(p_x'-p_x^0)\,\delta(p_y'-p_y^0)\,\delta(p_z'-p_z^0)\,e^{-iW't/h}.$$

If we use the transformation function*

$$(x'/p') = (2\pi h)^{-3/2} e^{i\Sigma_{xyz} p_x' x'/h},$$

and the transformation rule

$$\psi(x') = \int (x'/p')\,\psi(p')\,dp_x'\,dp_y'\,dp_z',$$

we obtain for the initial wave function in the co-ordinates x, y, z the value

$$a^0\,(2\pi h)^{-3/2}\,e^{i\Sigma_{xyz} p_x^0 x'/h}\,e^{-iW't/h}.$$

* The symbol x is used for brevity to denote x, y, z.

This corresponds to an initial distribution of $|a^0|^2(2\pi h)^{-3}$ electrons per unit volume. Since their velocity is P^0c^2/E^0, the number per unit time striking a unit surface at right-angles to their direction of motion is $|a^0|^2 P^0c^2/(2\pi h)^3 E^0$. Dividing this into the expression (24) we obtain, with the help of (25),

$$4\pi^2(2\pi h)^2 \frac{E'E^0}{c^4} |v(p'; p^0)|^2 \frac{P'}{P^0} \sin\theta' \, d\theta' \, d\phi'. \tag{26}$$

This is the effective area that must be hit by an electron in order that it shall be scattered in the solid angle $\sin\theta' \, d\theta' \, d\phi'$ with the energy E'. This result differs by the factor $(2\pi h)^2/2mE' \cdot P'/P^0$ from Born's.* The necessity for the factor P'/P^0 in (26) could have been predicted from the principle of detailed balancing, as the factor $|v(p'; p^0)|^2$ is symmetrical between the direct and reverse processes.†

§ 6. *Application to Light-Quanta.*

We shall now apply the theory of § 4 to the case when the systems of the assembly are light-quanta, the theory being applicable to this case since light-quanta obey the Einstein-Bose statistics and have no mutual interaction. A light-quantum is in a stationary state when it is moving with constant momentum in a straight line. Thus a stationary state r is fixed by the three components of momentum of the light-quantum and a variable that specifies its state of polarisation. We shall work on the assumption that there are a finite number of these stationary states, lying very close to one another, as it would be inconvenient to use continuous ranges. The interaction of the light-quanta with an atomic system will be described by a Hamiltonian of the form (20), in which $H_P(J)$ is the Hamiltonian for the atomic system alone, and the coefficients v_{rs} are for the present unknown. We shall show that this form for the Hamiltonian, with the v_{rs} arbitrary, leads to Einstein's laws for the emission and absorption of radiation.

The light-quantum has the peculiarity that it apparently ceases to exist when it is in one of its stationary states, namely, the zero state, in which its momentum, and therefore also its energy, are zero. When a light-quantum is absorbed it can be considered to jump into this zero state, and when one is emitted it can be considered to jump from the zero state to one in which it is

* In a more recent paper (' Nachr. Gesell. d. Wiss.,' Gottingen, p. 146 (1926)) Born has obtained a result in agreement with that of the present paper for non-relativity mechanics, by using an interpretation of the analysis based on the conservation theorems. I am indebted to Prof. N. Bohr for seeing an advance copy of this work.

† See Klein and Rosseland, ' Z. f. Physik,' vol. 4, p. 46, equation (4) (1921).

physically in evidence, so that it appears to have been created. Since there is no limit to the number of light-quanta that may be created in this way, we must suppose that there are an infinite number of light-quanta in the zero state, so that the N_0 of the Hamiltonian (20) is infinite. We must now have θ_0, the variable canonically conjugate to N_0, a constant, since

$$\dot{\theta}_0 = \partial F/\partial N_0 = W_0 + \text{terms involving } N_0^{-\frac{1}{2}} \text{ or } (N_0+1)^{-\frac{1}{2}}$$

and W_0 is zero. In order that the Hamiltonian (20) may remain finite it is necessary for the coefficients v_{r0}, v_{0r} to be infinitely small. We shall suppose that they are infinitely small in such a way as to make $v_{r0}N_0^{\frac{1}{2}}$ and $v_{0r}N_0^{\frac{1}{2}}$ finite, in order that the transition probability coefficients may be finite. Thus we put

$$v_{r0}\,(N_0+1)^{\frac{1}{2}}\,e^{-i\theta_0/h} = v_r, \quad v_{0r}N_0^{\frac{1}{2}}e^{i\theta_0/h} = v_r^*,$$

where v_r and v_r^* are finite and conjugate imaginaries. We may consider the v_r and v_r^* to be functions only of the J's and w's of the atomic system, since their factors $(N_0+1)^{\frac{1}{2}}\,e^{-i\theta_0/h}$ and $N_0^{\frac{1}{2}}e^{i\theta_0/h}$ are practically constants, the rate of change of N_0 being very small compared with N_0. The Hamiltonian (20) now becomes

$$F = H_P(J) + \Sigma_r W_r N_r + \Sigma_{r\neq 0}[v_r N_r^{\frac{1}{2}} e^{i\theta_r/h} + v_r^*(N_r+1)^{\frac{1}{2}}\,e^{-i\theta_r/h}]$$
$$+\,\Sigma_{r\neq 0}\Sigma_{s\neq 0}\,v_{rs}N_r^{\frac{1}{2}}\,(N_s+1-\delta_{rs})^{\frac{1}{2}}\,e^{i(\theta_r-\theta_s)/h}. \qquad (27)$$

The probability of a transition in which a light-quantum in the state r is absorbed is proportional to the square of the modulus of that matrix element of the Hamiltonian which refers to this transition. This matrix element must come from the term $v_r N_r^{\frac{1}{2}} e^{i\theta_r/h}$ in the Hamiltonian, and must therefore be proportional to $N_r'^{\frac{1}{2}}$ where N_r' is the number of light-quanta in state r before the process. The probability of the absorption process is thus proportional to N_r'. In the same way the probability of a light-quantum in state r being emitted is proportional to $(N_r'+1)$, and the probability of a light-quantum in state r being scattered into state s is proportional to $N_r'(N_s'+1)$. Radiative processes of the more general type considered by Einstein and Ehrenfest,† in which more than one light-quantum take part simultaneously, are not allowed on the present theory.

To establish a connection between the number of light-quanta per stationary state and the intensity of the radiation, we consider an enclosure of finite volume, A say, containing the radiation. The number of stationary states for light-quanta of a given type of polarisation whose frequency lies in the

† 'Z. f. Physik,' vol. 19, p. 301 (1923).

range ν_r to $\nu_r + d\nu_r$ and whose direction of motion lies in the solid angle $d\omega_r$ about the direction of motion for state r will now be $A\nu_r^2 d\nu_r d\omega_r/c^3$. The energy of the light-quanta in these stationary states is thus $N_r' . 2\pi h\nu_r . A\nu_r^2 d\nu_r d\omega_r/c^3$. This must equal $Ac^{-1}I_r d\nu_r d\omega_r$, where I_r is the intensity per unit frequency range of the radiation about the state r. Hence

$$I_r = N_r' (2\pi h)\nu_r^3/c^2, \tag{28}$$

so that N_r' is proportional to I_r and $(N_r' + 1)$ is proportional to $I_r + (2\pi h)\nu_r^3/c^2$. We thus obtain that the probability of an absorption process is proportional to I_r, the incident intensity per unit frequency range, and that of an emission process is proportional to $I_r + (2\pi h)\nu_r^3/c^2$, which are just Einstein's laws.* In the same way the probability of a process in which a light-quantum is scattered from a state r to a state s is proportional to $I_r [I_s + (2\pi h)\nu_r^3/c^2]$, which is Pauli's law for the scattering of radiation by an electron.†

§7. *The Probability Coefficients for Emission and Absorption.*

We shall now consider the interaction of an atom and radiation from the wave point of view. We resolve the radiation into its Fourier components, and suppose that their number is very large but finite. Let each component be labelled by a suffix r, and suppose there are σ_r components associated with the radiation of a definite type of polarisation per unit solid angle per unit frequency range about the component r. Each component r can be described by a vector potential κ_r chosen so as to make the scalar potential zero. The perturbation term to be added to the Hamiltonian will now be, according to the classical theory with neglect of relativity mechanics, $c^{-1} \Sigma_r \kappa_r \dot{X}_r$, where X_r is the component of the total polarisation of the atom in the direction of κ_r, which is the direction of the electric vector of the component r.

We can, as explained in §1, suppose the field to be described by the canonical variables N_r, θ_r, of which N_r is the number of quanta of energy of the component r, and θ_r is its canonically conjugate phase, equal to $2\pi h\nu_r$ times the θ_r of §1. We shall now have $\kappa_r = a_r \cos \theta_r/h$, where a_r is the amplitude of κ_r, which can be connected with N_r as follows:—The flow of energy per unit area per unit time for the component r is $\frac{1}{2}\pi c^{-1} a_r^2 \nu_r^2$. Hence the intensity

* The ratio of stimulated to spontaneous emission in the present theory is just twice its value in Einstein's. This is because in the present theory either polarised component of the incident radiation can stimulate only radiation polarised in the same way, while in Einstein's the two polarised components are treated together. This remark applies also to the scattering process.

† Pauli, 'Z. f. Physik,' vol. 18, p. 272 (1923).

per unit frequency range of the radiation in the neighbourhood of the component r is $I_r = \frac{1}{2}\pi c^{-1} a_r^2 \nu_r^2 \sigma_r$. Comparing this with equation (28), we obtain $a_r = 2(h\nu_r/c\sigma_r)^{\frac{1}{2}} N_r^{\frac{1}{2}}$, and hence

$$\kappa_r = 2(h\nu_r/c\sigma_r)^{\frac{1}{2}} N_r^{\frac{1}{2}} \cos\theta_r/h.$$

The Hamiltonian for the whole system of atom plus radiation would now be, according to the classical theory,

$$F = H_P(J) + \Sigma_r (2\pi h\nu_r) N_r + 2c^{-1}\Sigma_r (h\nu_r/c\sigma_r)^{\frac{1}{2}} \dot{X}_r N_r^{\frac{1}{2}} \cos\theta_r/h, \tag{29}$$

where $H_P(J)$ is the Hamiltonian for the atom alone. On the quantum theory we must make the variables N_r and θ_r canonical q-numbers like the variables J_k, w_k that describe the atom. We must now replace the $N_r^{\frac{1}{2}} \cos\theta_r/h$ in (29) by the real q-number

$$\tfrac{1}{2}\{N_r^{\frac{1}{2}} e^{i\theta_r/h} + e^{-i\theta_r/h} N_r^{\frac{1}{2}}\} = \tfrac{1}{2}\{N_r^{\frac{1}{2}} e^{i\theta_r/h} + (N_r+1)^{\frac{1}{2}} e^{-i\theta_r/h}\}$$

so that the Hamiltonian (29) becomes

$$F = H_P(J) + \Sigma_r (2\pi h\nu_r) N_r + h^{\frac{1}{2}} c^{-\frac{3}{2}} \Sigma_r (\nu_r/\sigma_r)^{\frac{1}{2}} \dot{X}_r \{N_r^{\frac{1}{2}} e^{i\theta_r/h} + (N_r+1)^{\frac{1}{2}} e^{-i\theta_r/h}\}. \tag{30}$$

This is of the form (27), with

$$v_r = v_r^* = h^{\frac{1}{2}} c^{-\frac{3}{2}} (\nu_r/\sigma_r)^{\frac{1}{2}} \dot{X}_r \tag{31}$$

and
$$v_{rs} = 0 \qquad (r, s \neq 0).$$

The wave point of view is thus consistent with the light-quantum point of view and gives values for the unknown interaction coefficient v_{rs} in the light-quantum theory. These values are not such as would enable one to express the interaction energy as an algebraic function of canonical variables. Since the wave theory gives $v_{rs} = 0$ for $r, s \neq 0$, it would seem to show that there are no direct scattering processes, but this may be due to an incompleteness in the present wave theory.

We shall now show that the Hamiltonian (30) leads to the correct expressions for Einstein's A's and B's. We must first modify slightly the analysis of § 5 so as to apply to the case when the system has a large number of discrete stationary states instead of a continuous range. Instead of equation (21) we shall now have

$$ih\,\dot{a}(\alpha') = \Sigma_{\alpha''} V(\alpha'\alpha'')\, a(\alpha'').$$

If the system is initially in the state α^0, we must take the initial value of $a(\alpha')$ to be $\delta_{\alpha'\alpha^0}$, which is now correctly normalised. This gives for a first approximation

$$ih\,\dot{a}(\alpha') = V(\alpha'\alpha^0) = v(\alpha'\alpha^0)\, e^{i[W(\alpha')-W(\alpha^0)]t/h},$$

which leads to

$$ih\,a(\alpha') = \delta_{\alpha'\alpha^0} + v(\alpha'\alpha^0)\,\frac{e^{i[W(\alpha')-W(\alpha^0)]t/h} - 1}{i[W(\alpha')-W(\alpha^0)]/h},$$

corresponding to (22). If, as before, we transform to the variables W, γ_1, $\gamma_2 \cdots \gamma_{u-1}$, we obtain (when $\gamma' \neq \gamma^0$)

$$a\,(\mathrm{W}'\gamma') = v\,(\mathrm{W}', \gamma'\,;\ \mathrm{W}^0, \gamma^0)\,[1 - e^{i(\mathrm{W}'-\mathrm{W}^0)t/h}]/(\mathrm{W}' - \mathrm{W}^0).$$

The probability of the system being in a state for which each γ_k equals γ_k' is $\Sigma_{\mathrm{W}'}\,|\,a\,(\mathrm{W}'\ \gamma')|^2$. If the stationary states lie close together and if the time t is not too great, we can replace this sum by the integral $(\Delta\mathrm{W})^{-1}\int |\,a\,(\mathrm{W}'\gamma')\,|^2\,d\mathrm{W}'$, where ΔW is the separation between the energy levels. Evaluating this integral as before, we obtain for the probability per unit time of a transition to a state for which each $\gamma_k = \gamma_k'$

$$2\pi/h\Delta\mathrm{W}\ .\ |\,v\,(\mathrm{W}^0, \gamma'\,;\ \mathrm{W}^0, \gamma^0)\,|^2. \tag{32}$$

In applying this result we can take the γ's to be any set of variables that are independent of the total proper energy W and that together with W define a stationary state.

We now return to the problem defined by the Hamiltonian (30) and consider an absorption process in which the atom jumps from the state J^0 to the state J$'$ with the absorption of a light-quantum from state r. We take the variables γ' to be the variables J$'$ of the atom together with variables that define the direction of motion and state of polarisation of the absorbed quantum, but not its energy. The matrix element $v\,(\mathrm{W}^0, \gamma'\,;\ \mathrm{W}^0, \gamma^0)$ is now

$$h^{1/2}c^{-3/2}\,(\nu_r/\sigma_r)^{1/2}\,\dot{\mathrm{X}}_r\,(\mathrm{J}^0\mathrm{J}')\mathrm{N}_r{}^0,$$

where $\dot{\mathrm{X}}_r\,(\mathrm{J}^0\mathrm{J}')$ is the ordinary $(\mathrm{J}^0\mathrm{J}')$ matrix element of $\dot{\mathrm{X}}_r$. Hence from (32) the probability per unit time of the absorption process is

$$\frac{2\pi}{h\Delta\mathrm{W}} \cdot \frac{h\nu_r}{c^3\sigma_r}\,|\,\dot{\mathrm{X}}_r\,(\mathrm{J}^0\mathrm{J}')\,|^2\mathrm{N}_r{}^0.$$

To obtain the probability for the process when the light-quantum comes from any direction in a solid angle $d\omega$, we must multiply this expression by the number of possible directions for the light-quantum in the solid angle $d\omega$, which is $d\omega\ \sigma_r\Delta\mathrm{W}/2\pi h$. This gives

$$d\omega\,\frac{\nu_r}{hc^3}\,|\,\dot{\mathrm{X}}_r\,(\mathrm{J}_0\mathrm{J}')\,|^2\,\mathrm{N}_r{}^0 = d\omega\,\frac{1}{2\pi h^2 c\nu_r{}^2}\,|\,\dot{\mathrm{X}}_r\,(\mathrm{J}^0\mathrm{J}')\,|^2\,\mathrm{I}_r$$

with the help of (28). Hence the probability coefficient for the absorption process is $1/2\pi h^2 c\nu_r{}^2\,.\,|\,\dot{\mathrm{X}}_r\,(\mathrm{J}^0\mathrm{J}')\,|^2$, in agreement with the usual value for Einstein's absorption coefficient in the matrix mechanics. The agreement for the emission coefficients may be verified in the same manner.

The present theory, since it gives a proper account of spontaneous emission, must presumably give the effect of radiation reaction on the emitting system, and enable one to calculate the natural breadths of spectral lines, if one can overcome the mathematical difficulties involved in the general solution of the wave problem corresponding to the Hamiltonian (30). Also the theory enables one to understand how it comes about that there is no violation of the law of the conservation of energy when, say, a photo-electron is emitted from an atom under the action of extremely weak incident radiation. The energy of interaction of the atom and the radiation is a q-number that does not commute with the first integrals of the motion of the atom alone or with the intensity of the radiation. Thus one cannot specify this energy by a c-number at the same time that one specifies the stationary state of the atom and the intensity of the radiation by c-numbers. In particular, one cannot say that the interaction energy tends to zero as the intensity of the incident radiation tends to zero. There is thus always an unspecifiable amount of interaction energy which can supply the energy for the photo-electron.

I would like to express my thanks to Prof. Niels Bohr for his interest in this work and for much friendly discussion about it.

Summary.

The problem is treated of an assembly of similar systems satisfying the Einstein-Bose statistical mechanics, which interact with another different system, a Hamiltonian function being obtained to describe the motion. The theory is applied to the interaction of an assembly of light-quanta with an ordinary atom, and it is shown that it gives Einstein's laws for the emission and absorption of radiation.

The interaction of an atom with electromagnetic waves is then considered, and it is shown that if one takes the energies and phases of the waves to be q-numbers satisfying the proper quantum conditions instead of c-numbers, the Hamiltonian function takes the same form as in the light-quantum treatment. The theory leads to the correct expressions for Einstein's A's and B's.

AMPLITUDE AND PHASE UNCERTAINTY RELATIONS

W. H. LOUISELL
Bell Telephone Laboratories, Incorporated, Murray Hill, New Jersey

Received 17 September 1963

In the discussion of narrow band radiation it is often extremely useful to describe the wave by an amplitude and phase function. Thus the voltage associated with a wave may be written as

$$v = v(t) \cos [\omega_0 t + \varphi(t)] . \quad (1)$$

It may also, of course, be written in terms of its in-phase and quadrature components as

$$v = v_1(t) \cos \omega_0 t + v_2(t) \sin \omega_0 t . \quad (2)$$

In these expressions the frequency ω_0 is always some more or less arbitrary reference frequency against which the signal is compared. Usually ω_0 is taken as the centre of the band to which the wave is restricted.

One would like to be able to carry these descriptions of waves into quantum mechanics. However, in quantum mechanics one is much more concerned with the operations of one's measuring apparatus than is the case classically. Thus one must inquire how one measures the various descriptive quantities $v_1(t)$, $v_2(t)$, $v(t)$, $\varphi(t)$. In the case of $v_1(t)$ and $v_2(t)$ the situation is clear enough. One merely beats v against a local oscillator signal of frequency ω_0 (this may be done in several stages) whose amplitude is much larger than v, and measures v_1 and v_2 by appropriate setting of the phase of the local oscillator. Correspondingly, one finds that the formulation of quantum operator equivalents of v_1 and v_2 is straightforward. On the other hand, the measurement of $v(t)$ and $\varphi(t)$ is not so obvious. A receiver which is power sensitive measures $v^2(t)$ but does not measure $\varphi(t)$. There does not seem to be any receiver which measures $\varphi(t)$ directly. Rather one always seems to infer a value of $\varphi(t)$ from measurements of $v_1(t)$ and $v_2(t)$. Correspondingly, a quantum mechanical operator equivalent to $\varphi(t)$ does not seem to exist. We show below that one can define operators equivalent to $\sin \varphi$ and $\cos \varphi$, but even then their value in describing physical situations does not seem to be great. Thus, we conclude that φ should *not* be used as a descriptive variable in quantum mechanics. Nonetheless, approximate significance can be ascribed to φ, and as the quantum description approaches the classical description in the limit of large numbers of photons (the correspondence principle) this significance becomes more and more precise.

The often accepted commutator relation [1,2] between the number of quanta, N, and the phase, φ,

$$[N, \varphi] = -i , \quad (3)$$

leads to certain difficulties when one attempts to evaluate the matrix elements of φ in a representation in which N is diagonal. In this representation $N|n\rangle = n|n\rangle$ where $\langle n|l\rangle = \delta_{nl}$. If we take the matrix elements of (3) between states $\langle l|$ and $|n\rangle$ we find that

$$\langle l|\varphi|n\rangle = -\frac{\delta_{nl}}{(n-l)} . \quad (4)$$

The difficulty is now apparent. If the eigenvalues of N are integers then the right side of (4) is undefined *. Only in the limit of large values of n and l (correspondence limit) does (4) take on a definite meaning.

Thus, if (3) has a meaning only in the limit of many quanta, the oft-quoted uncertainty relation [3].

$$\Delta N \, \Delta\varphi \geq \tfrac{1}{2} \quad (5)$$

derived from (3) also has a meaning only in the limit of many quanta **.

A self-consistent way out of the difficulty seems to be to take as the fundamental commutation relation

$$[N, f(\varphi)] = -\frac{\partial f(\varphi)}{\partial \varphi}, \tag{6}$$

and restrict $f(\varphi)$ to periodic functions of period 2π. This will rule out (3) as a valid commutator relation. We may then take $\cos\varphi$ and $\sin\varphi$ as Hermitian operators which satisfy

$$[N, \cos\varphi] = i\sin\varphi, \quad [N, \sin\varphi] = -i\cos\varphi. \tag{7}$$

It is easy to show that matrix elements of $\sin\varphi$ and $\cos\varphi$ do not run into the difficulty encountered above with φ. Accordingly, the uncertainty relations become [3)]

$$\Delta N \Delta \cos\varphi \geqslant \tfrac{1}{2}|\langle\sin\varphi\rangle|, \quad \Delta N \Delta \sin\varphi \geqslant \tfrac{1}{2}|\langle\cos\varphi\rangle|, \tag{8}$$

where $\langle\ \rangle$ signifies the expectation value for state $|\psi\rangle$ and $\Delta A = \sqrt{\langle A^2\rangle - \langle A\rangle^2}$. This represents a somewhat different approach than that used in ref. [4)] and could also be adapted to the angle and angular momentum case.

1) P.A.M.Dirac, Quantum mechanics, second ed. (Oxford University Press, Great Britain, 1935) p.137.
2) W.Heitler, Quantum theory of radiation, second ed. (Oxford University Press, Great Britain, 1944) p.68.
3) D. Bohm, Quantum theory, (Prentice-Hall, Inc., New York, 1951) p. 205.
4) D. Judge, Physics Letters 5 (1963) 189.
D. Judge and J.T. Lewis, Physics Letters 5 (1963) 190.

* In the case of matrix elements of q in a representation in which p is diagonal $p|p'\rangle = p'|p'\rangle$, $\langle p'|p''\rangle = \delta(p'-p'')$ taking the matrix element of $[q,p] = i\hbar$ between states $\langle p'|$ and $|p''\rangle$ leads to

$$\langle p'|q|p''\rangle = i\hbar\frac{\delta(p'-p'')}{p'-p''} = -i\hbar\delta'(p'-p'')$$

which is defined as the derivative of the δ-function since the eigenvalues of p and q are continuous.

** A similar difficulty for the angle, φ, and angular momentum, L_z, has been pointed out 4).

ON THE UNCERTAINTY RELATION FOR L_z AND φ

D. JUDGE
University College, Dublin, and
Dublin Institute for Advanced Studies, Dublin

Received 6 June 1963

Because L_z, the z-component of orbital angular momentum, can be represented in wave mechanics by the differential operator $-i\hbar\,\partial/\partial\varphi$, it is easy to assume, by analogy with the (p_x, x) case, that L_z and φ obey the uncertainty relation

$$\Delta L_z\,\Delta\varphi \geqslant \tfrac{1}{2}\hbar\,. \qquad (1)$$

However, a moment's thought shows that this cannot be correct, since the maximum value of $\Delta\varphi$ is $\pi/\sqrt{3}$, corresponding to a uniform distribution in φ. Thus as ΔL_z tends to zero, the left-hand side of (1) must tend to zero, and a contradiction results. The problem is, then, how to modify (1) while maintaining the spirit of the uncertainty principle.

First it is necessary to define the uncertainties involved. We assume that $\psi^*\psi$ is periodic in φ with period 2π, and define ΔL_z in the obvious way, which is independent of the choice of initial line. An appropriate definition of $\Delta\varphi$, which is also independent of the choice of initial line, is the following: let

$$V(\gamma) = \int_{-\pi}^{\pi} \psi^*(\varphi+\gamma)\,\varphi^2\,\psi(\varphi+\gamma)\,d\varphi\,,$$

then $(\Delta\varphi)^2$ is the minimum value of $V(\gamma)$ for γ in $[-\pi, \pi]$.

Next, using the Schwartz inequality as in the usual derivation of the (p_x, x) uncertainty relation (e.g. ref. 1)), one can show that

$$(\Delta L_z)^2\,V(\gamma) \geqslant \tfrac{1}{4}\hbar^2\,[1 - 2\pi\psi^*(\gamma+\pi)\,\psi(\gamma+\pi)]^2\,.$$

From this, an elementary argument can be given which shows that

$$\Delta L_z\,\frac{\Delta\varphi}{1 - 3(\Delta\varphi)^2/\pi^2} \geqslant 0.15\,\hbar\,. \qquad (2)$$

The particular factor 0.15 arises merely because of the simple type of proof used. It seems very probable that, for an optimum result, this factor should be replaced by $\frac{1}{2}$; but a proof of this seems to need much more delicate methods than those which yield (2). In any case, a relation of the form (2) has the chief features one demands of a satisfactory replacement for (1). It shows that, in conformity with the uncertainty principle, as $\Delta\varphi$ tends to zero, ΔL_z tends to infinity, while as ΔL_z tends to zero, $\Delta\varphi$ tends to $\pi/\sqrt{3}$, corresponding to a uniform distribution of φ.

It should be pointed out that the reason why (1) is incorrect is that the operator $-i\hbar\,\partial/\partial\varphi$ is self-adjoint only when restricted to a certain domain, namely, the functions whose values are equal at both ends of the range of integration. If $\psi(\varphi)$ is such, then $\varphi\psi(\varphi)$ is not (excluding the case $\psi(\pi) = 0$); and it is here that the analogy with the (p_x, x) case breaks down.

A full account of this work will be given elsewhere.

I wish to express my gratitude to Dr. J. T. Lewis for pointing out the problem and for some valuable discussions, and to Dr. D. ter Haar for a useful suggestions.

1) L. Schiff, Quantum mechanics (McGraw-Hill, New York, 1949) p. 54.

ON THE COMMUTATOR $[L_z, \varphi]_-$

D. JUDGE * and J. T. LEWIS **
Dublin Institute for Advanced Studies, Dublin

Received 6 June 1963

If A, B are self-adjoint operators, then, as Pauli pointed out [1], an application of Schwarz's inequality yields

$$(A\psi,A\psi)\,(B\psi,B\psi) \geq \tfrac{1}{4}\,|(\psi,[A,B]_-\psi)|^2 . \quad (1)$$

It follows that if

$$[A,B]_- = -\,i\hbar I , \quad (2)$$

then for all ψ

$$\Delta A\,\Delta B \geq \tfrac{1}{2}\hbar . \quad (3)$$

We wish to point out that (2), and hence (3), fail to hold for the case $A = L_z$, $B = \varphi$. For, expanding ψ in a Fourier series

$$\psi = \sum_{-\infty}^{\infty} C_n\, e^{in\varphi} / \sqrt{2\pi} , \quad (4)$$

we find that the matrix representation of $[L_z,\varphi]_-$ has elements

$$([L_z,\varphi]_-)_{mn} = -\,i\hbar\,[\delta_{mn} - (-1)^{m-n}] . \quad (5)$$

Use of this commutator on the right-hand side of (1) leads to the uncertainty relation given in a previous paper by one of us [2].

One can obtain some insight into the meaning of the commutation relation (5) by noting that the matrix with elements given by (5) acts as $-\,i\hbar I$ (and so agrees with (2)) on the subset of $L_2(-\pi,\pi)$ consisting of those functions for which $\Sigma\,(-1)^n C_n = 0$, i.e., those which vanish at $+\pi$ and $-\pi$. This is only to be expected, since $-\,i\hbar\partial/\partial\varphi$ is self-adjoint with spectrum consisting of the points $m\hbar$, m integral, if and only if the domain is taken to be sufficiently smooth functions *** which satisfy $\psi(\pi) = \psi(-\pi)$. Thus if ψ is in this domain, $\varphi\psi$ is in it if and only if $\psi(\pi) = 0$.

It may be thought that the real source of the failure of (3) in this case is due in some way to having restricted φ to lie in a finite range. But in fact it is necessary to take as angle observable in quantum mechanics multiplication, not by φ, the usual angle of trogonometry which runs from $-\infty$ to ∞, but instead by

$$Y(\varphi) \equiv \varphi - 2\pi \sum_{n=0}^{\infty} H[\varphi - (2n+1)\pi] + 2\pi \sum_{n=0}^{\infty} H[-\varphi - (2n+1)\pi] , \quad (6)$$

where $H(x)$ is the unit step function. That is, $Y(\varphi)$ is just φ (mod 2π) in the range $[-\pi,\pi]$. We must clearly take the range of integration in the scalar product to be the range of this observable, i.e., $[-\pi,\pi]$. The usual representation of L_z by $-\,i\hbar\partial/\partial\varphi$ is still valid. Since $Y(\varphi)$ and any functions of it are periodic, they are in the domain of L_z as a self-adjoint operator. (It may be necessary to define the end points of the range of integration with some care as, e.g., $-\pi+$ and $\pi+$, because of the ambiguity of $H(x)$ at $x = 0$.)

From (6) we see directly that

$$[L_z, Y(\varphi)]_- = -\,i\hbar\,[1 - 2\pi \sum_{-\infty}^{\infty} \delta\{\varphi - (2n+1)\pi\}] . \quad (7)$$

This is the commutation relation for L_z and $Y(\varphi)$ in the $Y(\varphi)$ representation. It may be noted that this is just $i\hbar$ times the classical Poisson bracket of L_z and $Y(\varphi)$. (7) is completely equivalent to (5).

The difference between the classical and quantum variables φ and $Y(\varphi)$ arises from the fact that a classical trajectory, implying continuous observation, is normally an inadmissable concept in quantum mechanics. Of a classical particle, it is reasonable to ask "What angle was turned through by the radius vector as the particle moved from its original to its present position?" - this angle might be many multiples of 2π. In quantum mechanics, such a question is in general meaningless, though, of course, it must become meaningful in the classical limit.

* And University College, Dublin.

** On sabbatical leave of absence from Brasenose College, Oxford.

*** Those which are absolutely continuous on $[-\pi,\pi]$, and whose derivatives belong to $L_2(-\pi,\pi)$.

References

1) H. Weyl, The theory of groups and quantum mechanics (Dover, New York) p. 77.
2) D. Judge, Physics Letters 5 (1963) 191.

QUANTUM MECHANICAL PHASE AND TIME OPERATOR

LEONARD SUSSKIND*
Laboratory of Nuclear Studies, Cornell University, Ithaca, New York

and

JONATHAN GLOGOWER

(*Received* 13 *May* 1964)

Abstract

The phase operator for an oscillator is shown not to exist. It is replaced by a pair of non-commuting sin and cos operators which can be used to define uncertainty relations for phase and number. The relation between phase and angle operators is carefully discussed. The possibility of using a phase variable as a quantum clock is demonstrated and the states for which the clock is most accurate are constructed.

I. Introduction

We shall discuss quantities which are special cases of time operators, that is, variables conjugate to the energy. Bohm and Aharanov[1] have attempted to use such operators in their discussion of the energy-time uncertainty principle. They suppose that an operator exists which is an appropriate quantum variable describing the generalized clock. In fact, Bohm and Aharanov assert that such operators exist for all quantum systems. For example, they consider a clock composed of a single free particle, whose position measures the time. The suggested time operator is

$$t_c = \frac{1}{2}M(XP_x^{-1} + P_x^{-1}X)$$

The questions of existence, Hermiticity, eigenvectors, etc., were not considered.

What we shall do is consider an example of a generalized time operator, the phase of a quantum oscillator, and give answers to these questions for the system of an oscillator clock.

Another reason to study the phase of the quantum oscillator is its possible relevance to the quantum theory of coherence.

The following convention is used. A caret will be used above a quantity to denote an operator. A quantity without a caret is a c number. Also $\hat{\phi}$ will be used as an operator with θ its eigen-values.

II. Definition of Phase Variables

We are going to study the quantum oscillator described by the Hamiltonian

$$\hat{H} = \hat{P}^2 + \frac{1}{4}w^2\hat{X}^2 \tag{1}$$

with special emphasis on the time phase variable.

*National Science Foundation Predoctoral Fellow.

Classically the solution to this Hamiltonian is the well-known

$$X = Ae^{i\phi} + A^*e^{-i\phi}, \qquad \phi = wt \tag{2}$$

$$P = i/2[Awe^{i\phi} - A^*we^{-i\phi}] \tag{2'}$$

The A's may be chosen real

$$X = 2A \cos \phi \tag{3}$$

$$P = -Aw \sin \phi \tag{3'}$$

In quantum mechanics the equivalent equations are

$$\hat{X} = \hat{a}^+ + \hat{a}^- \tag{4}$$

$$\hat{P} = -i/2w(\hat{a}^+ - \hat{a}^-) \tag{4'}$$

where we think of $\hat{a}^+$ and $\hat{a}^-$ as the positive and negative frequency components of $\hat{x}$. In analogy with (2) and (2′) we will consider a^+ to be the product of 2 operators, one Hermitian representing the amplitude of oscillation and one complex representing the phase. Ideally we should like to be able to express $\hat{a}^+$ and $\hat{a}^-$ in the form $\hat{R}e^{+i\hat{\phi}}$ and $e^{-i\hat{\phi}}R$ where R is Hermitian and $e^{i\hat{\phi}}$ is a unitary operator defining a Hermitian $\hat{\phi}$. This is what Heitler [2] and Dirac [3] try to do. We shall find, however, that their arguments are not correct. Since the number operator, n, is given by $\hat{a}^+\hat{a}^- = \hat{R}e^{-i\hat{\phi}}e^{i\hat{\phi}}\hat{R}$, where $e^{i\hat{\phi}}$ is unitary, then

$$\hat{n} = \hat{R}^2 \tag{5}$$

Suppose R is Hermitian. Then its representation in some basis is

$$\hat{R} = \begin{pmatrix} R_1 & & & \\ & R_2 & & \\ & & \cdot & \\ & & & \cdot \end{pmatrix} \qquad \hat{R}^2 = \begin{pmatrix} R_1^2 & & & \\ & R_2^2 & & \\ & & \cdot & \\ & & & \cdot \end{pmatrix} \tag{6}$$

But n is diagonal in but one basis system. Hence the system in which R is diagonal is the number representation and therefore

$$\hat{R} = \sqrt{\hat{n}}, \qquad \hat{a}^+ = \sqrt{\hat{n}}\, e^{-i\hat{\phi}}, \qquad \hat{a}^- = e^{i\hat{\phi}}\sqrt{\hat{n}} \tag{7}$$

These equations can be used to obtain

$$[\hat{n}, e^{i\hat{\phi}}] = e^{i\hat{\phi}} \tag{8}$$

This plus the supposed unitarity of $e^{i\hat{\phi}}$ is used to obtain

$$[\hat{n}, \hat{\phi}] = -i, \qquad \dot{\hat{\phi}} = -\omega \tag{8'}$$

However, taking matrix elements of this commutation relation we get

$$(n_1 - n_2)\langle n_1|\hat{\phi}|n_2\rangle = -i\delta_{n_1 n_2} \tag{9}$$

This is obviously impossible since when $n_1 = n_2$ the left side is zero while the right side is -1. The difficulty is not a basic one and lies in not properly taking account of the periodic nature of $\hat{\phi}$. This will be discussed in detail with angle operators.

A more important objection is that $e^{i\hat{\phi}}$ defined above is not unitary and does not define an Hermitian $\hat{\phi}$. Consider

$$\hat{a}^+ = \sqrt{\hat{n}}\, e^{-i\hat{\phi}}$$

Taking matrix elements in the n representation

$$\delta_{n,m+1}\sqrt{m+1} = \sqrt{n} <n|e^{-i\hat{\phi}}|m>. \tag{10}$$

For $n = 0$ both sides are identically zero. For $n \neq 0$ we get $<n|e^{-i\hat{\phi}}|m> = \delta_{n,m+1}$. The arbitrariness of $<0|e^{-i\hat{\phi}}|m>$ is due to n having zero eigenvalues and therefore no inverse.

Let

$$<0|e^{-i\hat{\phi}}|m> = r_m$$

Then

$$e^{-i\hat{\phi}} = \begin{pmatrix} r_0 & r_1 & r_2 & \cdots \\ 1 & 0 & 0 & \cdots \\ 0 & 1 & 0 & \cdots \\ 0 & 0 & 1 & \cdots \\ \cdot & \cdot & \cdot & \cdots \end{pmatrix}, \qquad e^{i\hat{\phi}} = \begin{pmatrix} r_0^* & 1 & 0 & 0 & \cdots \\ r_1^* & 0 & 1 & 0 & \cdots \\ r_2^* & 0 & 0 & 1 & \cdots \\ \cdot & \cdot & \cdot & \cdot & \cdots \end{pmatrix}$$

$$e^{-i\hat{\phi}}e^{i\hat{\phi}} = \begin{pmatrix} \sum r_i r_i^* & r_0 & r_1 & \cdots \\ r_0^* & 1 & 0 & \cdots \\ r_1^* & 0 & 1 & \cdots \\ \cdot & \cdot & \cdot & \cdots \end{pmatrix} \tag{11}$$

This cannot equal unity for any choice of the r_i. Therefore we shall cease to denote these operators by $e^{i\hat{\phi}}$ but rather call them $\hat{e}^{i\phi}$, to indicate that ϕ is not an operator. We can restrict the choice of r_i by demanding $\hat{e}^{i\phi}\,\hat{e}^{-i\phi} = 1$ which gives $r_i = 0$. In fact by so restricting r_i we can replace our definition of $\hat{e}^{i\phi}$ by

$$\hat{e}^{-i\phi} = \hat{a}^+(\hat{n}+1)^{-1/2}, \qquad \hat{e}^{i\phi} = (\hat{n}+1)^{-1/2}\hat{a}^- \tag{12}$$

This definition gives the same operators as the previous one. In fact $\hat{e}^{i\phi}$ is just the lowering operator

$$\hat{e}^{i\phi}|m> = |m-1>, \qquad \hat{e}^{-i\phi}|m> = |m+1> \tag{13}$$

It must be kept in mind that $\hat{e}^{i\phi}$ and $\hat{e}^{-i\phi}$ are symbolic expressions and do not represent some exponential function of a Hermitian phase. We can use these operators, however, to define trigonometric functions of phase and investigate their properties especially in the classical limit. We shall show that these operators are observables and that they do become the classical functions of phase in the limit of large amplitudes. Furthermore, they provide a set of quantities for measuring time. Consider

$$\cos\hat{\phi} \equiv \tfrac{1}{2}[\hat{e}^{i\phi} + \hat{e}^{-i\phi}]$$
$$\sin\hat{\phi} \equiv \frac{1}{2i}[\hat{e}^{i\phi} - \hat{e}^{-i\phi}] \tag{14}$$

Clearly they are Hermitian operators. According to the principles of quantum theory they are observable dynamical variables if they possess a complete set of eigenvectors. Therefore we shall proceed to investigate their eigenvalue spectrum and eigenvectors.

Since

$$[\cos\hat{\phi}, \sin\hat{\phi}] = \begin{pmatrix} 1 & 0 & 0 & 0 & . \\ 0 & 0 & 0 & 0 & . \\ 0 & 0 & 0 & 0 & . \\ . & . & . & . & . \end{pmatrix} \neq 0$$

the eigenstates of $\cos\hat{\phi}$ and $\sin\hat{\phi}$ will be distinct. That $[\cos\phi, \sin\phi] \neq 0$ is not too surprising since one is essentially the time derivative of the other. In fact, it is the non-commuting nature of $\cos\phi$ and $\sin\phi$ which makes $\hat{e}^{i\phi}$ not unitary.

The non-commuting of $\cos\hat{\phi}$ and $\sin\hat{\phi}$ prevents us from constructing states of known ϕ. In fact, we shall show that eigenstates of $\cos\hat{\phi}$ can be considered superpositions of states with $\pm\phi$, and $\sin\phi$ states are superpositions of states with ϕ and $\pi - \phi$.

We now find the eigenvalues and eigenvectors of $\cos\hat{\phi}$.

Let $|\cos\theta\rangle = \sum_n C_n|n\rangle$. Remembering that $\hat{e}^{i\phi}$ is a lowering operator and $\hat{e}^{-i\phi}$ a raising operator we obtain

$$2\cos\hat{\phi}|\cos\theta\rangle = \sum_n [C_n|n+1\rangle + C_n|n-1\rangle] = 2\lambda\sum_n C_n|n\rangle \tag{15}$$

for the eigenvalue equation. Hence

$$C_1 = 2\lambda C_0, \qquad C_n + C_{n+2} = 2\lambda C_{n+1} \tag{16}$$

Apart from an over all multiplicative factor this recursion relation has only one solution for each value of λ. The general solution to the second equation is

$$C_n = Ap^n + Bp^{-n} \tag{17}$$

where $p + 1/p = 2\lambda$. Then if $|\cos\theta\rangle$ is to be a normalizable vector $|p|$ must equal 1. Let $p = e^{i\theta}$. Then $\lambda = \cos\theta$ and we find that $\cos\phi$ has the eigenvalue spectrum -1 to $+1$. Inserting this into equation (16) we get

$$C_n = \sin(n+1)\theta$$

The eigenstate of $\cos\hat{\phi}$ with eigenvalue $\cos\theta$ is

$$\sum_n \sin(n+1)\theta\,|n\rangle \tag{18}$$

Notice that for each value of $\cos\theta$ there is only one state and not two.

Now $\cos\theta$ is a superposition of $\sum_n e^{in\theta}|n\rangle$ and $\sum_n e^{-in\theta}|n\rangle$. These states are the states that one might think should be the eigenstates of the conjugate to the $\hat{n}$ operator in analogy with the usual situation for an operator and its conjugate. Thus it might be suspected that $\sum_n e^{in\theta}|n\rangle$ provides a good definition of the eigenstates of ϕ. Might we have done much better in our definitions of phase by considering an operator ϕ which multiplies each state $\sum_n e^{in\theta}|n\rangle$ by θ?

For such an operator to be an observable at all, the states

$$|\theta\rangle = \sum_n e^{in\theta}|n\rangle$$

must form an orthonormal complete set. Consider the inner product

$$\langle\theta|\theta'\rangle = \sum_0^\infty e^{in(\theta'-\theta)}$$

Were the sum from $-\infty$ to $+\infty$ this would equal $\delta(\theta' - \theta)$. But as it stands it is not the delta function and the states are not orthogonal [4]. Thus if the operator exists at all it is not Hermitian. Furthermore the superposition of states

$$\int d\theta\, e^{im\theta} \sum_n e^{in\theta}|n\rangle = 0 \tag{19}$$

if m is a positive integer. The states are not even all linearly independent. Hence ϕ was not defined in a consistent manner.

Before going on we shall directly prove the orthogonality and completeness of the $|\cos\theta\rangle$

$$\langle\cos\theta|\cos\theta'\rangle = \sum_n \sin(n+1)\theta \sin(n+1)\theta'$$

$$= \sum_n \{\cos(n+1)(\theta-\theta') - \cos(n+1)(\theta+\theta')\} = \delta(\theta-\theta') - \delta(\theta+\theta') \tag{20}$$

Now we have found that for each value of θ there exists only one state. That is $|\cos\theta\rangle = -|\cos(-\theta)\rangle$. We can put the $\cos\phi$ states in 1-to-1 correspondence with the angles only if we restrict θ to half its range, say from 0 to π. Then one of the delta functions is spurious. The states $|\cos\theta\rangle$ are then orthogonal.

The remaining question is whether the $\cos\theta$ states are complete. If we can expand an arbitrary $|n\rangle$ state in $|\cos\theta\rangle$ states then the $|\cos\theta\rangle$ states are indeed complete. Hence we try

$$|n\rangle = \sum_n \int_0^\pi f(\theta) \sin(m+1)\theta|m\rangle \tag{22}$$

Let $f(\theta) = \frac{2}{\pi}\sin(n+1)\theta$. Then

$$\sum_m \int_0^\pi [\sin(m+1)\theta][\sin(n+1)\theta]|m\rangle = \frac{\pi}{2}\sum_m \delta_{nm}|m\rangle = \frac{\pi}{2}|n\rangle \tag{23}$$

Summarizing the results for the $\cos\hat{\phi}$ operator, it is an observable with spectrum -1 to 1 and the eigenvalues are nondegenerate. Each $|\cos\theta\rangle$ state can be thought of as a superposition of $|+\theta\rangle$ and $|-\theta\rangle$ states. The sense in which the $|\theta\rangle$ states can be thought of as states of known phase will be described later.

Similar results can be obtained for the $\sin\hat{\phi}$ operator. The recursion relation for the eigenstates becomes

$$C_m - C_{m+2} = 2i\mu C_{m+1}, \qquad C_1 = -2i\mu C_0 \tag{24}$$

where $C_m = \langle m|\sin\theta\rangle$ where μ is the eigenvalue of $\sin\hat{\phi}$.

We find the non-increasing solutions to be

$$|\sin\theta\rangle = \sum_m \{e^{-i(m+1)\theta} - e^{-i(m+1)(\pi-\theta)}\}|m\rangle \tag{25}$$

Hence the state $|\sin\theta\rangle$ is the same as $|\sin(\pi-\theta)\rangle$.

To put these states into 1-to-1 correspondence with angles we use only those angles in the first and fourth quadrant. The proof that $\langle\sin\theta|\sin\theta'\rangle = \delta(\theta-\theta')$ and the completeness of the $|\sin\theta\rangle$ states follows in much the same way as the corresponding proof for the cosines.

Thus the observables $\sin\hat{\phi}$ and $\cos\hat{\phi}$ exist but don't commute or define an operator $\hat{\phi}$. In particular the equation $\sin^2\hat{\phi} + \cos^2\hat{\phi} = 1$ must be replaced by

$$\sin^2\hat{\phi} + \cos^2\hat{\phi} = 1 - i[\cos\hat{\phi}, \sin\hat{\phi}] \tag{26}$$

We would like to show that $\langle\cos\theta|\sin\theta'\rangle$ is large only if $\sin^2\theta' + \cos^2\theta = 1$. It turns out that

$$\langle\cos\theta|\sin\theta'\rangle = f(\theta-\theta') - f(\theta+\theta') + f(\theta-\theta'-\pi) - f(\theta+\theta'-\pi)$$

where $f(\theta) = \Sigma_n e^{in\theta}$ which is very sharply peaked at $\theta = 0$.[4] Thus if θ is in the first quadrant the function is peaked at $\theta' = +\theta$. If θ is in the second quadrant the function is peaked at $\theta' = \theta - \pi$ and $\pi - \theta$. As expected $\langle\cos\theta|\sin\theta'\rangle$ is peaked at $\sin^2\theta' + \cos^2\theta = 1$.

III. The Oscillator Clock

Since $[\cos\hat{\phi}, \sin\hat{\phi}] \neq 0$ and $\sin\hat{\phi}$ is essentially the time derivative of $\cos\hat{\phi}$ we can expect spreading of the wave packet in $\cos\theta$. That is to say if we prepare a state $|\cos\theta\rangle$ at time $t = 0$, at time $t = \frac{2n\pi}{\omega}$ the state is not $|\cos(\theta \pm \omega t)\rangle$ but some wave packet in cos space. If we define an operator $\cos(\hat{\phi}+\epsilon)$ by $2\cos(\hat{\phi}+\epsilon) = e^{i\epsilon}\hat{e}^{i\phi} + e^{-i\epsilon}\hat{e}^{-i\phi}$ which incidentally does not commute with $\cos\hat{\phi}$, we would find the system in an eigenstate of $\cos(\hat{\phi} - \omega t)$ at time t. In this sense there is no spreading. However for any fixed choice of ϵ, wave packets in cos space spread with time. We might think that this spreading would prevent us from using the oscillator as a clock since a clock parameter must increase in time with essentially no spread in the possible values of the time which is measured. However, we shall show later that the oscillator can be used as a clock which can measure time with arbitrary accuracy if it is prepared in the correct states.

We have not yet exhibited states with a small uncertainty in both $\cos\hat{\phi}$ and $\sin\hat{\phi}$ simultaneously. The existence of such states is crucial for the existence of a classical limit. It might be questioned whether such states do exist since every $|\cos\theta\rangle$ state is a superposition of $|\sin\theta\rangle$ and $|\sin -\theta\rangle$ states. We must show that for every angle θ there exist states such that $\langle\cos\hat{\phi}\rangle = \cos\theta$ and $\langle\sin\hat{\phi}\rangle = \sin\theta$. Since the states $|\theta\rangle = \Sigma_n e^{im\theta}|m\rangle$ are the states of $\hat{\phi}$ suggested by a naive theory, we might expect them to have the desired properties. Consider first the inner product

$$\begin{aligned}\langle\theta|\cos\theta'\rangle = &-\frac{i}{2}e^{i(\theta+\theta')}e^{i\theta'}\Sigma_n e^{im(\theta+\theta')}\\ &-\frac{i}{2}e^{i(\theta-\theta')}\end{aligned} \tag{27}$$

This function is so strongly peaked at $\theta = \pm\theta'$ that the uncertainty in $\cos\hat{\phi}$ is zero. To show this we calculate the expectation value $\langle\theta|\cos\hat{\phi}|\phi\rangle$.

$$\begin{aligned}\langle\theta|\cos\hat{\phi}|\theta\rangle &= \Sigma_{mn}\langle m|e^{-im\theta}\cos\hat{\phi}\, e^{in\theta}|n\rangle/\Sigma_{mn}\langle m|e^{i(n-m)\theta}|n\rangle\\ &= \frac{\Sigma_n \cos\theta - e^{i\theta/2}}{\Sigma_n 1} = \cos\theta\end{aligned} \tag{28}$$

The normalization swamps out the $e^{i\theta}$ term. Similarly $\langle\sin\hat{\phi}\rangle = \sin\theta$. The uncertainty in $\cos\phi$ can be defined as $\langle\cos\hat{\phi}\rangle^2 - \langle\cos^2\phi\rangle$.

$$\langle\cos^2\hat{\phi}\rangle = \frac{1}{4}\frac{[\Sigma_n(e^{2i\theta} + e^{-2i\theta} + 2)] + \text{finite terms}}{\Sigma_n 1}$$

$$= \cos^2\theta \tag{29}$$

since the normalization again swamps out the finite terms. Hence

$$\langle\cos^2\rangle - \langle\cos\rangle^2 = 0 \tag{30}$$

In fact, one can show

$$\langle(\cos\phi - \langle\cos\phi\rangle)^n\rangle = 0 \tag{31}$$

If an oscillator is in a state $|\theta\rangle$ and $\sin\hat{\phi}$ or $\cos\hat{\phi}$ are measured, it is overwhelmingly probably that the measurements will yield $\cos\theta$ and $\sin\theta$.

Thus, although the states $|\theta\rangle$ are not eigenstates of a Hermitian $\hat{\phi}$, they are states for which the uncertainties in the non-commuting operators $\sin\hat{\phi}$ and $\cos\hat{\phi}$ are zero!

We can now see to what extent an oscillator can be used as a clock. We start by preparing the oscillator in the state

$$\sum_m e^{im\theta}|m\rangle = |\theta\rangle$$

After time t the state will have become

$$|\theta - \omega t\rangle$$

Since the uncertainties in $\sin\hat{\phi}$ and $\cos\hat{\phi}$ are "zero" in such states, simultaneous measurements of both can be performed to any degree of accuracy. That is, given any $\delta > 0$ both $\sin\hat{\phi}$ and $\cos\hat{\phi}$ can be measured within accuracy δ. But from what was said before, it is *overwhelmingly* probably that the measurement will yield $\cos(\theta + \omega t)$ and $\sin(\theta + \omega t)$ within accuracy δ which uniquely tell us the value of t. And, in fact, any quantum oscillator can be used as an arbitrarily accurate clock if its $\cos\hat{\phi}$ and $\sin\hat{\phi}$ can be measured.

IV. Periodic Systems

Now why does the phase operator have such strange properties? Why can we not define the conjugate to the $\hat{n}$ operator? The reason is that the $\hat{n}$ operator does not have all integer eigenvalues but only positive values. To understand this better it is instructive to examine the conjugate to an operator with such a full integer spectrum from $-\infty$ to $+\infty$.

The system is the angle-angular momentum system which has many formal similarities with the phase-energy system. Consider a bead on a circular wire. The position of the bead will be denoted by θ, and the momentum conjugate to θ is L, the angular momentum. The allowable wave functions are functions of θ defined between $\theta = -\pi$ and $\theta = +\pi$. In order to handle the difficulties that arise at the endpoints of the interval the functions are usually continued past the ends periodically in θ. Hence we consider the space of periodic functions of θ, where θ can be considered the rotation parameter. Letting $\hat{L} = -i\frac{\partial}{\partial\theta}$ we get

$$-i\frac{\partial}{\partial\theta}f_m = mf_m \qquad \text{and} \qquad f_m(\theta) = e^{im\theta}$$

and as usual m must be an integer to insure the periodic character of $f_m(\theta)$.

The commutation relations for θ and $\hat{L}$ are usually taken to be

$$[\hat{L}, \hat{\theta}] = i$$

However, this is clearly inconsistent. Taking matrix elements

$$\begin{aligned} \langle m|\hat{L}\hat{\theta} - \hat{\theta}\hat{L}|n\rangle &= i\delta_{mn} \\ (m - n)\langle m|\hat{\theta}|n\rangle &= i\delta_{mn} \end{aligned} \tag{33}$$

which is impossible for $m = n$.

The difficulty lies in not treating the periodic nature of θ correctly. θ is taken to be the rotation parameter and as such takes on values from $-\infty$ to $+\infty$. Clearly θ operating on a periodic $f(\theta)$ gives a non-periodic function. Hence multiplication by θ is not a good operator in the space of periodic functions. To remedy this we introduce an operator $\hat{\phi}$ which is periodic in θ:

$$\hat{\phi} f(\theta) = \theta f(\theta), \qquad -\pi < \theta < \pi \tag{34}$$

and is periodic. The operator $\hat{\phi}$ is represented in Figure 1.

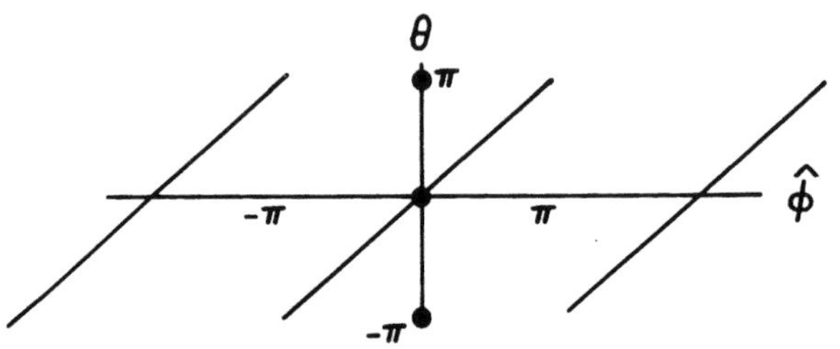

FIGURE 1

But $[\hat{L}, \hat{q}]$ is $i\dfrac{d\hat{q}}{d\theta}$ and not $i\dfrac{d\hat{q}}{d\hat{\phi}}$. In particular

$$[\hat{L}, \hat{\phi}] = i\frac{d\hat{\phi}}{d\theta} = i(1 - 2\pi\delta(\hat{\phi} - \pi)) \tag{35}$$

The choice of the delta function is somewhat arbitrary. It signifies that in a full 2π rotation the value of ϕ must somewhere jump through 2π. We conventionally choose this point to be $\phi = \pi$. Evaluating this commutator assuming $\langle m|\theta\rangle = \dfrac{1}{2\pi} e^{im\theta}$, where $|\theta\rangle$ is an eigenvector of $\hat{\phi}$, we obtain

$$\langle r|[\hat{L}, \hat{\phi}]|s\rangle = i\delta_{rs} - ie^{-i(r-s)\pi} \tag{36}$$

This is the general relation between a momentum and periodic coordinates and one finds that any argument based on the c number character of $[\hat{L}, \hat{\phi}]$ must be revised.

Let us suppose there exist a pair of hermitian operators $\hat{L}$ and $\hat{\phi}$ such that $[\hat{L}, \hat{\phi}] = i(1 - 2\pi\delta(\hat{\phi} - \pi))$, $\hat{L}$ with a discrete integer spectrum and $\hat{\phi}$ with a continuous spectrum from $-\pi$ to π. Can we infer anything about the spectrum $\hat{L}$? To answer this we introduce an operator $\hat{\mathscr{L}} = -i\dfrac{d}{d\phi}$ with the restriction that $\hat{\mathscr{L}} f(\theta) = -i\dfrac{df}{d\theta}$ if $f(\pi) = f(-\pi)$ and $\hat{\mathscr{L}} f(\theta) = -i\dfrac{df}{d\theta} - i[f(-\pi) - f(\pi)]\delta(\theta - \pi)$ otherwise. Now one can prove that $[\hat{\mathscr{L}}, \hat{\phi}] = [\hat{L}, \hat{\phi}]$ and that $[\hat{\mathscr{L}} - \hat{L}, \hat{\phi}] = 0$. This means that $\hat{L} = \hat{\mathscr{L}} + g(\hat{\phi})$ if $\hat{\phi}$ is to have a complete set of eigenvectors. The $g(\hat{\phi})$ can always be removed by unitary transformation, hence we lose no generality in assuming $\hat{L} = \hat{\mathscr{L}}$ [5]. But now the eigenstates of $\hat{\mathscr{L}}$ are clearly $\int d\theta e^{im\theta}|\theta>$ and its spectrum, m goes from $-\infty$ to $+\infty$. That is to say if $\hat{L}$ and $\hat{\phi}$ obey the commutation rules of conjugate variables it is possible to show that there exist functions of θ which are eigenfunctions of $\hat{L}$ with all integer eigenvalues. Clearly angular momentum has the required spectrum.

This sheds light on the phase operator. We see that since $\hat{n}$ goes from 0 to $+\infty$ it is impossible to find any phase operator that will be periodic in time and have the properties required by a conjugate of $\hat{n}$. The point is that because there are no negative energy eigenvalues the phase operator can't exist. We have proved a generalization of this theorem for arbitrary systems. A time operator cannot exist for any system with a lowest energy state.

V. Phase Difference of Two Oscillators

To further understand the nature of phase and its relation to interferences, consider the $\cos(\hat{\phi}_1 - \hat{\phi}_2)$ $\sin(\hat{\phi}_1 - \hat{\phi}_2)$ operators for a pair of oscillators. We define them in a manner similar to the $\cos\hat{\phi}$ and $\sin\hat{\phi}$ for a single oscillator. Suppose $\hat{e}^{i\phi_1}$, $\hat{e}^{-i\phi_1}$, $\hat{e}^{i\phi_2}$, $\hat{e}^{-i\phi_2}$ are the exponential phase variables for the separate 1 and 2 oscillators. Then define $\hat{e}^{i(\phi_1-\phi_2)} = \hat{e}^{i\phi_1}\hat{e}^{-i\phi_2}$ (remember that they commute), $\hat{e}^{-i(\phi_1-\phi_2)} = \hat{e}^{i\phi_2}\hat{e}^{-i\phi_1}$

$$
\begin{aligned}
2\cos(\hat{\phi}_1 - \hat{\phi}_2) &= \hat{e}^{i(\phi_1-\phi_2)} + \hat{e}^{-i(\phi_1-\phi_2)} \\
2i\sin(\hat{\phi}_1 - \hat{\phi}_2) &= \hat{e}^{i(\phi_1-\phi_2)} - \hat{e}^{-i(\phi_1-\phi_2)}
\end{aligned} \tag{37}
$$

The eigenvalue spectrum of $\cos(\hat{\phi}_1 - \hat{\phi}_2)$ and $\sin(\hat{\phi}_1 - \hat{\phi}_2)$ has some interesting properties. Let $|\psi>$ be an eigenstate of $\cos(\hat{\phi}_1 - \hat{\phi}_2)$.

$$\cos(\hat{\phi}_1 - \phi_2)|\psi> = \lambda|\psi> \tag{38}$$

Let $|\psi> = \sum_{mn} a_{mn}|m>|n)$ where $|m>$ is a number eigenstate of oscillator 1 and $|n)$, a state of oscillator 2. Then

$$
\begin{aligned}
2\cos(\hat{\phi}_1 - \hat{\phi}_2)|\psi> &= \sum_{m,n=1}^{\infty} (a_{m-1,n+1} + a_{m+1,n-1})|m>|n) \\
&\quad + \sum_{m=1}^{\infty} a_{m-1,1}|m>|0) + \sum_{n=1}^{\infty} a_{1,n-1}|0>|n) \\
&= \lambda \sum_{mn} a_{mn}|m>|n>
\end{aligned} \tag{39}
$$

which gives

$$
\begin{aligned}
a_{m-1,n+1} + a_{m+1,n-1} &= \lambda a_{mn} \\
a_{m-1,1} &= \lambda a_{m,0} \\
a_{1,n-1} &= \lambda a_{0,n}
\end{aligned} \tag{40}
$$

Notice that each term of each equation is an amplitude for a total number of excitations $n + m$. This means that we may separately write the recursion relations involving a total of R excitations. Let R be the total number of excitations and m, the excitation in oscillator 1. Denote the amplitude $a_{n,R-m}$ by R_m. Then

$$R_{m-1} + R_{m+1} = \lambda R_m \qquad \text{for} \qquad m \neq 0, R$$
$$R_{R-1} = \lambda R_R \tag{41}$$
$$R_1 = \lambda R_0$$

Each eigenstate of $\cos(\hat{\phi}_1 - \hat{\phi}_2)$ determined by these equations is an eigenstate of $m + n$. This is simply a consequence of the commutation relation

$$[\cos(\hat{\phi}_1 - \hat{\phi}_2), \hat{n}_1 + \hat{n}_2] = 0$$

As we have seen, the general solution to the first of equations (41) is $R_n = a p^n + b p^{-n}$ where $\lambda = p + \frac{1}{p}$. The second 2 relations determine the allowable values of p. The values of p are plotted on a complex plane in Figure 2. Note that the state with $p = 1/p'$ is the same as the state with $p = p'$. Hence only 1/2 the

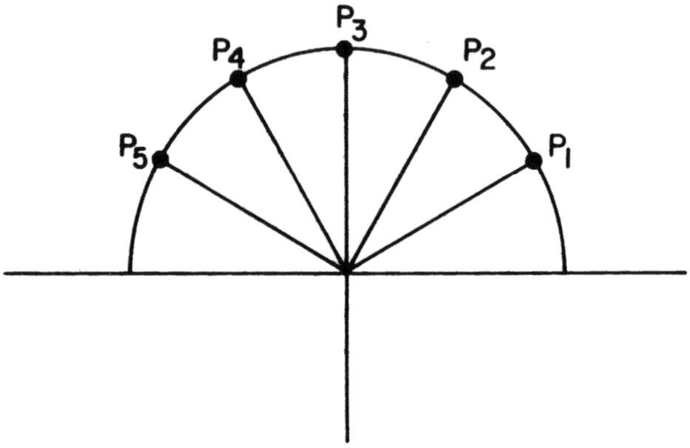

FIGURE 2

plane need be used. We shall use the top half. Also, the states with $p = \pm 1$ do not exist. Altogether then there are $R + 1$ orthonormal eigenstates of $\cos(\hat{\phi}_1 - \hat{\phi}_2)$, all of which are eigenstates of $\hat{m} + \hat{n}$. But there are $R + 1$ linearly independent eigenstates of $\hat{m} + \hat{n}$ and hence the eigenstates of $\cos(\hat{\phi}_1 - \hat{\phi}_2)$ form a complete set for the Hilbert space of R excitations. Considering the total of all eigenstates of $\cos(\hat{\phi}_1 - \hat{\phi}_2)$ of this type, they form a complete set and $\cos(\hat{\phi}_1 - \hat{\phi}_2)$ can have no other eigenstates or eigenvalues.

The eigenvalues of $\cos(\hat{\phi}_1 - \hat{\phi}_2)$[6] do not include all the points from -1 to $+1$ but only a countable infinity of them, namely $\cos\theta$, where θ is a rational multiple of π. We do not expect the phase difference variables to have many significant analogies with classical phase variables for states of low excitation since the eigenvalue spectrum is very unlike the classically allowed values of $\cos(\theta_1 - \theta_2)$ and $\sin(\theta_1 - \theta_2)$. However, let us look at states with a large number of quanta for which the cos, sin spectrum is dense.

Consider the "states with phase difference θ"

$$|R, \theta\rangle = \sum_0^R e^{im\theta}|m\rangle|R - m\rangle$$

and the amplitudes

$$<\cos\,\theta',\,R|R,\,\theta> \;=\; e^{i\theta'}\sum_{0}^{R} e^{in(\theta-\theta')} + e^{-i\theta'}\sum_{0}^{R} e^{in(\theta-\theta')} \tag{42}$$

Now if R is large, this function is very strongly peaked at $\theta = \theta'$ and $\theta = -\theta'$ or equivalently $\cos\theta = \cos\theta'$. Similarly for $\sin\hat{\phi}$.

The expectation value of $\cos\hat{\phi}$ in state $|R,\,\theta>$ is $\dfrac{R}{R+1}\cos\theta$ and

$$<\cos\hat{\phi}>^2 = \frac{R^2}{(R+1)^2}\cos^2\theta \tag{43}$$

Also

$$<\cos^2\hat{\phi}> = \frac{R-1}{R+1}\cos^2\theta \tag{44}$$

and the uncertainty in $\cos\hat{\phi}$ is

$$<\cos^2> - <\cos>^2 = \frac{1-2R}{(R+1)^2}\cos^2\theta \tag{45}$$

which goes as $1/R$ for large R. This justifies our use of $|R,\theta>$ as states of well known phase difference when R is large.

VI. Uncertainty Relations

The primary reason for Bohm and Aharonov's investigation of time operators was to demonstrate the energy-time uncertainty relations. We have shown that an oscillator's phase can be used as a clock variable, but we have not investigated limitations on the clock's accuracy due to the uncertainty principle. What is a good measure of the uncertainty in the phase? We should like to find some measure of probability for phase which reflects the probability distribution for its measurable trigonometric functions, and use this to define the uncertainty relations. Now the phase must be measured by measuring its trigonometric functions. Suppose our apparatus is designed to measure the quantity $\cos(\hat{\phi}+\epsilon)$. Denote the probability that the measurement yield $\cos(\hat{\phi}+\epsilon) = \cos(\theta+\epsilon)$ by $P[\cos(\hat{\phi}+\epsilon) = \cos(\theta+\epsilon)]$. Then classically

$$P[\cos(\hat{\phi}+\epsilon) = \cos(\theta+\epsilon)] = P[\hat{\phi} = \theta] + P[\hat{\phi} = -\theta - 2\epsilon] \tag{46}$$

where $P[\hat{\phi} = \theta]$ is the probability that $\hat{\phi}$ has value θ.

Another way of saying the same thing is that if the original state is displaced through phase angle ϵ then the probability, P_ϵ, that in the new state $\cos\hat{\phi} = \cos(\theta+\epsilon)$ is the probability, P_0, that in the old state $\hat{\phi} = \theta$ plus the probability that in the old state $\hat{\phi} = -\theta - 2\epsilon$.

$$P_\epsilon[\cos\hat{\phi} = \cos(\theta+\epsilon)] = P_0(\hat{\phi} = \theta) + P_0(\hat{\phi} = -\theta - 2\epsilon) \tag{47}$$

Now quantum mechanically a measurement of $\cos\hat{\phi}$ does not distinguish between $+\phi$ and $-\phi$. Because of this we must replace probabilities by corresponding probability amplitudes.

$$e^{i\epsilon}A_\epsilon[\cos\hat{\phi} = \cos(\theta+\epsilon)] = A_0(\hat{\phi} = \theta) + e^{i\alpha}A_0(\hat{\phi} = -\theta - 2\epsilon) \tag{48}$$

where $e^{i\epsilon}$ compensates for the rotation through angle ϵ and $e^{i\alpha}$ is an arbitrary undetermined phase. This is the equation we will use to define $A_0(\hat{\phi} = \theta)$. This definition is not trivial because for different values of ϵ the measurements of $\cos\hat{\phi}$ are not compatible and it is not clear, *a priori*, that a function $A_0(\hat{\phi} = \theta)$ can be found which obeys (48) independently of the value of ϵ.

Suppose now that equation (48) does define an amplitude A_0 for the phase taking on various values. Then we can be sure that for any value of ϵ the projections of a state onto an eigenstate of $\cos(\hat{\phi} + \epsilon)$ with eigenvalue $\cos(\theta + \epsilon)$ will be large only if either $A_0(\hat{\phi} = \theta)$ or $A_0(\hat{\phi} = -\theta - 2\epsilon)$ is large. Furthermore, suppose $A_0(\hat{\phi} = \theta)$ is large. For any given ϵ, $A\epsilon(\cos\hat{\phi} = \cos(\theta + \epsilon))$ may not be large. But we can always find some value of ϵ for which it will be large. For example, suppose $A_0(\hat{\phi} = \theta')$ is small. Then let $\epsilon = -(\theta - \theta')/2$. Thus if $\cos(\hat{\phi} + \epsilon)$ is measured in a state in which it is probable that it equals a given value, then one of the two angles which could yield this value must be in the region where $A(\hat{\phi} = \theta)$ is large. Also, for some method of measuring the phase (measuring $\cos(\hat{\phi} + \epsilon)$ the probability will be large that $\cos(\theta + \epsilon)$ is obtained for every angle in this region. We conclude that the width of an A function obeying (48) is a true measure of the uncertainty in phase, taking into account both the ordinary quantum mechanical uncertainty in measuring the trigonometric functions of phase and the incompatibility of the different methods of measuring it. Now it remains to construct a suitable A.

Consider any state $|\psi\rangle$ which can be expanded in the form

$$\sum_n \int d\bar{\theta}\, \rho(\bar{\theta})\, e^{im\bar{\theta}} |m\rangle$$

such that the Fourier expansion of $\rho(\bar{\theta})$ contains only exponentials with negative integers multiplying $\bar{\theta}$.

$$\rho(\bar{\theta}) = \sum_n a_m e^{-im\bar{\theta}} \tag{49}$$

To find $A\epsilon[\cos\hat{\phi} = \cos(\theta + \epsilon)]$ we rotate $|\psi\rangle$ through angle ϵ and take its projection on cos states

$$A\epsilon[\cos\hat{\phi} = \cos(\theta + \epsilon)] = \langle\cos(\theta + \epsilon)| \sum_m \int d\bar{\theta}\, \rho(\bar{\theta})\, e^{im\bar{\theta}} e^{im\epsilon} |m\rangle$$

$$= e^{i(\theta+\epsilon)} \sum_m e^{im(\theta+2\epsilon)} a_m - e^{-i(\theta+\epsilon)} \sum_m e^{-im\theta} a_m \tag{50}$$

Now since ρ has Fourier projections only on negative integer exponentials we can write this as

$$e^{i(\theta+\epsilon)} \rho(-\theta - 2\epsilon) - e^{-i(\theta+\epsilon)} \rho(\theta) = e^{-i\epsilon}[e^{-i(-\theta-2\epsilon)} \rho(-\theta - 2\epsilon) - e^{-i\theta} \rho(\theta)) \tag{51}$$

But this is the same form as equation (48) with $A(\hat{\phi} = \theta) = e^{-i\theta} \rho(\theta)$. The factor $e^{-i\theta}$ can be absorbed into the relative phase factor $e^{i\alpha}$. Actually the particular choice of phase can be made plausible by a self-consistency argument but, since our only use of A will be to define an uncertainty for ϕ, the question of the relative phase in unimportant. Thus for such states ρ can be used for A.

To show that any state can be expanded in the manner described, we make use of a special property of the $|\theta\rangle$ states. Although they are overcomplete they resolve the identity in the same way that a basis does.

$$\int |\theta\rangle\langle\theta|\, d\theta = I \tag{52}$$

Any state can be expanded in $|\theta\rangle$.

$$|\psi\rangle = \int d\theta \langle\theta|\psi\rangle |\theta\rangle \tag{53}$$

The over completeness of the $|\theta\rangle$ means that $|\psi\rangle$ may be expanded with many weighting factors other than $\langle\theta|\psi\rangle$ but only $\langle\theta|\psi\rangle$ has projections only on the negative integer exponents.

$$<\theta|\psi> = \sum_m e^{-im\theta}<m|\psi> \tag{54}$$

Also, if

$$\int [<\theta|\psi> + f(\theta)]|\theta> = |\psi>,$$

then

$$\int f(\theta)|\theta> = 0.$$

But then

$$\sum_m \int d\theta f(\theta) e^{im\theta}|m> = 0$$

The $|m>$ are linearly independent so that

$$\int d\theta f(\theta) e^{im\theta} = 0.$$

And f is totally composed of positive integer exponentials. Hence for any state $|\psi>$ we may simply choose A to be $<\theta|\psi>$.

The uncertainty relations become statements about the widths of functions

$$\rho(\theta) = \sum_0^\infty e^{-im\theta} a_m$$

with a_m being spread over Δ_m. Such functions have the property that their width, $\Delta\theta$ times Δ_m, is greater than 1 when $\Delta\theta$ becomes small. As the spread in m becomes small, $\Delta\theta$ must increase until $\rho(\theta)$ is spread over 2π. More exact statements could be derived, but our purpose here is only to show such uncertainty relations can be defined by the rigorous use of well defined quantum operator trigonometric functions of phase.

References

1. D. BOHM and Y. AHARONOV, *Phys. Rev.* **122**, 1649 (1961).
2. W. HEITLER, *Quantum Theory of Radiation* Chap. II. Oxford University Press (1954).
3. P. DIRAC, *Quantum Theory of Emission and Absorption in Quantum Electrodynamics* (Edited by J. SCHWINGER). Dover Publications, New York (1958).
4. Actually this function is $\delta(\theta \cdot \theta') + i \cot \frac{1}{2}(\theta - \theta')$.
5. This argument is the same as the one used by Dirac for momentum and position. See P. DIRAC, *The Principles of Quantum Mechanics* (4th ed.) Chap. 4. Oxford University Press (1958).
6. $\cos\hat{\phi}$ and θ refer to the phase difference of the two oscillators for the rest of this section.

States of the quantized radiation field

THE quantization of the radiation field is accomplished, as in the preceding chapter, by the association of a quantum-mechanical harmonic oscillator with each field mode. It is natural in this derivation to work with the states $|n_{\mathbf{k}}\rangle$ in which the cavity mode **k** contains an excitation of $n_{\mathbf{k}}$ photons. The total-field states $|\{n_{\mathbf{k}}\}\rangle$ defined in eqn (6.102) form a convenient complete set for the cavity electromagnetic field. There is, however, no reason to suppose that the electromagnetic field for real beams of light can be described by the assumption that a particular one of the states $|\{n_{\mathbf{k}}\}\rangle$ is excited. A more general state of the system involves some linear superposition of the basic states $|\{n_{\mathbf{k}}\}\rangle$. It is found that the light beams of physical interest correspond to such linear superpositions; the preparation, in practice, of a cavity state where the number of photons in mode **k** has a definite eigenvalue $n_{\mathbf{k}}$ is a formidable experimental problem, although it is possible in principle.

The task of the present chapter is to examine various types of field state which can be envisaged. Some parallels can be drawn between the kinds of light beam which occur in classical and quantum-mechanical theory. For example, it was shown in Chapter 5 that there are two completely different types of classical light beam. For the classical stable wave illustrated in Fig. 5.10, it was possible to specify the electric field of the light wave as having a definite magnitude, given in eqn (5.39), at each position and time. By contrast, the electric field of the light beam from a chaotic source could only be specified in terms of a certain probability that the field would have a given amplitude and phase at some observation position and time.

There is a similar distinction in quantum mechanics between two varieties of light beam which can be set up. A beam which has the photons in some definite state, expressible as a linear combination of the $|\{n_{\mathbf{k}}\}\rangle$, is said to be a *pure state* of the radiation field. However, if all that can be specified is a set of probabilities that the photons will be found in several states, corresponding to different linear combinations of the $|\{n_{\mathbf{k}}\}\rangle$, then the state of the system is called a *statistical mixture*. It will be shown that the classical stable wave and the chaotic beam are classical limits of photon states which in quantum-mechanics belong, respectively, to the categories of pure states and statistical mixtures.

We begin the chapter with a consideration of the pure states; for most of the derivations it is sufficient to consider just a single mode of the radiation field. Statistical distributions are introduced into quantum mechanics by means of the density operator, whose general properties are briefly discussed before consideration of its application to statistical mixtures of the radiation field.

It will be seen that quantization introduces characteristic quantum-mechanical effects into the properties of the electromagnetic field. For example, it is not possible to envisage a wave whose amplitude and phase are simultaneously precisely defined. As is common in quantum mechanics, accuracy of measurement is limited by uncertainty relations.

The photon phase operators

For the earlier part of the chapter we restrict attention to a particular mode $\mathbf{k}$ of the electromagnetic field, having specified polarization. With this understanding, the subscript $\mathbf{k}$ can be dropped from the variables which describe the mode under consideration, and the field quantities $\mathbf{A}$, $\mathbf{E}$, and $\mathbf{H}$ can be written as scalars.

In the classical theory of light waves, it is convenient to write the complex wave amplitude as a product of a real amplitude and a phase factor, as in eqn (5.17). If this procedure is followed for the classical vector potential of eqn (6.46), we can put

$$A_{\mathbf{k}} = A_0 \exp(\mathrm{i}\phi), \tag{7.1}$$

and the single-mode contribution is

$$A = A_0\{\exp(-\mathrm{i}\omega t + \mathrm{i}\mathbf{k}.\mathbf{r} + \mathrm{i}\phi) + \exp(\mathrm{i}\omega t - \mathrm{i}\mathbf{k}.\mathbf{r} - \mathrm{i}\phi)\}. \tag{7.2}$$

It is similarly convenient in quantum mechanics to make a separation into amplitude and phase factors, analogous to eqn (7.1). To do this, it is necessary to introduce the concept of phase into the quantum-mechanical description of the field.

The single-mode quantum-mechanical vector-potential operator obtained from eqn (6.105) is

$$\hat{A} = (\hbar/2\epsilon_0 V\omega)^{\frac{1}{2}}\{\hat{a}\exp(-\mathrm{i}\omega t + \mathrm{i}\mathbf{k}.\mathbf{r}) + \hat{a}^{\dagger}\exp(\mathrm{i}\omega t - \mathrm{i}\mathbf{k}.\mathbf{r})\}. \tag{7.3}$$

The analogue of eqn (7.1) is thus a separation of $\hat{a}$ into a product of amplitude and phase operators. There is, in fact, no exact prescription for the way in which the separation should be accomplished in quantum mechanics and there is a corresponding degree of arbitrariness in the definition of the quantum-mechanical phase operator. The main considerations are that the quantum-mechanical phase should have the same significance as the classical phase in the appropriate limit, and that the phase should be associated with Hermitian operators so that it is (at any rate in principle) an observable quantity.

Consider the phase operator $\hat{\phi}$ defined by the relation[1]

$$\hat{a} = (\hat{n}+1)^{\frac{1}{2}} \exp(i\hat{\phi}), \tag{7.4}$$

where $\hat{n}$ is the number operator defined in eqn (6.64). The Hermitian conjugate relation is

$$\hat{a}^{\dagger} = \exp(-i\hat{\phi})(\hat{n}+1)^{\frac{1}{2}}. \tag{7.5}$$

The operator $\hat{\phi}$ defined in this way will be adopted as the quantum-mechanical phase of the electromagnetic field, analogous to the classical phase, whose properties it will later be shown to approach in a suitable limit.

The basic properties of the phase operator can be calculated from the known properties of the creation, destruction, and number operators. According to eqns (7.4) and (7.5)

$$\exp(i\hat{\phi}) = (\hat{n}+1)^{-\frac{1}{2}}\hat{a}. \tag{7.6}$$

and

$$\exp(-i\hat{\phi}) = \hat{a}^{\dagger}(\hat{n}+1)^{-\frac{1}{2}}. \tag{7.7}$$

Since, according to eqns (6.62) and (6.64),

$$\hat{a}\hat{a}^{\dagger} = \hat{n}+1, \tag{7.8}$$

it follows from eqns (7.6) and (7.7) that

$$\exp(i\hat{\phi})\exp(-i\hat{\phi}) = 1. \tag{7.9}$$

Note, however, that the reverse-order product of the exponential operators is not equal to unity.

The results of applying $\exp(i\hat{\phi})$ and $\exp(-i\hat{\phi})$ to the states $|n\rangle$ can readily be calculated with the help of eqns (6.81), (6.86), and (6.87),

$$\begin{aligned} \exp(i\hat{\phi})\,|n\rangle = (\hat{n}+1)^{-\frac{1}{2}}n^{\frac{1}{2}}\,|n-1\rangle &= |n-1\rangle \quad \text{for } n \neq 0 \\ &= 0 \quad \text{for } n = 0, \end{aligned} \tag{7.10}$$

$$\exp(-i\hat{\phi})\,|n\rangle = \hat{a}^{\dagger}(n+1)^{-\frac{1}{2}}\,|n\rangle = |n+1\rangle. \tag{7.11}$$

The two exponential phase operators thus have non-vanishing matrix elements

$$\langle n-1|\exp(i\hat{\phi})\,|n\rangle = 1 \tag{7.12}$$

and

$$\langle n+1|\exp(-i\hat{\phi})\,|n\rangle = 1, \tag{7.13}$$

and all other types of matrix element are zero. These results are similar to the matrix elements of eqns (6.88) and (6.89) of the destruction and creation operators except that eqns (7.12) and (7.13) do not include the normalization factors of these latter matrix elements.

The above equations display the formal properties of the phase operators which arise in the decompositions of $\hat{a}$ and $\hat{a}^{\dagger}$ into amplitude and phase contributions. It is seen from eqns (7.12) and (7.13) that $\exp(\mathrm{i}\hat{\phi})$ and $\exp(-\mathrm{i}\hat{\phi})$ do not satisfy the relation (6.90); therefore they are not Hermitian operators, and they cannot represent observable properties of the electromagnetic field. They can, however, be combined to produce another pair of operators

$$\cos\hat{\phi} = \tfrac{1}{2}\{\exp(\mathrm{i}\hat{\phi})+\exp(-\mathrm{i}\hat{\phi})\} \tag{7.14}$$

and

$$\sin\hat{\phi} = \frac{1}{2\mathrm{i}}\{\exp(\mathrm{i}\hat{\phi})-\exp(-\mathrm{i}\hat{\phi})\}, \tag{7.15}$$

whose non-vanishing matrix elements are

$$\langle n-1|\cos\hat{\phi}\,|n\rangle = \langle n|\cos\hat{\phi}\,|n-1\rangle = \tfrac{1}{2} \tag{7.16}$$

and

$$\langle n-1|\sin\hat{\phi}\,|n\rangle = -\langle n|\sin\hat{\phi}\,|n-1\rangle = 1/2\mathrm{i}. \tag{7.17}$$

These matrix elements do satisfy eqn (6.90) and the operators $\cos\hat{\phi}$ and $\sin\hat{\phi}$ are Hermitian. We adopt them as the quantum-mechanical operators which represent the observable phase properties of the electromagnetic field.

Problem 7.1. Prove the commutation relation

$$[\cos\hat{\phi}, \sin\hat{\phi}] = \{\hat{a}^{\dagger}(\hat{n}+1)^{-1}\hat{a}-1\}/2\mathrm{i}, \tag{7.18}$$

and hence show that all matrix elements of the commutator are zero except for the diagonal ground-state matrix element

$$\langle 0|\,[\cos\hat{\phi}, \sin\hat{\phi}]\,|0\rangle = -1/2\mathrm{i}. \tag{7.19}$$

Problem 7.2. Prove the commutation relations

$$[\hat{n}, \hat{a}] = -\hat{a} \tag{7.20}$$

and

$$[\hat{n}, \hat{a}^{\dagger}] = \hat{a}^{\dagger} \tag{7.21}$$

and hence show that

$$[\hat{n}, \exp(\mathrm{i}\hat{\phi})] = -\exp(\mathrm{i}\hat{\phi}) \tag{7.22}$$

and

$$[\hat{n}, \exp(-\mathrm{i}\hat{\phi})] = \exp(-\mathrm{i}\hat{\phi}), \tag{7.23}$$

and that

$$[\hat{n}, \cos\hat{\phi}] = -\mathrm{i}\sin\hat{\phi} \tag{7.24}$$

and

$$[\hat{n}, \sin\hat{\phi}] = \mathrm{i}\cos\hat{\phi}. \tag{7.25}$$

The above commutation relations show that the number and phase operators do not commute and it is therefore not possible, in principle, to set up states of the radiation field which are simultaneous eigenstates of two of the operators. The amplitude of an electromagnetic wave, associated with $\hat{n}$, and the phase, associated with $\cos\hat{\phi}$ or $\sin\hat{\phi}$, cannot both be precisely specified. The results of measurements of the amplitude and phase are governed by uncertainty relations derived from eqns (7.24) and (7.25) in the usual way[2,3]

$$\Delta n\,\Delta\cos\phi \geqslant \tfrac{1}{2}|\langle\sin\phi\rangle| \tag{7.26}$$

$$\Delta n\,\Delta\sin\phi \geqslant \tfrac{1}{2}|\langle\cos\phi\rangle|. \tag{7.27}$$

Here Δ indicates the root-mean-square deviation of a series of measurements of the quantity which follows it, for a state of the field in which $\cos\hat{\phi}$ and $\sin\hat{\phi}$ have the respective expectation values $\langle\cos\phi\rangle$ and $\langle\sin\phi\rangle$. Some examples of the operation of these photon number-phase uncertainty relations are given later in the chapter.

We have therefore shown that the separations (eqns (7.4) and (7.5)) of the creation and destruction operators into amplitude and phase contributions, analogous to the classical result (7.1), lead to phase operators which do not commute with the amplitude operator. The resulting uncertainty relations are a characteristic feature of the quantized radiation field.

States of well-defined photon phase

The simultaneous harmonic-oscillator eigenstates $|n\rangle$ of the energy and of the number operator $\hat{n}$ have already been discussed. We now consider the nature of the states $|\phi\rangle$ which are eigenstates of the phase operators.

We have introduced two operators, $\cos\hat{\phi}$ and $\sin\hat{\phi}$, to represent the phase properties of the radiation field. In the usual concept of an electromagnetic wave, the phase is a single quantity and it seems unnecessary to represent it by two different operators in quantum mechanics. It would be possible to choose one of the phase operators and drop the other at this point in the calculations. On the other hand, the two operators form a symmetrical pair and little additional work is caused by the retention of both operators.

A striking feature of the phase operators is their failure to commute, according to eqn (7.18). It follows that it is impossible to form states which are simultaneous eigenstates of $\cos\hat{\phi}$ and $\sin\hat{\phi}$. However, owing to the circumstance that only one of the infinity of matrix elements of the commutator is non-vanishing, as given by eqn (7.19), it proves to be possible to form states which are simultaneous eigenstates of $\cos\hat{\phi}$ and $\sin\hat{\phi}$ in a certain limiting sense.

Consider the state $|\phi\rangle$ defined by

$$|\phi\rangle = \operatorname*{Lt}_{s\to\infty}(s+1)^{-\frac{1}{2}}\sum_{n=0}^{s}\exp(in\phi)\,|n\rangle. \tag{7.28}$$

The state is a linear superposition of all the number eigenstates $|n\rangle$ each weighted by a phase factor $\exp(in\phi)$. The normalization and orthogonality of the $|n\rangle$ ensure that $|\phi\rangle$ is normalized,

$$\langle\phi \mid \phi\rangle = 1. \tag{7.29}$$

Note that each energy eigenstate $|n\rangle$ has the same amplitude $(s+1)^{-\frac{1}{2}}$, which tends to zero as s tends to infinity, in the summation of eqn (7.28).

With the help of eqns (7.10), (7.11), and (7.14),

$$\begin{aligned}\cos\hat{\phi}\,|\phi\rangle &= \tfrac{1}{2}\operatorname*{Lt}_{s\to\infty}(s+1)^{-\frac{1}{2}}\left(\sum_{n=1}^{s}\exp(in\phi)\,|n-1\rangle + \sum_{n=0}^{s}\exp(in\phi)\,|n+1\rangle\right)\\ &= \cos\phi\,|\phi\rangle + \tfrac{1}{2}\operatorname*{Lt}_{s\to\infty}(s+1)^{-\frac{1}{2}}\times\\ &\quad\times\left[\exp(is\phi)\,|s+1\rangle - \exp\{i(s+1)\phi\}\,|s\rangle - \exp(-i\phi)\,|0\rangle\right].\end{aligned} \tag{7.30}$$

The state $|\phi\rangle$ thus fails to be a strict eigenstate of $\cos\hat{\phi}$ because of the contributions of the terms in the large bracket of eqn (7.30). However, the magnitudes of these contributions tend to zero in the limit $s\to\infty$, and eqn (7.30) then takes on the aspect of an eigenvalue equation.

The diagonal matrix element of $\cos\hat{\phi}$ derived from eqn (7.30) is

$$\langle\phi|\cos\hat{\phi}\,|\phi\rangle = \cos\phi\left\{1-\operatorname*{Lt}_{s\to\infty}(s+1)^{-1}\right\} = \cos\phi. \tag{7.31}$$

It can be shown by a similar procedure that

$$\langle\phi|\sin\hat{\phi}\,|\phi\rangle = \sin\phi\left\{1-\operatorname*{Lt}_{s\to\infty}(s+1)^{-1}\right\} = \sin\phi. \tag{7.32}$$

The matrix elements of the squares of the phase operators can also be easily calculated and, in the $s\to\infty$ limit, give the results

$$\langle\phi|\cos^2\hat{\phi}\,|\phi\rangle = \cos^2\phi, \qquad \langle\phi|\sin^2\hat{\phi}\,|\phi\rangle = \sin^2\phi. \tag{7.33}$$

These results show that the state $|\phi\rangle$ defined in eqn (7.28) behaves for most computations like a simultaneous eigenstate of $\cos\hat{\phi}$ and $\sin\hat{\phi}$, with ϕ as the observable phase angle. In a strict interpretation of the mathematics, it would be more correct to say that $|\phi\rangle$ is not a rigorous eigenstate of the phase operators, but is one in which, according to eqns (7.31), (7.32), and (7.33), the uncertainties in the measured values of $\cos\hat{\phi}$ and $\sin\hat{\phi}$ are simultaneously zero.[4] The non-existence of strict simultaneous eigenstates of $\cos\hat{\phi}$ and $\sin\hat{\phi}$ is a consequence of the non-vanishing of the commutator of eqn (7.18).

Problem 7.3. Consider the overlap $\langle\theta \mid \phi\rangle$ for two different phase states $|\theta\rangle$ and $|\phi\rangle$ defined as in eqn (7.28). Prove that $\langle\theta \mid \phi\rangle$ tends to zero in the $s\to\infty$ limit when $\theta \neq \phi$.

Physical properties of the single-mode number states

The number state $|n\rangle$ and the phase state $|\phi\rangle$ represent excitations of the electromagnetic field in a single mode of an optical cavity. We now determine the physical nature of such excitations and compare them with the types of excitation envisaged in classical electromagnetic theory.

Consider first the single-mode state where exactly n photons are excited. For such a state $|n\rangle$, the uncertainty in the number operator is zero,

$$\Delta n = 0. \qquad (7.34)$$

For the phase operators, use of eqns (7.14) and (7.15) gives

$$\langle n|\cos\hat{\phi}\,|n\rangle = \langle n|\sin\hat{\phi}\,|n\rangle = 0 \qquad (7.35)$$

and

$$\langle n|\cos^2\hat{\phi}\,|n\rangle = \langle n|\sin^2\hat{\phi}\,|n\rangle = \tfrac{1}{2}\text{ for } n \neq 0$$
$$= \tfrac{1}{4}\text{ for } n = 0. \qquad (7.36)$$

Thus, if the $n = 0$ state is excluded from consideration, the phase uncertainties are

$$\Delta\cos\phi = \Delta\sin\phi = 2^{-\frac{1}{2}}. \qquad (7.37)$$

Note that the results (7.34), (7.35), and (7.37) are consistent with the uncertainty relations (7.26) and (7.27).

The above results show that physically, the electromagnetic wave which corresponds to state $|n\rangle$ has a definite amplitude, but the phase angle ϕ determined by eqns (7.35) and (7.36) is equally likely to have any random value between 0 and 2π.

These properties can be illustrated pictorially in terms of the electric field distribution associated with the excitation $|n\rangle$. The single-mode electric field operator from eqn (6.106) is

$$\hat{E} = \mathrm{i}(\hbar\omega/2\epsilon_0 V)^{\frac{1}{2}}\{\hat{a}\exp(-\mathrm{i}\omega t+\mathrm{i}\mathbf{k}.\mathbf{r})-\hat{a}^\dagger\exp(\mathrm{i}\omega t-\mathrm{i}\mathbf{k}.\mathbf{r})\}. \qquad (7.38)$$

The expectation values of $\hat{E}$ and $\hat{E}^2$ for state $|n\rangle$ are

$$\langle n|\,\hat{E}\,|n\rangle = 0 \qquad (7.39)$$

and

$$\langle n|\,\hat{E}^2\,|n\rangle = (\hbar\omega/\epsilon_0 V)(n+\tfrac{1}{2}), \qquad (7.40)$$

in agreement with eqn (6.111). The root-mean-square deviation of the electric field is thus

$$\Delta E = (\hbar\omega/\epsilon_0 V)^{\frac{1}{2}}(n+\tfrac{1}{2})^{\frac{1}{2}}. \qquad (7.41)$$

Fig. 7.1 shows a pictorial representation of the single-mode excitation for state $|n\rangle$. The vertical axis represents the electric field at some fixed point in the cavity as a function of time. The field oscillates like a sine wave of known frequency ω. The above analysis shows that the electromagnetic

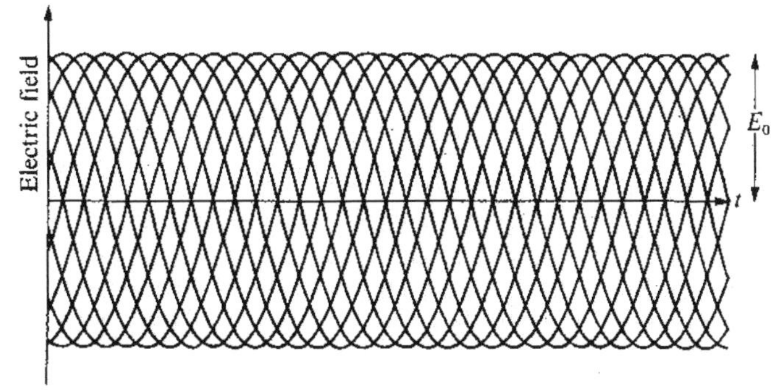

FIG. 7.1. Pictorial representation of the variation of electric field with time at a fixed point in a cavity mode excited to state $|n\rangle$. The sine waves should more accurately form a horizontal continuum. The amplitude E_0 is defined in the text.

wave amplitude can be represented by a quantity

$$E_0 = (2\hbar\omega/\epsilon_0 V)^{\frac{1}{2}}(n+\tfrac{1}{2})^{\frac{1}{2}}, \tag{7.42}$$

whose dispersion can be shown to be small for $n \gg 1$. However, the lateral position of the wave along the horizontal axis is completely undetermined owing to the complete uncertainty in the phase angle. This is indicated in the figure by the inclusion of several waves, all of the same amplitude and frequency, but with their nodes progressively displaced along the axis. In fact, the possible horizontal positions of the wave form a continuum, and the field at any time can take a continuous range of values between $-E_0$ and E_0.

Any pictorial representation of a quantum-mechanical state must be treated with caution. However, it can be seen that Fig. 7.1 is a correct representation of state $|n\rangle$, at least to the extent that it reproduces the various uncertainties shown in eqns (7.34), (7.37), and (7.41), and also ensures the vanishing of the expectation values of eqns (7.35) and (7.39).

Physical properties of the single-mode phase states

Now consider the single-mode state where a wave of precisely defined phase ϕ is excited. For such a state $|\phi\rangle$, discussed earlier in the chapter, the uncertainties in the phase operators are zero in the limit $s \to \infty$,

$$\Delta \cos \phi = \Delta \sin \phi = 0. \tag{7.43}$$

However, the number of photons present in the cavity mode is now uncertain. From the definition (7.28) of $|\phi\rangle$ in terms of the $|n\rangle$,

$$\langle\phi|\, \hat{n}\, |\phi\rangle = \underset{s\to\infty}{\mathrm{Lt}}\, (s+1)^{-1} \sum_{n=0}^{s} n = \underset{s\to\infty}{\mathrm{Lt}}\, \tfrac{1}{2}s. \tag{7.44}$$

and

$$\langle\phi|\,\hat{n}^2\,|\phi\rangle = \mathop{\mathrm{Lt}}_{s\to\infty}(s+1)^{-1}\sum_{n=0}^{s} n^2 = \mathop{\mathrm{Lt}}_{s\to\infty}\tfrac{1}{6}s(2s+1). \tag{7.45}$$

The photon-number expectation values are thus infinite and so also is the number uncertainty Δn. The ratio of this uncertainty to the mean number of photons is, however, finite,

$$\Delta n/\langle\phi|\,\hat{n}\,|\phi\rangle = 3^{-\frac{1}{2}}. \tag{7.46}$$

The left-hand sides of the uncertainty relations (7.26) and (7.27) are found to be infinite for the state $|\phi\rangle$, if the $s \to \infty$ limit is taken after the products of uncertainties are formed.

The electric-field expectation-value derived with the help of eqn (7.38) is

$$\langle\phi|\,\hat{E}\,|\phi\rangle = -2(\hbar\omega/2\epsilon_0 V)^{\frac{1}{2}}\sin(\mathbf{k}.\mathbf{r}-\omega t+\phi)\mathop{\mathrm{Lt}}_{s\to\infty}(s+1)^{-1}\sum_{n=0}^{s}(n+1)^{\frac{1}{2}}. \tag{7.47}$$

The summation diverges as $s^{\frac{1}{2}}$ for large s and leads to a divergent electric-field expectation-value. The uncertainty in electric field is also found to be infinite. However, the electric field oscillations have a definite phase ϕ, as expected for the excitation $|\phi\rangle$.

Fig. 7.2 is a pictorial representation of the single-mode excitation $|\phi\rangle$. The vertical axis again represents the electric field at some fixed point in

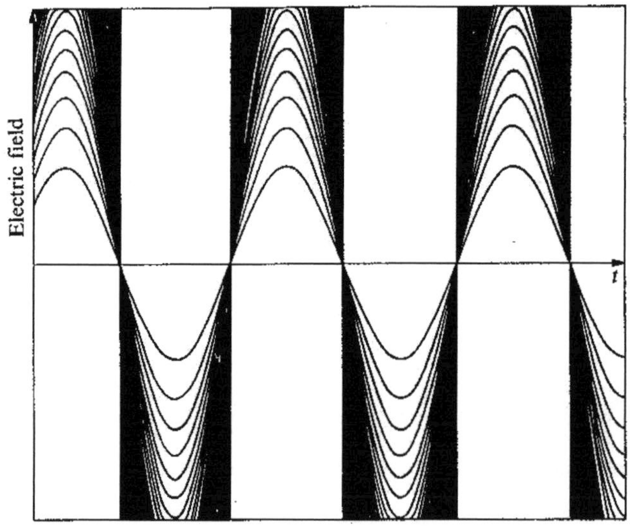

FIG. 7.2. Pictorial representation of the variation of electric field with time at a fixed point in a cavity mode excited to state $|\phi\rangle$. The net effect of adding together the contributions shown is an electric field whose expectation value changes abruptly between $+\infty$ and $-\infty$ every half cycle.

the cavity. The wave is a superposition of an infinite number of waves of different amplitude corresponding to all the different values of n. However, each subsidiary wave has the same frequency ω and phase ϕ so that the nodes of the composite wave occur at well-defined positions along the horizontal axis. The number of photons in the cavity mode is completely undetermined, in accordance with eqns (7.44) and (7.46), and the wave is entirely smeared out in the vertical direction.

The phase state $|\phi\rangle$ is something of an oddity. Eqn (7.44) shows that the mean energy of excitation of the state must be infinite, and a cavity mode could not be excited to state $|\phi\rangle$ in a practical experiment. The reason for considering state $|\phi\rangle$ is its role as one extreme case of the related uncertainties in photon number and phase. Figs. 7.1 and 7.2 illustrate the opposite extremes in the ranges of values for Δn and $\Delta \cos \phi$, or equivalently $\Delta \sin \phi$; the choice of zero uncertainty for one of the parameters causes a complete uncertainty in the other. The situation is analogous to that which results from the better-known position-momentum uncertainty relation in the quantum-mechanics of particles, where the eigenstates of the position and momentum operators correspond, respectively, to complete uncertainties in the particle momentum and position.

The coherent photon states

Neither of the single-mode states $|n\rangle$ or $|\phi\rangle$ illustrated in Figs. 7.1 and 7.2 bears any resemblance to the classical time-dependence expected for the electromagnetic wave of stable amplitude and fixed phase shown in Fig. 5.10. It is of interest to examine the states of the quantized radiation field which have properties similar to those of a classical electromagnetic wave in the limit of large amplitudes. Such states will be denoted $|\alpha\rangle$, and they are called coherent states of the radiation field[5] for reasons which will become apparent when the quantum theory of coherence is discussed in Chapter 9. The coherent states $|\alpha\rangle$ are important, not only because they are the closest quantum-mechanical approach to a classical electromagnetic wave, but also because the laser (discussed in Chapter 10) generates a coherent-state excitation of a cavity mode when operated well above threshold.

The coherent state $|\alpha\rangle$ is intermediate between $|n\rangle$ and $|\phi\rangle$ in the sense that neither the amplitude nor the phase of the state is precisely defined, but both quantities exhibit an uncertainty spread with a modest root-mean-square deviation. In particle mechanics, the quantum state most similar to a classical particle is a wave packet which has an analogous uncertainty spread in both its position and momentum.

Consider the properties of the state $|\alpha\rangle$ defined as the following linear superposition of the number states[5]

$$|\alpha\rangle = \exp(-\tfrac{1}{2}|\alpha|^2)\sum_n \frac{\alpha^n}{(n!)^{\frac{1}{2}}}|n\rangle. \tag{7.48}$$

In this expression α can be any complex number, and the coherent states so defined form a double continuum corresponding to the continuous ranges of values of the real and imaginary parts of α. It is easily verified that $|\alpha\rangle$ is normalized,

$$\langle\alpha\mid\alpha\rangle = \exp(-|\alpha|^2)\sum_n \frac{\alpha^{*n}\alpha^n}{n!} = 1. \tag{7.49}$$

The states are not however orthogonal, since for two different complex numbers α and β,

$$\begin{aligned}\langle\alpha\mid\beta\rangle &= \exp(-\tfrac{1}{2}|\alpha|^2-\tfrac{1}{2}|\beta|^2)\sum_n \frac{\alpha^{*n}\beta^n}{n!}\\ &= \exp(-\tfrac{1}{2}|\alpha|^2-\tfrac{1}{2}|\beta|^2+\alpha^*\beta).\end{aligned} \tag{7.50}$$

Thus

$$|\langle\alpha\mid\beta\rangle|^2 = \exp(-|\alpha-\beta|^2). \tag{7.51}$$

It is apparent from the definition eqn (7.48) that there are many more coherent states $|\alpha\rangle$ than there are states $|n\rangle$. The $|\alpha\rangle$ form an overcomplete set of states for the harmonic oscillator, and their lack of orthogonality is a consequence of this. Note, however, from eqn (7.51) that the states $|\alpha\rangle$ and $|\beta\rangle$ become approximately orthogonal if $|\alpha-\beta|$ is much greater than unity.

The coherent states are eigenstates of the destruction operator, since

$$\hat{a}\,|\alpha\rangle = \exp(-\tfrac{1}{2}|\alpha|^2)\sum_n \frac{\alpha^n}{(n!)^{\frac{1}{2}}}n^{\frac{1}{2}}|n-1\rangle = \alpha\,|\alpha\rangle. \tag{7.52}$$

The complex number α which labels the coherent state is thus the eigenvalue of the destruction operator. Note, however, that the state $|\alpha\rangle$ is not an eigenstate of the creation operator since the summation over n analogous to that in eqn (7.52) cannot be rearranged to reproduce the coherent state in the case of $\hat{a}^\dagger|\alpha\rangle$. An alternative approach to the coherent state is to take the eigenvalue eqn (7.52) as its definition; the expansion of eqn (7.48) can then be derived as one of the consequences.[5]

The relation (6.91) can be used to rewrite the definition (7.48) of the coherent state as

$$|\alpha\rangle = \exp(-\tfrac{1}{2}|\alpha|^2)\sum_n \frac{(\alpha\hat{a}^\dagger)^n}{n!}|0\rangle = \exp(\alpha\hat{a}^\dagger-\tfrac{1}{2}|\alpha|^2)\,|0\rangle. \tag{7.53}$$

Problem 7.4. Prove that the creation operator $\hat{a}^\dagger$ has no normalizable eigenstates.

Problem 7.5. The result

$$\exp(\hat{c})\exp(\hat{d}) = \exp(\hat{c}+\hat{d}+\tfrac{1}{2}[\hat{c},\hat{d}]) \tag{7.54}$$

can be proved[6] for any pair of operators $\hat{c}$ and $\hat{d}$ which commute with their commutator

$$[\hat{c},[\hat{c},\hat{d}]] = [\hat{d},[\hat{c},\hat{d}]] = 0. \tag{7.55}$$

With the help of this expression, cast eqn (7.53) into the compact form

$$|\alpha\rangle = \exp(\alpha\hat{a}^\dagger - \alpha^*\hat{a})\,|0\rangle. \tag{7.56}$$

The exponential operator in this equation is equivalent to a creation operator for the coherent state.

It is sometimes convenient to write α in terms of an amplitude and a phase. We take θ as a notation for the phase angle, arg α, and write

$$\alpha = |\alpha|\,e^{i\theta}. \tag{7.57}$$

Physical properties of the single-mode coherent states

The properties of a cavity mode excited to a coherent state $|\alpha\rangle$ can be determined by the methods applied earlier in the chapter to the number and phase states $|n\rangle$ and $|\phi\rangle$. The expectation values for the number operator are

$$\langle\alpha|\,\hat{n}\,|\alpha\rangle = \exp(-|\alpha|^2)\sum_n \frac{(\alpha^*\alpha)^n}{n!}\,n = |\alpha|^2, \tag{7.58}$$

$$\begin{aligned}\langle\alpha|\,\hat{n}^2\,|\alpha\rangle &= \exp(-|\alpha|^2)\sum_n \frac{(\alpha^*\alpha)^n}{n!}\,n^2 \\ &= \exp(-|\alpha|^2)\sum_n \frac{|\alpha|^{2n}}{n!}\{n(n-1)+n\} = |\alpha|^4+|\alpha|^2,\end{aligned} \tag{7.59}$$

and the root-mean-square deviation is therefore

$$\Delta n = |\alpha|. \tag{7.60}$$

The fractional uncertainty in the number of photons in the cavity mode is

$$\Delta n/\langle\alpha|\,\hat{n}\,|\alpha\rangle = |\alpha|^{-1}. \tag{7.61}$$

These results show that $|\alpha|^2$ is the mean number of photons in the cavity mode, and that the uncertainty spread about the mean is equal to the square root of the mean number of photons. The fractional uncertainty in photon number given by eqn (7.61) is proportional to the inverse square root of the mean photon number, and decreases with increasing excitation of the cavity mode. This result can be compared with the zero uncertainty in photon number (eqn (7.34)) for the state $|n\rangle$, and the constant fractional uncertainty (eqn (7.46)) for the state $|\phi\rangle$.

The definition (7.48) shows that the probability of finding n photons in the mode is

$$|\langle n \mid \alpha\rangle|^2 = \exp(-|\alpha|^2)\frac{|\alpha|^{2n}}{n!}. \tag{7.62}$$

This is a Poisson probability distribution (see reference 7 and Fig. 9.4) about a mean $|\alpha|^2$, and eqn (7.60) is the usual result for the spread of such a distribution.

The expectation values of the phase operators are given by somewhat complicated expressions for the coherent states.[8] We restrict attention to the cosine operator, and find

$$\begin{aligned}\langle\alpha|\cos\hat{\phi}\,|\alpha\rangle &= \tfrac{1}{2}\exp(-|\alpha|^2)\sum_n \frac{\alpha^{*n+1}\alpha^n+\alpha^{*n}\alpha^{n+1}}{\{(n+1)!\,n!\}^{\frac{1}{2}}}\\ &= |\alpha|\cos\theta\exp(-|\alpha|^2)\sum_n \frac{|\alpha|^{2n}}{n!\,(n+1)^{\frac{1}{2}}},\end{aligned} \tag{7.63}$$

where eqns (7.12), (7.13), (7.48), and (7.57) have been used. The expectation value of $\cos\hat{\phi}$ is thus proportional to the cosine of the phase of α. Similarly,

$$\begin{aligned}\langle\alpha|\cos^2\hat{\phi}\,|\alpha\rangle &= \tfrac{1}{2}-\tfrac{1}{4}\exp(-|\alpha|^2)\\ &\quad+|\alpha|^2(\cos^2\theta-\tfrac{1}{2})\exp(-|\alpha|^2)\sum_n \frac{|\alpha|^{2n}}{n!\,\{(n+1)(n+2)\}^{\frac{1}{2}}}.\end{aligned} \tag{7.64}$$

It is unfortunately not possible to evaluate the summations in eqns (7.63) and (7.64) analytically. There are, however, some simplifications in the limit where the mean number of photons in the mode is much larger than unity. The following asymptotic expansions,[8] which we quote without proof, can then be used,

$$\sum_n \frac{|\alpha|^{2n}}{n!\,(n+1)^{\frac{1}{2}}} = \frac{\exp(|\alpha|^2)}{|\alpha|}\left(1-\frac{1}{8|\alpha|^2}+\dots\right), \quad |\alpha|^2 \gg 1, \tag{7.65}$$

$$\sum_n \frac{|\alpha|^{2n}}{n!\,\{(n+1)(n+2)\}^{\frac{1}{2}}} = \frac{\exp(|\alpha|^2)}{|\alpha|^2}\left(1-\frac{1}{2|\alpha|^2}-\dots\right), \quad |\alpha|^2 \gg 1. \tag{7.66}$$

Thus for large mean numbers of photons, the phase expectation values become

$$\langle\alpha|\cos\hat{\phi}\,|\alpha\rangle = \cos\theta\left(1-\frac{1}{8|\alpha|^2}+\dots\right) \quad |\alpha|^2 \gg 1, \tag{7.67}$$

$$\langle\alpha|\cos^2\hat{\phi}\,|\alpha\rangle = \cos^2\theta-\frac{\cos^2\theta-\frac{1}{2}}{2|\alpha|^2}-\dots, \quad |\alpha|^2 \gg 1, \tag{7.68}$$

and the phase uncertainty is therefore

$$\Delta \cos \phi = \sin \theta/2|\alpha|, \qquad |\alpha|^2 \gg 1. \tag{7.69}$$

From eqns (7.60) and (7.69), the product of uncertainties is

$$\Delta n \, \Delta \cos \phi = \tfrac{1}{2} \sin \theta, \qquad |\alpha|^2 \gg 1. \tag{7.70}$$

It is possible to derive a result analogous to eqn (7.67) which shows that the expectation value of sin $\hat{\phi}$ is equal to sin θ for large $|\alpha|^2$. Thus for a large mean photon-number, the coherent state $|\alpha\rangle$ has the minimum uncertainty product allowed by the general relation (7.26). It is seen from eqns (7.61) and (7.69) that the fractional uncertainty in photon number and the phase uncertainty both vary like $|\alpha|^{-1}$; as the mean photon-number is increased, the wave becomes correspondingly better defined both in amplitude and in phase angle.

These properties can be seen more clearly by evaluation of the expectation value of the electric field in the coherent state. With the help of eqn (7.38)

$$\begin{aligned}\langle\alpha| \hat{E} |\alpha\rangle &= \mathrm{i}(\hbar\omega/2\epsilon_0 V)^{\frac{1}{2}}\{\alpha \exp(-\mathrm{i}\omega t+\mathrm{i}\mathbf{k}.\mathbf{r})-\alpha^* \exp(\mathrm{i}\omega t-\mathrm{i}\mathbf{k}.\mathbf{r})\} \\ &= -2(\hbar\omega/2\epsilon_0 V)^{\frac{1}{2}} |\alpha| \sin(\mathbf{k}.\mathbf{r}-\omega t+\theta),\end{aligned} \tag{7.71}$$

where we have used

$$\langle\alpha| \hat{a}^\dagger |\alpha\rangle = \alpha^*. \tag{7.72}$$

Similarly,

$$\langle\alpha| \hat{E}^2 |\alpha\rangle = (\hbar\omega/2\epsilon_0 V)\{4 |\alpha|^2 \sin^2(\mathbf{k}.\mathbf{r}-\omega t+\theta)+1\}, \tag{7.73}$$

and the root-mean-square deviation in electric field is therefore

$$\Delta E = (\hbar\omega/2\epsilon_0 V)^{\frac{1}{2}}. \tag{7.74}$$

Note that the results of the present paragraph apply irrespective of the amplitude of the wave, unlike the previous two paragraphs where large-amplitude approximations were made.

The time dependence of the electric field at a fixed point in the cavity is illustrated in Fig. 7.3 for coherent states $|\alpha\rangle$ all having the same phase angle, but three different values of the mean photon number. Since the uncertainty ΔE given by eqn (7.74) is independent of the wave amplitude, the definition of the waves improves with increasing $|\alpha|^2$. It is clear from the figure that the electric-field variation resembles that of a classical stable wave more and more closely as the mean number of photons in the coherent state is increased (compare Fig. 5.10).

It is possible to show that the coherent state $|\alpha\rangle$ is the only type of quantum-mechanical state of the electromagnetic field which can be directly related to the classical stable electromagnetic wave (see reference 9 for a detailed discussion of the correspondence). The coherent state does, of course, exhibit the characteristic quantum-mechanical uncertainty properties discussed above, but these effects all become relatively unimportant in the classical limit, when $|\alpha|^2$ is much larger than unity. Other types of cavity-mode excitation, for example the number states $|n\rangle$, do not tend towards the classical form when the photon number is much larger than unity.

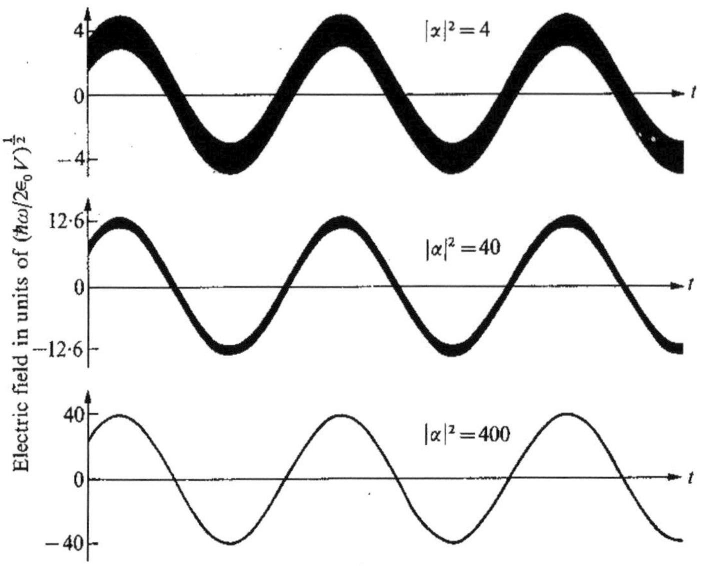

FIG. 7.3. Pictorial representation of the variation of electric field with time at a fixed point in a cavity mode excited to state $|\alpha\rangle$. Three different values of the mean photon number $|\alpha|^2$ are illustrated, the vertical scale being different for the three cases. The uncertainty spreads in field values are indicated by the vertical widths $2\Delta E$ of the sine waves. These spreads can also be regarded as due to a combination of the amplitude uncertainty associated with Δn and the phase uncertainty associated with $\Delta \cos\phi$.

The classical electric-field variation obtained from the vector potential A in eqn (7.2) is

$$E = -2\omega A_0 \sin(\mathbf{k}.\mathbf{r}-\omega t+\phi). \tag{7.75}$$

By comparison with eqn (7.71), it is seen that in the classical limit, the coherent state $|\alpha\rangle$ corresponds to a classical wave whose phase angle is equal to the phase of α, and whose electric field amplitude is related to $|\alpha|$ by

$$-2\omega A_0 = -2(\hbar\omega/2\epsilon_0 V)^{\frac{1}{2}}\,|\alpha|. \tag{7.76}$$

Despite the apparent uncertainty in the various observable characteristics of the coherent state, it will be seen from a consideration of the quantum mechanics of the detection processes for light beams in Chapter 9 that these states have perfect coherence properties.

The density operator

All the calculations so far in the chapter have been concerned with pure states of the radiation field in a single mode of an optical cavity, that is, states expressible as some definite linear combination of the basic number states $|n\rangle$. No difficulty is presented by the generalization to pure states of the complete radiation field in the cavity; the state of the total electromagnetic field is formed by a product of the states of the individual modes in the usual way, and any calculation for the multimode field reduces to a series of calculations for the single-mode fields taken one at a time.

A more radical generalization of the above theory is required to treat statistical-mixture states of the radiation field, where there is no precise specification of the state of the field, but only of the probabilities that the field will be observed to be in a range of possible states. It is convenient to proceed to a consideration of the statistical-mixture states before any discussion of the multimode states of the field. The pure state is a special case of the statistical mixture and both types of state can be treated within a common theoretical framework.

Let us consider a cavity electromagnetic field for which there is a known probability P_R that the field is in state $|R\rangle$. Here R is a label which runs over a set of pure states sufficient to describe the field. For a single cavity mode the states $|R\rangle$ could be the number states $|n\rangle$, the phase states $|\phi\rangle$ or the coherent states $|\alpha\rangle$ or they could be some other type of pure state. For the complete cavity field, the states $|R\rangle$ would be all possible products of the single-mode states, with one state for each mode of the cavity included in each of the basic states $|R\rangle$. The states given by eqn (6.102) are a possible set $|R\rangle$, based on the single-mode number states. The state described by the probability P_R is a statistical mixture; the magnitudes of the P_R for a given set of pure states $|R\rangle$ contain all the available information about the state.

The chaotic light beam is an example of a field which must be treated as a statistical mixture in quantum mechanics. It is evident from the classical treatment in Chapter 5 that the nature of a chaotic light source excludes the possibility of any definite prediction of the state of the emitted field. A probabilistic description is all that is possible in both classical and quantum-mechanical theories of chaotic light.

Another example of a statistical mixture is provided by the thermal excitation of photons in a cavity mode, derived in Chapter 1 as a step on the way to Planck's law. The most detailed description which can be given in this case is the probability P_n in eqn (1.42) that n photons are excited at temperature T. The distribution P_n is an example of the general probability P_R introduced above.

The result of an experiment which can be carried out on a beam of light generally depends on an ensemble average of some observable quantity.

For example, the fluctuation in photon number derived in eqn (1.71) requires a knowledge of the ensemble average of n^2, and the fringe intensity in eqn (5.44) requires the evaluation of an ensemble average of a product of electric fields. Consider some observable O which is represented by a quantum-mechanical operator $\hat{O}$. The average value of the observable for the pure state $|R\rangle$ is $\langle R|\hat{O}|R\rangle$, and hence the ensemble average of the observable for the statistical mixture specified by P_R is

$$\langle O\rangle = \sum_R P_R \langle R|\hat{O}|R\rangle. \tag{7.77}$$

It will be assumed that P_R is a normalized probability distribution

$$\sum_R P_R = 1. \tag{7.78}$$

Eqn (7.77) is the basic expression which enables useful predictions to be made for known statistical mixtures of the radiation field. It is, however, convenient to cast eqn (7.77) into a different form, which is usually easier to work with and leads to more elegant expressions for the ensemble averages. Let $|S\rangle$ represent some complete set of states for the field considered, S being a label which can take a series of values. Then according to the closure theorem of eqn (4.100)

$$\sum_S |S\rangle\langle S| = 1. \tag{7.79}$$

Insertion of this unit quantity immediately after the operator $\hat{O}$ in eqn (7.77) gives

$$\begin{aligned}\langle O\rangle &= \sum_R P_R \sum_S \langle R|\hat{O}|S\rangle\langle S|R\rangle \\ &= \sum_R \sum_S P_R \langle S|R\rangle\langle R|\hat{O}|S\rangle.\end{aligned} \tag{7.80}$$

The density operator $\hat{\rho}$ is defined to be[10]

$$\hat{\rho} = \sum_R P_R |R\rangle\langle R|, \tag{7.81}$$

and the average value of O from eqn (7.80) can be written

$$\langle O\rangle = \sum_S \langle S|\hat{\rho}\hat{O}|S\rangle. \tag{7.82}$$

The density operator contains exactly the same information as the probability distribution P_R, and $\hat{\rho}$ is determined once the P_R are specified for a given set of pure states $|R\rangle$. The advantage of the density operator lies in the simplicity of the manipulations by which it produces average values of the various observables.

The general result (7.82) is independent of the particular complete set of states $|S\rangle$ chosen for the evaluation. This is apparent from the way in which the states $|S\rangle$ were introduced, but can also be proved directly. Let the $|T\rangle$

be some other complete set of states for the radiation field. Use of the closure theorem twice in eqn (7.82), both before and after the operator $\hat{\rho}\hat{O}$, gives

$$\langle O \rangle = \sum_S \sum_T \sum_{T'} \langle S \mid T \rangle \langle T | \hat{\rho}\hat{O} | T' \rangle \langle T' \mid S \rangle, \tag{7.83}$$

where T and T' run over the same complete set of states. Rearrangement gives

$$\begin{aligned} \langle O \rangle &= \sum_S \sum_T \sum_{T'} \langle T' \mid S \rangle \langle S \mid T \rangle \langle T | \hat{\rho}\hat{O} | T' \rangle \\ &= \sum_T \langle T | \hat{\rho}\hat{O} | T \rangle, \end{aligned} \tag{7.84}$$

where eqn (7.79) has been used and orthonormality of the $|T\rangle$ has been assumed.

Since $\langle O \rangle$ is independent of the particular complete set of states used to evaluate the average, eqn (7.84) can be written

$$\langle O \rangle = \text{Trace}(\hat{\rho}\hat{O}), \tag{7.85}$$

where the trace of an operator (henceforth abbreviated to Tr) is the sum of its diagonal matrix elements for any complete set of states. This result is entirely equivalent to the basic expression (7.77).

The probability distribution P_R contains the physical information about the radiation field needed to evaluate the ensemble averages. It is important to choose carefully the particular complete set of pure states $|R\rangle$ to be used in defining the probability distribution. This choice must be made in such a way as to preserve all the available information about the state of the system, and the considerations involved are best appreciated by specific examples given later in the chapter.

It should be noted that the matrix elements of $\hat{\rho}$ itself have different properties for different complete sets of states. For the defining set $|R\rangle$ used in the construction of $\hat{\rho}$, only the diagonal matrix elements are non-zero,

$$\langle R' | \hat{\rho} | R'' \rangle = \sum_R P_R \langle R' \mid R \rangle \langle R \mid R'' \rangle = P_{R'}\, \delta_{R'R''}. \tag{7.86}$$

However, a typical matrix element for some other complete set of states is

$$\langle T | \hat{\rho} | T' \rangle = \sum_R P_R \langle T \mid R \rangle \langle R \mid T' \rangle, \tag{7.87}$$

and there are no general constraints on the $|T\rangle$ and $|T'\rangle$ for which the right-hand side is non-zero. The only general result is

$$\text{Tr}(\hat{\rho}) = \sum_T \langle T | \hat{\rho} | T \rangle = \sum_R P_R = 1, \tag{7.88}$$

which can be regarded as a special case of eqn (7.85) with $\hat{O}$ a pure number. For any state of the radiation field, there is always some complete set of states $|R\rangle$ for which the density operator has only diagonal matrix elements as in eqn (7.85).

Density operators for pure states

A pure state can be regarded as a special case of a statistical mixture in which one of the probabilities P_R is equal to unity, and all the remaining P_R are zero. From eqn (7.81), the density operator is

$$\hat{\rho} = |R\rangle\langle R|. \tag{7.89}$$

The radiation field is definitely in a particular pure state $|R\rangle$ in this case, and the statistical description becomes somewhat redundant. However, the concept of the density operator remains valid for the pure state, and some of its simpler properties can be conveniently illustrated in this limit. One result which holds only for a pure-state density operator is

$$\hat{\rho}^2 = \hat{\rho}, \tag{7.90}$$

easily proved from eqn (7.89).

The density operators for the various types of single-mode pure state considered earlier in the chapter are easily constructed. For a field in one of the number states $|n\rangle$, where n photons are definitely present, the density operator is simply

$$\hat{\rho} = |n\rangle\langle n|. \tag{7.91}$$

The only non-vanishing matrix element for the number states is

$$\langle n|\, \hat{\rho}\, |n\rangle = 1, \tag{7.92}$$

and the average value of an observable represented by operator $\hat{O}$ according to eqn (7.85) is

$$\langle O\rangle = \mathrm{Tr}(|n\rangle\langle n|\, \hat{O}) = \langle n|\, \hat{O}\, |n\rangle, \tag{7.93}$$

where the complete set of number states is used to evaluate the trace. The expression on the right of eqn (7.93) is the same as ordinarily used to compute an average value for state $|n\rangle$. The density operator formalism leads to all the usual results for the special case of a pure state.

The density operator for one of the states $|\phi\rangle$ of definite phase or for one of the coherent states $|\alpha\rangle$ can be similarly constructed,

$$\hat{\rho} = |\phi\rangle\langle\phi|, \tag{7.94}$$

$$\hat{\rho} = |\alpha\rangle\langle\alpha|. \tag{7.95}$$

Let us consider the coherent-state density operator in more detail, as being of greater physical interest.

It is evident from the normalization of the coherent states as in eqn (7.49) that the density operator of eqn (7.95) satisfies

$$\langle\alpha|\, \hat{\rho}\, |\alpha\rangle = 1, \tag{7.96}$$

analogous to the corresponding result (7.92) for the case of the number states. However, it follows that since different coherent states are not orthogonal, as shown by eqn (7.50), $\langle\alpha|\, \hat{\rho}\, |\alpha\rangle$ is not the only non-zero matrix element of $\hat{\rho}$ given by eqn (7.95). Indeed, every coherent-state matrix element of $\hat{\rho}$ is non-vanishing. This peculiar feature results from the over-completeness of the states $|\alpha\rangle$, mentioned earlier. Some of the normal theory of the density operator requires careful extension[11] to cover the use of coherent states, and some of the associated theorems must be generalized. These extensions and generalizations are not required for the calculations to be given below, which mainly employ density operators expressed in terms of the number states $|n\rangle$.

Consider now a general matrix element of the pure coherent-state density operator of eqn (7.95) for the number states. We have

$$\langle n|\,\hat{\rho}\,|n'\rangle = \exp(-|\alpha|^2)\,\frac{\alpha^n\alpha^{*n'}}{(n!\,n'!)^{\frac{1}{2}}}, \tag{7.97}$$

where eqn (7.48) has been used. The density operator for the pure coherent state thus has non-vanishing off-diagonal matrix elements for the number states. This is an example of the importance of choosing the appropriate states $|R\rangle$ in which to express the density operator, so that no information about the state of the system is lost. It is not possible to describe the pure coherent state $|\alpha\rangle$ fully in terms of a diagonal density operator based on the number states $|n\rangle$; such a density operator would have zero off-diagonal matrix elements $\langle n|\,\hat{\rho}\,|n'\rangle$ and the information contained in eqn (7.97) for $n \neq n'$ could not be reproduced.

The off-diagonal matrix elements of the density operator are particularly important in the calculation of average values of operators $\hat{O}$ which themselves have non-zero off-diagonal matrix elements $\langle n|\,\hat{O}\,|n'\rangle$. Consider, for example, the single-mode electric-field operator given in eqn (7.38). The average electric field determined by eqn (7.85) with the trace evaluated in terms of the number states is

$$\begin{aligned}\langle E\rangle &= \sum_n \langle n|\,\hat{\rho}\hat{E}\,|n\rangle\\ &= \mathrm{i}(\hbar\omega/2\epsilon_0 V)^{\frac{1}{2}}\sum_n \{\langle n|\,\hat{\rho}\,|n-1\rangle n^{\frac{1}{2}}\exp(-\mathrm{i}\omega t+\mathrm{i}\mathbf{k}.\mathbf{r})-\\ &\quad -\langle n|\,\hat{\rho}\,|n+1\rangle(n+1)^{\frac{1}{2}}\exp(\mathrm{i}\omega t-\mathrm{i}\mathbf{k}.\mathbf{r})\}.\end{aligned} \tag{7.98}$$

When n is replaced by $n+1$ in the first term of the large bracket, the average, becomes

$$\langle E\rangle = -2(\hbar\omega/2\epsilon_0 V)^{\frac{1}{2}}\sum_n (n+1)^{\frac{1}{2}}\,\mathrm{Im}\{\langle n+1|\,\hat{\rho}\,|n\rangle\exp(-\mathrm{i}\omega t+\mathrm{i}\mathbf{k}.\mathbf{r})\}. \tag{7.99}$$

It is evident that a state of the radiation field can have a non-vanishing average electric vector only if the density operator has non-zero off-diagonal matrix elements of the type $\langle n+1|\,\hat{\rho}\,|n\rangle$. For the pure coherent state $|\alpha\rangle$, eqns (7.57) and (7.97) give

$$\langle n+1|\,\hat{\rho}\,|n\rangle = \exp(-|\alpha|^2+i\theta)\frac{|\alpha|^{2n+1}}{n!\,(n+1)^{\frac{1}{2}}}, \tag{7.100}$$

and substitution into eqn (7.99) yields the same result as previously derived in eqn (7.71).

The diagonal number-state matrix element of the density operator, $\langle n|\,\hat{\rho}\,|n\rangle$, is the probability that n photons are excited in the state of the field described by $\hat{\rho}$. The diagonal matrix element for the pure coherent-state density operator obtained from eqn (7.97) reproduces the result (7.62) found earlier for the photon-number probability.

Pure states of the complete radiation field are treated in a similar manner. The state $|R\rangle$ in eqn (7.89) now represents all the modes of the cavity. If, for example, each cavity mode has a definite number of photons excited, the state $|R\rangle$ is one of the states $|\{n_{\mathbf{k}}\}\rangle$ defined in eqn (6.102) and the density operator is

$$\begin{aligned}\hat{\rho} &= |\{n_{\mathbf{k}}\}\rangle\langle\{n_{\mathbf{k}}\}| \\ &= |n_{\mathbf{k}_1}\rangle\,|n_{\mathbf{k}_2}\rangle\,|n_{\mathbf{k}_3}\rangle \ldots \langle n_{\mathbf{k}_3}|\,\langle n_{\mathbf{k}_2}|\,\langle n_{\mathbf{k}_1}|.\end{aligned} \tag{7.101}$$

Another example would be a state of the field in which each cavity mode is excited to a definite coherent state. The state of the total field in this case can be written

$$|\{\alpha_{\mathbf{k}}\}\rangle = |\alpha_{\mathbf{k}_1}\rangle\,|\alpha_{\mathbf{k}_2}\rangle\,|\alpha_{\mathbf{k}_3}\rangle \ldots, \tag{7.102}$$

and the corresponding density operator for the multimode coherent state is

$$\hat{\rho} = |\{\alpha_{\mathbf{k}}\}\rangle\langle\{\alpha_{\mathbf{k}}\}|. \tag{7.103}$$

Other pure states of the complete field are similarly treated. In each case, the matrix elements of $\hat{\rho}$ break up into products of matrix elements each of which refers to a single cavity mode, and no new physical principles are introduced by the generalization from one mode to many.

Statistical mixture states of the radiation field

The great utility of the density operator is made apparent in its application to the treatment of statistical-mixture states. As a first example, consider the thermal excitation of photons in a single mode of a cavity maintained at temperature T. This is the problem treated in Chapter 1 in the derivation of Planck's law, and the probability P_n that n photons are excited is given by eqn (1.42). Thus, according to eqn (7.81), the density operator based on the number states is

$$\begin{aligned}\hat{\rho} &= \sum_n P_n\,|n\rangle\langle n| \\ &= \{1-\exp(-\beta\hbar\omega)\}\sum_n \exp(-\beta n\hbar\omega)\,|n\rangle\langle n|,\end{aligned} \tag{7.104}$$

where

$$\beta = 1/k_B T. \tag{7.105}$$

The number states $|n\rangle$ are the correct basis for the density operator in this case because the thermal probability distribution gives information only on the probabilities of finding a system in its various energy eigenstates; the number states are the energy eigenstates of the photon system.

The density operator for the thermal photon distribution has only diagonal number-state matrix elements and thus by eqn (7.99), which holds equally for pure states and statistical mixtures, the average electric field is always zero. The density operator can be written in an alternative form if the mean number $\bar{n}$ of thermally excited photons is introduced.

According to eqn (7.85),

$$\bar{n} = \langle n \rangle = \mathrm{Tr}(\hat{\rho}\hat{a}^\dagger \hat{a}). \tag{7.106}$$

Problem 7.6. Use eqns (7.104) and (7.106) to evaluate the mean photon number $\bar{n}$ and show that the density operator can be re-expressed as

$$\hat{\rho} = \sum_n \frac{\bar{n}^n}{(1+\bar{n})^{1+n}} |n\rangle\langle n|. \tag{7.107}$$

Note that the calculation is algebraically similar to that of eqn (1.44).

Problem 7.7. Show that the density operator of eqn (7.104) for single-mode thermally excited light can be written in the equivalent form

$$\hat{\rho} = \{1 - \exp(-\beta\hbar\omega)\}\exp(-\beta\hbar\omega\hat{a}^\dagger\hat{a}), \tag{7.108}$$

where the exponential is defined by its usual power-series expansion.

Now consider the thermal excitation of all the cavity modes. The number states $|\{n_{\mathbf{k}}\}\rangle$ for the totality of modes are formed from products of number states for the individual cavity modes as in eqn (6.102). Since the different field modes are independent, the combined density operator is a product of the contributions of the different modes. Thus the density operator, expressed in general as

$$\hat{\rho} = \sum_{\{n_{\mathbf{k}}\}} P_{\{n_{\mathbf{k}}\}} |\{n_{\mathbf{k}}\}\rangle\langle\{n_{\mathbf{k}}\}|, \tag{7.109}$$

is obtained for the thermal case by multiplying together the factors $(\bar{n}_{\mathbf{k}})^{n_{\mathbf{k}}}/(1+\bar{n}_{\mathbf{k}})^{1+n_{\mathbf{k}}}$ for all the modes, to obtain a grand probability

$$P_{\{n_{\mathbf{k}}\}} = \prod_{\mathbf{k}} \frac{(\bar{n}_{\mathbf{k}})^{n_{\mathbf{k}}}}{(1+\bar{n}_{\mathbf{k}})^{1+n_{\mathbf{k}}}}. \tag{7.110}$$

In these expressions $\bar{n}_{\mathbf{k}}$ is the mean number of photons excited in mode $\mathbf{k}$, and the symbols $\{n_{\mathbf{k}}\}$ denote a set of numbers $n_{\mathbf{k}_1}, n_{\mathbf{k}_2}, n_{\mathbf{k}_3}, \ldots$, etc, of photons excited in every cavity mode. The summation in eqn (7.109) embraces every possible set of numbers $\{n_{\mathbf{k}}\}$.

The density operator for the radiation field in a thermal cavity obtained from eqns (7.109) and (7.110) is therefore

$$\hat{\rho} = \sum_{\{n_{\mathbf{k}}\}} |\{n_{\mathbf{k}}\}\rangle\langle\{n_{\mathbf{k}}\}| \prod_{\mathbf{k}} \frac{(\bar{n}_{\mathbf{k}})^{n_{\mathbf{k}}}}{(1+\bar{n}_{\mathbf{k}})^{1+n_{\mathbf{k}}}}. \tag{7.111}$$

Problem 7.8. For the density operator given by eqn (7.111), prove the normalization requirement (7.88) explicitly,

$$\mathrm{Tr}(\hat{\rho}) = 1, \tag{7.112}$$

and show that the mean number of photons in the cavity is

$$\bar{n} = \sum_{\mathbf{k}} \bar{n}_{\mathbf{k}}. \tag{7.113}$$

The mean photon numbers $\bar{n}_{\mathbf{k}}$ in eqn (7.111) are related to the mode frequency $\omega_{\mathbf{k}}$ and the temperature by eqn (1.45),

$$\bar{n}_{\mathbf{k}} = \{\exp(\beta\hbar\omega_{\mathbf{k}})-1\}^{-1}. \tag{7.114}$$

With this expression for the $\bar{n}_{\mathbf{k}}$, the density operator (7.111) contains essentially the same information as the results of the thermal probability calculation employed in the derivation of Planck's radiation law. However, eqn (7.111) applies not only to the thermal photon distribution but also to a wide range of excitations in which the statistical properties of the light generation are suitably random. For example, it is shown in Chapter 10 that the light generated by a source in which atoms are kept at an excitation level higher than that in thermal equilibrium has the same form of density operator as thermal radiation. This does not imply that the spectral distribution of the radiation is the same as that in thermal equilibrium and, of course, the mean photon numbers $\bar{n}_{\mathbf{k}}$ are not given by eqn (7.114) in this more general case, but are determined by the nature of the random field excitation.

The density operator (eqn (7.111)) applies in particular to the light beam emitted by a chaotic source and provides the tool by which the classical discussions of chaotic light given in Chapter 5 can be recast in quantum-mechanical form. For example, $\hat{\rho}$ in eqn (7.111) is the correct density operator for a beam of Lorentzian frequency distribution emitted by a chaotic source if the magnitude of $\bar{n}_{\mathbf{k}}\omega_{\mathbf{k}}$ is taken to have a Lorentzian dependence on $\omega_{\mathbf{k}}$.

Problem 7.9. Show that eqn (5.64) can be recast to give the mean number of photons excited in the modes of the quantized radiation field by a beam of Lorentzian frequency distribution as

$$\bar{n}_k = \frac{\bar{I}a}{\hbar\omega_k} \frac{\gamma}{(\omega_0-\omega_k)^2+\gamma^2}, \tag{7.115}$$

where a is the beam cross-sectional area. This result is used in Chapter 9 to compute the quantum-mechanical coherence of a Lorentzian beam.

It is clear from the above remarks that the density operator of eqn (7.111) for the complete radiation field has wide applicability for chaotic light. It is, of course, possible to set up density operators for total fields on the basis of other complete sets of states for the cavity modes, or on the basis of the over-complete set $|\{\alpha_k\}\rangle$ defined in eqn (7.102). However, the physical problems to be treated in subsequent chapters involve either multimode chaotic light, where eqn (7.111) is appropriate, or single-mode excitations, where the density operator simplifies, and we do not consider the other possible forms of density operator for the total field.

The single-mode density operator of eqn (7.107), for a random excitation of photons of the type associated with chaotic light, does not give any indication of the time scale of the fluctuations in photon number. The quantum-mechanical averages, which can be obtained with the use of the density operator, are similar to the classical ensemble averages described in Chapter 5, and there are similar conditions on the type of measurement which must be carried out in order to make comparisons with theoretical predictions. In general, the experimental averages must be computed from a series of measurements which extend over a time long compared to the fluctuation time, but each individual measurement must be completed in a time short compared to the period of the fluctuations.

The classical discussion of the dependence on cavity length of the excitation of normal modes by a Lorentzian emission line, given in connection with Fig. 5.12, is still valid for the quantized field, except for some obvious changes in wording. The conclusions regarding the magnitude of the single-mode fluctuation time, for cavities which are long or short compared with the coherence length, still hold. In particular, for a long cavity as specified by eqn (5.56), where many normal modes are excited, the fluctuation rate of each mode is determined by the mode spacing $\Delta\omega$ defined in eqn (5.51). In this case, the multimode density operator given by eqn (7.111) with eqn (7.115) substituted for $\bar{n}_k$ depends on γ and therefore contains information on the coherence time τ_c $(= 1/\gamma)$ of the beam as a whole.

In summary, it is seen that although the density operator for each individual mode of a cavity provides no information on the time dependence of the mode excitation, the multimode density operator for a beam whose frequency components spread across many modes does contain the magnitude of the coherence time of the beam. In both classical and quantum-mechanical theory, a light beam can be completely characterized by means of the normal-mode distribution of field intensity or photon number only if the normal modes are densely spaced with respect to the spread in the frequency components of the light.

Problem 7.10. Consider the quantum-mechanical analogues of the classical excitations of a pair of cavity modes envisaged in problems 5.6 and 5.7. If subscripts 1 and 2 denote the states of the two modes, show that the quantum-mechanical excitation most closely analogous to that described classically by eqn (5.81) is a pure state with density operator

$$\hat{\rho} = |\alpha_1\rangle\, |\alpha_2\rangle\langle\alpha_2|\, \langle\alpha_1|. \tag{7.116}$$

Derive the relation between α_1, α_2 and the classical fields E_1 and E_2 such that the quantum field has eqn (5.81) as its classical limit.

For the random excitation of the modes described in problem 5.7 show that the quantum-mechanical density operator is

$$\hat{\rho} = \sum_{n_1=0}^{\infty} \sum_{n_2=0}^{\infty} |n_1\rangle\, |n_2\rangle\langle n_2|\, \langle n_1|\, \frac{\bar{n}^{n_1+n_2}}{(1+\bar{n})^{2+n_1+n_2}}, \tag{7.117}$$

where $\bar{n}$ is the mean number of photons in each mode, and the validity of eqn (7.111) for any random excitation has been assumed.

References

1. SUSSKIND, L. and GLOGOWER, J. (1964). *Physics* **1**, 49.
2. See, for example, LANDAU, L. D. and LIFSHITZ, E. M. (1965). *Quantum mechanics*. Pergamon, Oxford. p. 46.
3. LOUISELL, W. H. (1963). *Phys. Lett.* **7**, 60.
4. See reference 1 for a rigorous account of the properties of the various states which can be defined in connection with the phase operators.
5. GLAUBER, R. J. (1963). *Phys. Rev.* **131**, 2766.
6. MESSIAH, A. (1964). *Quantum mechanics*, Vol. I. North Holland Pub. Co., Amsterdam. p. 442.
7. KITTEL, C. (1969). *Thermal physics*. Wiley, New York. Appendix G, p. 395.
8. CARRUTHERS, P. and NIETO, M. M. (1965). *Phys. Rev. Lett.* **14**, 387.
9. VINSON, J. F. (1969). *Cohérence optique, classique et quantique*, Dunod, Paris, Part 2, Chapter 1.
10. ZIMAN, J. M. (1969). *Elements of advanced quantum theory*. University Press, Cambridge. p. 94; LOUISELL, W. H. (1964). *Radiation and noise in quantum electronics*. McGraw-Hill, New York. Chap. 6.
11. See, for example, reference 5, Chap. 4 of reference 9, and GLAUBER, R. J. (1970). *Quantum optics* (edited by S. M. Kay and A. Maitland). Academic Press, London. p. 53.

Phase in quantum optics

S M Barnett and D T Pegg†

Optics Section, Blackett Laboratory, Imperial College of Science and Technology, London SW7 2BZ, UK

Received 6 March 1986

Abstract. Dirac's prescription for quantisation does not lead to a unique phase operator for the electromagnetic field. In this paper we consider the commonly employed phase operators due to Susskind and Glogower and their extension to unitary exponential phase opertors. However, we find that phase measuring experiments respond to a different operator. We discuss the form of the measured phase operator and its properties.

1. Introduction

Dirac (1927) (see also Heitler 1954) postulated the existence of Hermitian, canonical number and phase variables in his description of the quantised electromagnetic field. Comparison with classical equations of motion led Dirac to assume that the number and phase operators obey the canonical commutation relation

$$[\hat{\phi}_{\mathrm{D}}, \hat{N}] = -\mathrm{i} \tag{1.1}$$

where the hat denotes an operator and the subscript D denotes the Dirac phase operator. This commutation relation leads to difficulties when one attempts to calculate the matrix elements of $\hat{\phi}_{\mathrm{D}}$ in the representation in which $\hat{N}$ is diagonal (the photon number states) (Louisell 1963). The matrix elements between the states $\langle n'|$ and $|n\rangle$ are undefined:

$$(n-n')\langle n'|\hat{\phi}_{\mathrm{D}}|n\rangle = -\mathrm{i}\delta_{nn'}. \tag{1.2}$$

Dirac was aware that there were problems associated with his descripton of phase. However, he pointed out that the difficulties do not arise if the phase operator only appears together with the number operator in a polar decomposition of the field creation and annhilation operators (see, for example, Schweber 1984). Louisell (1963) suggested that the problem embodied in equation (1.2) could be overcome by considering periodic functions of the Dirac phase operator. (Judge and Lewis (1963) (see also Judge 1963) adopted a similar approach to the problem of angular momentum and rotation angle.) In particular, Louisell introduced the periodic operator functions $\cos\hat{\phi}_{\mathrm{D}}$ and $\sin\hat{\phi}_{\mathrm{D}}$ which satisfy the commutation relations

$$[\cos\hat{\phi}_{\mathrm{D}}, \hat{N}] = \mathrm{i}\sin\hat{\phi}_{\mathrm{D}} \tag{1.3a}$$

$$[\sin\hat{\phi}_{\mathrm{D}}, \hat{N}] = -\mathrm{i}\cos\hat{\phi}_{\mathrm{D}}. \tag{1.3b}$$

Susskind and Glogower (1964) considered a description of oscillator phase using exponential phase operators in a polar decomposition of the creation and annihilation

† Permanent address: School of Science, Griffith University, Nathan, Brisbane 4111, Australia.

0305-4470/86/183849+14$02.50

operators

$$\hat{a} = \hat{e}_{\mathrm{S}}^{\mathrm{i}\phi}\hat{N}^{1/2} \tag{1.4a}$$

$$\hat{a}^{\dagger} = \hat{N}^{1/2}\hat{e}_{\mathrm{S}}^{-\mathrm{i}\phi}. \tag{1.4b}$$

The exponential phase operators of Susskind and Glogower (1964) (which we denote by subscript S) are the normalised raising and lowering operators:

$$\hat{e}_{\mathrm{S}}^{\mathrm{i}\phi} = \sum_{n=0}^{\infty} |n\rangle\langle n+1| \tag{1.5a}$$

$$\hat{e}_{\mathrm{S}}^{-\mathrm{i}\phi} = \sum_{n=0}^{\infty} |n+1\rangle\langle n|. \tag{1.5b}$$

These operators do not comute and are not unitary. Therefore, the Susskind-Glogower formalism does not allow the existence of a unique Hermitian phase operator (Susskind and Glogower 1964, Carruthers and Nieto 1968, Loudon 1973 (p 140), Lévy-Leblond 1976). The operators of equations (1.5) cannot be considered as functions of a common phase operator. It is more natural to consider the Susskind-Glogower exponential phase operators $\hat{e}_{\mathrm{S}}^{\pm\mathrm{i}\phi}$ themselves as the fundamental phase-dependent operators.

The phase operators of Susskind and Glogower have been used in discussions of the properties of coherent states (Carruthers and Nieto 1965, 1968), squeezed states (Sanders *et al* 1986) and optical amplification processes (Matthys and Jaynes 1980, Loudon and Shepherd 1984). Number-phase uncertainty relations for the Susskind-Glogower operators and number-phase minimum uncertainty states have been considered by Carruthers and Nieto (1965, 1968), Jackiw (1968), Lévy-Leblond (1976) and Sanders *et al* (1986). Phase operators have also been used in the analysis of phase measurement experiments (Gerhardt *et al* 1973, 1974, Paul 1974, Nieto 1977, Shapiro and Wagner 1984, Walker and Carroll 1984).

In this paper we reconsider the definition of a phase operator for the quantised electromagnetic field. We find that there are many suitable candidates. In particular we discuss two new candidate phase operators: a unitary exponential phase operator, $\hat{e}_{\mathrm{u}}^{\mathrm{i}\phi}$, and a cosine phase operator, $\cos_{\mathrm{M}}\phi$, corresponding to the phase-dependent property measured in homodyne (Yuen and Chan 1983 and references therein) and prepared atom (Pegg 1981) experiments. In each case we compare the new phase operators with the conventional Susskind-Glogower operators.

In § 2 we review the general properties of phase operators. Using the requirement that phase operators reproduce classical results in the suitable limits, we find the general conditions that a phase operator must satisfy. We consider the Susskind-Glogower operators and our two new phase operators as special phase operators and compare their properties.

In § 3 we consider phase measurement experiments and in § 4 we redefine the phase operator in terms of the quantities usually measured in experiments. We find that the phase operators measured in experiments are proportional to the quadrature phase operators well known from discussions of squeezing (Slusher *et al* 1985). Finally, in § 5 we compare the phase operators and discuss our results.

2. Phase operators

In this section we identify appropriate operators for the quantum mechanical description of the phase of the radiation field by exploiting the well known correspondence between a single mode of the radiation field and a simple harmonic oscillator (see, for example, Loudon 1973, p 120) and considering the Poisson bracket formulation of the classical oscillator problem. The classical action (J) and angle (ϕ) variables for a simple harmonic oscillator are related to the position and momentum by the relations (see, for example, Goldstein 1980, Carruthers and Nieto 1968)

$$q=(2J/m\omega)^{1/2}\cos\phi \tag{2.1a}$$

$$p=(2Jm\omega)^{1/2}\sin\phi \tag{2.1b}$$

where m and ω are the oscillator mass and frequency. The classical Hamiltonian is

$$\begin{aligned} H&=(p^2/2m)+(m\omega^2q^2/2)\\ &=\omega J. \end{aligned} \tag{2.2}$$

In order to avoid the problem of multivaluedness of the phase angle it is natural to work with periodic functions of the phase ϕ. The time dependences of the phase variables $\sin\phi$ and $\cos\phi$ are given in terms of the Poisson brackets

$$(\mathrm{d}/\mathrm{d}t)\cos\phi=\{\cos\phi, H\}=\omega\sin\phi \tag{2.3a}$$

$$(\mathrm{d}/\mathrm{d}t)\sin\phi=\{\sin\phi, H\}=-\omega\cos\phi. \tag{2.3b}$$

Quantum mechanical operators whch reproduce the classical behaviour in the appropriate limit will be obtained if our operator commutators are related to the classical Poisson brackets according to the prescription (Dirac 1958)

$$[\hat{u}, \hat{v}]\leftrightarrow \mathrm{i}\hbar\{u, v\}. \tag{2.4}$$

The application of this technique to the problem of phase is due to Lerner (1968). Thus we look for Hermitian cosine and sine operators $\widehat{\cos}\,\phi$ and $\widehat{\sin}\,\phi$ obeying the commutation relations

$$[\widehat{\cos}\,\phi, \hat{N}]=\mathrm{i}\,\widehat{\sin}\,\phi \tag{2.5a}$$

$$[\widehat{\sin}\,\phi, \hat{N}]=-\mathrm{i}\,\widehat{\cos}\,\phi \tag{2.5b}$$

where the Hamiltonian operator is $\hat{H}=(\hat{N}+\frac{1}{2})\hbar\omega$. We also introduce the exponential phase operators $\hat{e}^{\pm\mathrm{i}\phi}$ by

$$\hat{e}^{\pm\mathrm{i}\phi}\equiv\widehat{\cos}\,\phi\pm\mathrm{i}\,\widehat{\sin}\,\phi \tag{2.6}$$

which from (2.5) will then obey the commutation relations

$$[\hat{e}^{\pm\mathrm{i}\phi}, \hat{N}]=\pm\hat{e}^{\pm\mathrm{i}\phi}. \tag{2.7}$$

The presumed Hermitian character of the cosine and sine operators implies that $\hat{e}^{\pm\mathrm{i}\phi}$ are the Hermitian conjugates of each other. In addition, our phase operators must reproduce the classically expected values for highly excited coherent ('classical') states:

$$\lim_{|\alpha|\to\infty}\langle\alpha|\,\widehat{\cos}\,\phi|\alpha\rangle=\cos\theta \tag{2.8a}$$

$$\lim_{|\alpha|\to\infty}\langle\alpha|\,\widehat{\sin}\,\phi|\alpha\rangle=\sin\theta \tag{2.8b}$$

where $\alpha=|\alpha|\,\mathrm{e}^{\mathrm{i}\theta}$. The Lerner criterion and classical correspondence are not sufficient, however, to define unique phase operators (Lerner 1968).

2.1. Susskind-Glogower phase operators

Dirac's (1927) original idea of radiation field phase was based upon a polar decomposition of the creation and annihilation operators into the product of an Hermitian amplitude operator and a unitary phase operator. Susskind and Glogower (1964) attempted to construct operators that were as close as possible to Dirac's conception. The resulting exponential phase operators are the bare raising and lowering operators satisfying the conditions

$$\hat{e}_{\mathrm{S}}^{\mathrm{i}\phi}|n\rangle = |n-1\rangle \tag{2.9a}$$

$$\hat{e}_{\mathrm{S}}^{-\mathrm{i}\phi}|n\rangle = |n+1\rangle. \tag{2.9b}$$

In addition, the Susskind-Glogower formalism requires the extra condition that

$$\hat{e}_{\mathrm{S}}^{\mathrm{i}\phi}|0\rangle = 0 \tag{2.10}$$

to avoid negative number states. The non-unitary character of the phase operators results from the termination of the eigenstates of $\hat{N}$ at the vacuum state $|0\rangle$. The Susskind-Glogower operators are 'one-sided unitary':

$$\hat{e}_{\mathrm{S}}^{\mathrm{i}\phi}\hat{e}_{\mathrm{S}}^{-\mathrm{i}\phi} = 1 \tag{2.11a}$$

$$\hat{e}_{\mathrm{S}}^{-\mathrm{i}\phi}\hat{e}_{\mathrm{S}}^{\mathrm{i}\phi} = 1 - |0\rangle\langle 0|. \tag{2.11b}$$

From these equations we can see that the non-commuting and non-unitary nature of $\hat{e}_{\mathrm{S}}^{\pm\mathrm{i}\phi}$ is only apparent for states of the radiation field that have a significant overlap with the vacuum

$$\langle\psi|[\hat{e}_{\mathrm{S}}^{\mathrm{i}\phi}, \hat{e}_{\mathrm{S}}^{-\mathrm{i}\phi}]|\psi\rangle = \langle\psi|0\rangle\langle 0|\psi\rangle. \tag{2.12}$$

The Susskind-Glogower phase operators obey the Lerner criterion

$$[\hat{e}_{\mathrm{S}}^{\pm\mathrm{i}\phi}, \hat{N}] = \pm\hat{e}_{\mathrm{S}}^{\pm\mathrm{i}\phi} \tag{2.13}$$

and have the required behaviour in the classical limit (equations (2.8)) (Carruthers and Nieto 1965, 1968). Extensive discussions of the Susskind-Glogower phase operators have been given by Carruthers and Nieto (1968) and Lévy-Leblond (1976).

In the quantum limit there are problems associated with interpreting the phase as described by the Susskind-Glogower operators. In particular we have

$$(\widehat{\cos_{\mathrm{S}}\phi})^2 + (\widehat{\sin_{\mathrm{S}}\phi})^2 \neq 1. \tag{2.14}$$

Also, the vacuum expectation values of $(\widehat{\cos_{\mathrm{S}}\phi})^2$ and $(\widehat{\sin_{\mathrm{S}}\phi})^2$ are $\frac{1}{4}$, not the $\frac{1}{2}$ which we would associate with a state of random phase. This implies that if it were possible to measure the Susskind-Glogower phase, a measurement of $(\widehat{\cos_{\mathrm{S}}\phi})^2$ would squeeze the vacuum. It should be noted that these considerations do not mean that the Susskind-Glogower formalism is inconsistent. However, it does mean that the phase operator expectation values are difficult to interpret and we require complicated uncertainty relations (Carruthers and Nieto 1965, 1968, Jackiw 1968, Lévy-Leblond 1976).

2.2. Unitary phase operators

In this subsection we introduce unitary commuting exponential phase operators (which we denote by a subscript u) $\hat{e}_{\mathrm{u}}^{\pm\mathrm{i}\phi}$. Our aim is to realise Dirac's idea of a polar decomposition of the creation and annihilation operators into Hermitian and unitary parts. We achieve our objective by extending the normal harmonic oscillator Hilbert space to include negative number states. We note that the derivation of the harmonic oscillator eigenstates merely requires a ground state that is annihilated by the annihilation operator so that the energy spectrum is bounded from below (see, for example, Merzbacher 1970). Negative energy states are not precluded, but they must be decoupled from the positive energy ground state that is annihilated by the annihilation operator. We shall see that the states containing a negative number of photons are inaccessible to a physical system so their mere existence in the formalism does not predict any new phenomena in quantum electrodynamics.

We define the unitary exponential phase operators as the normalised raising and lowering operators extending over the complete Hilbert space of all positive and negative photon number states:

$$\hat{e}_{\mathrm{u}}^{\mathrm{i}\phi} = \sum_{n=-\infty}^{\infty} |n\rangle\langle n+1| \tag{2.15a}$$

$$\hat{e}_{\mathrm{u}}^{-\mathrm{i}\phi} = \sum_{n=-\infty}^{\infty} |n+1\rangle\langle n|. \tag{2.15b}$$

With this extended basis of orthonormal oscillator states, the resolution for the identity becomes

$$\sum_{n=-\infty}^{\infty} |n\rangle\langle n| = 1. \tag{2.16}$$

The unitarity of $\hat{e}_{\mathrm{u}}^{\pm\mathrm{i}\phi}$ results from the absence of a cutoff in the summations defining the operators.

We retain the property that the number states are eigenstates of the number operator $\hat{N}$ with eigenvalues n

$$\hat{N}|n\rangle = n|n\rangle \tag{2.17}$$

for all positive and negative n. The polar decomposition of the creation and annihilation operators into an Hermitian amplitude and unitary phase then requires that

$$\hat{a} = \hat{e}_{\mathrm{u}}^{\mathrm{i}\phi}|\hat{N}^{1/2}| \tag{2.18a}$$

$$\hat{a}^{\dagger} = |\hat{N}^{1/2}|\hat{e}_{\mathrm{u}}^{-\mathrm{i}\phi} \tag{2.18b}$$

where $|\hat{N}^{1/2}|$ is the Hermitian amplitude operator

$$|\hat{N}^{1/2}| \equiv \sum_{n=-\infty}^{\infty} |n^{1/2}|\,|n\rangle\langle n|. \tag{2.19}$$

The expectation value of the commutator of $\hat{a}$ and $\hat{a}^{\dagger}$ depends upon whether the system is in a positive or negative photon number state:

$$[\hat{a}, \hat{a}^{\dagger}] = \sum_{n=0}^{\infty} |n\rangle\langle n| - \sum_{n=-\infty}^{-1} |n\rangle\langle n|. \tag{2.20}$$

For the creation and annihilation operators to correspond to the negative and positive frequency components of the free field, we require the free-field Hamiltonian to be

$$\hat{H} = (\hat{N} + \tfrac{1}{2})\hbar\omega \tag{2.21}$$

with both positive and negative energy eigenvalues.

The destruction of the vacuum state $|0\rangle$ by the annihilation operator and the destruction of the state $|-1\rangle$ by the creation operator are due to the Hermitian amplitude parts of the operators (2.18). Physical couplings to the radiation field take place via the creation and annihilation operators, the fundamental coupling being of the form

$$\hat{H}_I = (c \text{ numbers})(\hat{a} + \hat{a}^\dagger)(\text{current operators}). \tag{2.22}$$

It follows that the negative energy states cannot be coupled to any physical, i.e. positive energy, states because the Hamiltonian matrix elements between positive and negative energy states are always zero:

$$\langle n|\hat{H}_I|n'\rangle = 0 \tag{2.23}$$

if $|n\rangle$ is a negative energy state and $|n'\rangle$ is a positive energy state. Therefore, a field initially in a superposition of positive energy eigenstates can never evolve into a state containing a negative energy eigenstate. This lack of coupling between the positive and negative energy states means that quantum electrodynamical systems are restricted to the positive energy subspace where the familiar relations

$$[\hat{a}, \hat{a}^\dagger] = 1 \tag{2.24}$$

$$\sum_{n=0}^{\infty} |n\rangle\langle n| = 1 \tag{2.25}$$

are true.

The unitary phase operators obey the Lerner criterion if we employ the extended basis number and Hamiltonian operators

$$[\hat{e}_{\mathrm{u}}^{\pm i\phi}, \hat{N}] = \pm \hat{e}_{\mathrm{u}}^{\pm i\phi}. \tag{2.26}$$

The similarity between the unitary and the Susskind–Glogower phase operators ensures that the unitary phase operators have the required behaviour in the classical (positive energy) limit.

As with the Susskind–Glogower operators, there are problems associated with the unitary phase operators. In the unitary formalism the cosine and sine phase operators obey the trigonometric identity

$$(\widehat{\cos}_{\mathrm{u}}\, \phi)^2 + (\widehat{\sin}_{\mathrm{u}}\, \phi)^2 = 1 \tag{2.27}$$

and the vacuum state expectation values of $(\widehat{\cos}_{\mathrm{u}}\, \phi)^2$ and $(\widehat{\sin}_{\mathrm{u}}\, \phi)^2$ are $\frac{1}{2}$. However, these properties rely on the existence of unphysical negative photon number states. The unitary phase operators are unmeasurable because the fundamental interaction Hamiltonian (equation (2.22)) does not couple positive and negative number states together. Therefore no measuring device can be constructed that is sensitive to the negative number states in the definition of the unitary phase operators.

In addition, the Susskind–Glogower phase operators do not correspond to the quantities measured in homodyne (Yuen and Shapiro 1980) and prepared atom (Pegg 1981) experiments. We now turn our attention to defining different phase operators that correspond to these usual phase measuring experiments.

2.3. Measured phase operators

In the previous subsection we noted that the fundamental radiation–matter coupling in quantum electrodynamics is via the field creation and annihilation operators.

Therefore it seems natural to consider the creation and annihilation operators as fundamental and to construct phase operators in terms of $\hat{a}$ and $\hat{a}^\dagger$. We noted previously that the Lerner criterion does not define a unique phase operator; indeed the creation and annihilation operators themselves obey the Lerner criterion

$$[\hat{a}, \hat{N}] = \hat{a} \tag{2.28a}$$

$$[\hat{a}^\dagger, \hat{N}] = -\hat{a}^\dagger. \tag{2.28b}$$

In fact we can construct any operators of the general form

$$\hat{e}^{i\phi} = \hat{a}f(\hat{N}) + g(\hat{N})\hat{a} \tag{2.29a}$$

$$\hat{e}^{-i\phi} = f(\hat{N})\hat{a}^\dagger + \hat{a}^\dagger g(\hat{N}) \tag{2.29b}$$

where f and g are well behaved functions, which also satisfy the Lerner criterion. The Susskind–Glogower expressions (equations (1.4)) are just one particular member of this larger set.

The choice of phase operators is further constrained by the condition that they must reproduce the classically expected values for highly excited coherent states. Therefore the creation and annihilation operators themselves are not suitable as phase operators, although suitable phase operators of the form presented in equations (2.29) can be constructed. In particular Lerner (1968) has advocated the use of the symmerical expressions, either

$$\hat{e}^{i\phi} = \tfrac{1}{2}[(\hat{N}+\tfrac{1}{2})^{-1/2}\hat{a} + \hat{a}(\hat{N}+\tfrac{1}{2})^{-1/2}] \tag{2.30a}$$

$$\hat{e}^{-i\phi} = \tfrac{1}{2}[(\hat{N}+\tfrac{1}{2})^{-1/2}\hat{a}^\dagger + \hat{a}^\dagger(\hat{N}+\tfrac{1}{2})^{-1/2}] \tag{2.30b}$$

or

$$\hat{a} = \tfrac{1}{2}[(\hat{N}+\tfrac{1}{2})^{1/2}\hat{e}^{i\phi} + \hat{e}^{i\phi}(\hat{N}+\tfrac{1}{2})^{1/2}] \tag{2.31a}$$

$$\hat{a}^\dagger = \tfrac{1}{2}[(\hat{N}+\tfrac{1}{2})^{1/2}\hat{e}^{i\phi} + \hat{e}^{i\phi}(\hat{N}+\tfrac{1}{2})^{1/2}]. \tag{2.31b}$$

In this subsection we construct phase operators that correspond to the usual operational definition of a phase measurement. In § 3 we shall see that the quantity suggested by homodyne (Yuen and Shapiro 1980) and prepared atom (Pegg 1981) experiments is

$$\widehat{\cos}_M\,\phi = k(\hat{a} + \hat{a}^\dagger) \tag{2.32}$$

where k is a state-dependent c number (obtained by means of an independent experiment). We use the subscript M to denote these measured phase operators. The number k must be chosen so that

$$\lim_{|\alpha|\to\infty} \langle\alpha|k(\hat{a} + \hat{a}^\dagger)|\alpha\rangle = \cos\theta \tag{2.33}$$

where $\alpha = |\alpha|\,e^{i\theta}$. We also define $\widehat{\sin}_M\,\phi$ to be

$$\widehat{\sin}_M\,\phi = -ik(\hat{a} - \hat{a}^\dagger). \tag{2.34}$$

We give an analysis of the phase measurements that lead us to define our measured phase operators in § 3.

Our choice, for reasons discussed later, is to define

$$\widehat{\cos}_M\,\phi = \frac{\hat{a} + \hat{a}^\dagger}{2(\bar{n}+\tfrac{1}{2})^{1/2}} \tag{2.35a}$$

$$\widehat{\sin}_M\,\phi = \frac{\hat{a} - \hat{a}^\dagger}{2i(\bar{n}+\tfrac{1}{2})^{1/2}} \tag{2.35b}$$

where $\bar{n}$ is the mean photon number of the measured field. This choice is arrived at by placing the energy or intensity in the denominator of the classical phase measurement experiments. Here, in contrast to the operators of Lerner (1968, equations (2.30)), the energy denominator takes the form of an independently derived c number rather than an operator. The phase operators exhibit the classically expected property that

$$\langle(\widehat{\cos}_{\mathrm{M}}\,\phi)^2\rangle+\langle(\widehat{\sin}_{\mathrm{M}}\,\phi)^2\rangle=1. \tag{2.36}$$

The major problem associated with these phase operators is that their spectra are not bounded by the interval $(-1, 1)$. This may be demonstrated by considering the expectation values of the operators $(\cos_{\mathrm{M}}\,\phi)^m$. We address this problem in § 4. However, as these operators correspond to the quantities measured in conventional phase measurements it seems natural to adopt them as phase operators.

In table 1 we list some of the properties of the Susskind-Glogower, unitary and measured phase operators in order to highlight the differences between them.

Table 1. Some of the properties of the phase operators discussed in this paper.

	Susskind-Glogower	Unitary	Measured
$\hat{e}^{\mathrm{i}\phi}$	$\sum_{n=0}^{\infty} \|n\rangle\langle n+1\|$	$\sum_{n=-\infty}^{\infty} \|n\rangle\langle n+1\|$	$(\bar{n}+\frac{1}{2})^{-1/2}\hat{a}$
$\hat{e}^{-\mathrm{i}\phi}$	$\sum_{n=0}^{\infty} \|n+1\rangle\langle n\|$	$\sum_{n=-\infty}^{\infty} \|n+1\rangle\langle n\|$	$(\bar{n}+\frac{1}{2})^{-1/2}\hat{a}^\dagger$
$[\widehat{\cos}\,\phi, \widehat{\sin}\,\phi]$	$\frac{1}{2}\mathrm{i}\|0\rangle\langle 0\|$	0	$\frac{1}{2}\mathrm{i}(\bar{n}+\frac{1}{2})^{-1}$
$\langle 0\|(\widehat{\cos}\,\phi)^2\|0\rangle$	$\frac{1}{4}$	$\frac{1}{2}$	$\frac{1}{2}$
$\langle 0\|(\widehat{\sin}\,\phi)^2\|0\rangle$	$\frac{1}{4}$	$\frac{1}{2}$	$\frac{1}{2}$
$\langle 0\|(\widehat{\cos}\,\phi)^4\|0\rangle$	$\frac{1}{8}$	$\frac{3}{8}$	$\frac{3}{4}$
$\langle 0\|(\widehat{\sin}\,\phi)^4\|0\rangle$	$\frac{1}{8}$	$\frac{3}{8}$	$\frac{3}{4}$
$\langle(\widehat{\cos}\,\phi)^2\rangle+\langle(\widehat{\sin}\,\phi)^2\rangle$	$1-\frac{1}{2}\langle\|0\rangle\langle 0\|\rangle$	1	1

3. Phase measurements

Classically an absolute phase has no meaning and all measurements must be made relative to the phase of the reference system. The same can be expected for quantum mechnical systems, with the observable quantity being the phase difference between the quantum system and a reference oscillator. If the reference oscillator is in a highly excited coherent state then it has a well defined phase (equations (2.8)). In what follows we choose this reference phase to be precisley zero.

An observable is measured by means of an effect on the measuring apparatus. Electrodynamic fields interact with matter by means of an interaction Hamiltonian which, in the quantum mechanical case, will involve the creation and annihilation operators $\hat{a}^\dagger$ and $\hat{a}$. It is the properties of these operators which will predict the outcome of such measurements. We discuss briefly two physical processes where it is known that the phase of a classical field has a measurable effect on the measuring apparatus. We postulate that the corresponding effect produced by a quantum field will also be a phase effect, from which an operational means of defining the phase can be determined. A boundary condition which must be satisfied is that, when the measured field is also in a coherent state, with mean photon number $\bar{n}$, the result of a phase measurement must tend to the classically expected value as $\bar{n}$ increases.

The first measurement process involves homodyne detection (see, for example, Yuen and Shapiro 1980, Yuen and Chan 1983), i.e. mixing two fields of the same frequency and measuring the total intensity. Here the reference field will be a coherent field, whose intensity we shall allow to tend to infinity, and whose phase is defined to be zero. Classically this corresponds to a reference field which in the dipole approximation at the detector is $E_R \cos \omega t$. If this is mixed with a classical field $E_M \cos(\omega t + \phi)$ with a fluctuating phase ϕ, it is not difficult to show that for $I_R \gg I_M$, where I_R and I_M are the cycle-averaged intensities of the reference and measured fields

$$\cos \phi = \frac{I - I_R}{2(I_R I_M)^{1/2}}. \tag{3.1}$$

Here I is the total cycle-averaged intensity measured by the detector. Clearly the measurement needs to be made in a short time compared with the characteristic time of the fluctuations.

In the quantum mechanical case, where intensities are again measured, for example by photoelectron counts, a measurement of $\cos \phi$ could be defined in terms of quantities associated with those on the right-hand side of (3.1). By using suitable beam splitting and path differences a single measurement of $I - I_R$ could be made (Yuen and Shapiro 1980, Yuen and Chan 1983). Alternatively, because I_R is assumed to be without fluctuations, and thus the same at all times, a separate measurement of I_R could be made. It would be difficult in practice to measure $(I_M)^{1/2}$ at precisely the same time and place as the measurement of $I - I_R$ is performed. Also, in usual experiments a measurement of $(I_M)^{1/2}$ is not performed, either immediately before the measurement of $I - I_R$ or afterwards. Consequently, finding the operator counterpart of (3.1) involves replacing only the numerator with an operator $\hat{I} - \hat{I}_R$; the denominator will be a c number chosen to give correct dimensions and the correct limiting behaviour. The resulting operator will act on both the reference and measured fields. In order to obtain an operator which acts only on the measured field we use $\langle \beta | \hat{I} - \hat{I}_R | \beta \rangle$ in the numerator, where $|\beta\rangle$ is the state of the strong coherent reference field. This latter field is chosen to have phase zero and the eigenvalue β of its annihilation operator $\hat{b}$ is real. Writing $\hat{I}$ in terms of the combined field operators $\hat{b} + \hat{a}$ and its Hermitian conjugate, $\hat{I}_R$ in terms of the reference field operators $\hat{b}$ and $\hat{b}^\dagger$ and letting $\beta \to \infty$, we find the phase operator corresponding to the operation of homodyne measurement to be simply

$$\widehat{\cos}_M \phi = k(\hat{a} + \hat{a}^\dagger). \tag{3.2}$$

This operator is a normalised quadrature phase operator. The reduced fluctuations in one of the quadrature phases associated with squeezed states (Walls 1983 and references therein) imply reduced noise in the measured phase operators. We have used the fact that the c number representing the denominator in (3.1) must be proportional to β as $\beta \to \infty$ in order to obtain a finite expression, as one would with a classical reference field. Thus k is a c number which depends only on the measured field and has the dimensions of the inverse square root of a photon number. Also, if the measured field is in a coherent state, with mean photon number $\bar{n}$, k must approach $\frac{1}{2}(\bar{n})^{-1/2}$ for large

$\bar{n}$ in order to obtain the correct classical limit. Without loss of generality we can write

$$k = \tfrac{1}{2}(\bar{n} + F)^{-1/2} \tag{3.3}$$

where F is to be determined or defined subject to the condition that $F \ll \bar{n}$ for large $\bar{n}$.

The second method for the measurement of a phase associated property involves the interaction of the field with a two-level atom in a particular superposition state. It is well known that the occurrence of absorption or stimulated emission depends on the relative phases of the field and the prepared atomic state. It should be possible, therefore, to find a measure of the phase of the field by examining the initial change of the atomic state at the instant of interaction with the field. The atomic state could be prepared by a $\pi/2$ pulse from a reference field at an earlier time, in a similar manner to that described by Pegg (1981). To be specific, consider a two-level atom with excited and ground states $|e\rangle$ and $|g\rangle$ with transition frequency ω resonant with both the reference and measured field frequencies. The prior action of the intense reference field is equivalent to that of a classical field (with zero phase) so the Hamiltonian is

$$H = \omega|e\rangle\langle e| + \lambda E_{\mathrm{R}}(t)\cos\omega t(|e\rangle\langle g| + |g\rangle\langle e|) \tag{3.4}$$

where λ is the coupling constant. It is convenient to work in an interaction picture. We use the unitary operator

$$T = \exp(\mathrm{i}|e\rangle\langle e|\omega t) \tag{3.5}$$

to transform to a reference frame in which the field $E_{\mathrm{R}}(t)\cos\omega t$ becomes an effective field $\frac{1}{2}E_{\mathrm{R}}(t)$ when the rotating-wave approximation is made (see, for example, Knight and Allen 1983). In this frame the Hamiltonian is time independent:

$$H = \tfrac{1}{2}\lambda E_{\mathrm{R}}(t)(|e\rangle\langle g| + |g\rangle\langle e|). \tag{3.6}$$

The action of a $\pi/2$ pulse is to put an initially ground-state atom into a coherent superposition state which, in this frame, is

$$|A\rangle = 2^{-1/2}(|g\rangle - \mathrm{i}|e\rangle). \tag{3.7}$$

A more detailed discussion of the action of classical pulses on two-level atoms is given by Allen and Eberly (1975). The prepared atom retains the well defined phase information of the reference field in its dipole moment.

If a general fluctuating classical field $E_{\mathrm{M}}(t)\cos(\omega t - \phi(t))$ is applied to the prepared atom at time t_1 the value of $\langle\sigma_z\rangle$ will change, where $\sigma_z \equiv |e\rangle\langle e| - |g\rangle\langle g|$, in a manner dependent on the phase of the field. In the interaction picture in the rotating-wave approximation the Hamiltonian is

$$H(t) = \tfrac{1}{2}\lambda E_{\mathrm{M}}(t)(\mathrm{e}^{\mathrm{i}\phi(t)}|e\rangle\langle g| + \mathrm{e}^{-\mathrm{i}\phi(t)}|g\rangle\langle e|). \tag{3.8}$$

Under the action of this second field the initial rate of change of $\langle\sigma_z\rangle$ for the atom in the prepared atomic state is easily found from $\mathrm{i}\langle A|[H, \sigma_z](t_1)|A\rangle$ to be proportional to $\cos\phi(t_1)$, i.e.

$$\langle\dot{\sigma}_z(t_1)\rangle = \tfrac{1}{2}\lambda E_{\mathrm{M}}(t_1)\cos\phi(t_1). \tag{3.9}$$

In the quantum mechanical case the prepared atom is exposed at time t_1 to a quantised electric field. As in the semiclassical treatment above we expect the value of $\langle\sigma_z\rangle$ to change in a manner dependent on the phase of the field. In the interaction picture, and making the rotating-wave approximation, the Hamiltonian is

$$H = \tfrac{1}{2}\Lambda(\hat{a}|e\rangle\langle g| + \hat{a}^{\dagger}|g\rangle\langle e|) \tag{3.10}$$

where Λ is the fully quantum mechanical coupling constant. Under the action of the quantised field the initial rate of change of $\langle \sigma_z \rangle$ in the prepared atomic state is found from $\mathrm{i}\langle f|\langle A|[H, \sigma_z]|A\rangle|f\rangle$, where $|f\rangle$ is the field state at t_1, to be

$$\langle \dot{\sigma}_z(t_1) \rangle = \tfrac{1}{2}\Lambda \langle f|\hat{a} + \hat{a}^\dagger|f\rangle \tag{3.11}$$

which is proportional to $\langle \widehat{\cos}_\mathrm{M}\, \phi \rangle$. For a field in a strong coherent state with complex amplitude $\alpha = |\alpha|\mathrm{e}^{\mathrm{i}\theta}$ it is clear that $\langle \dot{\sigma}_z(t_1) \rangle$ is proportional to $\cos\theta$. For a general field, comparison of (3.9) and (3.11) shows that, as with the homodyne measurement, the phase operator corresponding to the operation of prepared atom phase measurement is $\widehat{\cos}_\mathrm{M}\, \phi$.

The corresponding sine phase operator, $\widehat{\sin}_\mathrm{M}\, \phi$, can be measured by altering the phase of the reference oscillator.

4. Choice of measured phase operators

From the preceding work it is clear that a definition of $\widehat{\cos}_\mathrm{M}\, \phi$ as

$$\widehat{\cos}_\mathrm{M}\, \phi = k(\hat{a} + \hat{a}^\dagger) \tag{4.1}$$

is at least in accord with measurable phase-dependent properties. Indeed, the original phase measurements of Gerhardt *et al* (1973, 1974) involved a homdyne technique. In this, the experiments appear to be measurements of the observable associated with $\widehat{\cos}_\mathrm{M}\, \phi$ rather than measurements of the Susskind–Glogower (1964) phase $\widehat{\cos}_\mathrm{S}\, \phi$ that they had intended to measure (Gerhardt *et al* 1974, Nieto 1977, Lévy-Leblond 1977). In classical homodyne experiments the maximum attainable value for the measured phase is $\cos\phi = 1$. However, in a quantum homodyne experiment the spectrum of $\widehat{\cos}_\mathrm{M}\, \phi$ is unbounded. This may be demonstrated by considering the expectation value of $(\widehat{\cos}_\mathrm{M}\, \phi)^{2m}$. If the spectrum of $(\widehat{\cos}_\mathrm{M}\, \phi)^{2m}$ is bounded then the spectrum of $(\widehat{\cos}_\mathrm{M}\, \phi)^{2m}$ will also be bounded and its expectation value will be finite for all m. This proves not to be the case. We calculate expectation values of the operator $(\widehat{\cos}_\mathrm{M}\, \phi)^{2m}$ by using the generating function $\hat{G}(x)$

$$\hat{G}(x) = \exp(x\, \widehat{\cos}_\mathrm{M}\, \phi). \tag{4.2}$$

We can write the generating function in a normally ordered form by using the Baker–Campbell–Hausdorf theorem (Louisell 1973, Hong and Mandel 1985)

$$\hat{G}(x) = \exp(x^2k^2/2) : \hat{G}(x) : \tag{4.3}$$

where the colons denote normal ordering. Expanding both sides of (4.3) as a series and equating coefficients of $x^{2m}/(2m)!$ we find a series for $(\widehat{\cos}_\mathrm{M}\, \phi)^{2m}$ in terms of normally ordered operators

$$\begin{aligned}(\widehat{\cos}_\mathrm{M}\, \phi)^{2m} = {} & :(\widehat{\cos}_\mathrm{M}\, \phi)^{2m}: + \tfrac{1}{2}k^2 2m(2m-1)/1!:(\widehat{\cos}_\mathrm{M}\, \phi)^{2m-2}: \\ & + (\tfrac{1}{2}k^2)^2 2m(2m-1)(2m-2)(2m-3)/2!:(\widehat{\cos}_\mathrm{M}\, \phi)^{2m-4}: + \ldots \\ & + (\tfrac{1}{2}k^2)^m (2m)!/m!. \end{aligned} \tag{4.4}$$

This series involves terms containing normally ordered even powers of $\widehat{\cos}_\mathrm{M}\, \phi$ with positive coefficients and a constant term $(k^2/2)^m(2m)!/m!$ which diverges as m increases. By choosing m large enough, we can make $\langle(\widehat{\cos}_\mathrm{M}\, \phi)^{2m}\rangle$ greater than any number we choose. This implies that the spectrum of $(\widehat{\cos}_\mathrm{M}\, \phi)^{2m}$, and therefore the spectrum of $\widehat{\cos}_M\,\phi$, is unbounded. In a classical homodyne experiment, the largest realisation of the intensity of the combined reference and signal fields can be chosen to correspond approximately to $\cos\phi = 1$. In this way the apparatus may be calibrated. However, the spectrum of $\widehat{\cos}_M\,\phi$ is unbounded and no such technique can be applied to calibrate a quantum phase measurement. There is no well defined method for using the largest realisation of the measured intensity to determine k

It then remains to choose a suitable value for k in the definition of $\widehat{\cos}_M\,\phi$. From the above discussion we cannot choose a value of k other than zero such that all experimentally measured values of $\widehat{\cos}_M\,\phi$ are less than one. However, guided by the relationship between k and $\frac{1}{2}(I_M)^{-1/2}$ in the classical expression for $\cos\phi$ (equation (3.1)) and remembering that c numbers can be obtained in quantum mechanics from expectation values, a reasonable choice for k is $\frac{1}{2}(\bar{n}+\frac{1}{2})^{-1/2}$. This can be compared with the operator expressions of Susskind and Glogower (1964) where, for example the operator $(\hat{N}+1)^{-1/2}$ is on the left of $\hat{a}$ or $\hat{N}^{-1/2}$ is on the right of $\hat{a}$ in

$$\hat{e}_S^{i\phi} = (\hat{N}+1)^{-1/2}\hat{a} = \hat{a}\hat{N}^{-1/2} \tag{4.5}$$

and the symmetrical expressions of Lerner (equations (2.30) and (2.31)). In the definition of $\widehat{\cos}_M\,\phi$, because $\bar{n}$ commutes with $\hat{a}$, the corresponding expressions have $(\bar{n}+\frac{1}{2})^{-1/2}$ on the right or left of $\hat{a}$. Our operator definition is thus

$$\widehat{\cos}_M\,\phi = (\hat{a}+\hat{a}^\dagger)/2(\bar{n}+\tfrac{1}{2})^{1/2}. \tag{4.6}$$

By altering the phase of the reference field by $\pi/2$ another measurement can be performed which, in the classical limit, behaves as $\sin\phi$. Following the same procedure as above gives the corresponding definition of $\widehat{\sin}_M\,\phi$ as

$$\widehat{\sin}_M\,\phi = (\hat{a}-\hat{a}^\dagger)/2i(\bar{n}+\tfrac{1}{2})^{1/2}. \tag{4.7}$$

It is apparent that the phase measurements of Gerhardt *et al* (1973, 1974) and, for example, Walker and Carroll (1984) correspond more closely to $\widehat{\cos}_M\,\phi$ and $\widehat{\sin}_M\,\phi$ than to $\widehat{\cos}_S\,\phi$ and $\widehat{\sin}_S\,\phi$. In the experiments of Gerhardt *et al* the phase measurements were normalised by using the largest occurring intensity. Thus the normalisation constant k will only be approximately the same as that above.

5. Discussion

In this paper we have investigated appropriate operators for the quantum mechanical description of the phase of the radiation field. In particular we have considered the Susskind–Glogower, unitary and measured phase operators. All the phase operators exhibit non-classical behaviour for quantum states but reproduce classical phase properties in the appropriate limit.

The original motivation for the introduction of phase operators was to describe the electric field in terms of polar (amplitude and phase) variables. Dirac (1927) suggested that the single mode creation and annihilation operators could be factorised into Hermitian amplitude and unitary exponential phase operators. Susskind and Glogower (1964) demonstrated that the exponential phase operators obtained from such a factorisation are only one-sided unitary. Of the number of different possible exponential phase operators which satisfy the Lerner criterion, and which involve only positive energy states, the Susskind–Glogower operator is that which is closest to being unitary. A nearly unitary operator is, nevertheless, still non-unitary. If the region of interest is that in which the non-unitarity is most apparent, this near unitarity is not a great advantage. In particular, fields with very small mean photon numbers, where non-classical effects might be expected to be observable, represent such a region. If a unitary operator is required, we see no alternative to using the complete Hilbert space which includes the negative energy states, but this requires an infinite set of unmeasurable states which are inaccessible for any physical system. On the other hand, if unitarity is not a necessary requirement, we prefer to abandon it entirely and to define phase operators in terms of the quantities actually measured in usual phase measuring experiments. This leads us to consider the single mode creation and annihilation operators ($\hat{a}^\dagger$ and $\hat{a}$) as the fundamental field operators and to define measured quantities in terms of them. The analysis presented in § 3 of this paper demonstrated that usual phase measuring experiments correspond to measurements of operators proportional to $\hat{a}+\hat{a}^\dagger$. Therefore we define our measured cosine phase operator to be proportional to $\hat{a}+\hat{a}^\dagger$. We note that in squeezing experiments it is in fact the fluctuations in the measured cosine and sine phase operators which can be said to be squeezed (Walls 1983 and references therein). Indeed, we can equally well discuss squeezed states in terms of the measured phase operators or the usual quadrature phases which are identical apart from a normalisation factor.

In conclusion we suggest the adoption of $\widehat{\cos}_M\,\phi$ and $\widehat{\sin}_M\,\phi$ as an operational definition of phase measurement.

Acknowledgments

We thank P L Knight, S Buckle, B J Dalton, B Piraux and B C Sanders for helpful and enlightening discussions. SMB thanks the Science and Engineering Research Council for the award of a postdoctoral fellowship. DTP thanks the British Council for a travel grant.

References

Allen L and Eberly J H 1975 *Optical Resonance and Two-Level Atoms* (New York: Wiley)
Carruthers P and Nieto M M 1965 *Phys. Rev. Lett.* **14** 387
—— 1968 *Rev. Mod. Phys.* **40** 411
Dirac P A M 1927 *Proc. R. Soc.* A **114** 243
—— 1958 *The Principles of Quantum Mechanics* (Oxford: Oxford University Press) 4th edn, p 84
Gerhardt H, Büchler U and Liftin G 1974 *Phys. Lett.* **49A** 119
Gerhardt H, Welling H and Frölich D 1973 *Appl. Phys.* **2** 91
Goldstein H 1980 *Classical Mechanics* (Reading, MA: Addison-Wesley) 2nd edn, p 457
Heitler W 1954 *The Quantum Theory of Radiation* (Oxford: Oxford University Press) 3rd edn, p 64
Hong C K and Mandel L 1985 *Phys. Rev.* A **32** 974
Jackiw R 1968 *J. Math. Phys.* **9** 339
Judge D 1963 *Phys. Lett.* **5** 189
Judge D and Lewis J T 1963 *Phys. Lett.* **5** 190
Knight P L and Allen L 1983 *Concepts of Quantum Optics* (Oxford: Pergamon) p 62
Lerner E C 1968 *Nuovo Cimento* B **56** 183
Lévy-Leblond J M 1976 *Ann. Phys., NY* **101** 319
—— 1977 *Phys. Lett.* **64A** 159
Loudon R 1973 *The Quantum Theory of Light* (Oxford: Oxford University Press)
Loudon R and Shepherd T J 1984 *Opt. Acta* **31** 1243
Louisell W H 1963 *Phys. Lett.* **7** 60

Louisell W H 1973 *Quantum Statistical Properties of Radiation* (New York: Wiley) p 137
Matthys D R and Jaynes E T 1980 *J. Opt. Soc. Am.* **70** 63
Merzbacher E 1970 *Quantum Mechanics* (New York: Wiley) 2nd edn, p 356
Nieto M M 1977 *Phys. Lett.* **60A** 401
Paul H 1974 *Fortschr. Phys.* **22** 657
Pegg D T 1981 *Opt. Commun.* **37** 353
Sanders B C, Barnett S M and Knight P L 1986 *Opt. Commun.* **58** 290
Schweber S S 1984 *Relativity, Groups and Topology* vol II, ed B S DeWitt and R Stora (Amsterdam: North-Holland) p 62
Shapiro J H and Wagner S S 1984 *IEEE J. Quantum Electron.* **QE-20** 803
Slusher R E, Hollberg L W, Yurke B, Mertz J C and Valley J F 1985 *Phys. Rev. Lett.* **55** 2409
Susskind L and Glogower J 1964 *Physics* **1** 49
Walker N G and Carroll J E 1984 *Electron. Lett.* **20** 981
Walls D F 1983 *Nature* **306** 141
Yuen H P and Shapiro J H 1980 *IEEE Trans. Inform. Theor.* **IT-26** 78
Yuen H P and Chan V W S 1983 *Opt. Lett.* **8** 177

2

The Phase Operator

The review in the previous chapter shows the general nature of the problem associated with defining a consistent quantum phase operator as it appeared in the late 1980's. The solution given by extending the Hilbert space to negative energy eigenstates (Fain 1967, Newton 1980, Barnett and Pegg 1986 **Paper 1.7**, Stenholm 1993), while in itself mathematically rigourous, lacks the physical appeal of a Hilbert space of states containing only physically meaningful elements. It is perhaps ironic that the final solution of the problem arose by considering the counter situation: spaces Ψ_s of finite but arbitrary large dimension $s+1$.

The utility of the finite-dimensional spaces Ψ_s stems from their simplicity. The preceding chapter suggests that the *unitary* exponential phase operator $\exp(i\hat{\phi})$ should be a ladder operator for the energy eigenstates. To be unitary it cannot 'annihilate' any state $|f\rangle$ in the sense that $\exp(i\hat{\phi})|f\rangle = 0$ but rather it must map every state into another state of the same norm. This mapping has a simple construction on the finite dimensional Hilbert space Ψ_s. Indeed, although not widely known for nearly 20 years, Popov and Yarunin had recognised this point in 1973. However, the crucial issue when working with finite dimensional subspaces is the limiting procedure used to generate a formalism to deal with states of the form $|f\rangle = \sum_{n=0}^{\infty} c_n|n\rangle$. The limiting procedure adopted by Popov and Yarunin leads to a formalism which yields *non-unitary* exponential phase operators. It was the special method of calculating limiting values, taken of expectation values and not operators themselves, that led to the *Pegg-Barnett phase formalism*.[1] This formalism has become a well-established method for theoretical analysis as well as forming the basis of experimental investigations of the phase properties of optical fields. In this chapter we briefly review the Pegg-Barnett phase formalism and discuss its salient features. The details can be found in the reprinted papers (**Papers 2.1 – 2.6**).

The elements of the formalism and its special limiting procedure were introduced in Pegg and Barnett (1988a **Paper 2.1**). The formalism begins by first considering a $(s+1)$-dimensional space Ψ_s which is spanned by the energy eigenstates $\{|n\rangle : n = 0, 1, \cdots s\}$. An alternate basis for Ψ_s, which more suitable for discussing phase, is given by a set of *phase states* $|\theta_m\rangle$ defined as

$$|\theta_m\rangle = \frac{1}{\sqrt{s+1}} \sum_{n=0}^{s} e^{in\theta_m}|n\rangle. \tag{2.1}$$

Here the phase angle is defined by $\theta_m = \theta_0 + m2\pi/(s+1)$ for $m = 0, 1, \cdots m$, and θ_0 is an arbitrary real phase angle analogous to the origin in the position representation. The set

[1] The phase formalism described here and the associated phase operator are now universally referred to in this way and we fear that not to do so might cause unnecessary confusion. We trust that the reader will not interpret this as an indication of any immodesty on behalf of one of the authors.

$\{|\theta_m\rangle : m = 0, 1, \cdots s\}$ is a complete orthonormal basis for Ψ_s, i.e.

$$\langle\theta_{m'}|\theta_m\rangle = \frac{1}{s+1}\sum_{n=0}^{s} e^{in(\theta_m-\theta_{m'})} = \delta_{m,m'}, \tag{2.2}$$

$$\sum_{m=0}^{s} |\theta_m\rangle\langle\theta_m| = \hat{1}. \tag{2.3}$$

More particularly, the sets $\{|n\rangle : n = 0, 1, \cdots s\}$ and $\{|\theta_m\rangle : m = 0, 1, \cdots s\}$ are two mutually unbiased bases for Ψ_s as $|\langle n|\theta_m\rangle|^2 = 1/(s+1)$ for $n, m = 0, 1, \cdots, s$. The exponential phase operator is defined in terms of the phase state basis as

$$e^{i\hat{\phi}_\theta} = \sum_{m=0}^{s} e^{i\theta_m}|\theta_m\rangle\langle\theta_m|, \tag{2.4}$$

or equivalently, in terms of the energy eigenstates as

$$e^{i\hat{\phi}_\theta} = \sum_{n=0}^{s-1} |n\rangle\langle n+1| + e^{i(s+1)\theta_0}|s\rangle\langle 0|. \tag{2.5}$$

The subscript θ serves as a reminder that the values of phase variable θ_m lie in the range from θ_0 to $\theta_0 + 2\pi$. This construction ensures that the phase states are the eigenstates of $e^{i\hat{\phi}_\theta}$, i.e.,

$$e^{i\hat{\phi}_\theta}|\theta_m\rangle = e^{i\theta_m}|\theta_m\rangle. \tag{2.6}$$

The exponential phase operator is unitary, because

$$(e^{i\hat{\phi}_\theta})^\dagger e^{i\hat{\phi}_\theta} = e^{i\hat{\phi}_\theta}(e^{i\hat{\phi}_\theta})^\dagger = \sum_{n=0}^{s} |n\rangle\langle n| = \hat{1}_s \tag{2.7}$$

where $\hat{1}_s$ is the identity operator on Ψ_s. The Hermitian phase operator representing the phase angle itself was introduced in (Pegg and Barnett 1988b **Paper 2.2**) and (Barnett and Pegg 1989 **Paper 2.3**) as simply

$$\hat{\phi}_\theta = \sum_{m=0}^{s} \theta_m|\theta_m\rangle\langle\theta_m|. \tag{2.8}$$

The operator on the left hand sides of Eqs. (2.4) and (2.5) is the exponential of $i\hat{\phi}_\theta$. Moreover, the operator corresponding to an arbitrary function of the phase angle $f(\theta)$ was introduced in Barnett and Pegg (1992 **Paper 3.3**) and also discussed in Pegg and Barnett (1997a **Paper 2.6**) as

$$f(\hat{\phi}_\theta) = \sum_{m=0}^{s} f(\theta_m)|\theta_m\rangle\langle\theta_m|. \tag{2.9}$$

It follows that the Hermitian conjugate of $e^{i\hat{\phi}_\theta}$ is the operator

$$(e^{i\hat{\phi}_\theta})^\dagger = \sum_{m=0}^{s} e^{-i\theta_m}|\theta_m\rangle\langle\theta_m| = e^{-i\hat{\phi}_\theta}. \tag{2.10}$$

This construction on the finite dimensional space Ψ_s elegantly sidesteps the problems faced with the infinite dimensional Hilbert space discussed in the previous chapter. Its simplicity is particularly appealing. The question remains, however, of how to deal with states such as $|f\rangle = \sum_{n=0}^{\infty} c_n|n\rangle$ which reside in the infinite dimensional Hilbert space. For each value of s, there exists a phase operator defined on Ψ_s; this leads naturally to sequences of operators, such as e.g. $\hat{\phi}_\theta|_{s=S}, \hat{\phi}_\theta|_{s=S+1}, \hat{\phi}_\theta|_{s=S+2}\cdots$ which are phase angle operators defined on the sequence of spaces $\Psi_S, \Psi_{S+1}, \Psi_{S+2}, \cdots$ for arbitrarily large integer S. The space Ψ_s contains a state $|f\rangle_s$ which approximates $|f\rangle$ to arbitrary precision for a sufficiently large value of s, such that the modulus of the overlap $\langle f|f\rangle_s$ is as close to unity as desired. Thus the issue of including states such as $|f\rangle$ entails dealing with such sequences as s tends to infinity.

Popov and Yarunin's approach (1973, 1992a, 1992b) to this question was to take the infinite-s limit of the operator sequences themselves. For example, the phase angle operator has the following expansion in the energy eigenbasis (Popov and Yarunin 1992a, Barnett and Pegg 1989 **Paper 2.3**):

$$\langle n|\hat{\phi}_\theta|n\rangle = \frac{\theta_0+\theta_s}{2}, \tag{2.11}$$

$$\langle m|\hat{\phi}_\theta|n\rangle = \frac{\exp[i(m-n)\theta_0]}{\exp[i(m-n)2\pi/(s+1)]-1} \quad \text{for } n \neq m. \tag{2.12}$$

In the limit $s \to \infty$ the matrix elements for fixed values of n and m are

$$\langle n|\hat{\phi}_\infty|n\rangle \equiv \lim_{s\to\infty}\langle n|\hat{\phi}_\theta|n\rangle = \frac{\theta_0+\theta_s}{2}, \tag{2.13}$$

$$\langle m|\hat{\phi}_\infty|n\rangle \equiv \lim_{s\to\infty}\langle m|\hat{\phi}_\theta|n\rangle = \frac{i}{m-n}\exp[i(m-n)\theta_0] \quad \text{for } n \neq m, \tag{2.14}$$

where the operator $\hat{\phi}_\infty$ is defined by these expressions. The operator $\hat{\phi}_\infty$ is Hermitian and acts on the infinite dimensional Hilbert space. Similarly, for any normalised states $|f\rangle, |g\rangle$,

$$\lim_{s\to\infty}\langle f|e^{i\hat{\phi}_\theta}|g\rangle = \sum_{n=0}^{\infty}\langle f|n\rangle\langle n+1|g\rangle \tag{2.15}$$

and so (Popov and Yarunin 1992a)

$$\lim_{s\to\infty}\langle f|e^{i\hat{\phi}_\theta}|g\rangle = \langle f|\widehat{e^{i\phi}}|g\rangle \tag{2.16}$$

where $\widehat{e^{i\phi}}$ is the Susskind-Glogower exponential phase operator discussed in the previous chapter. As the operator $\widehat{e^{i\phi}}$ is not unitary it cannot be expressed as the exponentiation of a Hermitian operator, and in particular $\widehat{e^{i\phi}}$ is *not the exponential of* $i\hat{\phi}_\infty$. Thus Popov and Yarunin's approach of taking the limits of the operator sequences leads to the same difficulties encountered by earlier workers. The appeal of the Ψ_s operators is subsequently lost in their approach.

The breakthrough presented in Pegg and Barnett (1988a **Paper 2.1**) was to treat the $s \to \infty$ limit in a different way. A careful analysis of the limiting procedure is given in Barnett and Pegg (1992 **Paper 3.3**). The calculation of all expectation values etc. should be made on Ψ_s for each value of s; this generates sequences of expectation values such as $\langle\hat{\phi}_\theta\rangle|_{s=S}$, $\langle\hat{\phi}_\theta\rangle|_{s=S+1}, \langle\hat{\phi}_\theta\rangle|_{s=S+2}\cdots$ and $\langle\hat{\phi}_\theta^2\rangle|_{s=S}, \langle\hat{\phi}_\theta^2\rangle|_{s=S+1}, \langle\hat{\phi}_\theta^2\rangle|_{s=S+2}\cdots$ where S is an arbitrarily large integer. The limit points of the sequences of expectation values are the values representing expectation values of phase operators. The limits do not exist for all states on the infinite dimensional Hilbert space. However they do exist for physically meaningful operators for a class of states $\mathcal{P}$ called *physically accessible or preparable* states which were introduced in Pegg and Barnett (1988a **Paper 2.1**) and later explored in more detail in Vaccaro and Pegg (1990b **Paper 4.4**). States belonging to $\mathcal{P}$, which hereafter we shall refer to simply as *physical states*, have the form $|p\rangle = \sum_{n=0}^{\infty} p_n|n\rangle$ where $\langle p|\hat{N}^m|p\rangle = \sum_{n=0}^{\infty}|p_n|^m < \infty$ for any given integer value of m, that is, all moments of the excitation number are finite. These states can be prepared from the vacuum state by coupling the oscillator to a finite energy source for a finite time by means of a finite interaction.

On first sight, this method of calculating expectation values may appear to be a major departure from the dogma of quantum physics. Indeed it does represent a major innovation in the calculation of quantum phase properties. However, the underlying procedure is analogous to well-established techniques in distribution theory and the related theory of generalised functions.[2] A distribution $T[\varphi]$ is a linear and continuous functional of smooth functions φ, that is, of functions φ which are differentiable to all orders: $|\partial\varphi^n(x)/\partial x^n| < \infty$ for any given value of n . The functional can be written in terms of a scalar product as $T[\varphi] = \int \varphi(x)^* T(x) dx = \langle\varphi|T\rangle$. The space of distributions $\{|T\rangle\}$ is larger than the Hilbert space of square-integrable functions L_2, which itself is larger than the space of smooth functions $\{|\varphi\rangle\}$. As a particular example, consider the Dirac "δ function" $\delta(x - x_0)$ which represents the eigenstate of position in the momentum representation, and which does not lie in the Hilbert space L_2 of square integrable functions. Actually this object is not a function at all but rather a distribution: indeed the end result of all calculations involving this object are scalar products such as $\delta_{x_0}[\varphi] = \int \varphi^*(x)\delta(x - x_0)dx$. To be strictly rigourous, we should refer to the "$\delta_{x_0}[\varphi]$ distribution" rather than the "δ function". The distribution $\delta_{x_0}[\varphi]$ has many equivalent definitions.[3] One is that it is the limit point as $\lambda \to \infty$ of the distribution $\delta_{x_0,\lambda}[\varphi] = \int \varphi(x)^* \delta_{x_0,\lambda}(x) dx$ where $\delta_{x_0,\lambda}(x) = \frac{\sin[\lambda(x-x_0)]}{\pi(x-x_0)}$. Calculations involving the $\delta_{x_0}[\varphi]$ distribution essentially entail performing all scalar products for the distributions $\delta_{x_0,\lambda}[\varphi]$, and then taking the $\lambda \to \infty$ limit of the results. The inclusion of distributions in quantum mechanics is necessary for incorporating the eigenstates of the position and momentum observables in a mathematically rigourous fashion.

The Pegg-Barnett formalism can be compared to distribution theory in the following way. The restricted class of physical states in the Pegg-Barnett phase formalism is analogous to the class of smooth functions in distribution theory. Taking the $s \to \infty$ limit of expectation values in the Pegg-Barnett phase formalism is analogous to taking the limit $\lambda \to \infty$ of scalar products in distribution theory. The Pegg-Barnett formalism allows the rigourous treatment of eigenstates of the phase observable which are analogous to the eigenstates of the position and momentum observables. Moreover, just as there is a formal (or symbolic) representation of $\delta_{x_0,\lambda}[\varphi]$ as a regular function $\delta(x - x_0)$ in the infinite λ limit, so too is there a formal representation of the Pegg-Barnett formalism in the infinite-s limit. Thus the solution to the phase problem given by the Pegg-Barnett formalism is analogous to Dirac's solution to the problem of defining the spectrum of the position and momentum observables. Discussion of further mathematical properties of the Pegg-Barnett formalism will be given in the chapter that follows. For now we discuss the formal and physically intuitive aspects of the Pegg-Barnett formalism.

The reason the class of physical states $\mathcal{P}$ was introduced in Pegg and Barnett (1988a **Paper 2.1**) is that various operator quantities have simplified and natural forms with respect to this class. In particular, the commutator of the annihilation ($\hat{a}$) and creation ($\hat{a}^\dagger$) operators is essentially unity for these states. These operators are defied on Ψ_s as

$$\hat{a} = e^{i\hat{\phi}_\theta}\sqrt{\hat{N}}\,, \tag{2.17}$$

$$\hat{a}^\dagger = \sqrt{\hat{N}}e^{-i\hat{\phi}_\theta}\,, \tag{2.18}$$

and their commutator is given by

$$[\hat{a}, \hat{a}^\dagger] = \hat{\mathbb{1}}_s - (s+1)|s\rangle\langle s|\,. \tag{2.19}$$

[2] A brief review of distribution theory is given in Appendix A of Messiah (1961) and Chapter 2 of Lighthill (1958) gives an introduction to generalised functions.

[3] More precisely, the distribution $\delta_{x_0}[\varphi]$ is defined as an equivalence class. The sequence of distributions given by $\delta_{x_0,\lambda}(x) = \frac{\sin[\lambda(x-x_0)]}{\pi(x-x_0)}$ for $\lambda \to \infty$ is one element of the equivalence class.

The difference in form in comparison with the more familiar commutator is an inevitable consequence of working in a finite-dimensional state space in which all commutators necessarily have zero trace. The matrix elements of arbitrary powers of the commutator, e.g.

$$\langle p'|[\hat{a},\hat{a}^\dagger]^n|p\rangle = \langle p'|\left[\mathbb{1}_s + [(-s)^n - 1)|s\rangle\langle s|\right]|p\rangle\,, \tag{2.20}$$

is unity for states $|p\rangle$ and $|p'\rangle$ belonging to $\mathcal{P}$ in the infinite-s limit and so, *formally*, the commutator

$$[\hat{a},\hat{a}^\dagger]_{\mathcal{P}} = \hat{\mathbb{1}} \tag{2.21}$$

is all that is needed when dealing with physical states $\mathcal{P}$. Here, and in the following, we use the subscript $\mathcal{P}$ to represent symbolically the fact that the operator $(\hat{Q}_{\mathcal{P}})^n$ has the same matrix elements as $\hat{Q}^n$ for states belonging to $\mathcal{P}$ in the infinite-s limit for arbitrary values of n.

The exponential phase operator $e^{i\hat{\phi}_\theta}$ and its Hermitian conjugate have the following simplified forms for physical states:

$$(e^{i\hat{\phi}_\theta})_{\mathcal{P}} = \sum_{n=0}^{\infty} |n\rangle\langle n+1|, \tag{2.22}$$

$$(e^{-i\hat{\phi}_\theta})_{\mathcal{P}} = \sum_{n=0}^{\infty} |n+1\rangle\langle n|. \tag{2.23}$$

Moreover

$$(e^{i\hat{\phi}_\theta}e^{-i\hat{\phi}_\theta})_{\mathcal{P}} = (e^{-i\hat{\phi}_\theta}e^{i\hat{\phi}_\theta})_{\mathcal{P}} = \hat{\mathbb{1}}. \tag{2.24}$$

One sees immediately that the appeal of the Ψ_s operators is retained in the Pegg-Barnett formalism. In particular, the difficulties associated with the Susskind-Glogower operators are elegantly circumvented using the limiting procedure and the subspace of physical states.

We now turn to an essential difference between *physical implications* of the Pegg-Barnett formalism and other approaches. By "physical implications" we mean the phase properties attributed to particular states. These properties are easily drawn out using the phase probability density, which was first introduced in Barnett and Pegg (1992 **Paper 3.3**) and further explored in Pegg and Barnett (1997a **Paper 2.6**). The phase probability distribution $Pr(\cdot)$ on Ψ_s for arbitrary state $|f\rangle$ is given by

$$Pr(\theta_m) = |\langle\theta_m|f\rangle|^2 = \frac{1}{s+1}\left|\sum_{n=0}^{s} e^{-in\theta_m}\langle n|f\rangle\right|^2. \tag{2.25}$$

Expectation values of an arbitrary function of phase $F(\hat{\phi}_\theta)$ can be written as

$$\langle f|F(\hat{\phi}_\theta)|f\rangle = \langle f|\left(\sum_{m=0}^{s} F(\theta_m)|\theta_m\rangle\langle\theta_m|\right)|f\rangle = \sum_{m=0}^{s} Pr(\theta_m)F(\theta_m)\,, \tag{2.26}$$

or, in terms of a probability *density* $P(\cdot)$, as

$$\langle f|F(\hat{\phi}_\theta)|f\rangle = \sum_{m=0}^{s} P(\theta_m)F(\theta_m)\Delta \tag{2.27}$$

where $\Delta = \frac{2\pi}{s+1}$ and

$$P(\vartheta) = \frac{s+1}{2\pi}Pr(\vartheta) = \frac{1}{2\pi}\sum_{n,m=0}^{s} e^{i(m-n)\vartheta}\langle n|f\rangle\langle f|m\rangle\,. \tag{2.28}$$

For physical state $|p\rangle \in \mathcal{P}$, the phase probability density converges in the infinite-s limit to (Barnett and Pegg 1992 **Paper 3.3**)

$$P_{\mathcal{P}}(\vartheta) = \frac{1}{2\pi} \sum_{n,m=0}^{\infty} e^{i(m-n)\vartheta} \langle n|p\rangle\langle p|m\rangle \tag{2.29}$$

and the sum in Eq. (2.27) becomes a Riemann integral over a 2π interval; in consequence the expectation value has the particularly simple form

$$\langle F(\hat{\phi}_\theta)\rangle_{\mathcal{P}} = \int_{2\pi} P_{\mathcal{P}}(\theta) F(\theta) d\theta. \tag{2.30}$$

These results follow directly from the construction of phase operators on the Ψ_s spaces in the infinite-s limit. A quite different approach to deriving Eq. (2.29) is given in the tutorial review (Pegg and Barnett 1997a **Paper 2.6**). There it is shown that Eq. (2.29) is the only suitable phase probability density for a harmonic oscillator based on the *physical motivation* that a time delay of Δt should produce a phase shift of $-\omega\Delta t$, that is, it should translate the phase probability density as $P_{\mathcal{P}}(\vartheta) \mapsto P_{\mathcal{P}}(\vartheta - \omega\Delta t)$, where ω is the angular frequency of the oscillator.

The equivalence between a time delay and a translation in phase implies that the stationary states of the Hamiltonian have a uniform phase probability density, because for these states $P_{\mathcal{P}}(\vartheta)$ is constant in time and so $P_{\mathcal{P}}(\vartheta - \omega\Delta t) = P_{\mathcal{P}}(\vartheta)$. The set of energy eigenstates are therefore a crucial testing ground for the physical implications of any phase description. It is easy to show that in the Pegg-Barnett formalism the phase probability density is

$$P_{\mathcal{P}}(\theta) = \frac{1}{2\pi} \tag{2.31}$$

for the energy eigenstate $|n\rangle$ for $n = 0, 1, 2, \cdots$. That is, every energy eigenstate has a uniform phase distribution; thus the phase observable for these states is purely random with maximal entropy. This result follows from the mutually unbiased nature of the two bases $\{|n\rangle : n = 0, 1, \cdots s\}$ and $\{|\theta_m\rangle : m = 0, 1, \cdots s\}$ of Ψ_s mentioned above.

These results differ from those of a number of other approaches. In particular the Susskind-Glogower (Susskind and Glogower 1964 **Paper 1.5**) and Garrison-Wong (Garrison and Wong 1970) approaches attribute a non-random phase to the energy eigenstates, that is, they assign higher probabilities to some phase angles compared than others. D'Ariano and Paris (1993) showed this explicitly for the Susskind-Glogower case. In these approaches the vacuum state, in particular, has *preferred phase angles*. However the preferred angles have the same values irrespective of the setting of the zero in the phase angle coordinate. This suggests that the preferred angles are a *mathematical artifact* arising from the approaches themselves and do not have physical relevance. Moreover the generation of the vacuum state for two separated oscillators does not require any mutual interaction between the oscillators. Yet in these approaches every oscillator in the vacuum state shares the same preferred phases. This further supports the conclusion that the preferred angles are of mathematical rather than physical origin. In contrast, in the Pegg-Barnett formalism, the vacuum state of all oscillators has random phase and these issues do not arise.

The existence of the phase probability density suggests that there is an associated probability operator measure (POM). Indeed, for physical states we identify a continuous probability operator measure (POM) as the set $\{d\hat{\Pi}(\theta)\}$ with (Vaccaro and Pegg 1993 **Paper 3.4**)

$$d\hat{\Pi}(\theta) = |\theta\rangle_\infty\langle\theta|d\theta \tag{2.32}$$

where

$$|\theta\rangle_\infty\langle\theta| = \frac{1}{2\pi}\sum_{n,m=0}^{\infty} e^{i(n-m)\theta}|n\rangle\langle m|, \tag{2.33}$$

$$\int_{2\pi} d\hat{\Pi}(\theta) = \hat{\mathbb{1}}, \tag{2.34}$$

$$P_{\mathcal{P}}(\theta)d\theta = \langle d\hat{\Pi}(\theta)\rangle. \tag{2.35}$$

The same POM was derived by Shapiro and Shepard (1991) using a quite different approach based on the Susskind-Glogower operators and quantum estimation theory. However, Shapiro and Shepard find that the right-hand side of Eq (2.35) is not consistent with a quantum phase observable on H_∞. Instead they associate the probability density given by Eq (2.35) with the measurement statistics of a *classical random variable* ϕ in place of a quantum phase observable. Nevertheless, as Shapiro and Shepard use the same phase probability density as $P_{\mathcal{P}}(\theta)$, their description of phase agrees with the Pegg-Barnett formalism. More will be said about this agreement in the following chapter.

The commutator between the phase angle operator and the excitation number operator is interesting not only for its departure from Dirac's version (Dirac 1927), but also for the way it manifests the cyclic nature of the phase angle. For physical states the commutator is given by (Pegg and Barnett 1989 **Paper 2.4**)

$$[\hat{N}, \hat{\phi}_\theta]_{\mathcal{P}} = i\left[\hat{\mathbb{1}} - 2\pi|\theta_0\rangle_\infty\langle\theta_0|\right] \tag{2.36}$$

where the projector $|\theta_0\rangle_\infty\langle\theta_0|$ is given by Eq. (2.33) with $\theta = \theta_0$. This differs markedly from Dirac's commutator $[\hat{N}, \hat{\phi}_\theta] = i\hat{\mathbb{1}}$. Dirac derived his commutator using the commutator-Poisson bracket correspondence and a non-cyclic variable for the phase angle. That is, in Dirac's theory phase angles differing by multiples of 2π are different; this is an unphysical characteristic for an isolated harmonic oscillator for which one period is equivalent to any other. In the Pegg-Barnett formalism, however, the phase angle is cyclic due to the cyclic property of the phase states themselves since from Eq. (2.1) we find

$$|\theta_m + 2n\pi\rangle = |\theta_m\rangle. \tag{2.37}$$

Indeed the expectation value of the time derivative of the phase angle, given by the Heisenberg equation of motion, for the physical state $|p\rangle \in \mathcal{P}$ is

$$\frac{d}{dt}\langle p|\hat{\phi}_\theta|p\rangle = \langle p|i[\omega\hat{N}, \hat{\phi}_\theta]_{\mathcal{P}}|p\rangle = -\omega + \omega 2\pi\, P(\theta_0) \tag{2.38}$$

where ω is the angular frequency of the oscillator. This can be compared with the corresponding expression for a classical cyclic phase variable $\phi(t)$ as follows. $\phi(t)$ is a "sawtooth" function of time in that it decreases linearly with time with slope $-\omega$ until it reaches a value of θ_0 where upon it undergoes a jump of 2π, i.e., for $t \geq 0$

$$\phi(t) = \phi_0 - \omega t + \sum_{n=0}^{\infty} 2\pi\, H([\theta_0 - 2n\pi - (\phi_0 - \omega t)]\,, \tag{2.39}$$

where $H(\cdot)$ is the Heaviside step function; the initial ($t = 0$) value of the phase is ϕ_0 which is, of course, in the range $\theta_0 \leq \phi_0 < \theta_0 + 2\pi$. Consider an ensemble of oscillators whose initial phase ϕ_0 is described by the probability density $P_{\text{cl}}(\phi_0)$, that is, $P_{\text{cl}}(\phi_0)d\phi_0$ is the probability that a given oscillator has an initial phase in the interval ϕ_0 and $\phi_0 + d\phi_0$. In Pegg and Barnett

(1989 **Paper 2.4**) it is shown that the average rate of change of $\phi(t)$ over the ensemble is given by

$$\frac{d}{dt}\langle\phi(t)\rangle = -\omega + 2\pi\omega P_{\rm cl}(\theta_0) \tag{2.40}$$

which agrees precisely with Eq. (2.38). The second term on the right-hand side arises from the derivative of the Heaviside step function in Eq. (2.39) and thus is due to the periodic nature of the phase variable. It is clear then, that the departure of the excitation number-phase commutator in the Pegg-Barnett formalism in Eq. (2.36) from the so-called *canonical* commutation relation "$[\hat{N}, \hat{\phi}_\theta] = i\hat{1}$" is due to the cyclic nature of the phase observable, which is an essential feature of a harmonic oscillator. This result resolved a long-standing problem associated with quantum phase.

Another mechanical system which is described by a periodic angle is given by the plane rotator. This system is an idealisation of, for example, a bead which is free to move on a circular wire. To be specific, let the wire lie in the $x-y$ plane so the bead is free to rotate about the z axis. The state space of the rotator is then spanned by the eigenstates of the azimuthal angular momentum observable $\hat{L}_z$. This observable has the eigenvalue spectrum $m\hbar$ for $m = 0, \pm 1, \pm 2, \cdots$, that is, the spectrum is unbounded above as well as below. Given that one of the hurdles in defining a consistent Hermitian phase operator of the harmonic oscillator was due to the fact that the energy spectrum of the harmonic oscillator is bounded below (i.e. there are no negative eigenvalues), one may expect that the definition of the rotation angle not require the same care as with the phase operator. However this is not the case. Barnett and Pegg (1990a **Paper 2.5**) show that inconsistencies arise when dealing with a periodic rotation operator of the type advocated by Judge and Lewis (1963 **Paper 1.4**) and are manifest in the commutator in Eq. (1.10). In particular, in the Appendix of their paper Barnett and Pegg show that this commutator is inconsistent with the eigenvalue spectrum of the angular momentum observable $\hat{L}_z$. To resolve these problems Barnett and Pegg (1990a **Paper 2.5**) applied the limiting procedure they invented for the harmonic oscillator to the plane rotator. Their plane rotator formalism can be described in very brief terms as follows.

The analogy of the $(s+1)$-dimensional Hilbert space of the harmonic oscillator is the $(2l+1)$-dimensional Hilbert space spanned by the eigenstates $|m\rangle$ of the $\hat{L}_z$ operator for $m = -l, \ldots, -1, 0, 1, \ldots, l$. The set of $2l+1$ eigenstates of the rotation angle are defined as

$$|\theta_n\rangle = \frac{1}{\sqrt{2l+1}} \sum_{m=-l}^{l} e^{-im\theta_n} |m\rangle, \tag{2.41}$$

where $\theta_n = 2\pi n/(2l+1)$ and $n = 0, 1, \ldots, 2l$. The rotation angle is defined as

$$\hat{\phi}_\theta = \sum_{n=0}^{2l} \theta_n |\theta_n\rangle\langle\theta_n| \,. \tag{2.42}$$

In place of the $s \to \infty$ limit for the harmonic oscillator, here the limit as $l \to \infty$ is taken of expectation values. The formalism of the plane rotator then follows closely that of the harmonic oscillator. The details can be found in Barnett and Pegg (1990a **Paper 2.5**). In particular the rotation angle-angular momentum commutator for physical states is

$$[\hat{\phi}_\theta, \hat{L}_z]_{\mathcal{P}} = i\hbar[1 - (2l+1)|\theta_0\rangle\langle\theta_0|] \tag{2.43}$$

where, in analogy with the harmonic oscillator, the physical states are defined as having finite moments of $\hat{L}_z$. This commutator is consistent with the commutator of Judge and Lewis (1963 **Paper 1.4**) in Eq. (1.10). However it arises in the Pegg-Barnett formalism only for physical states and does not apply, in particular, when acting on the eigenstates of the rotation angle. The inconsistencies mentioned above do not occur in the Pegg-Barnett formalism.

Unitary Phase Operator in Quantum Mechanics.

D. T. PEGG (*) and S. M. BARNETT (**)

(*) *School of Science, Griffith University, Nathan, Brisbane 4111, Australia*
(**) *Theoretical Physics Division, Harwell Laboratory*
Oxfordshire OX11 ORA, England
Wolfson College, Oxford OX2 6UD, England

(received 5 February 1988; accepted in final form 4 May 1988)

PACS. 03.65 – Quantum theory; quantum mechanics.
PACS. 03.70 – Theory of quantized fields.
PACS. 42.50 – Quantum optics.

Abstract. – The difficulties in formulating a natural and simple operator description of the phase of a quantum oscillator or single-mode electromagnetic field have been known for some time. We present a unitary phase operator whose eigenstates are well-defined phase states and whose properties coincide with those normally associated with a phase. The corresponding phase eigenvalues form only a dense subset of the real numbers. A natural extension to the definition of a time-measurement operator yields a corresponding countable infinity of eigenvalues.

In his original description of the quantized electromagnetic field, Dirac [1] postulated the existence of a Hermitian phase operator $\hat{\phi}$. This proposed operator would exist in a unitary exponential form $\exp[\pm i\hat{\phi}]$ which, together with the square root of the number operator $\hat{N}$, would appear in a decomposition of the creation and annihilation operators. However, difficulties have been found with this postulate. The problem of multivaluedness was easily overcome [2], but Susskind and Glogower [3] have emphasized the difficulty in actually finding an exponential phase operator which is unitary. Indeed it has been considered for some time that such an operator, with all the simple and desirable properties that would make it acceptable as a quantum phase, may not even exist [4]. For a review of these problems see ref. [4, 5]. A unitary phase operator with simple properties has been found [6], but this operates in an extended Hilbert space which includes unphysical negative number states.

Hermitian combinations of nearly-unitary operators have been used for a definition of phase [5]. This formulation exhibits desirable phaselike properties for fields with high average photon number. However, for fields with small average photon number, some peculiar effects associated with the nonunitarity of these operators become apparent. For example, the vacuum expectation values of $\cos^2\phi$ and $\sin^2\phi$ are not 1/2. This makes it difficult to interpret the vacuum as a state of random phase. Also, the eigenstates of these «cosine» and «sine» operators do not evolve in the simple manner which one would associate with a phase [5]. The nearly unitary operators (and Hermitian combinations of them) exhibit the anomalies associated with their nonunitarity in the very region of most interest—the «quantum» region of small photon number. Therefore, it has been suggested that unitarity be abandoned entirely and phase operators be defined in terms of the actual phase measuring processes used [6].

On the other hand, even though there have been difficulties in finding a suitably behaved phase operator, there do exist *states* which can be associated with a well-defined phase [7]

$$|\theta\rangle = \lim_{s\to\infty}(s+1)^{-1/2}\sum_{n=0}^{s}\exp[in\theta]\,|n\rangle\,, \tag{1}$$

where $|n\rangle$ are the $(s+1)$ number states spanning the $(s+1)$-dimensional space. These states have a natural and simple time-development; at time t the state evolves to $|\theta-\omega t\rangle$, where ω is the frequency of the harmonic oscillator or single mode field under consideration. Such states would, therefore, provide an ideal description of phase if used as phase states [4]. The attractiveness of these states and the apparent absence of a unitary or Hermitian operator of which they are eigenstates have caused serious problems. It has even been proposed that a fundamental rule of quantum mechanics, which assigns a Hermitian operator to every physical property, should be abandoned [4]. This underlines the extent of the problem.

The phase state (1) is well defined in the $(s+1)$-dimensional space, but care must be exercised in taking the $s\to\infty$ limit. Indeed, we maintain that the failure to obtain well-behaved phase operators was a result of disregarding the limiting process in (1) and replacing s by ∞ in the summation. Therefore, we shall work with the $(s+1)$-dimensional linear space Ψ spanned by $|n\rangle$ where $n=0, 1, \ldots, s$, where s can be made arbitrarily large and will be allowed to tend to infinity when expectation values are calculated. This space has no topological structure and it is debatable [8] whether physical experiments can determine the precise topology to introduce into Ψ, or indeed whether or not any space other than Ψ is actually required [8], but we shall not pursue these issues here. Within this space Ψ we select a reference phase state

$$|\theta_0\rangle = (s+1)^{-1/2}\sum_{n=0}^{s}\exp[in\theta_0]\,|n\rangle\,, \tag{2}$$

and find the subset of states $|\theta_m\rangle$, defined by replacing θ_0 by θ_m in (2), which are orthogonal to this state. The overlap between $|\theta_0\rangle$ and $|\theta_m\rangle$ is

$$\langle\theta_m|\theta_0\rangle = (s+1)^{-1}\sum_{n=0}^{s}\exp[in(\theta_0-\theta_m)]\,. \tag{3}$$

Summation of this geometrical series shows that $|\theta_0\rangle$ and $|\theta_m\rangle$ are orthogonal if $\exp[i(s+1)(\theta_0-\theta_m)]$ is unity with $\theta_0\neq\theta_m$. Consequently, the $(s+1)$-states $|\theta_m\rangle$ with

$$\theta_m = \theta_0 + \frac{2m\pi}{s+1}\,(m=0, 1, \ldots s) \tag{4}$$

form a complete orthogonal basis spanning the state space. The value of θ_0 can be chosen as any point in the continuum, that is on the real line. The choice of θ_0 determines the particular basis of the space. We can expand a number state $|n\rangle$ in terms of the θ_m-phase state basis $|\theta_m\rangle$ as

$$|n\rangle = \sum_{m=0}^{s}|\theta_m\rangle\langle\theta_m|n\rangle = (s+1)^{-1/2}\sum_{m=0}^{s}\exp[-in\theta_m]\,|\theta_m\rangle\,. \tag{5}$$

Equations (2) and (5) show the relationship between the phase and number basis states. We see that a system in a number state is equally likely to be found in any state $|\theta_m\rangle$ and, as θ_0 can be given any value, all phase states are equally likely. It is also evident that a system prepared in a phase state is equally likely to be found in any number state.

We seek a unitary operator $\widehat{\exp_\theta[i\phi]}$ whose eigenstates are the phase states $|\theta_m\rangle$:

$$\widehat{\exp_\theta[i\phi]}\,|\theta_m\rangle = \exp[i\theta_m]\,|\theta\rangle\,, \tag{6}$$

$$\widehat{\exp_\theta[-i\phi]}|\theta_m\rangle = \exp[-i\theta_m]\,|\theta_m\rangle\,. \tag{7}$$

Here, the position of the caret is intended to indicate that the whole expression is an operator. It is straightforward to show that the required unitary phase operator is

$$\widehat{\exp_\theta[i\phi]} = |0\rangle\,\langle 1| + |1\rangle\,\langle 2| + \ldots + |s-1\rangle\,\langle s| + \exp[i(s+1)\theta_0]\,|s\rangle\,\langle 0|\,, \tag{8}$$

with $\widehat{\exp_\theta[-i\phi]}$ being the Hermitian conjugate of this. Remarkably we also find from (8) that this simple addition to the Susskind-Glogower-type operators [3-7] has rendered $\widehat{\exp_\theta[i\phi]}$ unitary:

$$\widehat{\exp_\theta[i\phi]}\,\widehat{\exp_\theta[-i\phi]} = \widehat{\exp_\theta[-i\phi]}\,\widehat{\exp_\theta[i\phi]} = 1\,. \tag{9}$$

The new phase operators $\widehat{\exp_\theta[\pm i\phi]}$ differ from the well-known Susskind-Glogower operators only by the presence of the last term in eq. (8) which converts «ladder» behaviour into «cycle» behaviour. It is only possible to consider such a term because we have not taken the limit $s \to \infty$ prematurely. We note that the Susskind-Glogower operator is equal to the average of $\widehat{\exp_\theta[i\phi]}$ over all values of θ_0, which range over the continuum, in the limit as $s \to \infty$. However, the Susskind-Glogower operator is not unitary because the average of different unitary operators is not unitary.

From the unitary phase operators we can form Hermitian cosine and sine combinations. These operators are better behaved than their counterparts formed from the conventional Susskind-Glogower operators [3-7], with properties that are much more desirable for a description of phase, even in the «quantum» region. In particular we find

$$[\widehat{\cos_\theta\phi}]^2 + [\widehat{\sin_\theta\phi}]^2 = 1, \tag{10}$$

$$[\widehat{\cos_\theta\phi},\, \widehat{\sin_\theta\phi}] = 0\,, \tag{11}$$

$$\langle n|\,[\widehat{\cos_\theta\phi}]^2\,|n\rangle = \langle n|\,[\widehat{\sin_\theta\phi}]^2\,|n\rangle = \frac{1}{2}, \qquad \forall n\,, \tag{12}$$

where

$$\widehat{\cos_\theta\phi} = \frac{1}{2}(\widehat{\exp_\theta[i\phi]} + \widehat{\exp_\theta[-i\phi]}) \quad \text{and} \quad \widehat{\sin_\theta\phi} = \frac{1}{2i}(\widehat{\exp_\theta[i\phi]} - \widehat{\exp_\theta[-i\phi]})\,.$$

Our new definition of the phase operator is consistent with the phase of the vacuum being random and (11) means that $\cos\phi$ and $\sin\phi$ are compatible observables.

Within the space Ψ we can also define the number operator $\hat{N}$ by its action on the $s+1$ number states:

$$\hat{N}|n\rangle = n|n\rangle\,. \tag{13}$$

We can now define creation and annihilation operators acting on Ψ in terms of the unitary phase operator and the number operator:

$$\hat{a} = \overline{\exp}_\theta[i\phi]\hat{N}^{1/2}, \tag{14}$$

$$\hat{a}^\dagger = \hat{N}^{1/2}\overline{\exp}_\theta[-i\phi]. \tag{15}$$

The action of these operators mimics that of the conventional creation and annihilation operators if our state of interest is orthogonal to highly excited number states $|n\rangle$ with $n \sim s$. Equations (14) and (15) express $\hat{a}$ and $\hat{a}^\dagger$ in terms of a *unitary operator and a Hermitian one* ($\hat{N}^{1/2}$). Moreover, these expressions are identical to those obtained using the nonunitary Susskind-Glogower operators. The creation and annihilation operators defined here become the conventional raising and lowering operators in the limit as $s \to \infty$. From eqs. (8), (14) and (15) we find the commutation relation

$$[\hat{a}, \hat{a}^\dagger] = 1 - (s+1)|s\rangle\langle s|. \tag{16}$$

Because of the unfamiliar extra term on the right, the trace of this commutator vanishes, as it must for a finite-dimensional space. To show the effect of this term, consider a general linear combination

$$|f\rangle = \sum_{n=0}^{\infty} c_n |n\rangle \tag{17}$$

for which

$$\langle f|\hat{N}^p|f\rangle = \sum_{n=0}^{\infty} n^p |c_n|^2. \tag{18}$$

We describe $|f\rangle$ as a physically accessible, or preparable, state if the series in (18) is convergent for any given finite integer p. Such a state represents the state of the field which can be prepared from $|0\rangle$ by interaction for a finite time with a source whose energy is bounded, for example one which comprises a finite number of excited atoms. (See ref. [8] for the connection between states such as this and those of the Rigged Hilbert space.) It is not difficult to show that when $[\hat{a}, \hat{a}^\dagger]^q$ acts on any physically accessible state $|f\rangle$, where q is any given integer, we can always choose a value of s large enough so that the action of $[\hat{a}, \hat{a}^\dagger]^q$ is as close as we please to the action of the unit operator. We thus recover from (16) the usual commutator when operating on physically accessible states such as, for example, the coherent states or single number states.

The unitary phase operators directly couple the states $|0\rangle$ and $|s\rangle$. Is it then possible that a field could evolve, by means of a suitable interaction, from the ground state to the extremely energetic or infinite energy state $|s\rangle$? Physical couplings to the electromagnetic field take place through $\hat{a}$ and $\hat{a}^\dagger$ operators [6], for example via products with current operators. Therefore, the interaction Hamiltonian matrix element $\langle 0|\mathcal{H}_I|s\rangle$ and its complex conjugate will be zero because both $\hat{a}|0\rangle$ and $\hat{a}^\dagger|s\rangle$ vanish. Hence the time-evolution operator will not couple $|0\rangle$ and $|s\rangle$ directly and any transitions must occur in the usual way, up or down the ladder of number states.

A final and useful check on the properties of any proposed phase operator is whether or not it satisfies the Lerner criterion [5, 6] for acceptable phase operators. This condition is based on Dirac's receipe for correspondence between classical Poisson brackets and the commutators. Explicitly we find

$$[\widetilde{\exp}_\theta[i\phi], \hat{N}] = \widetilde{\exp}_\theta[i\phi] - (s+1)\exp[i(s+1)\theta_0]\,|s\rangle\,\langle 0|\,. \qquad (19)$$

In the expectation value of this commutator, in any physically accessible state (17), the second term in (19) will vanish when we let s tend to infinity for the same reason as discussed previously for $[\hat{a}, \hat{a}^\dagger]$. Thus, the Lerner criterion will be satisfied in this limit.

It is beyond the scope of this paper to pursue the far-reaching consequences of the existence of the unitary phase operator, but we wish to indicate its importance. Cosine and sine combinations of the conventional Susskind-Glogower phase operators have different sets of eigenstates, which have been regarded as useful for studying conceptual aspects of phase, but not as useful for practical calculations as the number or coherent states [5]. However, the phase states $|\theta_m\rangle$, which are eigenstates of *both* $\widetilde{\cos}_\theta\,\phi$ and $\widetilde{\sin}_\theta\,\phi$ should be a useful alternative to the number states as a basis set (particularly with a convenient choice of θ_0 such as zero). However, care must be taken in the use of these phase states because they are not physically accessible states with a convergent energy series. Therefore, precise expressions such as (16) and (19) must be retained during the calculation with the limit as s tends to infinity taken at the end of the calculation.

The existence of a unitary phase operator has some important conceptual consequences. In particular, it now becomes meaningful to consider a Hermitian phase operator $\hat{\phi}$ which can be defined relative to a convenient reference point determined, for example, by setting $\theta_0 = 0$. This reopens the possibility of a time-measurement operator $\hat{t}$ being defined as $-\hat{\phi}/\omega$ relative to a corresponding reference time, $t = 0$. We note that if a similar Hermitian phase operator could have been defined from the conventional Susskind-Glogower operators, which it cannot [5], then the eigenvalues of $\hat{\phi}$ and therefore $\hat{t}$ would have been a continuous spectrum over the real line. However, the $s+1$ eigenvalues of our operator $\hat{\phi}$ are (with $\theta_0 = 0$) $2m\pi/(s+1)$, which do not include irrational fractions of 2π. Thus the possible results of a time measurement in the corresponding interval 0 to $2\pi/\omega$ will form only a dense subset of all the real values which the parameter t can take. Of course as we let s become larger, the separation between consecutive time eigenvalues becomes as small as we wish, but the irrational fractions of $2\pi/\omega$ will not be included. Even in the limit of large s, the time will be quantized with a *countable* infinity of eigenvalues which, between 0 and $2\pi/\omega$, are equal in number to the number of energy eigenstates.

* * *

DTP thanks H. P. W. Gottlieb and SMB thanks R. Loudon for helpful discussions. Some of this work was carried out as part of the longer-term research within the Underlying Programme of the UKAEA.

REFERENCES

[1] Dirac P. A. M., *Proc. R. Soc. London, Ser. A*, **114** (1927) 243.
[2] Louisell W. H., *Phys. Lett.*, 7 (1963) 60.
[3] Susskind L. and Glogower J., *Physics*, 1 (1964) 49.
[4] Lévy-Leblond J. M., *Ann. Phys. (N.Y.)*, **101** (1976) 319.
[5] Carruthers P. and Nieto M. M., *Rev. Mod. Phys.*, **40** (1968) 411.
[6] Barnett S. M. and Pegg D. T., *J. Phys. A*, **19** (1986) 3849.
[7] Loudon R., *The Quantum Theory of Light*, 1st edition (Oxford University Press, Oxford) 1973, p 143.
[8] Böhm A., *The Rigged Hilbert Space and Quantum Mechanics* (Springer-Verlag, Berlin) 1978

Hermitian Phase Operator $\hat{\phi}$ in the Quantum Theory of Light

D. T. Pegg and S. M. Barnett

Abstract

We introduce the Hermitian phase operator for a single-mode electromagnetic field. The existence of this operator and its form follow directly and uniquely from the existence of states of well-defined phase. This removes phase from its hitherto exceptional position of being a classical observable without a quantum counterpart.

It has long been believed that no Hermitian optical phase operator exists[1]. In this paper we introduce the Hermitian phase operator $\hat{\phi}_\theta$ and show that the derived unitary operators $\exp(\pm i\hat{\phi}_\theta)$ allow a polar decomposition of the photon annihilation operator, precisely as anticipated by Dirac. The existence of $\hat{\phi}_\theta$ means that it is now meaningful to construct fully quantum mechanical expectation values and variances of optical phases. The results of these calculations are identical to those obtained using well known semiclassical and phenomenological approaches in appropriate limits.

It is well known that states of precisely defined phase exist[7]:

$$|\theta\rangle = \operatorname*{Lt}_{s\to\infty} (s+1)^{-\frac{1}{2}} \sum_{n=o}^{s} \exp(in\theta)|n\rangle \qquad (1)$$

where $|n\rangle$ are the (s+1) number states spanning the (s+1)-dimensional space. These states have a natural and simple time-development; at time t the state evolves to $|\theta-\omega t\rangle$, where ω is the frequency of the single field mode or harmonic oscillator under consideration. Such states provide an ideal description of phase when used as phase states[7]. The phase state (1) is well defined in the (s+1)-dimensional space but caution must be exercised in taking the $s\to\infty$ limit. Indeed, the early failure to obtain well behaved phase operators was a result of disregarding the limiting process in (1) and prematurely replacing s by ∞ in the summation[8]. Therefore we shall work with the (s+1)-dimensional linear space Ψ spanned by $|n\rangle$ where $n = 0,1, \ldots s$, where s can be made arbitrarily large and will be allowed to tend to infinity when expectation values are calculated.

The states $|\theta\rangle$ form an overcomplete basis for Ψ. However, we construct a complete orthonormal basis by selecting a subset of these states. Such a subset[8] will consist of a reference phase state

$$|\theta_o\rangle = (s+1)^{-\frac{1}{2}} \sum_{n=o}^{s} \exp(in\theta_o)|n\rangle \quad , \tag{2}$$

together with the s states that are orthogonal to it (and to each other). The basis states are then $|\theta_m\rangle$ where

$$\theta_m = \theta_o + \frac{2\pi m}{s+1} \qquad (m = 0,1, \ldots s) \quad . \tag{3}$$

The value of θ_o can be chosen as any point in the continuum, that is on the real line. The choice of θ_o determines the particular basis of the space. We can expand a number state $|n\rangle$ in terms of the θ_m-phase state basis $|\theta_m\rangle$ as

$$|n\rangle = \sum_{m=o}^{s} |\theta_m\rangle\langle\theta_m|n\rangle$$

$$= (s+1)^{-\frac{1}{2}} \sum_{m=o}^{s} \exp(-in\theta_m)|\theta_m\rangle \quad . \tag{4}$$

Equations (1) and (4) show the relationship between the phase and number basis states. We see that a system in a number state is equally likely to be found in any state $|\theta_m\rangle$ and, as θ_o can be given any value, all phase states are equally likely. We also see that a system prepared in a phase state is equally likely to be found in any number state.

We define the Hermitian phase operator as

$$\hat{\phi}_\theta = \sum_{m=o}^{s} \theta_m |\theta_m\rangle\langle\theta_m|$$

$$= \theta_o + \sum_{m=o}^{s} \frac{2\pi m}{s+1} |\theta_m\rangle\langle\theta_m| \quad . \qquad (5)$$

Naturally, the phase operator depends on the choice of reference phase and will have eigenvalues that range from θ_o to $\theta_o + 2\pi s/(s+1)$. The subscript θ on the phase operator indicates this dependence. The necessity of choosing a reference phase exists even in classicial physics where we define the inverses of trigonometrical functions to lie within a specified 2π interval. Thus, the problem of multivalued phases in quantum mechanics may be resolved in the same fashion as the identical problem in classicial physics.

We can now construct the operator-function $\exp(i\hat{\phi}_\theta)$ which will be unitary because $\hat{\phi}_\theta$ is Hermitian. The properties of this operator are determined by considering its action on the number state $|n\rangle$:

$$\exp(i\hat{\phi}_\theta)|n\rangle = \exp\left[i \sum_{m=o}^{s} \theta_m |\theta_m\rangle\langle\theta_m|\right]|n\rangle$$

$$= (s+1)^{-\frac{1}{2}} \sum_{m=o}^{s} \exp[-i(n-1)\theta_m]|\theta_m\rangle \ , \qquad (6)$$

where we have used the expansion (4). For $n>0$ the resulting state is simply the number state $|n-1\rangle$. For $n=0$ we obtain the state

$$\langle n|\hat{\phi}_\theta^2|n\rangle - \langle n|\hat{\phi}_\theta|n\rangle^2 = \pi^2/3 \quad , \qquad (12)$$

for all number states $|n\rangle$ <u>including</u> the vacuum state. These results verify the common belief that a number state, especially the vacuum, should have a completely random phase. This contrasts sharply with the results obtained[4] using the Susskind-Glogower operators. These results are precisely those that would be obtained for a uniform classical probability distribution of phases in the interval θ_o to $\theta_o + 2\pi$. For a field in a strong coherent state ($|\alpha|^2 >> 1$) we find that

$$\langle\alpha|\hat{\phi}_\theta|\alpha\rangle = \arg(\alpha) \qquad (13)$$

$$\langle\alpha|\hat{\phi}_\theta^2|\alpha\rangle - \langle\alpha|\hat{\phi}_\theta|\alpha\rangle^2 = \{4|\alpha|^2\}^{-1} \quad , \qquad (14)$$

where $\arg(\alpha)$ is evaluated in the interval θ_o to $\theta_o+2\pi$.
It is rewarding to see that a proper quantum mechanical derivation of the coherent state expectation value and variance, of the Hermitian phase operator, are in accord with previous semiclassical and phenomenological treatments.

In conclusion, we have found the Hermitian operator corresponding to the phase observable of the electromagnetic field. This operator is entirely consistent with the unitary exponential phase operator[8] and <u>also</u> with the physically verified properties of the photon creation and annihilation operators. It is at last meaningful to discuss the quantum mechanical properties of optical phase.

Acknowledgement

We thank P.L. Knight for his helpful suggestions. This work was suported in part by the Australian Academy of Science, the Royal Society and the Underlying Programme of the UKAEA.

References

1. The Hermitian phase operator was originally postualated by Dirac[2]. The original and simple commutator proposed for the phase and photon number operators led to problems[3-5]. It was suggested[6] that the source of the difficulties was the multivalued nature of the phase. Unfortunately, early attempts suggested that it also was impossible to construct well-behaved operators that correspond to periodic functions of the phase[3-5].

2. P.A.M. Dirac, Proc.R.Soc.Lond. A114, 243 (1927).

3. L. Susskind and J. Glogower, Physics 1, 49 (1964).

4. P. Carruthers and M. M. Nieto, Rev.Mod.Phys. 40, 411 (1968).

5. S.M. Barnett and D.T. Pegg, J.Phys.A:Math.Gen. 19, 3849 (1986).

6. W.H. Louisell, Phys.Lett. 7, 60 (1963).

7. R. Loudon, The Quantum Theory of Light, first edition (Oxford University Press, Oxfrod, 1973) p.143.

8. D.T. Pegg and S.M. Barnett, Europhys.Lett. (submitted).

9. E. Merzbacher, Quantum Mechanics, second edition (Wiley, New York, 1970) p.330.

On the Hermitian optical phase operator

S. M. BARNETT†

Theoretical Physics Division, Harwell Laboratory, Oxfordshire OX11 0RA, England
and Wolfson College, Oxford OX2 6UD, England

and D. T. PEGG

School of Science, Griffith University, Nathan, Brisbane 4111, Australia

(*Received 14 June 1988 and accepted 17 June 1988*)

Abstract. It has long been believed that no Hermitian optical phase operator exists. However, such an operator can be constructed from the phase states. We demonstrate that its properties are precisely in accord with the results of semi-classical and phenomenological approaches when such approximate methods are valid. We find that the number–phase commutator differs from that originally postulated by Dirac. This difference allows the consistent use of the commutator for inherently quantum states. It also leads to the correct periodic phase behaviour of the Poisson bracket in the classical regime.

1. Introduction

We have recently shown that a Hermitian optical phase operator $\hat{\phi}_\theta$ exists [1]. This result contradicts the well established belief that no such operator can be constructed. The existence of this operator allows us to construct unitary operators $\exp(\pm i\hat{\phi}_\theta)$ and to perform a polar decomposition of the photon annihilation operator [2], precisely as anticipated by Dirac [3]. It is now possible to describe the quantum properties of optical phase *without* recourse to semi-classical or phenomenological methods.

In this paper we describe some of the properties of the optical phase operator and seek to explain why it has been so elusive. With this objective in mind, we begin with an historical perspective on the development of the quantum optical phase. It is known that, despite the difficulties in obtaining a phase operator, states of precisely defined phase exist [4]. The phase operator $\hat{\phi}_\theta$ follows naturally and *uniquely* from the existence of these states.

The Poisson-bracket–commutator correspondence [5] was the key in the construction of the original phase operator [3]. We pay particular attention to the classical limit of the phase operator and describe the relationship between the number–phase commutator and the classical Poisson bracket.

† Present address: Department of Engineering Science, Oxford University, Parks Road, Oxford OX1 3PJ, England.

2. Historical perspective

The idea that the optical phase should be described by a Hermitian phase-operator dates from the earliest work on quantum electrodynamics [3]. Comparison with the classical Poisson bracket led to the assumption that (for each mode of the field) the photon number operator $\hat{N}$ and the phase operator obey the canonical commutation relation†

$$[\hat{\phi}, \hat{N}] = -\mathrm{i}, \tag{1}$$

where the caret denotes an operator. We note that the Poisson-bracket–commutator correspondence is *not* a rigorous method of quantization and was only expected to yield the correct commutator in simple cases [11]. A straightforward application of Heisenberg's uncertainty principle to this commutator yielded the number–phase uncertainty relation [6]

$$\Delta N \Delta \phi \geqslant \tfrac{1}{2}. \tag{2}$$

The properties of quantum optical phase remained a largely academic problem until the invention of the maser, which allowed the coherent amplification of electromagnetic waves consisting of only a few quanta. Here was a device that could probe both quantum intensity *and* phase fluctuations [12–15]. Indeed, this early work suggested that the output from a maser could minimize the uncertainty product (2). This kindled renewed interest in the quantum description of optical phase.

Closer investigation of the number–phase commutator (1) led to problems. In particular, the matrix elements of $\hat{\phi}$ in a number state basis (as calculated from (1)) are undefined [16]

$$(n-n')\langle n'|\hat{\phi}|n\rangle = -\mathrm{i}\delta_{nn'}. \tag{3}$$

Dirac was aware that there were problems associated with this description of phase but noted that the difficulties did not arise if $\hat{\phi}$ only appeared as part of a decomposition of annihilation and creation operators [17]. It was suggested that the source of the problem embodied in (3) was the multivalued nature of $\hat{\phi}$ and that the solution would be to introduce periodic operator functions of the phase [16, 18, 19]. Accordingly, Louisell introduced the operators $\cos\hat{\phi}$ and $\sin\hat{\phi}$ which he suggested should obey the commutation relations

$$\left.\begin{aligned}[\cos\hat{\phi}, \hat{N}] &= \mathrm{i}\sin\hat{\phi}, \\ [\sin\hat{\phi}, \hat{N}] &= -\mathrm{i}\cos\hat{\phi}.\end{aligned}\right\} \tag{4}$$

We will see that the problem expressed in (3) was *not* due to the multivalued nature of $\hat{\phi}$, but to an injudicious application of the Poisson-bracket–commutator correspondence.

Louisell's suggestion did not seem to help matters as it proved difficult to obtain an explicit form for the cosine and sine operators. Susskind and Glogower [7] returned to the original idea of optical phase based on exponential phase operators in a polar decomposition of the creation and annihilation operators

$$\begin{aligned}\hat{a} &= \hat{\exp}_{\mathrm{S}}(\mathrm{i}\phi)\hat{N}^{1/2}, \\ \hat{a}^{\dagger} &= \hat{N}^{1/2}\hat{\exp}_{\mathrm{S}}(-\mathrm{i}\phi).\end{aligned} \tag{5}$$

† Dirac actually chose the commutator $[\hat{\phi}, \hat{N}] = \mathrm{i}$, but we follow the convention adopted by later workers [6–10].

These relations define the Susskind–Glogower operators up to an arbitrary projector from the vacuum state to any other state $|\psi\rangle$

$$\begin{aligned}\widehat{\exp}_S(i\phi) &= \sum_{n=0}^{\infty} |n\rangle\langle n+1|\{+|\psi\rangle\langle 0|\},\\ \widehat{\exp}_S(-i\phi) &= \sum_{n=0}^{\infty} |n+1\rangle\langle n|\{+|0\rangle\langle\psi|\}.\end{aligned} \tag{6}$$

In order to determine the form of their operators precisely, Susskind and Glogower imposed the *additional* condition that the undetermined projector $|\psi\rangle\langle 0|$ was absent [7]. The resulting operators do not commute and are not unitary. Therefore the Susskind–Glogower formalism does not allow the existence of a unique Hermitian phase operator [7, 8, 20] as the operators $\widehat{\exp}_S(\pm i\phi)$ cannot be functions of a common operator. Nevertheless, the Susskind–Glogower operators have been employed in a range of quantum optical problems (see [10] and [21] for references to this work).

The development of the quantum theory of optical coherence [22], and the use of coherent state techniques [23], meant that it was no longer necessary to employ a phase operator to describe phase-sensitive optical effects. Interest in the phase-sensitive fluctuations of squeezed light [24, 25] focused attention on the field quadratures. An analysis of phase measuring experiments suggested that the measured cosine and sine components of the field were actually proportional to these quadratures [10]. Yet the problem of describing the quantum optical-phase remained. We were left in the uncomfortable position of having a classical observable (the phase) without a quantum counterpart.

3. Optical phase states

The failure of the Poisson-bracket–commutator correspondence to provide a well behaved optical phase operator suggests that we should try a different approach. Our method stems from the fact that states of precisely defined phase exist [4]. The phase state $|\theta\rangle$ is defined by

$$|\theta\rangle = \lim_{s\to\infty} (s+1)^{-1/2} \sum_{n=0}^{s} \exp(in\theta)|n\rangle, \tag{7}$$

where $|n\rangle$ are the $(s+1)$ number states spanning the $(s+1)$-dimensional space. This state is a normalized superposition of all the photon number states, each weighted by a phase factor $\exp(in\theta)$. The phase states provide an ideal description of optical phase for two reasons [4]. Firstly, the phase states have a natural time-development; at time t the state evolves to $|\theta-\omega t\rangle$† where ω is the frequency of the single-field mode. Secondly, the electric field is $\pm\infty$ and has a divergent variance everywhere *except* when it is about to change sign. At these times, when $\theta-\omega t = l\pi$, the field is *precisely* zero. These periodic zeros of the electric field determine the phase of the optical field. The phase state (7) is well defined in the $(s+1)$-dimensional state space but caution must be exercised in taking the $s\to\infty$ limit. Indeed, the early failure to obtain well behaved phase operators was at least in part a result of disregarding the limiting process in (7) and prematurely replacing s by ∞ in the summation [2]. Therefore we work with the $(s+1)$-dimensional linear space Ψ spanned by $|n\rangle$ where $n = 0, 1, \ldots, s$.

†Apart from an arbitrary dynamical phase due to the choice of the zero of energy.

Here s can be arbitrarily large and will ultimately be allowed to tend to infinity when expectation values are calculated.

Distinguishable phase states $|\theta\rangle$ exist for all θ in the range θ_0 to $\theta_0+2\pi$, where θ_0 is a reference phase. Therefore there are an uncountable infinity of different phase states. Yet we know that the state space is spanned by the countable basis of number states and so the phase states form an overcomplete basis for Ψ. However, we can construct a complete orthonormal basis by selecting a subset of these states. Such a subset will consist of a reference phase state

$$|\theta_0\rangle=(s+1)^{-1/2}\sum_{n=0}^{s}\exp(in\theta_0)|n\rangle, \tag{8}$$

together with the s phase states that are orthogonal to it (and to each other). The basis states are then $|\theta_m\rangle$, where

$$\theta_m=\theta_0+\frac{2\pi m}{s+1}, \quad m=0,1,\ldots,s. \tag{9}$$

The value of θ_0 can be chosen as any point in the continuum that is on the real line. The choice of θ_0 determines the particular phase state basis for the space. We can expand the number state $|n\rangle$ in terms of the θ_m phase state basis as

$$\begin{aligned}|n\rangle&=\sum_{m=0}^{s}|\theta_m\rangle\langle\theta_m|n\rangle\\&=(s+1)^{-1/2}\sum_{m=0}^{s}\exp(-in\theta_m)|\theta_m\rangle.\end{aligned} \tag{10}$$

Equations (7) and (10) show the relationship between the phase and number states. We see that a system in a number state is equally likely to be found in any state $|\theta_m\rangle$ and, as θ_0 can be given any value, all phase states are equally likely. We also see that a system prepared in a phase state is equally likely to be found in any number state.

We are now in a position to construct the Hermitian optical phase operator.

4. The Hermitian optical phase operator

We require a Hermitian phase operator which has eigenstates that are phase states with corresponding eigenvalues that are equal to the 'phase' of the state. We can construct such an operator directly from the orthonormal phase states

$$\begin{aligned}\hat{\phi}_\theta&\equiv\sum_{m=0}^{s}\theta_m|\theta_m\rangle\langle\theta_m|\\&=\theta_0+\sum_{m=0}^{s}\frac{2\pi m}{s+1}|\theta_m\rangle\langle\theta_m|.\end{aligned} \tag{11}$$

This operator is clearly Hermitian and satisfies the required eigenvalue equation

$$\hat{\phi}_\theta|\theta_m\rangle=\theta_m|\theta_m\rangle. \tag{12}$$

Naturally, the phase operator depends on the choice of reference phase in that its eigenvalues will range from θ_0 to $\theta_0+2\pi s/(s+1)$. The subscript θ on the phase operator indicates this dependence. There is nothing strange about limiting the phase eigenvalues to a 2π interval determined by a reference phase. The necessity of choosing a reference phase exists even in classical physics where we define the inverses of trigonometrical functions to lie within a specified 2π interval. Thus the problem of multivalued phases in quantum mechanics is resolved in precisely the same fashion as the identical problem is classical physics.

The original phase operator suffered from the fact that its matrix elements in a number state basis (as given by (3)) are undefined [16]. The Hermitian optical-phase operator $\hat{\phi}_\theta$ *does not* suffer from such problems; its diagonal and off-diagonal number state matrix elements can be determined from (11) and (10)

$$\langle n|\hat{\phi}_\theta|n\rangle = \theta_0 + \pi\frac{s}{s+1} \tag{13}$$

and

$$\langle n'|\hat{\phi}_\theta|n\rangle = \frac{2\pi}{s+1}\frac{\exp[\mathrm{i}(n'-n)\theta_0]}{\exp[2\pi\mathrm{i}(n'-n)/(s+1)]-1}, \quad n' \neq n \tag{14}$$

respectively. In all practical problems we will be interested in number states $|n\rangle$ for which $n \ll s$. In this limit the off-diagonal matrix elements of $\hat{\phi}_\theta$ become

$$\langle n'|\hat{\phi}_\theta|n\rangle \approx \frac{\mathrm{i}}{n-n'}\exp[\mathrm{i}(n'-n)\theta_0]. \tag{15}$$

These well behaved number state matrix elements of $\hat{\phi}_\theta$ imply that the original number-phase commutator (1) is incorrect. We can use the definition of our phase operator (11) and the properties of the phase states to calculate the correct number-phase commutator. The diagonal matrix elements of this commutator in a photon number or a phase state are zero

$$\langle n|[\hat{N}, \hat{\phi}_\theta]|n\rangle = 0 = \langle \theta_m|[\hat{N}, \hat{\phi}_\theta]|\theta_m\rangle. \tag{16}$$

The off-diagonal matrix elements of the commutator are found to be non-zero

$$\langle n'|[\hat{N}, \hat{\phi}_\theta]|n\rangle = \frac{2\pi}{s+1}\frac{(n'-n)\exp[\mathrm{i}(n'-n)\theta_0]}{\exp[2\pi\mathrm{i}(n'-n)/(s+1)]-1}, \tag{17}$$

$$\langle \theta_{m'}|[\hat{N}, \hat{\phi}_\theta]|\theta_m\rangle = \frac{2\pi}{s+1}\frac{(m'-m)}{1-\exp[2\pi\mathrm{i}(m-m')/(s+1)]}. \tag{18}$$

These matrix elements are in stark contrast to those of the original commutator (1) whose only non-zero matrix elements were the diagonal ones. If we again restrict our interest to those number states $|n\rangle$ for which $n \ll s$ we find that

$$\langle n'|[\hat{N}, \hat{\phi}_\theta]|n\rangle \approx -\mathrm{i}(1-\delta_{n'n})\exp[\mathrm{i}(n'-n)\theta_0]. \tag{19}$$

It is hard to conceive two more different results than the two commutators (1) and (19). The original result (1) was derived by comparison with the classical Poisson bracket and the dissimilarity between our result and the original raises the question of a suitable classical limit to our commutator. However, we emphasize that our result is the unique consequence of the existence of the phase states (7). The commutator (19) does reduce to the classical Poisson bracket in the appropriate limit but in quite a subtle way. We discuss this problem more fully in section 6.

Dirac [3] originally perceived the optical phase operator as part of a polar decomposition of the photon annihilation operator. We are now in a position to realise this decomposition by constructing the operator function $\exp(\mathrm{i}\hat{\phi}_\theta)$. This operator will be unitary because $\hat{\phi}_\theta$ is Hermitian and it will commute with the operator $\exp(-i\hat{\phi}_\theta)$. These two exponential phase operators will share the phase states as their common eigenstates†

$$\exp(\pm i\hat{\phi}_\theta)|\theta_m\rangle = \exp(\pm i\theta_m)|\theta_m\rangle. \tag{20}$$

The properties of $\exp(i\hat{\phi}_\theta)$ are determined by considering its action on the photon number state $|n\rangle$

$$\begin{aligned}\exp(i\hat{\phi}_\theta)|n\rangle &= \exp\left[i\sum_{m=0}^{s}\theta_m|\theta_m\rangle\langle\theta_m|\right]|n\rangle \\ &= (s+1)^{-1/2}\sum_{m=0}^{s}\exp[-i(n-1)\theta_m]\,|\theta_m\rangle,\end{aligned} \tag{21}$$

where we have used the expansion (10). For $n>0$ the resulting state is simply the lowered number state $|n-1\rangle$. For $n=0$ we effectively obtain a state which it is tempting to label as $|-1\rangle$ (compare with [10]), but which here must be identified with the highly excited state $|s\rangle$

$$\begin{aligned}(s+1)^{-1/2}\sum_{m=0}^{s}\exp(i\theta_m)|\theta_m\rangle &= (s+1)^{-1/2}\exp[i(s+1)\theta_0]\sum_{m=0}^{s}\exp(-is\theta_m)|\theta_m\rangle \\ &= \exp[i(s+1)\theta_0]|s\rangle.\end{aligned} \tag{22}$$

Therefore, the number-state representation of $\exp(i\hat{\phi}_\theta)$ is

$$\exp(i\hat{\phi}_\theta) = |0\rangle\langle 1| + |1\rangle\langle 2| + \ldots + |s-1\rangle\langle s| + \exp[i(s+1)\theta_0]|s\rangle\langle 0|. \tag{23}$$

This operator is closely related to the Susskind–Glogower operator (6), but the arbitrary state $|\psi\rangle$ appearing in the Susskind–Glogower expression is here *uniquely* defined as $\exp[i(s+1)\theta_0]|s\rangle$. The *additional* constraint of Susskind and Glogower that no $|\psi\rangle\langle 0|$ projector existed is removed and the unitary nature of $\exp(i\hat{\phi}_\theta)$ is restored.

From the unitary phase operators $\exp(\pm i\hat{\phi}_\theta)$ we can form Hermitian cosine and sine combinations. These operators are better behaved than their counterparts formed from the conventional Susskind–Glogower operators [7, 8] and have properties that are much more desirable for a description of phase. In particular we find

$$\cos^2(\hat{\phi}_\theta) + \sin^2(\hat{\phi}_\theta) = 1, \tag{24}$$

$$[\cos(\hat{\phi}_\theta), \sin(\hat{\phi}_\theta)] = 0, \tag{25}$$

$$\langle n|\cos^2(\hat{\phi}_\theta)|n\rangle = \langle n|\sin^2(\hat{\phi}_\theta)|n\rangle = \tfrac{1}{2}. \tag{26}$$

Our new phase operator is consistent with the phase of the vacuum being random, while (24) and (25) follow directly from the fact that the cosine and sine operators are functions of a common Hermitian phase operator.

We define the creation and annihilation operators acting on the space Ψ in terms of the Hermitian phase and the number operator $\hat{N}$

$$\begin{aligned}\hat{a} &= \exp(i\hat{\phi}_\theta)\hat{N}^{1/2} \\ &= |0\rangle\langle 1| + \sqrt{2}|1\rangle\langle 2| + \ldots + \sqrt{s}|s-1\rangle\langle s|,\end{aligned} \tag{27}$$

† These exponential phase operators may be derived directly from the requirement that they obey (20) [2]. However, they are introduced here as operator-functions of $\hat{\phi}_\theta$ [1].

with $\hat{a}^\dagger$ being the Hermitian conjugate of $\hat{a}$. As s tends to infinity $\hat{a}$ becomes the conventional annihilation operator. Note that the Hermitian amplitide operator $\hat{N}^{1/2}$ removes the $|s\rangle\langle 0|$ projector from $\hat{a}$. It is for this reason that Susskind and Glogower were unable to obtain this term from a polar decomposition of $\hat{a}$. The $|s\rangle\langle 0|$ projector *never* occurs in any physical coupling involving the electromagnetic field as these take place via $\hat{a}$ and $\hat{a}^\dagger$ [2, 10].

The commutation relation between $\hat{a}$ and $\hat{a}^\dagger$ as derived from equation (20) has a slightly unfamiliar form

$$[\hat{a}, \hat{a}^\dagger] = 1 - (s+1)|s\rangle\langle s|. \tag{28}$$

The second term in this commutator renders its trace zero, as it must be for a finite dimensional state space. This resolves the long-standing paradox that the trace of $[\hat{a}, \hat{a}^\dagger]$ appears not to vanish [26]. To show the effect of this term, consider the state

$$|f\rangle = \sum_{n=0}^{\infty} C_n|n\rangle, \tag{29}$$

for which

$$\langle f|\hat{N}^p|f\rangle = \sum_{n=0}^{\infty} n^p|C_n|^2. \tag{30}$$

We describe the state $|f\rangle$ as a physically accessible, or preparable, state if the series (30) is convergent for any given finite integer p. Such a state represents the state of the field which can be prepared from the vacuum by interaction for a finite time with a source whose energy is bounded, for example, one which comprises a finite number of excited atoms. It is not difficult to show that when $[\hat{a}, \hat{a}^\dagger]^q$ acts on any physically accessible state $|f\rangle$, where q is any given finite integer, we can always choose a value of s large enough that the action of $[\hat{a}, \hat{a}^\dagger]^q$ is as close as we please to the action of the unit operator. We thus recover the usual commutator from (28) when operating on physically accessible states such as, for example, the coherent states or single number states.

5. Phase properties of single-mode fields

In the previous section we introduced the Hermitian optical phase operator and showed how it could be used to perform a polar decomposition of the creation and annihilation operators. The existence of the phase operator means that it is now meaningful to describe the phase properties of optical fields in a fully quantum mechanical fashion. We illustrate this ability by considering the phase properties of three types of field states. We will find that care must be taken in interpreting the results obtained using the phase operator.

5.1. *States of random phase*

The phase state expansion of the number state (10) suggests that the number states are states of random phase. This should also be true for states that are represented by a density matrix that is diagonal in the number state basis (for example the thermal or chaotic state). The expectation value of the phase operator in a number state is

$$\langle n|\hat{\phi}_\theta|n\rangle = \sum_{m=0}^{s} \theta_m|\langle\theta_m|n\rangle|^2 = \theta_0 + \pi\frac{s}{s+1}, \tag{31}$$

and the expectation value of its square is

$$\langle n|\hat{\phi}_\theta^2|n\rangle = \sum_{m=0}^{s} \theta_m^2 |\langle \theta_m|n\rangle|^2$$

$$= \theta_0^2 + 2\pi\theta_0 \frac{s}{s+1} + \frac{4\pi^2}{3} \frac{s(s+\frac{1}{2})}{(s+1)^2}. \quad (32)$$

These results become true expectation values in the limit as s tends to infinity. In this limit the mean and variance of the phase are

$$\langle \hat{\phi}_\theta \rangle = \theta_0 + \pi, \quad (33)$$

and

$$\Delta\phi_\theta^2 = \pi^2/3 \quad (34)$$

respectively. These results are characteristic of a state with a random phase, for they are precisely those that would be obtained for a uniform classical probability distribution of phases in the interval θ_0 to $\theta_0 + 2\pi$. For such a *classical* random phase distribution we find

$$\bar{\phi} = \frac{1}{2\pi} \int_{\theta_0}^{\theta_0+2\pi} \phi \, d\phi = \theta_0 + \pi, \quad (35)$$

$$\mathrm{Var}\,\phi = \frac{1}{2\pi} \int_{\theta_0}^{\theta_0+2\pi} (\phi - \bar{\phi})^2 \, d\phi = \pi^2/3. \quad (36)$$

The results are identical to those obtained quantum mechanically for the photon number states. It is worth noting that our analysis includes the ***vacuum state*** as a state of random phase; this was not the case with the earlier Susskind–Glogower operators [4].

States that are represented by a density matrix that is diagonal in the number state basis

$$\rho = \sum_n P_n |n\rangle\langle n| \quad (37)$$

will also be states of random phase with a mean phase $\theta_0 + \pi$ and a variance equal to $\pi^2/3$. This is in accord with the expected result that thermal or chaotic states are states of random phase.

5.2. *States of partially determined phase*

If the state of interest has a partially (but not precisely) determined phase then the choice of reference phase θ_0 will be reflected in the calculated properties of the state. We can illustrate this point with a simple (if not particularly experimentally practical) example. Consider a state that is an equally weighted superposition of all the phase states corresponding to phase angles within $\pm\delta\phi$ of some mean angle ξ. The resulting distribution of phases is shown in figure 1.

The choice of reference phase θ_0 determines the range of eigenvalues of $\hat{\phi}_\theta$ and will influence the results of the calculated phase properties of the state. In figure 2 (*a*) we have chosen a value for θ_0 and superimposed the window of eigenvalues of $\hat{\phi}_\theta$ onto the phase distribution from figure 1. The calculated mean value of the phase operator is ξ and its variance is $(\delta\phi)^2/3$. These results are precisely as we would have expected from a cursory inspection of figure 1. The expectation value of $\hat{\phi}_\theta$ does depend on the value of θ_0, for clearly an increase in θ_0 of 2π would change the expectation value of $\hat{\phi}_\theta$ to $\xi+2\pi$.

A more serious situation is depicted in figure 2 (*b*). Here, we have unfortunately chosen θ_0 to coincide with ξ. The components of the phase distribution corresponding to phases greater or less than ξ appear at opposite ends of the eigenvalue range of $\hat{\phi}_\xi$. In this case the expectation value of $\hat{\phi}_\xi$ and its variance will be

$$\langle\hat{\phi}_\xi\rangle=\xi+\pi, \tag{38}$$

$$\Delta\phi_\xi^2=\pi^2-\pi\delta\phi+\tfrac{1}{3}\delta\phi^2. \tag{39}$$

If $\delta\phi$ is small then $\Delta\phi_\xi^2$ tends to π^2, which is even larger than the variance associated with a state of random phase! We wish to stress that this effect is not of quantum origin, but arises even for classical phase distributions and is associated with the multivalued nature of the phase. The expectation value and variance given in equations (38) and (39) are correct given the choice of reference phase $\theta_0=\xi$. These results emphasize the need for caution, just as in classical physics, in interpreting the results obtained by employing the phase operator (with a particular choice of θ_0) to describe the phase properties of a state. The results are most easily interpreted with a choice of window that minimizes the variance (as in figure 2 (*a*)).

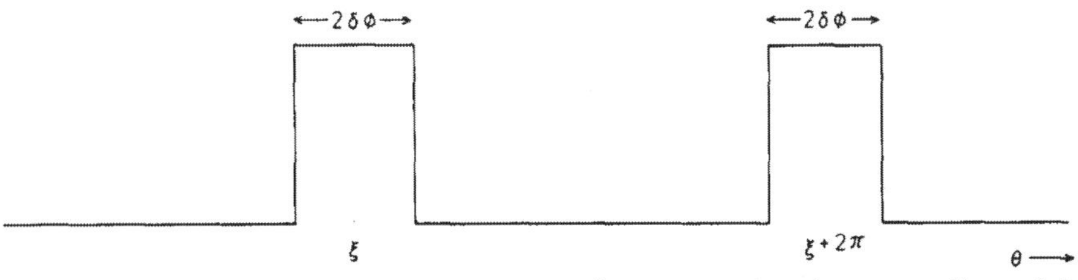

Figure 1. The phase probability distribution for a state that is an equally weighted superposition of all the phase states corresponding to the phase angles within $\pm\delta\phi$ of ξ.

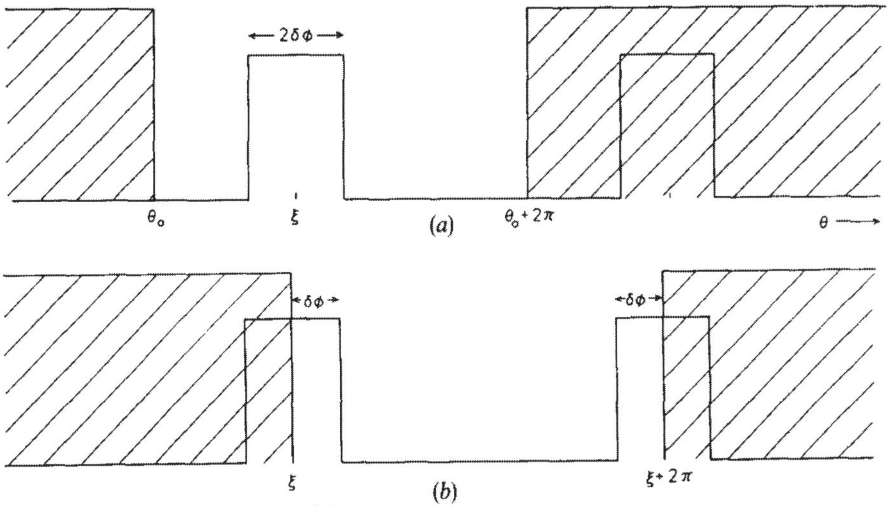

Figure 2. The spectrum of eigenvalues of (*a*) $\hat{\phi}_\theta$ with $\xi-2\pi+\delta\phi<\theta_0<\xi-\delta\phi$, (*b*) $\hat{\phi}_\theta$ with $\theta_0=\xi$ (i.e. $\hat{\phi}_\xi$) superimposed on the phase probability distribution depicted in figure 1.

5.3. *Coherent states*

Perhaps the most important states with a partially determined phase are the coherent states $|\alpha\rangle$ [23]. These states represent the ideal limit of the performance of a well stabilized laser operating above threshold. They are also important as the quantum states that most closely resemble the classical wave of well defined amplitude and phase. The coherent state may be expanded in a number state basis as

$$|\alpha\rangle = \exp(-r^2/2) \sum_n \frac{r^n}{\sqrt{(n!)}} \exp(in\xi)\,|n\rangle, \tag{40}$$

where $\alpha = r\exp(i\xi)$. We can use the number state expansion (10) to find the phase amplitude $\langle\theta_m|\alpha\rangle$

$$\langle\theta_m|\alpha\rangle = \exp(-r^2/2)(s+1)^{-1/2} \sum_n \frac{r^n}{\sqrt{(n!)}} \exp[i(\xi-\theta_m)n]. \tag{41}$$

From these amplitudes we can construct the phase probability distribution and calculate the expectation value and variance of the phase operator. In general, this procedure will require numerical summation of the sum in equation (41). However, we can obtain approximate analytical results for the highly excited coherent state ($r^2 \gg 1$).

In the limit of a large coherent amplitude the Poisson photon number distribution is well approximated by a continuous normal or Gaussian distribution [25]

$$\begin{aligned} P(n) &= \exp(-r^2)\, r^{2n}/n! \\ &\approx (2\pi r^2)^{-1/2} \exp\left[-\frac{(r^2-n)^2}{2r^2}\right]. \end{aligned} \tag{42}$$

This distribution is normalized so that

$$\int P(n)\,dn = 1.$$

We stress that there is no suggestion here of a continuous spectrum of field energy; the approximation in (42) is merely a convenient mathematical method of obtaining the approximate phase and photon-number probability distributions. We can now substitute the square root of $P(n)$ into equation (41) and perform the integration over n

$$\begin{aligned} \langle\theta_m|\alpha\rangle &\approx (s+1)^{-1/2}(2\pi r^2)^{-1/4} \int \exp[-(r^2-n)^2/4r^2] \exp[in(\xi-\theta_m)]\,dn \\ &= \left(\frac{2\pi}{s+1}\right)^{1/2} \left(\frac{4r^2}{2\pi}\right)^{1/4} \exp[-r^2(\xi-\theta_m)^2] \exp[ir^2(\xi-\theta_m)]. \end{aligned} \tag{43}$$

The resulting phase probability distribution is

$$\begin{aligned} P(\theta) &= |\langle\theta|\alpha\rangle|^2 \\ &\approx \frac{2\pi}{s+1} \left(\frac{4r^2}{2\pi}\right)^{1/2} \exp[-2r^2(\xi-\theta)^2], \end{aligned} \tag{44}$$

where we have dropped the subscript m on the phase angle θ. As with $P(n)$, this distribution is normalized so that

$$\int P(\theta)\frac{(s+1)}{2\pi}\,d\theta = 1, \tag{45}$$

where $(s+1)/2\pi$ is the density of the phase states.

The comments made earlier in section 5.2 still apply and calculation of the expectation value and phase of the coherent state will depend on the choice of θ_0. *However*, if θ_0 is sufficiently different from ξ for the whole of the narrow Gaussian peak in $P(\theta)$ to lie within the window of phase eigenvalues, then we find

$$\langle \hat{\phi}_\theta \rangle = \xi, \tag{46}$$

$$\Delta\phi_\theta^2 = \frac{1}{4\bar{n}}, \tag{47}$$

where $\bar{n} = r^2$ is the mean photon number. These results are in agreement with those obtained by Serber and Townes using phenomenological methods [12]. The variance in the photon number is

$$\Delta N^2 = \bar{n}, \tag{48}$$

and so the number-phase uncertainty product for this highly excited coherent state is

$$\Delta N \Delta\phi_\theta = \tfrac{1}{2}, \tag{49}$$

precisely as noted in early maser studies [12–14]. The difference between the early work and our analysis is that $\Delta\phi_\theta$ in equation (49) is the uncertainty in a well behaved Hermitian optical phase operator and *not* the result of a phenomenological analysis.

6. The classical limit

We have delayed discussion of the classical limit and the Poisson-bracket–commutator correspondence until now because the limiting process depends on the choice of a reference phase θ_0. The Poisson-bracket–commutator correspondence requires that the expectation value of the commutator in a classical-like state (that is a highly excited coherent state) should approach the form of the classical Poisson bracket. More specifically we require†

$$\lim_{s\to\infty} \langle \alpha | [\hat{\phi}, \hat{N}] | \alpha \rangle = i\hbar\{\phi, N\} = -i, \tag{50}$$

where $\{,\}$ denotes the classical Poisson bracket and N is the classical energy divided by $\hbar\omega$. Using the large s limit of the phase-number commutator (19) we find

$$\langle \alpha | [\hat{\phi}_\theta, \hat{N}] | \alpha \rangle = \sum_{n,n'} i(1-\delta_{nn'}) C_n C_{n'}^* \exp[i(n'-n)\theta_0], \tag{51}$$

where

$$C_n = \exp(-r^2/2)\frac{\alpha^n}{\sqrt{(n!)}}. \tag{52}$$

† The state $|\alpha\rangle$ must lie within the state space Ψ. Therefore we take the limit $s\to\infty$ before taking the large amplitude limit $r = |\alpha| \to \infty$.

We can re-express (51) as

$$\langle\alpha|[\hat{\phi}_\theta, \hat{N}]|\alpha\rangle = -\mathrm{i}+\mathrm{i}\left|\sum_n C_n \exp(-\mathrm{i}n\theta_0)\right|^2. \tag{53}$$

The first term $(-\mathrm{i})$ is precisely that which arises in the classical Poisson-bracket–commutator correspondence. The second term is apparently a quantum correction to this classical limit. We can calculate this correction in the large r limit by using the integral approximation to C_n described in the previous section. In this limit we find

$$\langle\alpha|[\hat{\phi}_\theta, \hat{N}]|\alpha\rangle = -\mathrm{i}+2\pi\mathrm{i}\left(\frac{4r^2}{2\pi}\right)^{1/2} \exp[-2r^2(\xi-\theta_0)^2]. \tag{54}$$

In the classical limit $(r\to\infty)$ this expectation value acquires the simple form

$$\lim_{r\to\infty} \langle\alpha|[\hat{\phi}_\theta, \hat{N}]|\alpha\rangle = -\mathrm{i}+2\pi\mathrm{i}\delta(\xi-\theta_0). \tag{55}$$

This is precisely the form expected from the classical Poisson bracket (50) *except* when $\xi=\theta_0$. *The reason for this special case is particularly interesting.* Our phase operator has eigenvalues that lie in the range θ_0 to $\theta_0+2\pi$. The phase corresponding to a small increase beyond $\theta_0+2\pi$ is identified with θ_0 plus a small amount. Therefore, there is a discontinuity in the phase at θ_0. The classical Poisson bracket assumes that the phase is a differentiable (continuous) function of the canonical coordinates. If we restrict the classical phase to values in the range θ_0 to $\theta_0+2\pi$ then the classical Poisson bracket will also exhibit a discontinuity at θ_0 *precisely* in agreement with (55). The commutator–Poisson-bracket correspondence is preserved in the classical limit.

The classical Poisson-bracket for the phase and Hamiltonian is the rate of change of the classical phase ϕ. The first term of (55) contributes a constant slope of $-\omega$ to $\phi(t)$. The second term is the derivative of a 2π step-function which suddenly steps the phase up by 2π when $\phi(t)$ reaches θ_0. Thus $\phi(t)$ has a saw-tooth, rather than a linearly decreasing, time dependence which keeps ϕ within the chosen 2π range. This follows automatically as a property of $\hat{\phi}_\theta$ and is not an *ad hoc* classical correction.

7. Conclusion

The existence and form of the Hermitian phase operator $\hat{\phi}_\theta$ follow directly and uniquely from the existence of states of well defined phase. Its elusive nature can now be seen to be associated with the premature replacement of s by infinity in the definition of the phase states. Our procedure involves finding the limit as s tends to infinity at a later stage of the calculation, usually *after* physical results such as expectation values are calculated.

We have examined the properties of $\hat{\phi}_\theta$ and have used it to describe the phase behaviour of randomly and partially phased states, including number states and coherent states. The last of these shows that the classical limit of the phase operator is well defined and also provides an interesting example of the problems associated with multi-valuedness *and* their solution. We have demonstrated that the correspondence between the commutator and the Poisson-bracket is preserved in the classical limit, but emphasize that this correspondence is not true in the quantum regime. Thus Dirac's expression (1), with suitable *ad hoc* compensation for periodicity [7], can be applied for strong laser light but *not* for manifestly quantum states. Consequently the use of $\hat{\phi}_\theta$ provides a rigorous (and fully quantum mechanical) justification of previous phenomenological descriptions of quantum optical phase which appear to be in accord with (1). However, the correct use of $\hat{\phi}_\theta$ solves the puzzling defect with Dirac's commutator which is illustrated by (3).

Finally, we feel that this work demonstrates that phase is no longer in the unique position of being a classical observable without a Hermitian quantum operator counterpart.

Note added in proof: We have recently shown that a correspondence between the phase-number commutator and Poisson-bracket may be established for *all* physical states of the field [27].

Acknowledgments

This work was supported in part by the Underlying Programme of the UKAEA, the Australian Academy of Science and the Royal Society.

References

[1] Pegg, D. T., and Barnett, S. M., 1988, UKAEA Harwell Report TP 1290.
[2] Pegg, D. T., and Barnett, S. M., 1988, *Europhys. Lett.*, **6**, 483.
[3] Dirac, P. A. M., 1927, *Proc. R. Soc. Lond.* A, **114**, 243.
[4] Loudon, R., 1973, *The Quantum Theory of Light* (first edition) (Oxford University Press), p. 143.
[5] Dirac, P. A. M., 1925, *Proc. R. Soc. Lond.* A, **109**, 642.
[6] Heitler, W., 1954, *The Quantum Theory of Radiation* (third edition) (Oxford University Press), p. 65.
[7] Susskind, L., and Glogower, J., 1964, *Physics*, **1**, 49.
[8] Carruthers, P., and Nieto, M. M., 1968, *Rev. mod. Phys.*, **40**, 411.
[9] Paul, H., 1974, *Fortschr. Phys.*, **22**, 657.
[10] Barnett, S. M., and Pegg, D. T., 1986, *J. Phys.* A, **19**, 3849.
[11] Dirac, P. A. M., 1957, *The Principles of Quantum Mechanics* (fourth edition) (Oxford University Press), p. 84.
[12] Serber, R., and Townes, C. H., 1960, *Quantum Electronics—A Symposium*, edited by C. H. Townes (New York: Columbia University Press), p. 233.
[13] Louisell, W., Yariv, A., and Siegman, A. E., 1961, *Phys. Rev.*, **124**, 1646.
[14] Hefner, H., 1962, *Proc. I.R.E.*, **50**, 1604.
[15] Haus, H. A., and Mullen, J. A., 1962, *Phys. Rev.*, **128**, 2407.
[16] Louisell, W. H., 1963, *Phys. Lett.*, **7**, 60.
[17] Schweber, S. S., 1984, *Relativity, Groups and Topology II*, edited by B. S. De Witt and R. Stora (Amsterdam: North-Holland), p. 62.
[18] Brunet, H., 1964, *Phys. Lett.*, **10**, 172.
[19] Harms, J., and Lorigny, J., 1964, *Phys. Lett.*, **10**, 173.
[20] Lévy-Leblond, J. M., 1976, *Ann. Phys., N.Y.*, **101**, 319.
[21] Sanders, B. C., Barnett, S. M., and Knight, P. L., 1986, *Optics Commun.*, **58**, 290.
[22] Glauber, R. J., 1963, *Phys. Rev.*, **130**, 2529.
[23] Glauber, R. J., 1963, *Phys. Rev.*, **131**, 2766.
[24] Walls, D. F., 1983, *Nature*, **306**, 141.
[25] Loudon, R., and Knight, P. L., 1987, *J. mod. Optics*, **34**, 709.
[26] Merzbacher, E., 1970, *Quantum Mechanics*, (second edition) (New York: Wiley), p. 330.
[27] Pegg, D. T., and Barnett, S. M., *Phys. Rev.* A (to be published).

Phase properties of the quantized single-mode electromagnetic field

D. T. Pegg
School of Science, Griffith University, Nathan, Brisbane 4111, Australia

S. M. Barnett
Department of Engineering Science, Oxford University, Parks Road, Oxford, OX1 3PJ, England
(Received 12 September 1988)

The usual mathematical model of the single-mode electromagnetic field is the harmonic oscillator with an infinite-dimensional state space, which unfortunately cannot accommodate the existence of a Hermitian phase operator. Recently we indicated that this difficulty may be circumvented by using an alternative, and physically indistinguishable, mathematical model of the single-mode field involving a finite but arbitrarily large state space, the dimension of which is allowed to tend to infinity *after* physically measurable results, such as expectation values, are calculated. In this paper we investigate the properties of a Hermitian phase operator which follows directly and uniquely from the form of the phase states in this space and find them to be well behaved. The phase-number commutator is not subject to the difficulties inherent in Dirac's original commutator, but still preserves the commutator–Poisson-bracket correspondence for physical field states. In the quantum regime of small field strengths, the phase operator predicts phase properties substantially different from those obtained using the conventional Susskind-Glogower operators. In particular, our results are consistent with the vacuum being a state of random phase and the phases of two vacuum fields being uncorrelated. For higher-intensity fields, the quantum phase properties agree with those previously obtained by phenomenological and semiclassical approaches, where such approximations are valid. We illustrate the properties of the phase with a discussion of partial phase states. The Hermitian phase operator also allows us to construct a unitary number-shift operator and phase-moment generating functions. We conclude that the alternative mathematical description of the single-mode field presented here provides a valid, and potentially useful, quantum-mechanical approach for calculating the phase properties of the electromagnetic field.

I. INTRODUCTION

The single-mode electromagnetic field is a well-known physical system which has been successfully modeled by the quantum harmonic oscillator. Indeed, the success of quantum electrodynamics, based on Dirac's approach,[1] is undeniable. For a long time, however, the nature of the phase of the quantized field has remained an enigma. The oscillator model produced a suitable Hermitian energy operator (or number operator[2] $\hat{N}$) but there was no corresponding Hermitian phase operator.[3–5] This placed the phase in the almost unique[6] position of being a classical observable without a corresponding Hermitian operator counterpart.[3]

While most experiments involved thermal or vacuum fields, the problems associated with quantum optical phase were not important. However, the advent of the maser and, more recently, work on squeezed light[8,9] has renewed interest in the problem.[10] It is tantalizing that a strong coherent field and even a suitably squeezed state with a large coherent amplitude[9] satisfy a phenomenological number-phase uncertainty relation:

$$\Delta N \Delta \phi \geq \tfrac{1}{2} \; . \tag{1.1}$$

This is precisely the relation that has been calculated[11] by using Dirac's quantum relation[1,12]

$$[\hat{\phi}, \hat{N}] = -i \; , \tag{1.2}$$

which is now known to be incorrect.[3–5]

Dirac obtained the commutator (1.2) by employing the correspondence between commutators and classical Poisson brackets.[13] The Hermitian number and phase operators were combined in a polar decomposition of the photon annihilation operator:

$$\hat{a} = \exp(i\hat{\phi})\hat{N}^{1/2} \; . \tag{1.3}$$

The assumed Hermiticity of $\hat{\phi}$ implied the unitarity of $\exp(i\hat{\phi})$. The difficulties with this approach, later realized by Dirac himself,[14] were clearly pointed out by Susskind and Glogower.[4] Firstly, the uncertainty relation (1.1) would imply that a well-defined number state would have a phase uncertainty of greater than 2π. This is a symptom of the fact that the commutator (1.2) does not take account of the periodic nature of the phase. Secondly, the commutator (1.2) gives rise to an inconsistency when matrix elements of the commutator are calculated in a number-state basis.[3] Finally, the "exponential" operator $[\exp(i\hat{\phi})]$ derived from this approach is not unitary and so does not define an Hermitian $\hat{\phi}$.[4] The failure of Dirac's approach means that it is now often accepted that a well-behaved Hermitian phase operator does not exist.[3–5,15]

The difficulties associated with the periodicity of the

phase operator also arise in the classical context. However, the apparent nonunitarity of $\exp(i\hat{\phi})$ [as derived from Eq. (1.3)] is more serious. The action of $\exp(i\hat{\phi})$ on a number state $|n\rangle$ can be determined from (1.3) except when $n=0$. The action of $\exp(i\hat{\phi})$ on the vacuum is indeterminate. The additional condition imposed in the Susskind-Glogower formalism is that

$$\exp(i\hat{\phi})|0\rangle=0 . \tag{1.4}$$

This condition immediately destroys the unitarity of $\exp(i\hat{\phi})$.

A well-behaved Hermitian phase operator $\hat{\phi}$ would lead to a unitary operator $\exp(i\hat{\phi})$ whose action on the vacuum state is well defined. There should be no need for additional constraints like (1.4). Moreover, the commutator of the number and phase operators should not suffer from the inconsistencies of Dirac's commutator (1.2). Because of the failure of Poisson–bracket-commutator correspondence and decomposition of the annihilation operator, an entirely new approach seems to be required.

We have recently indicated a new approach[16] which circumvents the difficulties discussed above. This involves describing the field mode in a finite but arbitrarily large state space of $s+1$ dimensions. Physical properties such as operator expectation values are evaluated in the limit as s tends to infinity. In this paper we explore the details and further consequences of our approach. We conclude that our approach and the conventional infinite state space model are *physically indistinguishable*. However, our method has the additional advantage of being able to incorporate a well-behaved Hermitian phase operator within the formalism. Our phase operator has properties which would normally be associated with phase, both in the classical regime and in the quantum regime of very low intensity fields. The resulting number-phase commutator does not lead to any inconsistencies yet satisfies the condition for Poisson–bracket-commutator correspondence. Periodic operator functions of the Hermitian phase operator can be defined. These are found to have very different properties from the conventional Susskind-Glogower operators[4,5] in the quantum region. For example, our phase operator is consistent with the vacuum being a state of random phase. The Susskind-Glogower operators are *not* consistent with the vacuum being a state of random phase.

II. CLASSICAL PHASE

It is not our intention to derive a phase operator from the commutator–Poisson-bracket correspondence. Nevertheless, it is essential that a meaningful phase operator should reproduce the classical phase in the appropriate limit. In this section we study the behavior of a classical oscillator phase ϕ, and pay particular attention to the problem of the multivalued nature of the phase.

The classical Hamiltonian for a unit mass harmonic oscillator (or single-mode field) is

$$H=\tfrac{1}{2}(p^2+\omega^2x^2), \tag{2.1}$$

where ω is the angular frequency of the oscillator. The phase of the oscillator is

$$\phi=\arctan[p/(\omega x)] \tag{2.2}$$

and is a multivalued property because the arctangent function only defines $\phi \bmod 2\pi$. If we allow ϕ to take a continuous range of values then the rate of change of ϕ has the standard form

$$\frac{d\phi}{dt}=\{\phi,H\}=-\omega , \tag{2.3}$$

where $\{\ \}$ denotes a Poisson bracket. However, it is more useful to restrict ϕ to lie within a specified 2π interval, $\theta_0\le\phi<\theta_0+2\pi$. A common choice of θ_0 might be 0 or $-\pi$, but we retain here the general case of arbitrary θ_0. We denote this choice of range by adding a subscript θ to the phase, thus ϕ_θ is restricted to lie in the range $\theta_0\le\phi_\theta<\theta_0+2\pi$. When the phase is restricted in this manner, the Poisson bracket equation of motion for the phase becomes

$$\frac{d\phi_\theta}{dt}=\{\phi_\theta,H\}=-\omega[1-2\pi\delta(\phi-\theta_0)] . \tag{2.4}$$

The phase ϕ_θ is a periodic sawtooth function of time: it decreases with slope $-\omega$ until it reaches the value θ_0, where it is immediately stepped by 2π and decreases towards θ_0 again.

Equation (2.4) describes the motion of a single classical oscillator. To describe the behavior of an ensemble of differently phased oscillators we use the phase probability function $P(\phi)$, where $P(\phi)d\phi$ is the probability of finding that the phase of a particular oscillator is in the range ϕ to $\phi+d\phi$. The phase probability function is normalized in the chosen 2π interval:

$$\int_{\theta_0}^{\theta_0+2\pi}P(\phi)d\phi=1 . \tag{2.5}$$

In the time interval between t and $t+\delta t$ the associated change $\delta\phi$ for a particular oscillator will have a contribution $-\omega\delta t$ from the first term in Eq. (2.4). If the value of ϕ_θ strays beyond θ_0 during δt, then the second Eq. (2.4) will step the phase up by 2π. The probability of the latter is $P(\theta_0)\omega\delta t$. Dividing by δt and taking the limit as $\delta t\to 0$ gives the rate of change of the expectation value of the phase

$$\frac{d}{dt}\langle\phi_\theta\rangle=-\omega+2\pi P(\theta_0)\omega , \tag{2.6}$$

where $P(\theta_0)$ will, in general, be a time-dependent quantity. Equation (2.6) emphasizes the significance of the choice of θ_0, that is, the particular choice of 2π interval in which ϕ_θ is defined.

We have seen that allowing the phase to have a continuous range or restricting it to a specified 2π interval leads to different Poisson brackets. It is not immediately obvious which, or if indeed either, should be used in conjunction with Dirac's commutator–Poisson-bracket correspondence. Dirac[1] originally favored the continuous phase, while Judge and Lewis[17] applied a restricted range to the problem of rotation angle. In view of the long history of difficulties we do not employ commutator–Poisson-bracket correspondence in order to obtain the phase operator. However, we will return to this problem of correspondence once we have derived our phase operator and photon-number-phase commutator by other means.

III. HERMITIAN PHASE OPERATOR

We have discussed some of the problems associated with formulating a Hermitian phase operator in the Introduction. Despite these difficulties, states of well-defined phase are known to exist.[18] These phase state are the starting point in our phase operator formalism. The phase state $|\theta\rangle$ is defined as

$$|\theta\rangle = \lim_{s\to\infty}(s+1)^{-1/2}\sum_{n=0}^{s}\exp(in\theta)|n\rangle \;, \qquad (3.1)$$

where $|n\rangle$ are the $(s+1)$ number states which span an $(s+1)$-dimensional state space Ψ. The state of zero phase has been chosen as the state in which all the number state amplitudes are equal. The limiting procedure is necessary in order to normalize the states. These states provide a good description in that their time development is such that at time t the state becomes $|\theta-\omega t\rangle$. Moreover, the expectation value of the electric field is $\pm\infty$ with a divergent variance at all times *except* when $\theta-\omega t$ is an integer multiple of π. At these times the field is precisely zero. These zeros (when the field changes sign) precisely determine the phase of the field.

The phase state is well defined in the space Ψ but care must be taken with the limiting process. As with all limiting procedures, errors can result if s is replaced by infinity prematurely. Our procedure, therefore, will be to work entirely with states and operators in the $(s+1)$-dimensional space Ψ (where s can be arbitrarily large) and then allow s to tend to infinity *after* physical results such as expectation values are calculated. The finiteness of our state space Ψ means that the operators involved may have slightly different properties than those of their infinite space counterparts. For example, the trace of all commutators in the finite space must be zero rather than being undefined.[19] However, we emphasize that such differences will not lead to detectable physical differences when the limit is eventually taken.[20] We shall not be letting s tend to infinity until a later stage and so dispense, for the present, with the limit notation in (3.1).

The parameter θ in the phase state (3.1) can take any real value, although distinct states will only occur for values of θ in a given 2π range. Therefore, there exists an uncountable infinity of different phase states, even in the finite but arbitrarily large state space Ψ. The phase states are necessarily overcomplete and are not in general orthogonal. However, it is not difficult to show that states with values of θ differing by integer multiples of $2\pi/(s+1)$ are orthogonal,[16,21] and consequently, given any reference state $|\theta_0\rangle$ we can find a complete set of $(s+1)$ orthonormal phase states given by

$$|\theta_m\rangle = \exp[i\hat{N}m2\pi/(s+1)]|\theta_0\rangle, \quad m=0,1,\ldots s \;. \qquad (3.2)$$

Here, we have used the unitary phase shift operator $\exp(i\hat{N}\gamma)$ which transforms $|\theta\rangle$ to $|\theta+\gamma\rangle$ (as can be seen from the definition of $|\theta\rangle$). If $\gamma=-\omega t$, then this unitary operator is the time-evolution operator and we recover the ideal time development of a phase state. We note that choosing $m\geq(s+1)$ in (3.2) reproduces the states $|\theta_m\rangle$ with values $0\leq m<s$.

The set of phase states $|\theta_m\rangle$ can be used as a basis to span Ψ. The freedom to choose an arbitrary value for θ_0 means that there is an uncountable infinity of such bases. We shall leave θ_0 as an arbitrary phase, allowing the flexibility to choose the most convenient basis to solve a particular problem. The $(s+1)$ values of θ_m are

$$\theta_m=\theta_0+\frac{2\pi m}{(s+1)} \;, \qquad (3.3)$$

which are spread evenly over the range $\theta_0\leq\theta_m<\theta_0+2\pi$. When s tends to infinity, these values correspond to θ_0 plus the rational fractions of 2π. In this limit they form a countable infinity of orthogonal states that exist in a one-to-one correspondence with the countable basis of number states.

The Hermitian phase operator is simply defined in terms of a suitable phase state basis as

$$\hat{\phi}_\theta\equiv\sum_{m=0}^{s}\theta_m|\theta_m\rangle\langle\theta_m|$$
$$=\theta_0+\frac{2\pi}{s+1}\sum_{m=0}^{s}m|\theta_m\rangle\langle\theta_m| \;. \qquad (3.4)$$

We note that the eigenvalues (θ_m) of $\hat{\phi}_\theta$ are single valued and depend on the chosen value of θ_0. This dependence is denoted by the subscript θ and we will show that the classical analogue of $\hat{\phi}_\theta$ is the single-valued ϕ_θ. The choice of reference state $|\theta_0\rangle$ determines the 2π range which the phase eigenvalues will occupy. This procedure is completely analogous to choosing a window in which to express the inverses of (classical) trigonometric functions as single-value numbers.

From the definition of the phase state (3.1), we can express the projector $|\theta_m\rangle\langle\theta_m|$ in terms of the number state basis:

$$|\theta_m\rangle\langle\theta_m|=(s+1)^{-1}\sum_{n,n'}\exp[i(n'-n)\theta_m]|n'\rangle\langle n| \;, \qquad (3.5)$$

with the summations running from 0 to s. Substituting this expression into (3.4) and performing the summation over m yields a number state expansion for the phase operator:

$$\hat{\phi}_\theta=\theta_0+\frac{s\pi}{s+1}+\frac{2\pi}{s+1}\sum_{n\neq n'}\frac{\exp[i(n'-n)\theta_0]|n'\rangle\langle n|}{\exp[i(n'-n)2\pi/(s+1)]-1} \;. \qquad (3.6)$$

We can also express the number operator in the phase space basis:

$$\hat{N}\equiv\sum_{n=0}^{s}n|n\rangle\langle n|$$
$$=\frac{s}{2}+\sum_{m\neq m'}\frac{|\theta_{m'}\rangle\langle\theta_m|}{\exp[-i(m'-m)2\pi/(s+1)]-1} \;, \qquad (3.7)$$

where we have used the result

$$|n\rangle=(s+1)^{-1/2}\sum_{m=0}^{s}\exp(-in\theta_m)|\theta_m\rangle\ , \tag{3.8}$$

which is easily derived from (3.1) using the orthonormality of the basis $\{|\theta_m\rangle\}$. Expressions (3.6) and (3.7) reveal a subtle symmetry between $\hat{\phi}_\theta$ and $\hat{N}$: both the phase and number operators consist of a constant corresponding to the middle of their eigenvalue range plus a sum of off-diagonal projectors. This symmetry is particularly striking if we compare the phase operator with $\theta_0=0$ and the scaled number operator:

$$\hat{\phi}_0=\frac{s\pi}{s+1}+\frac{2\pi}{s+1}\sum_{n\neq n'}\frac{|n'\rangle\langle n|}{\exp[-i(n'-n)2\pi/(s+1)]-1}\ , \tag{3.9a}$$

$$\frac{2\pi\hat{N}}{s+1}=\frac{s\pi}{s+1}+\frac{2\pi}{s+1}\sum_{m\neq m'}\frac{|\theta_{m'}\rangle\langle\theta_m|}{\exp[-i(m'-m)2\pi/(s+1)]-1}\ . \tag{3.9b}$$

IV. PHASE-NUMBER COMMUTATOR

In Sec. III we defined the Hermitian phase operator in terms of a complete set of orthonormal phase states. This technique is independent of commutator–Poisson-bracket correspondence and we are now in a position to *derive* the phase-number commutator. The phase-number commutator is easily calculated using expressions (3.6) and (3.7). Expressed in the number state basis the commutator is

$$[\hat{\phi}_\theta,\hat{N}]=\frac{2\pi}{s+1}\sum_{n\neq n'}\frac{(n-n')\exp[i(n'-n)\theta_0]|n'\rangle\langle n|}{\exp[-i(n'-n)2\pi/(s+1)]-1}\ , \tag{4.1}$$

while in terms of the phase state basis it has the form

$$[\hat{\phi}_\theta,\hat{N}]=\frac{2\pi}{s+1}\sum_{m\neq m'}\frac{(m'-m)|\theta_{m'}\rangle\langle\theta_m|}{\exp[-i(m-m')2\pi/(s+1)]-1}\ . \tag{4.2}$$

These expressions look very different from the Dirac relation (1.2). In particular, our commutator is traceless (as all commutators must be in Ψ). Moreover, it is clear that the vanishing trace follows directly from the fact that the expectation value of the commutator in a number state $|n\rangle$ or phase state $|\theta_m\rangle$ is zero. Our commutator does not suffer from the mathematical inconsistency associated with the Dirac commutator.

The matrix elements of the phase-number commutator in a number state basis are

$$\langle n|[\hat{\phi}_\theta,\hat{N}]|n\rangle=0\ , \tag{4.3a}$$

$$\langle n'|[\hat{\phi}_\theta,\hat{N}]|n\rangle=\frac{2\pi(n-n')\exp[i(n'-n)\theta_0]}{(s+1)\{\exp[i(n'-n)2\pi/(s+1)]-1\}}\ . \tag{4.3b}$$

These expressions are exact but complicated and for many purposes a simpler form can be found. With this aim in mind we define a physically accessible, or preparable, state as one which can be excited from the vacuum state by coupling the mode to a finite energy source for a finite time by means of a finite interaction. Such a physical state has the following properties:[16] firstly, the finite source ensures an upper bound to the number states which have any probability of being excited and secondly, the moments of the energy or photon number distribution $\langle n^q\rangle$ are bounded for any finite number q. The latter condition follows from the finite time and interaction strength and is a weaker condition than the former. Indeed, imposition of the former condition automatically ensures the latter. Examples of states obeying the latter condition include thermal states, coherent states, squeezed states, and single number states. We note that a phase state is *not* a physical state because the expectation value of the photon number diverges as $s\to\infty$. For a physical state of the field we can make s very much larger than the number n associated with any significant number state component $|n\rangle$ of the state. In this case (with $s\gg n,n'$) expression (4.3) reduces to

$$\langle n'|[\hat{\phi}_\theta,\hat{N}]|n\rangle\approx i(1-\delta_{nn'})\exp[i(n'-n)\theta_0]\ . \tag{4.4}$$

This approximate equality becomes exact for all finite n and n' in the limit as s tends to infinity. However, we stress that the exact expression (4.3) should be used in general with the limit as $s\to\infty$ being taken only when the final result has been obtained. If our final aim is to calculate the expectation value of the commutator for a physical state, then (4.4) can be used directly. The operator formed from the matrix elements (4.4) has the form

$$[\hat{\phi}_\theta,\hat{N}]_p=i\sum_{n,n'}|n'\rangle\langle n|(1-\delta_{nn'})\exp[i(n'-n)\theta_0]$$
$$=-i+i\sum_{n'}\exp(in'\theta_0)|n'\rangle\sum_{n}\exp(-in\theta_0)\langle n|\ , \tag{4.5}$$

where the subscript p distinguishes this commutator from the precise expression (4.1) and is a reminder that it is only valid when acting on physical states. The first term, which follows from the Kronecker delta, is the Dirac term. The presence of the second term ensures that the trace of the commutator vanishes. It is clear from the definition of the phase state (3.1) that the physical-state commutator reduces to

$$[\hat{\phi}_\theta,\hat{N}]_p=-i[(1-(s+1)|\theta_0\rangle\langle\theta_0|]\ . \tag{4.6}$$

The expectation value of the phase-number commutator in any physical state $|p\rangle$ is

$$\langle p|[\hat{\phi}_\theta,\hat{N}]|p\rangle=-i[1-(s+1)|\langle p|\theta_0\rangle|^2]\ , \tag{4.7}$$

where $|\langle p|\theta_0\rangle|^2$ is the probability that the phase of the state is θ_0. In the continuum limit (as $s\to\infty$) this may be expressed as $P(\theta_0)2\pi/(s+1)$, where $P(\theta_0)$ is the probability density and $(s+1)/(2\pi)$ is the density of states. With this substitution, the expectation value (4.7) becomes

$$\langle p|[\hat{\phi}_\theta,\hat{N}|p\rangle = -i[1-2\pi P(\theta_0)] \ . \tag{4.8}$$

Dirac's commutator–Poisson-bracket correspondence requires the form of the commutator and Poisson bracket to be related:[13]

$$[\hat{u},\hat{v}]\leftrightarrow i\hbar\{u,v\} \ . \tag{4.9}$$

The classical expression for the expectation value of the Poisson bracket is [cf. Eq. (2.6)]

$$\langle\{\phi_\theta,H\}\rangle = \int_{\theta_0}^{\theta_0+2\pi} P(\phi)\{\phi_\theta,H\}d\phi$$
$$= -\omega[1-2\pi P(\theta_0)] \ . \tag{4.10}$$

The quantum Hamiltonian is $(\hat{N}+\frac{1}{2})\hbar\omega$ and so the correspondence between the quantum (4.8) and classical (4.10) expressions is verified. This correspondence is *precise* for all physical states. If the phase probability distribution is very sharply peaked, as for example in the case of a highly excited coherent state,[22] then the expectation value of the commutator reproduces the classical δ-function Poisson bracket (2.4).

From the preceding discussion we see that it is essential for there to be a difference between the Dirac commutator (1.2) and that derived by our approach because the trace of the latter must vanish. This difference removes the inconsistency associated with the number state matrix elements of Dirac's commutator. It also produces the appropriate 2π step in the phase-number commutator for all physical states. This step maintains the phase eigenvalues within the chosen 2π range, that is, it automatically takes care of the periodicity problem. The classical result (2.6) has followed naturally from the quantum-mechanical description of phase applied to physical states. The category of physical states is extremely broad and includes practically all states used so far in quantum electrodynamics, with the phase states being the only notable exception. Nevertheless, (4.6) [and therefore (2.6)] is *not* universally applicable. It is now clear why Dirac's Poisson bracket recipe to extrapolate from (2.3) or even (2.6) to find an Hermitian phase operator, with well-behaved phase eigenstates, had little change of success.

The expectation value (4.8) is a measure of minimum uncertainty in that physical states must satisfy the relation

$$\Delta N\Delta\phi_\theta \geq \tfrac{1}{2}|1-2\pi P(\theta_0)| \ . \tag{4.11}$$

This uncertainty relation depends on θ_0, that is, the choice of range for the phase eigenvalues. This phenomenon persists in classical mechanics where the expectation value and variance of a classical phase distribution will depend on the range of phase values employed. The quantum (or classical) probability distributions $P(\theta)$ will be periodic with period 2π. If $P(\theta)$ is sharply peaked at and approximately symmetric about $\theta=\beta+2n\pi$, then a 2π window which totally encloses a peak [that is, chosen so that $P(\theta_0)$ is small] will yield a mean phase of approximately β and a small variance. If, however, the window is chosen such that $P(\theta_0)$ is large (for example, $\theta_0=\beta$) then the distribution in the 2π window has one peak at β and another at $2\pi+\beta$. In this case the calculated mean will be about $\beta+\pi$ and the variance will be large. This effect is explored in detail elsewhere,[22] but we note here that the mean and variance of the phase only have meaning if the particular window of phase eigenvalues is specified. For many distributions $P(\theta)$ is sufficiently small over a range of θ for the mean and variance to be reasonably insensitive to the precise choice of θ_0 [provided θ_0 is sufficiently different from the peak of $P(\theta)$].

V. CREATION AND ANNIHILATION OPERATORS

The phase operator was originally intended to combine with the square root of the number operator in a polar decomposition of the creation and annihilation operators.[1] However, the failure of early attempts to construct a Hermitian phase operator,[3] or even a unitary exponential phase operator,[4,5] suggested that this procedure was unsatisfactory. It seemed to be clear that the zero in the eigenvalue range of $\hat{N}$ precluded any possibility of constructing a unitary exponential phase operator $[\exp(\pm i\hat{\phi})]$.[23] In this section we reexamine the problem in using our Hermitian phase operator.

We can construct a unitary operator $\exp(i\hat{\phi}_\theta)$ from the Hermitian phase operator. This operator function may be defined by its series expansion and is guaranteed to be unitary by the Hermiticity of $\hat{\phi}_\theta$. the unitary operator (*and* its conjugate operator) will have the phase states as eigenstates:

$$\exp(\pm i\hat{\phi}_\theta)|\theta_m\rangle = \exp(\pm i\theta_m)|\theta_m\rangle \ . \tag{5.1}$$

The properties of the unitary operator are demonstrated by considering its action on the photon number states:

$$\exp(i\hat{\phi}_\theta)|n\rangle = \exp\left[i\sum_{m=0}^{s}\theta_m|\theta_m\rangle\langle\theta_m|\right]|n\rangle$$
$$=(s+1)^{-1/2}\sum_{m=0}^{s}\exp[-i(n-1)\theta_m]|\theta_m\rangle \ , \tag{5.2}$$

where we have used expression (3.8). For $n>0$, a comparison of (5.2) with (3.8) shows that

$$\exp(i\hat{\phi}_\theta)|n\rangle = |n-1\rangle \ . \tag{5.3}$$

For the vacuum state ($n=0$) the resulting state is

$$(s+1)^{-1/2}\sum_m \exp(i\theta_m)|\theta_m\rangle$$
$$=(s+1)^{-1/2}\exp[i(s+1)\theta_0]\sum_m\exp(-is\theta_m)|\theta_m\rangle$$
$$=\exp[i(s+1)\theta_0]|s\rangle \ . \tag{5.4}$$

Therefore, the number state representation of $\exp(i\hat{\phi}_\theta)$ is

$$\exp(i\hat{\phi}_\theta) = |0\rangle\langle 1| + |1\rangle\langle 2| + \cdots + |s-1\rangle\langle s|$$
$$+\exp[i(s+1)\theta_0]|s\rangle\langle 0| \ . \tag{5.5}$$

This operator resembles the one introduced by Susskind and Glogower,[4,5] but with the vital difference that the action on the vacuum state is not indeterminate and so cannot be arbitrarily set to zero. The result of acting on the vacuum with $\exp(i\hat{\phi}_\theta)$ is *uniquely and precisely determined* to be the state $\exp[i(s+1)\theta_0]|s\rangle$. The expansion (5.5) is fully consistent with $\exp(i\hat{\phi}_\theta)$ being a unitary operator. The conjugate operator $\exp(-i\hat{\varphi}_0)$ is also unitary and clearly commutes with $\exp(i\hat{\phi}_\theta)$.

We can define cosine and sine operators from the unitary operators. These will correspond to cosine and sine series in the Hermitian phase operator. These operators are more consistent than their counterparts formed from the Susskind-Glogower operators. In particular, we find

$$\cos^2\hat{\phi}_\theta+\sin^2\hat{\phi}_\theta=1 \ , \tag{5.6}$$

$$[\cos\hat{\phi}_\theta,\sin\hat{\phi}_\theta]=0 \ , \tag{5.7}$$

$$\langle n|\cos^2\hat{\phi}_\theta|n\rangle=\langle n|\sin^2\hat{\phi}_\theta|n\rangle=\tfrac{1}{2} \ . \tag{5.8}$$

The last of these results is consistent with the phase of the vacuum being random and is in marked contrast with the result obtained using the Susskind-Glogower operators. In the Susskind-Glogower formulation the vacuum expectation values of $\cos^2\hat{\phi}_\theta$ and $\sin^2\hat{\phi}_\theta$ are $\frac{1}{4}$. The major difference between our operators and those of Susskind and Glogower involves the action on the vacuum state. It should not be surprising that very different results will occur in the quantum regime involving field states with a significant vacuum component. Our operators are in *no sense* an approximation to the Susskind-Glogower operators and have similar properties to the latter only for fields with large energies where both sets of operators give results in accord with classical behavior. Moreover, our cosine and sine operators are derived as operator functions of an Hermitian phase operator. No analogous procedure is possible in the Susskind-Glogower formulation.

The creation and annihilation operators can now be constructed by definition:

$$\hat{a}\equiv\exp(i\hat{\phi}_\theta)\hat{N}^{1/2} \tag{5.9}$$

$$=|0\rangle\langle 1|+2^{1/2}|1\rangle\langle 2|+\cdots+s^{1/2}|s-1\rangle\langle s| \ , \tag{5.10}$$

with $\hat{a}^\dagger$ being the Hermitian conjugate of $\hat{a}$. As s tends to infinity the action of $\hat{a}$ becomes that of the conventional annihilation operator. Note that the Hermitian amplitude operator $\hat{N}^{1/2}$ removes the $|s\rangle\langle 0|$ projector term when acting on (5.5) to give (5.10). Thus, while $\exp(i\hat{\phi}_\theta)$ uniquely determines $\hat{a}$, it is now clear why the reverse procedure (involving a decomposition of $\hat{a}$) cannot produce the vital $|s\rangle\langle 0|$ projector term in (5.5). Such a procedure can at best only produce an indeterminate result, requiring an *extra* assumption as used by Susskind and Glogower. The $|s\rangle\langle 0|$ projector *never* occurs in any coupling involving the electromagnetic field as these take place via $\hat{a}$ and $\hat{a}^\dagger$.[16]

It is known[5] that an expression of the form (5.9) (as proposed by Dirac[1]) is inconsistent with the conventional commutation relation of $\hat{a}$ and $\hat{a}^\dagger$ if $\exp(i\hat{\phi}_\theta)$ is unitary. If it were possible, we would require that a unitary transformation $\exp(i\hat{\phi}_\theta)\hat{N}\exp(-i\hat{\phi}_\theta)$ would simply add the unit operator to $\hat{N}$. However, our approach does not suffer from such problems because of the finite (but arbitrarily large) nature of the state space. An extra term was necessary in Dirac's commutator (1.2) and we should also anticipate an additional term in the $\hat{a},\hat{a}^\dagger$ commutator, even if only to ensure that its trace vanishes. On calculating the commutator of $\hat{a}$ and $\hat{a}^\dagger$, as defined by (5.9), we indeed obtain a result that is traceless:

$$[\hat{a},\hat{a}^\dagger]=1-(s+1)|s\rangle\langle s| \ . \tag{5.11}$$

The last term exactly compensates the trace obtained from the first term in the same way as the additional part to Dirac's term in $[\hat{\phi}_0,\hat{N}]$. However, while the addition to the $\hat{\phi}_\theta,\hat{N}$ commutator has direct physical consequences, the additional term in (5.11) has no effect when $[\hat{a},\hat{a}^\dagger]$ acts on any physical state and so has no physically observable consequences. The physical-state commutator

$$[\hat{a},\hat{a}^\dagger]_p=1 \tag{5.12}$$

is sufficient when operating on physical states.

This is most easily seen for states excited from the vacuum by a finite energy source, but is also true for states which have no upper bound to $|n\rangle$ but which have finite energy moments. As a definite example, consider a coherent state $|\alpha\rangle$. If q is any chosen positive integer then we have

$$[\hat{a},\hat{a}^\dagger]^q=1-|s\rangle\langle s|[1-(-s)^q] \ . \tag{5.13}$$

The coherent state expectation value of this operator is

$$\langle\alpha|[\hat{a},\hat{a}^\dagger]^q|\alpha\rangle=1-\exp(-|\alpha|^2)\frac{|\alpha|^{2s}}{s!}[1-(-s)^q] \ . \tag{5.14}$$

As s tends to infinity the last term vanishes for any given (finite) q. For a more general physical state $|p\rangle$ we obtain

$$\langle p|[\hat{a},\hat{a}^\dagger]^q|p\rangle=1-|c_s|^2[1-(-s)^q] \ , \tag{5.15}$$

where $c_s=\langle s|p\rangle$. The requirement that the moment $\langle N^q\rangle$ be convergent as s tends to infinity ensures that $|c_s|^2s^q$ must tend to zero and the second term vanishes.

We emphasize again that when states other than the physical states (for example the phase states) are used in a calculation, then (5.11) must be used in place of (5.12) and the limit must be found at the end of the calculation.

VI. PHASE PROPERTIES OF A GENERAL STATE

For completeness and for future reference, we show how the phase operator can be used to examine the phase properties of a field state. Consider a general pure state of the field mode[24]

$$|f\rangle=\sum_{n=0}^{s}c_n|n\rangle \ . \tag{6.1}$$

This may be reexpressed in the phase state basis using (3.8):

$$|f\rangle=(s+1)^{-1/2}\sum_n\sum_m c_n\exp(-in\theta_m)|\theta_m\rangle \ . \tag{6.2}$$

The phase probability distribution is

$$|\langle \theta_m | f\rangle|^2=(s+1)^{-1}\left|\sum_n c_n \exp(-i\theta_m)\right|^2 , \tag{6.3}$$

with an expectation value and variance

$$\langle \hat{\phi}_\theta \rangle = \sum_m \theta_m |\langle \theta_m | f\rangle|^2 , \tag{6.4}$$

$$\Delta \hat{\phi}_\theta^2 = \sum_m (\theta_m - \langle \hat{\phi}_\theta \rangle)^2 |\langle \theta_m | f\rangle|^2 . \tag{6.5}$$

Any discussion of phase-number minimum uncertainty states requires the expectation value of the phase-number commutator:

$$\langle f|[\hat{\phi}_\theta, \hat{N}]|f\rangle = \frac{2\pi}{s+1}\sum_{n\neq n'} \frac{c_n^* c_{n'}(n-n')\exp[i(n-n')\theta_0]}{1-\exp[i(n-n')2\pi/(s+1)]} . \tag{6.6}$$

It is the purpose of this paper to look at phase properties of a single-mode field from as general a perspective as possible. Therefore, we shall not pursue the many applications of these formulas to specific field states. We note, however that when $|f\rangle$ is a single number state the expectation value and variance in the phase are $\theta_0+\pi$ and $\pi^2/3$, respectively (as $s\rightarrow\infty$). These values correspond to the mean and variance of a classical phase with a random value between θ_0 and $\theta_0+2\pi$. The details of this calculation will be presented elsewhere[22] along with other special cases including coherent states. In this paper we focus our attention on a class of states that we call partial phase states.

VII. PARTIAL PHASE STATES

The form of the phase state (3.1) and the phase probability (6.3) suggest that interesting phase properties are to be expected when the state is of the form

$$|b\rangle = \sum_{n=0}^{s} b_n \exp(in\beta)|n\rangle , \tag{7.1}$$

where b_n is real and positive. Obviously, the phase state is a special example of this with $b_n^2=(s+1)^{-1}$. The states $|b\rangle$, which we shall refer to as partial phase states, will not normally be eigenstates of phase. A very important subset of these states will be the physical partial phase states, of which the coherent state is a particular example. The phase states are themselves unphysical and so the best attempt at a physical phase measurement will only *project* the system into a physical partial phase state.

The phase probability distribution for a partial phase state is given [from (6.3) and (7.1)] by

$$|\langle \theta_m | b\rangle|^2 = (s+1)^{-1}\left|\sum_n b_n \exp[in(\beta-\theta_m)]\right|^2 = \frac{1}{s+1}+\frac{2}{s+1}\sum_{n>n'} b_n b_{n'}\cos[(n-n')(\beta-\theta_m)] . \tag{7.2}$$

The mean and variance of $\hat{\phi}_\theta$ will depend on the chosen value of θ_0. We note that all values are equally valid but choose θ_0 in the most convenient and physically transparent way. For the partial phase state $|b\rangle$ we set

$$\theta_0=\beta-\frac{\pi s}{s+1} \tag{7.3}$$

and define a new phase label

$$\mu = m-\frac{s}{2} . \tag{7.4}$$

From (3.3), Eq. (7.2) becomes

$$|\langle \theta_m | b\rangle|^2 = \frac{1}{s+1}+\frac{1}{s+2}\sum_{n>n'} b_n b_{n'}\cos[(n-n')\mu 2\pi/(s+1)] , \tag{7.5}$$

with μ ranging in integer steps from $-s/2$ to $s/2$. This distribution is symmetric in μ. Using Eqs. (7.3)–(7.5), we find that

$$\langle b|\hat{\phi}_\theta|b\rangle = \beta . \tag{7.6}$$

This is a very important and general result for partial phase states which can be applied immediately, for example, to any coherent state.

The choice of θ_0 means that the variance in the phase probability distribution has a particularly simple form:

$$\Delta\phi_\theta^2 = \frac{4\pi^2}{(s+1)^2}\sum_{\mu=-s/2} |\langle \theta_m | b\rangle|^2 \mu^2 . \tag{7.7}$$

The summation in (7.7) is most easily performed in the limit as s tends to infinity by transforming it into an integral. We replace $\mu 2\pi/(s+1)$ by θ, $2\pi/(s+1)$ by $d\theta$, and integrate from $-\pi$ to π to obtain

$$\Delta\phi_\theta^2 = (2\pi)^{-1}\int_{-\pi}^{\pi}\left[1+2\sum_{n>n'} b_n b_{n'}\cos[(n-n')\theta]\right]\theta^2 d\theta = \frac{\pi^2}{3}+4\sum_{n>n'} b_n b_{n'}(-1)^{(n-n')}(n-n')^{-2} . \tag{7.8}$$

We see that for the extreme case of a single number state (where only one of the b_n is nonzero) the last term vanishes and the variance is $\pi^2/3$. At the other extreme is the phase state [with $b_n=(s+1)^{-1/2}$]. Using the large s result that

$$\sum_{n>n'} \frac{(-1)^{(n-n')}}{(n-n')^2} \approx -s(1-\tfrac{1}{4}+\tfrac{1}{9}-\cdots) = -s\pi^2/12 , \tag{7.9}$$

we find that the phase variance vanishes as expected.

As a special example of a partial phase state, consider the "rectangular" state $|b,R\rangle$ for which the coefficients b_n equal a constant $(r^{-1/2})$ for $q\le n<q+r$ and are zero elsewhere. The photon number probability distribution is constant and nonzero between $|q\rangle$ and $|q+r-1\rangle$ and is zero outside this range—hence the name "rectangular" state. These states include as special cases both phase

states (when $q=0$ and $r=s+1$) and the single number state $|q\rangle$ when $r=1$. The phase probability distribution can be obtained by summing the series in (7.2) to give

$$|\langle\theta_m|b,R\rangle|^2=\frac{\sin^2[r(\beta-\theta_m)/2]}{r(s+1)\sin^2[(\beta-\theta_m)/2]} . \tag{7.10}$$

When s tends to infinity this gives a continuous phase probability distribution,

$$P(\theta_m)=\frac{\sin^2[r(\beta-\theta_m)/2]}{2\pi r\sin^2[(\beta-\theta_m)/2]} , \tag{7.11}$$

which is peaked and symmetric about $\theta_m=\beta$. If we let r increase to give a broad number state distribution, then the phase probability distribution approaches a δ function. With the choice of reference phase (7.3), the expectation value of the phase is β (as it must be for all partial phase states). From (7.8) we find the phase variance in the rectangular state to be

$$\begin{aligned}\Delta\phi_\theta^2&=\frac{\pi^2}{3}-(4/r)[(r-1)-(r-2)/4+(r-3)/9\\&\quad-\cdots+(-1)^r/(r-1)^2]\\&=\frac{\pi^2}{3}-4(1-\tfrac{1}{4}+\tfrac{1}{9}-\cdots)+(4/r)(1-\tfrac{1}{2}+\tfrac{1}{3}\ldots) .\end{aligned} \tag{7.12}$$

If r is reasonably large then we can replace the finite series by $\pi^2/12$ and ln2, respectively, giving the approximate result

$$\Delta\phi_\theta^2\approx(4/r)\ln 2 . \tag{7.13}$$

To examine the validity of our approximation, we have calculated (7.12) explicitly for values $r=1$, 2, 3, and 4. We find for the ratio $\Delta\phi_\theta^2/[(4/r)\ln 2]$ the results 1.19, 0.92, 1.03, and 0.98, respectively. It is clear that the approximation (7.13) is a good one for rectangular partial phase states with as low as three or perhaps even two different number state components.

The only rectangular partial phase state of interest to us which is not also a physical state is the phase state itself. We already know that the expectation value of the phase-number commutator vanishes for a phase state. We therefore restrict the following discussion of the minimum uncertainty properties to physical partial phase states. This restriction allows us to use expression (4.7) for the expectation value of the commutator in conjunction with (7.10) for a rectangular state. With our choice of θ_0 (7.3) and in the limit of large s we find

$$|\langle\theta_0|b,R\rangle|^2=\frac{1-(-1)^r}{2r(s+1)} , \tag{7.14}$$

with the result that the expectation value of the phase-number commutator is

$$\langle b,R|[\hat{\phi}_\theta,\hat{N}]|b,R\rangle=-i+\frac{i[1-(-1)^r]}{2r} . \tag{7.15}$$

For r even we regain the Dirac commutator expectation value. However, if r is odd then the situation is more complicated: the commutator expectation value vanishes if $r=1$ (single number state) and approaches the Dirac form for large r.

It is not difficult to calculate the photon number variance for the rectangular state $|b,R\rangle$:

$$\Delta N^2=(r^2-1)/12 . \tag{7.16}$$

This variance is zero for a single number state ($r=1$). In the limit of large r the number-phase uncertainty product becomes

$$\Delta N\Delta\phi_\theta\approx[(r\ln 2)/3]^{1/2} . \tag{7.17}$$

Comparison of this expression with one half the modulus of the commutator which for large r is $\frac{1}{2}$ [see (7.15)], shows that the rectangular partial phase states will *not* in general be number-phase minimum uncertainty states. The exception among the physical rectangular phase states is the number state (with $r=1$) for which both the uncertainty product and the commutator expectation value vanish.

It is beyond the scope of this paper to study further specific examples of partial phase states, but we conclude our discussion with a few general comments. Firstly, we emphasize the significance of the choice of reference phase θ_0. This phase can be assigned any value, but the choice $\theta_0=\beta-\pi$ is the natural (and most appealing) choice to make for partial phase states (7.1). With this choice of reference phase, the correction to the Dirac term in the expectation value of the phase-number commutator will be

$$\begin{aligned}(s+1)|\langle\theta_0|b\rangle|^2&=[(b_0+b_2+b_4+\cdots)\\&\quad-(b_1+b_3+b_5+\cdots)]^2 .\end{aligned} \tag{7.18}$$

If the variation of b_n with n is sufficiently smooth and the distribution includes a large number of nonzero b_n, then it follows that this term will be small. In this case the Dirac term alone will be a good approximation to the expectation value of the phase-number commutator. For such states, the phase-number minimum uncertainty states will satisfy the relation

$$\Delta N\Delta\phi_\theta=\tfrac{1}{2} . \tag{7.19}$$

The parallel between this expression and that for position-momentum minimum uncertainty states suggests that a high-intensity phase-number minimum uncertainty state would have a Gaussian distribution of number states. A detailed calculation for a high-intensity coherent state (approximated by a Gaussian photon number distribution) is presented elsewhere[22] and shows that the minimum uncertainty relation (7.19) is satisfied by such states. The Hermitian phase operator provides a fully quantum-mechanical explanation for what hitherto has been discussed only in phenomenological and semiclassical terms.[9]

Finally, we note that when $(s+1)|\langle\theta_0|b\rangle|^2$ is small the probability density $P(\theta_0)$ (4.8) is also small. Therefore, the chosen range of phase eigenvalues completely encloses the peak of the phase distribution. The mean and variance of the phase will be reasonably insensitive to variations in the precise choice of θ_0, *provided* such variations do not take θ_0 too close to β, that is, if $|\beta-\theta_0|$ is still much greater than the width of the phase distribution.

VIII. PHASE DIFFERENCES

The phase difference between two independent *classical* field modes, or oscillators, is well defined and is independent of time if the oscillators have the same frequency. This phase difference is simply the difference between the individual phases $(\phi_1-\phi_2)$. The conventional Susskind-Glogower exponential operators are not operator functions of a Hermitian phase operator and such a simple notion of phase difference is not possible in their approach.[4] Instead, it is necessary to define phase difference operators separately rather than as the difference between phase operators. The definition chosen was that which *would* apply if the series were genuine exponential series:[25]

$$\hat{\exp}_s[i(\phi_1-\phi_2)]\equiv\hat{\exp}_s(i\phi_1)\hat{\exp}_s(-i\phi_2) , \tag{8.1}$$

$$\hat{\exp}_s[-i(\phi_1-\phi_2)]\equiv\hat{\exp}_s(-i\phi_1)\hat{\exp}_s(i\phi_2) , \tag{8.2}$$

where the subscripts 1 and 2 refer to modes 1 and 2. Hermitian cosine and sine operators were then constructed from these in the same manner as they would be constructed if these were genuine unitary operators. This definition leads to cosine and sine operators $[\hat{\cos}_s(\phi_1-\phi_2)$ and $\hat{\sin}_s(\phi_1-\phi_2)]$ that commute with the total number operator $(\hat{N}_1+\hat{N}_2)$ but *not* with each other. These operators have the expected behavior in the classical limit, but have quite peculiar properties in the quantum regime where the vacuum is an important component of either of the two field states. The lowest energy eigenstate of $\hat{\cos}_s(\phi_1-\phi_2)$ is the double vacuum $|0,0\rangle$ with eigenvalue (labeled $\cos\theta$) of zero. This would correspond to a value of θ of $\pm\pi/2$. The next eigenstates are $2^{-1/2}\{|1,0\rangle\pm0,1\rangle\}$ with eigenvalues $\cos\theta=\pm\frac{1}{2}$. The next set of eigenstates are three orthogonal linear combinations of $|2,0\rangle$, $|0,2\rangle$, and $|1,1\rangle$ with three eigenvalues, and so on. Only in the limit of large total excitation number does the spectrum of $\hat{\cos}_s(\phi_1-\phi_2)$ become dense. These results would imply that a measurement of the phase difference between two modes, each in its vacuum state, *must* yield a phase difference of $\pm90°$. If the state of the system is $|0,1\rangle$ (that is, one mode in its vacuum state and one containing a single photon), then a measurement of phase difference must yield either $\pm60°$ or $\pm120°$. No other results are possible. The fundamental reason for these predictions is again the nonunitarity of $\hat{\exp}_s(i\phi)$, which imparts nonrandom phase properties to the vacuum.

However, we have demonstrated the existence of the Hermitian phase operator and there is nothing to prevent us from adopting the natural definition of phase difference. Our phase difference operator is simply $\hat{\phi}_{\theta_1}-\hat{\phi}_{\theta_2}$, where again the subscripts 1 and 2 refer to the individual modes. The eigenstates of this operator are just the products of the individual phase eigenstates:

$$(\hat{\phi}_{\theta_1}-\hat{\phi}_{\theta_2})|\theta_{m1}\rangle|\theta_{m2}\rangle=(\theta_{m1}-\theta_{m2})|\theta_{m1}\rangle|\theta_{m2}\rangle . \tag{8.3}$$

The spectrum of this operator is dense (in the limit as s_1 and s_2 tend to infinity) *even for states with a low total excitation number.* A phase difference measurement can lead to a countably infinite number of results regardless of total excitation number. For example, the two-mode vacuum state is

$$|0,0\rangle=(s_1+1)^{-1/2}(s_2+1)^{-1/2}\sum_{m1=0}^{s_1}\sum_{m2=0}^{s_2}|\theta_{m1}\rangle|\theta_{m2}\rangle \tag{8.4}$$

and the system is equally likely to be found in any of the $(s_1+1)(s_2+1)$ phase difference eigenstates. In the limit, as s_1 and s_2 tend to infinity, there will be a countable infinity of possible values which, depending on the choice of $\theta_{01}-\theta_{02}$ would ensure that all phase differences are between -2π and 2π. Our phase difference operator is entirely consistent with two modes, in their respective vacuum states, having uncorrelated and random phases.

IX. UNITARY TRANSFORMATIONS AND MOMENTS

It follows from the definition of the number state (3.1) that the number operator is the generator of increments in the phase:

$$\exp(i\hat{N}\gamma)|\theta\rangle=|\theta+\gamma\rangle . \tag{9.1}$$

Integer multiples of 2π can be added to or subtracted from $\theta+\gamma$ without altering the state, in order to keep the value of $\theta+\gamma$ inside the chosen 2π window. The operator shifts the phase of the state by γ (mod 2π). If γ is an integer multiple of $2\pi/(s+1)$, then the action of $\exp(i\hat{N}\gamma)$ on an eigenstate of $\hat{\phi}_\theta$ produces another eigenstate of $\hat{\phi}_\theta$. When this is not so, we have a unitary operator which allows us to transform to a different phase state which is not an eigenstate of $\hat{\phi}_\theta$.

The existence of the Hermitian phase operator allows us to construct a general unitary operator $\exp(-i\lambda\hat{\phi}_\theta)$. When $\lambda=1$, this is just the "down-shift" operator introduced earlier (5.5). From expression (5.5) it follows that $\exp(ij\hat{\phi}_\theta)$ shifts the number state to a new number outside the "window" between $|0\rangle$ and $|s\rangle$, then the shift is by an amount j plus a suitable multiple of $(s+1)$, including the appropriate phase factor. For example, the unitary operator $\exp(i4\hat{\phi}_\theta)$ lowers the photon number by four so that:

$$\begin{aligned}&\exp(i4\hat{\phi}_\theta)|6\rangle=|2\rangle ,\\&\exp(i4\hat{\phi}_\theta)|1\rangle=\exp[i(s+1)\theta_0]|s-2\rangle .\end{aligned} \tag{9.2}$$

If λ is not an integer, then the state $\exp(-i\lambda\hat{\phi}_\theta)|n\rangle$ will *not* be an eigenstate of the number operator. Nevertheless, this state will be one of a complete orthonormal set of $(s+1)$ basis states which can be used to span the state space Ψ. These states are the noninteger number states which we label $|n+\lambda\rangle$. Access to these states by means of a unitary transformation generated by $\hat{\phi}_\theta$ may be useful in solving future problems.

The unitary operators described here have an application in that they allow us to construct phase- and number-moment generating functions. Given the density matrix for the field ρ, we can define a phase-moment generating function or characteristic function:

$$\chi_\theta(\lambda)=\mathrm{Tr}[\rho\exp(i\lambda\hat{\phi}_\theta)] . \tag{9.3}$$

The moments of $\hat{\phi}_\theta$ are given by differentiation with respect to λ:

$$\langle\hat{\phi}_\theta^k\rangle=\left.\left[\frac{-i\partial}{\partial\lambda}\right]^k\chi_\theta(\lambda)\right|_{\lambda=0} . \tag{9.4}$$

The moment generating function is related by Fourier transform to the phase probability density introduced in Sec. IV:

$$\begin{aligned}P(\theta)&=\int_{-\infty}^{\infty}\frac{d\lambda}{(s+1)}e^{-i\lambda\theta}\chi_\theta(\lambda)\\&=\frac{2\pi}{(s+1)}\mathrm{Tr}[\rho\delta(\hat{\phi}_\theta-\theta)] .\end{aligned} \tag{9.5}$$

Evaluating the trace shows that this is indeed the phase distribution

$$P(\theta)=\frac{2\pi}{(s+1)}\sum_m\langle\theta_m|\rho|\theta_m\rangle\delta(\theta_m-\theta) \tag{9.6}$$

and that it is correctly normalized

$$\begin{aligned}\int_{\theta_0}^{\theta_0+2\pi}\frac{(s+1)}{2\pi}P(\theta)d\theta&=\mathrm{Tr}\rho\\&=1 .\end{aligned} \tag{9.7}$$

A parallel analysis involving the operator $\exp(i\hat{N}\gamma)$ will generate a number-moment generating function.

X. CONCLUSION

We have presented a mathematical model of the single-mode electromagnetic field which involves a finite but arbitrarily large state space Ψ. The dimensionality of Ψ is allowed to tend to infinity only *after* calculation of physical results, such as expectation values, are made. Our model and the usual harmonic-oscillator model are equally valid and are physically indistinguishable. The advantage of our approach is that it permits the existence of an Hermitian phase operator, thus removing phase from its hitherto rather unique position as a classical observable without a quantum Hermitian operator counterpart.

We have described how the existence and form of the phase operator follow directly and uniquely from the states of well-defined phase. The physical state expectation value of the resulting phase-number commutator corresponds precisely with the classical Poisson bracket of the single-valued phase with the energy. The commutator contains Dirac's term and an additional contribution which resolves the anomalies associated with Dirac's commutator. However, the *exact form* of the phase-number commutator (which must be used for unphysical states such as phase states) is such as to preclude a direct extrapolation from the Poisson bracket.

A vital term in the unitary operator function $\exp(i\hat{\phi}_\theta)$ vanishes when combined with $\hat{N}^{1/2}$ in order to form the annihilation operator. This is the source of the difficulty in attempting to define a phase operator from the annihilation operator. The genuine unitarity of $\exp(i\hat{\phi}_\theta)$ gives it very different properties than the conventional Susskind-Glogower operator $\widehat{\exp}_s(i\phi)$. This is particularly evident when operating on field states with a significant vacuum component. Our unitary operator is in no sense an approximation to the Susskind-Glogower operator. Indeed, the phase properties of the vacuum state, and particularly the phase difference between two fields, are dramatic illustrations of the difference between our formulation and that of Susskind-Glogower.

The derived phase properties of the partial phase states, which include the coherent states, are consistent with those obtained by phenomenological methods[9] where the latter are valid. The phase operator allows us to construct a continuous unitary transformation between the number states. This transformation also allows us to access new basis sets of noninteger number states which are *not* eigenstates of $\hat{N}$. The continuous unitary transformation is of utility in constructing moment generating functions for the phase operator.

We conclude that our model of an electromagnetic field mode is not only physically indistinguishable from the conventional mathematical model involving the infinite Hilbert space harmonic operator, but is also more useful in that it allows us to define a well-behaved phase operator. Optical phase can at last be treated within the framework of quantum electrodynamics.

ACKNOWLEDGMENTS

This work was supported in part by the Australian Academy of Science and the Royal Society. S.M.B. thanks GEC Research Limited and the Fellowship of Engineering for support.

---

[1]P. A. M. Dirac, Proc. R. Soc. London, Ser. A **114**, 243 (1927).
[2]We denote all quantum operators by a caret.
[3]W. H. Louisell, Phys. Lett. 7, 60 (1963).
[4]L. Susskind and J. Glogower, Physics **1**, 49 (1964).
[5]P. Carruthers and M. M. Nieto, Rev. Mod. Phys. **40**, 411 (1968), and references therein.
[6]There are similar problems associated with the quantum description of rotation angle. We shall discuss quantum angle variables elsewhere.
[7]For a recent review of attempted quantum descriptions of

phase, see S. M. Barnett and D. T. Pegg, J. Phys. A **19**, 3849 (1986).

[8]R. E. Slusher, L. W. Hollberg, B. Yurke, J. C. Mertz, and J. F. Valley, Phys. Rev. Lett. **55**, 2409 (1985); D. F. Walls, Nature **306**, 141 (1983); S. M. Barnett and C. R. Gilson, Eur. J. Phys. (to be published).

[9]R. Loudon and P. L. Knight, J. Mod. Opt. **34**, 709 (1987), and references therein.

[10]B. C. Sanders, S. M. Barnett, and P. L. Knight, Opt. Commun. **58**, 290 (1986); R. Lynch, J. Opt. Soc. Am B **4**, 1723 (1987).

[11]W. Heitler, *The Quantum Theory of Radiation,* 3rd ed. (Oxford University Press, Oxford, 1954), p. 65.

[12]Dirac actually chose the commutator $[\hat{\phi},\hat{N}]=i$, but we follow the convention adopted by later workers.

[13]P. A. M. Dirac, Proc. R. Soc. London, Ser. A **109**, 642 (1925); P. A. M. Dirac, *The Principles of Quantum Mechanics,* 4th ed. (Oxford University Press, Oxford, 1958), p. 84.

[14]S. S. Schweber, in *Relativity, Groups and Topology II,* edited by B. S. Dewitt and R. Stora (North-Holland, Amsterdam, 1984), p. 62.

[15]R. Jackiw, J. Math. Phys. **9**, 339 (1968); H. Paul, Fortschr. Phys. **22**, 657 (1974); I. Bialynicki-Birula, and Z. Bialynicka-Birula, Phys. Rev. A **14**, 1101 (1976); J. M. Levy-Leblond, Ann. Phys. (New York) **101**, 319 (1976).

[16]D. T. Pegg and S. M. Barnett, Europhys. Lett. **6**, 483 (1988).

[17]D. Judge and J. T. Lewis, Phys. Lett. **5**, 190 (1963).

[18]R. Loudon, *The Quantum Theory of Light,* 1st ed. (Oxford University Press, Oxford, 1973), p. 143; E. C. Lerner, H. W. Huang, and G. E. Walters, J. Math. Phys. **11**, 1679 (1970).

[19]E. Merzbacher, *Quantum Mechanics,* 2nd ed. (Wiley, New York, 1970), p. 330.

[20]The impossibility of distinguishing, by physical experiments, the difference between an infinite state space and one of finite but arbitrarily large dimensions is discussed by A. Böhm, *The Rigged Hilbert Space and Quantum Mechanics* (Springer-Verlag, Berlin, 1978), p. 4.

[21]Clearly, states with values differing by integer multiples of 2π are identical and so are not orthogonal.

[22]S. Barnett and D. T. Pegg, J. Mod. Opt. (to be published).

[23]E. C. Lerner, Nuovo Cimento **56B**, 183 (1968); S. Stenholm, Ann. Acad. Sci. Fenn. Ser. A **6**, 339 (1970).

[24]The generalization to mixed states may be straightforwardly achieved using the density matrix.

[25]We follow the notation of our earlier work (Ref. 7) and denote the Susskind-Glogower operators as $\widehat{\exp}_s(\pm i\phi)$.

Quantum theory of rotation angles

Stephen M. Barnett
Department of Engineering Science, University of Oxford, Parks Road, Oxford OX1 3PJ, England

D. T. Pegg
School of Science, Griffith University, Nathan, Brisbane, Queensland 4111, Australia
(Received 25 September 1989)

The formulation of the quantum description of the rotation angle of the plane rotator has been beset by many of the long-standing problems associated with harmonic-oscillator phases. We apply methods recently developed for oscillator phases to the problem of describing a rotation angle by a Hermitian operator. These methods involve use of a finite, but arbitrarily large, state space of dimension $2l+1$ that is used to calculate physically measurable quantum properties, such as expectation values, as a function l. Physical results are then recovered in the limit as l tends to infinity. This approach removes the indeterminacies caused by working directly with an infinite-dimensional state space. Our results show that the classical rotation angle observable does have a corresponding Hermitian operator with well-determined and reasonable properties. The existence of this operator provides deeper insight into the quantum-mechanical nature of rotating systems.

I. INTRODUCTION

Classical simple-harmonic motion may be described as the projection into one dimension of a two-dimensional uniform circular motion. The phase of the simple-harmonic motion is the rotation angle ϕ of the corresponding circular motion, and both of these quantities are multivalued classical observables. The phase and rotation angle often appear as the inverse trigonometric functions and are usually chosen to lie in a specified 2π range, that is, ϕ can take values from θ_0 up to, but not including, $\theta_0+2\pi$. The choice of θ_0 is arbitrary, but commonly used values are 0 and $-\pi$. If we take this restriction on the values of rotation angle (or phase) seriously, then the evolution of the rotation angle will not be smooth, but will jump by 2π when its value reached the edge of the allowed range. However, if we wish to avoid this discontinuity in the evolution, we can still choose an initial rotation angle and then allow it to evolve without a bound to its allowed values. Naturally, the physical properties of the system under investigation should be independent of the manner in which we deal with the multivalued nature of angle or phase. The quantum description of phase and angle is more complicated than its classical counterpart and has met with considerable difficulties since Dirac[1] first postulated the existence of an operator for the phase of an electromagnetic-field mode. These difficulties have received much theoretical interest and have been described by a number of authors.[2–5]

We have recently obtained the Hermitian operator corresponding to the phase of a single mode of the electromagnetic field.[6–8] A complete description of the field requires an infinite basis of number states, and thus an infinite limit must be involved in any theory. The crucial feature which distinguishes our procedure from previous approaches is the stage at which this limit is taken. We begin with a finite, but arbitrarily large, state space of $s+1$ dimensions and calculate measurable quantities, such as expectation values and variances, as a function of s. The limit as s tends to infinity is then taken only *after* these expectation values and variances are calculated and is thus simply the limit of a sequence of real numbers. This procedure avoids the indeterminacies associated with approaches in which the limit is taken at an earlier, intermediate stage, for example, by embedding the finite space in an infinite Hilbert space.

In this paper we extend our method to derive the form of the Hermitian rotation-angle operator (henceforth referred to as an angle operator). This operator corresponds to the angular position of a plane rotator, that is, a body in uniform circular motion—for example, a bead on a circular wire. The angle operator will be applicable to a wider range of problems than the bead on a wire, but we initially restrict our attention to this system because of its simplicity and because it has featured in earlier attempts to describe quantum angle variables.[2-4] Moreover, it is the natural analog of the oscillator with its associated phase. We note that finite state spaces have been used in some earlier discussions of phase and angle variables.[9,10] However, where a transition has been attempted to an infinite (unbounded) system, these approaches have involved limiting procedures that enforce a return to the original problems noted by Susskind and Glogower.[11] By delaying the taking of limits until the final stage of the calculation after physical results, that is, real numbers, are obtained, we circumvent these difficulties. As noted by Merzbecher,[12] by confining ourselves to a complex linear vector space of finite dimensions, we succeed in avoiding questions which concern the convergence of sums over infinitely many terms, the interchangeability of several such summations, and the legitimacy of certain limiting procedures.

One of the problems inherent in the oscillator phase problem was the existence of a cutoff in the spectrum of the number operator, which excludes the negative integers. For the plane rotator, however, the corresponding angular momentum operator has a spectrum that includes both positive and negative integers. This suggests that our approach, which delays taking the limit of the dimensions of the state space, may not be necessary for the derivation of the angle operator. We show, however, that direct use of an infinite state space can lead to problems and that these may be understood and overcome by using our limiting procedure. The infinite state space may only be used with extreme care.

II. CLASSICAL ROTATION ANGLES AND SIMPLE QUANTIZATION

We begin our discussion with a description of classical angles and their quantization by application of the correspondence principle. For simplicity and definiteness we restrict our investigation to a bead constrained to move on a circular wire whose axis is aligned in the z direction. The classical z component of angular momentum and the azimuthal rotation angle of the bead can be expressed in terms of the Cartesian coordinates and momenta as

$$L_z = xp_y - yp_x \,, \tag{2.1}$$

$$\phi = \arctan(y/x) \,. \tag{2.2}$$

The angle is defined as the inverse of a trigonometric function and may be defined to lie within a chosen 2π range or to be assigned an initial value and then evolve as a continuous and unbounded variable. If we treat ϕ as a continuous variable, then the Poisson bracket for the angular momentum and the angle has the form

$$\{\phi, L_z\} = 1 \,. \tag{2.3}$$

Direct application of the correspondence between Poisson brackets and commutators suggests that the angular momentum and angle operators obey a commutator of the form

$$[\hat{\phi}, \hat{L}_z] = i\hbar \,. \tag{2.4}$$

If we represent an angular momentum operator as

$$\hat{L}_z = -i\hbar \frac{\partial}{\partial\phi} \tag{2.5}$$

and the angle operator as multiplication by ϕ, then the commutator (2.4) is satisfied. However, this representation of the angle operator causes problems:[13] if $u(\phi)$ is a periodic wave function, then $\phi u(\phi)$ will not be and is therefore outside the angular momentum state space. Judge and Lewis realized that the eigenvalues of a well-behaved angle operator would have to be restricted to a 2π interval. Their solution was to modify the angle operator so that it corresponded to multiplication by ϕ plus a series of step functions. These step functions sharply change the angle by 2π at appropriate points.[14] The resulting commutation relation between this operator and $\hat{L}_z$ has a δ-function term in addition to the $i\hbar$ term from the commutator (2.4). The Judge-Lewis commutator corresponds to the classical Poisson bracket of L_z and a single-valued angle variable.[15] Another approach is to avoid the problem of multivaluedness by not dealing with an Hermitian angle operator at all, but rather only periodic functions of the angle operator.[2-4] Naturally, this approach does not allow us to investigate the properties of the angle operator itself.

There is a further difficulty associated with the proposed commutator (2.4). This problem was originally discovered in association with Dirac's phase operator,[16] but is readily extended to the present situation. The difficulty arises when we take matrix elements of the proposed commutator (2.4) in the angular momentum basis

$$\langle m | [\hat{\phi}, \hat{L}_z] | m' \rangle = i\hbar\delta_{mm'} \,, \tag{2.6}$$

where the states $|m\rangle$ are eigenstates of $\hat{L}_z$ with eigenvalue m. This expression implies that the matrix elements of $\hat{\phi}$ are undefined in the angular momentum basis

$$(m' - m)\langle m | \hat{\phi} | m' \rangle = i\hbar\delta_{mm'} \,. \tag{2.7}$$

Consideration of the diagonal matrix elements in this equation clearly demonstrates the problem. A similar problem occurs if we use this commutator in an attempt to construct an angle-state representation of the angular momentum operator.

The above difficulties are only partially resolved by employing a single-valued operator obeying a commutation relation similar to that proposed by Judge and Lewis.[13] We shall see in Sec. V that such a commutator arises naturally if we work directly in an infinite angular momentum state space. The resulting matrix elements of the angle operator in the angular momentum basis states are well defined and correct. However, the angle-state matrix elements of the angular momentum operator are not. We shall show how these problems associated with the angle operator can be resolved by methods previously used for the treatment of optical phase.[6-8]

III. ANGLE STATES AND THE HERMITIAN ANGLE OPERATOR

By analogy with our phase operator approach, we work in a $(2l+1)$-dimensional state space Ψ spanned by the $\hat{L}_z$ eigenvectors $|m\rangle$, with $m=-l,\ldots,-1,0,1,\ldots,l$. Later, and only *after* physical results such as expectation values are calculated, we shall let l tend to infinity.

A sensible rotation-angle state will have $\hat{L}_z$ as a generator, that is, it will obey

$$\exp(-i\eta\hat{L}_z/\hbar)|\phi\rangle=|\phi+\eta\rangle\ , \tag{3.1}$$

which can be achieved by defining

$$|\phi\rangle\equiv\exp(-i\phi\hat{L}_z/\hbar)|\alpha_0\rangle\ , \tag{3.2}$$

where $|\alpha_0\rangle$ is the zero angle state. If an Hermitian angle operator $\hat{\phi}$ exists in a conjugate relationship with $\hat{L}_z$, we would expect it to be the generator of an angular momentum shift,

$$\exp(in\hat{\phi})|m\rangle=|m+n\rangle \tag{3.3}$$

for all integers n. We can use these shift properties to obtain the form of the angle states by operating with $\exp(in\hat{\phi})$ on both sides of the expansion

$$|\alpha_0\rangle=\sum_m c_m|m\rangle\ . \tag{3.4}$$

If the zero angle state is an eigenstate of $\hat{\phi}$ with eigenvalue zero, we obtain

$$|\alpha_0\rangle=\sum_m c_m|m+n\rangle\ . \tag{3.5}$$

Comparison of (3.5) and (3.4) shows that the coefficients c_m should be independent of m.[17] The space Ψ has dimensional $2l+1$; thus we normalize the coefficients c_m to be $(2l+1)^{-1/2}$. Such a normalization cannot be achieved in an infinite state space. This analysis leads unambiguously to the form of the angle state $|\phi\rangle$,

$$|\phi\rangle=(2l+1)^{-1/2}\sum_{m=-l}^{l}\exp(-im\phi)|m\rangle\ . \tag{3.6}$$

The form of this state is similar to that of the optical phase states.[18]

We see from (3.6) that the angle states have a periodic structure: $|\phi+2\pi\rangle$ is the same state as $|\phi\rangle$. All distinct angle states can be specified by the points on the real line between some value θ_0 and up to, but not including, $\theta_0+2\pi$. These states, however, are overcomplete and are not all mutually orthogonal:

$$\begin{aligned}\langle\phi'|\phi\rangle&=(2l+1)^{-1}\sum_{m=-l}^{l}\exp[-im(\phi'-\phi)]\\&=(2l+1)^{-1}\frac{\sin[(2l+1(\phi-\phi')/2]}{\sin[(\phi-\phi')/2]}\ .\end{aligned} \tag{3.7}$$

If we were to ignore the factor $(2l+1)^{-1}$ and take the limit of the second factor as l tends to infinity, we would obtain a sum of δ functions with peaks separated by 2π. However, it makes little sense to take the limit of the second term without including the normalization factor. We see from (3.7) that two angles states $|\phi\rangle$ and $|\phi'\rangle$ are orthogonal if $\phi-\phi'$ is a nonzero multiple of $2\pi/(2l+1)$. We can thus form a complete orthonormal basis of $(2l+1)$ angle states $|\theta_n\rangle$ be selecting values of θ_n as

$$\theta_n=\theta_0+\frac{2\pi n}{2l+1}\quad(n=0,1,\ldots,2l)\ . \tag{3.8}$$

The choice of θ_0 is arbitrary and determines the particular basis set. Choosing the basis beginning with θ_0 to span the space Ψ corresponds to the classical procedure of choosing a particular 2π window in which to express the value of $\arctan(y/x)$. We have already noted a similar correspondence in our analysis of the optical phase operator.[7]

Angle states in different basis sets will not be orthogonal and will be eigenstates of *different noncommuting* angle operators. It is therefore necessary to attach a label to the angle operator in order to specify which basis set forms its eigenstates. We label the angle operator $\hat{\phi}_\theta$ to indicate that its eigenvalues are θ_n as given in (3.8):

$$\hat{\phi}_\theta\equiv\sum_{n=0}^{2l}\theta_n\ \theta_n\rangle\langle\theta_n| \tag{3.9}$$

$$=\theta_0+\sum_{n=0}^{2l}\frac{2\pi n}{2l+1}|\theta_n\rangle\langle\theta_n|\ . \tag{3.10}$$

The matrix elements of $\hat{\phi}_\theta$ in the angular momentum basis $|m\rangle$ are

$$\begin{aligned}\langle m'|\hat{\phi}_\theta|m\rangle&=\sum_{n=0}^{2l}\theta_n\langle m'|\theta_n\rangle\langle\theta_n|m\rangle\\&=(2l+1)^{-1}\sum_{n=0}^{2l}\theta_n\exp[i(m-m')\theta_n]\ ,\end{aligned} \tag{3.11}$$

which gives

$$\langle m|\hat{\phi}_\theta|m\rangle=\theta_0+\frac{2\pi l}{2l+1} \tag{3.12}$$

and

$$\langle m'|\hat{\phi}_\theta|m\rangle=\frac{2\pi\exp[i(m-m')\theta_0]}{(2l+1)\{\exp[i(m-m')2\pi/(2l+1)]-1\}} \tag{3.13}$$

for the diagonal and off-diagonal elements, respectively. These allow us to express the angle operator in the angular momentum basis as

$$\hat{\phi}_\theta=\theta_0+\frac{2\pi l}{2l+1}+\frac{2\pi}{2l+1}\sum_{\substack{m,m'\\m\neq m'}}\frac{\exp[i(m-m')\theta_0]|m'\rangle\langle m|}{\exp[i(m-m')2\pi/(2l+1)]-1}\ . \tag{3.14}$$

Of particular interest are the physical states for which all the moments of $\hat{L}_z$ are finite.[19] The states may be approximated to any desired accuracy by an expansion

$\sum_m b_m |m\rangle$, where all b_m are zero for $|m|>M$, with the bound M being as large as necessary, but always less than l. If we restrict the domain of $\hat{\phi}_\theta$ to these physical states, we can employ a "physical" angle operator obtained from (3.14) in the limit of large l:

$$(\hat{\phi}_\theta)_p=\theta_0+\pi-i\sum_{\substack{m,m'\\ m\neq m'}}\frac{\exp[i(m-m')\theta_0]}{m-m'}|m'\rangle\langle m| . \quad (3.15)$$

Here the label p is a reminder that this simplified form can only replace $\hat{\phi}_\theta$ when operating on physical states. Nonphysical states include the angle states themselves, for which the exact form (3.14) must be used.

From (3.14) we obtain the commutator

$$[\hat{\phi}_\theta,\hat{L}_z] = \frac{2\pi\hbar}{2l+1} \times \sum_{\substack{m,m'\\ m\neq m'}}\frac{(m-m')\exp[i(m-m')\theta_0]|m'\rangle\langle m|}{\exp[i(m-m')2\pi(2l+1)]-1} . \quad (3.16)$$

In the special case $\theta_0=0$, this reduces to the finite-space commutator obtained by Sunthanam.[10]

We can use the expansion

$$|m\rangle=\sum_{n=0}^{2l}|\theta_n\rangle\langle\theta_n|m\rangle = (2l+1)^{-1/2}\sum_{n=0}^{2l}\exp(im\theta_n)|\theta_n\rangle \quad (3.17)$$

to express $\hat{L}_z$ in the angle-state basis

$$\hat{L}_z=-i\frac{\hbar}{2}\sum_{\substack{n,n'\\ n\neq n'}}\frac{(-1)^{n-n'}|\theta_{n'}\rangle\langle\theta_n|}{\sin[(n-n')\pi/(2l+1)]} . \quad (3.18)$$

It is now straightforward to express the angle–angular-momentum commutator in the angle-state basis

$$[\hat{\phi}_\theta,\hat{L}_z]=i\frac{\hbar\pi}{2l+1} \times\sum_{\substack{n,n'\\ n\neq n'}}\frac{(n-n')(-1)^{n-n'}|\theta_{n'}\rangle\langle\theta_n|}{\sin[(n-n')\pi/(2l+1)]} . \quad (3.19)$$

The commutator clearly has well-defined matrix elements. In particular, the diagonal elements are

$$\langle m|[\hat{\phi}_\theta,\hat{L}_z]|m\rangle=0 , \quad (3.20)$$

$$\langle\theta_n|[\hat{\phi}_\theta,\hat{L}_z]|\theta_n\rangle=0 . \quad (3.21)$$

The commutator does not suffer from the difficulties discussed earlier in Sec. II.

When operating on physical state, the commutator (3.16) can be replaced by

$$[\hat{\phi}_\theta,\hat{L}_z]_p=-i\hbar\sum_{\substack{m,m'\\ m\neq m'}}\exp[i(m-m')\theta_0]|m'\rangle\langle m| , \quad (3.22)$$

which is obtained by taking the large-l limit. However, we stress that it is not in general possible to take the limit before expectation values are calculated. If there is any doubt as to the validity of this procedure, the full expressions involving l should be used. The physical commutator has matrix elements given by

$$\langle m'|[\hat{\phi}_\theta,\hat{L}_z]_p|m\rangle=-i\hbar(1-\delta_{mm'})\exp[i(m-m')\theta_0] . \quad (3.23)$$

We note that we cannot make a similar approximation when using the angle-state basis (3.19), as values of $n-n'$ up to $2l+1$ are allowed for physical states such as an angular momentum eigenstate. The definition of the angle states allows us to express the physical commutator in the form

$$[\hat{\phi}_\theta,\hat{L}_z]_p=i\hbar[1-(2l+1)|\theta_0\rangle\langle\theta_0|] . \quad (3.24)$$

For any physical state $|p\rangle$, the expectation value of the commutator will be

$$\langle p|[\hat{\phi}_\theta,\hat{L}_z]|p\rangle=i\hbar[1-(2l+1)|\langle p|\theta_0\rangle|^2] . \quad (3.25)$$

The second term can be written as $2\pi P(\theta_0)$, where $P(\theta_0)\delta\theta$, with $\delta\theta=2\pi/(2l+1)$, is the probability that the system will be found within $\delta\theta$ of the value θ_0. In the limit as $2l+1$ tends to infinity, $P(\theta)$ will be the normalized probability density

$$\int_{\theta_0}^{\theta_0+2\pi}P(\theta)d\theta=1 . \quad (3.26)$$

The expectation value of the commutator (3.25) corresponds precisely to the classical Poisson bracket for a single-valued angle variable.[20] The effect of the second term is to step the angle by 2π at $\theta_0+2\pi$. If $P(\theta)$ is a δ-function distribution $\delta(\theta-\theta')$, corresponding to a physical state of quite well-defined angle θ', then (3.25) becomes

$$\langle p|[\hat{\phi}_\theta,\hat{L}_z]|p\rangle=i\hbar1-2\pi\delta(\theta_0-\theta')] , \quad (3.27)$$

which clearly displays the 2π jump at $\theta'=\theta_0$. This is precisely the behavior anticipated by Judge and Lewis[13] and Susskind and Glogower[2] for a well-behaved single-valued operator. However, the expressions (3.25) and (3.27), which may be used for physical state of quite well-defined angle, are *not* applicable to state of precisely defined angle, that is, the angle states. For these we must resort to the exact expressions. It is clear that the expectation value of the physical commutator (3.24) for a system in state $|\theta_n\rangle$ is $i\hbar[1-(2l+1)\delta_{n0}]$, which is nonzero for all values of n. Clearly, (3.24) cannot be used in place of the exact commutator $[\hat{\phi}_\theta,\hat{L}_z]$ if the angle states are to be used because the commutator must have zero expectation value for angle eigenstates.

IV. PERIODIC ANGLE OPERATORS

In line with our earlier approach to phase, we construct unitary angle operators from the Hermitian angle operator:[21]

$$\exp(\pm i\hat{\phi}_\theta)=\exp\left[\pm i\sum_{n=0}^{2l}\theta_n|\theta_n\rangle\langle\theta_n|\right] . \tag{4.1}$$

The unitarity of these operators follows directly from the Hermiticity of the angle operator. The unitary operators act as angular momentum raising or lowering operators:

$$\exp(\pm i\hat{\phi}_\theta)|m\rangle=\exp\left[\pm i\sum_{n=0}^{2l}\theta_n|\theta_n\rangle\langle\theta_n|\right]\times\sum_{n'}\frac{\exp(im\theta_{n'})}{\sqrt{(2l+1)}}|\theta_{n'}\rangle$$
$$=(2l+1)^{-1/2}\sum_n\exp[i(m\pm1)\theta_n|\theta_n\rangle \tag{4.2}$$
$$=|m\pm1\rangle . \tag{4.3}$$

Here the states are labeled modulo $2l+1$, so that, for example,

$$|\pm(l\pm1)\rangle\equiv\exp[\pm i(2l+1)\theta_0]|\pm(-l)\rangle . \tag{4.4}$$

The cyclic nature of $\exp(i\hat{\phi}_\theta)$ is made clear by writing the unitary operator in the angular momentum basis

$$\exp(i\hat{\phi}_\theta)=|-l+1\rangle\langle-l|+\cdots+|m+1\rangle\langle m| + \cdots+|l\rangle\langle l-1| +\exp[i(2l+1)\theta_0]|-l\rangle\langle l| . \tag{4.5}$$

The lowering operator is given by the Hermitian conjugate of this operator. These unitary operators are functions of a common angle operator and must therefore commute. The sine and cosine angle operators have well-behaved properties. In general, we find

$$[\cos\hat{\phi}_\theta,\sin\hat{\phi}_\theta]=0 , \tag{4.6}$$

$$\cos^2\hat{\phi}_\theta+\sin^2\hat{\phi}_\theta=1 , \tag{4.7}$$

$$\langle m|\cos^2\hat{\phi}_\theta|m\rangle=\langle m|\sin^2\hat{\phi}_\theta|m\rangle=\tfrac{1}{2} , \tag{4.8}$$

$$\langle m|\cos\hat{\phi}_\theta|m\rangle=\langle m|\sin\hat{\phi}_\theta|m\rangle=0 . \tag{4.9}$$

The last two of those are consistent with a state of precise angular momentum having a random orientation.

If the zero angle state is an eigenstate of ϕ_θ, then θ_0 must be an integer multiple of $2\pi/(2l+1)$ and the exponential factor on the right side of (4.5) is unity. In this case the action of $\exp(i\hat{\phi}_\theta)$ on $|l\rangle$ gives $|-l\rangle$. Use of the operator identity

$$\exp(in\hat{\phi}_\theta)\equiv[\exp(i\hat{\phi}_\theta)]^n \tag{4.10}$$

justifies our remarks concerning the consistency of the representation of the zero angle state in the $(2l+1)$ dimensional space.[17]

For completeness, we give here the commutation relations between the unitary angle operators and $\hat{L}_z$:

$$[\exp(\pm i\hat{\phi}_\theta),\hat{L}_z]=\pm\hbar\{-\exp(\pm i\hat{\phi}_\theta)+(2l+1)\exp[\pm i(2l+1)\theta_0]|\pm(-l)\rangle\langle\pm l|\} . \tag{4.11}$$

These commutators have been obtained previously in a finite space.[9]

V. IMPROPER VECTORS IN INFINITE STATE SPACE

The symmetry between the Hermitian rotation angle operator $\hat{\phi}_\theta$ and $\hat{L}_z$ is evident throughout our work so far. In the limit as l tends to infinity, both have a countable infinity of eigenstates related to each other by (3.6) and (3.17). The angular momentum and angle eignestates are equal in number and from alternative bases for the same state space. If the eigenvalues of $\hat{\phi}_\theta$ are mapped as points on a line from θ_0 up to but not including $\theta_0+2\pi$, then, in the limit of large l, they correspond to θ_0 plus all the rational fractions of 2π. That is, the eigenvalue spectrum of the angle operator is dense.

The angular momentum operator has both positive and negative eigenvalues. This is not true for the photon number operator and therefore we might hope that it is possible to construct an angle operator directly in an infinite state space without the necessity of employing our limiting procedure. We show, however, that there are inherent difficulties associated with using an infinite state space directly.

In an infinite state space we cannot normalize an angle state vector, but begin instead with the improper, unnormalizable state vector given by the linear superposition[22]

$$|\Theta\rangle\equiv(2\pi)^{-1/2}\sum_{m=-\infty}^{\infty}\exp(-im\Theta)|m\rangle , \tag{5.1}$$

where we use Θ to distinguish these from our previous proper state vectors labeled by θ_n. We note from (5.1) that the states form an uncountably infinite set. The scalar product of two of these angle states involves a factor $(2\pi)^{-1}$ in place of the factor $(2l+1)^{-1}$ and (3.7). Thus we have a δ-function normalization of these states:

$$\langle\Theta'|\Theta\rangle=\delta(\Theta'-\Theta)\quad\text{for } |\Theta'-\Theta|<2\pi . \tag{5.2}$$

We can obtain a resolution of the identity by integration over the angle states

$$\int_{\Theta_0}^{\Theta_0+2\pi}|\Theta\rangle\langle\Theta|d\Theta=1 . \tag{5.3}$$

If we define an infinite space angle operator as

$$\hat{\Phi}_\theta \equiv \int_{\Theta_0}^{\Theta_0+2\pi} \Theta|\Theta\rangle\langle\Theta|d\Theta \ , \tag{5.4}$$

then we see from (5.2) the *all* the angle states are angle eigenstates:

$$\hat{\Phi}_\Theta|\Theta'\rangle = \Theta'|\Theta'\rangle \ . \tag{5.5}$$

Substituting the expansion (5.1) into our expression for the angle operator (5.4) we find

$$\hat{\Phi}_\Theta = (2\pi)^{-1}\int_{\Theta_0}^{\Theta_0+2\pi}\Theta \times \sum_{m,m'} \exp[i(m-m')\Theta] \times |m'\rangle\langle m|d\Theta \ . \tag{5.6}$$

If we allow ourselves to interchange the order of the infinite summation and the integral, then we obtain the expression

$$\hat{\Phi}_\Theta = \Theta_0 + \pi - i \sum_{\substack{m,m' \\ m\neq m'}} \frac{\exp[i(m-m')\Theta_0]}{m-m'}|m'\rangle\langle m| \ . \tag{5.7}$$

This form of the angle operator implies that the angle–angular-momentum commutator is

$$[\hat{\Phi}_\Theta, \hat{L}_z] = i\hbar(1 - 2\pi|\Theta_0\rangle\langle\Theta_0|) \ . \tag{5.8}$$

This commutation relation is equivalent to that postulated by Judge and Lewis.[13] We seem therefore to have arrived at the result (3.24), which we have seen is only applicable for physical states and not, for example, when operating on the angle states themselves. However, use of this commutator leads to inconsistencies. In the Appendix we reveal one of these inconsistencies by using the angle-state matrix elements of this commutator to rederive the form of $\hat{L}_z$. The source of these difficulties is the apparently innocent procedure of interchanging the order of infinite summations and integrations in the derivation of $\hat{\Phi}_\Theta$. From the definition of integration, we have

$$\int_{\Theta_0}^{\Theta_0+2\pi} \Theta \exp(ik\Theta)d\Theta \equiv \lim(\delta\to 0) \sum_{n=0}^{N} \exp[ik(\Theta_0+n\delta)](\Theta_0+n\delta) \ , \tag{5.9}$$

where $\delta = 2\pi/(N+1)$ and k is a nonzero integer. If we follow the usual rules of integration, the left-hand side becomes

$$\int_{\Theta_0}^{\Theta_0+2\pi} \Theta \exp(ik\Theta)d\Theta = \frac{2\pi}{ik}\exp(ik\Theta_0) \ . \tag{5.10}$$

However, the right-hand side can be summed exactly to give

$$\sum_{n=0}^{N} \exp[ik(\Theta_0+n\delta)](\Theta_0+n\delta) = -\frac{\delta^2(N+1)\exp(ik\delta\Theta_0)}{1-\exp(ik\delta)} \ . \tag{5.11}$$

Now *if and only if* $k\delta$ approaches zero as δ tends to zero does the limit of (5.11) become equal to (5.10). In our case $k = m - m'$ and therefore

$$k\delta = \frac{2\pi(m-m')}{N+1} \ . \tag{5.12}$$

Clearly, $k\delta$ approaches zero as N tends to infinity only if $m - m'$ is finite, as, for example, in the case of physical states. In general, however, the summation must be over all m and m', including $m - m'$ tending to infinity, and it is not permissible to exchange the order of summation and integration, unless we restrict the domain of operation to physical states.

In summary, we are not allowed in general to interchange the orders of limits associated with infinite summations and the integration. Such problems, which are associated with the direct use of an infinite state space, do not occur in our approach where we delay allowing the dimensionality of the space to approach infinity until after physical quantities such as expectation values are calculated, in which case the limits are those of sequences of real numbers. The problems described above emphasize the value of our approach.

VI. OTHER ANGULAR MOMENTUM OPERATORS

In this paper we have been concerned primarily with the quantum mechanics of a bead on a circular wire. This system has only one degree of freedom, the azimuthal angle, and one component of angular momentum. In this section we discuss breifly how the angle operator may be applied to more general problems in which all three components of angular momentum may be present. We examine a system with fixed total angular momentum—for example, a spinning top.

The eigenstates of this system are fixed by two quantum numbers and are labeled $|j,m\rangle$, where j is the total angular momentum quantum number which, bearing in mind that we shall eventually let l tend to infinity, is very much less than l. We begin by defining angular momentum raising and lowering operators in the space Ψ in terms of the angle operator $\hat{\phi}_\theta$.

$$\hat{J}_+ \equiv \exp(i\hat{\phi}_\theta)\hat{S}_j \ , \tag{6.1a}$$

$$\hat{J}_- \equiv \hat{S}_j\exp(-i\hat{\phi}_\theta) \ , \tag{6.1b}$$

where $\hat{S}_j$ is the amplitude operator

$$\hat{S}_j \equiv \hbar[j(j+1) - \hat{L}_z(\hat{L}_z+1)]^{1/2} \ . \tag{6.2}$$

We note that a different operator will be required for each value of j and that this amplitude operator will be Hermitian only when acting on states with $|m| \leq j$. The z component of angular momentum is as defined earlier:

$$\hat{L}_z = \sum_{m=-l}^{l} \hbar m|j,m\rangle\langle j,m| \ . \tag{6.3}$$

The action of the angular momentum raising and lowering operators on states for which $|m| \leq j$ is as expected:

$$\hat{J}_\pm|j,m\rangle = \hbar\{j(j+1) \pm [-m(m+1)]\}^{-1/2}|j,m\pm 1\rangle \ . \tag{6.4}$$

We note that these operators will also act on states $|m\rangle$ for which $|m|>j$. However, these states are not physically accessible, because the action of $\hat{J}_+$ or $\hat{J}_-$, which will occur in an interaction Hamiltonian, cannot couple states with $|m|\le j$ to those with $|m|>j$:

$$\hat{J}_{\pm}|j,\pm(-j)\rangle=0\ , \quad (6.5a)$$

$$\hat{J}_{\pm}|j,\pm(-j-1)\rangle=0\ . \quad (6.5b)$$

Thus the physical region of the state space and the unphysical region are uncoupled, and a system which initially is in a superposition of states of the physical region must always remain in some superposition of these states.

We can use the definitions of $\hat{L}_z$, $\hat{J}_+$, and $\hat{J}_-$ to obtain the angular momentum commutation relations. From the definitions of $\hat{J}_+$ and $\hat{J}_-$ and the commutators we find eventually that

$$[\hat{J}_+,\hat{J}_-]=2\hbar\hat{L}_z\ , \quad (6.6)$$

$$[\hat{J}_{\pm},\hat{L}_z]=\pm\hbar\{-\hat{J}_{\pm}+(2l+1)\exp[\pm i(2l+1)\theta_0][j(j+1)-l(l+1)]^{1/2}|\pm(-l)\rangle\langle\pm l|\}\ . \quad (6.7)$$

The first term in (6.7) is familiar. The second term will have no effect because l is greater than j, and therefore it gives zero when acting on states within the physical realm, for which $|m|\le j$. Within the physical subspace we have the conventional commutation relations.

It may seem strange that our angle operator $\hat{\phi}_\theta$ acts on the whole $(2l+1)$-dimensional space, that is, it acts both outside as well as inside the physically accessible subspace. However, the accurate localization of the angle must involve states of extremely high angular momentum. An inability to access those states places fundamental limitations on the accuracy of angle measurements.

VII. ORIENTATION OF ROTATING SYSTEMS

The quantum properties of a rotating system are usually expressed in terms of its angular momentum. However, we can also use the angle states and angle operator to investigate the orientation, or more specifically the azimuthal coordinate, of the system.

A. States of random orientation

The angle-state expansion of the angular momentum eigenstates indicates that they are states of random orientation. This can be verified by calculating the expectation value and variance of the angle operator for these states. Remembering that, after these are calculated as a function of l, our procedure is then to allow l to tend to infinity, we obtain in the limit

$$\langle m|\hat{\phi}_\theta|m\rangle=\theta_0+\pi\ , \quad (7.1)$$

$$\Delta\phi_\theta^2=\frac{\pi^2}{3}\ . \quad (7.2)$$

These results are precisely the same as those obtained for the phase of a photon number state[7] and are characteristic of a state of random orientation. Any state that may be represented by a density matrix that is diagonal in the angular momentum basis will also be a state of random orientation.

B. Coherent angular momentum states

An interesting class of states are those states with partially, but not precisely, determined orientation. We note that, as with the optical phase operator, the interpretation of expectation values and variances is made more complicated by the arbitrary nature of the value of θ_0. However, a sensible choice of θ_0 will usually allow us to interpret these expectation values and variances simply. Perhaps the most important of these are the angular momentum spin, or atomic coherent states.[23,24] For a given total quantum number j these states are defined as[25]

$$|\zeta\rangle\equiv\exp(\zeta\hat{J}_+-\zeta^*\hat{J}_-)|-j\rangle$$
$$=(1+|\tau|^2)^{-j}\exp(\tau\hat{J}_+)|-j\rangle\ , \quad (7.3)$$

where $\tau=(\tan|\zeta|)\exp(i\arg\zeta)$. It is straightforward to obtain an angular momentum state expansion of this state:

$$|\zeta\rangle=\sum_{m=-j}^{j}\frac{\tau^{j+m}}{(1+|\tau|^2)^j}\left[\frac{(2j)!}{(j+m)!(j-m)!}\right]^{1/2}|m\rangle\ . \quad (7.4)$$

The angular momentum coherent state has a binomial distribution of angular momentum values m. The angle-states probability amplitudes for this state may be obtained by using the expansion of angle states in terms of the angular momentum basis (3.6):

$$\langle\theta_n|\zeta\rangle=(2l+1)^{-1/2}\times\sum_{m=-j}^{j}\exp(im\theta_n)\frac{\tau^{j+m}}{(1+|\tau|^2)^j}\times\left[\frac{(2j)!}{(j+m)!(j-m)!}\right]^{1/2}\ . \quad (7.5)$$

In general, evaluation of these amplitudes requires numerical summations. However, in cases where j is large (but still, of course very much less than l, which will ultimately tend to infinity) with[26]

$$(2j)^{-1}\ll|\tau|^2\ll 2j\ , \quad (7.6)$$

the binomial distribution of angular momentum states may be approximated by a normal or Gaussian distribution[27]

$$P(m)\approx\frac{(1+|\tau|^2)}{(4\pi j|\tau|^2)^{1/2}}\exp\left[-\left[j+m-\frac{2j|\tau|^2}{(1+|\tau|^2)}\right]^2\times\frac{(1+|\tau|^2)^2}{4j|\tau|^2}\right]\ . \quad (7.7)$$

This distribution is normalized so that

$$\lim_{l\to\infty} \int_{-l}^{l} P(m)dm = 1 \ . \tag{7.8}$$

There is no suggestion of a continuous spectrum of m values; the continuum approximation is simply a convenient mathematical maneuver. We can obtain the angle-state probability amplitudes by performing the Fourier transform of $P^{1/2}(m)$. The resulting approximate angle probability distribution is

$$P(\theta)=\frac{2\pi}{2l+1}\left[\frac{8j|\tau|^2}{2\pi(1+|\tau|^2)}\right]^{1/2} \times\exp\left[-\frac{4j|\tau|^2}{(1+|\tau|^2)^2}(\theta-\arg\zeta)^2\right] . \tag{7.9}$$

If we choose the reference angle θ_0 so that $P(\theta_0)$ is small, then the expectation value and variance of the angle operator are

$$\langle\zeta|\hat{\phi}_\theta|\zeta\rangle=\arg\zeta \ , \tag{7.10}$$

$$\Delta\phi_\theta^2=\frac{(1+|\tau|^2)^2}{8j|\tau|^2} \ . \tag{7.11}$$

This variance may be very small if j is large. In the region where the Gaussian approximation is good, the expectation value and variance of $\hat{L}_z$ are

$$\langle\zeta|\hat{L}_z|\zeta\rangle=\hbar\left[-j+\frac{2j|\tau|^2}{(1+|\tau|^2)}\right] , \tag{7.12}$$

$$\Delta L_z^2=\frac{2\hbar^2 j|\tau|^2}{(1+|\tau|^2)^2} \ . \tag{7.13}$$

We see that, to a good approximation, these states are angle–angular-momentum minimum uncertainty states with an uncertainty product given by

$$\Delta\phi_\theta\Delta L_z=\frac{\hbar}{2} \ . \tag{7.14}$$

VIII. CONCLUSION

It is well known that the premature replacement of a mathematical variable in a calculation by infinity can lead to an indeterminant form. This problem can sometimes be circumvented by performing the calculation first and then taking the limit. This latter approach is essentially that used in this paper and is closely related to our earlier work on the problem of optical phase. Instead of using an infinite-dimensional state space directly, we calcualte physical results, such as expectation values, in a space of $2l+1$ dimensions and *then* find the limit as l tends to infinity. By this method we obtain a well-behaved Hermitian rotation-angle operator, with a countable infinity of proper state vectors as its eigenstates. These eigenstates of the angle operator can be used as an alternative basis for the angular momentum state space. The properties of the angle operator are well determined and physically reasonable. They automatically resolve the problem of periodicity and the inconsistencies associated with the earlier expressions for the angle–angular–momentum commutator. In particular, these earlier expressions cannot be used consistently with the angle states themselves. The infinite state space may be used only with caution and may *not* be used in calculations involving the angle states.

We have shown that the angle operator fits in with the general properties of rotating systems and may be applied to more complicated problems than the simple bead on a wire. It can also be applied to investigate the orientation of a rotating system.

Finally, it is intriguing to note than an approach based on a finite but arbitrarily large state space, the dimensionality of which is only allowed to tend to infinity after the final calculations, can incorporate operators and states which the infinite state space is *too small* to accommodate. In particular, our approach has allowed us to introduce Hermitian operators for both optical phase and rotation angle.

ACKNOWLEDGMENTS

S. M. Barnett thanks GEC Research Limited and the Fellowship of Engineering for support. D. T. Pegg thanks J. Vaccaro for helpful discussions.

APPENDIX

In this appendix we highlight the dangers associated with the use of the infinite state space. The angle-state matrix elements of the commutator (5.8) are

$$\langle\Theta|[\hat{\Phi}_\Theta,\hat{L}_z]|\Theta'\rangle=i\hbar[\delta(\Theta-\Theta') -2\pi\delta(\Theta-\Theta_0)\delta(\Theta'-\Theta_0)] , \tag{A1}$$

which imply that

$$\langle\Theta|\hat{L}_z|\Theta'\rangle=i\hbar\frac{\delta(\Theta-\Theta')}{\Theta-\Theta'}[1-2\pi\delta(\Theta_0-(\Theta+\Theta')/2)] , \tag{A2}$$

where we have used an integral representation of the δ function to factorize the product of δ functions.[28] A δ function divided by its argument is simply minus the derivative of the δ function,[29] so we can write

$$\langle\Theta|\hat{L}_z|\Theta'\rangle=-i\hbar\delta'(\Theta-\Theta') \times[1-2\pi\delta(\Theta_0-(\Theta+\Theta')/2)] , \tag{A3}$$

where

$$\delta'(x)\equiv\frac{d\delta(x)}{dx} \ . \tag{A4}$$

We can use these elements together with the angle-state resolution of the identity to obtain the angular momentum matrix elements of $\hat{L}_z$:

$$\langle m|\hat{L}_z|m'= \int_{\Theta_0}^{\Theta_0+2\pi} d\Theta \int_{\Theta_0}^{\Theta_0+2\pi} d\Theta' \langle m|\Theta\rangle\langle\Theta|\hat{L}_z|\Theta'\rangle\langle\Theta'|m'\rangle$$

$$= -\frac{i\hbar}{2\pi}\int_{\Theta_0}^{\Theta_0+2\pi} d\Theta \int_{\Theta_0}^{\Theta_0+2\pi} d\Theta' \exp(-im\Theta)\exp(im'\Theta')\delta'(\Theta-\Theta')[1-2\pi\delta(\Theta_0-(\Theta+\Theta')/2)] \ . \quad \text{(A5)}$$

Introducing a change of variables

$$\Theta_- = \Theta - \Theta' \ , \quad \text{(A6a)}$$

$$\Theta_+ = \frac{\Theta+\Theta'}{2} \ , \quad \text{(A6b)}$$

we find that

$$\langle m|\hat{L}_z|m'\rangle = -\frac{i\hbar}{2\pi}\int_{\Theta_0}^{\Theta_0+2\pi} d\Theta_+ \int_{-Y}^{Y} d\Theta_- \exp[i\Theta_+(m'-m)]\exp[-i\Theta_-(m+m')/2]\delta'(\Theta_-)[1-2\pi\delta(\Theta_0-\Theta_+)] \ , \quad \text{(A7)}$$

where $Y=2(\Theta_+-\Theta_0)$ for $\Theta_+ < \Theta_0+\pi$ and is $2(\Theta_0+2\pi-\Theta_+)$ for $\Theta_+ \geq \Theta_0+\pi$. We integrate first with respect to Θ_-, dealing with the derivative of the δ function by integration by parts:

$$\int f(x)\delta'(x)dx = -f'(0) \ , \quad \text{(A8)}$$

provided the range of integration includes $x=0$. Subsequent integration with respect to Θ_+ then yields

$$\langle m|\hat{L}_z|m'\rangle = \hbar m \delta_{mm'} - \frac{\hbar(m+m')}{2}\exp[i\Theta_0(m'-m)] \ . \quad \text{(A9)}$$

The presence of the second term makes this clearly inconsistent with our starting point that $|m\rangle$ is an eigenstate of $\hat{L}_z$:

$$\langle m|\hat{L}_z|m'\rangle = \hbar m \delta_{mm'} \ .$$

---

[1] P. A. M. Dirac, Proc. R. Soc. London Ser. A **114**, 243 (1927).

[2] L. Susskind and J. Glogower, Physics (N.Y.) **1**, 49 (1964).

[3] P. Carruthers and M. M. Nieto, Rev. Mod. Phys. **40**, 411 (1968) and references therein.

[4] J. M. Levy-Leblond, Ann. Phys. (N.Y.) **101**, 319 (1979).

[5] S. M. Barnett and D. T. Pegg, J. Phys. A **19**, 3849 (1986).

[6] D. T. Pegg and S. M. Barnett, Europhys. Lett. **6**, 483 (1988).

[7] S. M. Barnett and D. T. Pegg, J. Mod. Opt. **36**, 7 (1989).

[8] D. T. Pegg and S. M. Barnett, Phys. Rev. A **39**, 1665 (1989).

[9] J. M. Levy-Leblond, Rev. Mex. Fis. **22**, 15 (1973); I. Goldhirsch, J. Phys. A **13**, 3479 (1980).

[10] T. S. Santhanam, Found. Phys. **7**, 121 (1977).

[11] T. S. Santhanam, Aust. J. Phys. **31**, 233 (1978).

[12] E. Merzbacher, *Quantum Mechanics*, 2nd ed. (Wiley, New York, 1970), p. 296.

[13] D. Judge, Phys. Lett. **5**, 189 (1963); Nuovo Cimento **31**, 332 (1964); D. Judge and J. T. Lewis, Phys. Lett. **5**, 190 (1963).

[14] See also the work of Susskind and Glogower (Ref. 2).

[15] We have discussed the properties of classical single-valued angles elsewhere in the context of quantum optical phase (Ref. 8).

[16] W. H. Louisell, Phys. Lett. **7**, 60 (1963).

[17] If $m+n>l$, for example, then consistency in representing the zero angle state by a constant c_m requires that, if this state is an eigenvalue of $\hat{\phi}$, the state labeled $|m+n\rangle$ must be one of the eigenstates of $\hat{L}_z$. We see in Sec. IV that it is indeed just the state $|m+n-2l-1\rangle$.

[18] R. Loudon, *The Quantum Theory of Light*, 1st ed. (Oxford University Press, Oxford, 1973), p. 143.

[19] This is simply a statement of the fact that physical systems rotate with finite angular velocity. We have used a similar criterion to delineate the physical states of the harmonic oscillator or single-mode electromagnetic field (Refs. 6–8).

[20] We have given a parallel and detailed discussion of the classical Poisson bracket for a single-valued phase elsewhere (Ref. 8).

[21] These operators have also been discussed by D. Elinas (private communication).

[22] States such as these have been discussed previously in relation to infinite-state-space phase and angle operators (Refs. 3 and 4).

[23] J. M. Radcliffe, J. Phys. A **4**, 313 (1971).

[24] F. T. Arecchi, E. Courtens, R. Gilmore, and H. Thomas, Phys. Rev. A **6**, 2211 (1972).

[25] We follow the notation of Arrechi *et al.* (Ref. 24).

[26] This restiction on τ ensures that the probability amplitudes for the states $|\pm j\rangle$ are small.

[27] We have employed a similar approximation in describing the phase properties of the coherent states of the single-mode electromagnetic field (Ref. 7).

[28] U. Fano, Phys. Rev. **124**, 1866 (1961).

[29] M. J. Lighthill, *Introduction to Fourier Analysis and Generalized Functions* (Cambridge University Press, Cambridge, 1970), p. 29.

Quantum optical phase

DAVID T. PEGG

Faculty of Science and Technology, Griffith University, Nathan, Brisbane 4111, Australia

and STEPHEN M. BARNETT

Department of Physics and Applied Physics, University of Strathclyde, Glasgow G4 0NG, Scotland

(*Received 4 June 1996*)

Abstract. The phase associated with a single mode of the electromagnetic field is complementary to the photon number. This simple idea leads us to construct a probability density for the phase. The phase operator cannot be represented exactly in the usual infinite Hilbert space but can be constructed in a subspace of it and provides a valid representation of phase if used together with a suitable limiting procedure. We introduce the Hermitian optical phase operator and describe some of its most important properties. We also provide brief discussions of some alternative approaches to the phase problem and to the question of phase measurements. The paper concludes with an extensive bibliography.

1. Introduction

The concept of the phase of an oscillator plays an important part in mechanics and optics and was introduced into quantum mechanics and electrodynamics at an early stage [1, 2]. Yet problems associated with identifying a suitable Hermitian operator to represent the phase long hindered serious development of the idea of the quantum-optical phase [3, 4]. There have been very many ingenious attempts to resolve the problem of the phase operator, some of which have been discussed in [5]. We shall not attempt to describe this extensive body of work here, but rather we shall draw upon it to present what we hope is a logical introduction to the phase problem and its solution. Our aim is to explain the quantum-optical phase using only elementary ideas familiar to advanced undergraduate students. For readers interested in tracing the development of the topic, an extensive bibliography of papers on phase and related problems is given at the end of this tutorial review.

The problem that concerns us is easily stated: what is the correct quantum description of the phase of a harmonic oscillator or, equivalently, a single mode of the electromagnetic field? In answering this we can be guided by the classical limit of large coherent amplitude but, perhaps not surprisingly, this alone is not enough to give a unique form for the quantum phase. To define a unique, and hence useful, quantum phase we require an additional criterion that we can apply in the quantum regime. How we select this condition is essentially a matter of choice but will determine the form of our quantum-optical phase. This idea might seem strange but 'phase' is only a word and we can use it as a label for any element of the theory we choose! In our view there are compelling, even unanswerable reasons why the phase of a field mode should be the property which is the complement of photon number. There are two simple reasons for this. Firstly, complementarity ensures that all the photon number states, including the vacuum, will be states of random phase. This is natural for, if number states are not states of random phase, then we must ask '*what is it that provides the number states with their preferred phases?*'. A second and more pressing reason arises in the action of phase-shifting devices. If we accept that retarding plates act to shift the phase in quantum optics in the same way as they do in classical optics, then, as we shall see, we are led directly to the requirement that phase and photon number must be complementary. All phase-dependent experiments in quantum optics employ this idea to shift the phase and hence imply phase–photon number complementarity.

We begin our analysis of optical phase by re-examining the phase of a classical oscillator, paying particular attention to the idea that the phase is a multivalued quantity. The quantum theory of phase can be derived simply on the basis of complementarity as based on the action of a phase shifter. This leads very naturally to a phase representation and to a function which we would like to associate with a phase probability density. The problem arises when we examine the phase representation of phase eigenstates. In section 5 we present a solution to this problem and introduce the Hermitian optical phase operator itself [6, 7]. The phase operator cannot be represented exactly in the usual infinite Hilbert space but requires a more subtle construction based on a limiting procedure. This should not be too surprising as there are other examples including the position and momentum operators and their eigenfunctions that do not fit within Hilbert space [8]. It is possible, however, to find an operator (or rather sequence of operators) within Hilbert space that provide an arbitrarily good approximation to the phase. One way of doing this is described in section 6. It is, of course, phase sums and differences that we might expect to have most relevance to real experiments. We give the explicit forms of the probability densities for the phase sum and difference for two field modes in section 7. The remainder of the paper presents a brief discussion of some of the other descriptions of phase and of phase measurements. The paper ends with an extensive bibliography of papers on quantum-optical phase provided for those readers wishing to look at the development of the theory.

2. Classical phase

The classical meaning of phase is well understood. Phase can be defined for any motion which is periodic in time in such a way that the phase angle changes linearly with time by an amount 2π over one period T of the motion. The state of a classical system undergoing periodic motion changes with time but the system returns to any particular state after one period. We can label the various states through which the system evolves by means of a phase angle θ, where $\theta = 2\pi t/T$ with t being the elapsed time from some suitably chosen origin. States with phase angles which vary by integer multiples of 2π are identical. As a consequence, if we try to measure the phase by examining the state, the result that we obtain will not necessarily be θ, but we shall obtain θ modulo 2π. For example, by determining the state of a classical oscillator, we can determine cos θ and sin θ, which does not allow us to distinguish between values of θ which differ by 2π. It is conventional therefore to choose some 2π range, or window, in which to express the result. Common ranges are from $-\pi$ to π and from 0 to 2π. Here we wish to maintain generality by allowing the range to be from θ_0 to $\theta_0 + 2\pi$. Thus, although θ can take values from $-\infty$ to $+\infty$, the values which correspond to distinct states will be in this 2π range.

The classical harmonic oscillator undergoes simple harmonic motion, which is the projection of circular motion onto an axis. A convenient way to define the phase angle of the oscillator is to equate it to the angle executed by the circular motion. This gives rise to a minor ambiguity in that there are two circular motions in opposite directions which have the same projection. Thus the phase angle can increase linearly with time or decrease linearly with time. It is unimportant which of these we adopt provided that we are consistent. There is also the arbitrariness connected with the choice of the origin of time. This is no more an obstacle, of course, than defining a coordinate x as the distance from an arbitrarily chosen spatial origin. A quantity which is independent of the time origin is a phase shift, which will be proportional to a time difference. For two oscillators with the same frequency, a phase difference can be defined as the phase shift which must be applied to one oscillator to bring it to the same state as the other oscillator. This phase difference is therefore also independent of the origin. Although it is these *relative* phase quantities which one expects to be measurable, we can always define the phase itself either in terms of a chosen origin or as the phase difference with another oscillator of the same frequency whose phase is defined to be zero at a particular time.

3. Quantum phase

The periodic system in which we are interested is the single-mode electromagnetic field as represented by a quantum harmonic oscillator. We expect a field in a strong coherent state to have a well defined phase because such a field approximates a classical oscillator with well defined amplitude and phase. This provides a classical boundary condition on any description of quantum phase. This condition alone is not sufficient to determine a unique quantum-mechanical description, however. For this it is necessary also to have some property of phase which is obeyed by extremely weak fields, that is fields in the quantum domain. The principle of the correspondence between the classical Poisson bracket and the quantum-mechanical commutator places an important restriction on potential quantum descriptions of phase. Here we do not apply this principle immediately but use a more physical approach. Later we show that the correspondence is satisfied. Our classical definition was of a phase angle proportional to time, that is, if a classical oscillator of angular frequency ω evolves from one state time t_1 to another state at t_2, then the shift $\Delta\theta$ in the phase angle θ experienced by the oscillator has a magnitude $\omega(t_2 - t_1)$. The nature of time does not change as we go from classical to quantum physics; it plays a similar role as a parameter in quantum mechanics as it does in classical physics and so the proportionality between phase shift and time difference will carry through to the quantum domain. In the quantum domain, however, we must allow for the fact that an oscillator in a particular state may not have a precisely defined value of a particular observable but will only have some associated probability distribution. This is so even if the oscillator is in a pure state, in which case the probability distribution will be related to the square of the modulus of a complex probability amplitude. Thus we should speak in terms of the phase probability distribution of the system in the pure state $|f_1\rangle$ at time t_1 being shifted by an amount $\omega(t_2 - t_1)$ to beome the phase probability distribution associated with state $|f_2\rangle$, the state to which $|f_1\rangle$ evolves in time $t_2 - t_1$. These states are related by the unitary time shift operator:

$$|f_2\rangle = \exp\left[-i\hbar^{-1}\hat{H}(t_2 - t_1)\right]|f_1\rangle. \tag{1}$$

We shall write the Hamiltonian $\hat{H}$ as $\hat{N}\hbar\omega$, rather than $(\hat{N} + \frac{1}{2})\hbar\omega$, where $\hat{N}$ is the number operator, as the retention of $\frac{1}{2}$ introduces only a trivial unmeasurable factor $\exp(-i\omega t/2)$ into the state $|f(t)\rangle$ of the system at time t. The removal of $\frac{1}{2}$ corresponds to a choice of the zero of energy. Writing equation (1) as

$$|f_2\rangle = \exp\left[-i\hat{N}\Delta\theta\right]|f_1\rangle \tag{2}$$

then shows that $\hat{N}$ is the phase-shift generator, that is it generates a shift of $\Delta\theta$ of magnitude $\omega(t_2 - t_1)$ in the phase probability distribution.

Before proceeding, we should examine the consequences of accepting the important quantum-mechanical property that phase shifts are generated by $\hat{N}$. The immediate consequence is that the standard experimental method for shifting the phase of a travelling field at some point in space by inserting an extra optical path length *still applies for fields even at very low intensities*. This is completely in accord with the widely accepted practice of shifting the phase by insertion of a phase-retarding plate, such as a slab of glass, or by moving mirrors. Indeed it would be possible to begin a study of phase by using this property as an *operational definition* of a phase shift. A second consequence is that states which are eigenstates of $\hat{N}$, for example the photon number states $|n\rangle$ of the field, must have a uniform phase probability distribution. This follows because the unitary phase shift operator will merely multiply such states by a factor of modulus unity and so the physical properties of the state, which will include the phase probability distribution, will be unaltered by the shift $\Delta\theta$ in the distribution. The only distribution which in invariant under all shifts $\Delta\theta$ is a uniform distribution†. This consequence is completely in accord with the notion of *complementarity* of photon number and phase. That is, if it is possible to specify the number of photons in the field precisely, then the field contains no phase information. Another way of expressing this is that number states have completely random phase. A third consequence can be found by expanding $|f_1\rangle$ in equation (2) in terms of number states. We then find that a phase shift of 2π does not alter the state. Thus the phase probability distribution must be *periodic* with period 2π. This periodicity of phase is a direct result of $\hat{N}$ having integer eigenvalues and is also an essential feature of phase.

The phase probability density $P(\theta)$ is defined in the usual way as the limit, as $\delta\theta$ tends to zero, of $Pr(\theta, \delta\theta)/\delta\theta$ where $Pr(\theta, \delta\theta)$ is the probability that the field has a

† If the phase distribution $P(\theta)$ is invariant under all phase shifts $\Delta\theta$, then we must have $P(\theta + \Delta\theta) = P(\theta)$ for all $\Delta\theta$, and therefore $P(\theta)$ must be a constant or uniform distribution (independent of θ). Normalization over a 2π interval leads to the requirement that the phase probability distribution for all number states must be $1/2\pi$.

phase value between $\theta - \delta\theta/2$ and $\theta + \delta\theta/2$. From the above discussion, $P(\theta)$ will be a periodic function of θ as θ extends from $-\infty$ to ∞ but, because it is only possible to determine physically distinct values of phase in a 2π range, normalisation will be over a 2π interval, that is

$$\int_{\theta_0}^{\theta_0+2\pi} P(\theta)\,\mathrm{d}\theta = 1. \tag{3}$$

Periodicity of $P(\theta)$ ensures that equation (3) is true for all possible choices of θ_0. In this paper we shall be interested predominantly in pure states; thus $Pr(\theta, \delta\theta)$ will be the square of the modulus of a complex amplitude and therefore, in the limit $\delta\theta \to 0$, we can work in terms of a phase `wavefunction' $\psi(\theta)$ defined such that

$$P(\theta) = |\psi(\theta)|^2. \tag{4}$$

Each pure state of the field mode will have a corresponding phase wavefunction or phase representation.

In order to study the properties of phase, it is most straightforward to work in the phase representation [9]. This is analogous to the usual coordinate representation in wave mechanics in which the wavefunctions are functions of x, y and z. Operators representing algebraic functions of x in this latter representation are simple multiplicative operators with Dirac delta functions as eigenfunctions, but momentum operators are differential operators [8]. In the momentum representation, on the other hand, the momentum operator is multiplicative and the coordinate operators are differential. In the phase representation algebraic functions of phase will be represented by simple multiplicative operators; so there will be no difficulties with Hermiticity or unitarity, and the problem of finding the phase properties of a superposition of number states reduces to finding the appropriate wavefunctions for these states. In the phase representation, the fundamental phase shift relation becomes

$$\exp(-\mathrm{i}\hat{N}\Delta\theta)\,\psi(\theta) = \psi(\theta + \Delta\theta). \tag{5}$$

Comparing with Taylor's theorem

$$\exp\left(\Delta\theta\frac{\partial}{\partial\theta}\right)\psi(\theta) = \psi(\theta + \Delta\theta), \tag{6}$$

we find the required differential form for the number operator in the phase representation:

$$\hat{N} = \mathrm{i}\frac{\partial}{\partial\theta}. \tag{7}$$

From equations (3), (4) and (7) we find that the phase wavefunction for an eigenstate $|n\rangle$ of $\hat{N}$ must have the form

$$\psi_n(\theta) = \frac{1}{(2\pi)^{1/2}}\exp\left[\mathrm{i}(\beta_n - n\theta)\right], \tag{8}$$

where β_n is real and can be chosen arbitrarily. It is not difficult to show that the functions $\psi_n(\theta)$ form an orthonormal set. The choice of values for β_n determines the particular number state basis used, with different such representations being related by a simple unitary transformation. It is convenient to choose $\beta_n = 0$. We return to this point below.

In the phase representation, the operator representing $\exp(-i\theta)$ will simply be the multiplicative operator $\exp(-i\theta)\times$. It is easy to verify from equation (8) with $\beta_n = 0$ that

$$\exp(-i\theta)\,\psi_n(\theta) = \psi_{n+1}(\theta) \tag{9}$$

and, remembering that $\psi_{n+1}(\theta)$ is an eigenstate of $\hat{N}$ and hence of $\hat{N}^{1/2}$, that

$$\hat{N}^{1/2}\exp(-i\theta)\,\psi_n(\theta) = (n+1)^{1/2}\psi_{n+1}(\theta). \tag{10}$$

This allows us to associate $\hat{N}^{1/2}\exp(-i\theta)$ with the creation operator $\hat{a}^\dagger$. This is equivalent to Dirac's polar decomposition of the creation operator into amplitude and phase factors and is in accord with the corresponding classical expression for the field or oscillator. Choosing $\beta_n = 0$ as above is equivalent to choosing the number state basis in which

$$\hat{a}^\dagger|n\rangle = (n+1)^{1/2}|n+1\rangle, \tag{11}$$

which is the basis normally used in quantum optics. From equation (8) with $\beta_n = 0$ the phase probability distribution for a field in a superposition state $\sum c_n|n\rangle$ is given by

$$P(\theta) = \frac{1}{2\pi}\left|\sum_{n=0}^{\infty} c_n \exp(-in\theta)\right|^2. \tag{12}$$

Equation (12) is a well defined quantity from which can be found the quantum phase properties of any state. For example the mean value of any function of phase $F(\theta)$ can be calculated from

$$\langle F(\theta)\rangle = \int_{\theta_0}^{\theta_0+2\pi} F(\theta)P(\theta)\,d\theta. \tag{13}$$

Equations (12) and (13), which we have derived very simply and directly, *are all that are needed to calculate the phase properties of light*. It is interesting, nevertheless, to investigate further quantum-mechanical aspects of phase in the phase representation. The phase angle operator will also be a multiplicative operator, but we must remember that the physically distinguishable values of phase will not range from $-\infty$ to $+\infty$ but will lie in a 2π range from θ_0 to $\theta_0 + 2\pi$. Thus we should not use the multiplicative operator $\theta\times$ with its broad eigenvalue range because distinguishing a state with eigenvalue θ from a state with eigenvalue $\theta + 2\pi$ should not be possible as it would violate the periodicity requirement. Instead we must use the multiplicative operator $\varphi_{\theta_0}\times$ where φ_{θ_0} is θ modulo 2π. If the value of θ is between $\theta_0 + 2p\pi$ and $\theta_0 + 2(p+1)\pi$ say, where p is an integer, then the value of φ_{θ_0} is $\theta - 2p\pi$. We can express φ_{θ_0} in terms of Heaviside step functions as

$$\varphi_{\theta_0} = \theta - 2\pi \sum_{p=1}^{\infty} \Theta(\theta - \theta_0 - 2p\pi) + 2\pi \sum_{p=0}^{\infty} \Theta(-\theta + \theta_0 - 2p\pi), \tag{14}$$

where $\Theta(x)$ is zero for $x < 0$ and unity for $x > 0$.

From equations (7) and (14), remembering that the derivative of a step function is a Dirac delta function, we can calculate the photon number–phase commutator as

$$[\hat{N}, \varphi_{\theta_0}] = \mathrm{i}[1 - 2\pi\delta_{2\pi}(\theta - \theta_0)], \tag{15}$$

where $\delta_{2\pi}(\theta - \theta_0)$ is the 2π-periodic delta function defined by

$$\delta_{2\pi}(x) \equiv \sum_{p=-\infty}^{\infty} \delta(x - 2p\pi). \tag{16}$$

It is interesting to note that using the operator $\theta\times$ instead of $\varphi_{\theta_0}\times$ yields Dirac's commutator

$$[\hat{N}, \theta] = \mathrm{i}, \tag{17}$$

which has the following difficulty [10]. Taking the expectation value of each side of equation (17) for a number state we find for the left-hand side

$$\int_{\theta_0}^{\theta_0+2\pi} \{[\hat{N}\psi_n(\theta)]^* \theta\psi_n(\theta) - \psi_n^*(\theta)\theta\hat{N}\psi_n(\theta)\}\, \mathrm{d}\theta = 0, \tag{18}$$

while for the right-hand side we find the conflicting result

$$\int_{\theta_0}^{\theta_0+2\pi} \psi_n^*(\theta)\mathrm{i}\psi_n(\theta)\, \mathrm{d}\theta = \mathrm{i}. \tag{19}$$

It is important to check therefore that equation (15) is satisfactory in this regard. Finding the expectation value for a general wavefunction $\psi(\theta)$ gives

$$\begin{aligned} \mathrm{i}\int_{\theta_0}^{\theta_0+2\pi} \psi^*(\theta)[1 - 2\pi\delta_{2\pi}(\theta - \theta_0)]\psi(\theta)\, \mathrm{d}\theta &= \mathrm{i} - \mathrm{i}2\pi \int_{\theta_0}^{\theta_0+2\pi} |\psi(\theta)|^2 \delta(\theta - \theta_0)\, \mathrm{d}\theta \\ &= \mathrm{i} - \mathrm{i}2\pi P(\theta_0). \end{aligned} \tag{20}$$

In calculating this, we included the point θ_0 in our 2π interval but not the point $\theta_0 + 2\pi$. Had we excluded the former and included the latter we would have obtained $P(\theta_0 + 2\pi)$ in place of $P(\theta_0)$, which of course makes no difference. For a number state the phase probability distribution is uniform and so $P(\theta_0) = 1/2\pi$, and thus the right-hand side vanishes as required. The difficulty with equation (17) is thus seen to arise through not limiting the eigenvalues of the phase operator to physically distinguishable values. The second term in equation (15) arises from the last two terms in equation (14) which limit the phase eigenvalues to a 2π range.

It is also useful to check that equation (15) is physically reasonable, which we can do as follows. We can easily find the commutator of the Hamiltonian with the phase operator by multiplying equation (15) by $\hbar\omega$ and then, from the Heisenberg equation of motion, we can find the time development of the mean phase for a general state of the field. From equation (20) this yields

$$\frac{d\langle\varphi_{\theta_0}\rangle}{dt} = -\omega\left[1 - 2\pi P(\theta_0)\right]. \tag{21}$$

If we had used Dirac's commutator we would have obtained an expression for the rate of change of $\langle\theta\rangle$ as $-\omega$, which is indeed correct, given that our convention has yielded a phase angle θ which decreases with time. The second term in the square brackets in equation (21) is exactly that required to ensure that the physically determinable phase angle φ_{θ_0} jumps back to $\theta_0 + 2\pi$ when it reaches θ_0 in order to ensure that it remains in the same 2π range. This is completely in accord with the correspondence principle, and is what should be expected from a study of the classical Poisson brackets [7]. Indeed the commutator of the Hamiltonian and the phase operator obtained from equation (15) is identical, apart from the factor $i\hbar$, with the corresponding classical Poisson bracket. Thus the fundamental principle referred to earlier is satisfied. For a strong coherent state, for which $P(\theta)$ can be approximated by a delta function, the switchback jump can be seen to take place at a well defined time and a graph of phase against time will have a sawtooth shape. The number–phase uncertainty relation obtained from equations (15) and (20) is

$$\Delta N \Delta\varphi_{\theta_0} \geq \tfrac{1}{2}\left|\left[1 - 2\pi P(\theta_0)\right]\right|. \tag{22}$$

Only for states with phase probability densities which are zero at the boundaries of the selected 2π range does this reduce to $\Delta N \Delta\varphi_{\theta_0} \geq \frac{1}{2}$, which is the analogue of the momentum–position uncertainty relation and is that obtainable from equation (17).

The equality of $\exp(-i\varphi_{\theta_0})\times$ and $\exp(-i\theta)\times$ follows from the 2π periodicity of the complex exponential function. Hence, from equation (9), φ_{θ_0} can be seen to be the generator of a shift in photon number. Again, this is precisely what is required by complementarity.

4. The quantum phase problem

The procedure thus far has been straightforward with results which are very reasonable physically, which obey the correspondence principle and the associated commutator–Poisson bracket relationship and which are exactly in accord both with phase–photon number complementarity and with the accepted operational procedure for shifting phase. Indeed all our derivation could have been done many years ago, prior to the introduction of bras and kets or of the Hilbert space to quantum mechanics. What therefore is the problem which beset phase for so long?

The problem arises when we examine the eigenfunctions $\psi_{\theta'}(\theta)$ of the phase operator φ_{θ_0}. In the phase representation, eigenstates of phase will simply be delta functions $\delta(\theta - \theta')$ where $\theta_0 \leq \theta, \theta' < \theta_0 + 2\pi$. Let us assume that we can write the phase eigenstate as a superposition of the number states so that the wavefunction has the form

$$\psi_{\theta'} = \sum_{n=0}^{\infty} c_n \psi_n(\theta). \tag{23}$$

Then

$$\int_{\theta_0}^{\theta_0+2\pi} \psi_n^*(\theta)\psi_{\theta'}(\theta)\,d\theta = c_n \tag{24}$$

follows from the orthonormality of the number state wavefunctions. Using equation (8) with β_n zero and $\delta(\theta - \theta')$ for the phase eigenfunction we find that

$$c_n = \frac{1}{(2\pi)^{1/2}}\exp(in\theta'). \tag{25}$$

All these coefficients have the same modulus, which is in accord with number–phase complementarity. It follows from equations (25) and (23) that the phase representation of the phase eigenstate is

$$\psi_{\theta'}(\theta) = \frac{1}{2\pi}\sum_{n=0}^{\infty}\exp[in(\theta' - \theta)]. \tag{26}$$

This is *not* a delta function, however; so our assumption that we can express an eigenstate of phase as equation (23) is incorrect. (In fact the sum given in equation (26) is undefined, but insertion of a conversion factor does allow us to obtain an explicit form for equation (26) which is in general non-zero for $\theta \neq \theta'$.) To obtain the required delta function, we would have to include exponentials with negative values of n which do not appear to be the wavefunctions for any known states [5]. Thus the usual infinite-dimensional Hilbert space spanned by the number states $|n\rangle$ with $n = 0, 1, \ldots$ *cannot accommodate phase states*. This is essentially the quantum phase problem. That the difficulty lies in trying to write the phase states as equation (23) is also apparent because functions given by equation (26) are not orthogonal for different values of θ', so they cannot be the eigenfunctions of any Hermitian operator. That the fault does not lie with our choice of zero for β_n can readily be checked by retaining it as a general value and still arriving at equation (26). Furthermore, if, in accord with complementarity, we were to write the wavefunction for the phase state as *any* equally weighted superposition of number eigenfunctions and then apply the phase shift operator to generate a phase state with a different eigenvalue, we would find that the wavefunctions for the two phase states are not orthogonal. We note here that equation (26) usually in its equivalent ket form is sometimes referred to as a phase state on the basis that it is a right eigenstate of the non-unitary Susskind–Glogower operator [3]. As the vacuum is another such eigenstate, however, this justification is not well founded.

The problem is therefore as follows. On the one hand, states of the field have mathematically consistent and well defined phase properties given by the phase probability distribution (12). These properties are completely in accord with complementarity of photon number and phase, and with the accepted operational method for shifting phase. On the other hand, there seem to be no sensible phase states, that is eigenstates of phase, which can be written as a superposition of the number states. It appears that, to obtain a sensible wavefunction for a phase state, we must include in the superposition functions of the type $\exp(-in\theta)$ with negative values of n. While these are perfectly respectable mathematically, they are not the wavefunctions of physical states. We can show, for example, that

$$\frac{1}{(2\pi)^{1/2}} \exp(\mathrm{i}\theta) \neq \sum_{n=0}^{\infty} c_n \psi_n(\theta), \tag{27}$$

by multiplying each side by $\exp(-\mathrm{i}\theta)$ and integrating over a 2π range. Thus the state with wavefunction $(2\pi)^{-1/2} \exp(\mathrm{i}\theta)$ is orthogonal to all states of the usual Hilbert space. Can we generate a state with this wavefunction from a physical state? It is easy to show that

$$\exp(\mathrm{i}\theta)\, \psi_n(\theta) = \psi_{n-1}(\theta) \qquad \text{for} \quad n \geqslant 1. \tag{28}$$

Further, for the vacuum

$$\exp(\mathrm{i}\theta)\, \psi_0(\theta) = \frac{1}{(2\pi)^{1/2}} \exp(\mathrm{i}\theta); \tag{29}$$

so a state with such a wavefunction can indeed be generated mathematically from the vacuum by the multiplicative operator $\exp(\mathrm{i}\theta)$. However, the bare intensity-independent operator $\exp(\mathrm{i}\theta)$ is unlikely to occur as a separate term in any physical interaction Hamiltonian [5]. At best, a function of the annihilation operator $\exp(\mathrm{i}\theta)\hat{N}^{1/2}$ will be present, but this simply destroys the vacuum. Thus it seems impossible to prepare a field with such a state as part of its superposition. As a measurement of an observable involves leaving the field in an eigenstate of the corresponding operator, a measurement of phase should leave the field in a superposition involving these apparently inaccessible states. This leads one to the conclusion that a precise measurement of phase is impossible.

5. A solution

The quantum phase problem can, of course, be avoided by redefining `phase' to mean some observable which is not the complement of excitation or photon number and which can therefore be accommodated within the usual infinite-dimensional Hilbert space H_∞. Naturally such an observable would need to have the appropriate properties in the classical limit. There are a number of such approaches in the literature but there is no one consistently accepted useful definition. In this section we present a solution to the phase problem which allows us to retain number–phase complementarity. This involves the use of a particular limiting procedure which circumvents the inadequacy of H_∞.

5.1. *The phase probability density as a limit*

It is not difficult to show that equation (8) for the number state wavefunction, which yields the distribution $P(\theta)$ in equation (12), is the only possible normalizable function which satisfies the phase shift relation (5). Now the phase shift relation leads directly to complementarity of phase and photon number and to phase periodicity. Indeed the requirement of 2π periodicity would seem to demand that the eigenvalues of the phase shift generator are integers. In addition, equation (5) is widely accepted as forming the basis for practical phase shifters. Thus, if we wish to retain the notion of a phase which is complementary to photon number, we must conclude that the distribution $P(\theta)$ represents phase. Any difficulty with $P(\theta)$ lies therefore, not in its representation of phase but in whether or not it represents a *genuine probability density*. A concern about this stems from the fact that, if we wish to represent $P(\theta)$ for a general state

$$|f\rangle = \sum_{n=0}^{\infty} c_n |n\rangle \tag{30}$$

in terms of a projection, we would have

$$P(\theta) = |\langle f|\theta\rangle|^2 \tag{31}$$

where

$$|\theta\rangle = \frac{1}{(2\pi)^{1/2}} \sum_{n=0}^{\infty} \exp(in\theta)\,|n\rangle. \tag{32}$$

The wavefunction representing $|\theta\rangle$ is just equation (26), and we have seen that these states are not orthogonal and are not the eigenstates of any observable. Orthogonality of eigenstates is necessary for outcomes of a measurement to be mutually exclusive, which is in turn essential for a genuine probability distribution. This problem, of course, can be regarded as being associated with the apparent impossibility of making a precise phase measurement. In the light of equations (31) and (32) the question that we must ask is whether or not $P(\theta)$ can represent the probability density of an observable. To examine this question we return to the definition of a probability density for the phase,

$$\lim_{\delta\theta \to \infty} \left(\frac{Pr(\theta, \delta\theta)}{\delta\theta} \right), \tag{33}$$

where $Pr(\theta, \delta\theta)$ is the probability that the phase lies somewhere in a bin of size $\delta\theta$ located at θ. To show that $P(\theta)$ is a probability density, we must show that it can be written in the form (33).

We proceed by first noting that for any physical state, that is a state with finite energy moments [6, 7, 11, 12], given in the form (30) we can choose a truncated state

$$|f_s\rangle = \sum_{n=0}^{s} c_n |n\rangle, \tag{34}$$

for which the phase distribution $P_s(\theta)$ is given by equation (12) with the upper summation limit replaced by s so that

$$P_s(\theta) = \frac{1}{2\pi} \sum_{n=0}^{s} \sum_{n'=0}^{s} c_n^* c_{n'} \exp\left[i(n - n')\theta\right]. \tag{35}$$

In the appendix we prove that we can always choose s to be sufficiently large so that, for any given non-zero error ϵ, however small,

$$|P(\theta) - P_s(\theta)| < \epsilon \tag{36}$$

for all θ.

Let us sample values of the curve $P_s(\theta)$ at equally spaced points in some 2π range. For the histogram of the sampled values to give a reasonable representation of $P_s(\theta)$, the distance between adjacent points should not be larger than the period of the highest frequency oscillation in $P_s(\theta)$. From equation (35), this is $2\pi/s$. It is useful for our purposes here to take the difference between adjacent points as

$$\delta\theta = \frac{2\pi}{s+1} \tag{37}$$

with our sampling positions being given by

$$\theta_m = \theta_0 + m\frac{2\pi}{s+1} \tag{38}$$

and $m = 0, 1, 2, \ldots, s$. By this means we can approximate the curve $P_s(\theta)$ by a histogram with values $P_s(\theta_m)$ which will approach the curve as the bin size $\delta\theta \to 0$. From equations (35) and (37) we have

$$P_s(\theta_m)\,\delta\theta = \frac{1}{s+1}\left|\sum_{n=0}^{s} c_n \exp(-in\theta_m)\right|^2, \tag{39}$$

which in turn can be written as

$$Pr(\theta_m, \delta\theta) = |\langle f|\theta_m\rangle|^2, \tag{40}$$

where

$$|\theta_m\rangle = \frac{1}{(s+1)^{1/2}}\sum_{n=0}^{s} \exp(in\theta_m)|n\rangle. \tag{41}$$

It is not difficult to show that $\langle\theta_m|\theta_{m'}\rangle = \delta_{mm'}$; so these states are *orthonormal*. Thus $Pr(\theta_m, \delta\theta)$ can represent a genuine probability associated with the bin centred at θ_m. Outcomes in different bins will then correspond to mutually exclusive events. It follows that, because the histogram with values $P_s(\theta_m)$ approaches the curve $P(\theta)$ as $\delta\theta \to 0$, we can write

$$P(\theta) = \lim_{\delta\theta\to 0}\left(\frac{Pr(\theta, \delta\theta)}{\delta\theta}\right), \tag{42}$$

where $Pr(\theta, \delta\theta) \equiv Pr(\theta_m, \delta\theta)$ with θ_m chosen so that $|\theta - \theta_m| \leqslant \delta\theta/2$. Equation (42) thus establishes $P(\theta)$ as a genuine probability distribution from the fundamental definition of a probability density.

5.2. *Phase operator*

We have shown by means of the limiting procedure used in the basic definition of a probability density not only that equation (12) is consistent with complementarity of phase and photon number and with the operation of practical phase shifters, but also that it represents a genuine probability distribution for physical states of light. There is now no reason not to accept equation (12) as the phase probability distribution. While this distribution is all that is necessary to calculate the phase properties of physical states of light, it is interesting to pursue the fundamental quantum mechanics a little further.

We remarked previously that the apparent difficulty with the distribution (12) might be associated with the apparent impossibility of performing a precise measurement of phase. Now that the difficulty with equation (12) has been resolved, it is useful to examine the impact of this resolution on the question of the measurability of phase and on the identification of the operator associated with the phase observable. A quantum-mechanical property which is in principle measurable can be represented by an Hermitian operator. The limiting process used to find the phase probability density involved the states $|\theta_m\rangle$ given by equation (41). Our aim is to find an Hermitian operator having these states as eigenstates and the values of θ_m, on which the phase histogram is based, as eigenvalues. The states $|\theta_m\rangle$ form an orthonormal set, but it is also useful to be able to treat them as a *complete* set. This can be done by working in the $(s+1)$-dimensional space Ψ_s which is spanned by them. In this space a resolution of the identity is

$$\sum_{m=0}^{s} |\theta_m\rangle\langle\theta_m| = 1, \tag{43}$$

from which we can show using equation (41) that the number states of Ψ_s, which also form a basis, are

$$\begin{aligned} |n\rangle &= \sum_{m=0}^{s} |\theta_m\rangle\langle\theta_m|n\rangle \\ &= \frac{1}{(s+1)^{1/2}} \sum_{m=0}^{s} \exp(-in\theta_m)|\theta_m\rangle, \end{aligned} \tag{44}$$

where $n = 0, 1, 2, \ldots, s$. We must at some stage include *all* the number states in the formalism; so it will be necessary eventually to take the limit as $s \to \infty$. From equation (37) we see that this is equivalent to letting the distance $\delta\theta$ between possible values of θ_m tend to zero, which is also necessary for a satisfactory description of phase. The method for taking the required limit is discussed below. We note that in the space Ψ_s the number states and the states $|\theta_m\rangle$ are equally weighted superpositions of each other, which is a manifestation of complementarity. Further, the number operator acts as the generator of shifts in the value θ_m. These complementarity and phase-shift properties lead us to call $|\theta_m\rangle$ the *phase states* of Ψ_s. Consider the operator

$$\hat{\varphi}_\theta = \sum_{m=0}^{s} \theta_m |\theta_m\rangle\langle\theta_m|. \tag{45}$$

We have dropped the subscript zero from $\hat{\varphi}_\theta$ both for convenience of notation and to distinguish it from the operator φ_{θ_0} used previously which is *not* a function of s. The operator $\hat{\varphi}_\theta$ has eigenstates $|\theta_m\rangle$ with eigenvalues θ_m and the associated probability histogram will be given by equation (39). Because of the complementarity of its eigenstates with the number states of the space Ψ_s we call $\hat{\varphi}_\theta$ the *phase operator* and remember that it acts on the space Ψ_s rather than on the usual infinite Hilbert space. We note that the phase operator is both Hermitian and self-adjoint.

The orthogonality of the phase states $|\theta_m\rangle$ allows us to write

$$\hat{\varphi}_\theta^p = \sum_{m=0}^{s} \theta_m^p |\theta_m\rangle\langle\theta_m| \tag{46}$$

and hence to write any function $F(\hat{\varphi}_\theta)$ which is expressible as a power series, such as an exponential for example, as

$$F(\hat{\varphi}_\theta) = \sum_{m=0}^{s} F(\theta_m)|\theta_m\rangle\langle\theta_m|. \tag{47}$$

The important question now arises as to how to take the limit as $s \to \infty$. If we can find the correct expectation values of functions of phase, we can calculate all measurable phase properties for any state. We could do this, for example, by constructing the periodic phase probability distribution as a Fourier series from the expectation values of the appropriate exponential functions of phase. There are two obvious procedures by which we might obtain expectation values in the infinite-s limit. The first is to calculate expectation values of the operators on Ψ_s as a function of s and then to find the limit as $s \to \infty$. The second is to find the infinite-s limits of the operators first and then to take the expectation values. Perhaps surprisingly, these different limiting procedures yield different results. The simplest method involves taking the limit of expectation values, which are c numbers, rather than the limit of the phase operator itself; so let us study the outcome of this procedure first. The limit as $s \to \infty$ of the expectation value of the general function $F(\hat{\varphi}_\theta)$ for the general physical state $|f\rangle$ in equation (30) is

$$\lim_{s\to\infty}\left(\sum_{m=0}^{s} F(\theta_m)\langle f|\theta_m\rangle\langle\theta_m|f\rangle\right)$$
$$= \lim_{s\to\infty}\left(\frac{1}{2\pi}\sum_{m=0}^{s}\sum_{n=0}^{s}\sum_{n'=0}^{s} F(\theta_m)c_n^* c_{n'} \exp\left[i(n-n')\theta_m\right]\delta\theta\right), \tag{48}$$

where we have substituted $(s+1)^{-1} = \delta\theta/2\pi$ from equation (37). From the definition of the Riemann integral this expression becomes

$$\frac{1}{2\pi}\int_{\theta_0}^{\theta_0+2\pi} F(\theta)\sum_{n=0}^{\infty}\sum_{n'=0}^{\infty} c_n^* c_{n'} \exp\left[i(n-n')\theta\right]\mathrm{d}\theta = \int_{\theta_0}^{\theta_0+2\pi} F(\theta)P(\theta)\,\mathrm{d}\theta, \tag{49}$$

with $P(\theta)$ being the phase probability density given by equation (12). From equation (13) we see that equation (49) is the correct expectation value of $F(\theta)$. Thus finding the infinite-s limit of the expectation values of operators representing functions of the phase on the space Ψ_s yields the correct results. It is not difficult to verify that this procedure also produces the correct results for functions of other operators such as $\hat{N}$ and $\hat{a}$. Furthermore, by using this procedure to find $\langle\exp(ip\theta)\rangle$ for integer p, we can construct $P(\theta)$ correctly from its Fourier components. These considerations show that finding the limits of expectation values gives results which are precisely in accord with the phase probability distribution. As a further check, we can use this limiting procedure to find the expectation value of the commutator $[\hat{N}, \hat{\varphi}_\theta]$ for a physical state and recover the result (20) which leads to equation (21) as required by the correspondence principle [7]. We discuss the alternative limiting procedure, involving finding the limits of operators first, in the next section and show that it does *not* yield consistent results.

Earlier we noted that there was a problem associated with equation (29) in that the action of $\exp(i\theta)$ on the vacuum produced an unphysical state. It is instructive therefore to examine the consequence of the action of $\exp(i\hat{\varphi}_\theta)$ on the number states. We have

$$\begin{aligned}\exp(i\hat{\varphi}_\theta)|n\rangle &= \frac{1}{(s+1)^{1/2}}\sum_{m=0}^{2}\exp[-i(n-1)\theta_m]|\theta_m\rangle \\ &= |n-1\rangle \end{aligned} \qquad (50)$$

which follows from equation (44). This is in accord with our previous result (28). For $n = 0$, it is not difficult to show [6, 7, 11] that

$$\exp(i\hat{\varphi}_\theta)|0\rangle = \exp[i(s+1)\theta_0]|s\rangle. \qquad (51)$$

The phase wavefunction for this state is

$$\psi_s(\theta_m) = \frac{1}{(2\pi)^{1/2}}\exp[i(s+1)\theta_0]\exp(-is\theta_m) \qquad (52)$$

which, by using equation (38), we can show is $(2\pi)^{-1/2}\exp(i\theta_m)$ which is in accord with equation (29). Thus we see that the state generated by the action of $\exp(i\hat{\varphi}_\theta)$ on the vacuum is (apart from a phase factor) the state $|s\rangle$ which is within the state space Ψ_s. In the limit as $s \to \infty$, the unphysical state so generated becomes the `infinite' photon number state.

To sum up this section, we can use the Hermitian operator (45) as a valid quantum-mechanical description of phase which incorporates complementarity provided that we find the limit as $s \to \infty$ of expectation values obtained from it and from functions of it. This limiting procedure also gives the correct results for functions of the common quantum-mechanical operators.

6. Operators of H_∞

The procedure of taking the infinite-s limit of the expectation values of the phase operators produces consistent results agreeing with our earlier discussion of complementarity of photon number and phase. This procedure does not invoke the conventional infinite-dimensional Hilbert space H_∞ at all as all operators act on Ψ_s. Measurements, however, correspond to operators acting on H_∞. Thus, to investigate whether or not phase is measurable, we should ask what operator on H_∞ best corresponds to the phase operator $\hat{\varphi}_\theta$. An obvious possibility is to attempt to construct such an operator as the infinite-s limit of $\hat{\varphi}_\theta$. This brings us to the second limiting procedure mentioned in the previous section. It turns out that $\hat{\varphi}_\theta$ converges only *weakly* [13] on the infinite-dimensional Hilbert space and thus only its weak limit exists. We shall not enter into the details of this here but we remark that the weak limits of operators do not preserve the operator algebra. For example, the square of the weak limit of $\hat{\varphi}_\theta$ is not the weak limit of the square of $\hat{\varphi}_\theta$. Consequently, if the limit of the operator is found before performing algebraic manipulations, quite different results are found from those obtained from the limits of expectation values of algebraic expressions involving phase operators. Taking the weak limit leads to the operators studied by Popov and Yarunin [14, 15] and Garrison and Wong [16]. It is not difficult to show the weak limit $\hat{\varphi}_{\theta \mathrm{w}}$ of $\hat{\varphi}_\theta$ does not yield results in accord with equation (12) and that its eigenstates are not equally weighted superpositions of number states as required by complementarity. Even more seriously, as a potential phase operator it does not even give self-consistent results. For example, $\langle \hat{\varphi}_{\theta \mathrm{w}} \rangle$ for the vacuum state is $\theta_0 + \pi$ for *any* choice of θ_0 [17]. The only phase probability distribution consistent with this is the uniform distribution, which of course is in accord with complementarity. This distribution must have a phase variance of $\pi^2/3$, but the variance found from $\langle \hat{\varphi}^2_{\theta \mathrm{w}} \rangle$ is only $\pi^2/6$ [17, 18] which is therefore not consistent with the value of $\langle \hat{\varphi}_{\theta \mathrm{w}} \rangle$. The convergence problem does not arise with the more familiar operators of quantum electrodynamics such as the photon number operator. This operator is strongly convergent [13] and the same results are obtained whether we take the infinite-s limit of the Ψ_s number operator before finding expectation values or find expectation values before taking the limit. For the weakly convergent phase operator, however, we must use the more general approach of finding the limits of expectation values of Ψ_s operators. Taking the limit of the expectation values $\langle \hat{\varphi}_\theta \rangle$ and $\langle \hat{\varphi}^2_\theta \rangle$ yield a phase mean and variance of $\theta_0 + \pi$ and $\pi^2/3$ respectively, which are consistent both with each other and with complementarity.

The difficulty with taking the weak limit can be traced to the fact that the uppermost state $|s\rangle$ of Ψ_s is lost. This difficulty is most apparent if we attempt to reconstruct a unitary phase operator on H_∞ by taking the weak limit of $\exp(\mathrm{i}\hat{\varphi}_\theta)$. In this limit the relations (50) still apply but equation (51) becomes

$$\left[\exp(\mathrm{i}\hat{\varphi}_\theta)\right]_\mathrm{w}|0\rangle = 0. \tag{53}$$

This weak limit of the exponential operator is, in fact, the operator introduced by Susskind and Glogower [3] (see also [4]) and takes us back to the original phase problem, namely that the action of $\exp(\mathrm{i}\theta)$ on the vacuum state does not produce a state in H_∞. From equation (53) we see that the Susskind–Glogower operator is not unitary and therefore cannot represent the complex exponential of an Hermitian operator and, in particular, $\left[\exp(\mathrm{i}\hat{\varphi}_\theta)\right]_\mathrm{w} \neq \exp(\mathrm{i}\hat{\varphi}_{\theta \mathrm{w}})$.

The fact that $\hat{\varphi}_{\theta \mathrm{w}}$ is not a satisfactory phase operator tends to support the view that there is no phase operator acting on H_∞, with the consequence that phase is not precisely measurable. This leads us to ask whether there exists an operator acting on H_∞ which gives a good *approximate* measure of phase for fields in the quantum domain. We should note that, because the use of $\hat{\varphi}_{\theta \mathrm{w}}$ and powers of $\hat{\varphi}_{\theta \mathrm{w}}$ violate complementarity most strongly for weak fields in the quantum domain, it will not give results which are even reasonably approximate for such fields of interest. If we want an operator to represent phase at least approximately on H_∞, it is better to take the H_∞ operator

$$\hat{\varphi}_R = \sum_{m=0}^{R} \theta_m |\theta_m\rangle\langle\theta_m|, \tag{54}$$

with

$$|\theta_m\rangle = \frac{1}{(R+1)^{1/2}} \sum_{n=0}^{R} \exp(in\theta_m)|n\rangle \tag{55}$$

and $m = 0, 1, \ldots, R$ where R is some reasonably large number. The operator (54) is *not* a phase operator because it is not the complement of $\hat{N}$ in the same space H_∞. It has eigenstates $|\theta\rangle$ where $0 \leqslant m \leqslant R$ with eigenvalues θ_m separated by $2\pi/(R+1)$ *and* eigenstates $|n\rangle$ where $n > R$ with eigenvalues of zero. The associated probability histogram for a field in state $|f\rangle$ of equation (30) will have a value at θ_m given by equation (39) with s replaced by R, with an extra probability for eigenvalue zero given by $\sum_{n=R+1}^{\infty} |c_n|^2$. This latter probability can be made as small as desired for any normalizable state by choosing R sufficiently large. Thus this histogram can approximate equation (12) to within any given non-zero error by a suitable choice of R and so $\hat{\varphi}_R$ is an approximate representation of phase on H_∞ with the degree of accuracy determined by R.

It is worth noting here that, if we want a precise operator representation of phase on an infinite-dimensional Hilbert space, we require a space on which $\hat{\varphi}_\theta$ converges strongly. Then the complementarity with the number operator is not lost when the limit is taken. This can be achieved by using Vaccaro's [19] space H_{sym}, which contains H_∞ as a subspace and also contains the state $\exp(i\hat{\varphi}_\theta)|0\rangle$ even in the limit. Using H_{sym} allows consistent results to be obtained by finding the expectation values of the strong limits of operators. The results of this approach are the same as those obtained by our formalism of taking the limits of expectation values.

7. Phase differences and sums

The state of a two-mode field can be written in general as

$$|ff\rangle = \sum_{n_1,n_2} c_{n_1 n_2} |n_1\rangle |n_2\rangle, \tag{56}$$

where $|n_1\rangle$ and $|n_2\rangle$ are number states associated with modes 1 and 2. Using two $(s+1)$-dimensional spaces, we can obtain the joint probability for finding the fields with phases θ_{m_1} and θ_{m_2} by projecting $|ff\rangle$ onto the phase states with these eigenvalues. If $|ff\rangle$ is a physical state, we can then take the $s \to \infty$ limit and obtain the joint probability density for phases θ_1 and θ_2 as

$$P(\theta_1, \theta_2) = \frac{1}{4\pi^2} \left| \sum_{n_1,n_2} c_{n_1 n_2} \exp[-i(n_1\theta_1 + n_2\theta_2)] \right|^2. \tag{57}$$

The surface represented by $P(\theta_1, \theta_2)$ is 2π periodic along both the θ_1 and the θ_2 axes. It is also normalized over any two 2π ranges:

$$\int_{2\pi}\int_{2\pi} P(\theta_1, \theta_2)\, \mathrm{d}\theta_1\, \mathrm{d}\theta_2 = 1. \tag{58}$$

The probability density for the fields to have a phase difference of $\theta_- = \theta_2 - \theta_1$ is

$$P(\theta_-) = \int_{2\pi} P(\theta_1, \theta_1 + \theta_-)\, \mathrm{d}\theta_1. \tag{59}$$

This function is also 2π periodic and is normalized over any 2π range:

$$\int_{2\pi} P(\theta_-)\, \mathrm{d}\theta_- = 1. \tag{60}$$

For the physical state (56) we find from equations (59) and (57) that

$$\begin{aligned} P(\theta_-) &= \frac{1}{4\pi^2} \sum_{n_1,n_2} \sum_{n_1',n_2'} c^*_{n_1 n_2} c_{n_1' n_2'} \exp\left[\mathrm{i}(n_2 - n_2')\theta_-\right] \int_{2\pi} \exp\left[\mathrm{i}(n_1 - n_1' + n_2 - n_2')\right] \theta_1\, \mathrm{d}\theta_1 \\ &= \frac{1}{2\pi} \sum_{n_1,n_2} \sum_{n_1'} c^*_{n_1 n_2} c_{n_1'(n_1 - n_1' + n_2)} \exp\left[\mathrm{i}(n_1' - n_1)\theta_-\right]. \end{aligned} \tag{61}$$

In a similar way, the phase-sum probability can be found to be

$$P(\theta_+) = \frac{1}{2\pi} \sum_{n_1,n_2} \sum_{n_1'} c^*_{n_1 n_2} c_{n_1'(n_1' - n_1 + n_2)} \exp\left[\mathrm{i}(n_1 - n_1')\theta_+\right]. \tag{62}$$

As a check that these expressions are sensible, consider the case in which at least one of the fields is in a number state. It is then easy to see from equations (61) and (62) that both $P(\theta_-)$ and $P(\theta_+)$ are equal to $1/2\pi$. That is, both the phase difference and the phase sum are completely random. This is quite natural since the phase for a single-mode number state is random. For fields which are entangled such that $c_{n_1 n_2}$ is zero for $n_1 \neq n_2$, for example as for the two-mode squeezed vacuum states, it follows from equation (61) that $P(\theta_-)$ is $1/2\pi$ so that the phase difference is random. In this case, however, the phase sum will not be random [20].

8. Other descriptions of phase

We have seen that it does seem to be the case that there is no phase operator in the usual infinite-dimensional Hilbert space. Of course, this does not mean that physical states do not have well defined phase properties. Indeed these properties can be derived from equation (12).

Another method for deriving the phase probability distribution (12) without the use of a precise phase operator acting on H_∞ is the probability-operator measure approach [21, 22]. As this yields equation (12), it is also a description of phase as the complement of photon number. In this approach there is a phase probability density but no operator representing the phase observable.

There have also been approaches to phase which are not consistent with number–phase complementarity. The most notable of these is that of Susskind and Glogower, who represent $\exp(\pm i\theta)$ by non-unitary non-commuting matrices and $\cos\theta$ and $\sin\theta$ as Hermitian combinations of these. A simple calculation shows that for the vacuum $\langle\cos^2\theta\rangle = \frac{1}{4}$ [4], rather than the value of $\frac{1}{2}$ associated with the random phase demanded by complementarity. We might note that the Susskind–Glogower operators also have difficulty with self-consistency. If we use these to calculate $\langle\exp(ip\theta)\rangle$ for the vacuum, we find a result of zero for all integers p except zero. This implies that the phase probability distribution should be uniform, which is inconsistent with the value obtained for $\langle\cos^2\theta\rangle$. A noteworthy class of other descriptions includes the operational or 'measured phase' operators [5, 23–27]. Here, instead of defining phase in terms of complementarity, it is defined as the quantity measured by a particular experiment which, if used with classical fields, would be a phase measurement. The advantage of such descriptions is that they describe something relating to phase which is measurable in a reasonably direct way. The disadvantage is that different experiments lead to different operators; so there is no universally applicable operational definition of phase. None of the operational quantities defined so far has a probability distribution (12); so corresponding measurements will not yield this distribution. For these approaches, the idea of a phase shifter shifting the phase of all fields which, as we have seen, leads directly to the probability distribution (12) may need re-examination, particularly where use of such a device is part of the experimental procedure involved in the operational definition. We can only conclude that the quantities obtained from these experiments are *phase dependent* rather than representing the phase itself. For a fuller discussion of this point together with phase-space-based descriptions of phase see [28].

9. Measuring phase

If a precise Hermitian phase operator acting on H_∞ does not exist, it might be concluded that in principle a precise measurement of phase is impossible. This is not as serious as it may first sound, as a precise measurement of phase would leave the field in an unphysical state with infinite mean energy anyway. It should also be remembered that a precise measurement of the position of a particle would give it an infinite variance in momentum, which is also unphysical, but position is still a useful and observable quantum-mechanical quantity. A more reasonable question to ask is: 'Is it possible, in principle at least, to measure phase to within any given non-zero error?' Here the answer is definitely in the affirmative, with the corresponding operator acting on H_∞ being given by $\hat{\varphi}_R$ in equation (54). The larger the value chosen for R in equation (54), the better will be the measure of phase given by $\hat{\varphi}_R$ and its associated probability histogram. As discussed earlier, we can always select a value of R large enough to ensure an error less than any given non-zero value, however small. In practice, one does not measure probability densities; one measures probabilities associated with bins of finite size. The smallest possible bin size commensurate with performing the experiment in a reasonable time is often chosen. Thus the Hermitian operator $\hat{\varphi}_R$ with its associated histogram with finite bin size $2\pi/(R+1)$ might even be considered to

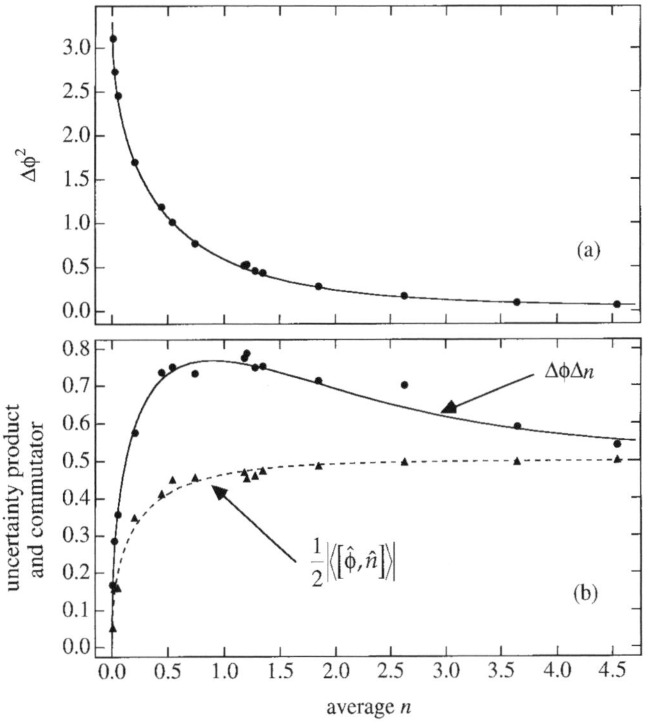

Figure (*a*) Variance of the phase plotted against the average number of photons in a coherent state. The points are the experimentally determined values and the theoretical solid curve is calculated using the phase operator. (*b*) The points are the experimentally determined values for the uncertainty product (●) and the commutator (▲). The curves are the theoretical values for the uncertainty product (——) and the commutator (– – –). (From Beck *et al.* [34].)

correspond more closely to an *actual* measurement than the zero bin size limit (12). The structure of $\hat{\varphi}_R$ may not immediately suggest how such a measurement should be made, but the existence of $\hat{\varphi}_R$ shows that such a measurement is possible in principle. As R is finite, the field will be projected onto a physical state. We know of no experiments which have been set up to measure equation (54) for $R > 2$. However, for some fields with sufficiently narrow phase distributions, such as coherent states with mean photon numbers of at least five, the simple balanced homodyne technique gives a reasonably good measurement of phase. What is actually being measured is the measured phase operator of [5], but for such fields this operator provides a reasonably good description of phase. Likewise other operational definitions such as those of Noh and co-workers [23–26] and of Raymer *et al.* [27] can provide a good approximate measure of the phase for some fields. We should emphasize here, because of the misunderstandings with some workers concerning the interpretation of the experiments of Noh and co-workers [23–26] (for example [29]) that these experimental results of course do not disprove *or even serve as a test of* the distribution (12). As Noh and co-workers were careful to point out [26], their measurement scheme and our formalism which results in equation (12) describe different things. Their measurements are not measurements of the phase distribution (12); they simply measure the operational phase operator constructed to describe the experiments. These experiments do not verify nor disprove any descriptions or definitions of phase. Experiments based on the measured phase operators of [5] and experiments to measure the operational phase of Raymer *et al.* [27] would yield different results still, as these measure neither the operational phase of Noh and co-workers nor the phase distribution (12).

Given a reproducible state, which is needed for the measurement of any probability distribution, there are other techniques which measure (12) to good accuracy even for coherent states with mean photon numbers as low as unity (see [30] and references therein). For fields with broader phase distributions where these techniques are not suitable, equation (12) can be experimentally determined by measuring c_n, that is by measuring the state of the field. This can be, and indeed has been, done by optical homodyne tomography [31–34]. The experimentally determined phase–variance and phase–number uncertainty product for weak coherent states (including the vacuum) are shown in the figure. The lines are calculated using equations (12), (15) and (20) together with the well known result that for the coherent states the variance in the photon number is equal to the mean. Again we should emphasize that, although the experimental results obtained are in excellent agreement with what is anticipated, these experiments do not verify or disprove that equation (12) is the phase probability density; they simply measure it. The disadvantage of tomographic techniques is that they are information expensive. They gather sufficient information to measure also the number state distribution. The question naturally arises as to whether it is possible to measure equation (12) with less information. This does seem to be the case as has been described elsewhere [35].

10. Conclusions

The phase of a quantum harmonic oscillator or single field mode is complementary to the excitation or photon number. This idea is a fundamental consequence of the fact that the Hamiltonian is the generator of time translations. We have shown that this also leads us to identify equation (12) with the probability distributions for the phase.

The quantum phase problem is simply that we cannot represent the phase eigenstate or the (exact) phase operator within the usual infinite Hilbert space spanned by the number states. Our solution to this problem is to work with a state space Ψ_s spanned by the first $s+1$ photon number states. Within this space we can construct the Hermitian optical phase operator $\hat{\varphi}_\theta$ which is complementary to the photon number operator. This gives the phase properties of a state by applying the limit $s \to \infty$ to the moments of the phase operator. If instead we were to apply the limit directly to the phase operator, then we would find results that are not consistent with the probability distribution (12) and hence with number–phase complementarity, and which also suffer self-consistency problems.

The phase probability distribution has been measured [33, 34]. It is important to emphasize, however, that not all phase-dependent measurements measure the phase itself.

Appendix

Here we provide a proof of the theorem (36). All physical states have finite moments of the photon number and hence for any positive integer q and positive number ϵ we can always find an integer L such that

$$\sum_{n=L+1}^{\infty} n^q |c_n|^2 < \epsilon, \tag{A 1}$$

where c_n is the coefficient of the number state $|n\rangle$ in the state under consideration. The inequality (A 1) implies that

$$|c_n|^2 < \frac{\epsilon}{n^q} \rightarrow |c_n| < \frac{\epsilon^{1/2}}{n^{q/2}}, \tag{A 2}$$

for all $n > L$. The phase representation of the state is

$$\psi(\theta) = \frac{1}{(2\pi)^{1/2}} \sum_{n=0}^{\infty} c_n \exp(-in\theta), \tag{A 3}$$

while, if we truncate the series at L, we find that

$$\psi_L(\theta) = \frac{1}{(2\pi)^{1/2}} \sum_{n=0}^{L} c_n \exp(-in\theta). \tag{A 4}$$

The phase probability density is given by the squared modulus equation (A 3) and its truncated approximation by the squared modulus of equation (A 4). The magnitude of the difference between these is then

$$\begin{aligned} \left| |\psi(\theta)|^2 - |\psi_L(\theta)|^2 \right| &= \left| |\psi(\theta)| - |\psi_L(\theta)| \right| (|\psi(\theta)| + |\psi_L(\theta)|) \\ &\leqslant |\psi(\theta) - \psi_L(\theta)| (|\psi(\theta)| + |\psi_L(\theta)|). \end{aligned} \tag{A 5}$$

It then follows from equation (A 1)–(A 4) that

$$|\psi(\theta) - \psi_L(\theta)| \leqslant (2\pi)^{-1/2} \sum_{n=L+1}^{\infty} |c_n| < \left(\frac{\epsilon}{2\pi}\right)^{1/2} \sum_{n=L+1}^{\infty} n^{-q/2}. \tag{A 6}$$

For $q > 2$ we can get an upper bound for the sum in equation (A 6)

$$\sum_{n=L+1}^{\infty} n^{-q/2} < \int_L^{\infty} x^{-q/2}\,\mathrm{d}x = \frac{2L^{(2-q)/2}}{q-2}. \tag{A 7}$$

Hence the magnitude of the difference in the two probability densities is bounded by

$$\begin{aligned} \left| |\psi(\theta)|^2 - |\psi_L(\theta)|^2 \right| &\leqslant \left(\frac{\epsilon}{2\pi}\right)^{1/2} \sum_{n=L+1}^{\infty} n^{-q/2} (|\psi(\theta)| - |\psi_L(\theta)| + 2|\psi_L(\theta)|) \\ &\leqslant \left(\frac{\epsilon}{2\pi}\right)^{1/2} \frac{2}{(q-2)L^{(2-q)/2}} \left[\left(\frac{\epsilon}{2\pi}\right)^{1/2} \frac{2}{(q-2)L^{(2-q)/2}} + 2|\psi_L(\theta)| \right]. \end{aligned} \tag{A 8}$$

We can also place an upper bound on $|\psi_L(\theta)|$ as

$$|\psi_L(\theta)| = (2\pi)^{-1/2}\left|\sum_{n=0}^{L} c_n \exp(-in\theta)\right| \leqslant (2\pi)^{-1/2}\sum_{n=0}^{L}|c_n| \leqslant \left(\frac{L+1}{2\pi}\right)^{1/2}. \quad \text{(A 9)}$$

The difference in the probabilities is then bounded by the inequality

$$\left||\psi(\theta)|^2 - |\psi_L(\theta)|^2\right| < \frac{2\varepsilon}{\pi(q-2)^2 L^{(2-q)}} + \frac{2\varepsilon^{1/2}(L+1)^{1/2}}{\pi(q-2)L^{(2-q)/2}}. \quad \text{(A 10)}$$

For $q > 2$, this can be made as small as we like. Hence, for L sufficiently large, the truncated phase probability distribution can be made arbitrarily close to the phase distribution (12).

Acknowledgments

Our work on optical phase has benefited greatly from a long-standing interaction with John Vaccaro. We are grateful to him and to Mike Raymer and Mark Beck for permission to reproduce their experimental data. We also thank Tony Chefles for his assistance with the time-consuming task of preparing the bibliography.

References

[1] London, F., 1926, *Z. Phys.*, **37,** 915.
[2] Dirac, P. A. M., 1927, *Proc. R. Soc.* A, **114,** 243.
[3] Susskind, L., and Glogower, J., 1964, *Physics*, **1,** 49.
[4] Carruthers, P., and Nieto, M. M., 1968, *Rev. mod. Phys.*, **40,** 411.
[5] Barnett, S. M., and Pegg, D. T., 1986, *J. Phys.* A, **19,** 3849.
[6] Barnett, S. M., and Pegg, D. T., 1989, *J. mod Optics*, **36,** 7.
[7] Pegg, D. T., and Barnett, S. M., 1989, *Phys. Rev.* A, **39,** 1665.
[8] Messiah, A., 1958, *Quantum Mechanics* (Amsterdam: North-Holland), p. 179.
[9] Biaynicki-Birula, I., and Van, C. L., 1980, *Acta Phys. Polonica* A, **57**, 599.
[10] Louisell, W. H., 1963, *Phys. Lett.*, **7,** 60.
[11] Pegg, D. T., and Barnett, S. M., 1988, *Europhys. Lett.*, **6,** 483.
[12] Vaccaro, J. A., and Pegg, D. T., 1990, *J. mod. Optics*, **37,** 17.
[13] Vaccaro, J. A., and Pegg, D. T., 1993, *Physica scripta* T, **48,** 22.
[14] Popov, V. N., and Yarunin, V. S., 1973, *Leningrad Univ. J.: Phys.*, **22,** 7.
[15] Povov, V. N., and Yarunin, V. S., 1992, *J. mod. Optics*, **39,** 1525.
[16] Garrison, J. C., and Wong, J., 1970, *J. math. Phys.*, **11,** 2242.
[17] Barnett, S. M., and Pegg, D. T., 1992, *J. mod. Optics*, **39,** 2121.
[18] Gantsog, T., Miranowicz, A., and Tanas, R., 1992, *Phys. Rev.* A, **46,** 2870.
[19] Vaccaro, J. A., 1995, *Phys. Rev.* A, **51,** 3309.
[20] Barnett, S. M., and Pegg, D. T., 1990, *Phys. Rev.* A, **42,** 6713.
[21] Helstrom, C. W., 1976, *Quantum Detection and Estimation Theory* (New York: Academic Press).
[22] Shapiro, J. H., and Shepard, S. R., 1991, *Phys. Rev.* A, **43,** 3795.
[23] Noh, J. W., Fougères, A., and Mandel, L., 1991, *Phys. Rev. Lett.*, **67,** 1426.
[24] Noh, J. W., Fougères, A., and Mandel, L., 1992, *Phys. Rev.* A, **45,** 424.
[25] Fougères, A., Noh, J. W., Grayson, T. P., and Mandel, L., 1994, *Phys. Rev.* A, **49,** 530.
[26] Noh, J. W., Fougères, A., and Mandel, L., 1993, *Phys. Rev. Lett.*, **71,** 2579.
[27] Raymer, M. G., Cooper, J., and Beck, M., 1993, *Phys. Rev.* A, **48,** 4617.

[28] BARNETT, S. M., and DALTON, B. J., 1993, *Physica scripta* T, **48**, 13.
[29] LYNCH, R., 1995, *Phys. Rep.*, **256**, 368.
[30] PEGG, D. T., VACCARO, J. A., and BARNETT, S. M., 1994, *Quantum Optics VI*, edited by D. F. Walls and J. D. Harvey (Berlin: Springer), p. 153.
[31] SMITHEY, D. T., BECK, M., RAYMER, M. G., and FARIDANI, A., 1993, *Phys. Rev. Lett.*, **70**, 1244.
[32] SMITHEY, D. T., BECK, M., COOPER, J., and RAYMER, M. G., 1993, *Phys. Rev.* A, **48**, 3159.
[33] SMITHEY, D. T., BECK, M., COOPER, J., RAYMER, M. G., and FARIDANI, A., 1993, *Physica scripta* T, **48**, 35.
[34] BECK, M., SMITHEY, D. T., COOPER, J., and RAYMER, M. G., 1993, *Optics Lett.*, **18**, 1259.
[35] BARNETT, S. M., and PEGG, D. T., 1996, *Phys. Rev. Lett.*, **76**, 4148.

Bibliography

The papers in this bibliography are presented in approximately chronological order of publication so that the reader can follow the development of the theory of phase. We have tried to be as complete as possible but, given the volume of work on the topic, some important omissions are inevitable.

LONDON, F., 1926, Über die Jacobischen Transformationen der Quantenmechanik, *Z. Phys.*, **37**, 915.
DIRAC, P. A. M., 1927, The quantum theory of the emission and absorption of radiation, *Proc. R. Soc.* A, **114**, 243.
LONDON, F., 1927, Winkelvariable und kanonische Transformationen in der Undulationmechanik, *Z. Phys.*, **40**, 193.
HEITLER, W., 1954, *The Quantum Theory of Radiation*, third edition (Oxford University Press).

1960–1964

SERBER, R., and TOWNES, C. H., 1960, Limits on electromagnetic amplification due to complementarity, *Quantum Electronics*, edited by C. H. Townes (New York: Columbia University Press).
LOUISELL, W. H., YARIV, A., and SIEGMAN, A. E., 1961, Quantum fluctuations and noise in parametric processes I, *Phys. Rev.*, **124**, 1646.
SENITZKY, I. R., 1962, Incoherence, quantum fluctuations and noise, *Phys. Rev.*, **128**, 2864.
HAUS, H. A., and MULLEN, J. A., 1962, Quantum noise in linear amplifiers, *Phys. Rev.*, **128**, 2407.
LOUISELL, W. H., 1963, Amplitude and phase uncertainty relations, *Phys. Lett.*, **7**, 60.
JUDGE, D., 1963, On the uncertainty relation for L_z and φ, *Phys. Lett.*, **5**, 189.
JUDGE, D., and LEWIS, J. T., 1963, On the commutator $[L_z, \varphi]$, *Phys. Lett.*, **5**, 190.
SUSSKIND, L., and GLOGOWER, J., 1964, Quantum mechanical phase and time operator, *Physics*, **1**, 49.
BRUNET, H., 1964, Quantum state of an ideal phase detector, *Phys. Lett.*, **10**, 172.
HARMS, J., and LORIGNY, J., 1964, Ideal phase detector quantum state, *Phys. Lett.*, **10**, 173.

1965–1969

CARRUTHERS, P., and NIETO, M. M., 1965, Coherent states and the number-phase uncertainty relation, *Phys. Rev. Lett.*, **14**, 387.
ANDERSON, P. W., 1966, Considerations on the flow of superfluid helium, *Rev. mod. Phys.*, **38**, 298.
LERNER, E. C., 1968, Harmonic-oscillator phase operators, *Nuovo Cim.* B, **56**, 183.
JACKIW, R., 1968, Minimum uncertainty product, number–phase uncertainty product, and coherent states, *J. math. Phys.*, **9**, 339.

Carruthers, P., and Nieto, M. M., 1968, Phase angle variables in quantum mechanics, *Rev. mod. Phys.*, **40**, 411.

Biatynicki-Birula, I., Mielnik, B., and Plebanski, 1969, Explicit solution of the continuous Baker–Campbell–Hausdorf problem and a new expression for the phase operator, *Ann. Phys. (N.Y.)*, **51**, 187.

1970–1974

Lerner, E. C., Huang, H. W., and Walters, G. E., 1970, Some properties of oscillator phase operators, *J. math. Phys.*, **11**, 1679.

Garrison, J. C., and Wong, J., 1970, Canonically conjugate pairs, uncertainty relations and phase operators, *J. math. Phys.*, **11**, 2242.

Stenholm, S., 1970, On the theory of non-Hermitean operators, *Ann. Acad. Scientiarum Fennicae* A, **6**, 339.

Ifantis, E. K., 1971, On the nature of the spectrum of generalized oscillator phase operators, *Lett. Nuovo Cim*, **2**, 1096.

Ifantis, E. K., 1971, Abstract formulation of the quantum measurement problem, *J. math. Phys.*, **12**, 1021.

Loudon, R., 1973, *The Quantum Theory of Light*, first edition (Oxford University Press).

Aharonov, Y., Lerner, E. C., Huang, H. W., and Knight, J. M., 1973, Oscillator phase states, thermal equilibrium and group representations, *J. math. Phys.*, **14**, 746.

Levy-Leblond, J. M., 1973, Azimuthal quantization of angular momentum, *Rev. Méx. Fis.*, **22**, 15.

Popov, V. N., and Yarunin, V. S., 1973, On the problem of the phase operator for a linear harmonic oscillator, *Leningrad Uni. J.: Phys.*, **22**, 7 (in Russian).

Gerhardt, H., Welling, H., and Frolich, D., 1973, Ideal laser amplifier as a phase measuring system of a microscope radiation field, *Appl. Phys.*, **2**, 91.

Paul, H., 1974, Phase of a microscopic electromagnetic field and its measurement, *Fdn. Phys.*, **22**, 657.

Provost, J. P., Rocca, F., and Vallee, G., 1974, Phase properties of some photon states with nonzero energy density, *J. math. Phys.*, **15**, 2079.

Gerhardt, H., Büchler, U., and Litfin, G., 1974, Phase measurement of a microscopic radiation field, *Phys. Lett.* A, **49**, 119.

1975–1979

Provost, J. P., Rocca, F., and Vallee, G., 1975, Coherent states, phase states, and condensed states, *Ann. Phys. (N.Y.)*, **94**, 307.

Biatynicki-Birula, I., and Bialynicka-Birula, Z., 1976, Quantum electrodynamics of intense photon beams. New approximation method, *Phys. Rev.* A, **14**, 1101.

Lévy-Leblond, J. M., 1976, Who is afraid of nonhermitian operators? A quantum description of angle and phase, *Ann. Phys. (N.Y.)*, **101**, 319.

Santhanam, T. S., and Tekumalla, A. R., 1976, Quantum mechanics in finite dimensions, *Fdn. Phys.*, **6**, 583.

Helstrom, C. W., 1976, *Quantum Detection and Estimation Theory* (New York: Academic Press).

Nieto, M. M., 1977, Phase-difference operator analysis of microscope radiation-field measurements, *Phys. Lett.* A, **60**, 401.

Lévy-Leblond, J. M., 1977, On the theoretical analysis of phase measurements for microscopic radiation fields, *Phys. Lett.* A, **64**, 159.

Santhanam, T. S., 1977, Quantum mechanics in discrete space and angular momentum, *Fdn. Phys.*, **7**, 121.

Santhanam, T. S., 1977, Does Weyl's commutation relation imply generalized statistics? *Lett. Nuovo Cim.*, **20**, 13.

Biatynicki-Birula, I., 1977, Classical limit of quantum electrodynamics, *Acta Phys. Austriaca Suppl.*, **18**, 111.

Damaskinsky, E. V., and Yarunin, V. S., 1978, Hermitian phase operator and Heisenberg representation of the canonical commutation relation, *Acad. Res. J. (Tomsk Univ.): Phys.*, **6**, 59 (in Russian).

SANTHANAM, T. S., and SINHA, K. B., 1978, Quantum mechanics in finite dimensions, *Aust. J. Phys.*, **31,** 233.

1980

BIATYNICKI-BIRULA, I., 1989, Phase representations of intense photon beams and its applications, *Foundations of Radiation Theory and Quantum Electrodynamics*, edited by A. O. Barut (New York: Plenum).

BIATYNICKI-BIRULA, I., and VAN, C. L., 1980, Energy levels of dressed atoms and resonance phenomena, *Acta Phys. Polonica* A, **57,** 599.

SANTHANAM, T. S., 1980, Generalized coherent states, *Symmetries in Science*, edited by B. Gruber and R. S. Millman (New York: Plenum).

GOLDHIRSCH, I., 1980, Phase operator and phase fluctuations of spins, *J. Phys.* A, **13,** 3479.

PERES, A., 1980, Measurement of time by quantum clocks, *Am. J. Phys.*, **48,** 552.

NEWTON, R. G., 1980, Quantum action-angle variables for harmonic oscillators, *Ann. Phys. (N.Y.)*, **124,** 327.

MATTHYS, D. R., and JAYNES, E. T., 1980, Phase-sensitive optical amplifier, *J. opt. Soc. Am.*, **70,** 263.

1981

ROGOVIN, D., NAGEL, J., and SCULLY, M., 1981, Quantum theory of the dc Josephson effect—current biased eigenstates and excitation spectrum, *Phys. Rev.* B, **23,** 1156.

GHEORGHE, V. N., and COLLINS, C. B., 1981, Photon-number and angular-momentum asymptotics, *Phys. Rev.* A, **24,** 927.

GOTO, T., YAMAGUCHI, K., and SUDO, N., 1981, On the time operator in quantum mechanics—3 typical examples, *Prog. theor. Phys.*, Osaka, **66,** 1525.

1982

YAMADA, K., 1982, Angular-momentum angle commutation relations and minimum uncertainty states, *Phys. Rev.* D, **25,** 3256.

CARSTOIU, F., DUMITRESCU, O., and FONDA, L., 1982, Semi-classical description of rotational states with inclusion of nuclear interactions, *Nuovo Cim.* A, **70,** 38.

ZWERGER, W., 1982, Influence of dissipation on phase tunneling in Josephson junctions, *Z. Phys.* B, **47,** 129.

1983

PARTOVI, H., 1983, Entropic formulation of uncertainty for quantum measurements, *Phys. Rev. Lett.*, **50,** 1883.

KOBE, D. H., 1983, Ehrenfest theorem for angular displacement, *Am. J. Phys.*, **51,** 912.

LEACOCK, R. A., and PADGETT, M. J., 1983, Hamilton–Jacobi action-angle quantum mechanics, *Phys. Rev.* D, **28,** 2491.

HOLEVO, A. S., 1983, Bounds for generalized uncertainty of the shift parameter, *Lect. Notes Math.*, **1021,** 243.

1984

DEOLIVEIRA, C. R., and MALTA, C. P., 1984, Action-angle variables for the harmonic oscillator—ambiguity spin X duplication formula, *Ann. Phys. (N.Y.)*, **155,** 447.

VAGLICA, A., and VERTI, G., 1984, Coherent spin states and energy-phase minimum uncertainty relation, *Optics Commun.*, **51,** 239.

GALINDO, A., 1984, Phase and number, *Lett. math. Phys.*, **8,** 495.

BOUNDURANT, R. S., and SHAPIRO, J. H., 1984, Squeezed states in phase-sensing interferometers, *Phys. Rev.* D, **30,** 2548.

LOUDON, R., and SHEPHERD, T. J., 1984, Properties of the optical quantum amplifier, *Optica Acta*, **31,** 1243.

SCHWEBER, S. S., 1984, Some chapters for a history of quantum field theory: 1938–1952, *Relativity, Groups and Topology II*, edited by B. de Witt and R. Stora (Amsterdam: Elsevier).

SHAPIRO, J. H., and WAGNER, S. S., 1984, Phase and amplitude uncertainties in heterodyne detection, *IEEE J. quant. Electron.*, **20,** 803.

Pedrotti, L. M., Sanders, V., and Scully, M. O., 1984, On the number-phase uncertainty relationship for a laser, Proceedings of the SPIE, Vol. 487 (Bellingham, Washington: SPIE), p. 39.

Walker, N. G., and Carroll, J. E., 1984, Simultaneous phase and amplitude measurements on optical signals using a multiport junction, *Electron. Lett.*, **20,** 981.

1985

Krinsky, S., 1985, Action-angle variables and fluctuation-dissipation relations for a driven quantum oscillator, *Phys. Rev.* A, **31,** 1267.

Breitenberger, E., 1985, Uncertainty measures and uncertainty relations for angle observables, *Fdn. Phys.,* **15,** 353.

Dawson, K. A., 1985, Application of action-angle variables to the calculation of the density matrix, *J. chem. Phys.,* **82,** 3222.

Biatynicki-Birula, I., 1985, Entropic uncertainty relations in quantum mechanics, *Lect. Notes in Math.,* **1136,** 90.

Stenholm, S., 1985, On the classical limit of quantum electrodynamics, *Ann. Phys. (Paris),* **10,** 817.

Imoto, N., Haus, H. A., and Yamamoto, Y., 1985, Quantum nondemolition measurement of the photon number via the optical Kerr effect, *Phys. Rev.* A, **32,** 2287.

1986

Sanders, B. C., Barnett, S. M., and Knight, P. L., 1986, Phase variables and squeezed states, *Optics Commun.*, **58,** 290.

Lynch, R., 1986, Phase operators for the electromagnetic field by squeezed-state theory, *J. opt. Soc. Am.* B, **3,** 1006.

Mandel, L., 1986, Non-classical states of the electromagnetic field, *Physica scripta* T, **12,** 34.

Kitagawa, M., and Yamamoto, Y., 1986, Number-phase minimum uncertainty state with reduced number uncertainty in a Kerr nonlinear interferometer, *Phys. Rev.* A, **34,** 3974.

Yamamoto, Y., Machida, S., and Nilsson, O., 1986, Amplitude squeezing in a pump noise suppressed laser oscillator, *Phys. Rev.* A, **34,** 4025.

Yamamoto, Y., and Haus, H. A., 1986, Preparation, measurement and information capacity of optical quantum states, *Rev. mod. Phys.*, **58,** 1001.

Walker, N. G., and Carroll, J. E., 1986, Multiport homodyne detection near the quantum noise limit, *Opt. quant. Electron.,* **18,** 355.

Barnett, S. M., and Pegg, D. T., 1986, Phase in quantum optics, *J. Phys.* A, **19,** 3849.

1987

Walker, N. G., 1987, Quantum theory of multiport optical homodyning, *J. mod. Optics*, **34,** 15.

Leacock, R. A., and Padgett, M. J., 1987, Quantum action-angle variable analysis of basic systems, *Am. J. Phys.*, **55,** 261.

Iwao, S., 1987, Set of supersymmetric oscillators, *Prog. theor. Phys., Osaka*, **77,** 798.

Yao, D. M., 1987, Phase properties of squeezed states of light, *Phys. Lett.* A, **122,** 77.

Santamura, M., Vaglica, A., and Vetri, G., 1987, Phase difference operators of 2 multi-spin systems, *Optics Commun.,* **62,** 317.

Kitagawa, M., Imoto, N., and Yamamoto, Y., 1987, Realization of number-phase minimum uncertainty states by quantum nondemolition measurement, *Phys. Rev.* A, **35,** 5270.

Grabowski, M., 1987, Entropic uncertainty relations for phase-number of quanta and time–energy, *Phys. Lett.* A, **124,** 19.

Leacock, R. A., 1987, The physical properties of linear and action-angle coordinates in classical and quantum mechanics, *Fdn. Phys.,* **17,** 799.

Helstrom, C. W., Charbit, M., and Bendjaballah, C., 1987, Cutoffs rate performance for phase-shift keying modulated coherent states, *Optics Commun.,* **64,** 253.

Lynch, R., 1987, Phase fluctuations in a squeezed state using measured phase operators, *J. opt. Soc. Am.* B, **10,** 1723.

Yamamoto, Y., Machida, S., Imoto, N., Kitagawa, M., and Bjork, G., 1987, Generation of number-phase minimum uncertainty states and number states, *J. opt. Soc. Am.* B, **4,** 1645.

1988

Fonda, L., Mankocborstnik, N., and Rosina, M., 1988, Coherent rotational states—their formulation and detection, *Phys. Rep.*, **158,** 160.

Dupertuis, M. A., 1988, Alternate orderings—a new tool for the study of phase and photon statistics, *Phys. Rev.* A, **37,** 4752.

Lynch, R., 1988, Phase fluctuations in the coherent light anharmonic oscillator system via measured phase operators, *Optics Commun.*, **67,** 67.

Bjork, G., and Yamamoto, Y., 1988, Generation of nonclassical photon states using correlated photon pairs and linear feedforward, *Phys. Rev.* A, **37,** 4229.

Pegg, D. T., and Barnett, S. M., 1988, Unitary phase operator in quantum mechanics, *Europhys. Lett.*, **6,** 483.

Agarwal, G. S., 1988, Nonclassical statistics of fields in pair coherent states, *J. opt. Soc. Am.* B, **5,** 1940.

Fan, H. Y., and Li, Y. P., 1988, Phase operators for atomic system in Schwinger boson representation and their properties, *Commun. theor. Phys.*, **9,** 341.

Watanabe, K., and Yamamoto, Y., 1988, Quantum correlation and state reduction of photon twins produced by a parametric amplifier, *Phys. Rev.* A, **38,** 3556.

Fan, H. Y., and Zaidi, H. R., 1988, An exact calculation of the expectation values of phase operators in squeezed states, *Optics Commun.*, **68,** 143.

Pimpale, A., and Razavy, M., 1988, Quantum mechanical phase space—a generalization of Wigner phase space formulation to arbitrary coordinate systems, *Phys. Rev.* A, **38,** 6046.

Saito, S., 1988, Minimum uncertainty in quantum mechanical cooperative phenomena, *Ann. Phys. (N.Y.)*, **187,** 249.

1989

Barnett, S. M., and Pegg, D. T., 1989, On the Hermitian optical phase operator, *J. mod. Optics*, **36,** 7.

Pegg, D. T., and Barnett, S. M., 1989, Phase properties of the quantized single-mode electromagnetic field, *Phys. Rev.* A, **39,** 1665.

Nath, R., and Kumar, P., 1989, Phase states and their statistical properties, *J. mod. Optics*, **36,** 1615.

Barnett, S. M., Stenholm, S., and Pegg, D. T., 1989, A new approach to optical phase diffusion, *Optics Commun.*, **73,** 314.

Gronbech-Jensen, N., Christiansen, P. L., and Ramanujam, P. S., 1989, Phase properties of squeezed states. *J. opt. Soc. Am.* B, **6,** 2423.

Vourdas, A., and Bishop, R. F., 1989, Phase-admixed states—coherence and incoherence, *Phys. Rev.* A, **39,** 214.

Ghosh, R., and Agarwal, G. S., 1989, Theory of two-photon squeezed laser-like oscillator, *Phys. Rev.* A, **39,** 1582.

Bergou, J., Orszag, M., Scully, M. O., and Wódkiewicz, K., 1989, Squeezing and quantum noise quenching in phase-sensitive optical system, *Phys. Rev.* A, **39,** 5136.

Schleich, W., Horowitz, R. J., and Varro, S., 1989, Bifurcation in the phase probability distribution of a highly squeezed state. *Phys. Rev.* A, **40,** 7405.

Grabowski, M., 1989, Spin phase, *Int. J. theor. Phys.*, **28,** 1215.

Shapiro, J. H., Shepherd, S. R., and Wong, N. C., 1989, Ultimate quantum limits on phase measurement, *Phys. Rev. Lett.*, **62,** 2377.

Shapiro, J. H., Shepherd, S. R., and Wong, N. C., 1989, Ultimate quantum limits on phase measurement—erratum, *Phys. Rev. Lett.*, **63,** 2002.

Barnett, S. M., and Pegg, D. T., 1989, Phase in quantum optics, *Dynamics of Non-linear Optical Systems*, edited by L. Pesquera and F. J. Bermejo (Singapore: World Scientific), p. 93.

PEGG, D. T., BARNETT, S. M., and VACCARO, J. A., 1989, Phase in quantum electrodynamics, *Quantum Optics V*, edited by J. D. Harvey and D. F. Walls (Berlin: Springer), p. 122.

1990

RICHARDSON, W. H., and SHELBY, R. M., 1990, Nonclassical light from a semiconductor laser operating at 4K, *Phys. Rev. Lett.*, **64**, 400.

PHOENIX, S. J. D., and KNIGHT, P. L., 1990, Periodicity, phase and entropy of two-photon resonance, *J. opt. Soc. Am.* B, **7**, 116.

VACCARO, J. A., and PEGG, D. T., 1990, Physical number phase intelligent and minimum uncertainty states of light, *J. mod. Opt.*, **37**, 17.

BANDILLA, A., 1990, Two-port homodyning of a coherent signal and the essence of shot noise, *J. mod. Optics*, **37**, 317.

VOURDAS, A., 1990, $SU(2)$ and $SU(1, 1)$ phase states, *Phys. Rev.* A, **41**, 1653.

HRADIL, Z., 1990, Noise minimum uncertainty states and the squeezing and antibunching of light, *Phys. Rev.* A, **41**, 400.

LYNCH, R., 1990, Fluctuation of the Pegg–Barnett phase operator in a coherent state, *Phys. Rev.* A, **41**, 2841.

SLUSHER, R. E., and YURKE, B., 1990, Squeezed light for coherent communications, *J. Lightwave Technol.*, **8**, 466.

CHAICHAN, M., and ELINAS, D., 1990, On the polar decomposition of the quantum $SU(2)$ algebra, *J. Phys.* A, **23**, L291.

GERRY, C. C., 1990, Phase fluctuation of coherent light in an anharmonic oscillator using the Hermitian phase operator, *Optics Commun.*, **75**, 168.

NATH, R., and KUMAR, P., 1990, Higher order coherence functions of phase states, *Optics Commun.*, **76**, 51.

BARNETT, S. M., and PEGG, D. T., 1990, Quantum theory of rotation angles, *Phys. Rev.* A, **41**, 3427.

VACCARO, J. A., and PEGG, D. T., 1990, Wigner function for number and phase, *Phys. Rev.* A, **41**, 5156.

GILLNER, L., BJORK, G., and YAMAMOTO, Y., 1990, Quantum noise properties of an injection-locked laser oscillator with pump-noise suppression and squeezed injection, *Phys. Rev.* A, **41**, 5053.

POSTELL, V., and UZER, T., 1990, Quantization of the asymmetric top using action-angle variables, *Phys. Rev.* A, **41**, 4035.

HRADIL, Z., 1990, Phase states in quantum optics and number phase minimum uncertainty, *Phys. Lett.* A, **146**, 1.

GERRY, C. C., and URBANSKI, K. E., 1990, Hermitian phase-difference operator analysis of microscopic radiation field measurements, *Phys. Rev.* A, **42**, 662.

SUMMY, G. S., and PEGG, D. T., 1990, Phase optimized quantum states of light, *Optics Commun.*, **77**, 75.

CIRAC, J. I., and SANCHEZ-SOTO, L. L., 1990, Population trapping in the Jaynes–Cummings model via phase coupling, *Phys. Rev.* A, **42**, 2851.

DALTON, B. J., and KNIGHT, P. L., 1990, Theory of photon counting for phase-dependent three-level detector systems, *Phys. Rev.* A, **42**, 3034.

LAKSHMI, P. A., and SWAIN, S., 1990, Phase in the correlated emission laser, *Phys. Rev.* A, **42**, 5632.

LU, N., 1990, Phase-diffusion coefficients and phase diffusion-rate, *Phys. Rev.* A, **42**, 5641.

LUKS, A., and PERINOVA, V., 1990, Compatibility of the cosine and sine phase operators, *Phys. Rev.* A, **42**, 5805.

YAMAMOTO, Y., MACHIDA, S., SAITO, S., IMOTO, N., YANAGAWA, T., KITAGAWA, M., and BJORK, G., 1990, Quantum mechanical limit in optical precision measurement and communication, *Prog. Optics*, **28**, 87.

PEGG, D. T., VACCARO, J. A., and BARNETT, S. M., 1990, Quantum optical phase and canonical conjugation, *J. mod. Optics*, **37**, 1703.

SAVAGE, C. M., 1990, Quantum optics with one atom in an optical cavity, *J. Mod. optics*, **37**, 1711.

BARNETT, S. M., and PEGG, D. T., 1990, Quantum theory optical phase correlations, *Phys. Rev.* A, **42**, 6713.

DUNG, H. T., TANAS, R., and SHUMOVKSY, A. S., 1990, Collapses, revivals and phase properties of the field in Jaynes–Cummings type models, *Optics Commun.*, **79**, 462.

ZHANG, X. L., 1990, Uncertainty relation for the generalized conjugate quantity pair, *Phys. Lett.* A, **150**, 219.

1991

EISELT, J., and RISKEN, H., 1991, Quasiprobability distributions for the Jaynes–Cummings model, *Phys. Rev.* A, **43**, 346.

ORSZAG, M., and SAAVEDRA, C., 1991, Phase fluctuations in a laser with atomic memory effects, *Phys. Rev.* A, **43**, 554.

ELLINAS, D., 1991, Phase operators via group contraction, *J. math. Phys.*, **32**, 135.

GANTSOG, T., and TANAS, R., 1991, Phase properties of the two-mode squeezed vacuum states, *Phys. Lett.* A, **152**, 251.

BAN, M., 1991, Number-phase quantization in ultra-small tunnel junctions, *Phys. Lett.* A, **152**, 223.

NATH, R., and KUMAR, P., 1991, Quasi-photon phase states, *J. mod. Optics*, **38**, 263.

BANDILLA, A., 1991, The broadening of the phase distribution due to linear amplification, *Optics Commun.*, **80**, 267.

XIN, M., and RHODES, W., 1991, Phase properties of the quantized single-mode electromagnetic field—comment, *Phys. Rev.* A, **43**, 2576.

PEGG, D. T., and BARNETT, S. M., 1991, Phase properties of the quantized single-mode electromagnetic field—reply, *Phys. Rev.* A, **43**, 2579.

ORSZAG, M., and SAAVEDRA, C., 1991, Phase difference fluctuations of the quantum beat laser, *Phys. Rev.* A, **43**, 2557.

CIBILS, M. B., CUCHE, Y., MARVULLE, V., and WRESZINSKI, W. F., 1991, Connection between Pegg–Barnett and Biarynicki–Birula phase operators, *Phys. Rev.* A, **43**, 4044.

SHAPIRO, J. H., and SHEPHERD, S. R., 1991, Quantum phase measurement—a system-theory perspective, *Phys. Rev.* A, **43**, 3795.

GANTSOG, T., TANAS, R., and ZAWODNY, R., 1991, Quantum phase fluctuations in parametric down conversion with quantum pump, *Optics Commun.*, **82**, 345.

GANTSOG, T., TANAS, R., and ZAWODNY, R., 1991, Quantum phase fluctuations in second harmonic generation, *Phys. Lett.* A, **155**, 1.

MEYSTRE, P., SLOSSER, J., and WILKENS, M., 1991, Cotangent states of the electromagnetic field—squeezing and phase difference, *Phys. Rev.* A, **43**, 4959.

KIM, C. H., and KUMAR, P., 1991, Tunable squeezed light generation from twin beams using an optical phase feedforward scheme, *Optics Lett.*, **16**, 755.

GANTSOG, T., and TANAS, R., 1991, Phase properties of pair coherent states, *Optics Commun.*, **82**, 145.

BAN, M., 1991, Phase operator and its eigenstates in Liouville space, *Phys. Lett.* A, **155**, 397.

MENG, H. X., and CHAI, C. L., 1991, Phase properties of coherent light in the Jaynes–Cummings model, *Phys. Lett.* A, **155**, 500.

GANTSOG, T., and TANAS, R., 1991, Phase properties of self-squeezed states generated by the anharmonic oscillator, *J. mod. Optics*, **38**, 1021.

BARNETT, S. M., and PHOENIX, S. J. D., 1991, Information theory, squeezing and quantum correlations, *Phys. Rev.* A, **44**, 535.

TANAS, R., GANTSOG, T., and ZAWODNY, R., 1991, Phase properties of second harmonics generated by different initial fields, *Optics Commun.*, **83**, 278.

NATH, R., and KUMAR, P., 1991, Phase properties of squeezed number states, *J. mod. Optics*, **38**, 1655.

GANTSOG, T., and TANAS, R., 1991, Phase properties of elliptically polarized light propagating in a Kerr medium, *J. mod. Optics*, **38**, 1537.

TANAS, R., GANTSOG, T., MIRANOWICZ, A., and KIELICH, S., 1991, Quasi-probability distribution $Q(\alpha,\alpha^*)$ versus phase distribution $P(\theta)$ in a description of superpositions of coherent states, *J. opt. Soc. Am.* B, 8, 1576.

HRADIL, Z., 1991, Extremal properties of near photon number eigenstate fields, *Phys. Rev.* A, **44**, 792.

GANTSOG, T., and TANAS, R., 1991, Phase properties of a damped anharmonic oscillator, *Phys. Rev.* A, **44**, 2086.

BERGOU, J., and ENGLERT, B. G., 1991, Operators of the phase—fundamentals, *Ann. Phys. (N.Y.)*, **209**, 479.

NOH, J. W., FOUGERES, A., and MANDEL, L., 1991, Measurement of the quantum phase by photon counting, *Phys. Rev. Lett.*, **67**, 1426.

GANTSOG, T., and TANAS, R., 1991, Phase properties of fractional coherent states, *Phys. Lett.* A, **157**, 330.

DUNG, H. T., TANAS, R., and SHUMOVSKY, A. S., 1991, Dynamic properties of the field phase in the Jaynes–Cummings Model, *J. mod. Optics*, **38**, 2069.

CIRAC, J. I., and SANCHEZ-SOTO, L. L., 1991, Population trapping in two-level models—spectral and statistical properties, *Phys. Rev.* A, **44**, 3317.

BAN, M., 1991, Relative number state representation and phase operator for physical systems, *J. math. Phys.*, **32**, 3077.

TANAS, R., 1991, Nonclassical effects in the model of anharmonic oscillator, *Optika Spektroskopya*, **70**, 637 (in Russian).

ELLINAS, D., 1991, Quantum phase angles and $su(\infty)$, *J. mod. Optics*, **38**, 2393.

GEA, BANACLOCHE, J., 1991, Atom state and field state evolution in the Jaynes–Cummings model for large initial fields, *Phys. Rev.* A, **44**, 5913.

ORSZAG, M., ROA, L., and RAMIREZ, R., 1991, Squeezing and bifurcation in the phase probability distribution in a two-photon micromaser, *Optics Commun.*, **86**, 147.

BAN, M., 1991, Phase variable and phase relaxation processes in Liouville space, *Physica* A, **179**, 103.

DOWLING, J. P., 1991, A quantum state with ultra-low phase noise, *Optics Commun.*, **86**, 119.

WILSON-GORDON, A. D., BUZEK, V., and KNIGHT, P. L., 1991, Statistical and phase properties of displaced Kerr states, *Phys. Rev.* A, **44**, 7647.

VOGEL, W., and SCHLEICH, W., 1991, Phase distribution of a quantum state without using phase states, *Phys. Rev.* A, **44**, 7642.

ADAM, P., JANSZKY, J., and VINOGRADOV, A. V., 1991, Amplitude squeezed and number-phase intelligent states via coherent state superposition, *Phys. Lett.* A, **160**, 506.

TANAS, R., and GANTSOG, T., 1991, Phase properties of elliptically polarized light propagating in a Kerr medium with dissipation. *J. opt. Soc. Am.* B, **8**, 2505.

LUKS, A., and PERINOVA, V., 1991, Extended number state basis and number-phase intelligent states of light field 1. Mapping and operator ordering approach to quantum phase problem, *Czech. J. Phys.*, **41**, 1205.

POPOV, V. N., 1991, Photon phase operator, *Theor. math. Phys.*, **89**, 1292.

BARANOV, L. Y., and LEVINE, R. D., 1991, On the complete sets of coherent and squeezed states, *Israel J. Chem.*, **31**, 403.

1992

NOH, J. W., FOUGERES, A., and MANDEL, L., 1992, Operational approach to the phase of a quantum field, *Phys. Rev.* A, **45**, 424.

MENG, H. X., CHAI, C. L., and ZHANG, Z. M., 1992, Phase dynamics for coherent light in the m-photon Jaynes–Cummings model, *Phys. Rev.* A, **45**, 2131.

VOURDAS, A., 1992, Analytic representations in the unit disk and applications to phase states and squeezing, *Phys. Rev.* A, **45**, 1943.

ELLINAS, D., 1992, Quantum phase and a q-deformed oscillator, *Phys. Rev.* A, **45**, 3358.

PENG, J. S., and LI, G. X., 1992, Phase fluctuations in the Jaynes–Cummings model without the rotating wave approximation, *Phys. Rev.* A, **45**, 3289.

THYLEN, L., GUSTAVSSON, M., KARLSSON, A., and GUSTAFSON, T. K., 1992, Phase noise in travelling-wave optical amplifiers, *J. opt. Soc. Am.* B, **9**, 369.

LOSS, D., and MULLEN, K., 1992, Commutation relations for periodic operators, *J. Phys.* A, **25**, L235.

TANAS, R., and GANTSOG, T., 1992, Quantum effects on the polarization of light propagating in a Kerr medium, *Optics Commun.*, **87**, 369.

LUKS, A., PERINOVA, V., and KREPELKA, 1992, Extended number state basis and number-phase intelligent states of light field 2. The reduced number-sine intelligent and Jackiw states, *Czech. J. Phys.*, **42**, 59.

DUNG, H. T., HUYEN, N. D., and SHUMOVSKY, A. S., 1992, Phase properties of a coherent field interacting with two two-level atoms, *Physica* A, **182**, 467.

VACCARO, J. A., BARNETT, S. M., and PEGG, D. T., 1992, Phase fluctuations and squeezing, *J. mod. Optics*, **39**, 603.

LUKS, A., and PERINOVA, V., 1992, Number-phase uncertainty products and minimizing states, *Phys. Rev.* A, **45**, 6710.

SCHLEICH, W., BANDILLA, A., and PAUL, H., 1992, Phase from Q-function via linear amplification, *Phys. Rev.* A, **45**, 6652.

BRUNE, M., HAROCHE, S., RAIMOND, J. M., DAVIDOVICH, L., and ZAGURY, N., 1992, Manipulation of photons in a cavity by dispersive atom–field coupling—quantum nondemolition measurements and generation of Schrödinger cat states, *Phys. Rev.* A, **45**, 5193.

TANAS, R., and GANTSOG, T., 1992, Phase properties of fields generated in a multiphoton down-converter, *Phys. Rev.* A, **45**, 5031.

GEA BANACLOCHE, J., 1992, A new look at the Jaynes–Cummings model for large fields—Bloch sphere evolution and detuning effects, *Optics Commun.*, **88**, 531.

CHIU, S. H., GRAY, R. W., and NELSON, C. A., 1992, The *q*-analog quantized radiation field and its uncertainty relations, *Phys. Lett.* A, **164**, 237.

AGARWAL, G. S., CHATURVEDI, S., TARA, K., and SRINIVASAN, V., 1992, Classical phase changes in nonlinear processes and their quantum counterparts, *Phys. Rev.* A, **45**, 4904.

BUZEK, V., WILSON-GORDON, A. D., KNIGHT, P. L., and LAI, W. K., 1992, Coherent states in a finite dimensional basis—their phase properties and relationship to coherent states of light, *Phys. Rev.* A, **45**, 8079.

SANDERS, B. C., 1992, Superpositions of distinct phase states by a nonlinear evolution, *Phys. Rev.* A, **45**, 7746.

BANDILLA, A., 1992, Strong linear amplification of quantum fields, *Annln Phys.*, **1**, 117.

ABE, S., 1992, Information entropic uncertainty in the measurements of photon number and phase in optical states, *Phys. Lett.*, A, **166**, 163.

HU, Z. F., 1992, Fluctuation of phase in the displaced number state, *J. mod. Optics*, **39**, 1381.

BAKASOV, A. A., and DENARDO, G., 1992, Quantum corrections to semiclassical laser theory, *Phys. Lett.* A, **167**, 37.

BAN, M., 1992, Phase operator formalism for a two-mode photon system, *J. opt. Soc. Am.* B, **9**, 1189.

BRECHA, R. J., PETERS, A., WAGNER, C., and WALTHER, H., 1992, Micromaser and separated oscillatory fields measurements, *Phys. Rev.* A, **46**, 567.

GUO, G. C., and CHAI, J. H., 1992, Generation of a photon number squeezed state by an optically pumped three-level atom system, *Chin. Phys.*, **12**, 264.

TSUI, Y. K., and REID, M. F., 1992, Unitary and Hermitian phase operators for the electromagnetic field, *Phys. Rev.* A, **46**, 549.

GALETTI, D., and PIZA, A. F. R. T., 1992, Discrete quantum phase spaces and the mod-*n* invariance, *Physica* A, **186**, 513.

LUKS, A., PERINOVA, V., and KREPELKA, J., 1992, Special states of the plane rotator relevant to the light field, *Phys. Rev.* A, **46**, 489.

SMITH, T. B., DUBIN, D. A., and HENNINGS, M. A., 1992, The Weyl quantization of phase angle, *J. mod. Optics*, **39**, 1603.

PENG, J. S., LI, G. X., and ZHOU, P., 1992, Phase properties and atomic coherent trapping in the system of a three-level atom interacting with a bimodal field, *Phys. Rev.* A, **46**, 1516.

RITZE, H. H., 1992, A proposal for the measurement of extremely small phase fluctuations, *Optics Commun.*, **92**, 127.

DROBNY, G., and JEX, I., 1992, Phase properties of field modes in the process of kth harmonic generation, *Phys. Lett.* A, **169**, 273.

POPOV, V. N., and YARUNIN, V. S., 1992, Quantum and quasi-classical states of the photon phase operator, *J. mod. Optics*, **39**, 1525.

NOH, J. W., FOUGERES, A., and MANDEL, L., 1992, Further investigations of the operationally defined quantum phase, *Phys. Rev.* A, **46**, 2840.

GANTSOG, T., MIRANOWICZ, A., and TANAS, R., 1992, Phase properties of real field states—the Garrison–Wong versus Pegg–Barnett predictions, *Phys. Rev.* A, **46**, 2870.

HRADIL, Z., 1992, Performance measures of quantum phase measurement, *Phys. Rev.* A, **46**, R2217.

BANDILLA, A., 1992, How to realize phase optimized quantum states, *Optics Commun.*, **94**, 273.

COHEN, D., BEN ARYEH, Y., and MANN, A., 1992, Phase variance of squeezed states, *Optics Commun.*, **94**, 227.

BAN, M., 1992, Phase state and phase distribution, *Optics Commun.*, **94**, 231.

BURAK, D., and WODKIEWICZ, K., 1992, Phase properties of quantum states of light, *Phys. Rev.* A, **46**, 2744.

FABRE, C., 1992, Squeezed states of light, *Phys. Rep.*, **219**, 215.

BARNETT, S. M., and PEGG, D. T., 1992, Limiting procedures for the optical phase operator, *J. mod. Optics*, **39**, 2121.

DUNG, H. T., and SHUMOVSKY, A. S., 1992, Quantum phase fluctuations in the Jaynes–Cummings model—effects of cavity dampling, *Phys. Lett.* A, **169**, 379.

BRAUNSTEIN, S. L., LANE, A. S., and CAVES, C. M., 1992, Maximum likelihood analysis of multiple quantum phase measurements, *Phys. Rev. Lett.*, **69**, 2153.

GHERI, K. M., WALLS, D. F., and MARTE, M. A., 1992, Systematic description of laser phase by linearized Ito equations, *Phys. Rev.* A, **46**, 6002.

GARRAWAY, B. M., and KNIGHT, P. L., 1992, Quantum phase distributions and quasidistributions, *Phys. Rev.* A, **46**, 5346.

WAGNER, C., BRECHA, R. J., SCHENZLE, A., and WALTHER, H., 1992, Phase diffusion and continuous quantum measurements in the micromaser, *Phys. Rev.* A, **46**, 5350.

GANTSOG, T., 1992, Collapses and revivals of phase fluctuations in parametric down-conversion with quantum pump, *Phys. Lett.* A, **170**, 249.

POPOV, V. N., and YARUNIN, V. S., 1992, Quantum states of the oscillator phase operator, *Soviet J. nucl. Phys.*, **55**, 1529.

BRAUNSTEIN, S. L., 1992, Quantum limits on precision measurement of phase, *Phys. Rev. Lett.*, **69**, 3598.

JEX, I., DROBNY, G., and MATSUOKA, M., 1992, Quantum phase properties of the *k*-photon down conversion with quantized pump, *Optics Commun.*, **94**, 619.

BAXTER, C., 1992, The Thomas–Reiche–Kuhn sum rule for an atom of *n* equally spaced energy levels, *J. Phys.* B, **25**, L589.

MOYSA-CESSA, H., and VIDIELLA-BARENCO, A., 1992, Interactions of squeezed light with two-level atoms. *J. mod. Optics*, **39**, 2481.

1993

TU, H. T., and GONG, C. D., 1993, Properties of the measured phase operators in the squeezed number states, *J. mod Optics*, **40**, 57.

FREYBERGER, M., and SCHLEICH, W., 1993, Photon counting, quantum phase and phase-space distributions, *Phys. Rev.* A, **47**, 30.

SMITHEY, D. T., BECK, M., RAYMER, M. G., and FARIDANI, A., 1993, Measurement of the Wigner distribution and the density matrix using optical homodyne tomography—application to squeezed states and the vacuum, *Phys. Rev. Lett.*, **70**, 1244.

JEX, I., and DROBNY, G., 1993, Phase properties and entanglement of the field modes in a two-mode coupler with intensity dependent coupling, *Phys. Rev.* A, **47**, 3251.

LAW, C. K., and EBERLY, J. H., 1993, Dynamics of a two-channel cavity QED model, *Phys. Rev.* A, **47**, 3195.

HRADIL, Z., 1993, Relation between ideal and feasible phase concepts, *Phys. Rev.* A, **47**, 2376.

MULLEN, K., LOSS, D., and STOOF, H. T. C., 1993, Resonant phenomena in compact and extended systems, *Phys. Rev.* B, **47**, 2689.

HONEGGER, R., 1993, On Heisenberg's uncertainty principle and the C. C. R., *Z. Phys. Nat.* A, **48**, 447.

LEONHARDT, U., and PAUL, H., 1993, Phase measurement and the *Q*-function, *Phys. Rev.* A, **47**, R2460.

LUIS, A., and SANCHEZ-SOTO, L. L., 1993, Alternative derivation of the Pegg–Barnett phase operator, *Phys. Rev.* A, **47,** 1492.

LYNCH, R., 1993, Comparison of the Vogel–Schleich phase theory with quantum phase measurements, *Phys. Rev.* A, **47,** 1576.

LANE, A. S., BRAUNSTEIN, S. L., and CAVES, C. M., 1993, Maximum likelihood statistics of multiple quantum phase measurements, *Phys. Rev.* A, **47,** 1667.

CHIZHOV, A. V., and MURZAKHMETOV, B. K., 1993, Photon statistics and phase properties of two-mode squeezed number states, *Phys. Lett.* A, **176,** 33.

BAN M., 1993, Phase operator in quantum optics, *Phys. Lett.* A, **176,** 47.

PEGG, D. T., 1993, Wave-function collapse in atomic physics, *Aust. J. Phys.*, **46,** 77.

FAN, A. F., 1993, Phase properties of a field in the Jaynes–Cummings model for nonresonant behaviour, *Optics Commun.*, **98,** 340.

WAGNER, C., BRECHA, R. J., SCHENZLE, A., and WALTHER, H., 1993, Phase diffusion, entangled states and quantum measurements in the micromaser, *Phys. Rev.* A, **47,** 5068.

TARA, K., AGARWAL, G. S., and CHATURVADI, S., 1993, Production of Schrödinger macroscopic superposition state in a Kerr medium, *Phys. Rev.* A, **47,** 5024.

COLLETT, M. J., 1993, Generations of number-phase squeezed states, *Phys. Rev. Lett.*, **70,** 3400.

VOGEL, W., and GRABOW, J., 1993, Statistics of difference events in homodyne detection, *Phys. rev.* A, **47,** 4227.

NOH, J. W., FOUGERES, A., and MANDEL, L., 1993, Operations approach to the phase of a quantum field—reply, *Phys. Rev.* A, **47,** 4535.

BARNETT, S. M., and PEGG, D. T., 1993, Phase measurements, *Phys. Rev.* A, **47,** 4537.

YU, Z. R., 1993, Q-photon phase operator and q-spin coherent state, *Phys. Lett.* A, **175,** 391.

BAN, M., 1993, Lie algebra methods in quantum optics—the Liouville space formulation, *Phys. Rev.* A, **47,** 5093.

TSUI, Y. K., 1993, Josephson tunneling between superconductors in the angle operator formalism, *Phys. Rev.* B, **47,** 12 296.

HALL, M. J. W., 1993, Phase resolutions and coherent phase states, *J. mod. Optics*, **40,** 809.

BECK, M., SMITHEY, D. T., COOPER, J., and RAYMER, M. G., 1993, Experimental determination of number–phase uncertainty relations, *Optics Lett.*, **18,** 1259.

QUANG, T., AGARWAL, G. S., BERGOU, J., SCULLY, M. O., WALTHER, H., VOGEL, K., and SCHLEICH, W. P., 1993, Calculation of the micromaser spectrum 1. Green function approach an approximate analytical techniques, *Phys. Rev.* A, **48,** 803.

IMAMOGLU, A., 1993, Quantum nondemolition measurements using dispersive atom–field coupling—Monte Carlo wave-function approach, *Phys. Rev.* A, **48,** 770.

LUIS, A., and SANCHEZ-SOTO, L. L., 1993, Canonical transformations to action and phase-angle variables and phase operators, *Phys. Rev.* A, **48,** 752.

AGARWAL, G. S., 1993, Eigenstates of the phase operator a la Dirac for a two mode field, *Optics Commun.*, **100,** 479.

HOLLAND, M. J., and BURNETT, K., 1993, Interferometric detection of optical phase shifts at the Heisenberg limit, *Phys. Rev. Lett.*, **71,** 1355.

YEOMAN, G., and BARNETT, S. M., 1993, Two-mode squeezed Gaussons, *J. mod. Optics*, **40,** 1497.

SHORE, B. W., and KNIGHT, P. L., 1993, The Jaynes–Cummings model, *J. mod. Optics*, **40,** 1195.

BECK, M., SMITHEY, D. T., and RAYMER, M. G., 1993, Experimental determination of quantum phase distributions using optical homodyne tomography, *Phys. Rev.* A, **48,** 890.

SCHIEVE, W. C., and MCGOWAN, R. R., 1993, Phase distribution in the micromaser, *Phys. Rev.* A, **48,** 2315.

DRUMMOND, P. D., SHELBY, R. M., FRIBERG, S., and YAMAMOTO, Y., 1993, Quantum solitons in optical fibres, *Nature*, **365,** 307.

FAN, H. Y., 1993, Generalization of Susskind–Glogower phase operators and inverse field operators to q-deformed case, *Commun. theor. Phys.*, **19,** 509.

DEB, B., GANGOPADHYAY, G., and RAY, D. S., 1993, Population trapping in a Raman-coupled model interacting with a two-mode quantized cavity field, *Phys. Rev.* A, **48,** 1400.

Nieto, M. M., 1993, Quantum phase and quantum phase operators: some physics and some history, *Physica scripta* T, **48**, 5.

Barnett, S. M., and Dalton, B. J., 1993, Conceptions of quantum optical phase, *Physica scripta* T, **48**, 13.

Vaccaro, J. A., and Pegg, D. T., 1993, Consistency of quantum descriptions of phase, *Physica scripta* T, **48**, 22.

Noh, J. W., Fougeres, A., and Mandel, L., 1993, Operational approach to phase operators based on classical optics, *Physica scripta* T, **48**, 29.

Smithey, D. T., Beck, M., Cooper, J., Raymer, M. G., and Faridani, A., 1993, Complete characterization of the quantum state of a light mode via the Wigner function and the density matrix: application to quantum phase distributions of vacuum and squeezed-vacuum states, *Physica scripta* T, **48**, 35.

Leonhardt, U., and Paul, H., 1993, Realistic measurement of phase, *Physica scripta* T, **48**, 45.

Bandilla, A., 1993, Strong local oscillator limit of the operational approach for quantum phase measurement, *Physica scripta* T, **48**, 49.

Tanas, R., Miranowicz, A., and Gantsog, T., 1993, Phase distributions of real field states, *Physica scripta* T, **48**, 53.

Herzog, U., Paul, H., and Richter, T., 1993, Wigner function for a phase state, *Physica scripta* T, **48**, 61.

Garraway, B. M., and Knight, P. L., 1993, Quantum superpositions, phase distributions and quasi-probabilities, *Physica scripta* T, **48**, 66.

Stenholm, S., 1993, Some formal properties of operator polar decomposition, *Physica scripta* T, **48**, 77.

Vourdas, A., 1993, Phase states: an analytic approach in the unit disc, *Physica scripta* T, **48**, 84.

Nienhuis, G., and van Enk, S. J., 1993, Spherical angle operators for quantum-mechanical angular momentum, *Physica scripta* T, **48**, 87.

Luks, A., and Perinova, V., 1993, Ordering of ladder operators, the Wigner function for number and phase, and the enlarged Hilbert space, *Physica scripta* T, **48**, 94.

Jones, K. R. W., 1993, Information theory and optimal phase measurement, *Physica scripta* T, **48**, 100.

Shapiro, J. H., 1993, Phase conjugate quantum communication with zero error probability at finite average photon number. *Physica scripta* T, **48**, 105.

Biatynicki-Birula, I., Freyberger, M., and Schleich, W., 1993, Various measures of quantum phase uncertainty: a comparative study, *Physica scripta* T, **48**, 113.

Daeubler, B., Miller, C., Risken, H., and Schoendorff, L., 1993, Quantum states with minimum phase uncertainty for the Süssmann measure, *Physica scripta* T, **48**, 119.

Collett, M. J., 1993, Phase noise in a squeezed state, *Physica scripta* T, **48**, 124.

Agarwal, G. S., Scully, M. O., and Walther, H., 1993, Phase narrowing a coherent state via repeated measures: only the no counts count, *Physica scripta* T, **48**, 128.

Buzek, V., Gantsog, T., and Kim, M. S., 1993, Phase properties of Schrödinger cat states of light decaying in phase-sensitive reservoirs, *Physica scripta* T, **48**, 131.

Drobny, G., and Jex, I., 1993, The system of N two-level atoms interacting with a field mode—entanglement and parametric approximation, *Optics Commun.*, **102**, 141.

Buzek, A., Kim, M. S., and Gantsog, T., 1993, Quantum phase distributions of amplified Schrödinger cat states, *Phys. Rev.* A, **48**, 3394.

Gou, S. C., 1993, Characteristic oscillations of phase properties for pair coherent states in the two-mode Jaynes–Cummings model dynamics, *Phys. Rev.* A, **48**, 3233.

Meng, H. X., Chai, C. L., and Zhang, Z. M., 1993, Phase and squeezing properties of two copropagating fields in a Kerr medium, *Phys. Rev.* A, **48**, 3219.

Smithey, D. T., Beck, M. Cooper, J., and Raymer, M. G., 1993, Measurement of number–phase uncertainty relations of optical fields, *Phys. Rev.* A, **48**, 3159.

Mizrahi, S. S., and Marchiolli, M. A., 1993, Pseudo-diffusion equation and information entropy of squeezed coherent states, *Physica* A, **199**, 96.

Noh, J. W., Fougeres, A., and Mandel, L., 1993, Measurements of the probability distribution of the operationally defined quantum phase difference, *Phys. Rev. Lett.*, **71**, 2579.

LIU, C. Y., CHEN, Q. M., and LI, Z. G., 1993, Radiation phase quantization and its application to the transport through a quantum-dot turnstile, *Chin. Phys. Lett.*, **10**, 393.

BAN, M., 1993, Relative-state formulation of quantum systems, *Phys. Rev.* A, **48**, 3452.

KUANG, L. M., 1993, A *q*-deformed harmonic oscillator in a finite-dimensional Hilbert space, *J. Phys.* A, **26**, L1079.

KUANG, L. M., WANG, F. B., and ZHOU, Y. G., 1993, Dynamics of a harmonic oscillator in a finite-dimensional Hilbert space, *Phys. Lett.* A, **183**, 1.

PERINOVA, V., and KREPEKA, J., 1993, Free and dissipative evolution of squeezed and displaced number states in the third order nonlinear oscillator, *Phys. Rev.* A, **48**, 3881.

KULAGA, A. A., and KHALILI, F. Y., 1993, Quantum states with phase minimum uncertainty, *Zh. eksp. I teore. Fiz.*, **104**, 3358.

RAYMER, M. G., COOPER, J., and BECK, M., 1993, Many-port homodyne detection of an optical phase, *Phys. Rev.* A, **48**, 4617.

LUIS, A., and SANCHEZ-SOTO, L. L., 1993, Phase difference operator, *Phys. Rev.* A, **48**, 4702.

D'ARIANO, G. M., and PARIS, M. G. A., 1993, Necessity of sine–cosine joint measurement, *Phys. Rev.* A, **48**, R4039.

1994

FOUGERES, A., NOH, J. W., GRAYSON, T. P., and MANDEL, L., 1994, Measurement of phase differences between two partially coherent fields, *Phys. Rev.* A, **49**, 530.

DROBNY, G., GANTSOG, T., and JEX, I., 1994, Phase properties of a field mode interacting with *n* two-level atoms, *Phys. Rev.* A, **49**, 622.

GARRAWAY, B. M., SHERMAN, B., MOYA-CESSA, H., KNIGHT, P. L., and KURIZKI, G., 1994, Generation and detection of nonclassical field states by conditional measurements following two-photon resonant interactions, *Phys. Rev.* A, **49**, 535.

SANCHEZ-SOTO, L. L., and LUIS, A., 1994, Quantum Stokes parameters and phase difference operator, *Optics Commun.*, **105**, 84.

HALL, M. J. W., 1994, Noise-dependent uncertainty relations for the harmonic oscillator, *Phys. Rev.* A, **49**, 42.

FAN, A. F., and WANG, Z. W., 1994, Phase, coherence properties and the numerical analysis of the field in the nonresonant Jaynes–Cummings model, *Phys. Rev.* A, **49**, 1509.

VACCARO, J. A., and PEGG, D. T., 1994, On measuring extremely small phase fluctuations, *Optics Commun.*, **105**, 335.

RIEGLER, P., and WÓDKIEWICZ, K., 1994, Phase space representation of operational phase operators, *Phys. Rev.* A, **49**, 1387.

FREUBERGER, M., and HERKOMMER, A. M., 1994, Probing a quantum state via atomic deflection, *Phys. Rev. Lett.*, **72**, 1952.

GERRY, C. C., 1994, Statistical properties of squeezed Kerr states, *Phys. Rev.* A, **49**, 2033.

KUANG, L. M., and CHEN, X., 1994, Coherent state formalism for the Pegg–Barnett Hermitian phase theory, *Phys. Lett.* A, **186**, 8.

D'ARIANO, G. M., and PARIS, M. G. A., 1994, Lower bounds on phase sensitivity in ideal and feasible measurements, *Phys. Rev.* A, **49**, 3022.

DAS, H. K., 1994, Phase distribution in the Yurke–Stoler states, *Physica scripta*, **49**, 606.

AGARWAL, G. S., GRAF, M., ORSZAG, M., SCULLY, M. O., and WALTHER, H., 1994, State preparation via quantum coherence and continuous measurement, *Phys. rev.* A, **49**, 4077.

KIS, Z., ADAM, P., and JANSZKY, J., 1994, Properties of states generated by excitations on the amplitude squeezed state, *Phys. Lett.* A, **188**, 16.

LIU, R. C., and YAMAMOTO, Y., 1994, Suppression of quantum partition noise in mesoscopic electron branching circuits, *Phys. Rev.* B, **49**, 10 520.

BIZARRO, J. P., 1994, Weyl–Wigner formalism for rotation angle and angular momentum variables in quantum mechanics, *Phys. Rev.* A, **49**, 3255.

Franson, J. D., 1994, Nonlocal measurements of the wave-function by quantum phase measurements, *Phys. Rev.* A, **49,** 3221.

Gangopadhyay, G., 1994, Coherent phase state and displaced phase state in a finite dimensional basis and their light field limits, *J. mod. Optics,* **41,** 525.

Khan, M. A., and Chaudry, M. A., 1994, Applicability of Vogel–Schleich phase theory to quantum phase measurements, *Optik,* **96,** 49.

Kryuchkyan, G. Y., and Kheruntsyan, K. V., 1994, Phase-dependent correlations and intense squeezed light spectra in above the threshold mode of four wave mixing, *Zh. eksp. theor. Fiz.,* **105,** 1161.

Herzog, U., and Richter, T., 1994, Phase destruction and quantum non-demolition measurement of the photon number by atomic beam deflection. *J. mod. Optics,* **41,** 553.

Vaccaro, J. A., and Pegg, D. T., 1994, Phase properties of optical linear amplifiers, *Phys. Rev.* A, **49,** 4985.

Freyberger, M., and Schleich, W., 1994, Phase uncertainties of a squeezed state, *Phys. Rev.* A, **49,** 5056.

Chumakov, S. M., Klimov, A. B., and Sanchez-Mondragon, J. J., 1994, General properties of quantum optical systems in a strong field limit, *Phys. Rev.* A, **49,** 4972.

Bandilla, A., and Ritze, H. H., 1994, Very small optimized phase uncertainties of two-photon coherent states, *Phys. Rev.* A, **49,** 4912.

Milburn, G. J., Chen, W. Y., and Jones, K. R., 1994, Hyperbolic phase and squeeze-parameter estimation, *Phys. Rev.* A, **50,** 801.

Luks, A., Perinova, V., and Krepelka, J., 1994, Rotation angle, phases of oscillators with definite circular polarizations and the composite ideal phase operator, *Phys. Rev.* A, **50,** 818.

Janszky, J., Vinogradov, A. V., Walmsley, I. A., and Mostowski, J., 1994, Competition between geometrical and dynamical squeezing during a Franck–Condon transition, *Phys. Rev.* A, **50,** 732.

Vaccaro, J. A., and Pegg, D. T., 1994, Nondiffusive phase dynamics from linear amplifiers and attenuators in the weak field regime, *J. mod. Optics,* **41,** 1079.

Lyra, M. L., and Gouveia Neto, A. S., 1994, Evolution of coherent states in a dispersionless fibre with saturable nonlinearity and the generation of macroscopic quantum superposition states, *J. mod. Optics,* **41,** 1361.

Kuang, L. M., Wang, F. B., and Zhou, Y. G., 1994, Coherent states of a harmonic oscillator in a finite-dimensional Hilbert space and their squeezing properties, *J. mod. Optics,* **41,** 1307.

Paul, H., and Leonhardt, U., 1994, Realistic measurement of phase, *Acta Phys. Polonica* A, **86,** 213.

Raymer, M. G., Smithey, D. T., Beck, M., and Cooper, J., 1994, Quantum states and number-phase uncertainty relations measured by optical homodyne tomography, *Acta Phys. Polonica* A, **86,** 71.

Meng, H. X., Guo, A. Q., and Xing, C. Z., 1994, Dependence of phase dynamics of light interacting with two-level atoms on the initial state of the atoms, *Phys. Lett.* A, **190,** 455.

Mendas, I., and Popovic, D. B., 1994, Number-phase uncertainty product for displaced number states, *Phys. Rev.* A, **50,** 947.

Zhu, J. Y., and Kuang, L. M., 1994, Squeezing properties of even coherent states in a finite-dimensional Hilbert space, *Chin. Phys. Lett.,* **11,** 424.

Andreev, V. A., Um, C. I., and Yeon, K. H., 1994, A photon cluster approach to multiphoton interactions, *J. Korean phys. Soc.,* **27,** 360.

Janszky, J., Vinogradov, A. V., Kobayashi, T., and Kis, Z., 1994, Vibrational Schrödinger cat states, *Phys. Rev.* A, **50,** 1777.

Baxter, C., 1994, Counterrotating terms and the rotating wave approximation in minimally coupled multilevel systems, *Ann. Phys. (N.Y.),* **234,** 404.

Ban, M., 1994, Unitary equivalence between ideal and feasible phases, *Phys. Rev.* A, **50,** 2785.

Agarwal, G. S., and Ghosh, R., 1994, Two-photon squeezed laser with long-lived atoms, *Phys. Rev.* A, **50,** 1950.

Brif, C., and Ben Aryeh, Y., 1994, Antinormal ordering of Susskind–Glogower quantum phase operators, *Phys. Rev.* A, **50,** 2727.

Zhu, J. Y., and Kuang, L. M., 1994, Even and odd coherent states of a harmonic oscillator in a finite-dimensional Hilbert space and their squeezing properties, *Phys. Lett.* A, **193,** 227.

Elion, W. J., Matters, M., Geigenmuller, U., and Mooij, J. E., 1994, Direct demonstration of Heisenberg's uncertainty principle in a superconductor, *Nature,* **371,** 594.

Miranowicz, A., Piatek, K., and Tanas, R., 1994, Coherent states in a finite-dimensional Hilbert space, *Phys. Rev.* A, **50,** 3423.

Brif, C., and Ben Aryeh, Y., 1994, Phase state representation in quantum optics, *Phys. Rev.* A, **50,** 3505.

Fougeres, A., Monken, C. H., and Mandel, L., 1994, Measurements of the probability distribution of the phase difference between two quantum fields, *Optics Lett.*, **19,** 1771.

Andreev, V. A., and Lerner, P. B., 1994, Photon clusters and supersymmetry of multiphoton transitions, *Nuovo Cim.* A, **107,** 1767.

Zagury, N., and Piza, A. F. R. D., 1994, Large correlation effects of small pertubations by preselection and postselection of states, *Phys. Rev.* A, **50,** 2908.

Schuēr, S., Freyberger, M., and Schleich, W., 1994, The birth of a phase cat, *J. mod. Optics,* **41,** 1765.

Basieia, B., Vyas, R., Danras, M. A., and Bagnato, V. S., 1994, Scattering of atoms by light-probing a quantum state and the variance of the phase operator, *Phys. Lett.* A, **194,** 153.

Gennaro, G., Leonardi, C., Lillo, F., Vaglica, A., and Vetri, G., 1994, Internal coherence and quantum phase difference between two EM fields, *Optics Commun.*, **112,** 67.

Hakioglu, T., Shumovsky, A. S., and Aytur, O., 1994, Operational approach to quantum limits on polarization, *Phys. Lett.* A, **194,** 304.

Opatrny, T., 1994, Mean value and uncertainty on optical phase—a simple mechanical analogy, *J. Phys.* A, **27,** 7201.

Kuang, L. M., and Chen, X., 1994, Phase coherent states and their squeezing properties, *Phys. Rev.* A, **50,** 4228.

Enderle, M., and Neumann, H., 1994, Embedding of the classical into the quantum description of photons, *Fdn. Phys.,* **24,** 1415.

Videilla-Barenco, A., and Roversi, J. A., 1994, Statistical and phase properties of the binomial states of the electromagnetic field, *Phys. Rev.* A, **50,** 5233.

Dubin, D. A., Hennings, M. A., and Smith, T. B., 1994, Quantization in polar coordinates and the phase operator, *Publ. Res. Inst. math. Sci.,* **30,** 479.

Biatynicka-Birula, Z., and Bialynicki-Birula, I., 1994, Reconstruction of the wave-function from the photon number and quantum phase distributions, *J. mod. Optics,* **41,** 2203.

Ban, M., 1994, Phase operator formalism in thermofield dynamics, *Physica* A, **212,** 327.

Perinova, V., and Luks, A., 1994, Quantum statistics of dissipative nonlinear oscillators, *Prog. Optics,* **33,** 129.

Pegg, D. T., Vaccaro, J. A., and Barnett, S. M., 1994, Quantum optical phase and its measurement, *Quantum Optics VI,* edited by D. F. Walls and J. D. Harvey (Berlin: Springer), p. 153.

1995

Busch, P., Grabowski, M., and Lahti, P. J., 1995, Who is afraid of POV measures—unified approach to quantum phase observables, *Ann. Phys. (N.Y.),* **237,** 1.

Leonhardt, U., Vaccaro, J. A., Bohmer, B., and Paul, H., 1995, Canonical and measured phase distributions, *Phys. Rev.* A, **51,** 84.

Pegg, D. T., and Vaccaro, J. A., 1995, Phase difference operator—comment, *Phys. Rev.* A, **51,** 859.

Luis, A., and Sanchez-Soto, L. L., 1995, Phase difference operator—reply, *Phys. Rev.* A, **51,** 861.

Vaccaro, J. A., 1995, New Wigner function for number and phase, *Optics Commun.*, **113**, 421.

Vaccaro, J. A., and Ben Aryeh, Y., 1995, Antinormal ordering of phase operators and the algebra of weak limits, *Optics Commun.*, **113**, 427.

Vaccaro, J. A., and Bonner, R. F., 1995, Pegg–Barnett operators of infinite rank, *Phys. Lett.* A, **198**, 167.

Leonhardt, U., and Paul, H., 1995, Measuring the quantum state of light, *Prog. quant. Electron.*, **19**, 89.

Leng, B., Du, S. D., and Gong, C. D., 1995, Phase properties of the displaced number state in a Kerr medium, *J. mod. Optics*, 42, 435.

Gerry, C. C., 1995, Two-mode squeezed pair coherent states. *J. mod. Optics,* **42**, 585.

Roy, A. K., and Mehta, C. L., 1995, Boson inverse operators and a new family of two-photon annihilation operators. *J. mod. Optics,* **42**, 707.

Perinova, V., Vrana, V., and Luks, A., Quantum statistics of displaced Kerr states, *Phys. Rev.* A, **51**, 2499.

Luis, A., Sanchez-Soto, L. L., and Tanas, R., 1995, Phase properties of light propagating in a Kerr medium—Stokes parameters versus Pegg–Barnett predictions, *Phys. Rev.* A, **51**, 1634.

Bandyopadhyay, A., and Rai, J., 1995, Uncertainties of Schwinger angular momentum operators for squeezed radiation in interferometers, *Phys. Rev.* A, **51**, 1597.

Ban, M., 1995, Quantum phase superoperator, *Phys. Rev.* A, **51**, 2469.

Wiseman, H. M., 1995, Using feedback to eliminate back-action in quantum measurements, *Phys. Rev.* A, **51**, 2459.

Nelson, C. A., and Fields, M. H., 1995, Number and phase uncertainties of the q-analog quantized field, *Phys. Rev.* A, **51**, 2410.

Hillery, M., Freyberger, M., and Schleich, W., 1995, Phase distributions and large amplitude states, *Phys. Rev. A*, **51**, 1792.

Ban, M., 1995, Quantum phase superoperator and the antinormal ordering of the Susskind–Glogower phase operators, *Phys. Lett.* A, **199**, 275.

Yang, Y. P., and Yu, Z. R., 1995, Q-phase operator in a finite-dimensional Hilbert space, *Mod. Phys. Lett.* A, **10**, 347.

Vaccaro, J. A., 1995, Phase operators on Hilbert space, *Phys. Rev.* A, **51**, 3309.

Fang, M. F., and Liu, H. E., 1995, Properties of entropy and phase of the field in the two-photon Jaynes–Cummings model with an added Kerr medium, *Phys. Lett.* A, **200**, 250.

Abe, S., 1995, The Pegg–Barnett phase operator formalism as a q-deformed theory—limiting procedure and weak deformation, *Phys. Lett.* A, **200**, 239.

Bagchi, B., and Roy, P. K., 1995, A new look at the harmonic oscillator problem in a finite-dimensional Hilbert space, *Phys. Lett.* A, **200**, 411.

Vaccaro, J. A., and Orlowski, A., 1995, Phase properties of Kerr media via variance and entropy as measures of uncertainty, *Phys. Rev.* A, **51**, 4172.

Gerry, C. C., and Grobe, R., 1995, Two-mode intelligent *s u* (1, 1) states, *Phys. Rev.* A, **51**, 4123.

Lynch, R., 1995, The quantum phase problem—a critical review, *Phys. Rep.*, **256**, 368.

Arancibiliabulnes, C. A., Moya-Cessa, H., and Sanchez-Mondragon, J. J., 1995, Purifying a thermal field in a lossless micromaser, *Phys. Rev.* A, **51**, 5032.

Braginsky, V. B., Khalili, F. Y., and Kulaga, A. A., 1995, Quantum non-demolition measurement of the phase, *Phys. Lett.* A, **202**, 1.

Klimov, A. B., and Chumakov, S. M., 1995, Semiclassical quantization of the evolution operator for a class of optical models, *Phys. Lett.* A, **202**, 145.

Shepard, 1995, A quantum theory of angle, *Ann. Phys. (N.Y.)*, **755**, 812.

Barone, V., and Penna, V., 1995, Group theory of the number-phase representation, *Mod. Phys. Lett.* B, **9**, 685.

Hu, X. A. M., 1995, Phase properties of a strong field interacting with many atoms with and without initial atomic coherences, *J. mod. optics*, **42**, 1505.

Abdel-Hafez, A. M., Obada, A. S. F., and Essawy, A. H., 1995, Phase properties of multiphoton Jaynes–Cummings model, *Annls. Phys. (Paris)*, **20**, 47.

Englert, B. G., Wodkiewicz, K., and Riegler, P., 1995, Intrinsic phase operator of the Noh–Fougeres–Mandel experiments, *Phys. Rev.* A, **52**, 1704.

Baseia, B., Delima, A. F., and Marques, G. C., 1995, Intermediate number-phase states of the quantized radiation field, *Phys. Lett.* A, **204**, 1.

Opatrny, T., Buzek, V., Bajer, J., and Drobny, G., 1995, Propensities in discrete phase spaces—Q-function of a state in a finite-dimensional Hilbert space, *Phys. Rev.* A, **52**, 2419.

Gonzalez, A. R., Vaccaro, J. A., and Barnett, S. M., 1995, Entropic uncertainty relations for canonically conjugate operators, *Phys. Lett.* A, **205**, 247.

Fujikawa, K., 1995, Phase operator for the photon field and an index theorem, *Phys. Rev.* A, **52**, 3299.

Vaccaro, J. A., and Barnett, S. M., 1995, Reconstructing the wave-function in quantum optics, *J. mod. Optics,* **42**, 2165.

Baseia, B., Marques, G. C., Chaba, A. N., Debrito, A. L., and Duarte, S. B., 1995, Jaynes–Cummings model with intensity-dependent interaction via the phase operators, *Mod. Phys. Lett.* B, **9**, 1199.

Dubin, D. A., Hennings, M. A., and Smith, T. B., 1995, Mathematical aspects of quantum phase, *Int. J. mod. Phys.* B, **9**, 2597.

Kar, T. K., and Bhaumik, D., 1995, The azimuthal angle operator for angular momentum and phase operators of oscillators, *Phys. Lett.* A, **207**, 243.

Vaccaro, J. A., 1995, Number-phase Wigner function on Fock space, *Phys. Rev.* A, **52**, 3474.

Fujikawa, K., Kwek, L. C., and Oh, C. H., 1995, Q-deformed oscillator algebra and an index theorem for the photon phase operator, *Mod. Phys. Lett.* A, **10**, 2543.

Mendas, I., and Popovic, D. B., 1995, Number–phase uncertainty product for generalized squeezed states arising from the Pegg–Barnett Hermitian phase operator formalism, *Phys. Rev.* A, **52**, 4356.

Opatrny, T., 1995, Number–phase uncertainty relations, *J. Phys.* A, **28**, 6961.

Hennings, M. A., Smith, T. B., and Dubin, D. A., 1995, Approximations to the quantum phase operator, *J. Phys.* A, **28**, 6809.

Hirayama, M., and Zhang, H. M., 1995, Phase operator associated with radiation field, *Prog. theor. Phys.*, Osaka, **94**, 989.

1996

Gantsog, T., and Tanas, R., 1996, Quantum phase properties of the field in a lossless micromaser cavity, *Phys. Rev.* A, **53**, 562.

Royer, A., 1996, Phase states and phase operators for the quantum harmonic oscillator, *Phys. Rev.* A, **53**, 70.

Luis, A., and Sanchez-Soto, L. L., 1996, Probability distributions for the phase difference, *Phys. Rev.* A, **53**, 495.

Bandilla, A., Drobny, G., and Jex, I., 1996, Nondegenerate parametric interactions and nonclassical effects, *Phys. Rev.* A, **53**, 507.

Nath, R., and Kumar, P., 1996, Phase properties of squeezed thermal states, *J. mod. Optics*, **43**, 7.

Perinova, V., Luks, A., Krepelka, J., Sibilia, C., and Bertolotti, M., 1996, Quantum phase properties of a cubically behaved second-order medium, *J. mod. Optics*, **43**, 13.

Barnett, S. M., and Pegg, D. T., 1996, Phase measurement by projection synthesis, *Phys. Rev. Lett.*, **76**, 4148.

Tanas, R., Miranowicz, A., and Gantsog, Ts., Quantum phase properties of nonlinear optical phenomena, Progress in Optics xxxv, edited by E. Wolf (Amsterdam: Elsevier).

3

Mathematical Elaborations

The development of the theory of quantum phase had been forestalled by the lack of a consistent and physically meaningful definition for the phase angle. However, once the Pegg-Barnett phase formalism was introduced this impasse was removed and the theoretical and experimental research into quantum phase began with vigour. From the outset there was a need to further develop the more mathematical aspects of the formalism. These included exploring a new Wigner representation of the state of a harmonic oscillator, the nature of the canonical conjugate relationship between $\hat{N}$ and $\hat{\phi}_\theta$, the implications of the special limiting procedure $s \to \infty$ and the consistency of the formalism with the traditional description of a harmonic oscillator. In this chapter we review the main results of the investigations into these issues.

We begin with a novel representation of the harmonic oscillator provided by the new formalism. In 1932 Wigner introduced a new kind of probability distribution, $W(q, p)$, in order to study quantum corrections to classical statistical mechanics (Wigner 1932, Hillery *et al.* 1984). For the sake of clarity we shall refer to $W(q, p)$ as the *position-momentum Wigner function*. It is given in the momentum basis $|p\rangle$ as

$$W(q, p) = \left\langle \frac{1}{\pi} \int_{-\infty}^{\infty} e^{-i2qy} |p + y\rangle\langle p - y| dy \right\rangle. \tag{3.1}$$

This function yields an alternate representation of quantum physics in the sense that the distribution gives a one-to-one representation of the state density operator from which any expectation value can be calculated. The function is not a probability distribution, as it is negative for a particular class of density operators, but rather it is one of a class of so-called *quasi-probability distributions* (Barnett and Radmore 1997). Nevertheless, the marginals of the function, i.e. the integrals $\int W(q, p)dp$ and $\int W(q, p)dq$, are the position and momentum distributions, respectively. This implies that the Wigner function is biased towards the canonically-conjugate position and momentum observables. The advent of the Pegg-Barnett phase formalism provided a new canonically-conjugate pair of observables and thus the possibility of a Wigner function for excitation number (or energy) and phase. Wigner functions have been explored previously for discrete variables (e.g. Bout 1974, Wootters 1987, Leonhardt 1995, and more recently Gibbons *et al.* 2004). Wootters' discrete Wigner function is defined on an N-dimensional Hilbert space and so provides a natural basis for constructing Wigner functions on the $(s + 1)$ dimensional space Ψ_s in the Pegg-Barnett formalism. Indeed, Vaccaro and Pegg (1990a **Paper 3.1**) introduced the *number-phase Wigner function* $W_{N,\phi}(n, \theta_m)$ for the photon number and phase observables based on Wootters' work. The new function is given in the phase state basis $\{|\theta_m\rangle : m = 0, 1 \ldots, s\}$ as

$$W_{N,\phi}(n, \theta_m) = \left\langle \frac{1}{s+1} \sum_{k=0}^{s} e^{-i2nk\Delta} |\theta_{m+k}\rangle\langle\theta_{m-k}| \right\rangle \tag{3.2}$$

where $\Delta = 2\pi/(s+1)$ and we have implicitly used the periodic property of the phase states, viz. $|\theta_m\rangle = |\theta_m \pm 2\pi\rangle = |\theta_{m\pm(s+1)}\rangle$. One can see a close formal relationship between Eqs. (3.1) and (3.2). The marginals of $W_{N,\phi}(n,\theta_m)$ are the photon number and phase probability distributions; i.e. $\sum_{m=0}^{s} W_{N,\phi}(n,\theta_m) = \langle n|\hat{\rho}|n\rangle$ and $\sum_{n=0}^{s} W_{N,\phi}(n,\theta_m) = \langle\theta_m|\hat{\rho}|\theta_m\rangle$, respectively, for state density operator $\hat{\rho}$. Thus, in contrast to the position-momentum Wigner function, $W_{N,\phi}(n,\theta_m)$ is biased towards the conjugate number and phase observables. Indeed, the $W_{N,\phi}(n,\theta_m)$ representation of the Fock state $|j\rangle$ is zero everywhere except for $n = j$, illustrating the characteristic properties of number states, i.e. well-defined photon number and random phase. One can compare this with the $W(q,p)$ representation of the same state which is proportional to $\exp(-r^2)L_j(2r^2)$ where $r^2 = q^2 + p^2$ and $L_n(\cdot)$ is the nth order Laguerre polynomial (Barnett and Radmore 1997). Moreover the $W_{N,\phi}(n,\theta_m)$ representation of the phase state $|\theta_j\rangle$ is zero everywhere except for $\theta_m = \theta_j$ which displays the well defined phase and random photon number nature of phase states. Further properties of this function are explored in Vaccaro and Pegg (1990a, **Paper 3.1**). One, perhaps unusual, property is that for physical states $W_{N,\phi}(n,\theta_m)$ first tends to decay with increasing n and then exhibits a kind of "revival" for $n > s/2$.

We saw in Chapter 2 that it is possible to construct a phase probability density $P_{\mathcal{P}}(\theta)$, as given in Eq. (2.29), on the conventionally used, infinitely-dimensional Hilbert space H_∞, where

$$H_\infty \equiv \left\{ \sum_{n=0}^{\infty} x_n|n\rangle \;:\; \hat{N}|n\rangle = n|n\rangle,\; \sum_{n=0}^{\infty} |x_n|^2 < \infty \right\}. \tag{3.3}$$

So it is perhaps not surprising to find that one can also construct a meaningful Wigner function for number and phase directly on H_∞. For example, Lukš and Peřinová (1993) defined the function $W_{n\varphi}(n,\varphi)$ over the set of values $n = 0, 1/2, 1, 3/2, \cdots$, that is, a set which includes half odd values. It is not difficult to show that $W_{n\varphi}(n,\varphi)$ is directly related to $W_{N,\phi}(n,\theta_m)$ above as follows:

$$W_{n\varphi}(n,\varphi) = \begin{cases} \frac{s+1}{2\pi} W_{N\phi}(n,\varphi) & \text{for } n = 0, 1, 2, \cdots \\ \frac{s+1}{2\pi} W_{N\phi}(n + \frac{s+1}{2}, \varphi) & \text{for } n = \frac{1}{2}, \frac{3}{2}, \frac{5}{2}, \cdots \end{cases} \tag{3.4}$$

for physical states in the $s \to \infty$ limit with $\theta_0 = 0$. The sequence of values of s are restricted to being odd only. The half odd values of n represent the "revival" in $W_{n\varphi}(n,\varphi)$ for $n > s/2$.

A quite different approach was taken in Vaccaro (1995a) and then further developed in Vaccaro (1995b, **Paper 3.6**) which avoids both the "revival" and the half-integer values of n. The guiding principle is to define a Wigner function $S_{N\phi}(n,\theta)$ on H_∞ which has all seven properties of Wigner's original function listed in Hillery *et al.* (1984) but in respect of the number and phase observables rather than position and momentum. While the seven properties uniquely define the position-momentum Wigner function, they are found to be insufficient to uniquely define $S_{N\phi}(n,\theta)$ and there remains some measure of choice. An additional property, which is responsible for the quantum interference fringes displayed by Wigner's function for Schrodinger cat states, is chosen to define $S_{N\phi}(n,\theta)$ uniquely as

$$S_{N\phi}(n,\theta) = \left\langle \frac{1}{2\pi} \int_{2\pi} e^{-i2n\phi}(1+e^{i\phi})|\theta+\phi\rangle\langle\theta-\phi|d\phi \right\rangle. \tag{3.5}$$

Alternate definitions are possible with different choices, and one example is given in footnote [53] of Vaccaro (1995b, **Paper 3.6**). For this reason $S_{N\phi}(n,\theta)$ is called the *special number-phase Wigner function*. It gives a graphical representation of the number and phase properties of states. Moreover it illustrates the Schrödinger cat nature for a coherent superposition state as interference fringes, where the states of the superposition have either distinct photon number or distinct phase values. Applications are explored in Vaccaro and Joshi (1998), Jie *et al.* (1998), Joshi *et al.* (1998), Joshi (2001) and elsewhere, and an extension of $S_{N\phi}(n, \theta)$ to the two-mode case is given by (Fan and Sun 2000).

We next look at how the concept of canonical conjugate observables is generalised by the new formalism. Prior to the new formalism the view held by many was that two observables, such as the position $\hat{x}$ and the momentum $\hat{p}_x$ along the x axis, are canonically conjugate if, and only if, their commutator is proportional to the identity

$$[\hat{x}, \hat{p}_x] = i\hbar\hat{\mathbb{1}}. \tag{3.6}$$

In contrast, recall from Eq. (2.36) that in the new formalism the commutator between the number and phase angle variables for physical states is (Pegg and Barnett 1989 **Paper 2.4**)

$$[\hat{N}, \hat{\phi}_\theta]_{\mathcal{P}} = i\left[\hat{\mathbb{1}} - 2\pi\,|\theta_0\rangle_\infty\langle\theta_0|\right] . \tag{3.7}$$

The observables $\hat{N}$ and $\hat{\phi}_\theta$ are *conjugate* in the sense (Bohr 1935a) that it is not possible to assign definite values to both. Indeed preparing a state with a definite value for $\hat{N}$ implies that $\hat{\phi}_\theta$ is completely uncertain, and vise versa. This property arises directly from the mutually-unbiased nature of the respective eigenbases: i.e. the modulus-squared overlap between the basis states $|\langle n|\theta_m\rangle|^2 = (s+1)^{-1}$ is independent of n and m for the sets of number states $\{|n\rangle : n = 0, 1, \cdots s\}$ and phase states $\{|\theta_m\rangle : m = 0, 1, \cdots s\}$. Moreover, it is not possible for the pointer variable of a measuring apparatus to simultaneously represent the values of two non-commuting observables as they do not have common eigenstates. Experiments to measure $\hat{N}$ and $\hat{\phi}_\theta$ are therefore mutually exclusive, and so following Bohr (1935b), we find $\hat{N}$ and $\hat{\phi}_\theta$ are also *complementary* observables.

The difference between the two forms of the commutators in Eqs. (3.6) and (3.7) suggests that the new formalism generalises the concept of canonical conjugation. It also raises the question of whether the new formalism can provide a unifying connection between these two canonical commutation relations. These issues were addressed in Pegg, Vaccaro and Barnett (1990, **Paper 3.2**) where it is shown how a single general method of defining conjugate observables gives rise to a family of canonical commutation relations, including Eqs. (3.6) and (3.7). The method begins with a general pair of observables defined on an $(s+1)$-dimensional Hilbert space and whose eigenbases are mutually unbiased. The mutually unbiased nature ensures that the pair of observables are *conjugate* (as definite values cannot be assigned simultaneously to both) and *complementary* (as measurements of their values require mutually exclusive experiments). However, it does not fix the commutation relation between the observables. The simplest, and hence *canonical*, way to fix the relation is to require each of the observables to generate a simple cyclic shift (i.e. translation) in the set of eigenstates of the other observable. This yields a generic canonical commutation relation between the observables. Finally one must specify both the step size between sequential eigenvalues as a function of s and the classes of physical states in the infinite-s limit. The resulting spectra of the observables are either discrete or continuous[1] depending on whether the step size has a non-zero or zero value in the infinite-s limit, respectively. This gives rise to two types of canonical conjugate observable pairs depending on the nature of the spectra of the pair of observables: either one is discrete and the other is continuous or both are continuous. (The case were both spectra are discrete does not occur in the infinite-s limit.) The first type corresponds to number and phase or equivalently to angular momentum and rotation angle, and the second to position and momentum. These results not only justify Eq. (3.7) as a valid canonical commutation relation but also *unify the concept of canonical conjugation across different kinds of observables.*

[1] Strictly speaking, the spectrum is *dense* in the case where the step size is zero in the infinite-s limit. However the difference between a dense spectrum and a continuous spectrum, obtained by its closure, is not discernable by physical means.

The requirement that phase is canonically conjugate to photon number was also used by Leonhardt *et al.* (1995, **Paper 3.5**) to investigate a family of phase probability distributions. The motivation of this work was to study phase properties and measurements without going beyond the traditionally used infinite-dimensional Hilbert space H_∞. The price paid for restricting the analysis to H_∞ is the lack of an observable representing phase. Nevertheless, provided one has a properly defined phase probability distribution, one can analyse phase properties despite this handicap. The advantage of this approach is that it allows phase properties to be examined within the conventional mathematical framework.

The objective of the approach is to derive a distribution $\Pr(\varphi)$ for a state density operator $\hat{\rho}$ where $\Pr(\varphi)d\varphi = \text{tr}[\hat{\rho} d\hat{\Pi}(\varphi)]$. Here, the set $\{d\hat{\Pi}(\varphi)\}$ represents a probability operator measure (POM) which, in general, could be mixed, i.e., its elements could be mixed states. Two conditions are imposed to ensure that $\Pr(\varphi)$ yields properties which are canonically conjugate to those of the photon number:

(A) a phase shifter shifts the phase distribution, and

(B) a number shifter does not shift the phase distribution.

These conditions are a *subset* of those that ensure a pair of observables are canonically conjugate, as reviewed above (Pegg, Vaccaro and Barnett 1990 **Paper 3.2**). The phase and number shifts are defined in the number state basis as follows: a phase shift of δ is generated by the operator $e^{i\hat{N}\delta}$ whereas a number shift is given by the mapping $|n\rangle \mapsto |n+1\rangle$. The resulting distribution $\Pr(\varphi)$ is found to be a convolution of a distribution $g(\phi)$ and the Pegg-Barnett phase distribution $P_{\mathcal{P}}(\theta)$ in Eq. (2.35), i.e.

$$\Pr(\varphi) = \int_{-\pi}^{\pi} \frac{d\phi}{2\pi} g(\phi) P_{\mathcal{P}}(\varphi - \phi) \,. \tag{3.8}$$

The convolution implies that the distribution $\Pr(\varphi)$ is a "noisy" representation of the phase properties. Thus it would correspond, for example, to the measured phase distribution in situations where some extra phase noise is present in a measurement apparatus. The distribution $\Pr(\varphi)$ is found to reduce to the canonical phase distribution $P_{\mathcal{P}}(\varphi)$ for the case of a pure state POM, i.e. in absence of extra noise. Finally the measured distributions arising in a number of experiments are found to be equivalent to noisy measurement of phase.

In Chapter 2 we compared the Pegg-Barnett phase formalism with Dirac's "delta function" distribution. Just as the introduction of the latter spurned the development of distribution theory in the 1950's, the introduction of the new phase formalism resulted in a number of studies which further elaborated its mathematical framework. These included investigations into the limiting procedure underlying the formalism, developing tools to assist its implementation as well as constructing faithful representations using different state spaces. We now review some of these studies.

The mathematical treatment of the quantum harmonic oscillator inevitably makes use of an unbounded excitation spectrum and this necessarily involves a limit of some sort. The paper (Barnett and Pegg 1992 **Paper 3.3**) discusses the various possibilities for taking the limit. The conventional treatment is to first build the infinite-dimensional Hilbert space H_∞ which essentially involves taking the limit of the state space[2] and then defining operators on H_∞. The shortcomings of this approach were reviewed in Chapter 1. Alternatively, if one

[2] The infinite-dimensional space H_∞ can be constructed from the finite-dimensional spaces Ψ_s in the following way (Böhm 1978). Consider the elements $|x\rangle_s = \sum_{n=0}^{s} x_n|n\rangle$ belonging to Ψ_s. The sequence $|x\rangle_0, |x\rangle_1, |x\rangle_2, \cdots$ converges to the limit point $|x\rangle = \sum_{n=0}^{\infty} x_n|n\rangle$ iff $\sum_{n=0}^{\infty} |x_n|^2 < \infty$. H_∞ is the closure of the state space $\Psi_\infty = \bigcup_{s=0}^{\infty} \Psi_s$, that is, it is the union of Ψ_∞ and the set of these limit points.

defines operators on finite dimensional *subspaces* of H_∞ the temptation is to take the limit points of those operators before calculating expectation values and moments, as Popov and Yarunin did (1973, 1992a, 1992b), but again similar problems arise as discussed in Chapter 2. It is only when one performs the analysis of the oscillator on the finite dimensional spaces Ψ_s and then takes limit of expectation values and moments that a consistent and meaningful description of quantum phase emerges.

Nonetheless, the phase formalism and the conventional approach do agree for many calculations. The extent of the consistency between the two approaches was established in Vaccaro and Pegg (1993 **Paper 3.4**). This paper draws attention to two different types of convergence for sequences of operators, strong and weak convergence: the sequence $\hat{Q}_0, \hat{Q}_1, \hat{Q}_2, \cdots$ converges strongly or weakly to $\hat{Q}_\infty$ on the Hilbert space h depending on whether $\langle g_n|g_n\rangle \to 0$ or $\langle f|\hat{Q}_n - \hat{Q}_\infty|g\rangle \to 0$, respectively, as $n \to \infty$ for all $|f\rangle, |g\rangle$ belonging to h and where $|g_n\rangle = (\hat{Q}_n - \hat{Q}_\infty)|f\rangle$ (Richtmyer 1978). We write this as $\hat{Q}_n \overset{s}{\to} \hat{Q}_\infty$ or $\hat{Q}_n \overset{w}{\to} \hat{Q}_\infty$, respectively, as $n \to \infty$. Note that if a sequence converges strongly, it also converges weakly to the same limit point, however the converse is not true, i.e. weak convergence does not imply strong convergence. Also, if two sequences converge strongly, i.e. $\hat{A}_n \overset{s}{\to} \hat{A}_\infty$ and $\hat{B}_n \overset{s}{\to} \hat{B}_\infty$ as $n \to \infty$, then the sequence of the products strongly converges to the product of the limits, namely $(\hat{A}_n\hat{B}_n) \overset{s}{\to} \hat{A}_\infty\hat{B}_\infty$ as $n \to \infty$. This means that the algebra of strongly converging sequences is preserved in the limit. For example, the H_∞-space annihilation and creation operators $\hat{a}_\infty$ and $\hat{a}_\infty^\dagger$ are the strong limits of the Ψ_s-space operators $\hat{a}$ and $\hat{a}^\dagger$ considered in Eqs. (2.17) and (2.18) in Chapter 2 for physical states $\mathcal{P}$, i.e. for the set of states $\mathcal{P} = \{|p\rangle : \langle p|\hat{N}^m|p\rangle < \infty \; \forall \; m\}$. The expectation value of the operator function $F(\hat{a}_\infty, \hat{a}_\infty^\dagger)$, representing a linear combination of powers of $\hat{a}_\infty$ and $\hat{a}_\infty^\dagger$ is therefore equivalent to that given in the Pegg-Barnett phase formalism by virtue of the strong limit, for the physical states $\mathcal{P}$. Hence those calculations in the conventional mathematical treatment that can be expressed in terms of operator functions of the form of $F(\hat{a}_\infty, \hat{a}_\infty^\dagger)$ have the same result as the corresponding calculations in the Pegg-Barnett phase formalism, for physical states. We conclude that the new formalism *encompasses* the conventional one in this regard.

However, for weakly converging sequences, $\hat{A}_n \overset{w}{\to} \hat{A}_\infty$ and $\hat{B}_n \overset{w}{\to} \hat{B}_\infty$ as $n \to \infty$, the sequence of their products is not necessarily equal to $\hat{A}_\infty\hat{B}_\infty$. Thus the algebra of the weakly converging sequences is not preserved in the limit, in general. In particular, the exponential phase operator $e^{i\hat{\phi}_\theta}$ in Eq. (2.4) converges only *weakly* to the Susskind-Glogower operator $\widehat{e^{i\phi}}$ in Eq. (1.14). The weak convergence of $e^{i\hat{\phi}_\theta}$ is manifest in Eq. (2.16). Whereas the products $(e^{i\hat{\phi}_\theta})(e^{-i\hat{\phi}_\theta})$ and $(e^{-i\hat{\phi}_\theta})(e^{i\hat{\phi}_\theta})$ both converge to the identity $\hat{1}_\infty$ on H_∞, we note that the product of the weak limits, $(\widehat{e^{i\phi}})(\widehat{e^{-i\phi}}) = \hat{1}_\infty$ and $(\widehat{e^{-i\phi}})(\widehat{e^{i\phi}}) = \hat{1}_\infty - |0\rangle\langle 0|$, are not both the identity. Thus the algebra of the phase operators is lost in the weak limit on H_∞. This also explains why Popov and Yarunin's (1973, 1992a, 1992b) phase angle operator $\hat{\phi}_\infty$ in Eqs. (2.13) and (2.14) does not obey the same algebra at the Pegg-Barnett phase angle operator $\hat{\phi}_\theta$ in Eq. (2.8), as the former is the weak limit of the latter. Again we find that the Pegg-Barnett formalism must be used in order to have a consistent and physically meaningful description of quantum phase. In other words the Pegg-Barnett phase formalism both *encompasses and extends the conventional approach based on* H_∞.

The consistency between Shapiro and Shepard's POM approach (1991) and the Pegg-Barnett phase formalism, which was briefly mentioned in Chapter 2, can also be understood in terms of weak limits. The expectation value of the function $F(\phi)$ of Shapiro and Shepard's *classical random phase variable* ϕ is given in their approach by

$$\langle F(\phi)\rangle = \int_{2\pi} P_{\mathcal{P}}(\theta)F(\theta)d\theta \tag{3.9}$$

where $P_{\mathcal{P}}(\theta)d\theta = \langle d\hat{\Pi}(\theta)\rangle$ and $\{d\hat{\Pi}(\theta)\}$ is the continuous POM in Eqs. (2.32) and (2.35). The value of $\langle F(\phi)\rangle$ is equivalent to the expectation value of the operator

$$\widehat{F(\phi)} = \int_{2\pi} F(\theta)d\hat{\Pi}(\theta). \tag{3.10}$$

However, $\widehat{F(\phi)}$ is *not* equal to $F(\hat{\phi})$, in general, where

$$\hat{\phi} = \int_{2\pi} \theta d\hat{\Pi}(\theta), \tag{3.11}$$

and so Eq. (3.10) does not give functions of a fixed phase operator on H_∞. Rather, we find that Eq. (3.10) generates the weak limit of the function of phase $F(\hat{\phi}_\theta)$ defined on Ψ_s (Vaccaro and Pegg 1993 **Paper 3.4**), that is

$$F(\hat{\phi}_\theta) \stackrel{w}{\rightarrow} \widehat{F(\phi)}, \tag{3.12}$$

where the limit is with respect to $s \to \infty$. We note that $\langle\hat{Q}_n\rangle \to \langle\hat{Q}_\infty\rangle$ for either $\hat{Q}_n \stackrel{s}{\rightarrow} \hat{Q}_\infty$ or $\hat{Q}_n \stackrel{w}{\rightarrow} \hat{Q}_\infty$ as $n \to \infty$, i.e. the limit has the same expectation value as the sequence of operators. Hence the agreement between the Shapiro-Shepard and the Pegg-Barnett calculations is due to the operators associated with the former being the weak limits of those of the latter. As the algebra is lost for the weak limits, there appears to be no fixed phase observable associated with the former. But, of course, the underlying phase observable is actually $\hat{\phi}_\theta$, defined in the Pegg-Barnett phase formalism and reduced to its weak limit on H_∞ in the Shapiro-Shepard approach.

Given that the weak limits give the correct expectation values, all one has to do in order to regain the analysis of the sequences is impose the algebra of the sequences on the weak limits. Indeed this is essentially the mechanism underlying the *anti-normal ordering* technique introduced by Lukš and Peřinová (1993) and developed extensively by Brif and Ben-Aryeh (1994) and Vaccaro and Ben-Aryeh (1995 **Paper 3.7**). The operator $\widehat{F(\phi)}$ in Eq. (3.10) can be written as a power series in terms of the Susskind-Glogower operators $\widehat{e^{i\phi}}$ and $\widehat{e^{-i\phi}}$ given in Eq. (1.14):

$$\widehat{F(\phi)} = \sum_{n=-\infty}^{\infty} f_n\widehat{e_n} \tag{3.13}$$

where $F(\cdot)$ has the Fourier expansion $F(\varphi) = \sum_{n=-\infty}^{\infty} f_n e^{in\varphi}$ and $\widehat{e_n}$ is defined by

$$\widehat{e_n} = \begin{cases} (\widehat{e^{i\phi}})^n & \text{for } n > 0 \\ \hat{1}_\infty & \text{for } n = 0 \\ (\widehat{e^{-i\phi}})^{|n|} & \text{for } n < 0. \end{cases} \tag{3.14}$$

Moreover, Brif and Ben-Aryeh (1994) show that if $F(\cdot) = G(\cdot)H(\cdot)$ then

$$\widehat{F(\phi)} = {}^*_*\widehat{G(\phi)}\widehat{H(\phi)}{}^*_* \tag{3.15}$$

where $\widehat{G(\phi)}$ and $\widehat{H(\phi)}$ are given by Eq. (3.10) with F replaced by G and H, respectively. The operator ${}^*_*\cdots{}^*_*$ on the right-hand side of Eq. (3.15) *antinormally orders* the Susskind-Glogower operators appearing in the expression "$\cdots$" by placing all powers of $\widehat{e^{i\phi}}$ to the right of all powers of $\widehat{e^{-i\phi}}$, leaving coefficients unchanged. That is,

$$ {}^*_*\widehat{G(\phi)}\widehat{H(\phi)}{}^*_* = {}^*_*\left[\sum_{n=-\infty}^{\infty} g_n\widehat{e_n}\right]\left[\sum_{m=-\infty}^{\infty} h_m\widehat{e_m}\right]{}^*_* \tag{3.16} $$

$$ = \sum_{n=0}^{\infty}\sum_{m=-\infty}^{\infty} g_n h_m \widehat{e_n}\widehat{e_m} + \sum_{n=-\infty}^{-1}\sum_{m=-\infty}^{\infty} g_n h_m \widehat{e_m}\widehat{e_n} \tag{3.17} $$

$$ = \sum_{n=-\infty}^{\infty}\sum_{m=-\infty}^{\infty} g_n h_m \widehat{e_{n+m}} \tag{3.18} $$

where we have used $\widehat{e_n}\widehat{e_m} = \widehat{e_{n+m}}$ for $n \geq 0$ and $\widehat{e_m}\widehat{e_n} = \widehat{e_{m+n}}$ for $n < 0$, and the Fourier expansions $G(\varphi) = \sum_{n=-\infty}^{\infty} g_n e^{in\varphi}$ and $H(\varphi) = \sum_{n=-\infty}^{\infty} h_n e^{in\varphi}$. Noting $f_n = \sum_{m=-\infty}^{\infty} g_{n-m}h_m$ shows that the last line is just $\widehat{F(\phi)}$ in accord with Eq. (3.15). Vaccaro and Ben-Aryeh (1995 **Paper 3.7**) also show rigourously that the weak limit itself generates antinormally-ordered limit points. Given that the expectation value of $\widehat{F(\phi)}$ is the expectation value of the function of phase $F(\hat{\phi}_\theta)$, this work shows that to obtain a consistent phase formalism one must antinormally order products of the operators given by Eq. (3.10), or equivalently, by the weak limit. Hence *the antinormal ordering operation essentially reimposes the algebra of the Pegg-Barnett phase operators on their weak limits.*

We have already noted in Chapter 2 that Dirac's "delta-function" $\delta(x)$ is actually a distribution. Distributions are generalised functions belonging to a vector space which includes, as a proper subspace, the Hilbert space of square integrable functions L_2 which, incidently, is isomorphic to H_∞. One may wonder if the phase observables in the Pegg-Barnett phase formalism are "generalised operators" acting on a vector space larger than H_∞. This question was addressed by Vaccaro and Bonner (1995 **Paper 3.8**). They showed that the Pegg-Barnett phase formalism essentially builds a linear vector space E, of which H_∞ is a proper subspace, on which act operators that are generalised versions of operators on H_∞. The elements of E are *infinite sequences of states* such as $|f\rangle = (|f\rangle_0, |f\rangle_1, |f\rangle_2, \ldots)$ where $|f\rangle_s$ belongs to Ψ_s, and the operators on E are *infinite sequences of operators* such as $\hat{A} = (\hat{A}_0, \hat{A}_1, \hat{A}_2, \ldots)$ where $\hat{A}_s$ operates on Ψ_s. Expectation values are defined by $\langle f|\hat{A}|f\rangle = \lim_{s\to\infty}({}_s\langle f|\hat{A}_s|f\rangle_s/{}_s\langle f|f\rangle_s)$. The conventional treatment based on H_∞ as well as the Pegg-Barnett phase formalism are faithfully represented in this system. The important point about this work is that it includes an infinite-rank phase angle operator which acts on the infinite-dimensional vector space E. In this sense it *establishes the phase angle observable as a bona fide infinite-rank operator*, albeit a generalised one, on par with other observables such as position and momentum.

Despite this success, the linear vector space E is not a Hilbert space as it is not equipped with an inner product. It would be far more attractive to find a Hilbert-space subspace of E on which the phase observable could act. This is essentially what was reported in Vaccaro (1995c **Paper 3.9**). Here an infinite-dimensional Hilbert space H_{sym} is constructed from the infinite set of spaces Ψ_s for $s = 0, 1, 2, \ldots$ The important properties of H_{sym} rest on the special relationship between the spaces Ψ_s for different values of s which we now briefly outline. We treat odd values of s only for brevity and we make use of two infinite sets of mutually orthonormal states $|0\rangle, |1\rangle, |2\rangle, \ldots$ and $|\widetilde{\infty}\rangle, |\widetilde{\infty}-1\rangle, |\widetilde{\infty}-2\rangle, \ldots$; the elements of one set are also orthogonal to the elements of the other. The reason for the use of the symbol "$\widetilde{\infty}$" will become apparent in the following. In effect the space Ψ_s is spanned by the sequence of $(s+1)$ orthogonal states $|0\rangle, |1\rangle, |2\rangle, \ldots, |\frac{1}{2}(s-1)\rangle, |\widetilde{\infty}-\frac{1}{2}(s-1)\rangle, \ldots, |\widetilde{\infty}-2\rangle, |\widetilde{\infty}-1\rangle, |\widetilde{\infty}\rangle$. Notice that as s increases, new states are added to the *middle* of the sequence, and not at the end, as implied in the standard Pegg-Barnett formalism. The Hilbert space H_{sym} is defined to be the closure of $\Psi_\infty = \bigcup_{s=0}^{\infty} \Psi_s$, that is, it is the union of Ψ_∞ and the set of limit points of all (strongly) convergent sequences on Ψ_∞. The result is that H_{sym} contains all the states defined on H_∞ such as $|h\rangle = \sum_{n=0}^{\infty} h_n|n\rangle$ as well as states of the form $|\mu\rangle = \sum_{n=0}^{\infty} h_n|\widetilde{\infty} - n\rangle$, where $\sum_{n=0}^{\infty} |h_n|^2 < \infty$. In this sense H_{sym} is symmetric. The definition of operators on Ψ_s is also special. For example, the number operator $\hat{N}_s$ on Ψ_s is given by

$$\hat{N}_s = \sum_{n=0}^{\frac{1}{2}(s-1)} n|n\rangle\langle n| + (s-n)|\widetilde{\infty} - n\rangle\langle\widetilde{\infty} - n| \tag{3.19}$$

and so $\hat{N}_s|\widetilde{\infty}\rangle = s|\widetilde{\infty}\rangle$. Thus, in the $s \to \infty$ limit the state $|\widetilde{\infty}\rangle$ in H_{sym} represents a state of *infinite* excitation; this underlies the use of the symbol "$\widetilde{\infty}$". This construction endows the Pegg-Barnett phase operators with *strong* convergence on H_{sym} and the corresponding limit points are ordinary operators on H_{sym} obeying the algebra of the Pegg-Barnett operators. Hence the Pegg-Barnett phase formalism is represented faithfully by infinite-rank operators acting on an infinite-dimensional Hilbert space. A related approach was also taken by Ozawa (1997) who used the theory of non-standard numbers to effectively deal with states such as $|\widetilde{\infty}\rangle$ in H_{sym}. The outcome is essentially the same. The approach described here has some similarity with others based on extended Hilbert spaces such as Fain (1967), Newton (1980), Barnett and Pegg (1986 **Paper 1.7**), and Stenholm (1993) as discussed in Chapter 1. However, these other approaches make use of states representing *negative* excitation number $|n\rangle$ for $n < 0$, whereas in the H_{sym} approach the states $|\widetilde{\infty} - n\rangle$ for $n \geq 0$ represent states of *infinite* excitation. Indeed, the construction of H_{sym} stems from a quite different physical interpretation of the extension of H_∞.

Wigner function for number and phase

John A. Vaccaro* and D. T. Pegg
Division of Science and Technology, Griffith University, Nathan, Brisbane 4111, Australia
(Received 30 November 1989)

Various quasiprobability distributions have been developed in the past using the Hilbert space of the single-mode light field. The development of a quasiprobability distribution associated with a phase operator has previously been impossible because of the absence of a unique Hermitian phase operator defined on the Hilbert space. Recently, however, Pegg and Barnett [Europhys. Lett. **6**, 483 (1988); Phys. Rev. A **39**, 1665 (1989)] and Barnett and Pegg [J. Mod. Optics **36**, 7 (1989)] introduced a new formalism that does allow the construction of a Hermitian phase operator and associated phase eigenstates. In this paper we develop a quasiprobability distribution associated with the number and phase operators of the single-mode light field in the new formalism. The new distribution, which we call the number-phase Wigner function, has properties analogous to the Wigner function. We also derive the number-phase Wigner representation of number states, phase states, general physical states, coherent states, and the squeezed vacuum. We find this new representation has features that are related to the number and phase properties of states. For example, the number-phase Wigner representation of a number state is nonzero only on a circle, while the representation of a phase state is only nonzero along a radial line.

I. INTRODUCTION

Various quasiprobability distributions (QPD's) have been defined using the Hilbert space of the single-mode light field. These functions display the statistical properties of field states and also give c-number formulations of the quantum dynamics of the field. Well-known examples of QPD's include the Q function[1] and the Wigner function.[3] These particular QPD's allow the statistical nature of the quadrature amplitude operators $\hat{X}$ and $\hat{Y}$ to be represented graphically. A case in point is Yuen's[3] Q representation of the ideal-squeezed states, which is a two-dimensional Gaussian function. This function gives a vivid picture of the "squeezing" of the uncertainty from one quadrature amplitude to the other that characterizes these states. The Q function and the Wigner function have also been used to illustrate the statistical nature of the phase of the field in two ways. Firstly, the so-called "measured-phase" operators[4] are proportional to the quadrature amplitude operators and so these QPD's display the measured-phase properties of states. Secondly, the polar angle of the Wigner function has been interpreted phenomenologically as the phase angle of large-amplitude fields.[5,6] However, both of these illustrations have the disadvantage that they are not based on a Hermitian phase operator corresponding to phase angle. This problem stems from the fact that no unique Hermitian phase operator has yet been found for the Hilbert space itself.[7-9]

Recently, Pegg and Barnett introduced a new quantum-mechanical formalism for describing the phase of a single-mode field.[7-9] This new formalism allows the construction of a Hermitian phase operator $\hat{\phi}_\theta$ with properties normally associated with a phase angle. The new phase formalism is based on a linear space Ψ spanned by the $(s+1)$ number states $|0\rangle, |1\rangle, \ldots, |s\rangle$. A complete description of the single-mode field involves an infinite set of number states and here this corresponds to the limit of infinite s. An essential feature of the new formalism is the method of taking this limit when calculating physical properties, such as expectation values. These properties are first calculated with s finite and only then is the limit of infinite s taken. The space Ψ is also spanned by the $(s+1)$ orthonormal phase states

$$|\theta_m\rangle \equiv (s+1)^{-1/2} \sum_{n=0}^{s} \exp(in\theta_m)|n\rangle , \qquad (1.1)$$

where

$$\theta_m \equiv \theta_0 + m\Delta$$

and

$$\Delta \equiv 2\pi/(s+1) ,$$

for $m = 0, 1, \ldots, s$. Here θ_0 is arbitrary and Δ is the step in phase between successive phase states in this basis. The Hermitian phase operator is defined as

$$\hat{\phi}_\theta \equiv \sum_{m=0}^{s} \theta_m |\theta_m\rangle\langle\theta_m| , \qquad (1.2)$$

and has eigenvalues in the interval $[\theta_0, \theta_0 + 2\pi]$, which is closed at the lower end. This operator is conjugate to the number operator $\hat{N}$ which is defined by

$$\hat{N} \equiv \sum_{n=0}^{s} n|n\rangle\langle n| .$$

We wish to develop a QPD that can display the field-phase properties associated with the Hermitian phase operator, as opposed to the quadrature-amplitude properties displayed by the usual QPD's. The question arises as to whether a QPD in the new formalism should be defined for infinite or finite s. Our approach is to define the QPD for the finite $(s+1)$-dimensional space Ψ. All calculations of physical properties can then be performed first with s finite before the infinite-s limit is taken. This approach is consistent with the Pegg-Barnett formalism, and it is doubtful whether a number-phase QPD based on an infinite-state space is possible.

Wootters[10] recently defined a QPD called the *discrete* Wigner function for systems whose state space is finite (e.g., spin systems). This function has properties analogous to the Wigner function defined on the infinite Hilbert space. It gives a representation of the statistical nature associated with an arbitrary pair of conjugate observables. In this paper we apply Wootters's discrete Wigner function to the single-mode field and construct a function associated with the number and phase operators in the new formalism. For convenience we call this function the *number-phase Wigner function.* We also examine the number-phase Wigner representation of various states with well-known properties to establish the way it displays their number and phase properties.

II. THE NUMBER-PHASE WIGNER FUNCTION

In this section we use Wootters's discrete Wigner function to define the number-phase Wigner function. The discrete Wigner function is defined on an N-dimensional space.[10] Although N can be arbitrary, the analysis of this function becomes particularly simple for the special cases where N is a prime number. In this paper, we take N to be prime. The discrete Wigner function is defined for N prime as

$$W_{n,m} \equiv N^{-1}\langle \hat{A}(n,m)\rangle \ , \qquad (2.1)$$

where

$$\langle k|\hat{U}\hat{A}(n,m)\hat{U}^{\dagger}|l\rangle \equiv \bar{\delta}_{2n,k+l}\exp[2\pi i m(k-l)/N] \qquad (2.2)$$

for $N>2$. Here $\hat{U}$ is an arbitrary unitary operator, and the states $|k\rangle$ for $k=0,1,\ldots,(N-1)$ form a complete orthonormal basis. In Eq. (2.2) n and m are integers and $\bar{\delta}_{i,j}$ is a *periodic* Kronecker δ which we define as

$$\bar{\delta}_{i,j} \equiv \bar{\delta}_{i,j+N} \equiv \bar{\delta}_{i+N,j}$$

for all values of i and j, and

$$\bar{\delta}_{i,j} \equiv \delta_{i,j}$$

for i and j in the range $[0,N-1]$. The overbar distinguishes $\bar{\delta}_{i,j}$ from the usual Kronecker $\delta_{i,j}$. Thus, for example, $\bar{\delta}_{N+3,2N-8}=\bar{\delta}_{3,N-8}=\delta_{3,N-8}=0$ for $N>11$.

For our case $N=s+1$. We choose $\hat{U}$ to be the phase-shift operator $\exp(-i\theta_0\hat{N})$ and define

$$\hat{A}_{N\phi}(n,m) \equiv \sum_{k=0}^{s}\sum_{l=0}^{s}\bar{\delta}_{2n,k+l}\exp[im(k-l)\Delta] \times \exp(i\theta_0\hat{N})|k\rangle\langle l|\exp(-i\theta_0\hat{N}) \ , \qquad (2.3)$$

where $\Delta=2\pi/(s+1)$. Using the result

$$\bar{\delta}_{2n,k+l}=(s+1)^{-1}\sum_{p=0}^{s}\exp[ip(k+l-2n)\Delta] \ , \qquad (2.4)$$

the expression $\theta_m=\theta_0+m\Delta$, and the definition of the phase states (1.1), we find

$$\hat{A}_{N\phi}(n,m)=(s+1)^{-1}\sum_{k=0}^{s}\sum_{l=0}^{s}\sum_{p=0}^{s}\exp(-ip2n\Delta)\exp(ik\theta_{m+p})|k\rangle\langle l|\exp(-il\theta_{m-p})$$
$$=\sum_{p=0}^{s}\exp(-ip2n\Delta)|\theta_{m+p}\rangle\langle\theta_{m-p}| \ . \qquad (2.5)$$

This operator is Hermitian: on substituting $p=s+1-q$ and using the periodic property of the phase states

$$|\theta_m\rangle=|\theta_m\pm 2\pi\rangle=|\theta_{m\pm(s+1)}\rangle \ ,$$

we obtain

$$\hat{A}_{N\phi}(n,m)=|\theta_m\rangle\langle\theta_m|+\sum_{q=1}^{s}\exp(i2nq\Delta)|\theta_{m-q}\rangle\langle\theta_{m+q}|$$
$$=\hat{A}_{N\phi}(n,m)^{\dagger} \ .$$

We define the number-phase Wigner function according to Eq. (2.1) by

$$W_{N\phi}(n,\theta_m) \equiv (s+1)^{-1}\langle\hat{A}_{N\phi}(n,m)\rangle \ , \qquad (2.6)$$

where the indices n,m range over $0,1,\ldots,s$. For our analysis we have taken N to be a prime number and here this corresponds to $(s+1)$ being an arbitrarily large prime number. To find physical results from $W_{N\phi}$ we let $(s+1)$ tend to infinity through the prime numbers.[11] We also note that $W_{N\phi}$ is real because $\hat{A}_{N\phi}$ is Hermitian. Equations (2.5) and (2.6) can be compared with Wigner's original function for the harmonic oscillator:

$$W(q,p)=\left\langle \pi^{-1}\int_{-\infty}^{\infty}\exp(-2iqy)|p+y\rangle\langle p-y|dy\right\rangle \ ,$$

where $|p\rangle$ are the momentum eigenstates.

The mathematical properties of the number-phase Wigner function are precisely those of the discrete Wigner function. The main property that we exploit in this paper concerns the probability interpretation of $W_{N\phi}$. We now derive this property here; the details of other properties can be found in Wootters's paper.[10] In analogy with *integrating* Wigner's original function $W(p,q)$ over either p or q to obtain a probability density, we *sum* $W_{N\phi}(n,\theta_m)$ over one of the indices n or m to produce a probability distribution. For instance, consider first the operator sum

$$(s+1)^{-1}\sum_{n=0}^{s}\hat{A}_{N\phi}(n,m)$$
$$=(s+1)^{-1}\sum_{n=0}^{s}\sum_{p=0}^{s}\exp(-2inp\Delta)|\theta_{m+p}\rangle\langle\theta_{m-p}| \, . \tag{2.7}$$

The sum over n of $\exp(-2inp\Delta)$ is $(s+1)\bar{\delta}_{2p,0}$. We note[12] that because $(s+1)$ is a prime number, $2p \bmod (s+1)$ is zero only for p zero or a multiple of $(s+1)$, and thus $\bar{\delta}_{2p,0}$ is unity for these values of p and zero otherwise. Hence the right-hand side of Eq. (2.7) reduces to $|\theta_m\rangle\langle\theta_m|$. Taking the expectation value of both sides of Eq. (2.7) yields, from Eq. (2.6),

$$\sum_{n=0}^{s}W_{N\phi}(n,\theta_m)=\langle(|\theta_m\rangle\langle\theta_m|)\rangle$$
$$=P(\theta_m)\, , \tag{2.8}$$

which is the probability that a measurement of the phase operator $\hat{\phi}_\theta$ will yield a value of θ_m. The sum of $(s+1)^{-1}\hat{A}_{N\phi}(n,m)$ over the other index m is found from Eq. (2.3) to be

$$\sum_{m=0}^{s}(s+1)^{-1}\hat{A}_{N\phi}(n,m)$$
$$=\sum_{k=0}^{s}\sum_{l=0}^{s}\bar{\delta}_{2n,k+l}\delta_{k,l}\exp(i\theta_0\hat{N})|k\rangle\langle l|\exp(-i\theta_0\hat{N})$$
$$=\sum_{k=0}^{s}\bar{\delta}_{2n,2k}\,|k\rangle\langle k| \, .$$

Because $(s+1)$ is prime and therefore odd, $\bar{\delta}_{2n,2k}$ is nonzero only for $k=n$, and thus the right-hand side is $|n\rangle\langle n|$. Taking the expectation value of both sides yields, therefore, from Eq. (2.6),

$$\sum_{m=0}^{s}W_{N\phi}(n,\theta_m)=\langle(|n\rangle\langle n|)\rangle=P_n\, , \tag{2.9}$$

which is the probability that a measurement of photon number will yield a value of n photons.

In analogy with the usual treatment of Wigner's original function we represent the number-phase Wigner function as $(s+1)^2$ discrete points in the three-dimensional space. The points in cylindrical coordinates (r,θ,z) are given by $(n,\theta_m,W_{N\phi}(n,\theta_m))$ for $n,m=0,1,\ldots,s$. The phase probability distribution $P(\theta_m)$ is, from Eq. (2.8), the sum of the z components of the $(s+1)$ points above the radial line $\theta=\theta_m$, while the number probability distribution P_n is, from Eq. (2.9), the sum of the z components of the $(s+1)$ points above the circle $r=n$. Thus the radial coordinate $r=n$ and the polar angle $\theta=\theta_m$ are associated with the number and phase operators, and hence the number and phase properties, respectively. In the following sections we derive the number-phase Wigner representation of various states, whose number and phase properties are well known, to establish the actual manner in which these properties are expressed.

III. THE $W_{N\phi}$ REPRESENTATION OF NUMBER AND PHASE STATES

The number-phase Wigner representations of number and phase states provide the clearest examples of how the number and phase properties of states are expressed by this function. The $W_{N\phi}$ representation of the number state $|k\rangle$ is found from Eqs. (2.6) and (2.3) to be

$$W_{N\phi}(n,\theta_m)=(s+1)^{-1}\bar{\delta}_{2n,2k}\, ,$$

or, because $(s+1)$ is odd,

$$W_{N\phi}(n,\theta_m)=(s+1)^{-1}\delta_{n,k}\, .$$

Thus $W_{N\phi}$ is only nonzero on the circle of radius $r=k$. This circle is illustrated in Fig. 1. $W_{N\phi}$ is defined at $(s+1)$ equidistant points on this circle; as $s\to\infty$ these points become dense. The value of $W_{N\phi}$ is $(s+1)^{-1}$ at each of these points and is zero elsewhere. Clearly $W_{N\phi}$ exhibits the characteristic properties of number states, that is, well-defined photon number and random phase.

The $W_{N\phi}$ representation of the phase state $|\theta_k\rangle$, where $0\le k\le s$, is found from Eqs. (2.6) and (2.5) to be

$$W_{N\phi}(n,\theta_m)=(s+1)^{-1}\sum_{p=0}^{s}\exp(-2inp\Delta)\bar{\delta}_{m+p,k}\bar{\delta}_{m-p,k}\, ,$$

where $0\le m\le s$. For $p=0$ the product $\bar{\delta}_{m+p,k}\bar{\delta}_{m-p,k}$ is equal to $\delta_{m,k}$. For all other values of p, k cannot be equal to both $m+p$ and $m-p$; nor can k equal both $m+p+j(s+1)$ and $m-p+l(s+1)$, where j and l are integers, because $(s+1)$ is odd and $p\le s$. Thus this product is zero for $p\ne 0$ and hence

$$W_{N\phi}(n,\theta_m)=(s+1)^{-1}\delta_{m,k}\, . \tag{3.1}$$

In Fig. 2 we again plot only those points in the x-y plane for which $W_{N\phi}$ is nonzero. We find that $W_{N\phi}$ is nonzero only on the radial line $\theta=\theta_k$ and this vividly illustrates the random photon number and the well-defined phase

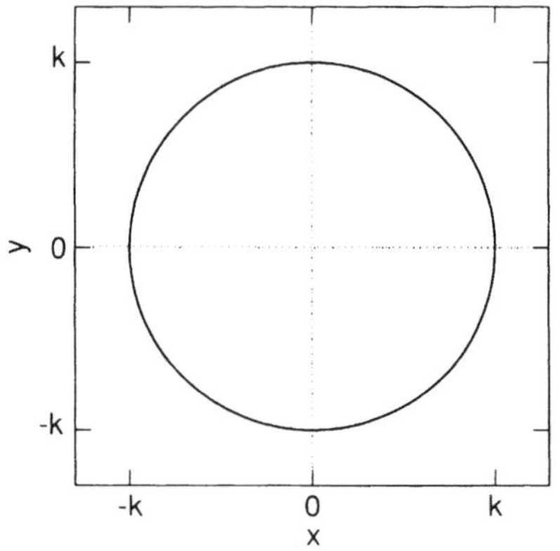

FIG. 1. Circle in the x-y plane on which the $W_{N\phi}$-representation of the number state $|k\rangle$ is nonzero.

properties of $|\theta_k\rangle$, which is an eigenstate of $\hat{\phi}_\theta$. Here when $s\to\infty$, the points in Fig. 2 are not dense but remain separated by the distance corresponding to a difference in photon number of unity, and the line of points extends to infinity.

Not all phase states[7–9] are necessarily eigenstates of $\hat{\phi}_\theta$ with eigenvalues between θ_0 and $\theta_0+2\pi$. By shifting the value of θ_0 by $\lambda\Delta$ where $0<\lambda<1$ we can create a new basis set of $(s+1)$ phase states $|\theta_j+\lambda\Delta\rangle$ with eigenvalues shifted from the eigenvalues of $\hat{\phi}_\theta$. The question now arises as to the nature of $W_{N\phi}$ for a field in a phase state $|\theta_j+\lambda\Delta\rangle$. Using the overlap

$$\langle\theta_j+\lambda\Delta|\theta_m\rangle=(s+1)^{-1}\sum_{k=0}^{s}\exp[ik(\theta_m-\theta_j-\lambda\Delta)] ,$$

we find

$$W_{N\phi}(n,\theta_m)=(s+1)^{-3}\sum_{p=0}^{s}\exp(-i2np\Delta)\sum_{k=0}^{s}\exp[ik(\theta_{m+p}-\theta_j-\lambda\Delta)]\sum_{l=0}^{s}\exp[-il(\theta_{m-p}-\theta_j-\lambda\Delta)] . \tag{3.2}$$

After some manipulation we obtain

$$W_{N\phi}(n,\theta_m)\Delta^{-1}=\frac{\sin(\lambda\pi)}{\pi(s+1)}[\sin(\lambda\pi+M\theta)-\cos(\lambda\pi+M\theta)\cot\theta] , \tag{3.3}$$

where $M\equiv 2n \bmod(s+1)$ and $\theta\equiv\theta_m-\theta_j-\lambda\Delta$. The details of this calculation are given in the Appendix. The singularities in the $\cot\theta$ factor ensure that the right-hand side is nonvanishing as s tends to infinity for $\theta=a\Delta$ and $\theta=a\Delta+\pi$, that is, $\theta_m=\theta_j+\lambda\Delta+a\Delta$ and $\theta_m=\theta_j+\lambda\Delta+a\Delta+\pi$, where a is a real number independent of s. For all other values of θ the right-hand side approaches zero as $s\to\infty$. Thus $W_{N\phi}(n,\theta_m)\Delta^{-1}$ has relatively large values only for $\theta_m\approx\theta_j$ and $\theta_m\approx\theta_j+\pi$. For arbitrarily large s, the relatively large values of $W_{N\phi}(n,\theta_m)\Delta^{-1}$ for $\theta_m\approx\theta_j$ indicate that the phase of the state $|\theta_j+\lambda\Delta\rangle$ is approximately θ_j. The question remains as to what information about the phase (if any) is conveyed by the relatively large values of $W_{N\phi}(n,\theta_m)\Delta^{-1}$ for $\theta_m\approx\theta_j+\pi$. To address this question we examine the phase probability density $P(\theta_m)\Delta^{-1}$,

$$P(\theta_m)\Delta^{-1}=\sum_{n=0}^{s}W_{N\phi}(n,\theta_m)\Delta^{-1} , \tag{3.4}$$

because it represents all the information available about the phase. Before evaluating the sum in Eq. (3.4) completely consider first the sum of the pair of terms $[W_{N\phi}(n,\theta_m)+W_{N\phi}(n+\frac{1}{2}s,\theta_m)]\Delta^{-1}$ for $n\le\frac{1}{2}s$. We find from Eq. (3.3) that this sum approaches zero in the region where $\theta_m\approx\theta_j+\pi$ as s tends to infinity. Thus the values of $W_{N\phi}(n,\theta_m)\Delta^{-1}$ and $W_{N\phi}(n+\frac{1}{2}s,\theta_m)\Delta^{-1}$ tend to cancel each other in the sum in Eq. (3.4) in this region. This leads us to conclude that the relatively large values of $W_{N\phi}(n,\theta_m)\Delta^{-1}$ for $\theta_m\approx\theta_j+\pi$ do not convey any information about the phase because they contribute no detail to $P(\theta_m)\Delta^{-1}$.

Evaluating completely the sum in Eq. (3.4) eventually gives

$$P(\theta_m)\Delta^{-1}=\frac{\sin^2(\lambda\pi)}{2\pi(s+1)\sin^2[\frac{1}{2}(\theta_m-\theta_j-\lambda\Delta)]} . \tag{3.5}$$

As $s\to\infty$ we find that for $\theta_m=a\Delta+\theta_j$, where a is independent of s,

$$P(\theta_m)\Delta^{-1}\to\frac{(s+1)\sin^2(\lambda\pi)}{2\pi^3(a-\lambda)^2} ,$$

while for $\theta_m=b+\theta_j$, where b is independent of s,

$$P(\theta_m)\Delta^{-1}\to 0 .$$

Evidently, in the infinite-s limit the density $P(\theta_m)\Delta^{-1}$ is zero for all θ_m except for an infinitesimal region near $\theta_m=\theta_j$ where the density is infinitely large. Furthermore, because the distribution $P(\theta_m)$ is normalized for all s then the density $P(\theta_m)\Delta^{-1}$ is also normalized in the limit $s\to\infty$ according to

$$1=\lim_{s\to\infty}\sum_{m=0}^{s}[P(\theta_m)\Delta^{-1}]\Delta .$$

This limit can be compared to the integral of a Dirac δ function. Moreover, the right-hand side of Eq. (3.5) is

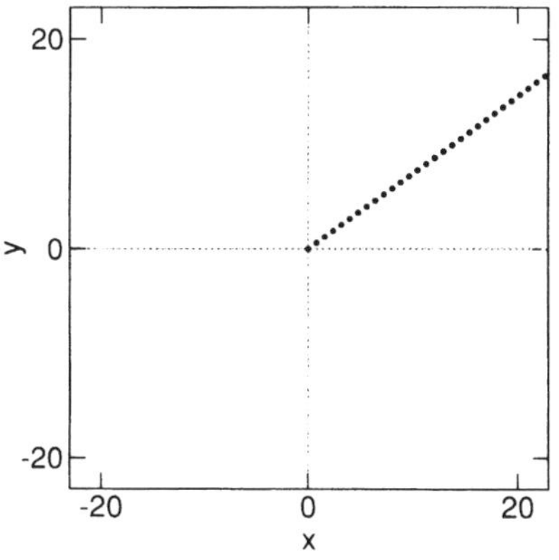

FIG. 2. Points on the x-y plane for which the $W_{N\phi}$-representation of the phase state $|\theta_k\rangle$ is nonzero. These points lie on the radial line that makes an angle of θ_k to the x axis.

periodic in θ_m with a period of 2π. Thus we conclude that the phase probability density $P(\theta_m)\Delta^{-1}$ for the phase state $|\theta_j+\lambda\Delta\rangle$ has periodic Dirac δ-function behavior in the infinite-s limit. This behavior clearly shows that the phase state $|\theta_j+\lambda\Delta\rangle$ gives well-defined values of phase when operated on by $\hat{\phi}_\theta$ even though it is not an eigenstate of $\hat{\phi}_\theta$.

IV. THE $W_{N\phi}$ REPRESENTATION OF PHYSICAL STATES

We illustrated in Sec. III the way in which the number and phase properties of the number and phase states are expressed by the $W_{N\phi}$ function. The number state belongs to a wide class of states called physical states. This class of states is important because it represents states that can be prepared physically. It includes nearly all states used in quantum optics, with the notable exception being the phase states themselves, for which the expectation value of energy is infinite. Formally, physical states $|p\rangle$,

$$|p\rangle\equiv\sum_{n=0}^{s}d_n|n\rangle\ , \tag{4.1}$$

are those states that have finite moments of the number operator $\langle\hat{N}^q\rangle$ for any given q in the limit of infinite s.[6–9] That is, states for which

$$\langle\hat{N}^q\rangle=\lim_{s\to\infty}\sum_{n=0}^{s}|d_n|^2n^q<B_q\ , \tag{4.2}$$

where B_q is some bound.

The $W_{N\phi}$ representation of physical states is found, by a similar calculation to that which leads to Eq. (A1) in the Appendix, to be

$$W_{N\phi}(n,\theta_m)=(s+1)^{-1}\sum_{k=0}^{M}d_k^*d_{M-k}\exp[i(2k-M)\theta_m]+(s+1)^{-1}\sum_{k=M+1}^{s}d_k^*d_{M-k+s+1}\exp[i(2k-M-s-1)\theta_m]\ , \tag{4.3}$$

where $M\equiv 2n\ \mathrm{mod}(s+1)$. From Eq. (4.2) it follows that

$$|d_n|^2<B_qm^{-q}$$

and hence

$$|d_k^*d_{l-k}|<B_q[k(l-k)]^{-(1/2)q}\ ,$$

for $l>k$. Using $k(l-k)\geq\frac{1}{2}l$ for $l>k$ and $k\neq0$ we get

$$|d_k^*d_{l-k}|<B_q(\tfrac{1}{2}l)^{-(1/2)q},$$

from which it follows that the magnitude of the second term on the right-hand side of Eq. (4.3) is less than $B_q[\frac{1}{2}(M+s+1)]^{-(1/2)q}$ for any given q. Thus, by choosing sufficiently large s, this term is negligible and we can approximate Eq. (4.3) to any desired accuracy by

$$W_{N\phi}(n,\theta_m)=(s+1)^{-1}\sum_{k=0}^{M}d_k^*d_{M-k}\exp[i(2k-M)\theta_m]\ . \tag{4.4}$$

This simple expression gives the number-phase Wigner function for physical states in general. We now examine two particular examples of physical states: the coherent state $|\alpha\rangle$ and the squeezed vacuum[13] $|0,\xi\rangle$. The phase properties of these states depend on the amplitude $|\alpha|$ and squeezing $|\xi|$ parameters. As these parameters are increased from zero the phase of each state transforms from being completely random to a more well-defined value.[8,14]

A. Coherent state

Substituting the number-state coefficients for $|\alpha\rangle$,

$$d_n=\alpha^n(n!)^{-1/2}\exp(-\tfrac{1}{2}|\alpha|^2)\ ,$$

into Eq. (4.4) yields eventually

$$W_{N\phi}(n,\theta_m)=(s+1)^{-1}\Lambda(n,r)\Phi(n,\theta_m,\phi)\ ,$$

where $\Lambda(n,r)\equiv r^M\exp(-r^2)/l!$,

$$\Phi(n,\theta_m,\phi)\equiv\sum_{k=0}^{M}\frac{l!\cos[(2k-M)(\theta_m-\phi)]}{(k!)^{1/2}[(M-k)!]^{1/2}}\ , \tag{4.5}$$

$\alpha=r\exp(i\phi)$, $M\equiv2n\ \mathrm{mod}(s+1)$, and l is the largest integer not exceeding $\frac{1}{2}M$. We notice here that $W_{N\phi}$ is the product of a constant $(s+1)^{-1}$, an amplitude-r-dependent factor $\Lambda(n,r)$ and a phase-ϕ-dependent factor $\Phi(n,\theta_m,\phi)$. This factorization provides an alternative viewpoint as to why more intense coherent states have more well-defined phases. The factor $\Lambda(n,r)$ is related to the photon-number probability distribution $P_n=r^{2n}\exp(-r^2)/n!$ by

$$\Lambda(n,r)=P_n,\quad 0\leq n\leq\tfrac{1}{2}s \tag{4.6a}$$

$$\Lambda(\tfrac{1}{2}s+n,r)=rP_{n-1},\quad 0<n\leq\tfrac{1}{2}s\ . \tag{4.6b}$$

Here we find that as n increases from zero, $\Lambda(n,r)$ decays to negligible values as n approaches $\frac{1}{2}s$ and then it undergoes a "revival" from $n=\frac{1}{2}s+1$. This should perhaps not be unexpected because in the definition of $W_{N\phi}(n,\theta_m)$, Eqs. (2.3) and (2.6), the dependence on n appears only as $\bar{\delta}_{2n,j}$, which has the property

$$\bar{\delta}_{2n+s,j}=\bar{\delta}_{2n-1,j}$$

for all n and j. Thus $\bar{\delta}_{2n,j}$ is *almost* cyclic in n and it is this property that gives rise to the revival in $\Lambda(n,r)$. We also find from Eq. (4.5) the symmetry property that $\Phi(n,\theta_m,\phi)$ equals $\Phi(n,\theta_m+\pi,\phi)$ for $n\leq\frac{1}{2}s$ and $-\Phi(n,\theta_m+\pi,\phi)$ for $n>\frac{1}{2}s$. In Fig. 3 we have plotted

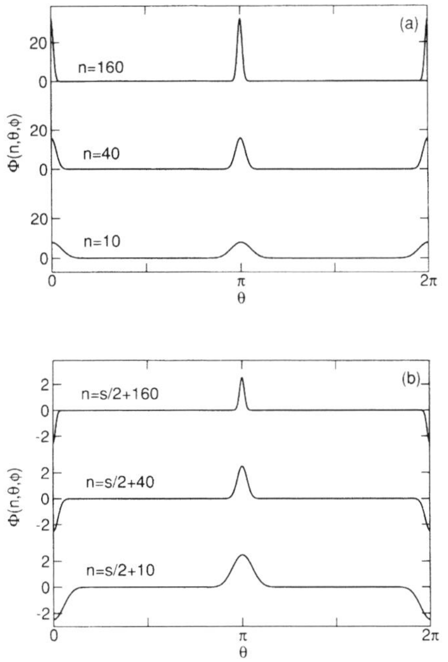

FIG. 3. Phase-dependent factor $\Phi(n,\theta,\phi)$ as a function of the continuous parameter θ for the coherent state $|\alpha\rangle$ with $\arg(\alpha)=\pi$ for (a) $n=10$, 40, and 160, and (b) $n=\frac{1}{2}s+10$, $\frac{1}{2}s+40$, and $\frac{1}{2}s+160$.

$\Phi(n,\theta,\phi)$ as a function of the continuous parameter θ for selected values of n for the coherent state with $\phi=\pi$ and with an arbitrarily large value of s. The $(s+1)$ values of $\Phi(n,\theta_m,\phi)$ for $m=0$ to s form a subset of discrete points on these curves. The periodic behavior of Φ is clearly visible here. The curves have a peak near $\theta=\pi$, which is the expected phase of the coherent state, and peaks (for $n\leq\frac{1}{2}s$) or troughs (for $n>\frac{1}{2}s$) near $\theta=0$. We also note that all peaks and troughs become progressively narrower as n increases from 0 to $\frac{1}{2}s$ and, separately, from $\frac{1}{2}s+1$ to s. The phase-probability distribution $P(\theta_m)$ is given by summing $W_{N\phi}(n,\theta_m)$ over n, i.e.,

$$P(\theta_m)=(s+1)^{-1}\sum_{n=0}^{s}\Lambda(n,r)\Phi(n,\theta_m,\phi)\ . \tag{4.7}$$

In this sum the amplitude-dependent function $\Lambda(n,r)$ acts as a weighting factor. According to Eqs. (4.6a) and (4.6b), $\Lambda(n,r)$ is related to a Poisson distribution with a mean of r^2, and so the most significant contribution to $P(\theta_m)$ will be from terms for which $n\approx r^2$ and $n\approx r^2+\frac{1}{2}s$. In the sum of these terms in Eq. (4.7) the peaks and troughs of $\Phi(n,\theta_m,\phi)$ near $\theta_m=0$ tend to cancel each other while peaks near $\theta_m=\pi$ reinforce each other. Thus $P(\theta_m)$ will exhibit a narrow peak near $\theta_m\approx\pi$ and will be relatively flat elsewhere. As r is increased, the sum in Eq. (4.7) will be dominated by terms involving $\Phi(n,\theta_m,\phi)$ with larger n, that is, terms with narrower peaks in θ_m, giving $P(\theta_m)$ a narrower peak near $\theta_m\approx\pi$. This explains why coherent states have more well-defined phases as the intensity r^2 is increased: the amplitude factor $\Lambda(n,r)$, which is related to the photon-number probability distribution, determines in part the phase-probability distribution through the heavier weighting of $\Phi(n,\theta_m\phi)$ with narrower peaks in Eq. (4.7).

B. Squeezed state

The squeezed vacuum $|0,\xi\rangle$ is also a physical state[6] and so its number-phase Wigner representation can also be found using the approximate expression (4.4). The number-state coefficients of $|0,\xi\rangle$ are[3,5]

$$d_{2n}=\frac{(-\tanh t)^n}{2^n n!}\left[\frac{(2n)!}{\cosh t}\right]^{1/2}\exp(in\eta)$$

and $d_{2n+1}=0$, where $\xi=t\exp(i\eta)$. After some manipulation we find $W_{N\phi}$ for $|0,\xi\rangle$ is approximately

$$W_{N\phi}=(s+1)^{-1}\Lambda'(n,t)\Phi'(n,\theta_m,\eta)\ ,$$

where

$$\Lambda'(n,t)\equiv\frac{(\tanh t)^n(2q)!}{\cosh t\ q!^2 2^{2q}},\quad n\leq\tfrac{1}{2}s$$

$$\Lambda'(n,t)\equiv 0,\quad n>\tfrac{1}{2}s\ , \tag{4.8}$$

$$\Phi'(n,\theta_m,\eta)\equiv\frac{q!^2(-1)^n}{(2q)!2^{n-2q}}\sum_{k=0}^{n}\frac{[(2k)!(2n-2k)!]^{1/2}}{k!(n-k)!}\times\cos[(n-2k)(\eta-2\theta_m)]\ , \tag{4.9}$$

and q is the largest integer not exceeding $\frac{1}{2}n$. Thus $W_{N\phi}$ for the squeezed vacuum also factorizes into three factors: a constant $(s+1)^{-1}$, a factor Λ' dependent on t, and a phase-η-dependent factor Φ'. The factor $\Lambda'(n,r)$ steadily decreases with increasing n. It is not difficult to show that Λ' is related to the photon-number probability distribution P_n for the squeezed vacuum by

$$\Lambda'(2n,t)=P_{2n},\quad 2n\leq\tfrac{1}{2}s$$

$$\Lambda'(2n+1,t)=(\tanh t)P_{2n},\quad (2n+1)\leq\tfrac{1}{2}s\ .$$

We note that, unlike Λ for the coherent state, Λ' does *not* undergo any revival for $n>\frac{1}{2}s$. This can be traced to the fact that the odd-number state coefficients of the squeezed vacuum are zero. On the other hand, we can see from Eq. (4.9) that Φ' possesses the following symmetries:

$$\Phi'(n,\theta_m,\eta)=\begin{cases}\Phi'(n,\theta_m+\frac{1}{2}\pi,\eta) & \text{for even } n,\\ -\Phi'(n,\theta_m+\frac{1}{2}\pi,\eta) & \text{for odd } n,\\ \Phi'(n,\theta_m+\pi,\eta) & \text{for all } n\ .\end{cases}$$

These symmetries are clearly visible in the plots of

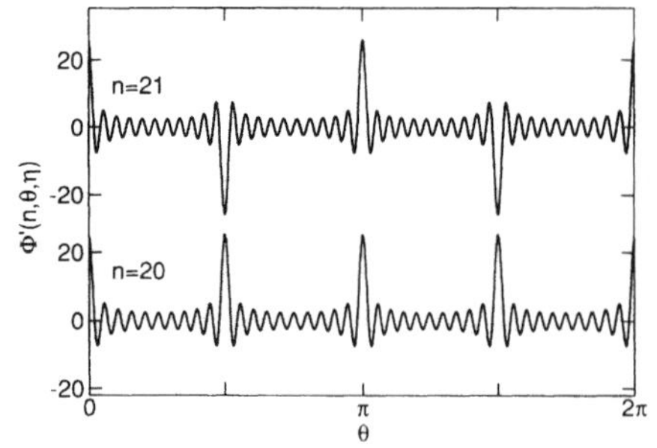

FIG. 4. Two curves illustrating the general properties of the phase-dependent factor $\Phi'(n,\theta,\eta)$ as a function of the continuous parameter θ for the squeezed vacuum $|0,\xi\rangle$ with $\arg(\xi)=\pi$. The $s+1$ values of $\Phi'(n,\theta_m,\eta)$ for $m=0$ to s form a discrete subset of points on these curves.

$\Phi'(n,\theta,\eta)$ for $n=20$ and 21 in Fig. 4 for the squeezed vacuum with $\eta=\pi$. These two curves illustrate the general properties of Φ'. Although not shown here, we find that the relatively large peaks at $\theta=0$ and π become narrower for larger n, with widths of the order of π/n. The peaks for n even and troughs for n odd at $\theta=\pi/2$ and $3\pi/2$ tend to cancel each other in the sum

$$P(\theta_m)=(s+1)^{-1}\sum_{n=0}^{s}\Lambda'(n,t)\Phi'(n,\theta_m,\eta)\ . \qquad (4.10)$$

Thus these peaks and troughs do not convey any significant phase information. Perhaps surprisingly, we find approximately 40 smaller oscillations in each curve in Fig. 4. In general, it turns out that there are approximately $2n$ such oscillations in the curve of $\Phi'(n,\theta,\eta)$. However, in the sum in Eq. (4.10), $\Phi'(n,\theta_m,\eta)$ is weighted by $\Lambda'(n,t)$, which decays relatively slowly with increasing n, and so many terms involving $\Phi'(n,\theta,\eta)$ of different frequencies contribute to $P(\theta_m)$. Thus the smaller oscillations tend to cancel each other and make no significant contribution to the phase information. Hence $P(\theta_m)$ has peaks at $\theta_m\approx 0$ and π only, and is relatively flat in other regions. Finally we note that as the squeezing parameter t increases, the ratios

$$\frac{\Lambda'(2n+2,t)}{\Lambda'(2n,t)}=\frac{\Lambda'(2n+3,t)}{\Lambda'(2n+1,t)}=\frac{2n+1}{2n+2}(\tanh t)^2$$

also increase and this implies $\Lambda'(n,t)$ decays more slowly with n. Thus the contribution to $P(\theta_m)$ from values of $\Phi'(n,\theta_m,\eta)$ with larger n becomes more significant with increased squeezing, and so the two peaks in $P(\theta_m)$ will become narrower. Hence, as the squeezing is increased the phase of the squeezed vacuum becomes more well defined at two values and this can be compared to the similar effect found for the single-peaked phase distribution of the coherent state.

In summary, we note that the coherent states and the squeezed vacuum exhibit a similar mechanism for reducing the phase uncertainty. We found that as the amplitude or the squeezing parameter is increased, values of $W_{N\phi}(n,\theta_m)$ with narrower peaks in θ_m contribute more significantly to the phase-probability distributions $P(\theta_m)$ for these states. Thus these distributions have narrower peaks, which indicates more well-defined phases. We also found that the $W_{N\phi}$ representation of these states has oscillations with θ_m, as illustrated in Figs. 3 and 4, that do not contribute any detail to $P(\theta_m)$. This can be compared to a similar effect we found in Sec. III for the phase state $|\theta_j+\lambda\Delta\rangle$.

V. CONCLUSION

In this paper we have developed the number-phase Wigner function $W_{N\phi}$ by using Wootters's discrete Wigner function to represent a single-mode light field in the Pegg-Barnett formalism. Unlike Wigner's original function, which is defined everywhere on a plane, the number-phase Wigner function $W_{N\phi}(n,\theta_m)$ is defined only on dense subsets of points on circles of integer radius centered on the origin. These points (n,θ_m) correspond to a polar representation of the eigenvalues of the number and phase operators, that is, n are the eigenvalues of $\hat{N}$, and θ_m are the eigenvalues of $\hat{\phi}_\theta$.

We have found that the number-phase Wigner function gives a picture of the number and phase properties of states. For example, $W_{N\phi}$ for the number state $|n\rangle$ is nonzero only along the circle of radius n, while for the phase eigenstate $|\theta_m\rangle$, $W_{N\phi}$ is nonzero only along the radial line, which is at an angle of θ_m to the x axis. A reasonably simple expression was obtained for the number-phase Wigner representation of a general physical state. This expression was used to find the $W_{N\phi}$ representation of the coherent state and the squeezed vacuum. The phase properties of these states were found to be expressed by their respective $W_{N\phi}$ representations as follows. Although $W_{N\phi}$ for these states may contain many oscillations with θ_m, only particular peaks convey the phase information. These particular peaks are narrower for larger n. As the amplitude (or squeezing) is increased, values of $W_{N\phi}(n,\theta_m)$ with larger n contribute more significantly to the phase-probability distribution $P(\theta_m)$, and thus the peaks in $P(\theta_m)$ become narrower, giving a more well-defined phase.

In conclusion, we have introduced the number-phase Wigner function and shown how it gives a new way of viewing the number and phase properties of states in the new formalism.

ACKNOWLEDGMENTS

J. A. V. acknowledges financial support from the Department of Employment, Education and Training.

APPENDIX

Here we give the details of the derivation of Eq. (3.3) from Eq. (3.2). Performing the sum over p in Eq. (3.2), using $\theta_{m+p}=\theta_m+p\Delta$ and Eq. (2.4), yields

$$W_{N\phi}(n,\theta_m)=(s+1)^{-2}\sum_{n=0}^{s}\sum_{l=0}^{s}\bar{\delta}_{k+l,2n}\exp[i(k-l)(\theta_m-\theta_j-\lambda\Delta)] .$$

We can replace $2n$ by $M\equiv 2n \bmod (s+1)$ in the index of the $\bar{\delta}$ function because $\bar{\delta}$ is periodic with a period of $(s+1)$. Separating the sum over k into two parts then gives

$$W_{N\phi}(n,\theta_m)=(s+1)^{-2}\left[\sum_{k=0}^{M}\sum_{l=0}^{s}\bar{\delta}_{k+l,M}\exp[i(k-l)\theta]+\sum_{k=M+1}^{s}\sum_{l=0}^{s}\bar{\delta}_{k+l,M}\exp[i(k-l)\theta]\right] ,$$

where for convenience we have written $\theta\equiv\theta_m-\theta_j-\lambda\Delta$. In the first sum $k\le M$ and so

$$\bar{\delta}_{k+l,M}=\bar{\delta}_{l,M-k}=\delta_{l,M-k} ,$$

while in the second sum $k>M$ and thus

$$\bar{\delta}_{k+l,M}=\bar{\delta}_{l,M-k}=\delta_{l,M-k+s+1} .$$

Hence we find

$$W_{N\phi}(n,\theta_m)=(s+1)^{-2}\left[\sum_{k=0}^{M}\exp[i(2k-M)\theta]+\sum_{k=M+1}^{s}\exp[i(2k-M-s-1)\theta]\right] . \tag{A1}$$

Replacing the dummy summation index k in the second sum with $q=k-M-1$ gives

$$W_{N\phi}(n,\theta_m)=(s+1)^{-2}\left[\sum_{k=0}^{M}\exp[i(2k-M)\theta]+\sum_{q=0}^{s-M-1}\exp[i(2q-s+M+1)\theta]\right] .$$

The right-hand side now contains two geometric series of the general form

$$\sum_{k=0}^{j}\exp(ik2\theta)\exp(-ij\theta)=\sin[(j+1)\theta]/\sin\theta ,$$

where j is an integer. Replacing the geometric series with their closed expressions then gives

$$\begin{aligned}W_{N\phi}(n,\theta_m)&=(s+1)^{-2}\{\sin[(M+1)\theta]/\sin\theta\\&\quad+\sin[(s-M)\theta]/\sin\theta\}\\&=2(s+1)^{-2}\sin[\tfrac{1}{2}(s+1)\theta]\\&\quad\times\cos\{[\tfrac{1}{2}(s-1)-M]\theta\}/\sin\theta .\end{aligned} \tag{A2}$$

We note that $\theta=\theta_m-\theta_j-\lambda\Delta=(m-j-\lambda)\Delta$ and $\Delta=2\pi/(s+1)$, and so

$$\sin[\tfrac{1}{2}(s+1)\theta]=-(-1)^{m-j}\sin(\lambda\pi) ,$$

$$\begin{aligned}\cos\{[\tfrac{1}{2}(s-1)-M]\theta\}&=(-1)^{m-j}\cos[\lambda\pi+(M+1)\theta]\\&=(-1)^{m-j}[\cos(\lambda\pi+M\theta)\cos\theta\\&\quad-\sin(\lambda\pi+M\theta)\sin\theta] .\end{aligned}$$

Substituting these two results into Eq. (A2) and then dividing by Δ yields Eq. (3.3):

$$W_{N\phi}(n,\theta_m)\Delta^{-1}=\frac{\sin(\lambda\pi)}{\pi(s+1)}[\sin(\lambda\pi+M\theta)-\cos(\lambda\pi+M\theta)\cot\theta] .$$

---

*Present address: Division of Commerce and Administration, Griffith University, Nathan, Brisbane 4111, Australia.

[1]P. D. Drummond, C. W. Gardiner, and D. F. Walls, Phys. Rev. A **24**, 914 (1981).
[2]E. P. Wigner, Phys. Rev. **40**, 749 (1932).
[3]H. P. Yuen, Phys. Rev. A **13**, 2226 (1976).
[4]S. M. Barnett and D. T. Pegg, J. Phys. A **19**, 3849 (1986).
[5]R. Loudon and P. L. Knight, J. Mod. Opt. **34**, 709 (1987).
[6]J. A. Vaccaro and D. T. Pegg, J. Mod. Opt. **37**, 17 (1990).
[7]D. T. Pegg and S. M. Barnett, Europhys. Lett. **6**, 483 (1988).
[8]S. M. Barnett and D. T. Pegg, J. Mod. Opt. **36**, 7 (1989).
[9]D. T. Pegg and S. M. Barnett, Phys. Rev. A **39**, 1665 (1989).
[10]W. K. Wootters, Ann. Phys. **176**, 1 (1987).
[11]We can show that odd $(s+1)$ is sufficient for the essential properties of the number-phase Wigner function. However, in keeping with Wootters's analysis we treat prime $(s+1)$ only in this paper.
[12](a mod b) is the remainder when a is divided by b; for example, (105 mod191) = 105, while (205 mod191) = 14.
[13]For reviews of squeezed light see, e.g., D. F. Walls, Nature **306**, 141 (1983); and also Ref. 5.
[14]J. A. Vaccaro and D. T. Pegg, Opt. Commun. **70**, 529 (1989)

Quantum-optical phase and canonical conjugation

D. T. PEGG, J. A. VACCARO

Division of Science and Technology, Griffith University, Nathan, Brisbane, 4111, Australia

and STEPHEN M. BARNETT

Department of Engineering Science, Oxford University, Parks Road, Oxford OX1 3PJ, England

Abstract. We use a new limiting procedure, developed to study quantum-optical phase, to examine canonically conjugate operators in general. We find that Dirac's assumption that photon number and phase should be canonically conjugate variables, similar to momentum and position, is essentially correct. The difficulties with Dirac's approach are shown to arise through use of a form of the canonical commutator which, although the only possible form in the usual infinite Hilbert space approach, is not sufficiently general to be used as a model for a number–phase commutator.The approach in this paper unifies the theory of conjugate operators, which include photon number and phase, angular momentum and angle, and momentum and position as particular cases. The usual position-momentum commutator is regained from a more generally applicable expression by means of a domain restriction which cannot be used for the phase–number commutator.

1. Introduction

In Dirac's original approach to the theory of quantum electrodynamics [1], he made the intuitive assumption that optical phase and photon number should be canonically conjugate variables, with a commutator

$$[\hat{\varphi},\hat{N}]=\mathrm{i}, \tag{1}$$

which is the direct analogue of the position–momentum commutator. Later, however, it was shown that expression (1) leads to inconsistencies [2, 3] which imply that an Hermitian phase operator cannot exist in the conventional infinite Hilbert space

Recently Pegg and Barnett [4] suggested a new limiting procedure for quantum mechanical calculations which allows the use of an Hermitian operator $\hat{\varphi}_\theta$ to represent quantum optical phase, while still predicting results of physical experiments identical to those predicted by the conventional approach. Instead of invoking the infinite Hilbert space from the outset, the new procedure involves the use of a finite but arbitrarily large state space Ψ of $s+1$ dimensions, with well behaved operators, to calculate c-numbers, such as means, variances and possible results of physical measurements, as a function of s. The final part of the procedure is to find the limit of these c-numbers as s tends to infinity. Nowhere is it necessary to find the limits of states or operators. The properties of $\hat{\varphi}_\theta$ have been investigated in greater detail [5, 6] with the conclusion that this operator is the appropriate quantum-mechanical representation of the phase observable.

In this paper we explore the concept of canonically conjugate operators in general from the perspective of the new limiting procedure. We find that it is possible to derive the canonical commutation relations for position and momentum, phase and number and angular momentum and angle from a single common expression. Differences only appear in the commutators when varying restrictions are placed on the states of Ψ on which the commutator operators are allowed to act. Dirac's assumption of equivalent commutators is thus intrinsically correct, with the inconsistencies arising only through use of an inappropriately restricted version of the position–momentum commutator as a model of a canonical commutator. The relations between phase and number and between position and momentum are not essentially different after all.

2. Conjugate operators in space Ψ

As discussed in the introduction, we consider a state space Ψ of $(s+1)$ dimensions, spanned by an orthonormal basis set of states $|n\rangle, n=0,1,\ldots s$. This space is also spanned by other orthonormal basis sets $|m\rangle, m=0,1,\ldots s$. We are interested in states $|m\rangle$ which are maximum uncertainty states in relation to $|n\rangle$, that is, for which a system in state $|m\rangle$ will have the same probability of being found in any of the states $|n\rangle$. Together with the normalization condition, this implies that

$$\langle n|m\rangle=(s+1)^{-1/2}\exp[\mathrm{i}f(m,n)], \tag{2}$$

where the function $f(m,n)$ is yet to be determined. Expression (2) also implies that $|n\rangle$ is a maximum uncertainty state in relation to $|m\rangle$.

Operators with uniformly spaced eigenvalue spectra can be defined as

$$\hat{N}=\sum_n n|n\rangle\langle n|, \tag{3}$$

$$\hat{M}=\sum_m m|m\rangle\langle m| \tag{4}$$

which have eigenvalues $0,1,\ldots s$. We generalize these operators to $\hat{N}+a\hat{1}$ and $\hat{M}+b\hat{1}$, where $\hat{1}$ is the unit operator and a and b are arbitrary. These operators have the same eigenstates and eigenvalue differences as $\hat{N}$ and $\hat{M}$ but with adjustable ranges of eigenvalues.

We now specify that $\hat{N}+a\hat{1}$ and $\hat{M}+b\hat{1}$ can generate shifts in each other's eigenstates according to

$$\exp[-\mathrm{i}(\hat{N}+a)j\delta]|m\rangle=|m+j\rangle, \tag{5}$$

$$\exp[\mathrm{i}(\hat{M}+b)k\delta]n\rangle=|n+k\rangle, \tag{6}$$

where j and k are integers such that $m+j$ and $n+k$ are less than or equal to s, and δ is yet to be determined. We note that we could have specified these relations with opposite signs of i, which would alter the sign of the commutator $[\hat{M},\hat{N}]$ which we determine later.

By expanding $|m\rangle$ in terms of $|n\rangle$ using (2), substituting in (5), and likewise expanding $|n\rangle$ in terms of $|m\rangle$ in (6) we obtain the functional equations

$$f(m,n)-(n+a)j\delta=f(m+j,n), \tag{7}$$

$$f(m,n)-(m+b)k\delta=f(m,n+k), \tag{8}$$

which are satisfied by

$$f(m, n) = -(m+b)(n+a)\delta. \tag{9}$$

By using the fact that the sum of all the $s+1$ complex $(s+1)$th roots of unity is zero, we find that orthogonality of $|m\rangle$ and $|m'\rangle$ can be satisfied for all $m \neq m'$ with

$$\delta = 2\pi/(s+1). \tag{10}$$

Our relation is finally

$$\langle n|m\rangle = (s+1)^{-1/2} \exp[-\mathrm{i}(n+a)(m+b)2\pi/(s+1)]. \tag{11}$$

For convenience, we shall continue to use the symbol δ as given by (10).

When a general state $|t\rangle$ is expanded in terms of $|n\rangle$ and $|m\rangle$, with coefficients $\langle n|t\rangle$ and $\langle m|t\rangle$ respectively, then $\langle m|t\rangle$ is the *discrete Fourier transform* [7] of $\langle n|t\rangle$, and $\langle n|t\rangle$ is the inverse transform of $\langle m|t\rangle$:

$$\langle m|t\rangle = (s+1)^{-1/2} \sum_{n=0}^{s} \exp[\mathrm{i}(n+a)(m+b)2\pi/(s+1)]\langle n|t\rangle, \tag{12}$$

$$\langle n|t\rangle = (s+1)^{-1/2} \sum_{m=0}^{s} \exp[-\mathrm{i}(n+a)(m+b)2\pi/(s+1)]\langle m|t\rangle. \tag{13}$$

By expanding $|m\rangle$ in terms of $|n'\rangle$ and performing the appropriate summations over m, we eventually find the commutator of $\hat{M}$ and $\hat{N}$ to be

$$\begin{aligned}[\hat{M}, \hat{N}] &= \sum_{n,m} nm(|m\rangle\langle m|n\rangle\langle n| - |n\rangle\langle n|m\rangle\langle m|) \\ &= \sum_{n \neq n'} \frac{(n-n')\exp[\mathrm{i}(n-n')b\delta]|n'\rangle\langle n|}{\exp[\mathrm{i}(n-n')\delta]-1}.\end{aligned} \tag{14}$$

Guided by (5) and (6) we define our conjugate operators $\hat{M}_\beta$ and $\hat{N}_\beta$ as

$$\hat{M}_\beta = (\hat{M} + b\hat{I})\delta/\beta, \tag{15}$$

$$\hat{N}_\beta = (\hat{N} + a\hat{I})\beta, \tag{16}$$

where β, for the present, is unspecified. $\hat{N}_\beta$ has eigenvalues $a\beta$, $(a+1)\beta, \ldots (a+s)\beta$, that is, the eigenvalue spectrum has a width of $(s+1)\beta$ and the difference between successive eigenvalues is β. The $\hat{M}_\beta$ eigenvalue spectrum likewise has a width of $(s+1)\delta/\beta = 2\pi/\beta$ and an eigenvalue difference of $\delta/\beta = 2\pi/[(s+1)\beta]$. With the adoption of the definitions (15) and (16), $\hat{N}_\beta$ can be interpreted, from (5) and (6), as the generator of a change in the observable associated with $\hat{M}_\beta$, that is, $\hat{N}_\beta j\delta/\beta$ in the exponent of a unitary transformation changes the value of the $\hat{M}_\beta$ observable by $j\delta/\beta$. Likewise $\hat{M}_\beta k\beta$ causes a shift of $k\beta$ along the $\hat{N}_\beta$ eigenvalue spectrum.

The commutator for these conjugate operators is

$$[\hat{M}_\beta, \hat{N}_\beta] = \delta[\hat{M}, \hat{N}], \tag{17}$$

which, from (14), is completely off-diagonal in $|n\rangle$ and looks nothing like the diagonal canonical commutator exemplified by (1). To proceed further we place a restriction on $\hat{N}_\beta$ and $\hat{M}_\beta$: we choose β in (16) and (17) so that $(s+1)\beta$ approaches infinity when s tends to infinity, that is, we limit our discussion to operators $\hat{N}_\beta$ with infinite eigenvalue spectra when s tends to infinity. Thus $\hat{N}_\beta$ might represent observables such as photon number, energy, angular momentum or momentum. Secondly we impose a restriction on states, which we call the *first physical condition*: we assume that only a finite portion of the infinite $\hat{N}_\beta$ eigenvalue spectrum is physically accessible, that is, we assume, for example, that we cannot prepare a state of infinite energy or momentum. Actually it is sufficient to restrict states to those with finite moments of $\hat{N}_\beta$, but we have shown elsewhere [8] that such states can be approximated to any desired accuracy by states $\Sigma c_n|n\rangle$ for which the coefficients c_n are non-zero only for n in the range from n_1 to n_2, where $|(n_2-n_1)\beta|$ is bounded, provided this bound is sufficiently large. We call such states physical states.

If we take the matrix elements of (14) between physical states, then the terms in the sum over n and n' which do not vanish involve coefficients c_n and $c_{n'}$ for which $|(n-n')\beta|<B$, where B is some upper bound. Combining this with the infinite range condition that $(s+1)\beta$ approaches infinity as s tends to infinity, we see that $(n-n')\delta$, which is just $2\pi(n-n')\beta/(s+1)\beta]$, can be made as small as we please by making s sufficiently large. Thus, provided we restrict the action of the commutator to operate only on physical states, we can replace (17) by the physical commutator

$$[\hat{M}_\beta,\hat{N}_\beta]_p=-\mathrm{i}\sum_{n\neq n'}\exp[\mathrm{i}(n-n')b\delta]|n'\rangle\langle n|. \tag{18}$$

whose physical state matrix elements are equal to those obtained by allowing s to tend to infinity in the physical state matrix elements of (17).

We can write (18) as

$$\begin{aligned}[\hat{M}_\beta,\hat{N}_\beta]_p&=\mathrm{i}(1-\sum_{n'}\exp(-\mathrm{i}n'b\delta)|n'\rangle\sum_{n}\exp(\mathrm{i}nb\delta)\langle n|\\&=\mathrm{i}[1-(s+1)|m=0\rangle\langle m=0|].\end{aligned} \tag{19}$$

where $|m=0\rangle$ is the eigenstate of $\hat{M}_\beta$ with minimum eigenvalue, that is, with eigenvalue $b\delta/\beta$. The restricted form (19) of the canonical commutator still differs from (1) by the presence of the second term. We discuss the meaning of this term in the next section.

We have restricted β, which may be a function of s, to vary in such a way that $(s+1)\beta$ approaches infinity as s tends to infinity. β also represents the difference in successive eigenvalues of the $\hat{N}_\beta$ spectrum so, in order for $\hat{N}_\beta$ to represent a sensible multistate physical system, β itself should not approach infinity as s tends to infinity. There are two possibilities: (a) β may approach a finite, non-zero, limit, or (b) β may approach zero as s tends to infinity. We deal with those two cases separately in the next sections.

3. Phase and angle operators

In the case where $(s+1)\beta$ approaches infinity and β approaches a non-zero finite limit as s tends to infinity, the width of the $\hat{N}_\beta$ eigenvalue spectrum becomes infinite and the successive eigenvalue difference β in non-zero. The associated $\hat{M}_\beta$ spectrum must have a finite width $2\pi/\beta$ and a successive eigenvalue difference of $2\pi/[(s+1)\beta]$ which approaches zero. The $\hat{M}_\beta$ eigenvalue spectrum is thus dense in the large s limit. The special case $\beta=1$ is of particular interest because the $\hat{M}_\beta$ spectrum is then dense and of width 2π, with eigenvalues ranging from $b2\pi/(s+1)$ to $(b+s)2\pi/(s+1)$. Thus $\hat{M}_\beta$ can represent phase or rotation angle. Indeed it is not difficult to show in this case that a shift of 2π in phase or angle, as generated by the action of $\exp(-\mathrm{i}\hat{N}_\beta 2\pi)$ on $|m\rangle$, yields an eigenstate of $\hat{M}_\beta$ with the same eigenvalue as $|m\rangle$.

This displays explicitly the 2π periodicity of phase or rotation angle. Also it follows because $\beta=1$, that the successive eigenvalue difference of $\hat{N}_\beta$ is unity. Thus $\hat{N}_\beta$ can represent the photon number operator or the z-component of angular momentum. To obtain the former, we choose the value of a in (16) to be zero, and for the latter we put $a=-s/2$. Since s can be either even or odd, we can let s approach infinity through even or odd integers, in which case the eigenvalues of $\hat{N}_\beta$ with $a=-s/2$ would be integers or integers plus one half respectively. The choice of b determines the particular 2π range of values which we specify to express the phase or angle eigenvalues, that is the range, or window, from θ_0 to $\theta_0+2\pi$ where $\theta_0=b2\pi/(s+1)$. This is a standard way of expressing the value of multi-valued functions [4, 5, 6].

The commutator (19) for phase and photon number becomes

$$[\phi_\theta, \hat{N}]_p = \mathrm{i}[1-(s+1)|\theta_0\rangle\langle\theta_0|] \tag{20}$$

which can be compared directly with (1). As discussed elsewhere [5, 6], the second term removes the inconsistencies associated with (1) and also has the essential effect of stepping the value of the phase through 2π when the field evolves to the state $|\theta_0\rangle$, that is, reaches the edge of the 2π window in which the value of the phase is expressed. A similar effect occurs for the rotation angle of a rotating particle [9]. We note that sign of i in (20) is the opposite of that of our previous number phase commutator [5, 6]. This is due to our choice of the sign of i (5) and (6), which corresponds to Dirac's original sign in (1).

Because the eigenvalue spectrum of $\hat{M}_\beta$ is dense in the large s limit, when a field in state $|t\rangle$ is expanded in number and phase states $|n\rangle$ and $|m\rangle$, the coefficients $\langle n|t\rangle$ and $\langle m|t\rangle$ are related in this limit by a ***finite Fourier transform*** pair [10]:

$$\Psi(\theta)=(2\pi)^{-1/2}\sum_{n=0}\langle n|t\rangle \exp[\mathrm{i}(n+a)\theta], \tag{21}$$

$$\langle n|t\rangle=(2\pi)^{-1/2}\int_{\theta_0}^{\theta_0+2\pi}\Psi(\theta)\exp[-\mathrm{i}(n+a)\theta]\,\mathrm{d}\theta, \tag{22}$$

where $\theta=\theta_0+m\delta$ and the wave function $\Psi(\theta)=[(s+1)/(2\pi)]^{1/2}\langle m|t\rangle$. Equation (22) follows from (13) and the mathematical definition of the integral as the large s limit of the corresponding sum. We have used $\mathrm{d}\theta=\delta=2\pi/(s+1)$. The modulus squared of $\Psi(\theta)$ is the probability density for finding the field with phase θ. For the number-phase case $a=0$. The same expressions apply to the angular momentum-rotation angle case with $a=-s/2$, but here it is probably more convenient to choose a new label $\bar{n}=n-s/2$ for the angular momentum states.

4. Momentum and position

We now examine the case for which $(s+1)\beta$ approaches infinity and β approaches zero as s tends to infinity. For example β might be inversely proportional to a power of $(s+1)$ which is between, but not including, zero and unity; however the precise form of β does not concern us. Here the widths of both the $\hat{N}_\beta$ and $\hat{M}_\beta$ eigenvalue spectra, $(s+1)\beta$ and $2\pi/\beta$ respectively, become infinite, and the eigenvalues differences β and $2\pi/[(s+1)\beta]$ approach zero. Thus both spectra become infinite and dense. To be definite, let us take the values of both a and b in (15) and (16) to be $-s/2$, to ensure that both spectra have infinite extension in both positive and negative directions when s tends to infinity, so that the operators can represent momentum and position components in a particular direction. We label the operators $\hat{N}_\beta$ and $\hat{M}_\beta$

as $\hat{p}$ and $\hat{q}$ with eigenvalues p and q given by $(n-s/2)\beta$ and $(m-s/2)2\pi/[(s+1)\beta]$ respectively.

For this case, for a system in state $|t\rangle$ the coefficients $\langle m|t\rangle$ and $\langle n|t\rangle$ are related, in the large s limit, by a ***Fourier transform*** pair:

$$[\beta(s+1)/(2\pi)]^{1/2}\langle m|t\rangle=(2\pi)^{-1/2}\int_{-\infty}^{\infty}\exp(\mathrm{i}pq)\,\beta^{-1/2}\langle n|t\rangle\,\mathrm{d}p, \tag{23}$$

$$\beta^{-1/2}\langle n|t\rangle=(2\pi)^{-1/2}\int_{-\infty}^{\infty}\exp(-\mathrm{i}pq)\,[\beta(s+1)/(2\pi)]^{1/2}\langle m|t\rangle\,\mathrm{d}q, \tag{24}$$

where again we have written the amplitudes as factors of wavefunctions. These follow from the large s limit of (12) and (13), with $\mathrm{d}p=\beta$ and $\mathrm{d}q=2\pi/[(s+1)\beta]$.

The commutator (19) becomes

$$[\hat{q},\hat{p}]_{\mathrm{p}}=\mathrm{i}[1-(s+1)|-s/2\rangle\langle -s/2|], \tag{25}$$

where $|-s/2\rangle$ is the minimum position eigenstate, whose eigenvalue is $-s\pi/[(s+1)\beta]$. Clearly this eigenvalue approaches negative infinity as s tends to infinity and β tends to zero, which yields the integration limit in (24). The action of the second term in (25) is precisely the same as the corresponding term in (20). *There is an important physical difference however*: the phase of a physical field, or the rotation angle of a localized particle, will reach θ_0, the finite-width window edge value, in a finite evolution time, which is less than or equal to the period of oscillation or rotation, even in the infinite s limit, whereas a particle with finite momentum, which is reasonably localized at a finite distance from the origin, must travel an infinite distance, and hence take an infinite time, to reach the edge of the infinite width position window. Thus it would seem that we can ignore the second term in (25) and regain the usual form of the canonical commuter as given by (1) for suitably localized states. Formally, we can replace (25) by

$$[\hat{q},\hat{p}]_{\mathrm{pp}}=\mathrm{i}, \tag{26}$$

if we impose a *second physical condition*: all systems, for example a particle, must be in a state with a finite extension in space, or at least with finite moments of position. The extra subscript in (26) is a reminder of this second physical condition on the position, which corresponds precisely to the first physical condition imposed on the momentum, or in general on the $\hat{N}_\beta$ observable. We note that the first physical condition means that (20), for example, cannot be used to operate on a system in a phase state [4–6] because this state has divergent energy moments, and similarly (25) should not operate on a position eigenstate because this has divergent momentum moments. The second physical condition means that (26) should not be used to operate on ***momentum*** eigenstates, because these have divergent position moments. Thus to use (26) we must restrict the domain of operation to exclude eigenstates of both momentum ***and*** position, that is, to exclude states of precisely defined momentum or position. At best we must have a distribution of momentum states, for example. No matter how sharp this distribution is made in order to define more closely the momentum of the state, it must, in the infinite s limit, contain an infinite number of momentum eigenstates. However, because the momentum eigenstate spectrum is dense in this limit, we can always choose a distribution sufficiently narrow to specify the momentum to within any given non-zero error, however small. Thus, although we cannot use a Kronecker delta type of distribution, involving only a single eigenstate, we can use a Dirac delta function distribution. The same applies for states of position. We note the clear correspondence with the usual infinite Hilbert space approach, in which we cannot describe position or momentum by proper vectors, that is, by vectors normalizable to unity, but only by improper vectors which are normalizable to the Dirac delta function. Our approach has the advantage that we can either use improper vectors with the usual commutator (26) *or* use proper vectors with the unrestricted commutator (17) and (14). We can even use an improper position vector with a proper momentum vector in conjunction with (25). This would correspond precisely to the photon number–phase case where we must be able to use proper number states, because eigenstates of photon number can actually be prepared.

4. Discussion and conclusion

The approach adopted in this paper has been based on procedure involving a $(s+1)$-dimensional state space Ψ with the limit as s tends to infinity being taken after c-number results are calculated as a function of s. As such, it is distinguished from other approaches to phase and angle which also involve finite state spaces [11]. In earlier work [4–6] we have illustrated the value of the approach for treating optical phase within the framework of quantum electrodynamics.

In this paper we have used our procedure to provide a unified approach to canonically conjugate variables in general. We find that conjugate operators $\hat{N}_\beta$ and $\hat{M}_\beta$ have the following properties. The $\hat{M}_\beta$ eigenvalue spectrum has a width of $2\pi/\beta$ where β is the difference between successive eigenvalues of the $\hat{N}_\beta$ spectrum. A $\hat{M}_\beta$ spectrum width of 2π, which can apply to phase or angle observables, thus requires a $\hat{N}_\beta$ eigenvalues difference of unity, as occurs for example for the photon number operator. If the periodicity of $\hat{M}_\beta$ were 4π, for example, the $\hat{N}_\beta$ eigenvalue difference would be $\frac{1}{2}$. If the width of the $\hat{M}_\beta$ spectrum approaches infinity, the $\hat{N}_\beta$ eigenvalue difference approaches zero. Further, if the width of the $\hat{N}_\beta$ spectrum approaches infinity, the $\hat{M}_\beta$ eigenvalue difference approaches zero. When these last two conditions are fulfilled the conjugate operators can represent momentum and position.

In all the cases of interest here, that is, where $\hat{N}_\beta$ represents photon number, the z component of angular momentum or a component of linear momentum, the width of the $\hat{N}_\beta$ eigenvalue spectrum is infinite and the $\hat{N}_\beta$ eigenvalue difference is finite or zero. This allows us to impose a physical condition which restricts the domain of operation of the canonical commutator to physically preparable states, and thus to replace it by a simpler commutator which has an extra term additional to that of the usual canonical commutator. Because the $\hat{M}_\beta$ eigenvalue spectrum is also of infinite width for position, we can likewise impose a second physical condition which further restricts the domain of operation of the commutator, allowing us to use the usual position–momentum commutator in this smaller domain, which does not include either the momentum or position eigenstates. For phase and angle, however, because of the finite width $\hat{M}_\beta$ spectrum, we cannot apply this second restriction, which is the *essential* reason why we cannot use Dirac's commutator (1). Thus the difference between phase–number and position–momentum relationship is not of a fundamental origin, it arises simply because we cannot make the second restriction to the domain of the phase–number commutator, which is useful, but not absolutely necessary for the position–momentum commutator. Indeed we can use the form (19) or (25) in a domain which includes the momentum eigenstates just as we can use (20) in a domain which includes the number, or energy, eigenstates. Further, we can use the exact forms (17) and (14) if we do not wish to restrict the domain at all, but include number and phase or momentum *and* position eigenstates. It is interesting that the $(s+1)$-dimensional state space approach, in conjunction with our limiting procedure, allows us to incorporate states, such as phase, angle, momentum and position eigenstates which cannot be incorporated into the usual infinite-dimensional Hilbert space approach.

As an added note, we should mention that Shapiro *et al.* [12] have recently used an analysis based on quantum estimation theory, from which they find that the optimum probability operator measure for phase is a projector involving the un-normalized phase state. This leads to a phase probability distribution identical to that obtainable, for example, from (21) and (22). The fact that these two independent approaches lead to the same results lends a measure of support to both as descriptions of phase. The major advantage of the limiting procedure described in this paper, however, is that it allows the use of a well behaved Hermitian phase operator to represent the classical phase observable, and thus promotes the concept of phase to the same status as other observables such as momentum, position and energy. On the other hand, in the approach of Shapiro *et al.*, because the infinite limit has effectively been taken at a much earlier stage, the only phase operator at their disposal is the Susskind–Glogower operator [2], with the well known difficulties associated with its non-unitarity [2, 3].

We conclude that our limiting procedure, which was successful for studying quantum optical phase, also gives a deeper insight into linear momentum and position. Dirac's intuitive assumption that the conjugate relationship between phase and number should be precisely the same as that between position and momentum is indeed correct. The difficulties with Dirac's approach arose simply through use of an overly restricted form of the canonical position–momentum commutator.

References

[1] Dirac, P. A. M., 1927, *Proc. R. Soc.* A, **114**, 243.
[2] Susskind, L., and Glogower, J., 1964, *Physics*, **1**, 49.
[3] Carruthers, P., and Nieto, M. M., 1968, *Rev. mod. Phys.*, **40**, 411.
[4] Pegg, D. T., and Barnett, S. M., 1988, *Europhys. Lett.*, **6**, 483.
[5] Barnett, S. M., and Pegg, D. T., 1989, *J. mod. Optics*, **36**, 7.
[6] Pegg, D. T., and Barnett, S. M. 1989, *Phys. Rev. A*, **39**, 1665.
[7] Brigham, E. O., 1974, *The Fast Fourier Transform* (London: Prentice-Hall), p. 98.
[8] Vaccaro, J. A., and Pegg, D. T., 1990, *J. mod. Optics*, **37**, 17.
[9] Barnett, S. M., and Pegg, D. T., 1990, *Phys. Rev.* A, **41**, 3427.
[10] Tranter, C. J., 1959, *Integral Transforms in Mathematical Physics* (London: Meuthuen), p. 77.
[11] Lévy-Leblond, 1973, *Revista Mexicana de Físca*, **22**, 15; Santhanam, T. S., 1978, *Aust. J. Phys.*, **31**, 233; Goldhirsh, I., 1980, *J. Phys.* A, **13**, 3479.
[12] Shapiro, J. H., Shepard, S. R., and Wong, N. C., 1989, *Phys. Rev. Lett.*, **62**, 2377; *ibid.*, **63**, 2002.

Limiting procedures for the optical phase operator

STEPHEN M. BARNETT
Department of Physics and Applied Physics,
University of Strathclyde, Glasgow G4 0NG, Scotland

and D. T. PEGG
Division of Science and Technology,
Griffith University, Nathan, Brisbane 4111, Australia

(*Received 22 May 1992 and accepted 31 May 1992*)

Abstract. We examine some of the attempts to describe the phase of a single field mode by a quantum operator acting in the conventional infinite Hilbert space. These operators lead to bizarre properties such as non-random phases for the number states and experience consistency difficulties when used to obtain a phase probability density. Moreover, in these approaches operator functions of phase are not simply functions of a phase operator. We show that these peculiarities do not arise when the Hermitian optical phase operator is employed. In our opinion, the problems associated with the descriptions of phase in conventional infinite Hilbert space arise from the nature of the limiting process.

1. Introduction

The introduction of the Hermitian optical phase operator [1–3] has led to renewed interest in the quantum description of phase. A range of new phase operator formulations has been proposed and some older ones reinvestigated [4–12]. The central difference between these approaches and our Hermitian optical phase operator is the manner in which limits are taken and phase moments calculated. In our approach, the dimension of the state space is allowed to tend to infinity only after expectation values and moments have been calculated [1–3]. The alternative formulations attempt to realise a consistent description of the phase in the conventional infinite Hilbert space.

In this paper we examine the implicit limiting procedures that have been applied to form phase operators in infinite Hilbert space [4–6, 9, 11, 12]. These formalisms lead to difficulties such as number state phase moments that are not characteristic of random phase and, in some cases at least, are not consistent with a periodic phase probability density. In addition operator functions of phase are not simply functions of a phase operator; for example the operator representing the exponential of the phase is not simply the exponential of the operator representing the phase itself. We show that all of these difficulties arise from the premature application of the infinite state space limit to the state space itself. If the evaluation of the limit is delayed until after c-number moments are calculated (as we advocate) then none of these problems arise.

2. Infinite Hilbert space approaches

Whenever the symbol ∞ appears in a mathematical expression there is the implicit requirement that a limit has to be evaluated. For many purposes, the quantity of interest does not depend on the exact manner in which the limit is taken, or indeed on the order in which multiple limits are taken. However, we should not forget that infinity is not a number but rather the signature of a limiting process. In some situations, the manner in which limits are evaluated is of crucial importance.

The approaches of Garrison and Wong [4], Popov, Yarunin and Damaskinsky [5, 6, 12], Santhanam and Sinha [9] and of Bergou and Englert [11] all involve the formulation of a phase operator in infinite Hilbert space. This means that there is an implicit infinite limit to be applied to the dimension of the state space. Each of these approaches is slightly different from the others but the central idea of applying the limit to the state space is the same. For the sake of brevity we will base this Section on the work of Yarunin *et al.* but will highlight common features with the other approaches.

The formalism of Yarunin *et al.* (and of Santhanam *et al.*) begins with a finite-dimensional state space spanned by r states $\{|n\rangle\}$ in which the matrix elements of a phase operator can be defined [12]:

$$\langle n|\hat{\phi}_r|n\rangle = \pi\frac{r-1}{r}, \tag{1 a}$$

$$\langle m|\hat{\phi}_r|n\rangle = -\frac{\pi}{r}\left[1+\mathrm{i}\cot\left(\pi\frac{m-n}{r}\right)\right], \quad (m \neq n). \tag{1 b}$$

These formulae are found to be equivalent to those that we have obtained for the Hermitian optical phase operator [2] if we make the substitution $s+1=r$ and choose the special value of $\theta_0=0$. However, as we will see, the formalisms are significantly different. It is also possible to form a unitary operator corresponding to the exponential of the phase multiplied by i. This operator has the form [9, 12]:

$$\hat{U}_r = \sum_{n=0}^{r-2} |n\rangle\langle n+1| + |r-1\rangle\langle 0|, \tag{2}$$

which again corresponds to a special case of our unitary phase operator [1–3].

Operators suitable for use in conventional infinite Hilbert space are obtained by evaluating the matrix elements of the operators (1) and (2) for finite number states and taking the limit as r tends to infinity. This procedure leads to new operators $\hat{\phi}$ and $\hat{V}$ (obtained as limits of $\hat{\phi}_r$ and $\hat{U}_r$ respectively) whose matrix elements are:

$$\langle n|\hat{\phi}|n\rangle = \pi, \tag{3 a}$$

$$\langle m|\hat{\phi}|n\rangle = \frac{\mathrm{i}}{n-m}, \quad (m \neq n), \tag{3 b}$$

$$\langle m|\hat{V}|n\rangle = \delta_{m,n-1}, \tag{4}$$

for all number states in the infinite Hilbert space. Unlike its finite space counterpart, the operator $\hat{V}$ is not unitary but rather isometric or one sided unitary. It is, in fact, the operator introduced by Susskind and Glogower in their attempt to realise a polar decomposition of the annihilation operator [13–15]. The operator $\hat{V}$ does not commute with its Hermitian conjugate and therefore these operators cannot be operator functions of a common phase operator. In particular, the operator $\hat{V}$ cannot be obtained by exponentiation of the operator $\hat{\phi}$. This is in marked contrast to the relationship between the original operators formed in the finite dimensional space for which [2, 3, 8]:

$$\bar{U}_r = \exp(\mathrm{i}\hat{\phi}_r). \tag{5}$$

The fact that this simple relationship is not preserved when we generate the new infinite Hilbert space operators is a consequence of the manner in which the limit has been taken. The infinite Hilbert space limit of an operator function of $\hat{\phi}_r$ is *not* the same as the operator function of the infinite Hilbert space limit of $\hat{\phi}_r$.

The lack of any functional relationship between the infinite Hilbert space operators leads to problems when trying to evaluate moments of the phase, or even in deciding what the correct expressions for these moments should be! If we describe the phase variance in terms of the operator $\hat{\phi}_r$ whose matrix elements are given in (3), then we find that the number states have phase variances given by:

$$\Delta\phi^2 \equiv \langle n|(\hat{\phi} - \langle\hat{\phi}\rangle)^2|n\rangle = \frac{\pi^2}{6} + \sum_{k=1}^{n} \frac{1}{k^2}. \tag{6}$$

This is in agreement with the result obtained by Bergou and Englert [11]. For large values of n this variance tends to the value $\pi^2/3$ which coincides with the expected value for a state of random phase [2]. However, for smaller values of n, the phase variance calculated on the basis of the operator $\hat{\phi}$ is less than that expected for a state of random phase. In the extreme case of the vacuum state the resulting phase variance is calculated to be only *half* of the value characteristic of a state of random phase. The implication is that if this formulation is to be believed then we must interpret the phase of the vacuum as being non-random. Popov and Yarunin have devised a means for calculating moments of the phase based on analytic representations of the matrix elements [5, 12]. It is difficult to arrive at explicit forms for moments of the phase within this formalism, even for the number states. However, it seems clear that these number state moments will be characteristic of random phase only for large n, with the biggest departure from randomness occurring for the vacuum state [12]. These results are at odds with those obtained directly from the operator $\hat{\phi}_r$ for which the phase variance for a number state is [2]:

$$\Delta\phi_r^2 \equiv \langle n|(\hat{\phi}_r - \langle\hat{\phi}_r\rangle)^2|n\rangle = \frac{\pi^2}{3}\left(\frac{r^2-1}{r^2}\right), \tag{7}$$

which is clearly independent of n and tends to the value $\pi^2/3$ as r tends to infinity. Again this difference is a consequence of the manner in which the limit has been taken.

3. Hermitian optical phase operator

The Hermitian optical phase operator $\hat{\phi}_\theta$ acts in a formally finite state space Ψ spanned by $s+1$ number states $\{|n\rangle\}$ [2, 3]. Evaluation of the properties of the phase operator requires taking the limit as s tends to infinity. However, the manner in which this is done is crucially different from that advocated by other authors [4–6, 9, 11, 12] in that the limit is applied only to c-number expectation values and moments and ***not directly to the state space itself***. Thus all states and operators are defined in an arbitrarily large although formally finite state space.

The $(s+1)$-dimensional state space, Ψ, can be spanned by a complete orthonormal basis of phase states:

$$|\theta_m\rangle=(s+1)^{-1/2}\sum_{n=0}^{s}\exp(in\theta_m)|n\rangle, \tag{8}$$

where

$$\theta_m=\theta_0+\frac{2\pi m}{s+1}, \qquad m=0,1,\ldots,s. \tag{9}$$

We can choose a particular phase basis by picking the value of θ_0. The Hermitian optical phase operator has the phase states as eigenstates with corresponding eigenvalues θ_m:

$$\hat{\phi}_\theta=\sum_{m=0}^{s}\theta_m|\theta_m\rangle\langle\theta_m|. \tag{10}$$

The phase and number operators are canonically conjugate in the state space Ψ [16]. The matrix elements of the phase operator in the number state basis are [2, 3]:

$$\langle n|\hat{\phi}_\theta|n\rangle=\theta_0+\pi\frac{s}{s+1}, \tag{11 a}$$

$$\langle n'|\hat{\phi}_\theta|n\rangle=\frac{2\pi}{s+1}\frac{\exp[i(n'-n)\theta_0]}{\exp[2\pi i(n'-n)/(s+1)]-1}, \qquad n'\neq n. \tag{11 b}$$

We note the similarity between these expressions and those obtained by Yarunin *et al.* but emphasize that the difference lies in the manner in which the limits are taken. The expectation value and variance of the phase operator (as well as higher-order moments) are evaluated in the state space Ψ and the dimension of the space is only then allowed to tend to infinity. For example, the expectation value and variance of the phase for a number state $|n\rangle$ is:

$$\begin{aligned}\langle n|\hat{\phi}_\theta|n\rangle&=\lim_{(s\to\infty)}\langle n|\sum_{m=0}^{s}\theta_m|\theta_m\rangle\langle\theta_m|n\rangle\\&=\theta_0+\pi,\end{aligned} \tag{12 a}$$

$$\begin{aligned}\Delta\phi_\theta^2&=\lim_{(s\to\infty)}\langle n|\left\{\sum_{m=0}^{s}\theta_m|\theta_m\rangle\langle\theta_m|-\langle n|\hat{\phi}_\theta|n\rangle\right\}^2|n\rangle\\&=\frac{\pi^2}{3}.\end{aligned} \tag{12 b}$$

These results suggest that all the number states are states of random phase [2] and are in marked contrast to the expressions obtained when the limit is applied directly to the states and operators themselves [11, 12].

The exponential of i multiplied by the Hermitian optical phase operator is a unitary operator which acts in the number state basis to shift the number:

$$\exp(i\hat{\phi}_\theta)=|0\rangle\langle 1|+|1\rangle\langle 2|+\ldots+|s-1\rangle\langle s|+\exp[i(s+1)\theta_0]|s\rangle\langle 0|. \tag{13}$$

This operator differs from the infinite Hilbert space operator $\hat{V}$ by the presence of a projector coupling the vacuum state $|0\rangle$ to the most highly excited state $|s\rangle$. The absence of this term is responsible for the non-unitarity of $\hat{V}$ and for the fact that it does not commute with its Hermitian conjugate. As with the Hermitian optical phase operator, moments of the unitary phase operator, of its conjugate operator and of the associated trigonometric operators are calculated in the space Ψ and only then is s allowed to tend to infinity.

There is one further pertinent feature of the Hermitian optical phase operator formalism, namely the existence of a phase probability distribution. The phase states form a complete orthonormal basis in the state space Ψ and therefore we can express the probability that the system is in a given phase state as:

$$P(\theta_m)=\langle\theta_m|\rho|\theta_m\rangle, \tag{14}$$

where ρ is the density matrix describing the state. Moments of the phase operator and of operator functions of the phase operator can be calculated using this phase probability distribution, for example the expectation value of the cosine of the phase is:

$$\langle\cos(\hat{\phi}_\theta)\rangle=\lim_{(s\to\infty)}\sum_{m=0}^{s}P(\theta_m)\cos(\theta_m). \tag{15}$$

For *physical states*† these expressions will converge to a simpler form involving a continuous probability density:

$$\begin{aligned}\langle f(\hat{\phi}_\theta)\rangle&=\lim_{(s\to\infty)}\sum_{m=0}^{s}P(\theta_m)f(\theta_m)\\&=\int_{\theta_0}^{\theta_0+2\pi}f(\theta)\Pi(\theta)\,\mathrm{d}\theta,\end{aligned} \tag{16}$$

where $f(\ldots)$ denotes a function of the phase and $\Pi(\theta)$ is the probability density defined in terms of the limit:

$$\begin{aligned}\Pi(\theta)&=\lim_{(s\to\infty)}\frac{s+1}{2\pi}P(\theta)\\&=\frac{1}{2\pi}\sum_{n=0}^{\infty}\sum_{n'=0}^{\infty}\langle n'|\rho|n\rangle\exp[\mathrm{i}(n-n')\theta].\end{aligned} \tag{17}$$

We emphasize that this description, while applicable for physical states, cannot be applied to all states of Ψ [17]. In particular, expressions (16) and (17) cannot be applied to analyse properties of the phase states themselves.

4. Significance of the limiting procedure

It might seem strange that anything as subtle as the distinction between applying a limit to operators and states and applying it to expectation values and moments can have a significant effect. Yet the calculations of the number state phase variances give very different results depending on the manner in which the limit is evaluated (equations (6) and (12 *b*)). In order to understand why this should be we examine the calculation of the phase variance for the vacuum state, $|0\rangle$, in some detail.

Within the state space Ψ the phase operator may be written as:

$$\hat{\phi}_\theta=\theta_0+\frac{2\pi}{s+1}\left\{\frac{s}{2}+\sum_{n\neq n'}\frac{\exp[\mathrm{i}(n'-n)\theta_0]|n'\rangle\langle n|}{\exp[2\pi\mathrm{i}(n'-n)/(s+1)]-1}\right\}, \tag{18}$$

† That is states for which finite moments of the number operator are bounded in the limit that s tends to infinity [2, 3, 17].

where the summation is over all off-diagonal projectors in the number state basis. Following Popov and Yarunin [12] we assume that $|n'-n|$ is very much less than s for all states of interest and take the limit of (18) as s tends to infinity. The resulting operator, $\hat{\phi}$ is Hermitian with diagonal matrix elements $\theta_0+\pi$ and off-diagonal matrix elements $\mathrm{i}\exp[\mathrm{i}(n'-n)\theta_0]/(n-n')$. These matrix elements correspond to those given in equation (3) for the special case $\theta_0=0$. If we calculate the expectation values of $\hat{\phi}$ and its square for the vacuum then we find:

$$\langle 0|\hat{\phi}|0\rangle=\theta_0+\pi, \tag{19a}$$

$$\begin{aligned}\langle 0|\hat{\phi}^2|0\rangle&=(\theta_0+\pi)^2+\sum_{n\neq 1}\langle 0|\hat{\phi}|n\rangle\langle n|\hat{\phi}|0\rangle\\&=(\theta_0+\pi)^2+\sum_{n=1}^{\infty}\frac{1}{n^2}.\end{aligned} \tag{19b}$$

The sum is equal to $\pi^2/6$ and therefore the vacuum phase variance is:

$$\Delta\phi^2=\frac{\pi^2}{6}. \tag{20}$$

For a state of random phase we would expect the phase variance to be $\pi^2/3$. Moreover, as proven in the Appendix, the results (19 *a*) and (20) cannot *both* be consistent with *any* 2π periodic phase density!

Applying the limit to the state space and obtaining operators acting in infinite Hilbert space requires us to *neglect* all matrix elements of the phase that involve highly excited number states in Ψ. The *assumption* is that these matrix elements cannot be significant and must vanish as the dimension of the space tends to infinity. This is *incorrect*, for if we examine the number state matrix elements of the Hermitian optical phase operator we find:

$$\langle s-n|\hat{\phi}_\theta|0\rangle=\exp[\mathrm{i}(s+1)\theta_0]\langle 0|\hat{\phi}_\theta|n+1\rangle, \tag{21}$$

which clearly shows that phase operator matrix elements between the vacuum state and the *lowest* lying number states are of *comparable magnitude* to those between the vacuum and the *highest* lying states in Ψ. It is true that the overlap between the most highly excited states and any physically accessible state must vanish in the limit as s tends to infinity but this does not mean that these states are unimportant in the representation of the phase. In particular, when using the completeness relation to calculate the expectation value of the square of the phase:

$$\langle 0|\hat{\phi}_\theta^2|0\rangle=\sum_{n=0}^{s}\langle 0|\hat{\phi}_\theta|n\rangle\langle n|\hat{\phi}_\theta|0\rangle, \tag{22}$$

we *cannot* ignore the contribution from number states with n comparable to s. From equation (21) it is apparent that the most highly excited states are *equally* as important as those around $n=0$. We can highlight the roles of these different sets of number states by separating the sum in (22) into two components, one from the lower lying number states and one from the higher:

$$\langle 0|\hat{\phi}_\theta^2|0\rangle=\sum_{n=0}^{M}\langle 0|\hat{\phi}_\theta|n\rangle\langle n|\hat{\phi}_\theta|0\rangle+\sum_{n'=M+1}^{s}\langle 0|\hat{\phi}_\theta|n'\rangle\langle n'|\hat{\phi}_\theta|0\rangle, \tag{24}$$

where M is the greatest integer less than or equal to $s/2$. Making the change of variables $n' = s - n + 1$ gives:

$$\langle 0|\hat{\phi}_\theta^2|0\rangle = (\theta_0 + \pi)^2 + \sum_{n=1}^{M} \frac{1}{n^2} + \sum_{n=1}^{s-M} \frac{1}{n^2}. \tag{25}$$

In the limit as s tends to infinity *each of these sums* contributes $\pi^2/6$ to the phase moment and the resulting phase variance is:

$$\Delta\phi_\theta^2 = \frac{\pi^2}{3}$$

$$= 2 \times \Delta\phi^2. \tag{26}$$

The premature application of the limiting procedure to the operator and state space leads to the loss of half the full expression for the vacuum phase variance.

We can also trace the isometric rather than unitary property of the operator $\hat{V}$ to the application of the limiting procedure to the state space. The non-vanishing matrix elements of the unitary exponential phase operator (13) have the same magnitude, in particular:

$$\langle 0|\exp(i\hat{\phi}_\theta)|1\rangle = \exp[-i(s+1)\theta_0]\langle s|\exp(i\hat{\phi}_\theta)|0\rangle, \tag{27}$$

and they are therefore *equally important* in the representation of the operator. If we neglect the matrix element linking the vacuum state with the state $|s\rangle$ then we are left with operators, $\hat{V}$ and its conjugate, that are no longer unitary.

5. Discussion

In this paper we have contrasted two approaches to the description of the phase of a single mode field by a quantum mechanical operator. The difference between them is the form of the state space in which the operator acts. This difference is manifest in the limiting procedures employed. As stated by Popov and Yarunin [12] the two approaches lead to different results, particularly for the vacuum. However, we *do not accept* that the procedure of Popov and Yarunin is correct and that ours is a mistake. Indeed from our earliest work [1] we have attributed the previous difficulties with the quantum description of phase to taking the limit as s tends to infinity prematurely, which removes the phase operator matrix elements between $|0\rangle$ and $|s\rangle$. The procedure of taking the limit only after expectation values are calculated as functions of s is *the essential feature* of our approach.

Our procedure *avoids* inconsistencies such as those described above in relation to the phase of the vacuum. It is reasonably simple to use for calculation the phase properties of various states of light and gives physically sensible and self-consistent results. Within the state space, Ψ, the phase operator is conjugate to the number operator [16]. Moreover, the results obtained using the Hermitian optical phase operator are consistent with those obtained using the system-theory approach to quantum measurement [18]. The formalism has also been extended to describe the properties of related systems such as rotation angle operators conjugate to angular momentum [19] and q-deformed oscillators [20]. We see no reason to regard our approach as being in any way approximate or inexact. However, there are very good physical reasons (notably in describing the vacuum and obtaining consistency with a 2π periodic phase probability density) for *not* employing the infinite Hilbert space approaches to describe phase.

Since submitting this manuscript we have received a preprint in which the Garrison-Wong formalism is compared with the Hermitian optical phase operator [21]. The authors of this paper reach similar conclusions to those presented here.

Acknowledgments

We thank Adam Miranowicz for interesting discussions and for preparing translations of the early papers by Yarunin and coworkers. S.M.B. thanks the Royal Society of Edinburgh for the award of a Research Fellowship.

Appendix

In this appendix we prove the following result: If for a 2π periodic phase probability distribution $[\Pi(\theta)=\Pi(\theta+2\pi)]$ the mean value of the phase for any given 2π range of phase values $[\theta_0, \theta_0+2\pi)$ is:

$$\langle\theta\rangle=\theta_0+\pi, \tag{A 1}$$

for *all* θ_0, then the phase distribution *must be flat:*

$$\Pi(\theta)=\frac{1}{2\pi}. \tag{A 2}$$

The proof of this is based on the basic theorem of calculus: If

$$F(t)=\int_{t_0}^{t} f(\tau)\,\mathrm{d}\tau, \tag{A 3}$$

then

$$\frac{\mathrm{d}F(t)}{\mathrm{d}t}=f(t). \tag{A 4}$$

From equation (A 1) we can write the mean value of the phase as:

$$\begin{aligned}\theta_0+\pi&=\int_{\theta_0}^{\theta_0+2\pi}\theta\Pi(\theta)\,\mathrm{d}\theta\\&=\int_{\theta_0}^{\theta_0+\alpha}\theta\Pi(\theta)\,\mathrm{d}\theta+\int_{\theta_0+\alpha}^{\theta_0+2\pi}\theta\Pi(\theta)\,\mathrm{d}\theta.\end{aligned} \tag{A 5}$$

We can use the assumption that (A 1) holds for any value of θ_0 and choose a new range of phase values by replacing θ_0 by $\theta_0+\alpha$ to give:

$$\begin{aligned}\theta_0+\pi+\alpha&=\int_{\theta_0+\alpha}^{\theta_0+2\pi+\alpha}\theta\Pi(\theta)\,\mathrm{d}\theta\\&=\int_{\theta_0+\alpha}^{\theta_0+2\pi}\theta\Pi(\theta)\,\mathrm{d}\theta+\int_{\theta_0+2\pi}^{\theta_0+2\pi+\alpha}\theta\Pi(\theta)\,\mathrm{d}\theta.\end{aligned} \tag{A 6}$$

Subtracting (A 5) from (A 6) we find:

$$\alpha=\int_{\theta_0+2\pi}^{\theta_0+\alpha+2\pi}\varphi\Pi(\varphi)\,\mathrm{d}\varphi-\int_{\theta_0}^{\theta_0+\alpha}\theta(\theta)\,\mathrm{d}\theta. \tag{A 7}$$

We can make the change of dummy variable $\varphi = \theta + 2\pi$ and use the assumed 2π periodicity of $\Pi(\theta)$ to rewrite (A 7) as:

$$\alpha = \int_{\theta_0}^{\theta_0+\alpha} (\theta + 2\pi)\Pi(\theta)\,d\theta - \int_{\theta_0}^{\theta_0+\alpha} \theta\Pi(\theta)\,d\theta$$

$$= 2\pi \int_{\theta_0}^{\theta_0+\alpha} \Pi(\theta)\,d\theta, \tag{A 8}$$

or

$$\frac{\alpha}{2\pi} = \int_{\theta_0}^{\theta_0+\alpha} \Pi(\theta)\,d\theta. \tag{A 9}$$

Therefore from the basic theorem of calculus we have:

$$\frac{d}{d(\theta_0+\alpha)}\left(\frac{\alpha}{2\pi}\right) = \Pi(\theta_0+\alpha), \tag{A 10}$$

which implies that:

$$\Pi(\theta_0+\alpha) = \frac{1}{2\pi} \tag{A 11}$$

for *any* α. The phase distribution is flat and a calculation of the variance of the phase for this distribution leads to the result [2]:

$$\Delta\theta^2 = \frac{\pi^2}{3}. \tag{A 12}$$

Thus any operator for which (A 1) holds (for *all* θ_0) but for which (A 12) does not *cannot* be consistent with a 2π periodic probability density and thus *cannot* represent phase.

References

[1] Pegg, D. T., and Barnett, S. M., 1988, *Europhys. Lett.*, **6**, 483.
[2] Barnett, S. M., and Pegg, D. T., 1989, *J. mod. Opt.*, **36**, 7.
[3] Pegg, D. T., and Barnett, S. M., 1989, *Phys. Rev.* A., **39**, 1665.
[4] Garrison, J. C., and Wong, J., 1970, *J. Math. Phys.*, **11**, 2243.
[5] Popov, V. N., and Yarunin, V. S., 1973, *The Leningrad University Journal: Physics*, **22**, 7.
[6] Damaskinsky, E. V., and Yarunin, V. S., 1978, *The Academic Research Journal (Tomsk University): Physics*, **6**, 59.
[7] Levy-Leblond, J.-M., 1973, *Revista Mexicana de Fisica*, **22**, 15.
[8] Santhanam, T. S., 1977, *Found. Phys.*, **7**, 121.
[9] Santhanam, T. S., and Shina, K. B., 1978, *Aust. J. Phys.*, **31**, 233.
[10] Goldhirsch, I., 1980, *J. Phys.* A., **13**, 3479.
[11] Bergou, J., and Englert, B.-G., 1991, *Ann. Phys.*, **209**, 479.
[12] Popov, V. N., and Yarunin, V. S., 1992, *J. mod. Optics*, **39**, 1525.
[13] Susskind, L., and Glogower, J., 1964, *Physics*, **1**, 49.
[14] Carruthers, P., and Nieto, M. M., 1968, *Rev. mod. Phys.*, **40**, 411.
[15] Levy-Leblond, J.-M., 1976, *Ann. Phys.*, **101**, 173.
[16] Pegg, D. T., Vaccaro, J. A., and Barnett, S. M., 1990, *J. mod. Optics*, **37**, 1703.
[17] Vaccaro, J. A., and Pegg, D. T., 1990, *J. mod. Optics*, **37**, 17.
[18] Shapiro, J. H., and Shepard, S. R., 1991, *Phys. Rev.* A, **43**, 3795.
[19] Barnett, S. M., and Pegg, D. T., 1990, *Phys. Rev.* A, **41**, 3427.
[20] Ellinas, D., 1992, *Phys. Rev.* A, **45**, 3358.
[21] Gantsog, Ts., Miranowicz, A., and Tanas, R., *Phys. Rev.* A (in press).

Consistency of Quantum Descriptions of Phase

J. A. Vaccaro* and D. T. Pegg

Faculty of Science and Technology, Griffith University, Brisbane 4111, Australia

Received October 6, 1992; accepted in revised form January 5, 1993

Abstract

In the Ψ-space limiting procedure limits of expectation values, rather than operator or states, are found as the state space dimensionality tends to infinity. This approach has been applied successfully to the calculation of the phase properties of various states of light, but its status as a valid quantum mechanical thory equivalent to the usual infinite Hilbert space (H-space) approach has not yet been fully accepted. Here we address this issue by investigating the formal relationship between the two approaches. We establish the consistency between the Ψ-space and H-space approaches for observables which are amenable to an H-space treatment. Such observables are represented in H by operators which are strong limits of Ψ-space operators and which obey the same algebra as the corresponding Ψ-space operators. The phase operator, however, exists in H only as a weak limit of a Ψ-space operator. For such limits the Ψ-space operator algebra is not preserved, which is the fundamental reason for the difficulties in constructing a consistent quantum description of phase in H. We show that for the phase observable the Ψ-space approach is consistent with the probability-operator measure (POM) method with the important distinction that, whereas the relation between non-orthogonal POMs and probability has to be accepted in the latter method as a postulate, the corresponding relation is derived in the Ψ-space approach. We conclude that the Ψ-space approach is not only equivalent to, but is also more fundamental than both the H-space and POM approaches.

1. Introduction

The conventional approach to calculating properties of the quantum electromagnetic field is to use operators and states of the infinite Hilbert space H. For example, the predicted probability distribution of measured values of an observable for a given state of the field can be determined by calculating the moments of the distribution as expectation values of powers of the associated operator. Acceptance of this approach leads to well-known difficulties in finding a physically reasonable quantum description of phase. The underlying reason for this can be appreciated by examining the construction of the space H from a more fundamental starting point, that is, as the completion of a finite-dimensional Hilbert space Ψ_s with respect to a particular topology [1]. The limiting procedure used for this is described in the next section, but the essential point is that one does not simply replace the dimensionality of Ψ_s by infinity. If this were done, the resulting space would no longer be a Hilbert space because it would contain non-normalizable states. The limiting procedure must also involve a restriction to yield a Hilbert space, and it is this restriction which precludes the existence of phase states and a phase operator in H. From this fundamental point of view, the conventional approach may be described as starting with Ψ_s, immediately applying the limiting procedure to obtain H, constructing states and operators of H and then calculating expectation values and probability distributions. While this prescription has the advantage that the limiting process has already been incorporated into the states and operators, it has the disadvantage that the necessary restriction associated with the limiting process reduces the number of physical quantities, such as phase, for which the approach is applicable.

An alternative procedure has been introduced for calculating quantum mechanical expectation values and hence probability distributions [2]. This procedure does not need to incorporate the above restriction. Here states and operators of Ψ_s are first used to calculate an expectation value $E(s)$ which, in general, is a function of s. The limiting process is then applied at this stage, in that the expectation value of an observable for any state, even those which contain an infinite number of coefficients, is found as the limit E of $E(s)$ as s tends to infinity. E is thus the limit point of a sequence of numbers on the real line, each of which corresponds to a particular finite dimension of the state space. E need not be one of the numbers in the sequence and therefore need not have a value corresponding to that obtained from the operators or states of a space of a finite dimension. We shall refer to this approach as the Ψ-space formalism to distinguish it from the more conventional H-space formalism.

The Ψ-space procedure has been used to investigate the quantum-mechanical nature of phase [2–5] and other observables [6, 7] which are not readily amenable to the H-space approach. It has been found useful for calculating the phase properties of various states of light belonging to H, but its precise relationship to the conventional H-space approach has not been fully established. In this paper we address this problem and examine the consistency between the two approaches by deriving the formal links. We then extend our analysis to show that the phase operators in the Ψ-space approach are consistent with operators constructed on H from the probability-operator measure approach but that the former approach is more fundamental in that the expression corresponding to the basic postulate of the latter is derived in the Ψ-space approach.

2. States and operators of Ψ

The limiting procedure by which the infinite space H is constructed is as follows [1]. The $(s+1)$-dimensional Hilbert space Ψ_s, spanned by the number states $|0\rangle, |1\rangle, \ldots, |s\rangle$ contains all states $|f\rangle_s$ with number state expansions $\sum_{n=0}^{s} c_n |n\rangle$. As we allow s to increase to arbitrarily large, but finite, values we contruct a sequence of spaces $\Psi_1, \Psi_2, \ldots$ where Ψ_n is a subspace of Ψ_{n+1}. Sequences of states in these

---

* Permanent address: Faculty of CAD, Griffith University, Brisbane 4111, Australia.

spaces include, for example, $|f\rangle_1, |f\rangle_2, \ldots$ where $|f\rangle_n \in \Psi_n$. In particular there are Cauchy sequences of states for which the norm of $|f\rangle_n - |f\rangle_m$ tends to zero as n and m tend to infinity but whose limit points do not lie within Ψ_s. To form H we adjoin these limit points, which correspond to states $|f\rangle$ with an infinite number of coefficients such that the norm

$$\||f\rangle\| = \left(\sum_{m=0}^{\infty} |c_m|^2\right)^{1/2}$$

is bounded. This restricts the states of H to those for which $|c_n| \to 0$ as $n \to \infty$. Thus, whereas we can make s large enough for Ψ_s to include any state with a finite number of coefficients c_n, whether or not c_n decrease in modulus as n increases, H does not contain all states with an infinite number of coefficients. For example a state with all coefficients of equal modulus, which might be expected for a phase state, does not belong to H. We also note that the limiting process which allows s in the number state $|s\rangle$ to tend to infinity is carried out in constructing H, and is thus already incorporated into the states and operators of H.

The states and operators of H are familiar. For example $|f\rangle$ could be a coherent state. In order to use the alternative Ψ-space formalism, we must define corresponding states and operators in Ψ_s. To relate states and operators of both approaches we define the Ψ-restrictions of states and operators as follows. $|f\rangle_s$ is the Ψ-restriction of $|f\rangle$ if

$$\langle n|f\rangle_s = \langle n|f\rangle \quad \text{for } n \leqslant s$$

$$\langle n|f\rangle_s = 0 \quad \text{for } n > s$$

and $\hat{A}_s$ is the Ψ-restriction of $\hat{A}$ if

$$\langle n|\hat{A}_s|m\rangle = \langle n|\hat{A}|m\rangle \quad \text{for } n \leqslant s \text{ and } m \leqslant s$$

$$\langle n|\hat{A}_s|m\rangle = 0 \quad \text{for } n > s \text{ or } m > s$$

where $|n\rangle$ and $|m\rangle$ are number states.

In the Ψ-space procedure operators on H are in general represented in Ψ_s by their restrictions, for example the number, annihilation and creation operators $\hat{N}_H$, $\hat{a}_H$ and $\hat{a}_H^\dagger$ are represented by $\hat{N}_s$, $\hat{a}_s$ and $\hat{a}_s^\dagger$ respectively. H-space states $|f\rangle$ are also represented by their renormalized restriction $k_s|f\rangle_s$ where k_s is a renormalization factor. As s is allowed to become arbitrarily large to form the sequence of expectation values $|k_s|^2 {}_s\langle f|\hat{A}_s|f\rangle_s$, the states $k_s|f\rangle_s$ will either eventually become exactly equal to their H-space counterparts if these have a finite number of number state coefficients, or will approach their counterparts if an infinite number of coefficients is involved. As we are interested only in the limiting value of the expectation and the renormalization factor k_s approaches unity as $s \to \infty$, we often ignore this factor and use instead the sequence ${}_s\langle f|\hat{A}_s|f\rangle_s$, where $|f\rangle_s$ are the unrenormalized restrictions, because this sequence has the same limit. From these Ψ-space states and operators we can find complementary states and conjugate operators in Ψ_s, the most notable of these are the phase states and the phase operator [2–5] but other conjugate observables have also been examined [6, 7]. Because states such as phase states are not restrictions of H-space states, it is necessary to retain the precise normalization factor for these states in the limiting procedure. Phase states on Ψ_s are normalized restrictions of states which have an infinite number of number-state coefficients of equal modulus and which belong to the state space isomorphic to C^∞, a much larger state space than H.

In addition to incorporating states of equal coefficients, the Hilbert space Ψ_s has mathematical advantages over the infinite-dimensional Hilbert space H. All the relevant operators in Ψ_s are bounded, every self-adjoint operator has a complete system of eigenstates and all associative algebraic operations with operators are allowed, with no questions arising concerning the domain of definition. We also note that, because the operators act on a finite-dimensional space, there is no need to distinguish between self-adjoint and Hermitian operators. For example the phase operator defined by eq. (16) is both self-adjoint and Hermitian. Irrespective of the terminology used, the important question is whether or not such operators can suitably represent physical quantities. The fundamental postulate of the probabilistic interpretation of quantum mechanics requires that the operator associated with a dynamical variable must yield real expectation values for any states of the system. In the Ψ-space formalism the use of self-adjoint operators acting on Ψ_s ensures that the expectation value is the limit of a sequence of points on the real line and hence must itself be real. The fundamental postulate is thereby satisfied.

3. Strongly convergent operators

As Ψ_s is a subspace of H, operators defined on Ψ_s also form a sequence on H as s is allowed to increase. In this section we explore the relationship between calculations performed on Ψ_s and H by examining the sequences of operators and their limit points on H.

For a sequence of operators to have a strong limit, the sequence of states generated by the operators must converge strongly. We note that the state $|f\rangle$ is the strong limit point of a Cauchy sequence of states $|f\rangle_1, |f\rangle_2, \ldots$ if, and only if, $\||f\rangle - |f\rangle_n\| \to 0$ as $n \to \infty$ [8]. The construction described in the Introduction ensures that all such limit points $|f\rangle$ belong to H. A sequence of operators $\hat{Q}_n$ converges *strongly* to a limit point $\hat{Q}$ on H if, and only if, the resultant state $\hat{Q}_n|f\rangle = |g\rangle_n$ converges strongly to the state $|g\rangle = \hat{Q}|f\rangle$ for all $|f\rangle$ in H [8]. The transformed state $|g\rangle$ also belongs to H and can be acted on by another operator. Clearly the operators $\hat{Q}_1, \hat{Q}_2, \ldots$ and $\hat{Q}$ must be bounded on H, that is, $\|\hat{Q}_n|f\rangle\| < \infty$ for all n and $\|\hat{Q}|f\rangle\| < \infty$ for all $|f\rangle \in$ H.

We first examine the convergence of Ψ-restrictions of all bounded H-space operators $\hat{Q}$. Firstly we note that $\hat{Q}$ operates only on states in its domain D($\hat{Q}$), defined as the set of states $|f\rangle \in$ H for which the resultant state $\hat{Q}|f\rangle$ belongs to H, that is $\|\hat{Q}|f\rangle\| < \infty$. Thus the convergence behaviour is in relation to states of D($\hat{Q}$). Let

$$\hat{A}_s|f\rangle = |g\rangle_s \tag{1}$$

$$\hat{A}|f\rangle = |g\rangle \tag{2}$$

for $|f\rangle \in$ D($\hat{A}$), where $\hat{A}_s$ is the Ψ-restriction of the H operator $\hat{A}$. Then

$$|g\rangle - |g\rangle_s = \sum_{n=0}^{\infty} \sum_{m=s+1}^{\infty} |n\rangle A_{nm} f_m + \sum_{n=s+1}^{\infty} \sum_{m=0}^{s} |n\rangle A_{nm} f_m \tag{3}$$

where $A_{nm} = \langle n|\hat{A}|m\rangle$ and $f_m = \langle m|f\rangle$. The square of the norm of this is

$$\||g\rangle - |g\rangle_s\|^2 = \left(\sum_{p=s+1}^{\infty}\sum_{n=0}^{\infty}\sum_{m=s+1}^{\infty} + \sum_{p=s+1}^{\infty}\sum_{n=s+1}^{\infty}\sum_{m=0}^{s} + \sum_{p=0}^{s}\sum_{n=s+1}^{\infty}\sum_{m=s+1}^{\infty} + \sum_{p=0}^{s}\sum_{n=s+1}^{\infty}\sum_{m=0}^{s}\right) \times A^*_{np}A_{nm}f^*_p f_m \tag{4}$$

We note that

$$\||g\rangle\|^2 = \sum_{n=0}^{\infty}\sum_{m=0}^{\infty}\sum_{p=0}^{\infty} A^*_{np}A_{nm}f^*_p f_m < \infty$$

and also that each term in eq. (4) involves at least one summation from $s+1$ to ∞. It follows from this that eq. (4) must vanish as $s \to \infty$ and thus $\||g\rangle - |g\rangle_s\|$ also vanishes in this limit. Hence the Ψ-restrictions of all bounded operaors defined on H converge strongly.

We now explore the relationship between the algebra of the bounded operators defined on H and their Ψ-restrictions defined on Ψ-space. Consider two Ψ-restrictions $\hat{A}_s$ and $\hat{B}_s$ of the bounded H-space operators $\hat{A}$ and $\hat{B}$, respectively. The states $|g\rangle_s = \hat{A}_s|f\rangle$ and $|h\rangle_s = \hat{B}_s|g\rangle$ will converge strongly to $|g\rangle = \hat{A}|f\rangle$ and $|h\rangle = \hat{B}|g\rangle$ where $|f\rangle \in D(\hat{A})$ and $|g\rangle \in D(\hat{B})$. We can find the strong limit of the product $\hat{B}_s\hat{A}_s$ as follows:

$$\hat{B}_s\hat{A}_s|f\rangle = \hat{B}_s|g\rangle_s \tag{5}$$

so

$$\hat{B}_s\hat{A}_s|f\rangle - \hat{B}\hat{A}|f\rangle = |\Delta h_s\rangle + \hat{B}_s|\Delta g_s\rangle \tag{6}$$

where

$$|\Delta g_s\rangle = |g\rangle_s - |g\rangle \quad \text{and} \quad |\Delta h_s\rangle = |h\rangle_s - |h\rangle.$$

The norm of the right-hand side of eq. (6) is, by the triangle inequality [8],

$$\||\Delta h_s\rangle + \hat{B}_s|\Delta g_s\rangle\| \leqslant \||\Delta h_s\rangle\| + \|\hat{B}_s|\Delta g_s\rangle\|. \tag{7}$$

As $s \to \infty$ the norm $\||\Delta h\rangle_s\| \to 0$ because $|h\rangle_s$ converges strongly to $|h\rangle$. After some algebraic manipulation we find that

$$\|\hat{B}_s|\Delta g_s\rangle\|^2 = \sum_{n=0}^{s}\sum_{q=s+1}^{\infty}\sum_{t=0}^{s}\sum_{p=s+1}^{\infty}\sum_{m=0}^{s} B^*_{nt}B_{nm}A^*_{tq}A_{mp}f^*_q f_p. \tag{8}$$

Noting that $\hat{B}_s$ is bounded:

$$\|\hat{B}_s|g\rangle\|^2 = \sum_{n=0}^{s}\sum_{q=0}^{\infty}\sum_{t=0}^{s}\sum_{p=0}^{\infty}\sum_{m=0}^{s} B^*_{nt}B_{nm}A^*_{tq}A_{mp}f^*_q f_p < \infty$$

and that the sums over q and p in eq. (8) are from $s+1$ to infinity, we find that $\|\hat{B}_s|\Delta g_s\rangle\| \to 0$ as $s \to \infty$. Thus the norm of the left side of eq. (6) vanishes and so $\hat{B}_s\hat{A}_s$ converges strongly to $\hat{B}\hat{A}$ as $s \to \infty$. We can symbolize this important result as

$$\left(\text{"}\lim_{s\to\infty}\text{"}\ \hat{B}_s\right)\left(\text{"}\lim_{s\to\infty}\text{"}\ \hat{A}_s\right)|f\rangle = \text{"}\lim_{s\to\infty}\text{"}\ (\hat{B}_s\hat{A}_s)|f\rangle.$$

Thus the algebra of all bounded operators defined on H is the *same* as that of the operators of Ψ_s which are their Ψ-restrictions.

We are now in a position to answer the following crucial question. Suppose $\hat{Q}$ is a bounded H-space operator and $|f\rangle$ is a state within its domain. Will calculations of the expectation value of $\hat{Q}$ for this state, firstly by means of the Ψ-space formalism using the Ψ-restriction $\hat{Q}_s$ and secondly by the H-space formalism, yield a consistent result? An example of this might be the calculation of the mean square energy of a coherent state. We obtain consistency if

$$\lim_{s\to\infty} {}_s\langle f|\hat{Q}_s|f\rangle_s = \langle f|\hat{Q}|f\rangle. \tag{9}$$

where $|f\rangle_s$ is the Ψ-restriction of $|f\rangle$. Since $\hat{Q}_s|n\rangle = 0$ and $\langle n|\hat{Q}_s = 0$ for $n > s$, we have ${}_s\langle f|\hat{Q}_s|f\rangle_s = \langle f|\hat{Q}_s|f\rangle$. Further, $\hat{Q}_s$ converges strongly to $\hat{Q}$, therefore the resultant state $\hat{Q}_s|f\rangle$ converges to the state $\hat{Q}|f\rangle$ as $s \to \infty$. It follows that the l.h.s. of eq. (9) can be written as

$$\lim_{s\to\infty} \langle f|\hat{Q}_s|f\rangle = \left\langle f\left|\lim_{s\to\infty}\hat{Q}_s\right|f\right\rangle = \langle f|\hat{Q}|f\rangle. \tag{10}$$

Our previous discusson then shows that if $\hat{Q}$ is an operator function $F(\hat{A}, \hat{B})$ of other bounded operators $\hat{A}$ and $\hat{B}$ each of which are the strong limits of Ψ-restrictions $\hat{A}_s$ and $\hat{B}_s$, where $F(x, y)$ is a power series in x and y, then

$$\lim_{s\to\infty} {}_s\langle f|F(\hat{A}_s, \hat{B}_s)|f\rangle_s = \langle f|F(\hat{A}, \hat{B})|f\rangle \tag{11}$$

and the two approaches are consistent.

As an example, the operators $\hat{a}_s$ and $\hat{a}_s^\dagger$ as defined in Ψ_s [2] turn out to be Ψ-restrictions of $\hat{a}_H$ and $\hat{a}_H^\dagger$ defined in H. The domain of the operator function $F(\hat{a}_H, \hat{a}_H^\dagger)$ which represents a linear combination of products of $\hat{a}_H$ and $\hat{a}_H^\dagger$ is the set of physical states $|\Phi\rangle$, that is, those for which $\langle\Phi|(\hat{a}_H^\dagger\hat{a}_H)^q|\Phi\rangle < \infty$ for all given q. This excludes, for instance, the H-state $(6^{1/2}/\pi)\sum^{\infty}(n+1)^{-1}|n\rangle$ because the action of $\hat{a}_H^\dagger\hat{a}_H + 1$ generates a state in the space isomorphic to C^∞ but not in the space H. $F(\hat{a}_H, \hat{a}_H^\dagger)$ includes most operators used in quantum electrodynamics. A special example of $F(\hat{a}_H, \hat{a}_H^\dagger)$ is the commutator $[\hat{a}_H, \hat{a}_H^\dagger]$ which is equivalent in both approaches to the unit operator when acting on $|\Phi\rangle$ or its Ψ-restriction.

4. Weakly convergent operators

There also exist sequences of operators defined on Ψ_s which do not converge strongly on H. We now consider operators which converge in a weaker mode called *weak* convergence. A sequence of operators $\hat{A}_s$ converges weakly to the weak limit $\hat{A}_W$ on H if

$$\langle g|\hat{A}_s|f\rangle \to \langle g|\hat{A}_W|f\rangle \quad \text{as } s \to \infty$$

for all $|g\rangle$ and $|f\rangle$ belonging to H [8]. If a sequence of operators is strongly convergent it is also weakly convergent, but there are weakly converging sequences which do not converge strongly. We now examine the algebra of the operators of the latter. We note that these will not be Ψ-restrictions of operators in H.

Let $\hat{C}_s$ and $\hat{G}_s$ be operators defined on Ψ_s which have weak limits $\hat{C}_W$ and $\hat{G}_W$ respectively on H:

$$\langle g|(\hat{C}_s - \hat{C}_W)|f\rangle \to 0 \quad \text{as } s \to \infty$$

$$\langle g|(\hat{G}_s - \hat{G}_W)|h\rangle \to 0 \quad \text{as } s \to \infty$$

for all $|f\rangle, |g\rangle, |h\rangle \in$ H. Further let $|h\rangle_s = \hat{C}_s|f\rangle$ not converge strongly to a state in H. An example of such a state is provided by the unitary phase operator, that is, $|h\rangle_s$ is given by

$$\exp(\mathrm{i}\hat{\phi}_\theta)_s|f\rangle = \sum_{n=1}^{s}|n-1\rangle\langle n|f\rangle + |s\rangle\langle 0|f\rangle \tag{12}$$

where we have taken $\theta_0 = 0$ for convenience [2]. The weak limit of $\exp(\mathrm{i}\hat{\phi}_\theta)_s$ is the Susskind–Glogower operator [9] $\hat{e}_{SG}$, for which $\hat{e}_{SG}|f\rangle = \sum_{n=1}^{\infty}|n-1\rangle\langle n|f\rangle \equiv |\mathrm{H}\rangle$. This is consistent with the general results that $\hat{C}_W|f\rangle$ is a state $|\mathrm{H}\rangle$ belonging to H (which can have zero norm). We note that $|h\rangle_s$ is *not* the Ψ-restriction of $|\mathrm{H}\rangle$, for otherwise it would converge strongly to $|\mathrm{H}\rangle$. The difference between the weak limit of the product $\hat{G}_s\hat{C}_s$ and the product of the weak limits $\hat{G}_W\hat{C}_W$ is illustrated by

$$\begin{aligned}\lim_{s\to\infty} \langle g|(\hat{G}_s\hat{C}_s - \hat{G}_W\hat{C}_W)|f\rangle \\ &= \lim_{s\to\infty}\langle g|(\hat{G}_s|h\rangle_s - \hat{G}_W|\mathrm{H}\rangle) \\ &= \lim_{s\to\infty}\langle g|\hat{G}_s(|h\rangle_s - |\mathrm{H}\rangle_s)\end{aligned} \tag{13}$$

where $|\mathrm{H}\rangle_s$ is the Ψ-restriction of $|\mathrm{H}\rangle$. Now $|h\rangle_s$ and $|\mathrm{H}\rangle_s$ can be quite different states, yielding a non-zero r.h.s. of eq. (13). As a concrete example, when $\hat{G}_s^\dagger = \hat{C}_s = \exp(\mathrm{i}\hat{\phi}_\theta)_s$ the r.h.s. of eq. (13) becomes $\langle g|0\rangle\langle 0|f\rangle$ which is non-zero for many $|f\rangle$ and $|g\rangle$. Thus, in general, the weak limit of the product, $(\hat{G}\hat{C})_W$, is not the product $\hat{G}_W\hat{C}_W$ of the weak limits.

We now address the following question. Suppose $\hat{R}$ is an H-space operator which is also the weak limit of an operator $\hat{R}_s$ on Ψ and $|f\rangle_s$ is a Ψ-restriction of an H-space state $|f\rangle$. Do the H-space and Ψ-space formalisms give the same expectation for this operator and state? The same result will be obtained only if

$$\lim_{s\to\infty} {}_s\langle f|\hat{R}_s|f\rangle_s = \langle f|\hat{R}|f\rangle. \tag{14}$$

Using again the fact that ${}_s\langle f|\hat{R}_s|f\rangle_s = \langle f|\hat{R}_s|f\rangle$ because $|f\rangle_s$ is a Ψ-restriction of $|f\rangle$, eq. (14) follows immediately from the definition of the weak limit. Suppose now that $\hat{R}_s$ is a power series function $F(\hat{A}_s, \hat{B}_s)$ of other operators $\hat{A}_s$ and $\hat{B}_s$; do we obtain a result analogous to eq. (11)? This time the answer is no, if the function involves the product of two operators which converge weakly but not strongly, because the weak limit of the product is not necessarily the product of the weak limits. Thus in general for such a case

$$\lim_{s\to\infty} {}_s\langle f|F(\hat{A}_s, \hat{B}_s)|f\rangle_s \neq \langle f|F(\hat{A}_W, \hat{B}_W)|f\rangle. \tag{15}$$

We use the phase operators $\exp(-\mathrm{i}\hat{\phi}_\theta)_s$ and $\exp(\mathrm{i}\hat{\phi}_\theta)_s$ to illustrate this result. The Hermitian combination $(\cos\hat{\phi}_\theta)_s$ has as a weak limit the Susskind–Glogower [9] operator $(\cos\hat{\phi}_\theta)_W$. Relation (14) shows that the expectation value of $\cos\phi$ for the state $|f\rangle$ can be calculated by using both formalisms with the same result. The weak limit of $(\cos^2\hat{\phi}_\theta)_s$, however, is $(\cos^2\hat{\phi}_\theta)_W$ which is *not* the same as $[(\cos\hat{\phi}_\theta)_W]^2$. Thus calculations of the expectation values of these two weak-limit operators in H will, in general, lead to different results. For example for the vacuum the first gives a value of $\frac{1}{2}$, which is both physically reasonable and consistent with the Ψ-space formalism, while the second gives only $\frac{1}{4}$.

If we wish to construct phase operators in H-space which obey the normal operator algebra we could, of course, choose a basic operator to represent a particular function of phase, for example ϕ, $\cos\phi$ or $\cos^2\phi$ and then define the other functions of phase in terms of normal algebraic functions of the chosen operator. If we do this, however, we still cannot obtain phase distributions consistent with number–phase complementarity. For example if we choose our basic operator such that $\langle\cos^2\phi\rangle$ is $\frac{1}{2}$ for a number state, which is consistent with a uniform phase probability distribution, the higher moments such as that of $\cos^4\phi \equiv (\cos^2\phi)^2$, will not be consistent with a uniform distribution. Clearly we *can* calculate consistent expectation values of a function of phase angle using operators in Hilbert space *provided* we calculate the H-space operator representing the function as a weak limit of the corresponding function in Ψ_s, and *not* calculate it as a function of weak limits. Thus, for example, there is a well-defined operator in H which can be used for the direct calculation of the mean phase angle and there is also a well-defined operator in H for the square of the phase angle which will allow, for example, the phase variance of states in H to be calculated, but the latter operator is not the square of the former. In general, the phase operator algebra of Ψ-space, which is the normal operator algebra, is not preserved on H for weak limits of phase operators.

5. H-space phase operators and POMs

From the previous section, we see that we can calculate the phase properties of any given quantum state of light in H either by using the full limiting procedure of the Ψ-formalism or by using H-space operators for which no limiting procedure is necessary. For the latter method, however, we would require a list of weak limits of various functions of phase. Is there an H-space procedure which is simpler than this? In this section we seek to establish a formula for the weak limit of a general function $F(\hat{\phi}_\theta)$. This is an H-space operator which can be used to calculate the phase properties of H-space states.

In Ψ-space the basic phase operator is [3, 4]

$$\hat{\phi}_\theta = \sum_{m=0}^{s}\theta_m|\theta_m\rangle\langle\theta_m|. \tag{16}$$

From the orthogonality of the phase states $|\theta_m\rangle$ it follows that we can write an operator function of phase as

$$F(\hat{\phi}_\theta) = \sum_{m=0}^{s}F(\theta_m)|\theta_m\rangle\langle\theta_m| \tag{17}$$

with an expectation value for state $|f\rangle$ of

$$\langle F(\hat{\phi}_\theta)\rangle = \sum_{m=0}^{s}F(\theta_m)P(\theta_m) \tag{18}$$

where $P(\theta_m)$ is the probability $\langle f|\theta_m\rangle\langle\theta_m|f\rangle$. We can calculate the probability Pr (Δ) for a value of phase somewhere in the interval Δ as the total probability $\sum_\Delta P(\theta_m)$ because the phase states are orthogonal, that is, mutually exclusive. We can therefore define an operator

$$\hat{\Pi}_s(\Delta) = \sum_\Delta|\theta_m\rangle\langle\theta_m| \tag{19}$$

and deduce that

$$\Pr(\Delta) = \langle \hat{\Pi}_s(\Delta) \rangle. \quad (20)$$

For mixed states this becomes $\mathrm{Tr}\,[\hat{\rho}\hat{\Pi}_s(\Delta)]$. $\hat{\Pi}_s(\Delta)$ is the function $F(\hat{\phi}_\theta)$ in eq. (17) for which $F(\theta)$ is the top-hat function with a value of unity for θ in the interval Δ and zero elsewhere. It is also the operator associated with a phase probability-operator measure (POM) [10]. Relation (20) is essentially equivalent to the fundamental postulate [10] concerning POMs, which in the case here is a direct consequence of the orthogonality of the phase states.

Let us now find the weak limit of eq. (17) as $s \to \infty$. From eq. (17)

$$\langle g | F(\hat{\phi}_\theta) | h \rangle = \sum_{p,q=0}^{s} \sum_{m=0}^{s} F(\theta_m) \times \exp\,[\mathrm{i}(p-q)\theta_m] g_p^* h_q/(s+1) \quad (21)$$

where g_p and h_q are number state coefficients $\langle p | g \rangle$ and $\langle q | h \rangle$, and we have used the phase state expansion [2] in number states

$$|\theta_m\rangle = (s+1)^{-1/2} \sum_{n=0}^{s} \mathrm{e}^{in\theta_m} |n\rangle, \quad (22)$$

where $\theta_m = 2\pi m/(s+1)$. We note that from the definition of weak limits the states $|g\rangle, |h\rangle \in \mathrm{H}$ must be fixed as $s \to \infty$ so that the coefficients g_p and h_q are not functions of s. This excludes, for example, $|g\rangle$ being a phase eigenstate. In order to avoid obscuring our essential argument with unnecessary mathematical detail at this stage, we further confine $|g\rangle$ and $|h\rangle$ to the subspace of states of H for which g_n and h_n are zero for $n > B$, where B is some number which can be made as large as desired but remains finite as we let $s \to \infty$. We shall relax this restriction later. For states in this subspace we can find the limit of eq. (21) by taking the limits of the three summations independently. From the Riemann integral expression where $\delta\theta = 2\pi/(s+1)$ we then obtain

$$\lim_{s\to\infty} \langle g | F(\hat{\phi}_\theta) | h \rangle = (2\pi)^{-1} \sum_{p,q=0}^{\infty} g_p^* h_q \int_{\theta_0}^{\theta_0+2\pi} F(\theta) \exp\,[\mathrm{i}(p-q)\theta]\, \mathrm{d}\theta \quad (23)$$

and thus the weak limit on this subspace is

$$[F(\hat{\phi}_\theta)]_\mathrm{W} = (2\pi)^{-1} \sum_{p,q} \int_{\theta_0}^{\theta_0+2\pi} F(\theta) \times \exp\,[\mathrm{i}(p-q)\theta] |p\rangle\langle q|\, \mathrm{d}\theta \quad (24)$$

$$= \int_{\theta_0}^{\theta_0+2\pi} F(\theta) |\theta\rangle\langle\theta|\, \mathrm{d}\theta \quad (25)$$

where $|\theta\rangle$ is the unnormalizable linear number state combination $(2\pi)^{-1/2} \sum_n \mathrm{e}^{in\theta} |n\rangle$, which of course is not a state of H. The expectation value of a function of phase of any state of this particular subspace can be found from the expectation value of the operator (24).

We have shown that eq. (25) is the weak limit of eq. (17) on a particular subspace of H which is, in fact, the subspace of physically preparable states [5]. We shall show below that this is also true on all of H for particular functions of θ. Before doing so, however, we examine the consequences of eq. (25) being the weak limit of eq. (17) at least on this subspace.

By taking the expectation value of eq. (25), we see that the phase probability density $\mathscr{P}(\theta)$ is just $\langle |\theta\rangle\langle\theta| \rangle$, from which it follows that

$$\Pr(\Delta) = \left\langle \left[\int_\Delta |\theta\rangle\langle\theta|\, \mathrm{d}\theta \right] \right\rangle. \quad (26)$$

The weak limit $\hat{\Pi}(\Delta)$ of the operator $\hat{\Pi}_s(\Delta)$ can be found by using the top-hat function for $F(\theta)$ in eq. (25), yielding

$$\hat{\Pi}(\Delta) = \int_\Delta |\theta\rangle\langle\theta|\, \mathrm{d}\theta. \quad (27)$$

From eqs (26) and (27) it follows that

$$\Pr(\Delta) = \langle \hat{\Pi}(\Delta) \rangle. \quad (28)$$

It is important to note that eq. (27) is just the phase POM operator [10, 11]. In the POM formalism, however, eq. (26) is obtained from eq. (27) by adopting relation (28) as the fundamental POM *postulate*. While this postulate is easy to justify for orthogonal POMs in the way eq. (20) follows from eq. (19), there is no way of justifying it in terms of probability theory and H-space entities alone for non-orthogonal POM operators such as eq. (27), which involve entities $|\theta\rangle$ that are *not orthogonal even in the Dirac δ-function sense*. By contrast, here expression (26) has been derived from eq. (25) which in turn follows from eq. (17), and eq. (17) is based on *orthogonal* states. Thus the Ψ-space approach not only predicts the same expectation values and probability distributions as the POM approach, but it also provides important insight into the fundamental POM postulate as it applies to phase. Essentially, in going from the Ψ-space POM operator (19) to the H-space POM operator (27) by means of the weak limit, the orthogonality is lost, but the probabilistic relationship (20) is preserved. Thus we do not require H-space POMs to be orthogonal, that is projection-valued, in order to find probability distributions from them.

The Ψ-space approach also allows us to understand the nature of phase operators defined in terms of POMs. The POM generated by

$$\mathrm{d}\hat{\Pi}(\theta) = |\theta\rangle\langle\theta|\, \mathrm{d}\theta \quad (29)$$

is not orthogonal, thus according to the POM approach an operator defined by

$$\hat{F}(\theta) = \int_{\theta_0}^{\theta_0+2\pi} F(\theta)\, \mathrm{d}\hat{\Pi}(\theta)$$

is *not* in general given by $F(\hat{\theta})$ [10]. In the Ψ-space approach the same result follows because eq. (25) is a *weak* limit and the weak limit of a function of $\hat{\phi}_\theta$ is not in general the same as the function of the weak limit of $\hat{\phi}_\theta$.

We now relax our previous restriction and turn our attention to the use of eq. (23) for states of H which are superpositions of an infinite number of number states. Let $F(\theta)$ have the infinite Fourier series

$$F(\theta) = \sum_{k=-\infty}^{\infty} a_k \exp(ik\theta). \quad (30)$$

We wish to find the conditions under which $|A(s)| \to 0$ as $s \to \infty$, where $A(s)$ is the difference between the r.h.s. of eqs

(21) and (23). Using eq. (30) we have

$$A(s) = A'(s) - \left(\sum_{p=0}^{\infty}\sum_{q=s+1}^{\infty} + \sum_{p=s+1}^{\infty}\sum_{q=0}^{s}\right)\sum_{k=-\infty}^{\infty} a_k(2\pi)^{-1} \times \int_{\theta_0}^{\theta_0+2\pi} \exp\,[i(k+p-q)\theta]\, d\theta g_p^* h_q \tag{31}$$

where

$$A'(s) = \sum_{p,q=0}^{s}\sum_{k=-\infty}^{\infty} a_k(s+1)^{-1} \times \sum_{m=0}^{s} \exp\,[i(k+p-q)\theta_m] g_p^* h_q - \sum_{p,q=0}^{s}\sum_{k=-\infty}^{\infty} a_k(2\pi)^{-1} \times \int_{\theta_0}^{\theta_0+2\pi} \exp\,[i(k+p-q)\theta]\, d\theta g_p^* h_q . \tag{32}$$

We limit ourselves to functions $F(\theta)$ for which the expression on the r.h.s. of eq. (23) is finite. With $F(\theta)$ as in eq. (30) this is

$$\sum_{p=0}^{\infty}\sum_{q=0}^{\infty}\sum_{k=-\infty}^{\infty} a_k(2\pi)^{-1} \int_{\theta_0}^{\theta_0+2\pi} \exp\,[i(k+p-q)\theta]\, d\theta g_p^* h_q . \tag{33}$$

Because this is finite, the last term of eq. (31) vanishes as $s \to \infty$ and thus $A(s)$ vanishes as $s \to \infty$ whenever $A'(s)$ vanishes. $A'(s)$ can be simplified by using the relations

$$\int_{\theta_0}^{\theta_0+2\pi} \exp\,[i(k+p-q)\theta]\, d\theta = 2\pi\delta_{0,k+p-q} \tag{34}$$

where $\delta_{i,j}$ is the Kronecker delta, and

$$\sum_{m=0}^{s} \exp\,[i(k+p-q)\theta_m] \frac{1}{s+1} = \exp\,[i(k+p-q)\theta_0] \sum_{r=-\infty}^{\infty} \delta_{0,k+p-q+r(s+1)} . \tag{35}$$

Consider functions $F(\theta)$ which can be represented by a finite Fourier series, such as linear combinations of exponential functions of θ. For such functions $a_k = 0$ for $|k| > L$ where L stays constant as $s \to \infty$. L of course can be made as large as desired. Since $L < s$, then for this restriction on k the Kronecker delta in eq. (35) is non-zero only for $r = 0, \pm 1$. When we substitute eqs (34) and (35) into eq. (32), the $r = 0$ contribution to the first term in eq. (32) cancels the second term in eq. (32). Consequently we obtain

$$A'(s) = \sum_{p,q=0}^{s} \{\exp\,[i(s+1)\theta_0] a_{q-p+(s+1)} + \exp\,[-i(s+1)\theta_0] a_{q-p-(s+1)}\} g_p^* h_q \tag{36}$$

where, because of the above restriction on k, $|q - p \pm (s+1)| \leqslant L$, that is

$$|q - p| \geqslant (s+1) - L. \tag{37}$$

There are only a finite number of values of q and p for which this is true. For example, for $L = 2$ the values of (p, q) are $(s, 0)$, $(s, 1)$, $(s-1, 0)$, $(0, s)$, $(1, s)$, $(0, s-1)$. In general either $p < L$ and $q > s - L$ or *vice versa*, so the summation in eq. (36) is over a fixed number of terms for which either $|g_p|$ or $|h_q| \to 0$ as $s \to \infty$ because for all states of H, $|g_p|$ and $|h_q|$ must eventually tend to zero. Thus $|A'(s)| \to 0$ and so $|A(s)| \to 0$ as $s \to \infty$. Hence a sufficient, but not necessary, condition for expression (23) to be valid for all states of H is that $F(\theta)$ is expressible as a finite Fourier series in eq. (30). Such functions include all powers of $\cos\theta$ and $\sin\theta$. Also, eq. (23) is valid for finite Fourier series approximations to other functions, such as powers of θ, where the number of terms in the series can be made as large as necessary for any desired degree of accuracy.

We have shown that eq. (23) is valid for all $F(\theta)$ for states which are superpositions of a finite number of number states and for all states on H if $F(\theta)$ can be represented as a finite Fourier series. Since the truncated superposition can include as many terms as necessary for the norm of the difference between it and any H state to be less than any given non-zero amount and, similarly, the finite Fourer series can approximate any infinite convergent series as accurately as desired, it follows that eq. (23) is applicable for all practical purposes. Nevertheless it is interesting to also consider whether eq. (23) is valid for functions $F(\theta)$ represented by infinite Fourier series in conjunction with states which are infinite superpositions of number states, and not just for extremely good approximations to such functions. The limit of eq. (21) with $F(\theta)$ replaced by eq. (30) is

$$\lim_{s\to\infty} \langle g|F(\hat{\phi}_\theta)|h\rangle = \lim_{s\to\infty} \sum_{p,q=0}^{s}\sum_{m=0}^{s} \lim_{L\to\infty}\sum_{k=-L}^{L} a_k \times \exp\,[i(k+p-q)\theta_m] g_p^* h_q/(s+1). \tag{38}$$

For this to be the limit of a sequence of finite-L approximations as L increases we must have

$$\lim_{L\to\infty}\lim_{s\to\infty}\sum_{p,q=0}^{s}\sum_{m=0}^{s}\sum_{k=-L}^{L}\cdots = \lim_{s\to\infty}\sum_{p,q=0}^{s}\sum_{m=0}^{s}\lim_{L\to\infty}\sum_{k=-L}^{L}\cdots$$

which requires the series over q, p and k to converge sufficiently rapidly. We can show that this is true at least for states of finite mean energy in conjunction with commonly-used functions of θ, such as θ, θ^2, ... and the top hat functions. The physical states belong to this subspace of states.

In summary, the range of applicability of eq. (23) includes all functions $F(\theta)$ in conjunction with H states with a finite cutoff, which can be as large as necessary to approximate any given state to any order of accuracy, and all H states in conjunction with functions $F(\theta)$ with a finite Fourier series (30). The latter include, for example, functions such as $\cos\theta$, $\cos^2\theta$, ... and also include approximations to functions such as θ, θ^2, ... to any desired degree of accuracy. While this is sufficient for all practical purposes, eq. (23) is in fact valid for a wider range of combinations of states and functions.

6. Conclusion

In this paper we have established the formal relationship between the Ψ-space limiting procedure and the more conventional H-space approach. Although the Ψ-space approach has already found much practical use for calculating the phase properties for various quantum states of light, we feel that an understanding of this relationship is important in helping to justify the Ψ-space approach. It also sheds light on the fundamental reason for the difficulties that the H-space formalism has encountered in trying to incorporate a consistent quantum description of phase.

When H-space operators are derived as infinite-s limits of their Ψ-space counterparts, where $s+1$ is the dimensionality of the space, some observables are represented in H-space by the strong limits of their Ψ-space operator counterparts, but other observables can only be represented on H-space by the weak limits of the Ψ-space operators. Notable examples of the former are power series functions of $\hat{a}$ and $\hat{a}^\dagger$, such as the energy operator, and examples of the latter are power series functions of the phase operator. The former obey the same associative operator algebra of their Ψ-space counterparts but the latter do not, for example the cosine of the weak limit of the phase operator is not equal to the weak limit of the cosine of the phase operator. Thus for functions of $\hat{a}$ and $\hat{a}^\dagger$ entirely consistent results are obtained by using the normal operator algebra either in H-space or in Ψ-space. If, however, this algebra is used for phase operators on both spaces, consistent results are not obtained. The option of using normal algebra for phase operators in H-space by begining with the weak limit of one arbitrarily chosen function of phase, such as the phase angle (Popov–Yarunin operator [12]) or the cosine of phase angle (Susskind–Glogower operator [9]) and so on, and then defining all the other functions of phase by means of the normal algebra would not only lead to a description of phase not consistent with descriptions based on other choices of initial function, but also none of these descriptions would be consistent with number–phase complementarity. On the other hand phase does have a consistent description in the Ψ-space formalism which is in accord with complementarity.

The derivation of a formula for the weak limits of functions of phase makes it possible to calculate the phase properties of states without use of the full Ψ-space limiting procedure. For example, we can calculate $\langle \cos^2 \theta \rangle$ simply by finding the expectation value of the H-space operator (24), with $F(\theta) = \cos^2 \theta$, in terms of the appropriate H-space states $|h\rangle$ or, even more simply, by using the eq. (23) directly with $g = h$. It is interesting that this is equivalent to calculating the expectation value by using the non-orthogonal phase POM defined by eq. (27) [10, 11]. A very important point, however, is that whereas the connection between expectation values and non-orthogonal POMs in the POM approach is made by postulating eq. (28), the corresponding result is derived in the Ψ-space approach from the orthogonality of the Ψ-space phase states. The Ψ-space approach therefore explains why the non-orthogonal POM postulate gives correct results, at least for phase. Essentially the non-orthogonal POM on H corresponds to the weak limit of the Ψ-space orthogonal POM, for which the connection with probablity is firmly established from the usual probabilistic interpretation of quantum mechanics. It is interesting also to note that not only does the unitary phase operator lose its unitarity when its limit on H-space is found, but also the orthogonal phase POM loses its orthogonality. Both the unitarity of the exponential operator and the orthogonality of phase states are fundamental properties which permit a consistent description of phase in the Ψ-space formalism.

In conclusion, we remark that we are so accustomed to study the quantum electromagnetic field by commencing with the infinite space H that it is natural to have reservations about adopting a different limiting procedure. An aim of this paper is to diminish such reservations. The price paid for fervent adherence to the customary prescription is forgoing a full understanding of the quantum nature of phase. This understanding requires a more fundamental approach. By starting with Ψ_s we are able to avoid invoking H with its limited applicability and still obtain an equivalent quantum description for the physical quantities which the H-space approach can accommodate. In addition, we also obtain a quantum description of other quantities such as phase. Although the restrictive nature of the H-space approach can also be overcome to some extent by introducing the non-orthogonal POM concept through an additional *ad hoc* postulate, we note that the Ψ-space approach achieves the same results without the need for such a postulate. Finally we are led to the conclusion that the insight into both the H-space approach and the POM approach offered by the Ψ-space formalism strengths its position as a more fundamental approach to quantum mechanics.

Acknowledgements

We thank Dr. S. M. Barnett for discussions and valuable comments on J.A.V.'s Ph.D. thesis from which the work in this paper has been developed. J.A.V. also thanks Dr. R. Bonner for helpful discussions concerning the properties of Hilbert space. We acknowledge communications with Dr. M. Hall.

References

1. Böhm, A., "The Rigged Hilbert Space and Quantum Mechanics" (Springer-Verlag, Berlin 1978).
2. Pegg, D. T. and Barnett, S. M., Europhysics Letters **6**, 483 (1988).
3. Barnett, S. M. and Pegg, D. T., J. Mod. Opt. **36**, 7 (1989).
4. Pegg, D. T. and Barnett, S. M., Phys. Rev. **A39**, 1665 (1989).
5. Vaccaro, J. A. and Pegg, D. T., J. Mod. Opt. **37**, 17 (1990).
6. Pegg, D. T., Vaccaro, J. A. and Barnett, S. M., J. Mod. Opt. **37**, 1703 (1990).
7. Barnett, S. M. and Pegg, D. T., Phys. Rev. **A41**, 3427 (1990).
8. Akhiezer, N. I. and Glazman, I. M., "Theory of Linear Operators in Hilbert Space" (Translated from the Russian by M. Nestell) (Frederick Ungar, New York 1961), vol. I.
9. Susskind, L. and Glogower, J., Physics **1**, 49 (1964).
10. Helstrom, C. W., "Quantum Detection and Estimation Theory" (Academic Press, New York 1976), p. 53.
11. Shapiro, J. H. and Shepard, S. R., Phys. Rev. **A43**, 3795 (1991).
12. Popov, V. N. and Yarunin, V. S., The Leningrad University Journal: Physics **22**, 7 (1973).

Canonical and measured phase distributions

U. Leonhardt, J. A. Vaccaro, B. Böhmer, and H. Paul
Arbeitsgruppe "Nichtklassische Strahlung" der Max-Planck-Gesellschaft an der Humboldt-Universität zu Berlin, Rudower Chaussee 5, 12484 Berlin, Germany
(Received 27 June 1994)

We derive relationships between canonical and measured phase distributions for quantum-oscillator states in the semiclassical regime. First, we extend the formalism for the canonical phase to include external measurement-induced uncertainty. We require that a phase shifter shifts a phase distribution while a number shifter does not change it. These axioms determine pure canonical phase distributions uniquely while a noisy distribution can be interpreted as a weighted average of pure phase distributions. As a second step, we show that measured phase distributions, i.e., s-parametrized phase distributions fulfill approximately the axioms of noisy canonical phase, and we derive simple analytical expressions for the corresponding weight functions. Our analysis thus bridges all three conceptions of quantum-optical phase (canonical phase, s-parametrized phase, phase from measurements) and provides important physical insight into the relationship between them.

PACS number(s): 03.65.Bz, 42.50.Dv

I. INTRODUCTION

Although the quantum-optical phase is still a controversial subject [1,2], recently significant progress has been made in unifying the various different conceptions of phase [3]. First, the results of different formalisms embodying the concept of phase as an observable canonically conjugate to the photon number have been shown to be physically equivalent. In particular, the phase distribution associated with the Helstrom-Shapiro-Shepard probability operator measure [4,5] is equivalent to that derived from the Pegg-Barnett formalism [6] for physical states in the infinite dimensional limit [5,7,8]. Also, the Newton-Barnett-Pegg formalism [9] where the Hilbert space has been doubled gives the same physical results as the formalism mentioned above [10]. Considered purely mathematically, these approaches are quite distinct. However, the phase-distribution function for a given physical state is the same irrespective from which formalism it has been derived. We call this common distribution a *canonical phase distribution* since the Pegg-Barnett formalism [6], for instance, is motivated by the requirement that the physical quantity *phase* be canonically conjugate to the *photon number* similar to definitions of the standard variables *position* and *momentum.*

A second conception of phase [3] is based on examining phase properties via s-parametrized quasiprobability distributions of position and momentum [11]. A phase distribution is obtained from integrating the quasiprobability distribution over the radial coordinate. This conception is not motivated by the complementarity of number and phase nor is it free from arbitrariness. Which s-parametrized phase distribution should be taken? Phase from the Wigner function [12] has at least the merit of coinciding with the canonical phase in the limit of very large intensities [13]. On the other hand, it has the drawback of yielding negative distribution functions for some physical states in the quantum regime [14]. However, it was recently shown [15,16] that s-parametrized phase distributions describe experimentally measured phase probability distributions [17,18] when the parameter s is less or equal to -1. Thus, the motivation for this concept comes *a posteriori* from experiment.

At first glance, the third conception [3], the operational approach to quantum phase [18,19,20] introduces an element of subjectivity into the definition of a physical quantity, especially for weak fields in the quantum regime. Possible experiments are different and depend on the intentions of their designers. So it is highly remarkable that most of them yield the same results under reasonable assumptions [21]. The distributions measured in these experiments [22] can be interpreted as being derived from the Q function [15] or a smoothed Q function in the case of inefficient detection [16]. We call them *measured phase distributions* [23].

As a last step, it remains to be shown how the *measured phase distributions* are related to the *canonical phase distribution.* It would be very surprising indeed if there did not exist a deeper relationship between both, since the experiments are designed as classical phase-measurement schemes and in the classical limit, measured and canonical phase coincides. We note that significant differences occur in the extreme quantum regime of low photon numbers. Here, we are interested in a semiclassical domain of relatively large photon numbers. What would we expect? Since the phase-measurement schemes realize simultaneous yet noisy measurements of position and momentum [24,25], we anticipate that the measured phase distributions should be somewhat broader than the canonical distribution. A comparison of phase variances [26] and calculations for particular states [27] support this assertion. In this paper, we quantify the general asymptotic relationship between canonical and measured phase distributions. Before we address this problem, we provide in Sec. II an alternative and more general theoretical approach to canonical phase distributions. It is not motivated by a quantum-estimation problem as in Shapiro's and Shepard's classic paper [5] nor by requiring the construction of Hermitian phase operators. Our approach is based purely on the complementarity of phase and photon number. It has the merit that it can describe phase as a variable influenced by external noise. This element of external statistics and the basic consequences of our axioms of complementarity will give us the key for relating canonical phase to measured phase. We will show in Sec. III that in the semiclassical domain measured phase distributions are averaged pure canonical distributions with respect to a distribution function of reference phases. The latter depends on the mean photon number and the s parameter in a simple way. In the classical limit, it reduces to a δ function, as we would expect. The results are summarized in Sec. IV.

II. CANONICAL PHASE DISTRIBUTIONS

A. Axioms for quantum phase

Preliminaries

In this approach to quantum phase, we consider phase distributions. Since a probability distribution of phase contains all statistical information on the phase properties for a given state it represents the physical quantity *phase* completely. Realistic measurements of phase [24] always involve extra noise beyond that due to the intrinsic quantum phase fluctuations described by the canonical phase distribution [4–6,9]. Consequently, we need a general method that allows the description of phase in the presence of noise. For this we use *probability operator measures* (POM's) [28], as in Shapiro's and Shepard's classic paper [5]. Within the POM formalism, we introduce a probability distribution $\Pr(\varphi)$, $\varphi\in(-\pi,\pi]$, as [29]

$$\Pr(\varphi)=\mathrm{Tr}\{\hat{\rho}\hat{\Pi}(\varphi)\} , \tag{1}$$

where $\hat{\rho}$ is the density matrix and $\hat{\Pi}(\varphi)$ denotes a set of suitable operators parametrized by the phase variable φ. Since probability distributions are real functions, $\hat{\Pi}(\varphi)$ must be Hermitian,

$$\hat{\Pi}(\varphi)^{\dagger}=\hat{\Pi}(\varphi) . \tag{2}$$

A probability distribution is normalized to unity. Consequently, the set of operators $\hat{\Pi}(\varphi)$ must be normalized as well,

$$\int_{-\pi}^{+\pi}d\varphi\,\hat{\Pi}(\varphi)=1 . \tag{3}$$

Lastly, a probability distribution is nonnegative

$$\Pr(\varphi)\geq 0 . \tag{4}$$

This implies that the eigenvalues of $\hat{\Pi}(\varphi)$ must be nonnegative as well. This property together with (2) and (3) is sufficient to identify $\hat{\Pi}(\varphi)$ as a density operator which is often laxly called a *phase state.* One important point is, however, that the operator $\hat{\Pi}(\varphi)$ might be unnormalizable, as the London phase states [30] are, although we note that such states can be represented in larger spaces, e.g., the rigged Hilbert space [31]. Another important point is that a phase state $\hat{\Pi}(\varphi)$ can be a mixed state. This simply means that the measure of φ is not precise and so the probability distribution $\Pr(\varphi)$ represents the results of a noisy measurement of φ. For this reason, we call $\Pr(\varphi)$ a *noisy phase distribution* and $\hat{\Pi}(\varphi)$ a *mixed phase state.*

The basic expression (1) for a probability distribution in quantum mechanics can be also expressed as an overlap of Wigner functions [32,33],

$$\Pr(\varphi)=2\pi\int_{-\infty}^{+\infty}dx\int_{-\infty}^{+\infty}dp\; W(x,p)W(x,p;\varphi) . \tag{5}$$

Here,

$$W(x,p)\equiv\int_{-\infty}^{+\infty}\frac{dy}{\pi}\,e^{2ipy}\langle x-y|\hat{\rho}|x+y\rangle , \tag{6}$$

and

$$W(x,p;\varphi)\equiv\int_{-\infty}^{+\infty}\frac{dy}{\pi}\,e^{2ipy}\langle x-y|\hat{\Pi}(\varphi)|x+y\rangle \tag{7}$$

are the Wigner functions of the quantum state $\hat{\rho}$ and the mixed phase state $\hat{\Pi}(\varphi)$, respectively [34]. The overlap relation (5) closely resembles the probability overlap in classical statistical physics and hence it helps us to understand quantum mechanics more intuitively. In addition, it is a quite useful tool for finding the semiclassical asymptotics of a given quantum-mechanical problem. Later on we will use the overlap relation for understanding the asymptotic relations between measured phase distributions.

So far, we have briefly summarized the general properties of POM's. Now, we turn to two specific requirements for a noisy quantum-phase distribution. Our goal is the definition of quantum-optical phase as canonically conjugate variable with respect to photon number. How can this be achieved? We wish to treat *number* and *phase* similarly to the basic canonically conjugate variables *position* and *momentum.* We could, for instance, extend the canonical commutation relation for position and momentum operators $\hat{q}$ and $\hat{p}$,

$$[\hat{q},\hat{p}]=i \tag{8}$$

to a commutation relation for number $\hat{n}=\hat{a}^{\dagger}\hat{a}$ and phase $\hat{\varphi}$. (As usual, $\hat{a}$ denotes the annihilation operator.) It is well known, however, that a Hermitian phase operator $\hat{\varphi}$ does not exist on the Fock space. Instead, we are considering not phase operators but phase distributions. In defining *phase* as canonically conjugate to *photon number,* we must translate some typical properties of position and momentum distributions into the language of number and phase and regard them as being fundamental. We obtain from the canonical commutation relation that the operator $\hat{S}(q_0)=\exp(-iq_0\hat{p})$ shifts position eigenstates by the amount q_0 while $\hat{T}(p_0)=\exp(ip_0\hat{q})$ shifts momentum eigenstates by p_0, see, for instance, Ref. [35]. Hence the momentum distribution $\mathrm{Tr}\{\hat{\rho}|p\rangle\langle p|\}$ is shifted by p_0 when $\hat{T}(p_0)$ is applied to the quantum state $\hat{\rho}$. On the other hand, the momentum distribution is not changed when $\hat{S}(q_0)$ is applied to $\hat{\rho}$, i.e., when the position is shifted. Position and momentum are strictly independent since shifting one variable does not affect the other. We may regard this mutual independence of the canonically conjugate variables as being fundamental and require the same for *number* and *phase.*

Axioms

We require that a phase-distribution function of a single mode satisfies the following axioms.

(A) A phase shifter shifts the phase distribution.

(B) A number shifter does not change the phase distribution (complementarity). To be explicit, a phase shifter is represented by the unitary transformation operator,

$$\hat{U}(\phi) \equiv \exp(i\phi\hat{a}^\dagger\hat{a}) \ . \tag{9}$$

Axiom (A), thus, means that

$$\begin{aligned} \mathrm{Pr}'(\varphi) &\equiv \mathrm{Tr}\{\hat{U}(\phi)\hat{\rho}\hat{U}(\phi)^\dagger\hat{\Pi}(\varphi)\} \\ &= \mathrm{Pr}(\varphi-\phi) \\ &= \mathrm{Tr}\{\hat{\rho}\hat{\Pi}(\varphi-\phi)\} \ . \end{aligned} \tag{10}$$

A number shifter is expressed by the operator

$$\hat{E} \equiv \sum_{n=0}^{\infty} |n+1\rangle\langle n| \ . \tag{11}$$

It shifts the photon-number distribution up by one step. ($|n\rangle$ denotes a Fock state.) The operator $\hat{E}$ is nothing but the Susskind-Glogower exponential phase operator $\widehat{\exp}(-i\varphi)$ [36]. As is well known, it is neither a unitary nor a strictly invertible operator, since $\hat{E}^\dagger$ annihilates the vacuum state. However, we are not concerned about the problems of $\hat{E}$ as an exponential phase operator here. We only use the number-shifter property of $\hat{E}$. Axiom (B) thus requires

$$\mathrm{Pr}'(\varphi) \equiv \mathrm{Tr}\{\hat{E}\hat{\rho}\hat{E}^\dagger\hat{\Pi}(\varphi)\} = \mathrm{Pr}(\varphi) \ . \tag{12}$$

Comments

Both axioms (A) and (B) together determine a phase distribution. What do they mean physically? Axiom (A) is almost trivial. We only require that the phase distribution should indeed reflect the basic feature of quantum phase, i.e., that a phase shifter is a phase-distribution shifter. Naturally, many phase-sensitive quantities have the property (A). Axiom (B) is more specific [37]. It means that the distribution function $\mathrm{Pr}(\varphi)$ contains the properties of quantum phase and nothing else. It must not reflect any properties of the canonically conjugate variable, the photon number. Hence (B) means that phase should be complementary to photon number. We also note, however, that if a particular distribution function $\mathrm{Pr}(\varphi)$ satisfies the axioms then so does the weighted average $p_1\mathrm{Pr}(\varphi)+p_2\mathrm{Pr}(\varphi+\delta)$ of this function and the phase-shifted distribution $\mathrm{Pr}(\varphi+\delta)$, which describes uncertainty in the reference phase. We interpret this as the axioms allow for a noisy measure of phase. The nature of this noise is very special in that the resulting distribution still satisfies the axioms of complementarity. Thus our approach here contains, in essence, the basic prescription for describing a noisy measurement of phase without contamination from the complementary observable, photon number. Now we consider the detailed consequences of both axioms.

B. Consequences

Fock representation

We express the noisy phase probability distribution in the Fock basis,

$$\mathrm{Pr}(\varphi) = \sum_{n,m=0}^{\infty} \langle m|\hat{\Pi}(\varphi)|n\rangle\langle n|\hat{\rho}|m\rangle \ . \tag{13}$$

We use the axiom (A),

$$\begin{aligned} \mathrm{Pr}(\varphi) &= \mathrm{Pr}[0-(-\varphi)] \\ &= \sum_{n,m=0}^{\infty} \langle m|\hat{\Pi}(0)|n\rangle\langle n|\hat{U}(-\varphi)\hat{\rho}\hat{U}(\varphi)|m\rangle \\ &= \sum_{n,m=0}^{\infty} \langle m|\hat{\Pi}(0)|n\rangle e^{i(m-n)\varphi}\langle n|\hat{\rho}|m\rangle \ , \end{aligned} \tag{14}$$

define the coefficients

$$B_{n,m} \equiv 2\pi\langle m|\hat{\Pi}(0)|n\rangle \ , \tag{15}$$

and obtain

$$\mathrm{Pr}(\varphi) = \frac{1}{2\pi}\sum_{n,m=0}^{\infty} B_{n,m}e^{i(m-n)\varphi}\langle n|\hat{\rho}|m\rangle \ . \tag{16}$$

Since the operator $\hat{\Pi}(\varphi)$ is Hermitian, the matrix $B_{n,m}$ must be Hermitian as well,

$$B_{n,m} = B^*_{m,n} \ . \tag{17}$$

Expressions of the type (16) have been known for several phase-dependent distributions for a long time (cf., for instance, Ref. [19]). As we have seen here, the root of these formulas lies in the phase-shifter axiom (A). Now, we consider the consequences of axiom (B):

$$\begin{aligned} \mathrm{Pr}(\varphi) &= \frac{1}{2\pi}\sum_{n,m=0}^{\infty} B_{n,m}e^{i(m-n)\varphi}\langle n|\hat{E}\hat{\rho}\hat{E}^\dagger|m\rangle \\ &= \frac{1}{2\pi}\sum_{n,m=1}^{\infty} B_{n,m}e^{i(m-n)\varphi}\langle n-1|\hat{\rho}|m-1\rangle \\ &= \frac{1}{2\pi}\sum_{n,m=0}^{\infty} B_{n+1,m+1}e^{i(m-n)\varphi}\langle n|\hat{\rho}|m\rangle \ . \end{aligned} \tag{18}$$

Consequently, the B coefficients should have the number-shift invariance as well,

$$B_{n+1,m+1} = B_{n,m} \ . \tag{19}$$

This simple relation will provide us with the key for relating canonical and measured phase distributions.

Convolution

Because of the invariance relation (19) and the Hermitian condition (17) the B-coefficients depend on a single row of free parameters,

$$B_{n,m}=b_{m-n}, \quad b_\nu=\begin{cases} B_{0,\nu} & \text{for } \nu\geq 0, \\ b^*_{-\nu} & \text{for } \nu<0. \end{cases} \tag{20}$$

These parameters characterize all possible noisy phase distributions satisfying both axioms (A) and (B). Using the definition (20) and the Fock expansion (16) for a noisy phase distribution, we find

$$\Pr(\varphi)=\sum_{\nu=-\infty}^{+\infty} e^{i\nu\varphi}b_\nu c_\nu , \tag{21}$$

with

$$c_\nu=\begin{cases} \frac{1}{2\pi}\sum_{n=0}^{\infty}\langle n|\hat\rho|n+\nu\rangle & \text{for } \nu\geq 0, \\ c^*_{-\nu} & \text{for } \nu<0. \end{cases} \tag{22}$$

Here the noisy phase distribution $\Pr(\varphi)$ is expressed as a Fourier series. According to the convolution theorem, we obtain

$$\Pr(\varphi)=\int_{-\pi}^{+\pi}\frac{d\phi}{2\pi}g(\phi)\Pr_p(\varphi-\phi) , \tag{23}$$

with

$$g(\phi)\equiv\sum_{\nu=-\infty}^{+\infty} e^{i\nu\phi}b_\nu \tag{24}$$

and

$$\Pr_p(\varphi)\equiv\sum_{\nu=-\infty}^{+\infty} e^{i\nu\varphi}c_\nu=\frac{1}{2\pi}\sum_{n,m=0}^{\infty} e^{i(m-n)\varphi}\langle n|\hat\rho|m\rangle . \tag{25}$$

The function $\Pr_p(\varphi)$ is nothing but the Helstrom-Shapiro-Shepard phase distribution [4,5] and the Pegg-Barnett phase distribution [6] for physical states in the infinite-dimensional limit. We call $\Pr_p(\varphi)$ a *pure canonical phase distribution.* The function $g(\phi)$ is real because of the definition (20) of the b coefficients. Moreover, it must be nonnegative and normalized to unity since the distributions $\Pr(\varphi)$ and $\Pr_p(\varphi)$ are nonnegative and normalized for all states. Hence we can interpret $g(\phi)$ as a probability distribution. Our result (23) thus means that any noisy phase distribution $\Pr(\varphi)$ satisfying both axioms (A) and (B) consists of pure canonical phase distributions $\Pr_p(\varphi)$ averaged with respect to a certain probability distribution $g(\phi)$ of reference phases ϕ which represents the noise.

Pure phase distributions

Finally, it remains to be proved that the pure canonical phase distributions $\Pr_p(\varphi)$ are the only ones that deserve the designation "pure," in the sense that they correspond to pure phase states $\hat\Pi(\varphi)=|\varphi\rangle\langle\varphi|$. In fact, they are the only distributions having both properties (A) and (B) and a coefficient matrix $B_{n,m}$ that factorizes according to

$$B_{n,m}=B^*_n B_m . \tag{26}$$

The proof is rather simple. Because of the invariance principle (19), we have

$$B^*_{n+\nu}B_{m+\nu}=B^*_n B_m . \tag{27}$$

Setting $n=m$, we obtain $|B_n|^2=|B_0|^2$, and because of the normalization of the phase distribution $|B_n|^2=1$. We express B_n as $\exp(-i\beta_n)$ and obtain

$$B^*_n B_1=e^{i(\beta_n-\beta_1)}=e^{i(\beta_{n-1}-\beta_0)} \tag{28}$$

and, consequently,

$$e^{i\beta_n}=e^{i\beta_{n-1}}e^{i\phi}, \quad \phi=\beta_1-\beta_0 . \tag{29}$$

Applying this relation n times, we get, finally,

$$B_m=e^{-i\beta_n}=e^{-i\beta_0}e^{-in\phi} . \tag{30}$$

Hence, the phase distribution $\Pr(\varphi)$ is

$$\Pr(\varphi)=\frac{1}{2\pi}\sum_{n,m=0}^{\infty} e^{i(m-n)(\varphi-\phi)}\langle n|\hat\rho|m\rangle$$
$$=\Pr_p(\varphi-\phi) . \tag{31}$$

It corresponds to the well-known phase-distribution first introduced by London [30]. Hence up to a reference phase our basic axioms (A) and (B) determine a canonical phase distribution uniquely when we consider a pure distribution, while, in general, any noisy phase distribution satisfying the complementarity axioms can be seen as a statistical mixture of pure canonical phase distributions.

III. MEASURED PHASE DISTRIBUTIONS

A. Q-phase distribution

The phase distribution measured in the Noh-Fougères-Mandel experiment [18] is the radius-integrated Q function (Q phase) [15], provided a strong local oscillator and perfect detectors have been employed [21]. Also, in other operational approaches to quantum phase [19,20], the Q-phase distribution is measured. In this Sec. we show that we can interpret this distribution as a smoothed canonical phase distribution. The Q-phase distribution is defined as

$$\Pr_Q(\varphi)=\int_0^\infty dr\,\frac{r}{\pi}\langle re^{i\varphi}|\hat\rho|re^{i\varphi}\rangle . \tag{32}$$

($|re^{i\varphi}\rangle$ is a coherent state of amplitude r and phase φ). It reads in terms of the POM formalism,

$$\Pr_Q(\varphi)=\mathrm{Tr}\{\hat\rho\hat\Pi_Q(\varphi)\} , \tag{33}$$

with

$$\hat\Pi_Q(\varphi)\equiv\int_0^\infty \mathrm{d}r\,\frac{r}{\pi}|re^{i\varphi}\rangle\langle re^{i\varphi}| . \tag{34}$$

This type of phase state for the Q phase has been studied in detail by Paul [19].

Wigner function

To compare the Q-phase distribution with the pure canonical phase distribution, we can use the Wigner-function-overlap relation [32], i.e., according to Eq. (5), we express $\Pr_Q(\varphi)$ in terms of Wigner functions,

$$\Pr_Q(\varphi)=2\pi\int_{-\infty}^{+\infty}dx\int_{-\infty}^{+\infty}dp\; W(x,p)W_Q(x,p;\varphi)\ , \quad (35)$$

with

$$W_Q(x,p;\varphi)=\int_0^{\infty}dr\frac{r}{\pi}W_{\mathrm{coh}}(x,p)\ . \quad (36)$$

Here, $W_{\mathrm{coh}}(x,p)$ denotes the Wigner function of the coherent state $|re^{i\varphi}\rangle$

$$W_{\mathrm{coh}}(x,p)=\frac{1}{\pi}\exp[-(x-2^{1/2}r\cos\varphi)^2 -(p-2^{1/2}r\sin\varphi)^2]\ . \quad (37)$$

We substitute

$$x_\varphi=x\cos\varphi+p\sin\varphi,\quad p_\varphi=-x\sin\varphi+p\cos\varphi\ , \quad (38)$$

and obtain

$$W_Q(x,p;\varphi)=\frac{1}{(2\pi)^2}e^{-p_\varphi^2}\{e^{-x_\varphi^2}+x_\varphi\pi^{1/2}[1+\mathrm{erf}(x_\varphi)]\}\ . \quad (39)$$

The Wigner function associated with the Q-phase distribution grows linearly in the phase direction φ. It has a Gaussian profile which originates physically from the vacuum noise involved in realistic measurements of phase [24], see Fig. 1. In contrast, the Wigner function of a pure phase state (cf. Fig. 3 in Ref [34]) grows quadratically in the phase direction. It is much narrower, shows characteristic oscillations, and becomes negative in certain regions. Here, on the other hand, the Wigner function of the Q-phase state is always positive which already indicates that it represents a statistical mixture. (Only Gaussian pure states have non-negative Wigner functions [33].)

Fock representation

In Fock representation, the Q-phase distribution reads

$$\Pr_Q(\varphi)=\frac{1}{2\pi}\sum^{\infty} B_{n,m}e^{i(m-n)\varphi}\langle n|\hat{\rho}|m\rangle\ , \quad (40)$$

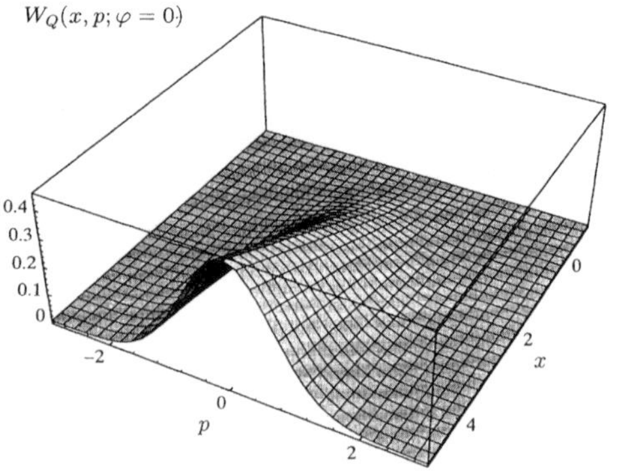

FIG. 1. Wigner function (39) for the Q-phase state (34).

tion. On the other hand, we note that the B coefficients tend to unity in the limit $n,m\to\infty$. Thus, the Q phase coincides with the pure canonical phase for very large photon numbers. Hence, the number-shift invariance cannot be exact. Here, we are interested in an intermediate regime where n and m are relatively large. We will now quantify the asymptotics of the Q-phase distribution.

Convolution

For relatively narrow photon-number distributions compared to the mean photon number, we can approximate

$$B_{n,n+|\nu|}\approx B_{N,N+|\nu|}\ , \quad (43)$$

with N being the mean photon number

$$N=\mathrm{Tr}\{\hat{\rho}\hat{a}^\dagger\hat{a}\}\ . \quad (44)$$

Consequently, the Q-phase distribution reads

$$\Pr_Q(\varphi)=\sum_{\nu=-\infty}^{+\infty}e^{i\nu\varphi}B_{N,N+|\nu|}c_\nu\ , \quad (45)$$

$$\ln B_{N,N+|\nu|}\approx\left[N+\frac{|\nu|}{2}+\tfrac{1}{2}\right]\ln\left[N+\frac{|\nu|}{2}+1\right]$$
$$-\tfrac{1}{2}(N+\tfrac{1}{2})\ln(N+1)$$
$$-\tfrac{1}{2}(N+|\nu|+\tfrac{1}{2})\ln(N+|\nu|+1)$$
$$=-\frac{3+2N}{16(N+1)^2}\nu^2+O(|\nu|^3)\ . \quad (48)$$

Hence, we get for the distribution function

$$g(\phi;N)=\int_{-\infty}^{+\infty}d\nu\exp\left[i\nu\phi-\frac{3+2N}{16(N+1)^2}\nu^2\right] \quad (49)$$

and, finally,

$$g(\phi;N)=2\pi\left[\frac{4(N+1)^2}{\pi(2N+3)}\right]^{1/2}\exp\left[-\frac{4(N+1)^2}{2N+3}\phi^2\right]. \quad (50)$$

This simple formula describes the asymptotics of the reference-phase distribution $g(\phi;N)$. It is a Gaussian with a width depending roughly inversely on the mean photon number N which reflects the extra noise involved in phase measurements [24]. The distribution $g(\phi;N)$ gets narrower with increasing N and finally tends to a δ function for very large N, since with increasing intensity the influence of this noise decreases. In the macroscopic regime the measurement-induced noise is negligible. This is readily understood from the particular source of extra noise being present in an experimental setup that allows measurement of the Q phase. It is either vacuum noise introducing the apparatus via the unused port of a beam splitter [15,20] or amplification noise [19]. In both cases, the noise becomes negligible when the intensity of the initial field is high. On the other hand, a distribution function $g(\phi)$ that is independent of the field can most easily be interpreted in taking the term "reference-phase distribution" literally, i.e., in identifying the noise with phase instabilities in the reference beam, e.g., a high-intensity laser beam. In contrast, the N-dependent distribution $g(\phi;N)$ shows that the Q-phase distribution is also an (approximative) noisy measure of the pure canonical phase distribution but now the noise relative to the field intensity decreases as the latter increases. In particular, if the field is in a coherent state then both the relative measurement-induced noise and the intrinsic quantum-phase fluctuations of the field vanish at roughly equal rates as the intensity increases. In this respect, the Q phase remains a good measure of the pure canonical phase distribution even for lower values of N. This is, however, a special result for coherent states. For states with narrower phase distributions, e.g., phase-optimized states [40], a reference-phase distribution that decreases at a faster rate is required to preserve the relative accuracy of the noisy phase distribution [26].

Numerical tests

We tested the accuracy of our treatment of the Q phase as a noisy canonical phase for squeezed states [41] having quadrature wave functions,

$$\psi_{sq}(x)\equiv\langle x|\psi_{sq}\rangle=(\zeta/\pi)^{1/4}\exp[-(\zeta/2)(x-\sqrt{2}\alpha)^2]\ , \quad (51)$$

see Ref. [12], Eq. (3). Here, ζ is a real and positive parameter which characterizes the squeezing. (It is equivalent to Schleich's s [12]. In order to avoid confusion with the quasiprobability parameter, we denote it by ζ.) The real parameter α characterizes the coherent amplitude of the squeezed state. The mean photon number N of the state $|\psi_{sq}\rangle$ is given by [41]

$$N=|\alpha|^2+\tfrac{1}{4}(\zeta+\zeta^{-1}-2)\ . \quad (52)$$

We calculated numerically the pure canonical phase distribution (25) using the photon-number probability amplitudes $\langle m|\psi_{sq}\rangle$ given by Ref. 12, Eq. (4). According to Eq. (46) this distribution was convoluted numerically with the weight function $g(\phi;N)$ of Eq. (50) and compared with the exact Q-phase distribution for squeezed states, as found in Ref. [27], Eqs. (32), (33) with $\mu=\zeta,\alpha_0=\alpha$. Figure 2 shows the exact versus the approximate Q-phase distributions for some squeezed states with mean photon numbers $N=25$. We see that the approximate curve fits quite well the exact one. We observed that for lower mean photon numbers the agreement between both curves becomes worse while for higher intensities they become almost indistinguishable.

B. *s*-phase distribution

Due to losses in overall detection efficiency the phase from an s-parametrized quasiprobability distribution (s phase) is measured in realistic experiments [16]. The parameter s is, in general, less than -1, which means that the quasiprobability distribution is a smoothed Q function. The measured phase distribution is given by

$$\Pr_s(\varphi)\equiv\int_0^{\infty}dr\,rW(r\cos\varphi,r\sin\varphi;s)\ . \quad (53)$$

Here, $W(r\cos\varphi,r\sin\varphi;s)$ denotes an s-parametrized quasiprobability distribution [11,16],

$$W(r\cos\varphi,r\sin\varphi;s)=\int_{-\infty}^{+\infty}dx\int_{-\infty}^{+\infty}dp\,W(x,p)\times W_{DT}(x,p)\ , \quad (54)$$

with

$$W_{DT}(x,p)=\frac{-1}{\pi s}\exp\left[\frac{1}{s}(x-r\cos\varphi)^2+\frac{1}{s}(p-r\sin\varphi)^2\right]. \quad (55)$$

The function $W_{DT}(x,p)$ can be interpreted as the Wigner function of a displaced thermal state $\hat{\rho}_{DT}$. Using some standard expressions for thermal states $\hat{\rho}_T$ and Ref. [42], we easily obtain the corresponding density operator,

$$\hat{\rho}_{DT}=\frac{2}{1-s}\exp\left[-2\,\mathrm{arcoth}(-s)\left[\hat{a}-\frac{r}{\sqrt{2}}e^{i\varphi}\right]^{\dagger}\times\left[\hat{a}-\frac{r}{\sqrt{2}}e^{i\varphi}\right]\right]. \quad (56)$$

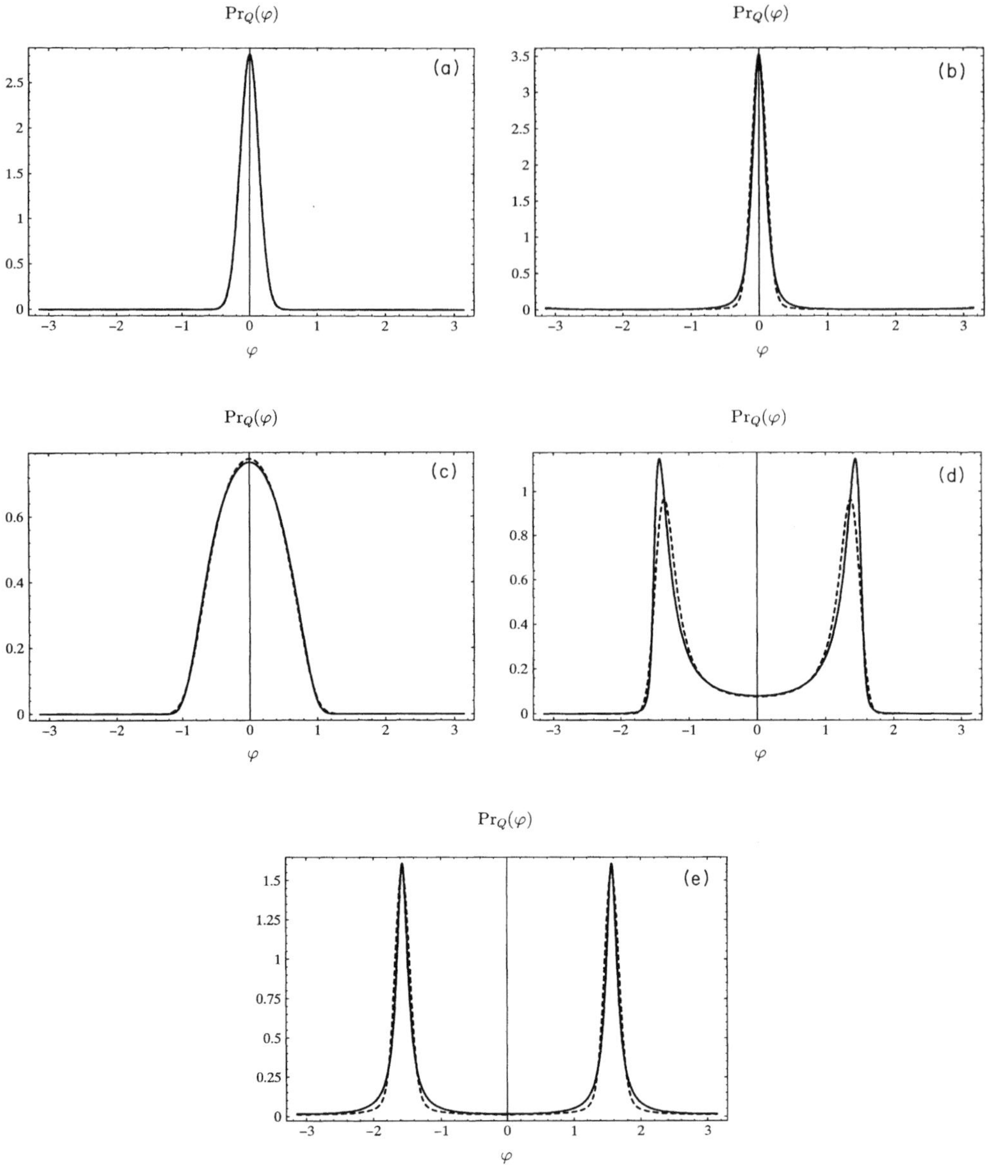

FIG. 2. Comparison of exact Q-phase distributions for squeezed states (51) with the results of our treatment of the Q phase as a noisy canonical phase, i.e., as convolutions (46) of pure canonical phase distributions (25) with weight functions (50). The exact Q-phase distribution (line) and the approximation (dashed line) is plotted for (a) a coherent state with $\alpha=5$ and $\zeta=1$ (no difference between the exact and the approximative distribution is visible); (b) a phase-squeezed state with $\alpha=4.5$ and $\zeta=0.047\,727\,5$; (c) an amplitude-squeezed state with $\alpha=4.5$ and $\zeta=20.9523$ [the inverse squeezing parameter as in (b)]; (d) an amplitude-squeezed state with $\alpha=1$ and $\zeta=97.9898$ showing bifurcation of the phase distribution [12]; and (e) a squeezed-vacuum state with $\alpha=0$ and $\zeta=101.99$. In all cases the mean photon number N was fixed to $N=25$.

According to the Wigner-function-overlap relation [32], we can write Eq. (54) as

$$W(r\cos\varphi, r\sin\varphi; s) = \frac{1}{2\pi}\,\mathrm{Tr}\{\hat{\rho}\hat{\rho}_{\mathrm{DT}}\}\ . \tag{57}$$

We substitute this result in Eq. (53), replace $r2^{-1/2}$ by r, and obtain, finally, the POM form for the s-phase distribution,

$$\mathrm{Pr}_s(\varphi) = \mathrm{Tr}\{\hat{\rho}\hat{\Pi}_s(\varphi)\}\ , \tag{58}$$

with

$$\hat{\Pi}_s(\varphi) = \frac{2}{\pi(1-s)}\int_0^\infty \mathrm{d}r\, r\exp[-2\,\mathrm{arcoth}(-s)(\hat{a} - re^{i\varphi})^\dagger \times(\hat{a} - re^{i\varphi})]\ . \tag{59}$$

In the following paragraphs, we relate the s-phase distribution to the pure canonical phase distribution. First, we calculate the Wigner function for the "s-phase state" $\hat{\Pi}_s(\varphi)$ and then, we apply a similar procedure as for the Q phase. We verify that the number-shift invariance (19) is fulfilled approximately, and calculate the asymptotics of the related reference-phase distribution.

Wigner function

Similar to the Wigner function associated with the Q phase, we obtain from Eqs. (53)–(55) the Wigner function for the "s-phase state" $\hat{\Pi}_s(\varphi)$,

$$W_s(x,p;\varphi) = \frac{-1}{2s\pi^2}\int_0^\infty dr\, r\exp\left[\frac{1}{s}(x - r\cos\varphi)^2 + \frac{1}{s}(p - r\sin\varphi)^2\right]. \tag{60}$$

Evidently, $W_s(x,p;\varphi)$ is simply a scaled Q-phase Wigner function,

$$W_s(x,p;\varphi) = \frac{1}{-s}W_Q\left[\frac{x}{\sqrt{-s}}, \frac{p}{\sqrt{-s}};\varphi\right]. \tag{61}$$

(Note that $s < -1$.) This indicates that the s-phase distribution is broader than the Q-phase distribution, as we would expect since extra detection noise is involved in inefficient phase-measurement schemes [16]. We now derive an approximative relation which links all s-phase distributions for $s \le 0$. Writing the expression (39) for the Q-phase Wigner function $W_Q(x,p;\varphi)$ in polar coordinates, we find that approximately

$$W_Q(r\cos\theta, r\sin\theta;\varphi) \approx \frac{r}{2\pi^{3/2}}\exp[-r^2(\varphi-\theta)^2] \tag{62}$$

holds for $r \gg 1$. According to the overlap relation (5) written in polar coordinates, and utilizing Eq. (61), we obtain for the s-phase distribution

$$\begin{aligned}\mathrm{Pr}_s(\varphi) &= \int_0^\infty dr\, r\int_{-\pi}^{+\pi} d\theta\, W(r\cos\theta, r\sin\theta) \times\frac{r}{\sqrt{-\pi s}}\exp\left[\frac{r^2}{s}(\varphi-\theta)^2\right]\\ &= \int_{-\pi}^{+\pi} d\theta\int_0^\infty dr\, rW[r\cos(\varphi-\theta), r\sin(\varphi-\theta)] \times\frac{r}{\sqrt{-\pi s}}\exp\left[\frac{r^2}{s}\theta^2\right].\end{aligned} \tag{63}$$

We approximate

$$\mathrm{Pr}_s(\varphi) \approx \int_{-\pi}^{+\pi} d\theta\frac{R}{\sqrt{-\pi s}}\exp\left[\frac{R^2}{s}\theta^2\right] \times\int_0^\infty dr\, rW[r\cos(\varphi-\theta), r\sin(\varphi-\theta)]\ , \tag{64}$$

where R is a typical radius of the Wigner function $W(x,p)$ for the particular physical state $\hat{\rho}$, for instance, the mean radius. The integral with respect to r yields the phase distribution derived from the Wigner function ($s=0$). Thus, we find that the relation

$$\mathrm{Pr}_s(\varphi) = \int_{-\pi}^{+\pi}\frac{d\phi}{2\pi}G_s(\phi;R)\mathrm{Pr}_0(\varphi-\phi) \tag{65}$$

between an s-phase distribution and the Wigner-phase distribution holds approximately in the semiclassical regime. Hence, we can interpret $\mathrm{Pr}_s(\varphi)$ as consisting of Wigner-phase distributions averaged with respect to a reference-phase distribution given by

$$G_s(\phi;R) = \left[\frac{4\pi^2 R}{-s}\right]^{1/2}\exp\left[\frac{R^2}{s}\phi^2\right]. \tag{66}$$

(Note that $s < 0$.) Since the Wigner phase approaches the pure canonical phase in the limit of very large intensities [13], we anticipate that an s-phase distribution is approximately a noisy canonical phase distribution. We may already guess what the reference-phase distribution looks like.

Fock representation

Now, we investigate how accurately the axioms (A) and (B) are fulfilled. Tanás, Miranowicz, and Gantsog [27] found the Fock representation of an s-parametrized phase distribution to be given by

$$\mathrm{Pr}_s(\varphi) = \frac{1}{2\pi}\sum_{n,m=0}^{\infty} B_{n,m}e^{i(m-n)\varphi}\langle n|\hat{\rho}|m\rangle\ , \tag{67}$$

with

$$B_{n,m} = \left[\frac{2}{1-s}\right]^{(n+m)/2}(n!m!)^{1/2} \times\sum_{k=0}^{\min(n,m)}\frac{[\frac{1}{2}(n+m)-k]!}{k!(n-k)!(m-k)!}\left[-\frac{1+s}{2}\right]^k. \tag{68}$$

Evidently, the phase-shifter axiom (A) for a canonical phase distribution is satisfied. How accurately is the number-shift invariance (B) fulfilled? In order to test axiom (B) it is convenient to write the expression (68) for the B coefficients in a different way. We recall the definition [39] [Vol. 2, Eq. 10.8(12)] of the Jacobi polynomials,

$$P_n^{(\alpha,\beta)}(x)=2^{-n}\sum_{k=0}^{n}\begin{pmatrix} n+\alpha \\ k\end{pmatrix}\begin{pmatrix} n+\beta \\ n-k\end{pmatrix} \times(x-1)^{n-k}(x+1)^k , \tag{69}$$

and find

$$B_{n,m}=\left[\frac{m-n}{2}\right]!\left[\frac{2}{1-s}\right]^{(m-n)/2}\left[\frac{s+1}{s-1}\right]^n\left[\frac{n!}{m!}\right]^{1/2} \times P_n^{[m-n,-(m+n)/2]}[1-4(s+1)^{-1}] \tag{70}$$

for $m \geq n$. (Otherwise, n and m should be interchanged.) We utilize a property [39] [Vol. 2, Eq. 10.8(33)] of the Jacobi polynomials and obtain

$$B_{n+1,m+1}=\frac{\frac{1}{2}(n+1+m+1)}{(n+1)^{1/2}(m+1)^{1/2}}B_{m,m}(1+\varepsilon), \tag{71}$$

with ε given by

$$\varepsilon=\frac{m-n}{n+1+m+1}\left[\frac{s+1}{s-1}\right]\Theta \tag{72}$$

and

$$\Theta=\frac{P_n^{[m-n,-(m+n)/2-1]}[1-4(s+1)^{-1}]}{P_n^{(m-n,-(m+n)/2)}[1-4(s+1)^{-1}]} . \tag{73}$$

The expression (71) looks quite similar to the approximate number-shift invariance (42) of the Q-phase distribution. Apart from the factor $(1+\varepsilon)$, the shifted coefficients $B_{n+1,m+1}$ differ from the initial coefficients $B_{n,m}$ by the ratio of the arithmetic and the geometric mean of $n+1$ and $m+1$, as the Q-phase coefficients do. Evidently, for the Q phase where s equals -1, the correction ε is zero. Now, we prove that ε is small for $s<-1$ [43]. Because of

$$\begin{pmatrix} n+\beta+1 \\ n-k\end{pmatrix} > \begin{pmatrix} n+\beta \\ n-k\end{pmatrix} , \tag{74}$$

we obtain from the definition (69) of the Jacobi polynomials that for $x>1$,

$$0<P_n^{(\alpha,\beta)}(x)<P_n^{(\alpha,\beta+1)}(x) . \tag{75}$$

For $s<-1$, we get $x=1-4(s+1)^{-1}>1$, and, hence,

$$0<\Theta<1 . \tag{76}$$

According to Eq. (72) this implies that the correction ε in the number-shift relation (71) is indeed small for states having narrow photon distributions compared to the mean photon number. Consequently, the number-shift invariance (19) holds approximately in the semiclassical regime. Thus, measured phase distributions, i.e., s-phase distributions with $s \leq -1$, can be regarded as noisy canonical phase distributions.

Convolution

Hence, we can interpret $\mathrm{Pr}_s(\varphi)$ as pure canonical phase distributions $\mathrm{Pr}_p(\varphi)$ averaged with respect to a weight function,

$$\mathrm{Pr}_s(\varphi)=\int_{-\pi}^{+\pi}\frac{d\phi}{2\pi}g_s(\phi;N)\mathrm{Pr}_p(\varphi-\phi) , \tag{77}$$

with

$$g_s(\phi;N)=\sum_{\nu=-\infty}^{+\infty}e^{i\nu\phi}B_{N,N+|\nu|} . \tag{78}$$

As before, N denotes the mean photon number. Similar to the case of the Q phase, we derive an asymptotic expression for $g_s(\phi;N)$. We replace the Fourier series by an integral and use the saddle-point method for evaluating it. It is shown in the Appendix that

$$B_{N,N}=1, \quad \left.\frac{\partial B_{N,N+|\nu|}}{\partial \nu}\right|_{\nu=+0}\to 0 , \quad \left.\frac{\partial^2 B_{N,N+|\nu|}}{\partial \nu^2}\right|_{\nu=+0}\to\frac{s}{4N} \tag{79}$$

for large mean photon numbers N. Hence, the saddle point lies at $\nu=0$ and we can approximate $B_{N,N+|\nu|}$ by

$$B_{N,N+|\nu|}\approx \exp\left[\frac{s\nu^2}{8N}\right] \tag{80}$$

in the Fourier series (78). Replacing the series by an integral, we obtain, finally,

$$g_s(\phi;N)=2\pi\left[-\frac{2N}{\pi s}\right]^{1/2}\exp\left[\frac{2N}{s}\phi^2\right] . \tag{81}$$

(Note that $s<-1$.) This expression describes the intensity-dependent probability distribution of reference phases for inefficiently measured phase distributions considered as noisy canonical phase distributions. Our expression (81) is very similar to Eq. (66), which was motivated by the overlap relation (5). Note that the general formula (81) is less accurate than the specific expression (50) for the Q phase since some more approximations are involved. However, when we set $s=-1$, both formulas converge to the same expression for large mean photon numbers N. The s dependence in Eq. (81) is easy to understand. The more inefficient the phase measurement is the larger is the modulus of the s parameter and the broader is the weight function. We also note that as in the case of the Q phase, the reference-phase distribution for the s-parametrized phase gets narrower with increasing intensity. It approaches a δ function in the macroscopic limit. We tested numerically the accuracy of our treatment of s-parametrized phase as a noisy canonical phase. For $N/|s|>20$, we found a similar accuracy as in the case of the Q phase.

IV. SUMMARY

We have extended the treatment of canonical quantum-optical phase within the formalism of probability distributions to include external measurement-induced uncertainty. We require that quantum phase satisfies two elementary axioms motivated by the complementarity of phase and photon number: *A phase shifter shifts a phase distribution* while *a number shifter does not change it* [37]. These requirements determine a pure canonical phase distribution uniquely as being the Helstrom-Shapiro-Shepard phase distribution [4,5] or the Pegg-Barnett phase distribution [6] for physical states when the limit of infinite Hilbert-space dimension has been taken. A noisy canonical phase distribution can be interpreted as a weighted average of pure canonical phase distributions where the weight function represents uncertainty in the reference phase.

As a second step, we have linked recent *phase measurements* [18] with *canonical phase* in the semiclassical regime. Under reasonable assumptions [21], measured phase distributions [22] are s-parametrized phase distributions, i.e., integrals of smoothed Q functions over the radial coordinate [15,19,20,16]. We have shown that these distributions fulfill approximately the basic axioms of canonical quantum phase for states having a narrow photon distribution compared to the mean photon number. In this case, we can interpret a measured phase distribution as a weighted average of pure canonical phase distributions. The important point is that the measurement is not contaminated (at least to the level of the approximation) by the conjugate variable, photon number. The weight function depends on the s-parameter which comprises the overall detection noise in phase measurements [16]. In contrast to an axiomatically defined noisy phase distribution, the weight function depends weakly on the mean photon number as well. It tends to a δ function for very large intensities since the extra noise involved in phase measurements [24] becomes negligible in the classical domain. Numerical tests illustrate that our treatment of measured phase distributions as noisy canonical phase distributions is well justified in the semiclassical regime for mean photon numbers exceeding roughly twenty times the modulus of the s parameter. Our analysis, thus, bridges all three conceptions of quantum-optical phase [3] (canonical phase, s-parametrized phase, phase from measurements) and provides important physical insight into the relationship between them.

ACKNOWLEDGMENTS

J.A.V. thanks D. T. Pegg and S. M. Barnett for previous discussions regarding quantum-optical phase.

APPENDIX

We wish to derive asymptotic expressions for $B_{N,N}$,

$$B'_{N,N} \equiv \left. \frac{\partial B_{N,N+|\nu|}}{\partial \nu} \right|_{\nu=+0}, \tag{A1}$$

$$B''_{N,N} \equiv \left. \frac{\partial^2 B_{N,N+|\nu|}}{\partial \nu^2} \right|_{\nu=+0}, \tag{A2}$$

in the limit of large N for the B coefficients being defined by Eq. (68). Since $P_N^{(0,-N)}(x)=[\frac{1}{2}(x+1)]^N$, we obtain immediately from Eq. (70)

$$B_{N,N}=1 . \tag{A3}$$

Now, we calculate the first derivative $B'_{N,N}$ of $B_{N,N+|\nu|}$ with respect to ν at $\nu=+0$. We get from the number-shift relation (71) the recurrence relation,

$$B'_{N+1,N+1}=B'_{N,N}+\varepsilon' , \tag{A4}$$

with

$$\varepsilon' \equiv \left. \frac{\partial}{\partial \nu} \frac{\nu/2}{N+\frac{\nu}{2}+1} \left[\frac{s+1}{s-1} \right] \Theta \right|_{\nu=0} \tag{A5}$$

$$= \frac{1}{2(N+1)} \left[\frac{s+1}{s-1} \right] \Theta_0 . \tag{A6}$$

Here, Θ_0 is given by

$$\Theta_0 = \Theta|_{\nu=0} = \frac{P_N^{(0,-N-1)}\left[1-\frac{4}{s+1}\right]}{P_N^{(0,-N)}\left[1-\frac{4}{s+1}\right]} = \left[\frac{s+1}{s-1} \right]^N . \tag{A7}$$

Hence, the recurrence relation for the $B'_{N,N}$ reads

$$B'_{N+1,N+1}=B'_{N,N}+\frac{1}{2(N+1)} \left[\frac{s+1}{s-1} \right]^{N+1} . \tag{A8}$$

According to the definition (68) of the B coefficients, we have

$$B_{0,\nu} = \left[\frac{2}{1-s} \right]^{\nu/2} \frac{\left[\frac{\nu}{2} \right]!}{\sqrt{\nu !}} \tag{A9}$$

and, hence,

$$B'_{0,0}=\tfrac{1}{2} \ln \left[\frac{2}{1-s} \right] . \tag{A10}$$

Solving the recurrence relation (A8) and using the initial value (A10), we obtain

$$B'_{N,N}=\frac{1}{2} \left[\ln \left[\frac{2}{1-s} \right] + \sum_{k=1}^{N} \frac{1}{k} \left[\frac{s+1}{s-1} \right]^k \right] . \tag{A11}$$

For $s=-1$, thus, considering the Q phase, the coefficients $B'_{N,N}$ are always zero. For $s<-1$, they approach zero in the limit of large N,

$$B'_{N,N} \to 0 , \tag{A12}$$

because [44] (Vol. 1, Eq. 5.2.4.4)

$$\sum_{k=1}^{\infty} \frac{1}{k} \left[\frac{s+1}{s-1} \right]^k = -\ln \left[\frac{2}{1-s} \right] . \tag{A13}$$

Now, we derive an asymptotic expression for the second derivative $B''_{N,N}$ of $B_{N,N+|\nu|}$ at $\nu=+0$. First, we obtain from the definition (68) of the B coefficients the differential equation,

$$(1-s)\frac{\partial B_{N,N+|\nu|}}{\partial s}=\left[N+\frac{\nu}{2}\right]B_{N,N+|\nu|}-[N(N+|\nu|)]^{1/2}B_{N-1,N-1+|\nu|}\ . \qquad (A14)$$

Consequently,

$$\begin{aligned}(1-s)\frac{\partial B''_{N,N}}{\partial s}&=N(B''_{N,N}-B''_{N-1,N-1})+B'_{N,N}-B'_{N-1,N-1}+\frac{1}{4N}\\&=N(B''_{N,N}-B''_{N-1,N-1})+\frac{1}{2N}\left[\frac{s+1}{s-1}\right]^{N}+\frac{1}{4N}\ ,\end{aligned} \qquad (A15)$$

where the recurrence relation (A8) for $B'_{N,N}$ has been used. Motivated by the result (66) derived using the overlap relation (5), we make the ansatz

$$B''_{N,N}=a_N s\ . \qquad (A16)$$

Inserting the ansatz (A16) into the differential equation (A15) we obtain, neglecting the $(2N)^{-1}(s+1)^N(s-1)^{-N}$ term,

$$(1-s)a_N=N(a_N-a_{N-1})s+\frac{1}{4N}\ . \qquad (A17)$$

Comparing the powers of s on the left and the right side of this equation, we get

$$a_N=\frac{1}{4N} \qquad (A18)$$

and

$$(N+1)a_N=Na_{N-1}\ . \qquad (A19)$$

Equations (A18) and (A19) do not contradict each other for large N. Hence the ansatz (A16) is justified, and we obtain, finally,

$$B''_{N,N}\to\frac{s}{4N}\ . \qquad (A20)$$

[1] The recent special issue of Phys. Scr. **T48** (1993), edited by S. M. Barnett and W. Schleich, nicely illustrates the different opinions people have on quantum-optical phase.

[2] The interpretation of the recent stimulating phase measurements [18] was debated in a series of Comments and Replies, see Z. Hradil, Phys. Rev. A **47**, 4532 (1993); J. W. Noh, A. Fougères, and L. Mandel, *ibid.* **47**, 4535 (1993); S. M. Barnett and D. T. Pegg, *ibid.* **47**, 4537 (1993); J. W. Noh, A. Fougères, and L. Mandel, *ibid.* **47**, 4541 (1993); Z. Hradil and J. Bajer, *ibid.* **48**, 1717 (1993); J. W. Noh, A. Fougères, and L. Mandel, *ibid.* **48**, 1719 (1993).

[3] S. M. Barnett and B. J. Dalton, Phys. Scr. **T48**, 13 (1993).

[4] C. W. Helstrom, Int. J. Theor. Phys. **11**, 357 (1974); J. W. Shapiro, S. R. Shepard, and N. C. Wong, Phys. Rev. Lett. **62**, 2377 (1989).

[5] J. W. Shapiro and S. R. Shepard, Phys. Rev. A **43**, 3795 (1991).

[6] D. T. Pegg and S. M. Barnett, Europhys. Lett. **6**, 483 (1988); Phys. Rev. A **39**, 1665 (1989); S. M. Barnett and D. T. Pegg, J. Mod. Opt. **36**, 7 (1989); **39**, 2121 (1992).

[7] M. J. W. Hall, Quantum Opt. **3**, 7 (1991).

[8] J. A. Vaccaro and D. T. Pegg, Phys. Scr. **T48**, 22 (1993).

[9] R. G. Newton, Ann. Phys. (N.Y.) **124**, 327 (1980); S. M. Barnett and D. T. Pegg, J. Phys. A **19**, 3849 (1986).

[10] J. A. Vaccaro (unpublished).

[11] K. E. Cahill and R. J. Glauber, Phys. Rev. **177**, 1882 (1969).

[12] W. Schleich, R. J. Horowicz, and S. Varro, Phys. Rev. A **40**, 7405 (1989).

[13] The reason is that the Wigner function for a London phase state [30] approaches a δ function in regions far from the origin in phase space [34]. According to the overlap relation [32], the radius integrated Wigner function thus approaches the pure canonical phase distribution.

[14] B. M Garraway and P. L. Knight, Phys. Rev. A **46**, 5346 (1992); Phys. Scr. **T48**, 66 (1993).

[15] U. Leonhardt and H. Paul, Phys. Rev. A **47**, R2460 (1993); M. Freyberger, K. Vogel, and W. Schleich, Phys. Lett. A **176**, 41 (1993); see, also, N. G. Walker, J. Mod. Opt. **34**, 15 (1987); Y. Lai and H. A. Haus, Quantum Opt. **1**, 99 (1989).

[16] U. Leonhardt and H. Paul, Phys. Rev. A **48**, 4598 (1993).

[17] H. Gerhardt, H. Welling, and D. Frölich, Appl. Phys. **2**, 91 (1973); H. Gerhardt, U. Büchler, and G. Litfin, Phys. Lett. **49A**, 119 (1974).

[18] J. W. Noh, A. Fougères, and L. Mandel, Phys. Rev. Lett. **67**, 1426 (1991); Phys. Rev. A **45**, 424 (1992); **46**, 2840 (1992).

[19] Linear amplification as a phase-measurement scheme for fields in the quantum regime was proposed by A. Bandilla and H. Paul, Ann. Phys. (Leipzig) **23**, 323 (1969); **24**, 119 (1970); and studied in detail by H. Paul, Fortschr. Phys. **22**, 657 (1974). Experiments are reported in Ref. [17]. W. Schleich, A. Bandilla, and H. Paul, Phys. Rev. A **45**, 6652 (1992), pointed out that the *Q*-phase distribution is measured here.

[20] Phase measurement via heterodyning was proposed by J. H. Shapiro and S. S. Wagner, IEEE J. Quantum Electron. **QE-20**, 803 (1984).

[21] The assumption for the Noh-Fougéres-Mandel experiment [18] is that the local oscillator be strong with respect to the signal. This means that the phase difference with respect to a physical state of well-defined phase is measured. In more recent experiments, however, the operationally-defined phase difference of two weak fields was measured while simultaneously avoiding the problems highlighted in the Comments in Ref. [2], see J. W. Noh, A. Fougères, and L. Mandel, Phys. Rev. Lett. **71**, 2579 (1993); Phys. Scr. **T48**, 29 (1993). Phase measurements of fields with variable coherence are reported in A. Fougères, J. R. Torgerson, and L. Mandel, Opt. Commun. **105**, 199 (1994); A. Fougères, J. W. Noh, T. P. Grayson, and L. Mandel, Phys. Rev. A **49**, 530 (1994).

[22] Note that we are considering direct phase measurements here. Phase distributions can be determined, however, indirectly using *optical homodyne tomography,* see D. T.

Smithey, M. Beck, M. G. Raymer, and A. Faridani, Phys. Rev. Lett. **70**, 1244 (1993); Phys. Scr. **T48**, 35 (1993). By this method the density matrix for a given quantum state of light is reconstructed from homodyne measurements. Knowing the density matrix any physical quantity and, in particular, arbitrarily defined phase properties can be calculated, see M. Beck, D. T. Smithey, J. Cooper, and M. G. Raymer, Opt. Lett. **18**, 1259 (1993); M. Beck, D. T. Smithey, and M. G. Raymer, Phys. Rev. A **48**, R890 (1993); D. T. Smithey, M Beck, J. Cooper, and M. G. Raymer, *ibid.* **48**, 3159 (1993).

[23] We also note that the Ban-formalism [M. Ban, Phys. Lett. A **152**, 223 (1991); **152**, 397 (1991); J. Math. Phys. **32**, 3077 (1991); J. Opt. Soc. Am. B **9**, 1189 (1992); Phys. Lett. A **176**, 47 (1993)] for a two-mode bosonic system is equivalent to the phase-measurement schemes [15,19,20]. See A. Lukš, V. Peřinová, and J. Křepelka, Phys. Rev. A **50**, 818 (1994). One mode corresponds to the investigated system and the other to the extra noise involved in phase measurements [24].

[24] U. Leonhardt and H. Paul, Phys. Scr. **T48**, 45 (1993).

[25] U. Leonhardt and H. Paul, J. Mod. Opt. **40**, 1745 (1993), or see E. Arthurs and J. L. Kelly, Jr., Bell Syst. Tech. J. **44**, 725 (1965); S. L. Braunstein, C. M. Caves, and G. J. Milburn, Phys. Rev. A **43**, 1153 (1991); nicely reviewed by S. Stenholm, Ann. Phys. (N.Y.) **218**, 233 (1992).

[26] A. Bandilla and H.-H. Ritze, Phys. Rev. A **49**, 4912 (1994).

[27] R. Tanaś, A. Miranowicz, and Ts. Gantsog, Phys. Scr. **T48**, 53 (1993).

[28] C. W. Helstom, *Quantum Detection and Estimation Theory* (Academic, New York, 1976). For an excellent introduction to POM's see Ref. [5], Sec. II A.

[29] Probability operator measures are usually denoted as *measures* $d\mu = \mathrm{Tr}\{\hat{\rho}\, d\hat{\mu}\}$. However, here we drop the differential for simplicity. Strictly speaking, this means we are only considering the operators $d\hat{\mu}/d\varphi = \hat{\Pi}(\varphi)$ as ordinary Hilbert-space operators. Alternatively, the full mathematical generality of *measures* can be restored by interpreting all integrals involving $\hat{\Pi}(\varphi)$ as appropriate Stieltjes integrals. In terms of simple physics, we are considering probability densities instead of probabilities.

[30] F. London, Z. Phys. **40**, 193 (1927), Eq. (13).

[31] A. Böhm, *The Rigged Hilbert Space and Quantum Mechanics* (Springer, Berlin, 1978).

[32] Usually, the overlap relation (5) is formulated for pure states as

$$|\langle \psi_1|\psi_2\rangle|^2 = 2\pi \int_{-\infty}^{+\infty} dx \int_{-\infty}^{+\infty} dp\; W_1(x,p) W_2(x,p) \; .$$

See, for instance, R. F. O'Connell and E. P. Wigner, Phys. Lett. **83A**, 145 (1981). However, it is easily generalized for mixed states. We express the density operators $\hat{\rho}_1$ and $\hat{\rho}_2$ as $\hat{\rho}_1 = \sum_a p_a |a\rangle\langle a|$ and $\hat{\rho}_2 = \sum_b p_b |b\rangle\langle b|$, and obtain

$$\begin{aligned} \mathrm{Tr}\{\hat{\rho}_1\hat{\rho}_2\} &= \sum_{ab} p_a p_b |\langle a|b\rangle|^2 \\ &= 2\pi \sum_{ab} p_a p_b \int_{-\infty}^{+\infty} dx \int_{-\infty}^{+\infty} dp\; W_a(x,p) W_b(x,p) \\ &= 2\pi \int_{-\infty}^{+\infty} dx \int_{-\infty}^{+\infty} dp\; W_1(x,p) W_2(x,p) \; . \end{aligned}$$

[33] For review articles on the Wigner representation, see, V. I. Tatarskii, Usp. Fiz. Nauk. **139**, 587 (1983) [Sov. Phys. Usp. **26**, 311 (1983)]; M. Hillery, R. F. O'Connell, M. O. Scully, and E. P. Wigner, Phys. Rep. **106**, 121 (1984).

[34] U. Herzog, H. Paul, and Th. Richter, Phys. Scr. **T48**, 61 (1993).

[35] C. Cohen-Tannoudji, B. Diu, and F. Laloë, *Quantum Mechanics* (Wiley, New York, 1977); Chap. II, Complement *E*.

[36] L. Susskind and J. Glogower, Phys. **1**, 49 (1964); P. Carruthers and M. M. Nieto, Rev. Mod. Phys. **40**, 411 (1968).

[37] Axiom A, *a phase shifter is a phase-distribution shifter*, has been used previously by D. T. Pegg, J. A. Vaccaro, and S. M. Barnett, *Quantum Optical Phase and its Measurement in Quantum Optics VI*, edited by J. D. Harvey and D. F. Walls (Springer, Berlin, 1994). As a second axiom, they required that strong coherent states have well-defined phases, i.e., their phase-distributions approach δ functions. This condition, however, is not sufficient to determine pure canonical phase distributions uniquely.

[38] The replacement of the geometric by the arithmetic mean in the Fock representation was also the basic approximation in analytic expressions for the wave function and the Wigner function of a London phase state [30], see H. Paul, in *International Conference on Quantum Electronics Technical Digest Series 1992* (Institut für Nachrichtentechnik der Tu Wien, Vienna, 1992), Vol. 9, p. 274; H. Paul and W. Schleich (unpublished). Excellent agreement of the approximate with the numerically calculated exact Wigner function was pointed out in Ref. [34].

[39] A. Erdélyi, *Higher Transcendental Functions* (McGraw-Hill, New York, 1955).

[40] Phase-optimized states are introduced and analyzed in G. S. Summy and D. T. Pegg, Opt. Commun. **77**, 75 (1990); A. Bandilla, H. Paul, and H.-H. Ritze, Quantum Opt. **3**, 267 (1991).

[41] R. Loudon and P. L. Knight, J. Mod. Opt. **34**, 709 (1987).

[42] C. W. Gardiner, *Quantum Noise* (Springer, Berlin, 1991), Chap. 4.4.

[43] We note that the number-shift invariance does not hold approximately for *s*-phase distributions with $s > -1$. This is connected with the fact that these distributions are not measurable as probability distributions in phase-measurement schemes.

[44] A. P. Prudnikov, Yu. A. Brychkov, and O. I. Marichev, *Integrals and Series* (Gordon and Breach, New York, 1992).

Number-phase Wigner function on Fock space

John Vaccaro*
Arbeitsgruppe "Nichtklassische Strahlung" der Max-Planck-Gesellschaft an der Humboldt–Universität zu Berlin, Rudower Chaussee 5, 12484 Berlin, Germany
(Received 21 March 1995)

We define a quasiprobability distribution $S_{\mathrm{NP}}(n,\theta)$ which describes the quantum statistics of the photon number and phase observables of a single-mode field (or, equivalently, a harmonic oscillator). The properties of $S_{\mathrm{NP}}(n,\theta)$ are the photon number and phase analogies of the properties of Wigner's original function, which describes the position and momentum observables. For example, the marginals of $S_{\mathrm{NP}}(n,\theta)$ are the continuous phase and the discrete photon-number probability distributions. We give examples of the $S_{\mathrm{NP}}(n,\theta)$ representation of various states and show, in particular, that $S_{\mathrm{NP}}(n,\theta)$ displays the quantum interference associated with Schrödinger cat states. We also describe how $S_{\mathrm{NP}}(n,\theta)$ can be determined from quantities that are, in principle, measurable.

PACS number(s): 03.65.Ca, 42.50.−p

I. INTRODUCTION

It is now over 60 years since Wigner introduced his celebrated quasiprobability distribution for illustrating the difference between classical and quantum statistics [1]. Wigner's function expresses the quantum statistics of a pair of canonically conjugate observables that have continuous spectra. Over the intervening years a number of related quasiprobability distributions have been introduced mainly for quantum observables with discrete spectra and associated with finite-dimensional systems. For example, Agarwal [2] (see also [3]) introduced quasiprobability distributions for finite-dimensional systems based on the atomic coherent state formalism of Arecchi *et al.* [4]. Agarwal's quasiprobability distributions are functions of continuous variables in contrast to the discrete Wigner function defined by Wootters [5], which is a function of discrete variables for the same finite-dimensional systems. The discrete nature of Wootters's function reflects the discrete nature of the spectra of the underlying pair of general canonically conjugate observables. Mukunda [6] introduced a Wigner function for the canonically conjugate rotation angle and angular momentum observables; recently this function was further justified and studied comprehensively by Bizarro [7]. An interesting point about the Mukunda-Bizarro function is that one variable is continuous (angle) and the other is discrete (angular momentum) and so it strides both discrete and continuous domains.

Another pair of canonically conjugate observables having both a discrete and a continuous spectrum are the phase and photon-number observables of a single-mode field (or, equivalently, a harmonic oscillator). The definition of the phase observable has been studied extensively in recent years [8] and there are now a number of different formalisms for describing it that yield identical physical results; these are the Newton, Pegg-Barnett, Ban, Helstrom-Shiparo-Shepard probability-operator measure (POM), and the H_{sym} formalisms [9]. Each of the these formalisms is based on a different state space. Newton [10] uses a Hilbert space that contains negative as well as positive photon-number states. Pegg and Barnett [11–14] use a Hilbert space of finite dimension and a special procedure for taking the infinite-dimensional limit. It has been shown recently [15] that the special limiting procedure essentially builds a vector space E that is larger than the conventionally used Fock space. Ban's approach [16,17] employs a tensor product of two Fock spaces. In the POM approach of Helstrom [18] and Shapiro and Shepard [19] (see also [20,21]) only the conventionally used Fock space is required and phase operators are not considered explicitly. Instead only the phase probability distribution is defined and from this expectation values of a stochastic phase variable are found. The H_{sym} formalism [22] is based on an infinite-dimensional Hilbert space H_{sym}, which is an extension of the conventionally used Fock space and contains vectors representing infinite photon number; H_{sym} has the distinction of supporting the strong limits of the Pegg-Barnett phase operators.

In this paper we wish to define a quasiprobability distribution that has properties analogous to those of Wigner's function, but is associated with the canonically conjugate [23] phase and photon-number observables rather than the position and momentum observables treated by Wigner [24]. For this we require a state space on which phase and photon-number probability distributions are defined. Fortunately, all the previously mentioned phase formalisms agree with respect to these distributions for physically relevant states. Moreover, it turns out that we do not require a phase operator explicitly for the definition of the quasiprobability distribution and so we have some latitude in our choice of the state space. Since the Fock space conventionally used to represent the state of a single-mode field does support the phase and photon-number probability distributions it provides a sufficient state space for our analysis. Thus we confine our attention to defining a quasiprobability distribution on the conventionally used Fock space in this paper.

In a previous work [25] we defined a quasiprobability distribution $W_{N\phi}(n,\theta)$ associated specifically with the Pegg-Barnett phase formalism. This function, which we shall call the *discrete number-phase Wigner function*, is based on

*Present address: Physics Department, The Open University, Walton Hall, Milton Keynes MK7 6AA, United Kingdom.

Wootters's definition of a discrete Wigner function for a finite-dimensional space (e.g., of dimension $s+1$). In accordance with the Pegg-Barnett formalism, the limit of infinite s is taken only after expectation values have been calculated. An unusual feature of $W_{N\phi}(n,\theta)$ is that for some states of the field (including coherent states) it undergoes a revival as the number variable n is increased beyond the value representing half the dimension $s/2$. Thus there appears to be no simple way of taking the infinite-dimensional limit of the Wigner function itself for these states. Lukš and Peřinová [26] attempted to overcome this problem by defining an analogous function $W_{n\varphi}(n,\theta)$ at half integer values of the photon-number variable. Although the function $W_{n\varphi}(n,\theta)$ at the fractional values of n has no direct physical relevance, nevertheless, the values of $W_{n\varphi}(n,\theta)$ at these values of n are required for normalization.

We take a fresh look at the problem in this paper by adopting a *first-principles* approach. We require the number-phase Wigner function to have the analogous properties of Wigner's original function as described by Hillery *et al.* [27]. The revivals found for the discrete number-phase Wigner function do not appear and there is no need for fractional values of n as used by Lukš and Peřinová. However, even after ensuring that the number-phase Wigner function has properties analogous to those of Wigner's original function as listed by Hillery *et al.*, we find that the number-phase Wigner function is not uniquely defined. We enlist an extra property, which is responsible for the quantum interference fringes displayed by Wigner's function for Schrödinger cat states, to lead us to the definition of the *special number-phase Wigner function* $S_{\rm NP}(n,\theta)$. A brief preliminary description of $S_{\rm NP}(n,\theta)$ was given recently in [28]. Here we give a more-detailed analysis. We illustrate $S_{\rm NP}(n,\theta)$ for various states and show, in particular, that it displays interference fringes for Schrödinger cat states in a manner similar to that of Wigner's original function. We also show that $S_{\rm NP}(n,\theta)$ can be determined from quantities that are, in principle, measurable.

The format of this paper is as follows. In Sec. II we define the number-phase Wigner function $S_{\rm NP}(n,\theta)$. In Sec. III we illustrate $S_{\rm NP}(n,\theta)$ for various states and in Sec. IV we discuss the $S_{\rm NP}$ representation of arbitrary operators. Then, in Sec. V we show how $S_{\rm NP}(n,\theta)$ for an unknown state of the field can be determined from measurable quantities. We end with a discussion in Sec. VI.

II. SPECIAL NUMBER-PHASE WIGNER FUNCTION $S_{\rm NP}$

We wish to give the special number-phase Wigner function $S_{\rm NP}(n,\theta)$ properties analogous to those of the position-momentum Wigner function $W(x,p)$ (for convenience we call these functions the NP-Wigner function and the PM-Wigner function, respectively). To this end we shall take the list of properties given by Hillery *et al.* [27] for the PM-Wigner function and transform them into analogous properties for the NP-Wigner function. We note that Wigner [29] has shown that the PM-Wigner function is uniquely determined by just five of these properties and also that O'Connell and Wigner [30] have shown that the PM-Wigner function is also uniquely determined by a different set of five properties. Our use of a superset of seven analogous defining properties will therefore yield a function that is acceptable as the Wigner function for the photon-number and phase observables. We use the symbol $S_{\rm NP}$ and the adjective "special" to distinguish the NP-Wigner function defined here from the discrete NP-Wigner function and the NP-Wigner function defined by Lukš and Peřinová [26].

A. Seven basic properties

We begin by ensuring that $S_{\rm NP}(n,\theta)$ is a bilinear functional of the state vector by specifying that it is the expectation value of the NP-Wigner operator $\hat{S}_{\rm NP}(n,\theta)$, which we represent as

$$\hat{S}_{\rm NP}(n,\theta)=\frac{1}{2\pi}\sum_{p,q=0}^{\infty}\Omega_{p,q}(n,\theta)|p\rangle\langle q| \tag{1}$$

in the Fock basis for $n=0,1,2,\ldots$ and for a real value of θ. Our task is to specify the elements $\Omega_{p,q}(n,\theta)$ of the matrix $\mathbf{\Omega}(n,\theta)$ that give $S_{\rm NP}$ its required properties. Most of these properties will be expressed in terms of expectation values; such expressions can be put into the form

$$\sum_{p,q=0}^{\infty} M_{p,q}f_p^*f_q=0 ,$$

where $M_{p,q}$ are the matrix elements of an operator $\hat{M}$ in the Fock basis and $f_n=\langle n|f\rangle$ are the Fock state coefficients of an arbitrary vector $|f\rangle$ in the Fock space. Since $|f\rangle$ is arbitrary it follows that the operator $\hat{M}$ itself vanishes. In the following, this allows us to translate the required properties of $S_{\rm NP}$ into requirements for the NP-Wigner operator $\hat{S}_{\rm NP}(n,\theta)$ and the associated matrix elements $\Omega_{p,q}(n,\theta)$. We now specify seven properties of $S_{\rm NP}$ using the same numbering scheme as Hillery *et al.* to aid a comparison of $S_{\rm NP}$ with the PM-Wigner function $W(x,p)$.

(i) $S_{\rm NP}(n,\theta)$ should be real and so $\hat{S}_{\rm NP}(n,\theta)$ should be Hermitian. Thus

$$\Omega_{p,q}(n,\theta)^*=\Omega_{q,p}(n,\theta) . \tag{2}$$

(ii) The marginal distributions of $S_{\rm NP}$ should be the (normalized) number and phase probability distributions

$$\int_{2\pi} S_{\rm NP}(n,\theta)d\theta=\langle|n\rangle\langle n|\rangle , \tag{3}$$

$$\sum_{n=0}^{\infty} S_{\rm NP}(n,\theta)=\langle|\theta\rangle\langle\theta|\rangle . \tag{4}$$

The phase state $|\theta\rangle\equiv(2\pi)^{-1/2}\Sigma_{n=0}^{\infty}\exp(in\theta)|n\rangle$ in the expression (4) belongs to a rigged Hilbert space [31]. It yields the now-well-established [10–13,15–22,32–34] definition of the phase probability distribution $P(\theta)=\langle|\theta\rangle\langle\theta|\rangle$ for states belonging to the conventionally used infinite-dimensional Fock space. Thus we require the NP-Wigner operator to satisfy

$$\int_{2\pi} \hat{S}_{\rm NP}(n,\theta)d\theta = |n\rangle\langle n| \ ,$$

$$\sum_{n=0}^{\infty} \hat{S}_{\rm NP}(n,\theta) = |\theta\rangle\langle\theta| \ .$$

These expressions imply that

$$\frac{1}{2\pi}\int_{2\pi} \Omega_{p,q}(n,\theta)d\theta = \delta_{p,n}\delta_{q,n} \tag{5}$$

and

$$\sum_{n=0}^{\infty} \Omega_{p,q}(n,\theta) = e^{i(p-q)\theta} \ , \tag{6}$$

respectively, where $\delta_{n,m}$ is the Kronecker delta function.

(iii) $S_{\rm NP}$ should be Galilei invariant in the sense that shifts in phase $\langle\theta|f\rangle \mapsto \langle\theta+\Delta|f\rangle$ and photon number $\langle n+1|f\rangle \mapsto \langle n|f\rangle$ should produce the corresponding shifts $S_{\rm NP}(n,\theta) \mapsto S_{\rm NP}(n,\theta+\Delta)$ and $S_{\rm NP}(n+1,\theta) \mapsto S_{\rm NP}(n,\theta)$, respectively, in $S_{\rm NP}$. A phase shift of Δ is generated by the operator $\exp(i\hat{N}\Delta)$, where $\hat{N}$ is the photon-number operator. Thus we require that

$$e^{i\hat{N}\Delta}\hat{S}_{\rm NP}(n,\theta)e^{-i\hat{N}\Delta} = \hat{S}_{\rm NP}(n,\theta+\Delta) \ .$$

Expressing the NP-Wigner operator in terms of the coefficients given by Eq. (1) and equating matrix elements in the Fock basis yields

$$\Omega_{p,q}(n,\theta)e^{i(p-q)\Delta} = \Omega_{p,q}(n,\theta+\Delta)$$

and thus we have

$$\Omega_{p,q}(n,\theta) = \Omega_{p,q}(n)e^{i(p-q)\theta} \ , \tag{7}$$

where we have defined

$$\Omega_{p,q}(n) \equiv \Omega_{p,q}(n,0) \ . \tag{8}$$

Substituting $\Omega_{p,q}(n,\theta)$ from Eq. (7) into Eqs. (2), (5), and (6) gives

$$\Omega_{p,q}(n)^* = \Omega_{q,p}(n) \ , \tag{9}$$

$$\Omega_{p,p}(n) = \delta_{p,n} \ , \tag{10}$$

$$\sum_{n=0}^{\infty} \Omega_{p,q}(n) = 1 \ , \tag{11}$$

respectively.

The shift in photon number $\langle n+1|f\rangle \mapsto \langle n|f\rangle$ is produced by the operator, $\widehat{e^{-i\phi}} = \sum_{n=0}^{\infty}|n+1\rangle\langle n|$, i.e., $\langle n+1|\widehat{e^{-i\phi}}|f\rangle = \langle n|f\rangle$. (We note that $\widehat{e^{-i\phi}}$ is the Susskind-Glogower [35] exponential phase operator, which is well known to be nonunitary. However, we need only the "upward" number-shifting property of $\widehat{e^{-i\phi}}$; the nonunitary nature of $\widehat{e^{-i\phi}}$ is of no consequence here because the operation $|f\rangle \mapsto \widehat{e^{-i\phi}}|f\rangle$ preserves the norm.) Thus, for the number-shift invariance property we require that

$$\widehat{e^{i\phi}}\hat{S}_{\rm NP}(n+1,\theta)\widehat{e^{-i\phi}} = \hat{S}_{\rm NP}(n,\theta) \ ,$$

where $\widehat{e^{i\phi}} = \widehat{e^{-i\phi}}^\dagger$, from which we find that

$$\Omega_{p,q}(n) = \Omega_{p+1,q+1}(n+1) \tag{12}$$

for non-negative integers p,q,n. This translational property allows a more-convenient representation of the matrix elements $\Omega_{p,q}(n)$ as follows: we extend the definition of $\Omega_{p,q}(0)$ to negative integer values of p and q according to

$$\Omega_{p-n,q-n}(0) = \Omega_{p,q}(n) \tag{13}$$

for $n>p,q$. That this gives a unique definition of $\Omega_{p,q}(0)$ for negative integers p and q is easily proved from Eq. (12). Thus, specifying $\Omega_{p,q}(0)$ for all integers p and q is equivalent to specifying $\Omega_{p,q}(n)$ for all non-negative integers p,q,n. Equation (11) can now be written as

$$\sum_{n=0}^{\infty} \Omega_{p-n,q-n}(0) = 1 \ , \tag{14}$$

which holds for all non-negative integers p,q.

(iv) $S_{\rm NP}$ should be invariant with respect to a reflection in time and a phase shift of π rad. That is, if $\langle n|f\rangle \mapsto \langle n|f\rangle^*$, then $S_{\rm NP}(n,\theta) \mapsto S_{\rm NP}(n,-\theta)$, and if $\langle\theta|f\rangle \mapsto \langle\theta+\pi|f\rangle$, then $S_{\rm NP}(n,\theta) \mapsto S_{\rm NP}(n,\theta+\pi)$.

In the former invariance, the transformation is a time reflection in the sense that if $f_n(t) \equiv \langle n|f(t)\rangle = \langle n|\hat{U}(t)|f(0)\rangle = e^{-in\omega t}f_n(0)$, where $\hat{U}(t) = e^{-i\hat{N}\omega t}$ is the time evolution operator at time t, i.e., $|f(t)\rangle = \hat{U}(t)|f(0)\rangle$, then $|f(t)^*\rangle \equiv \Sigma f_n(t)^*|n\rangle = \hat{U}(-t)|f(0)^*\rangle$. This invariance implies that

$$\sum_{p,q=0}^{\infty} \Omega_{p,q}(n,\theta)f_pf_q^* = \sum_{p,q=0}^{\infty} \Omega_{p,q}(n,-\theta)f_p^*f_q$$

for arbitrary vector $|f\rangle$ and so

$$\Omega_{p,q}(n,\theta) = \Omega_{q,p}(n,-\theta) \ .$$

Combining this result with Eq. (7) yields

$$\Omega_{p,q}(n) = \Omega_{q,p}(n) \tag{15}$$

and so from Eq. (9) we obtain

$$\Omega_{p,q}(n) = \Omega_{p,q}(n)^* \ , \tag{16}$$

that is, $\mathbf{\Omega}(n)$ is a real symmetric matrix. We can now express the matrices $\mathbf{\Omega}(n)$ in a more convenient form by identifying the diagonals of the matrix $\mathbf{\Omega}(0)$ as follows: let $\mathbf{\Lambda}(r)$ be the rth diagonal of $\mathbf{\Omega}(0)$, where

$$\Lambda_p(r) \equiv \Omega_{p,p-r}(0) = \Omega_{p-r,p}(0) \tag{17}$$

for integers p,q and $r=0,1,2,\ldots$, and thus, from Eq. (13),

$$\Omega_{p-n,q-n}(0) = \Omega_{p,q}(n) = \begin{cases} \Lambda_{p-n}(p-q) & \text{for } p\geq q \\ \Lambda_{q-n}(q-p) & \text{for } p<q \end{cases} \tag{18}$$

for non-negative integers p, q, and n.

The latter invariance under a phase shift of π rad is the analogy of the invariance of the PM-Wigner function to a spatial reflection, i.e., where $W(x,p)\mapsto W(-x,-p)$, which is a rotation by π rad in the x-p plane about the origin. However, this invariance supplies no extra restriction on $S_{\rm NP}$ as the invariance under phase shifts follows automatically from the Galilei invariance property (iii).

(v) The equation of motion of $S_{\rm NP}$ for a free oscillator should be the classical one. Under free evolution the wave function experiences the phase shift $\langle\theta|f\rangle\mapsto\langle\theta+\omega t|f\rangle$ and so from the Galilei invariance property (iii) we find that $S_{\rm NP}(n,\theta,t)\mapsto S_{\rm NP}(n,\theta+\omega t,0)$, which is the expected evolution of the corresponding classical phase-space distribution. [Here $S_{\rm NP}(n,\theta,t)$ is the Wigner function at time t.] Thus the classical equation of motion follows automatically from the Galilei invariance under phase shifts.

(vi) $S_{\rm NP}$ should have the overlap property

$$2\pi\int_{2\pi}d\theta\sum_{n=0}^{\infty}S_{\rm NP}(n,\theta)S'_{\rm NP}(n,\theta)=|\langle f|g\rangle|^2 ,$$

where $S_{\rm NP}(n,\theta)$ and $S'_{\rm NP}(n,\theta)$ are the NP-Wigner functions for the pure states $|f\rangle$ and $|g\rangle$, respectively. This property should also extend to mixed states. We note that because this property is a special case of the next property (vii) we need only consider the latter more-general property.

(vii) $S_{\rm NP}$ should give the trace of a product of general operators as

$$\mathrm{tr}(\hat A\hat B)=2\pi\int_{2\pi}d\theta\sum_{n=0}^{\infty}S_{\rm NP}^{(A)}(n,\theta)S_{\rm NP}^{(B)}(n,\theta) , \tag{19}$$

where $S_{\rm NP}^{(A)}$ and $S_{\rm NP}^{(B)}$ are the NP-Wigner representations

$$S_{\rm NP}^{(A)}(n,\theta)=\mathrm{tr}[\hat S_{\rm NP}(n,\theta)\hat A] , \tag{20}$$

$$S_{\rm NP}^{(B)}(n,\theta)=\mathrm{tr}[\hat S_{\rm NP}(n,\theta)\hat B] \tag{21}$$

of any two linear operators $\hat A$ and $\hat B$ on the Fock space. We have introduced a different notation in these expressions to distinguish the Wigner function $S_{\rm NP}$, which is the expectation value of the NP-Wigner operator $\hat S_{\rm NP}$ and has all the properties discussed in this section, from the Wigner representation $S_{\rm NP}^{(A)}$ of an arbitrary linear operator $\hat A$, which is given by Eq. (20) and only need satisfy the trace property of Eq. (19). In the special case where $\hat A$ and $\hat B$ are density operators, Eqs. (20) and (21) are identical to the definition of the Wigner functions given by Eq. (1) and so we can drop the superscripts (A) and (B); thus Eq. (19) extends property (vi) to the general situation, which includes mixed states. Substituting for $\hat S_{\rm NP}$ in Eqs. (20), and (21), evaluating the trace in the Fock basis, making use of Eq. (7) and performing the integral over θ yields

$$\sum_{p,q=0}^{\infty}A_{p,q}B_{q,p}=\sum_{p,q=0}^{\infty}\sum_{p',q'=0}^{\infty}\sum_{n=0}^{\infty}\Omega_{p,q}(n)\Omega_{p',q'}(n)\times\delta_{p-q+p'-q',0}A_{q,p}B_{q',p'} ,$$

where $A_{p,q}\equiv\langle p|\hat A|q\rangle$ and $B_{p,q}\equiv\langle p|\hat B|q\rangle$. Since $\hat A$ and $\hat B$ are arbitrary operators we may choose the matrix elements $A_{p,q}$ and $B_{p,q}$ at will; thus choosing $A_{j,k}$ and $B_{r,s}$ as the only nonzero matrix elements gives

$$A_{j,k}B_{r,s}\delta_{k,r}\delta_{j,s}=\sum_{n=0}^{\infty}\Omega_{k,j}(n)\Omega_{s,r}(n)\delta_{k-j+s-r,0}A_{j,k}B_{r,s}$$

and thus

$$\delta_{k,r}\delta_{j,s}=\sum_{n=0}^{\infty}\Omega_{k,j}(n)\Omega_{s,r}(n)\delta_{k-j,r-s} . \tag{22}$$

Setting $k=j+r-s$ and $r\geqslant s$ and making use of Eq. (18) gives

$$\delta_{j,s}=\sum_{n=0}^{\infty}\Lambda_{j+r-s-n}(r-s)\Lambda_{r-n}(r-s) . \tag{23}$$

The expression for $s>r$ is obtained on interchanging r and s.

Hillery *et al.* also listed an eighth property concerning a symmetry between the position and momentum representations of the PM-Wigner function. This property is, however, unsuitable for defining the NP-Wigner function because of the asymmetry between the unbounded discrete photon-number and bounded continuous-phase spectra. Rather, we define the NP-Wigner function by another criterion in Sec. II B and then check *a posteriori* the presence of this property.

Collecting our results we find from Eqs. (1), (7), and (18) that

$$\hat S_{\rm NP}(n,\theta)=\frac{1}{2\pi}\left\{\sum_{p=0}^{\infty}\Lambda_{p-n}(0)|p\rangle\langle p|+\left[\sum_{p=0}^{\infty}\sum_{q=0}^{p-1}\Lambda_{p-n}(p-q)\times e^{i(p-q)\theta}|p\rangle\langle q|+\text{H.c.}\right]\right\} , \tag{24}$$

where $\{\boldsymbol\Lambda(n)\}_{n=0,1,2,\ldots}$ is a set of vectors with the following properties. The elements of $\boldsymbol\Lambda(n)$ are given by $\Lambda_m(n)$ for integer m and, according to Eqs. (10), (11), and (16) with Eq. (18), satisfy

$$\Lambda_{p-n}(0)=\delta_{p,n} , \tag{25}$$

$$\sum_{m=0}^{\infty}\Lambda_{p-m}(r)=1 , \tag{26}$$

$$\Lambda_{p-n}(r)=\Lambda_{p-n}(r)^* , \tag{27}$$

respectively, for non-negative integers n and p and for $0\leqslant r\leqslant p$. The elements also satisfy Eq. (23), which can be rewritten as

$$\delta_{p,q}=\sum_{n=0}^{\infty}\Lambda_{p-n}(r)\Lambda_{q-n}(r) \tag{28}$$

for non-negative p and q and for $0 \leqslant r \leqslant p,q$.

It is worthwhile to trace the origins of these expressions if only to keep track of the analysis so far: the Hermitian form of Eq. (24) arises from property (i); the factorization of the θ dependence in Eq. (24) is due to the invariance to shifts in phase (iii) and the classical equation of motion (v); Eqs. (25) and (26) arise from the marginal distribution requirements (ii) for the photon-number and phase probability distributions, respectively; Eq. (27) is due to the invariance to time reflections (iv) and Eq. (28) arises from the overlap (vi) and trace (vii) requirements. The invariance to shifts in photon number, property (iii), is responsible for the translational property of Eq. (12), which allows $\Omega_{p,q}(n)$ to be written in terms of the vectors $\mathbf{\Lambda}(r)$ as given by Eq. (18).

Any set of vectors $\{\mathbf{\Lambda}(n)\}$ satisfying these equations will give rise to a NP-Wigner operator, which has all seven of the properties considered. Let us examine what sort of vectors they are. Equation (25) specifies uniquely the vector $\mathbf{\Lambda}(0)$; it has only one nonzero element i.e., $\Lambda_0(0)=1$. Setting $p=r,r+1,r+2,\ldots$ successively in Eq. (26) reveals that

$$\Lambda_m(r)=0 \quad \text{for } m>r \tag{29}$$

for all vectors $\mathbf{\Lambda}(r)$ and so Eq. (26) implies that the sum of all the elements of any given vector is unity. We now show that only one element of each vector is nonzero. Choose any vector except $\mathbf{\Lambda}(0)$, say, $\mathbf{\Lambda}(r')$, with $r'>0$, and for clarity relabel its elements as $d_n=\Lambda_{r'-n}(r')$ for $n=0,1,2,\ldots$, where d_n are the elements of the vector $\boldsymbol{d}$. The remaining elements of $\mathbf{\Lambda}(r')$ are zero according to Equation (29). Equation (27) implies that the d_n are real and Eq. (28) with $p=q=r=r'$ shows that the vector $\boldsymbol{d}$ is normalized as $\Sigma_{n=0}^{\infty}(d_n)^2=1$. The periodic function $f(\theta)$ defined as

$$f(\theta)=\frac{1}{2\pi}\left|\sum_{n=0}^{\infty} d_n e^{in\theta}\right|^2$$

has Fourier coefficients f_m, which are given by

$$f_m=\int_{2\pi} d\theta\ f(\theta)e^{im\theta}$$

$$=\sum_{n=|m|}^{\infty} d_n d_{n-|m|}=\sum_{n=|m|}^{\infty} \Lambda_{r'-n}(r')\Lambda_{r'+|m|-n}(r')=\delta_{m,0}$$

for an integer m. We arrived at the last two lines by making use of Eq. (28) with $p=r'$, $q=r'+|m|$, and $r=r'$ and noting that $\Lambda_{r'+|m|-n}(r')=0$ for $n=0,1,2,\ldots,|m|-1$ according to Eq. (29). Reconstructing $f(\theta)$ from its Fourier components yields

$$f(\theta)=\frac{1}{2\pi}\sum_{m=-\infty}^{\infty} f_m e^{-im\theta}=\frac{1}{2\pi},$$

which is independent of θ. Hence *only one element* of $\boldsymbol{d}$, and thus of $\mathbf{\Lambda}(r')$, is nonzero and, from Eq. (26), the nonzero value is unity. But there are no further restrictions on the vectors $\mathbf{\Lambda}(r)$. Apart from satisfying Eqs. (25) and (29), the unit values can otherwise occur at any position in the vector $\mathbf{\Lambda}(r)$ and so $\hat{S}_{\mathrm{NP}}$ is not uniquely specified at this stage. Thus, even though only five of the properties listed by Hillery *et al.* [27,29,30] are sufficient to uniquely define the PM-Wigner function $W(x,p)$, all seven of the analogous properties considered here are not sufficient to define the NP-Wigner function uniquely.

B. An additional property

We need an additional property to further restrict the vectors $\mathbf{\Lambda}(n)$. Which extra property of the PM-Wigner function $W(x,p)$ should we use? The interference fringes [36–38] displayed by $W(x,y)$ have been a valuable tool in the study of coherent superpositions (e.g. Schrödinger cat states [39–41]) and the loss of coherence in noisy environments [42]. Giving S_{NP} an analogous property would be a great advantage. The interference fringes in $W(x,p)$ arise as a result of the wave function interfering with itself; that is, $W(x,p)$ is the Fourier transform of the "interfering" product $f(x-y)f^*(x+y)$, where $f(x)$ is the wave function. This property is expressed in terms of the PM-Wigner operator matrix elements as a "skew diagonal" form, i.e.,

$$\langle\mu|\hat{W}(x,p)|\nu\rangle=\frac{1}{\pi}\int_{-\infty}^{\infty} dy\ e^{2ipy}\langle\mu|x+y\rangle\langle x-y|\nu\rangle$$

$$=\frac{1}{\pi}e^{2ip(\mu-x)}\delta(\mu+\nu-2x), \tag{30}$$

where the bras and kets in this expression are position eigenstates and $\delta(x)$ is the Dirac distribution. $W(x,p)$ is given by the expectation value of $\hat{W}(x,p)$.

A first attempt at translating this form to one suitable for the number-phase Wigner function might be to require that $\langle p|\hat{S}_{\mathrm{NP}}(n,\theta)|q\rangle$ be proportional to the Kronecker delta $\delta_{p+q,2n}$. However, we would then find that the resulting vector $\mathbf{\Lambda}(p-q)$, for $p\geqslant q$, would not satisfy Eq. (26) since the left-hand side of Eq. (26) would be zero for odd values of $(p+q)$. Similarly, requiring $\langle p|\hat{S}_{\mathrm{NP}}(n,\theta)|q\rangle$ to be proportional to $\delta_{p+q,2n-1}$ results in a vector $\mathbf{\Lambda}(p-q)$, for $p\geqslant q$, which gives a zero on the left-hand side of Eq. (26) for even $(p+q)$. The solution to this problem is to require $\langle p|\hat{S}_{\mathrm{NP}}(n,\theta)|q\rangle$ to be proportional to $\delta_{p+q,2n}+\delta_{p+q,2n-1}$, i.e.,

$$\Lambda_{p-n}(p-q)=\delta_{p+q,2n}+\delta_{p+q,2n-1} \tag{31}$$

for $p\geqslant q$. It is easily checked that this expression for $\mathbf{\Lambda}(p-q)$ satisfies all the requirements made above. Thus we *define* the special number-phase Wigner function in the Fock basis as [28]

$$\hat{S}_{\mathrm{NP}}(n,\theta)\equiv\frac{1}{2\pi}\sum_{p,q=0}^{\infty} e^{i(p-q)\theta}(\delta_{p+q,2n}+\delta_{p+q,2n-1})|p\rangle\langle q|, \tag{32}$$

which can be rearranged as

$$\hat{S}_{\rm NP}(n,\theta)=\frac{1}{2\pi}\left(\sum_{p=-n}^{n} e^{i2p\theta}|n+p\rangle\langle n-p| + \sum_{p=-n}^{n-1} e^{i(2p+1)\theta}|n+p\rangle\langle n-p-1|\right) \quad (33)$$

provided we take the second sum as being zero for $n=0$. $S_{\rm NP}$ can be expressed in terms of the phase state basis as

$$\hat{S}_{\rm NP}(n,\theta)=\frac{1}{2\pi}\int_{2\pi} d\phi\, e^{-2in\phi}(1+e^{i\phi})|\theta+\phi\rangle\langle\theta-\phi| \ . \quad (34)$$

One can check this last step by evaluating the Fock state matrix elements $\langle p|\hat{S}_{\rm NP}(n,\theta)|q\rangle$ from Eq. (34) and comparing them with the corresponding elements in Eq. (32) or (33).

C. Comparison with other Wigner functions

It is interesting to compare $S_{\rm NP}$ with the PM-Wigner function

$$W(x,p)=\frac{1}{\pi}\left\langle\int_{-\infty}^{\infty} dy\ e^{2ipy}|x+y\rangle\langle x-y|\right\rangle \ ,$$

where the bras and kets are position eigenstates, or, equivalently,

$$W(x,p)=\frac{1}{\pi}\left\langle\int_{-\infty}^{\infty} dy\ e^{-2ixy}|p+y\rangle\langle p-y|\right\rangle, \quad (35)$$

where the bras and kets are momentum eigenstates. The expressions on the right-hand sides of Eqs. (34) and (35) share a high degree of similarity. The main formal difference between the operators in Eqs. (35) and (34), apart from the different limits of integration, is the extra factor $\frac{1}{2}[1+\exp(i\phi)]$ in Eq. (34). This factor is a direct result of the *two* Kronecker δ's on the right-hand side of Eq. (31) whose presence can be traced to the discrete nature of the photon-number spectrum as follows. If the photon number had a continuous spectrum, then the variables p,q,n in Eq. (31) would be continuous and there would exist solutions to $p+q=2n$ for every value of p and q; thus we would need only one term on the right-hand side of Eq. (31) and so the factor $\frac{1}{2}[1+\exp(i\phi)]$ in Eq. (34) would be replaced with 1. The upshot of this hypothetical continuous n case is that Eq. (34) would then be in exact formal agreement with Eq. (35). Hence, we conclude the main formal difference between $W(x,p)$ and $S_{\rm NP}(n,\theta)$ is due to the discrete nature of the photon-number spectrum.

It is also interesting to compare $S_{\rm NP}$ with previous definitions of number-phase Wigner functions. We consider first the discrete number-phase Wigner function $W_{N\phi}(n,\theta)$ for physical states. We find that

$$S_{\rm NP}(n,\theta)=\lim_{s\to\infty}\frac{s+1}{2\pi}[W_{N\phi}(n,\theta)+W_{N\phi}(n+s/2,\theta)] \ ,$$

where $W_{N\phi}(n,\theta)$ is defined on a $(s+1)$-dimensional Hilbert space [see Ref. [25], Eq. (4.4)]. This shows how $S_{\rm NP}$ takes care of the approximate revival in the n dependence of $W_{N\phi}$ for $n>s/2$. For example, in the case of a coherent state the second peak in $W_{N\phi}$ is simply superimposed on the first peak. Comparing $S_{\rm NP}$ now with the function $W_{n\varphi}$ defined by Lukš and Peřinová [26], we find that at the half-odd photon-number values of n, $W_{n\varphi}$ corresponds to the expectation value of the second term in Eq. (33). In fact,

$$S_{\rm NP}(n,\theta)=W_{n\varphi}(n,\theta)+W_{n\varphi}(n-1/2,\theta)$$

for $n=0,1,2,\ldots$, where each term on the right-hand side is equal to the expectation value of the corresponding operator on the right-hand side of Eq. (33). Hence, whereas Lukš and Peřinová take the values of $W_{N\phi}(n+s/2,\theta)$ as the half-odd photon-number values of their function, here we avoid the introduction of unphysical half-odd photon numbers altogether by, in effect, simply adding $W_{N\phi}(n+s/2,\theta)$ to $W_{N\phi}(n,\theta)$.

D. Approximate symmetry property

We now consider the symmetry (or lack of it) between the expressions for $S_{\rm NP}$ in the Fock and phase state bases; this symmetry is analogous to the eighth property of the PM-Wigner function listed by Hillery *et al.* [27]. Let us rewrite Eq. (32) in the following way:

$$\hat{S}_{\rm NP}(n,\theta)=\frac{1}{2\pi}\sum_{p,q=0}^{\infty} e^{2i(n-q)\theta} \times(\delta_{p+q,2n}+e^{-i\theta}\delta_{p+q,2n-1})|p\rangle\langle q|.$$

We wish to interchange n with θ and swap the Fock state with the phase state, Kronecker's delta $\delta_{k,j}$ with the periodic Dirac delta $\delta(k-j)=\Sigma_{n=-\infty}^{\infty}\exp[in(k-j)]/(2\pi)$, the photon-number variables p,q with the continuous phase variables ϕ,φ, and the sums $\Sigma_{p=0}^{\infty}$, $\Sigma_{q=0}^{\infty}$ with the integrals $\int_{2\pi}d\phi, \int_{2\pi}d\varphi$, respectively, in the expression on the right-hand side. To be consistent with the symmetry of $W(x,y)$ we must also swap c numbers with their complex conjugates. We note that the two Kronecker deltas $\delta_{p+q,2n}$ and $\delta_{p+q,2n-1}$ differ by a single step in the photon-number variable. If, in this process of interchanging and swapping, we adopt the principle that a single step in the photon number corresponds to an infinitesimally small (or zero) step in phase, then both Kronecker δ's become the same periodic Dirac δ distribution and we obtain

$$\frac{1}{2\pi}\int_{2\pi} d\phi\int_{2\pi} d\varphi\ e^{-2i(\theta-\varphi)n} \times\delta(\phi+\varphi-2\theta)(1+e^{in})|\phi\rangle\langle\varphi| \ .$$

Performing the integral over φ gives

$$\frac{1}{2\pi}\int_{2\pi} d\phi\ e^{-2in\phi}(1+e^{in})|\theta+\phi\rangle\langle\theta-\phi|, \quad (36)$$

which is quite similar to the right-hand side of Eq. (34). The only difference is that Eq. (34) contains the factor $\frac{1}{2}[1+\exp(i\phi)]$, whereas Eq. (36) contains the factor $\frac{1}{2}[1+\exp(in)]$. Hence this shows that there is an approximate

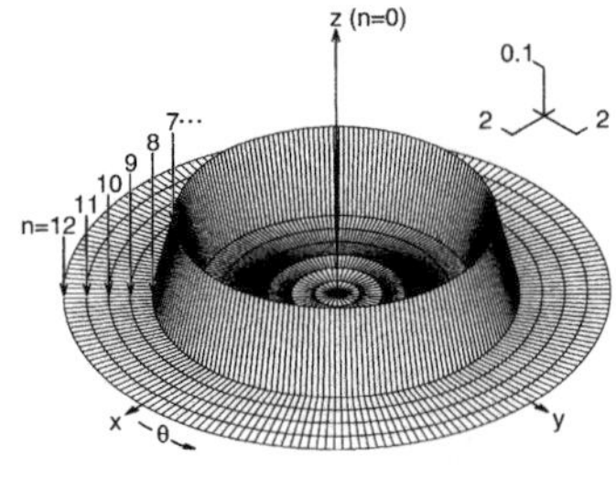

FIG. 1. Three-dimensional polar diagram illustrating S_{NP} for the single Fock state $|7\rangle$. The function $S_{NP}(n,\theta)$ is depicted as an interpolated surface at a height of $z=S_{NP}(n,\theta)$ above the point $(n\cos\theta, n\sin\theta)$ in the x-y plane, i.e., above the point (n,θ) in polar coordinates, for $n=0,1,2,\ldots$ and θ in $(0,2\pi)$. The surface is interpolated linearly between the integral values of n. On the surface radial lines indicate lines of constant θ and circles indicate lines of constant n. The triad in the top right corner gives the units for each axis. The raised ring at $n=7$ illustrates the uncertain phase and sharp photon number of a Fock state.

symmetry between the two forms of S_{NP} in Eqs. (33) and (34), which is analogous to the symmetry of the PM-Wigner function under the transformation from the position to the momentum representation and vice versa. We noted earlier that the reason the factor $\frac{1}{2}[1+\exp(i\phi)]$ in Eq. (34) differs from unity can be attributed to the discrete nature of the photon-number spectrum. This suggests that the lack of an *exact* symmetry is due to the asymmetry between the discrete photon-number and continuous phase variables.

III. S_{NP} FOR VARIOUS STATES

The Fock and Glauber coherent states are two of the most familiar states in quantum optics. The S_{NP} representation of the general Fock state $|m\rangle$

$$S_{NP}(n,\theta)=\frac{1}{2\pi}\delta_{n,m} \tag{37}$$

is a "raised ring" of radius m as illustrated in Fig. 1. The phase is completely uncertain whereas the photon number is sharp.

For the coherent state $|\alpha\rangle$ we find [28]

$$S_{NP}(n,\theta)=\frac{e^{-|\alpha|^2}}{2\pi}\left[|\alpha|^{2n}\sum_{p=-n}^{n}\frac{e^{i2p(\theta-\varphi)}}{\sqrt{(n+p)!(n-p)!}}+|\alpha|^{2n-1}\sum_{p=-n}^{n-1}\frac{e^{i(2p+1)(\theta-\varphi)}}{\sqrt{(n+p)!(n-p-1)!}}\right], \tag{38}$$

where $\alpha=|\alpha|\exp(i\varphi)$. Let us look first at the weak-field regime. In the limit as $\alpha\to 0$ the coherent state $|\alpha\rangle$ becomes the vacuum state, which is represented by a raised point (i.e., a raised ring of zero radius) at the origin. For small, but nonvanishing, values of α we find that $S_{NP}(0,\theta)\approx 1/(2\pi)$, $S_{NP}(1,\theta)\approx|\alpha|\cos(\theta-\varphi)/\pi$, and $S_{NP}(n,\theta)\approx 0$ for $n>1$ to first order in $|\alpha|$. Thus the raised point of the vacuum state broadens as $|\alpha|$ increases. This is illustrated in Fig. 2 for $\alpha=0.1$. It is perhaps a little surprising that $S_{NP}(1,\theta)$ is negative for $\theta\leq\varphi-\pi/2$ and $\theta\geq\varphi+\pi/2$ in view of the fact that the PM-Wigner function is positive for all coherent states. Nevertheless, the phase probability distribution $P(\theta)$ obtained as a marginal of $S_{NP}(n,\theta)$ is always positive as expected, i.e., $P(\theta)\approx[1+2|\alpha|\cos(\theta-\varphi)]/(2\pi)$ to first order in $|\alpha|$. For a slightly higher intensity with $\alpha\approx 0.64$ the picture shown in Fig. 3 is similar. The highest point is now situated on the $n=1$ curve along the $\theta=0$ direction.

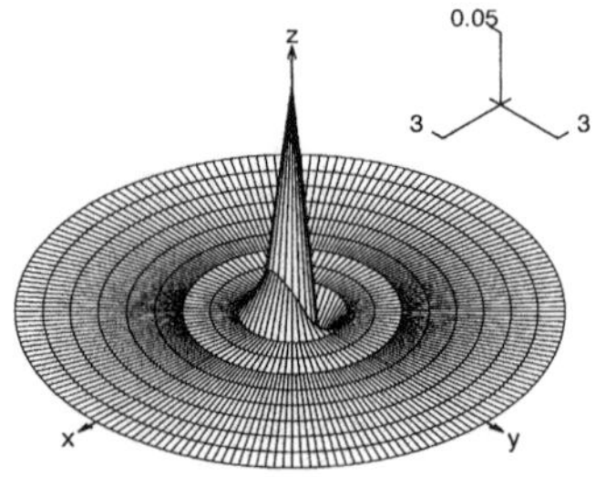

FIG. 2. S_{NP} for the weak coherent state $|\alpha\rangle$ with $\alpha=0.1$. The largest value of n is 10. The raised point of the vacuum state has broadened slightly.

For relatively intense coherent states we can approximate the Poisson photon-number distribution of the coherent state with a Gaussian of the same mean and variance. We find eventually that

$$S_{NP}(n,\theta)\approx\frac{1}{2\pi}\{G_P(\theta)[G_N(n)+G_N(n-1/2)]+G_P(\theta+\pi)\times[G_N(n)-G_N(n-1/2)]\}, \tag{39}$$

where $G_N(n)\equiv\exp[-(n-\bar{n})^2/(2\bar{n})]$ and $G_P(\theta)=\exp[-2\bar{n}(\theta-\varphi)^2]$ for $\bar{n}=|\alpha|^2\gg 1$. In this intense-field regime the fluctuations in number and phase decouple and both the number and phase dependence of S_{NP} become approximately Gaussian. These features are evident even in the plot in Fig. 4 of

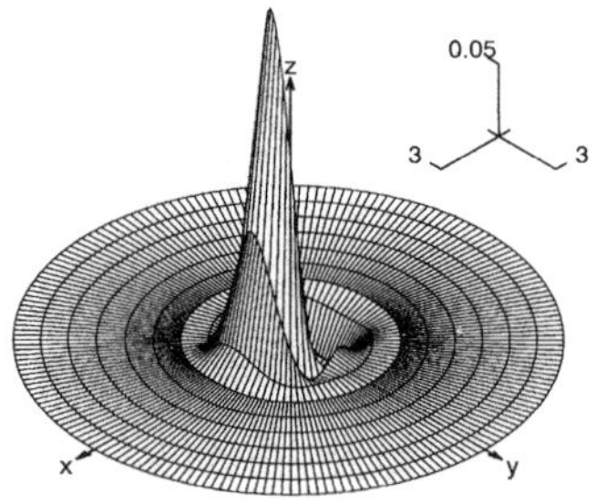

FIG. 3. S_{NP} for a coherent state $|\alpha\rangle$ of a slightly higher intensity with $\alpha=0.64$. The largest value of n is 10. The peak in Fig. 2 has moved to a point above the curve $n=1$ in the $\theta=0$ direction.

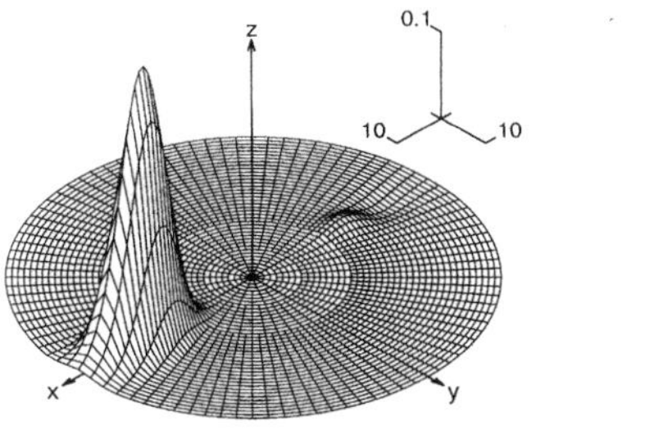

FIG. 4. S_{NP} for the coherent state $|\alpha\rangle$ with $\alpha=5$. The largest value of n is 40. The hill is centered on the point $n=25, \theta=0$.

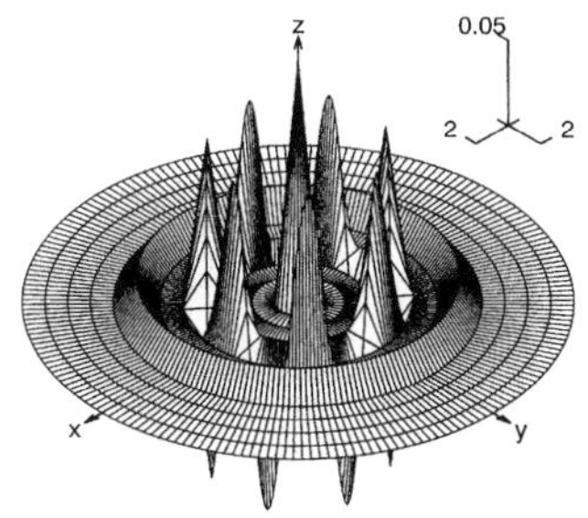

FIG. 5. S_{NP} for the Fock cat $\cos(\eta)|0\rangle+\sin(\eta)|7\rangle$ with $\eta=\pi/10$. The largest value of n is 12. The interference ring lies halfway between the raised point at the origin and the raised ring at $n=7$.

S_{NP} for the coherent state $|\alpha\rangle$ with $\alpha=5$ where a relatively "smooth" hill shape is centered on the point $n=25$, $\theta=0$. Interestingly, there is also a small "wave" along $\theta=\pi$ in Fig. 4. This feature, however, has little effect on the phase distribution, which is obtained by summing S_{NP} radially. It arises from the relatively small difference between the two Gaussians $G_N(n)$ and $G_N(n-1/2)$ in Eq. (39). This difference vanishes as $\bar{n}$ increases and so in the limit of large $\bar{n}$ we obtain

$$S_{\mathrm{NP}}(n,\theta)\approx\frac{1}{\pi}G_P(\theta)G_N(n) \ , \tag{40}$$

which is a two-dimensional Gaussian in the polar coordinates n,θ centered on the point $n=|\alpha|^2$, $\theta=\phi$.

To compare this result with the PM-Wigner representation $W(x,p)$ of coherent states we need to change coordinates. First note that $S_{\mathrm{NP}}(n,\theta)$ is vanishingly small unless n and θ differ by relatively small amounts from their mean values $\bar{n}$ and φ, respectively. The transformation from the Cartesian coordinates (x,p) to the polar coordinates (r,θ), where $x=r\cos\theta$ and $p=r\sin\theta$, is characterized by the metric $dx^2+dp^2=dr^2+r^2d\theta^2$. Here the radial coordinate r corresponds to amplitude in the $W(x,p)$ phase space whereas n in Eq. (40) represents intensity. Transforming to an intensity coordinate n with $r=\sqrt{2n}$ gives $dx^2+dp^2=dn^2/(2n)+2nd\theta^2$. [The $\sqrt{2}$ factor arises from the fact that the mean of x^2+p^2 for $W(x,p)$ is $\langle 2\hat{N}+1\rangle$.] Replacing differentials with small deviations from mean values gives

$$(x-\bar{x})^2+(p-\bar{p})^2\approx(n-\bar{n})^2/(2\bar{n})+2\bar{n}(\theta-\varphi)^2 \ ,$$

where $x=\sqrt{2n}\cos(\theta)$, $p=\sqrt{2n}\sin(\theta)$. Hence the expression for $S_{\mathrm{NP}}(n,\theta)$ in Eq. (40) becomes approximately

$$\frac{1}{\pi}\exp[-(x-\bar{x})^2-(p-\bar{p})^2] \ ,$$

which is the PM-Wigner function $W(x,p)$ for the same coherent state where $\bar{x}=\sqrt{2\bar{n}}\cos(\varphi)$, $\bar{p}=\sqrt{2\bar{n}}\sin(\varphi)$ with $\bar{n}=|\alpha|^2$ and $\alpha=|\alpha|\exp(i\varphi)$. Thus $S_{\mathrm{NP}}(n,\theta)$, under the change of variables $n,\theta\mapsto x,p$, converges asymptotically to $W(x,p)$ for large $\bar{n}$.

There has been quite a deal of attention given recently to the study of Schrödinger cats states [39]. These states may be defined generally as superpositions of macroscopically distinguishable states [40,41]. The position-momentum Wigner function has been found to exhibit interference fringes for such states [36–38,42]. In [28] we gave the S_{NP} representation of the simplest "Fock cat," which is an equally weighted superposition of two Fock states. Here we shall give the S_{NP} representation of the most general two-state Fock cat $\cos(\eta)|m\rangle+e^{i\phi}\sin(\eta)|m+r\rangle$. This state has a S_{NP} representation given by

$$S_{\mathrm{NP}}(n,\theta)=\frac{1}{2\pi}[\delta_{n,m}\cos^2\eta+\delta_{n,m+r}\sin^2\eta + \delta_{n,m+k}\sin(2\eta)\cos(r\theta-\varphi)] \ ,$$

where k is the largest integer not exceeding $(r+1)/2$. A plot for $m=0$, $r=7$, $\phi=0$, and $\eta=\pi/10$ is given in Fig. 5. $S_{\mathrm{NP}}(n,\theta)$ consists of two raised rings at radii $n=m$ and $n=m+r$, corresponding to the individual Fock states, and an interference ring at $n=m+k$. The relative height of each ring corresponds to the relative probability of finding the Fock cat in each Fock state. The number of oscillations in the interference ring depends on the distance r between the other two rings and the orientation of the interference ring depends on the phase factor ϕ. A similar dependence is evident in the interference displayed for Schrödinger cats by the PM-Wigner function. We note also that the interference ring is not present for the S_{NP} function of the corresponding mixture of two Fock states given by the density operator $\cos^2(\eta)|m\rangle\langle m|+\sin^2(\eta)|m+r\rangle\langle m+r|$.

It is now well known that Schrödinger cat states involving a discrete superposition of coherent states can be produced by a Kerr medium [40], by transferring atomic coherences to a cavity field [36] and by the atom-cavity interaction at the half atomic-inversion revival time in the Jaynes-Cummings model [43]. Let us look at the so-called odd and even coherent states as typical examples. We give a plot of the even coherent state $c(|\alpha\rangle+|-\alpha\rangle)$ for $\alpha=0.83$, where c is a normalization constant, in Fig. 6. This value of α gives approximately the same mean photon number as the coherent state in Fig. 3. In Fig. 6 there are ridges along the $\theta=0$ and $\theta=\pi$ directions where we would expect to see the hill shape

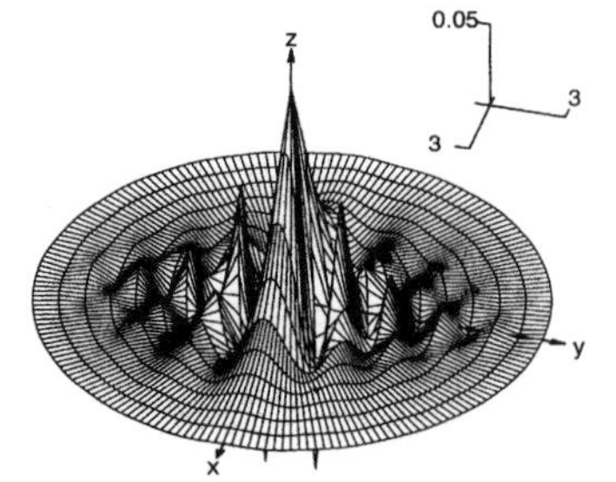

FIG. 9. $S_{\rm NP}$ for the squeezed vacuum $|0,\xi\rangle$ with $\xi=-0.6$. The largest value of n is 10. There is a ridge along the $\theta=0$ and $\theta=\pi$ directions. Interference fringes occur in between the ridges and reach a maximum along the $\theta=-\pi/2$ and $\theta=\pi/2$ directions. The relatively large photon-number spread of this state is the reason why the interference fringes are relatively spread out. The mean photon number of this state is approximately the same as for the even coherent state in Fig. 6.

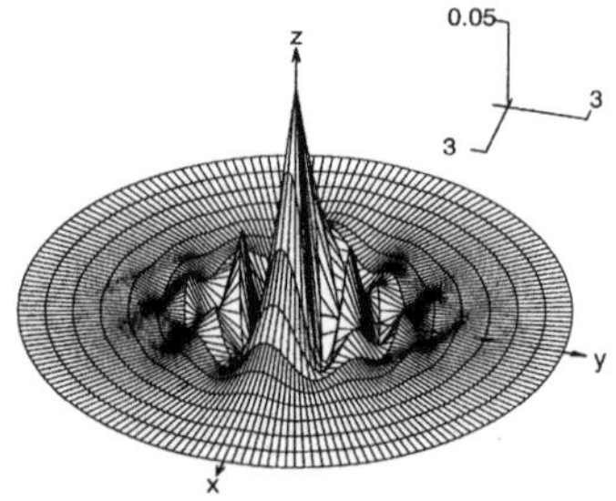

FIG. 10. $S_{\rm NP}$ for the squeezed vacuum of Fig. 9 after having interacted with a noisy environment and suffered a loss of 15% of its energy. The largest value of n is 10. The interference fringes are reduced only slightly compared to the fringes in Fig. 9. This shows that the interference fringes associated with the squeezed vacuum are relatively robust.

ing the squeezed vacuum as a phase cat can help explain the features of the figure. The ridge oriented along $\theta=0$ and $\theta=\pi$ in Fig. 9 corresponds to the two states of the phase cat, one with a mean phase of 0 and the other π, whereas the oscillations along $\theta=-\pi/2$ and $\theta=\pi/2$ are the interference fringes associated with the superposition of the two states of the cat.

It is interesting to compare Fig. 9 with the $S_{\rm NP}$ representation of the even coherent state in Fig. 6. The states in both figures have a mean photon number of approximately 0.41. Although the main features of the even coherent state are confined to a much smaller region, nevertheless, the interference fringes in both figures do share common features such as oscillations along $\theta=-\pi/2$ and $\theta=\pi/2$.

The picture of the squeezed vacuum given by $S_{\rm NP}$ in Fig. 9 contrasts markedly with the corresponding PM-Wigner representation, which is a smooth Gaussian and positive everywhere. For this reason the squeezed vacuum is perhaps the most interesting state considered in this section. Moreover, it appears that the $S_{\rm NP}$ representation of the squeezed vacuum will give a *picture of quantum interference in phase space* for an experimentally determined quantum state. Indeed, the experiments of optical quantum-state determination by Raymer's group [50,51] involved the squeezed vacuum and just recently the PM-Wigner representation of a state with approximately the same degree of squeezing as the field represented in Fig. 9 was determined experimentally [52]. We note that the $S_{\rm NP}$ representation can be determined from any representation of the quantum state, including the PM-Wigner representation, as the expectation value of $\hat{S}_{\rm NP}(n,\theta)$ using, for example, Eq. (33). Furthermore, the squeezing referred to in the latter case is the squeezing detected with nonideal photodetectors; the actual squeezing in the field is much greater [52] and, correspondingly, the field itself would have a $S_{\rm NP}$ representation with more pronounced interference fringes.

It is well known that Schrödinger cat states can quickly decohere in noisy environments [40]. One might wonder then how robust the interference fringes in Fig. 9 are. To see the effects of a noisy environment (or detection losses) on the fringes, consider the following. Let the squeezed vacuum represented in Fig. 9 be the initial state of the field and let the field interact linearly with the environment. We can model this interaction by imagining that the field is mixed with the vacuum at a beam splitter that has a nonunity transmittance [40]. Let the transmittance be some reasonable value, say, 0.85. (This transmittance value, in fact, models the detection losses in the first optical quantum-state determination experiment by Raymer's group [50].) The state of the field at one of the output ports of the beam splitter corresponds to the mixed state of the field after it has interacted with the noisy environment and suffered losses. Figure 10 shows the $S_{\rm NP}$ representation of the field at this point. The interference fringes are clearly still evident. We conclude that the interference fringes associated with *the squeezed vacuum are relatively robust* in this case and it is quite likely that they will be seen in experimental data.

IV. THE $S_{\rm NP}$-REPRESENTATION OF ARBITRARY OPERATORS

The NP-Wigner function also supplies a representation of operators. Each linear operator $\hat{A}$ on the Fock space has a corresponding unique Wigner representation $S_{\rm NP}^{(A)}(n,\theta)$ given by Eq. (20). The uniqueness can be proved by showing that every nonzero operator has a nonzero Wigner representation and then using the linear property of the operators and the Wigner representations to show that different operators have different representations. In particular, this means that the special number-phase Wigner function represents uniquely the number and phase properties of any quantum state $\hat{\rho}$.

Conversely, an operator $\hat{A}$ can be reconstructed from its NP-Wigner representation $S_{\rm NP}^{(A)}(n,\theta)$ by

$$\hat{A}=2\pi\sum_{n=0}^{\infty}\int_{2\pi}d\theta\,\hat{S}_{\rm NP}(n,\theta)S_{\rm NP}^{(A)}(n,\theta)\ . \tag{41}$$

This can be proved easily by taking matrix elements in the Fock basis of both sides of Eq. (41). However, this mapping is not unique and there is more than one function $A(n,\theta)$, which maps to $\hat{A}$ according to

$$\hat{A}=2\pi\sum_{n=0}^{\infty}\int_{2\pi}d\theta\,\hat{S}_{\rm NP}(n,\theta)A(n,\theta)\quad . \tag{42}$$

Let us call the set of functions $\{A_i(n,\theta)\}_i$ a set of *$\hat{A}$-equivalent functions* if all elements map to the same operator $\hat{A}$ by Eq. (42). For example, adding $e^{ik\theta}\delta_{m,n}$, where $|k|>2m$ and m is a non-negative integer, to $A(n,\theta)$ does not alter the left-hand side of Eq. (42); in other words, $A(n,\theta)$ and $A(n,\theta)+e^{ik\theta}\delta_{m,n}$ are $\hat{A}$-equivalent functions. The set of $\hat{A}$-equivalent functions for any given operator $\hat{A}$ contains a unique element $S_{\rm NP}^{(A)}(n,\theta)$, which is given by Eq. (20). We can always obtain this unique element by taking an arbitrary element of the set, say $A(n,\theta)$, determining the operator $\hat{A}$ by Eq. (42) and then using $\hat{A}$ to produce $S_{\rm NP}^{(A)}(n,\theta)$ via Eq. (20). This procedure can be represented as a single operation on $A(n,\theta)$ as

$$S_{\rm NP}^{(A)}(n,\theta)=\mathrm{tr}\left[\hat{S}_{\rm NP}(n,\theta)2\pi\sum_{m=0}^{\infty}\int_{2\pi}d\varphi\;\hat{S}_{\rm NP}(m,\varphi)A(m,\varphi)\right]$$

$$=\frac{1}{2\pi}\int_{2\pi}d\varphi\left[\sum_{p=-n}^{n}e^{i2p(\theta-\varphi)}+\sum_{p=-n}^{n-1}e^{i(2p+1)(\theta-\varphi)}\right]A(n,\varphi)\quad .$$

Thus, given any function $A(n,\theta)$ one can calculate the unique $S_{\rm NP}$ representation of the associated operator $\hat{A}$ [53].

What then are the operators $\hat{A}$ associated with the functions $A(n,\theta)=(1/2\pi)\exp(im\theta)$? A simple calculation reveals that $\hat{A}$ is $(\widehat{e^{i\phi}})^m$ or $(\widehat{e^{-i\phi}})^{|m|}$ for positive and negative m, respectively, where $\widehat{e^{i\phi}}=\Sigma_{n=0}^{\infty}|n\rangle\langle n+1|\ =\widehat{e^{-i\phi}}^{\dagger}$ is the nonunitary Susskind-Glogower exponential phase operator. This result should not be unexpected for the following reasons. Our use of the phase probability distribution $P(\theta)$ and the infinite-dimensional Fock space implies that we are in the domain of the infinite-s limit of the Pegg-Barnett formalism and so it follows that the exponential phase operators represented by $S_{\rm NP}$ will be the corresponding Susskind-Glogower operators as these are the weak limits [34] of the Pegg-Barnett unitary phase operators. Moreover, the restriction of our analysis to the Fock space (instead of a larger space) implies that the phase operators of the Newton, Ban, and $H_{\rm sym}$ formalisms, which operate operate on larger spaces, will be represented here by their projection onto operators on the Fock space, that is, by the corresponding Susskind-Glogower operators.

Taking this one step further, the operator given by Eq. (42) with $A(n,\theta)=(1/2\pi)\cos^2(\theta)$ is not the square of the operator represented by $(1/2\pi)\cos\theta$,

$$\tfrac{1}{4}(\widehat{e^{i\phi}}+\widehat{e^{-i\phi}})^2=\tfrac{1}{4}(\widehat{e^{i\phi}}^{2}+\widehat{e^{-i\phi}}^{2}+2-|0\rangle\langle 0|)\quad ,$$

but rather

$${}^{*}_{*}\tfrac{1}{4}(\widehat{e^{i\phi}}+\widehat{e^{-i\phi}})^2\,{}^{*}_{*}=\tfrac{1}{4}(\widehat{e^{i\phi}}^{2}+\widehat{e^{-i\phi}}^{2}+2), \tag{43}$$

where ${}^{*}_{*}\cdots{}^{*}_{*}$ represents the antinormal ordering operation introduced by Lukš and Peřinová [54,55]. This operation places all positive powers of $\widehat{e^{i\phi}}$ to the left of all positive powers of $\widehat{e^{-i\phi}}$. Similarly the operator corresponding to $A(n,\theta)=(1/2\pi)\sin^2(\theta)$ is ${}^{*}_{*}-\tfrac{1}{4}(\widehat{e^{i\phi}}-\widehat{e^{-i\phi}})^2\,{}^{*}_{*}\ =\tfrac{1}{4}(-\widehat{e^{i\phi}}^{2}-\widehat{e^{-i\phi}}^{2}+2)$. Adding this to the right-hand side of Eq. (43) gives unity: that is, the $S_{\rm NP}$-representation gives the operator equivalent of the mathematical identity $\cos^2(\theta)+\sin^2(\theta)=1$. Thus the Wigner-Weyl correspondence between Wigner functions and operators, embodied here by Eqs. (41) and (42), leads to a consistent set of phase operators provided we adopt Lukš and Peřinová's antinormal ordering.

Let us now look at the relationship between the photon-number and phase operators given by Eq. (42). Setting $A(n,\theta)$ in Eq. (42) alternatively to $(1/2\pi)n$ and $(1/2\pi)e^{in}$ gives the photon-number operators $\hat{N}$ and $e^{i\hat{N}}$, respectively. Using the fact that the phase operators are antinormally ordered we find that

$${}^{*}_{*}\widehat{e^{i\phi}}e^{i\hat{N}}\widehat{e^{-i\phi}}\,{}^{*}_{*}={}^{*}_{*}e^{i(\hat{N}+1)}\widehat{e^{i\phi}}\widehat{e^{-i\phi}}\,{}^{*}_{*}=e^{i(\hat{N}+1)}\widehat{e^{i\phi}}\widehat{e^{-i\phi}}$$
$$=e^{i}e^{i\hat{N}}\quad ,$$

$${}^{*}_{*}\widehat{e^{-i\phi}}e^{i\hat{N}}\widehat{e^{i\phi}}\,{}^{*}_{*}={}^{*}_{*}e^{i(\hat{N}-1)}\widehat{e^{-i\phi}}\widehat{e^{i\phi}}\,{}^{*}_{*}=e^{i(\hat{N}-1)}\widehat{e^{i\phi}}\widehat{e^{-i\phi}}$$
$$=e^{-i}e^{i\hat{N}}\quad ,$$

where we have used the fact that $\widehat{e^{i\phi}}e^{i\hat{N}}=e^{i(\hat{N}+1)}\widehat{e^{i\phi}}$, $\widehat{e^{-i\phi}}e^{i\hat{N}}=e^{i(\hat{N}-1)}\widehat{e^{-i\phi}}$ and $\widehat{e^{i\phi}}\widehat{e^{-i\phi}}=1$. This shows that the number and phase observables given by Eq. (42) are canonically conjugate in the Weyl sense [56].

Finally, an important point, which should be stressed, is that the presence here of the Susskind-Glogower operators does not mean that the description of phase is that given by Susskind and Glogower. For example, consider the vacuum state whose representation is simply $S_{\rm NP}(n,\theta)=(1/2\pi)\delta_{n,0}$. This representation yields a uniform phase probability distribution and hence attributes the vacuum state with a random phase in contrast to the nonrandom phase description of Susskind and Glogower.

V. DETERMINATION OF THE QUANTUM STATE VIA $S_{\rm NP}$

An important feature of the PM-Wigner function $W(x,p)$ is that it can be determined experimentally [57,58,50,51]. Since $W(x,p)$ represents uniquely the density operator of the system this allows experimenters to determine (or, in a sense, to "measure") the quantum state. Can the quantum state be determined using the special number-phase Wigner function? The results of the preceding section show that the density operator $\hat{\rho}$ can be uniquely determined from knowledge of $S_{\rm NP}$. We now consider whether $S_{\rm NP}$ can be determined experimentally. For this we require the expectation value of the Wigner operator $\hat{S}_{\rm NP}(n,\theta)$ for each value of n and θ. These expectation values can be calculated from the probabilities of finding the field mode in the eigenstates of the Hermitian operator $\hat{S}_{\rm NP}(n,\theta)$, i.e.,

$$\hat{S}_{NP}(n,\theta)=\sum_i \lambda_i(n,\theta)|\psi_i(n,\theta)\rangle\langle\psi_i(n,\theta)| \ ,$$

where $\lambda_i(n,\theta)$ and $|\psi_i(n,\theta)\rangle$ is an eigenvalue and an eigenvector, respectively, of $\hat{S}_{NP}(n,\theta)$. Thus the task reduces to finding the probabilities $\langle\psi_i(n,\theta)|\hat{\rho}|\psi_i(n,\theta)\rangle$, which can be determined experimentally, in principle. This procedure of determining the Wigner function by diagonalizing the Wigner operator is quite general and, in fact, can be applied to the PM-Wigner function.

It turns out, however, that the diagonalization operation in the present case can be simplified considerably by adopting a slightly modified procedure. Instead of diagonalizing $\hat{S}_{NP}(n,\theta)$ we diagonalize the two terms $\hat{F}(n,\theta)$ and $\hat{G}(n,\theta)$ in Eq. (33) separately, where

$$\hat{F}(n,\theta)=\frac{1}{2\pi}\sum_{p=-n}^{n} e^{i2p\theta}|n+p\rangle\langle n-p| \ ,$$

$$\hat{G}(n,\theta)=\frac{1}{2\pi}\sum_{p=-n}^{n-1} e^{i(2p+1)\theta}|n+p\rangle\langle n-p-1| \ .$$

Solving the eigenvalue equation $\hat{F}(n,\theta)|f\rangle=\lambda|f\rangle$ shows that $\langle m|f\rangle=0$ for $m>2n$ and

$$e^{i2m\theta}\langle n-m|f\rangle=2\pi\lambda\langle n+m|f\rangle$$

for $0\le m\le n$. From this we find that for $n=0$ there is a single eigenvalue and eigenvector $\lambda=1/(2\pi)$ and $|f\rangle=|0\rangle$, whereas for $n>0$ there are just two degenerate eigenvalues $\lambda=\pm 1/(2\pi)$ and many different sets of eigenvectors. A particularly simple set of eigenvectors for $n>0$ is given in the Fock basis by

$$|f_m^+(n,\theta)\rangle=\frac{1}{\sqrt{2}}(|n-m\rangle+|n+m\rangle e^{i2m\theta}) \ ,$$

$$|f_m^-(n,\theta)\rangle=\frac{1}{\sqrt{2}}(|n-m\rangle-|n+m\rangle e^{i2m\theta}) \ ,$$

$$|f_n^+\rangle=|n\rangle$$

for $m=1,2,\ldots,n$, where the superscripts refer to the sign of the eigenvalue. Thus we now have

$$\hat{F}(n,\theta)=\frac{1}{2\pi}\Big[|n\rangle\langle n|+\sum_{m=1}^{n}|f_m^+(n,\theta)\rangle\langle f_m^+(n,\theta)| - |f_m^-(n,\theta)\rangle\langle f_m^-(n,\theta)|\Big]$$

with it being understood that the first sum is zero for the $n=0$ case. In a similar way we also find that $\hat{G}(n,\theta)$ can be diagonalized as

$$\hat{G}(n,\theta)=\frac{1}{2\pi}\sum_{m=1}^{n}|g_m^+(n,\theta)\rangle\langle g_m^+(n,\theta)|-|g_m^-(n,\theta)\rangle \times\langle g_m^-(n,\theta)| \ ,$$

where

$$|g_m^+(n,\theta)\rangle=\frac{1}{\sqrt{2}}(|n-m\rangle+|n+m-1\rangle e^{i(2m-1)\theta}) \ ,$$

$$|g_m^-(n,\theta)\rangle=\frac{1}{\sqrt{2}}(|n-m\rangle-|n+m-1\rangle e^{i(2m-1)\theta})$$

for $m=1,2,\ldots,n$. Thus the expectation values of $\hat{F}(n,\theta)$ and $\hat{G}(n,\theta)$ can be calculated from knowledge of the probabilities of finding the field mode in the eigenstates $|f_m^{\pm}\rangle,|g_m^{\pm}\rangle$ for $m=1,2,\ldots,n$ and the Fock state $|n\rangle$. In other words, the special number-phase Wigner function can be determined from knowledge of the photon-number probability distribution and the probabilities of finding the field mode in the states $|m\rangle\pm|n\rangle\exp[i(n-m)\varphi]$ for all $n>m$ and all φ.

These probabilities are, in principle, measurable quantities. Indeed, probabilities of this type can be determined experimentally via an application of homodyne detection [59]. We now describe briefly an alternate scheme based directly on photodetection. Let the field in question, which we shall call the signal field, be in the state $|\psi\rangle$. For brevity we only treat the pure state case here; the extension to the more-general mixed state case is, however, straightforward. The signal field is mixed with a probe field, which is in the specially prepared state $|\phi\rangle$, at an ideal beam splitter as shown in Fig. 11. The output fields of the beam splitter are directed onto photodetectors of known quantum efficiency ϵ. We assume that the signal and probe fields are pulsed synchronously and that the detectors, field intensities, and pulse duration are chosen such that they allow the counting of the individual photoelectron events produced in the detectors in a manner similar to the operational-phase measurements of Mandel *et al.* [60]. The joint probability distribution giving the probability of finding p and q photons in the output fields 1 and 2, respectively, is given by

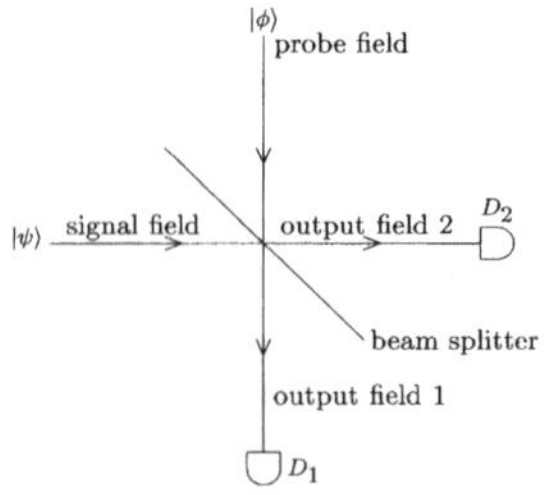

FIG. 11. Experimental setup for the determination of S_{NP}. The signal and probe fields are mixed at a beam splitter and then directed onto the two photodetectors D_1 and D_2. The S_{NP}-representation of the signal field is obtained from the joint photocount distributions for a given set of probe field states.

$$P(p,q)=\left|\sum_{k=0}^{p+q}\psi_k\phi_{p+q-k}C_p(k,p+q-k)\right|^2 \ , \qquad (44)$$

where ψ_n and ϕ_m are the Fock state coefficients $\psi_n=\langle n|\psi\rangle$ and $\phi_m=\langle m|\phi\rangle$. The coefficients $C_k(n,m)$ are the complex numbers [61,62]

$$C_k(n,m)=(-1)^n\sqrt{\frac{k!(n+m-k)!}{n!m!}}e^{i\varphi_\tau(n-k)}e^{i\varphi_\rho(m-k)}$$
$$\times\sum_{p=0}^{n}\sum_{q=0}^{m}(-1)^p\binom{n}{p}\binom{m}{q}$$
$$\times\sqrt{\tau^{m+p-q}\rho^{n-p+q}}\delta_{n+m-k,p+q}\ ,$$

where $\delta_{n,m}$ is the Kronecker delta, $\binom{n}{m}$ is the binomial coefficient $n!/m!(n-m)!$, τ, and ρ are the transmittance and reflectance, respectively, of the beam splitter, and φ_τ and φ_ρ are phase factors as defined by Campos *et al.* [62]. It follows from Eq. (44) that if the probe field is prepared in the state

$$|\phi\rangle=c(a|n\rangle+b|m\rangle e^{i\eta})\ , \quad (45)$$

where $c=1/\sqrt{[a|]^2+[b|]^2}$ is a normalization constant, $a=C_{p'}(p'+q'-n,n)^{-1}$, $b=C_{p'}(p'+q'-m,m)^{-1}$, and η is a phase parameter, then the particular value of $P(p,q)$ at $p=p'$ and $q=q'$ is

$$P(p',q')=2|c|^2|\langle\psi|\Theta\rangle|^2$$

for $p'+q'\geqslant n,m$, where the state $|\Theta\rangle=1/\sqrt{2}(|p'+q'-n\rangle+|p'+q'-m\rangle e^{-i\eta})$ is in the form of the eigenstates $|f_m^\pm\rangle$ and $|g_m^\pm\rangle$ of $\hat{F}$ and $\hat{G}$. Thus, by choosing appropriate values of p', q', m, n, and η for the probe field state $|\phi\rangle$ in Eq. (45) we can determine the required probabilities of finding the signal field in the eigenstates $|f_m^\pm\rangle$ and $|g_m^\pm\rangle$. Using these probabilities and the corresponding eigenvalues we can determine the expectation value of $\hat{S}_{\mathrm{NP}}(n,\theta)=\hat{F}(n,\theta)+\hat{G}(n,\theta)$, which is the S_{NP} representation of the signal field $S_{\mathrm{NP}}(n,\theta)$.

Of course, real photodetectors have quantum efficiencies ϵ that are less than unity and so the relative frequency $M(p,q)$ of counting p and q photoelectron events at detectors 1 and 2, respectively, is not equal to the joint photon-number probability distribution $P(p,q)$. By modeling the loss in each detector as the loss from a beam splitter of transmittance ϵ placed in front of an ideal detector it is not difficult to show that $M(p,q)$ and $P(p,q)$ are related by

$$M(m,n)=\sum_{p=n}^{\infty}\binom{p}{n}\epsilon_n(1-\epsilon)^{p-n}$$
$$\times\sum_{q=m}^{\infty}\binom{q}{m}\epsilon_m(1-\epsilon)^{q-m}P(p,q)\ .$$

This expression is in the form of a double Bernoulli transform which can be inverted [63] to give

$$P(p,q)=\sum_{n=p}^{\infty}\binom{n}{p}\left(1-\frac{1}{\epsilon}\right)^{n-p}\epsilon^{-p}$$
$$\times\sum_{m=q}^{\infty}\binom{m}{q}\left(1-\frac{1}{\epsilon}\right)^{m-q}\epsilon^{-q}M(n,m)\ .$$

As the values of n and m increase, the moduli of the factors

$$\binom{n}{p}\left(1-\frac{1}{\epsilon}\right)^{n}\ ,\ \binom{m}{q}\left(1-\frac{1}{\epsilon}\right)^{m}$$

decay exponentially for $\frac{1}{2}<\epsilon<1$ and diverge otherwise. Thus the experimental determination of the photon-number probability distribution $P(p,q)$, and hence the determination of the S_{NP} representation of the signal field, can be obtained using realistic detectors provided the quantum efficiency is greater than $\frac{1}{2}$.

Experimentally the most difficult part of the scheme is the preparation of the probe field in the states given by Eq. (45) for a continuous range of η values and all two-state superpositions of Fock states. Nevertheless, it may be possible in the near future [64] to produce such states containing Fock state components near the vacuum and so this scheme may find a use in determining the S_{NP} representation of very weak fields [65]. We conclude that the special number-phase Wigner function $S_{\mathrm{NP}}(n,\theta)$ *can be determined from quantities that are, in principle, measurable.* The quantum state of the field mode can then be found by evaluating the density operator $\hat{\rho}$ according to

$$\hat{\rho}=2\pi\sum_{n=0}^{\infty}\int_{2\pi}d\theta\ \hat{S}_{\mathrm{NP}}(n,\theta)S_{\mathrm{NP}}(n,\theta)\ .$$

VI. SUMMARY AND CONCLUSION

We have analyzed the problem of defining a Wigner function $S_{\mathrm{NP}}(n,\theta)$ associated with the photon-number and phase observables on the infinite-dimensional Fock space, which is conventionally used to represent the state of a single-mode field. We began by requiring $S_{\mathrm{NP}}(n,\theta)$ to be a bilinear functional of the wave function and then we specified several properties that the function must exhibit. For this we used the properties of Wigner's original function for position and momentum observables, as listed by Hillery *et al.* [27], as a basis. These properties are sufficient to define the position-momentum Wigner function uniquely. Translated into analogous properties for the photon-number and phase observables, they require that the number-phase Wigner function be real, give the discrete photon-number and continuous phase probability distributions as marginals, be Galilei invariant under phase and photon-number shifts (and thus automatically satisfy the classical equation of motion), be invariant under time reflections, have the overlap property, and give the trace property for the associated representation of arbitrary operators. We found, however, that these properties are not sufficient to define the number-phase Wigner function uniquely. An extra property of the position-momentum Wigner function was used to define the special number-phase Wigner function $S_{\mathrm{NP}}(n,\theta)$. This property, which is a skew diagonal form of the position-momentum Wigner function, generates the quantum interferences fringes that are characteristic of Schrödinger cat states.

The outcome of this procedure is that $S_{\mathrm{NP}}(n,\theta)$ is a quasiprobability distribution that has properties analogous to those of the Wigner's original function but with the distinctive feature in that it gives a representation of states that displays their underlying photon-number and phase properties. For Schrödinger cat states, in particular, it exhibits interference fringes that reveal the coherent superpositions of states with either different mean photon number or different mean phase. Also, every operator (including every density operator) has a unique S_{NP} representation. $S_{NP}(n,\theta)$ can be determined from quantities that are, in principle, measurable. We conclude that $S_{NP}(n,\theta)$ is a bona fide Wigner function, which should be useful in the study of the phase and photon-number properties of quantum optical systems.

ACKNOWLEDGMENTS

I would like to thank D.T. Pegg, S.M. Barnett, P.L. Knight, and A. Orłowski for comments and encouragement. I am particularly grateful to O. Steuernagel for entertaining discussions generally, to U. Leonhardt for discussions concerning recent experimental work on the determination of the PM-Wigner function and also to H. Paul for his hospitality at the Arbeitsgruppe "Nichtklassische Strahlung."

---

[1] E. P. Wigner, Phys. Rev. **40**, 749 (1932).
[2] G. S. Agarwal, Phys. Rev. A **24**, 2889 (1981).
[3] J. Dowling, G. S. Agarwal, and W. P. Schleich, Phys. Rev. A **49**, 4101 (1994).
[4] F. T. Arecchi, E. Courtens, R. Gilmore, and H. Thomas, Phys. Rev. A **6**, 2211 (1972).
[5] W. K. Wootters, Ann. Phys. **176**, 1 (1987).
[6] N. Mukunda, Am. J. Phys. **47**, 182 (1979).
[7] J. P. Bizarro, Phys. Rev. A **49**, 3255 (1994).
[8] A comprehensive survey of work in this field is given in S. M. Barnett and B. J. Dalton, Phys. Scr. **T48**, 13 (1993).
[9] We should mention that other approaches for describing quantum optical phase have been taken such as the operational approach of Mandel's group [60], earlier theoretical work based on defining operators, e.g., by Susskind and Glogower [35], Carruthers and Nieto [66], Paul [67], Garrison and Wong [68], Galindo [69], Lévy-Leblond [70], Popov and Yarunin [71], and more recent work such as the phase-space overlap approach [72] and the Wigner-Weyl quantization method by Smith *et al.* [73]. However, we are interested here with the approaches that define the phase observable as being canonically conjugate to the number operator and that attribute the vacuum state as having a random phase.
[10] R. G. Newton, Ann. Phys. (N.Y.) **124**, 327 (1980); see also [74].
[11] D. T. Pegg and S. M. Barnett, Europhys. Lett. **6**, 483 (1988).
[12] S. M. Barnett and D. T. Pegg, J. Mod. Opt. **36**, 7 (1989).
[13] D. T. Pegg and S. M. Barnett, Phys. Rev. A **39**, 1665 (1989).
[14] See also D. Ellinas, J. Math. Phys. **32**, 135 (1991) for a restatement of the Pegg-Barnet limiting procedure as a contractive limit.
[15] J. A. Vaccaro and R. F. Bonner, Phys. Lett. A **198**, 167 (1995).
[16] M. Ban, J. Math. Phys. **32**, 3077 (1991).
[17] M. Ban, Opt. Commun. **94**, 231 (1992).
[18] C. W. Helstrom, Int. J. Theor. Phys. **11**, 357 (1974).
[19] J. H. Shapiro and S. R. Shepard, Phys. Rev. A **43**, 3795 (1991).
[20] I. Białynicki-Birula and Z. Białynicka-Birula, Phys. Rev. A **14**, 1101 (1976).
[21] U. Leonhardt, J. A. Vaccaro, B. Böhmer, and H. Paul, Phys. Rev. A **51**, 84 (1995).
[22] J. A. Vaccaro, Phys. Rev. A, **51** 3309, (1995).
[23] Since all the phase formalisms in Refs. [10–22] give identical physical results, such as identical probability distributions and expectation values for physically accessible states, the underlying conjugacy relationship between photon-number and phase in one formalism automatically carries, as far as the physical interpretation is concerned, to them all. Thus we can say that the photon number and phase observables are canonically conjugacy in two senses. First, we find from the Pegg-Barnett formalism that they are canonically conjugate in the sense that the photon number and phase operators are the generators of phase and photon number shifts, respectively, and that the overlap of any eigenstate of the photon-number operator with any eigenstate of the phase operator is the same constant. (See Ref. [75] for a general discussion of canonical conjugacy of this type.) Second, the photon number and phase observables have also been shown in the H_{sym} formalism [22] to be canonically conjugate in the sense of Weyl [56].
[24] Schleich *et al.* [49] gave an intuitive approach to the problem of defining a quasiprobability distribution that has the photon number and phase distributions as marginals for squeezed states.
[25] J. A. Vaccaro and D. T. Pegg, Phys. Rev. A **41**, 5156 (1990).
[26] A. Lukš and V. Peřinová, Phys. Scr. **T48**, 94 (1993).
[27] M. Hillery, R. F. O'Connell, M. O. Scully, and E. P. Wigner, Phys. Reports **106** 121 (1984).
[28] J. A. Vaccaro, Opt. Commun. **113**, 421 (1995).
[29] E. P. Wigner, in *Perspectives in Quantum Theory*, edited by W. Yourgrau and A. van der Merwe (Dover, New York, 1979).
[30] R. F. O'Connell and E. P. Wigner, Phys. Lett. **83A**, 145 (1981).
[31] A concise introduction to the rigged Hilbert space is given in L.E. Ballentine, *Quantum Mechanics* (Prentice-Hall, Englewood Cliffs, NJ, 1990).
[32] F. London, Z. Phys. **37**, 915 (1926); **40**, 193 (1927).
[33] R. Loudon, *The Quantum Theory of Light*, 1st ed. (Oxford University Press, Oxford, 1973).
[34] J. A. Vaccaro and D. T. Pegg, Phys. Scr. **T48**, 22 (1993).
[35] L. Susskind and J. Glogower, Physics **1**, 49 (1964).
[36] M. Brune, S. Haroche, J. M. Raimond, L. Davidovich, and N. Zagury, Phys. Rev. A **45**, 5193 (1992).
[37] A. Vidiella Barranco, V. Bužek, P. L. Knight, and W. K. Lai, in *Quantum Measurements in Optics*, edited by P. Tombesi (Plenum, New York, 1991).
[38] W. Schleich, M. Pernigo, and F. Le Kien, Phys. Rev. A **44**, 2172 (1991).
[39] E. Schrödinger, Naturwissenschaften **23**, 812 (1935); see also *Quantum Theory and Measurement*, edited by J. A. Wheeler and W. H. Zurek (Princeton University Press, Princeton, 1983), p. 152.
[40] B. Yurke and D. Stoler, Phys. Rev. Lett. **57**, 13 (1986).
[41] W. H. Zurek, Phys. Today **44**(10), 36 (1991).
[42] B. M. Garraway and P. L. Knight, Phys. Rev. A **50**, 2548 (1994).

[43] V. Bužek, H. Moya-Cessa, P. L. Knight, and S. J. D. Phoenix, Phys. Rev. A **45**, 8190 (1992).
[44] D. F. Walls, Nature **324**, 210 (1986).
[45] H. P. Yuen, Phys. Rev. A **13**, 2226 (1976).
[46] R. Loudon and P.L. Knight, J. Mod. Opt. **34** 709 (1987).
[47] J. A. Vaccaro and D. T. Pegg, Opt. Commun. **70**, 529 (1989).
[48] J. A. Vaccaro, S. M. Barnett, and D. T. Pegg, J. Mod. Opt. **39**, 603 (1992).
[49] W. Schleich, R. J. Horowicz, and S. Varro, in *Quantum Optics V*, edited by J.D. Harvey and D.F. Walls (Springer-Verlag, Berlin, 1989).
[50] D. T. Smithey, M. Beck, M. G. Raymer, and A. Faridani, Phys. Rev. Lett. **70**, 1244 (1993).
[51] U. Leonhardt and H. Paul, Phys. Rev. A **48**, 4598 (1993).
[52] U. Leonhardt (private communication).
[53] It is possible to define a NP-Wigner function $\mathcal{A}(n,\theta)$ that does give a one-to-one mapping between functions $A(n,\theta)$ and operators $\hat{A}$. The vectors $\mathbf{\Lambda}(r)$ for this have the form $\Lambda_p(r)=\delta_{r,p}$ for all non-negative integers r. These vectors satisfy all properties described in Sec. II A and give an alternate NP-Wigner function as the expectation value of

$$\hat{\mathcal{A}}(n,\theta)\equiv\frac{1}{2\pi}\left[|n\rangle\langle n|+\sum_{p=n+1}^{\infty}e^{i(p-n)\theta}|p\rangle\times\langle n|+e^{-i(p-n)\theta}|n\rangle\langle p|\right].$$

However, the price for this is the loss of the skew-diagonal property described in Sec. II B with the result that $\mathcal{A}(n,\theta)$ displays quantum interferences in quite a different manner to $W(x,p)$ and $S_{\mathrm{NP}}(n,\theta)$.
[54] A. Lukš and V. Peřinová, Phys. Scr. **T48**, 94 (1993).
[55] J. A. Vaccaro and Y. Ben-Aryeh, Opt. Commun. **113**, 427 (1995).
[56] H. Weyl, *The Theory of Groups and Quantum Mechanics* (Dover, New York, 1950), Chap. IV, Sec. 14.
[57] A. Royer, Phys. Rev. Lett. **55**, 2745 (1985).
[58] K. Vogel and H. Risken, Phys. Rev. A **40**, 2847 (1989).
[59] U. Leonhardt, Phys. Rev. A **48**, 3265 (1993).
[60] See, e.g., J. W. Noh, A. Fougères, and L. Mandel, Phys. Scr. **T48**, 29 (1993), and references therein.
[61] W. Brunner, H. Paul, and G. Richter, Ann. Phys (Leipzig) **15**, 17 (1965); H. Paul, W. Brunner, and G. Richter, **17**, 262 (1966).
[62] R. A. Campos, B. E. A. Saleh, and M. Teich, Phys. Rev. A **40**, 1371 (1989).
[63] C. T. Lee, Phys. Rev. A **48**, 2285 (1993).
[64] A. S. Parkins, P. Marte, P. Zoller, O. Carnal, and H. J. Kimble, Phys. Rev. A **51**, 1578 (1995).
[65] This suggested scheme for determining the quantum state using photoncounting by first determining the S_{NP} representation is not the most efficient [76]. It does show, however, that S_{NP} can be determined from quantities that are, in principle, measurable.
[66] P. Carruthers and M. M. Nieto, Rev. Mod. Phys. **40**, 411 (1968).
[67] H. Paul, Fortschr. Phys. **22**, 657 (1974).
[68] J. C. Garrison and J. Wong, J. Math. Phys. **11**, 2242 (1970).
[69] A. Galindo, Lett. Math. Phys. **8**, 495 (1984); **9**, 263 (1985).
[70] J. M. Lévy-Leblond, Ann. Phys. (N.Y.) **101**, 319 (1976); Phys. Lett. **64A**, 159 (1977).
[71] V. N. Popov and V. S. Yarunin, J. Mod. Opt. **39**, 1525 (1992).
[72] See, e.g., W. Vogel and W. Schleich, Phys. Rev. A **44**, 7642 (1991).
[73] T. B. Smith, D. A. Dubin, and M. A. Hennings, J. Mod. Opt. **39**, 1603 (1992); D. A. Dubin, M. A. Hennings, and T. B. Smith, Publ. Res. Inst. Math. Sci. Kyoto Univ. **30**, 479 (1994).
[74] S. M. Barnett and D. T. Pegg, J. Phys. A **19**, 3849 (1986).
[75] D. T. Pegg, J. A. Vaccaro, and S. M. Barnett, J. Mod. Opt. **37**, 1703 (1990).
[76] O. Steuernagel and J. A. Vaccaro Phys. Rev. Lett. (to be published).

Antinormally ordering of phase operators and the algebra of weak limits

John A. Vaccaro [a], Y. Ben-Aryeh [b]

[a] Arbeitsgruppe "Nichtklassische Strahlung" der Max-Planck-Gesellschaft an der Humboldt-Universität zu Berlin, Rudower Chaussee 5, 12484 Berlin, Germany

[b] Department of Physics, Technion, Haifa 32000, Israel

Received 22 September 1994

Abstract

We show that the antinormal ordering introduced for the Susskind–Glogower phase operators can be justified formally from the Pegg–Barnett Ψ-space formalism. The Susskind–Glogower operators are the weak limits on the infinite-dimensional Hilbert space H of the corresponding operators in the Pegg–Barnett formalism. Because the weak limit of a product of two sequences of operators is not necessarily the product of the respective weak limits of the two sequences it follows that the algebra of the Pegg–Barnett operators will not be preserved by their weak limits on H. However, we show that the antinormal ordering of the Susskind–Glogower operators is just what is required to preserve the algebra for the weak limits on H.

1. Introduction

The last few years has yielded a number of different formalisms for describing the quantum mechanical phase properties of a harmonic oscillator or, equivalently, a single mode of the field [1]. A sound understanding of the underlying nature of quantum mechanical phase can only be achieved by establishing the relationships between the different formalisms. Indeed, a number of studies have already been directed at this very topic. In particular it has been shown [2–4] that the approach based on the Susskind–Glogower [5] probability operator measure (SG-POM) [2,6] is consistent with the Ψ-space formalism of Pegg and Barnett [7]. Also recently Lukš and Peřinová [8] introduced an interesting "antinormal ordering" technique which allows the calculation of physically reasonable results using the Susskind–Glogower operators. Their justification for the technique is based on an analogy with coherent states and the antinormal ordering of the creation and annihilation operators for the Q function. Recently, Brif and Ben-Aryeh [10] showed that use of the technique is equivalent to defining the vacuum as a state of random phase. In this paper we give another, more formal justification of the technique by showing that the antinormal ordering of the Susskind–Glogower operators is required because these operators are the weak limits of Pegg–Barnett operators and the algebra of the weak limits can only be made consistent with the Pegg–Barnett formalism by antinormally ordering the weak limits.

2. Antinormal ordering

Let us first define the operators of the SG-POM formalism. The phase probability density is postulated to be given by [2,6]

$$P(\theta)\,d\theta = \langle|\theta\rangle\langle\theta|\rangle\,d\theta,$$

where $|\theta\rangle$ is the unnormalizable London state $|\theta\rangle = \sum_{n=0}^{\infty}\exp(in\theta)|n\rangle/\sqrt{2\pi}$ [9] and $|n\rangle$ is a Fock state. The probability $P(\theta)$ describes the statistics of a stochastic variable θ for which the expectation value of a function of phase $G(\theta)$ is given by

$$\int_{2\pi} G(\theta)P(\theta)\,d\theta\,.$$

The same expectation value is obtained from the operator

$$\hat{G}_{SG} = \int_{2\pi} G(\theta)|\theta\rangle\langle\theta|\,d\theta\,. \tag{1}$$

However, it is well known that because the kets $|\theta\rangle$ are not orthogonal $\hat{G}_{SG}\hat{H}_{SG} \neq \widehat{(GH)}_{SG}$ in general where

$$\widehat{(GH)}_{SG} = \int_{2\pi} G(\theta)H(\theta)|\theta\rangle\langle\theta|\,d\theta\,.$$

A particularly important example is given by the angle operators for which $\widehat{\phi_{SG}}^2 \neq \widehat{\phi^2_{SG}}$ where $\widehat{\phi_{SG}}$ and $\widehat{\phi^2_{SG}}$ are defined by Eq. (1) with $G(\theta)$ replaced by θ and θ^2, respectively.

A solution to this problem has been given recently by Lukš and Peřinová [8]. They introduced an antinormal ordering operation, which has been represented [10] by the symbol ${}^*_*\cdots{}^*_*$, such that

$${}^*_*\hat{G}_{SG}\hat{H}_{SG}{}^*_* = \widehat{(GH)}_{SG}\,.$$

This antinormal ordering operation is defined as follows. First the expansions of $\hat{G}_{SG}$ and $\hat{H}_{SG}$ are found in terms of the Susskind–Glogower exponential phase operators $\hat{e}_k$:

$$\hat{e}_k \equiv \int_{2\pi} \exp(ik\theta)|\theta\rangle\langle\theta|\,d\theta,$$

where $k = 0, \pm1, \pm2, \cdots$. For example

$$\hat{G}_{SG} = \sum_{k=-\infty}^{\infty} a_k\hat{e}_k, \tag{2}$$

where a_k are the Fourier coefficients of the function $G(\theta)$:

$$G(\theta) = \sum_{k=-\infty}^{\infty} a_k\exp(ik\theta)\,.$$

Next the antinormal ordered product of $\hat{G}_{SG}$ and $\hat{H}_{SG}$ is defined as [8]

$${}^*_*\hat{G}_{SG}\hat{H}_{SG}{}^*_* = {}^*_*\sum_{k=-\infty}^{\infty} a_k\hat{e}_k \sum_{j=-\infty}^{\infty} b_j\hat{e}_j{}^*_*$$

$$\equiv \sum_{k=0}^{\infty}\sum_{j=-\infty}^{\infty} a_kb_j\hat{e}_k\hat{e}_j + \sum_{k=-1}^{-\infty}\sum_{j=-\infty}^{\infty} a_kb_j\hat{e}_j\hat{e}_k,$$

where b_j are the Fourier coefficients of $H(\theta)$. For example, the Susskind–Glogower exponential phase operators themselves satisfy

$${}^*_*\hat{e}_1\hat{e}_{-1}{}^*_* = {}^*_*\hat{e}_{-1}\hat{e}_1{}^*_* = \hat{1},$$

whereas it is well known that $\hat{e}_1\hat{e}_{-1} = \hat{1}$ and $\hat{e}_{-1}\hat{e}_1 = \hat{1} - |0\rangle\langle 0|$. Hence with this clever technique Lukš and Peřinová [8] succeed in overcoming the shortcomings of the Susskind–Glogower formalism. Indeed, taking the measure of phase variance [8] to be $\langle{}^*_*\widehat{\phi^2_{SG}}{}^*_*\rangle - \langle{}^*_*\widehat{\phi_{SG}}{}^*_*\rangle^2$, where $\widehat{\phi_{SG}}$ and $\widehat{\phi^2_{SG}}$ are given by Eq. (2) with a_k replaced with the Fourier coefficients of θ and θ^2, respectively, gives physically reasonable results and, e.g., attributes the vacuum state with random phase properties.

However, while this technique may give physically reasonable results the question remains as to whether it can be justified on formal grounds. We already know that the Susskind–Glogower operators are the weak limits of the Pegg–Barnett operators and that the weak limits do not obey the same algebra as the Pegg–Barnett operators [4]. This follows from the property that the product of weak limits is not necessarily the weak limit of a product of sequences of operators. We now show that the Pegg–Barnett algebra is preserved in the weak limit only if we use the antinormal ordering technique of Lukš and Peřinová [8] .

We recall that the operator $\hat{Q}$ is the weak limit of the sequence of operators $\hat{Q}_0, \hat{Q}_1, \hat{Q}_2, \cdots$ if, and only if,

$$\langle f|\hat{Q}_s|g\rangle \to \langle f|\hat{Q}|g\rangle,$$

for all $|f\rangle, |g\rangle \in H$ as $s \to \infty$. We shall write this as $\hat{Q}_s \xrightarrow{w} \hat{Q}$. It has already been shown that [4]

$$\exp(ik\hat{\phi}_\theta)_s \xrightarrow{w} \hat{e}_k,$$

where the Pegg–Barnett unitary phase operator on the left-hand side operates on the $(s+1)$-dimensional space Ψ spanned by the Fock states $|0\rangle, |1\rangle, |2\rangle, \cdots, |s\rangle$ and k is any integer. We shall first show that

$$\exp(ip\hat{\phi}_\theta)_s \exp(im\hat{\phi}_\theta)_s \xrightarrow{w} {}^*_*\hat{e}_p\hat{e}_m{}^*_*,$$

for integer p, m. For each value of s the two operators on the left-hand side commute and can be written in any order. Thus the weak limit of the left hand side is

$$\sum_{n=0}^{\infty} |n\rangle\langle n+p+m|, \quad \text{for } p+m \geq 0,$$

$$\sum_{n=0}^{\infty} |n-p-m\rangle\langle n|, \quad \text{for } p+m < 0,$$

regardless of the ordering chosen. In contrast, these expressions are only given for the product of the operators $\hat{e}_p$ and $\hat{e}_m$ when put in the antinormal order with the factor (if any) which has a negative index on the right, i.e. $\hat{e}_p\hat{e}_m$ if $m < 0$ and $\hat{e}_m\hat{e}_p$ otherwise. Thus we have established that *the weak limit of a product of Pegg–Barnett unitary phase operators is the antinormally ordered product of the respective weak limits.*

Turning now to more complicated situations, we consider the general polynomial functions of $\exp(i\theta)$. Let $U_T(x, y)$ be a polynomial in x and y, i.e.,

$$U_T(x, y) = \sum_{n,m=0}^{T} U_{n,m} x^n y^m ,$$

where T is the highest power and $U_{n,m}$ are complex coefficients, and consider the weak limit of the operator $U_T(\exp(i\hat{\phi}_\theta)_s, \exp(-i\hat{\phi}_\theta)_s)$. From the last result we find

$$\begin{aligned} &U_T(\exp(i\hat{\phi}_\theta)_s, \exp(-i\hat{\phi}_\theta)_s) \\ &\quad = \sum_{n,m} U_{n,m} \exp(in\hat{\phi}_\theta)_s \exp(-im\hat{\phi}_\theta)_s \\ &\quad \xrightarrow{w} \sum_{n,m} U_{n,m}\hat{e}_n\hat{e}_{-m} = {}^*_*U_T(\hat{e}_1, \hat{e}_{-1}){}^*_* . \end{aligned}$$

It also follows that

$$\begin{aligned} &U_T(\exp(i\hat{\phi}_\theta)_s, \exp(-i\hat{\phi}_\theta)_s) \\ &\quad \times V_{T'}(\exp(i\hat{\phi}_\theta)_s, \exp(-i\hat{\phi}_\theta)_s) \\ &\quad \xrightarrow{w} {}^*_*U_T(\hat{e}_1, \hat{e}_{-1}) V_{T'}(\hat{e}_1, \hat{e}_{-1}){}^*_*, \end{aligned}$$

where $V_{T'}$ is a polynomial in x and y of degree T'.

Finally consider an arbitrary bounded function of phase $G(\hat{\phi}_\theta)_s$ defined as follows:

$$G(\hat{\phi}_\theta)_s = \sum_{m=0}^{s} G(\theta_m)|\theta_m\rangle\langle\theta_m|,$$

where $|G(\theta)| \leq B$ for all θ and B is a positive bound, and where $\theta_m = \theta_0 + 2\pi m/(s+1)$ and $|\theta_m\rangle$ is a phase eigenstate on the $(s+1)$-dimensional space Ψ [7]. In the appendix we show that $G(\hat{\phi}_\theta)_s$ converges weakly to the operator $\hat{G}_{SG}$,

$$G(\hat{\phi}_\theta)_s \xrightarrow{w} \sum_{k=-\infty}^{\infty} a_k\hat{e}_k \equiv \hat{G}_{SG},$$

on the subspace $\overline{H}$ containing vectors satisfying

$$\sum_{n=0}^{\infty} |\langle f|n\rangle| < \infty,$$

where

$$a_k = \int_{\theta_0}^{\theta_0+2\pi} G(\theta) \exp(-ik\theta)\, d\theta/2\pi .$$

(This result extends an earlier analysis of the weak limits of Ψ-space operators on the infinite dimensional Hilbert space in Ref. [4] where the weak convergence was examined on the subspace of H containing vectors which are a superposition of only a finite number of photon number states.) We find that the antinormal ordering technique also applies to products of such operators as follows. Let $H(\hat{\phi}_\theta)_s$ be another arbitrary bounded function of phase

$$H(\hat{\phi}_\theta)_s = \sum_{m=0}^{s} H(\theta_m)|\theta_m\rangle\langle\theta_m|,$$

with

$$b_k = \int_{\theta_0}^{\theta_0+2\pi} H(\theta) \exp(-ik\theta)\, d\theta/2\pi,$$

and so from the appendix $H(\hat{\phi}_\theta)_s \xrightarrow{w} \hat{H}_{SG}$ where $\hat{H}_{SG} \equiv \sum b_k\hat{e}_k$. Also let

$$F(\hat{\phi}_\theta)_s = G(\hat{\phi}_\theta)_s H(\hat{\phi}_\theta)_s .$$

From the orthogonality of the phase states $|\theta_m\rangle$ we find

$$F(\hat{\phi}_\theta)_s = \sum_{m=0}^{s} F(\theta_m)|\theta_m\rangle\langle\theta_m|,$$

where

$$F(\theta) = G(\theta)H(\theta) = \sum_{k=-\infty}^{\infty} c_k \exp(ik\theta),$$

and thus $c_k = \sum_n a_n b_{k-n}$. Again using the result in the appendix we find

$$F(\hat{\phi}_\theta)_s \xrightarrow{w} \sum_{m,n=-\infty}^{\infty} a_n b_m \hat{e}_{m+n} \,. \tag{3}$$

We also note that $\hat{e}_{n+m} = {}^*_*\hat{e}_n\hat{e}_m{}^*_*$ and so we find the right-hand side of Eq. (3) is the antinormally ordered product of $\hat{G}_{SG}$ and $\hat{H}_{SG}$. This shows that

$$G(\hat{\phi}_\theta)_s H(\hat{\phi}_\theta)_s \xrightarrow{w} {}^*_*\hat{G}_{SG}\hat{H}_{SG}{}^*_*$$

on the subspace $\overline{H}$ and so we find that *the algebra of the Pegg–Barnett operators on the Ψ-space is preserved in the weak limit if the weak limits are antinormally ordered.* This means that physically-reasonable results can be obtained using using this technique with the Susskind–Glogower operators and also the general operators associated with the SG-POM formalism. For example,

$$(\hat{\phi}_\theta)_s \xrightarrow{w} \widehat{\phi_{SG}},$$

$$(\hat{\phi}_\theta)_s^2 \xrightarrow{w} {}^*_*(\widehat{\phi_{SG}})^2{}^*_* = \widehat{\phi^2_{SG}},$$

where

$$\widehat{\phi_{SG}} = \sum_{k=-\infty}^{\infty} d_k \hat{e}_k,$$

$$\widehat{\phi^2_{SG}} = \sum_{k=-\infty}^{\infty} h_k \hat{e}_k,$$

$$d_k = \int_{\theta_0}^{\theta_0+2\pi} \theta \exp(-ik\theta)\, d\theta/2\pi,$$

$$h_k = \int_{\theta_0}^{\theta_0+2\pi} \theta^2 \exp(-ik\theta)\, d\theta/2\pi,$$

and hence

$$\langle {}^*_*\widehat{\Delta\phi_{SG}}^2{}^*_* \rangle = \langle \Delta\hat{\phi}_\theta{}^2 \rangle,$$

where the right-hand side is the Pegg–Barnett result. The antinormally ordered Susskind–Glogower operators therefore give the same results as the Pegg–Barnett operators. For example, both give the phase variance of the vacuum as $\pi^2/3$ which is consistent with that of a random phase. Thus the Lukš-Peřinová antinormal ordering technique overcomes the original shortcoming of the Susskind–Glogower operators which attribute unusual phase properties to the field such as a non-random phase to the vacuum state.

3. Discussion

An explanation of why the antinormal ordering technique is necessary can be found by considering a specific example. The sequence of Pegg–Barnett exponential phase operators $\hat{A}_0, \hat{A}_1, \hat{A}_2, \cdots$, where $\hat{A}_s \equiv \exp(i\hat{\phi}_\theta)_s$, converges only weakly on H. It maps a vector $|f\rangle$, for which $\langle 0|f\rangle \neq 0$, to a sequence of vectors $|g\rangle_0, |g\rangle_1, |g\rangle_2, \cdots$, which does not converge strongly to another vector in H. In fact it can be shown that the infinite sequence of vectors is represented by a vector $|g)$ that belongs to a space E which is larger than H [11]. It is possible, nevertheless, to represent the sequence of vectors by its weak limit, $|g\rangle = \sum_{n=0}^{\infty} |n\rangle\langle n+1|f\rangle$, in H. Effectively this replaces the sequence $\hat{A}_0, \hat{A}_1, \hat{A}_2, \cdots$, by its weak limit $\hat{e}_1$ which is the Susskind–Glogower phase operator. This replacement, however, leads to a problem. Consider another sequence of Pegg–Barnett operators given by $\hat{B}_0, \hat{B}_1, \hat{B}_2, \cdots$, where $\hat{B}_s = \exp(-i\hat{\phi}_\theta)_s$. This operator sequence maps a vector in H into a sequence of vectors which does have a strong limit point in H and so the operator sequence converges strongly. The limit of the operator sequence is the Susskind–Glogower operator $\hat{e}_{-1}$. Applying the operator sequence to the state $|g)$ yields a sequence of vectors which converges to $|f\rangle - |0\rangle\langle 0|f\rangle$. (The same final state is also reached by applying the limit $\hat{e}_{-1}$ on $|g\rangle$.) The fact that the final state is not $|f\rangle$ lies at the root of the non-unitarity problem of the Susskind–Glogower operators $\hat{e}_1$, $\hat{e}_{-1}$. On the other hand the product $\hat{C}_s \equiv \hat{B}_s\hat{A}_s = \sum_{n=0}^{s} |n\rangle\langle n|$ converges strongly to $\hat{C} = \sum_{n=0}^{\infty} |n\rangle\langle n|$ which is the identity operator for H. Thus $\hat{C}$ maps $|f\rangle$ onto itself. By considering the action of its factors we see that $\hat{C}$ effectively maps $|f\rangle$ to $|g\rangle$ and then $|g\rangle$ back to $|f\rangle$. Clearly the non-unitarity problem arises from replacing $|g\rangle \in E$ with $|g\rangle \in H$. This problem can be avoided by ensuring that the operator expressions never map vectors in H to a larger space and back to H. Since the phase operators commute in the Pegg–Barnett formalism we can reorder the product $\hat{B}_s\hat{A}_s$ to $\hat{A}_s\hat{B}_s$ which also converges strongly to $\hat{C}$. The important point now is that this product effectively maps $|f\rangle$ to another vector in H and then back to $|f\rangle$ without ever leaving H. The final result is the same as that obtained from the Pegg–Barnett formalism. The general rule for this is to arrange the order of the exponential phase operators so that positive powers of $\exp(\mathrm{i}\hat{\phi}_\theta)$ appear on the left of positive powers of $\exp(-\mathrm{i}\hat{\phi}_\theta)$. Only in this way can we insure that a vector in H is not mapped to a larger space and back to H again. This is the essence of why the antinormal ordering technique of Lukš and Peřinová works.

In conclusion we note that the antinormally ordering technique simultaneously overcomes the problems associated with the non-orthogonality of the phase states $|\theta\rangle$ (i.e. the difference $\widehat{\phi_{\mathrm{SG}}}^2 \neq \widehat{\phi^2_{\mathrm{SG}}}$) and the topology of H (which forbids unitary phase operators on H) in the SG phase formalism and is able to yield the same physical results as the Pegg–Barnett Ψ-space formalism for which the phase eigenstates are orthogonal and the restriction of H is not a problem. Our analysis not only gives a formal justification of the antinormal ordering technique of Lukš and Peřinová from the Pegg–Barnett phase formalism, but in addition, it provides further insight into why the technique works.

Acknowledgements

J.A.V. thanks Prof. D.T. Pegg and Dr. S.M. Barnett for encouragement and helpful discussions. This work was supported by the Max-Planck-Gesellschaft.

Appendix

We prove here that an arbitrary bounded function of phase

$$G(\hat{\phi}_\theta)_s = \sum_{m=0}^{s} G(\theta_m)|\theta_m\rangle\langle\theta_m|,$$

where $|G(\theta)| \leq B$ for all θ and $B > 0$, converges weakly as follows

$$G(\hat{\phi}_\theta)_s \xrightarrow{w} \sum_{k=-\infty}^{\infty} a_k\hat{e}_k \equiv \hat{G}_{\mathrm{SG}},$$

where

$$a_k = \int_{\theta_0}^{\theta_0+2\pi} G(\theta)\exp(-ik\theta),$$

on the subspace $\overline{H}$ containing vectors satisfying

$$\sum_{n=0}^{\infty} |\langle f|n\rangle| < \infty .$$

Let $f_s(\theta)$ and $f(\theta)$ be defined as

$$f_s(\theta) \equiv \sum_{n=0}^{s}\langle n|f\rangle \exp(-in\theta)/\sqrt{2\pi}, \tag{4}$$

$$f(\theta) \equiv \sum_{n=0}^{\infty}\langle n|f\rangle \exp(-in\theta)/\sqrt{2\pi}, \tag{5}$$

where $|f\rangle \in \overline{H}$. It can be shown that $f_s(\theta)$ converges uniformly to $f(\theta)$ for all θ. Similarly, $g_s(\theta)$ converges uniformly to $g(\theta)$ for all θ where $g_s(\theta)$ and $g(\theta)$ are given by Eqs. (4, 5), respectively, on replacing $|f\rangle$ with $|g\rangle \in \overline{H}$. It follows that $g_s^*(\theta)f_s(\theta)$ converges uniformly to $g^*(\theta)f(\theta)$ for all θ. Now consider

$$\langle g|G(\hat{\phi}_\theta)_s|f\rangle = \sum_{m=0}^{s} G(\theta_m)g_s^*(\theta_m)f_s(\theta_m)2\pi/(s+1) .$$

Because of the uniform convergence of $g_s^*(\theta)f_s(\theta)$ we can always choose s sufficiently large to make

$$\left|\langle g|G(\hat{\phi}_\theta)_s|f\rangle - \sum_{m=0}^{s} G(\theta_m)g^*(\theta_m)f(\theta_m)2\pi/(s+1)\right|,$$

less than any $\epsilon > 0$, however small. Since the second term converges to a Riemann integral we can write

$$\lim_{s\to\infty} \langle g|G(\hat{\phi}_\theta)_s|f\rangle = \int_{2\pi} G(\theta) g^*(\theta) f(\theta)\, \mathrm{d}\theta$$

$$= \sum_{n,p=0}^{\infty} a_{n-p} \langle g|n\rangle \langle p|f\rangle,$$

where, in the second line, we have expanded $f(\theta)$ and $g(\theta)$ according to Eq. (5) and performed the integral. On further rearranging the right-hand side becomes

$$\sum_{k=0}^{\infty}\sum_{p=0}^{\infty} a_k \langle g|k+p\rangle\langle p|f\rangle + \sum_{k=-1}^{-\infty}\sum_{n=0}^{\infty} a_k \langle g|n\rangle\langle n-k|f\rangle$$

$$= \langle g| \sum_{k=-\infty}^{\infty} a_k \hat{e}_k |f\rangle \equiv \langle g|\hat{G}_{\mathrm{SG}}|f\rangle,$$

which shows that $G(\hat{\phi}_\theta)_s$ converges weakly on $\overline{H}$ to $\hat{G}_{\mathrm{SG}}$. This completes the proof.

References

[1] See, e.g. various approaches taken in: W.P. Schleich and S.M. Barnett, eds., Phys. Scripta T 48 (1993).
[2] J.H. Shapiro and S.R. Shepard, Phys. Rev. A 43 (1991) 3795.
[3] M.J.W. Hall, Quantum Optics 3 (1991) 7.
[4] J.A. Vaccaro and D.T. Pegg, Phys. Scripta T 48 (1993) 22.
[5] L. Susskind and J. Glogower, Physics 1 (1964) 49.
[6] C.W. Helstrom, Int. J. Theor. Phys. 11 (1974) 357.
[7] D.T. Pegg and S.M. Barnett, Europhys. Lett. 6 (1988) 483; Phys. Rev. A 39 (1989) 1665;
S.M. Barnett and D.T. Pegg, J. Mod. Optics 36 (1989) 7.
[8] A. Lukš and V. Peřinová, Phys. Scripta T 48 (1993) 94.
[9] F. London, Z. Phys. 40 (1927) 193.
[10] C. Brif and Y. Ben-Aryeh, Phys. Rev. A 50 (1994) 2727.
[11] J.A. Vaccaro and R.F. Bonner, in preparation (1994).

Pegg–Barnett phase operators of infinite rank

John A. Vaccaro [a], Richard F. Bonner [b]

[a] *Arbeitsgruppe "Nichtklassische Strahlung" der Max-Planck-Gesellschaft an der Humboldt-Universität zu Berlin, Rudower Chaussee 5, 12484 Berlin, Germany*

[b] *Faculty of CAD, Griffith University, Nathan 4111, Australia*

Received 12 December 1994; revised manuscript received 16 January 1995; accepted for publication 16 January 1995

Communicated by P.R. Holland

Abstract

We extend the Pegg–Barnett formalism, which describes the phase of a single-mode light field and is based on finite-dimensional Hilbert spaces, to the infinite-dimensional case. The infinite-dimensional versions of the Pegg–Barnett phase operators operate on a linear vector space which contains the conventionally used infinite-dimensional Hilbert space as a subspace.

1. Introduction

The quantum-mechanical description of phase was first investigated by London and Dirac in the 1920s [1,2]. It has been known now for some thirty years that the infinite-dimensional Hilbert space H conventionally used for describing a single field-mode does not support consistent unitary or Hermitian phase operators [3–5]. A number of different approaches, which nevertheless give equivalent physical descriptions, have been introduced to circumvent this problem (see e.g. Refs. [6–11]). In particular, the limitations of H are avoided in the Pegg–Barnett formalism [11] by the use of $(s+1)$-dimensional spaces Ψ_s and a special limiting procedure in which the infinite-dimensional limit $s \to \infty$ is taken of expectation values only. A great deal of work has been performed using the Pegg–Barnett formalism to examine the phase properties of the field [12].

The limiting procedure in the Pegg–Barnett formalism can be avoided by choosing the value of s sufficiently large to allow any physical result to be calculated to arbitrary precision for any given physical situation [11]. However, the conventional and generally accepted view in quantum optics (and the one adopted in this Letter) is that the photon number spectrum is unbounded above and so, for example, the photon number operator is of infinite rank and, correspondingly, the states of the field are represented in an infinite-dimensional space. Thus in this conventional view the limiting procedure is necessary.

This brings us to the task of finding the infinite-rank operators and infinite-dimensional space associated with the Pegg–Barnett formalism. One may ask is it possible to simply take the limit points on H of the vectors in Ψ_s as the dimension s tends to infinity? To show that the answer is "no" in general we first note that the Pegg–Barnett formalism allows the treatment of states such as the phase states which, in the infinite-dimensional limit, have no representation in H [14]. For example, any given sequence of phase states does *not* converge *strongly* to a vector in H as $s \to \infty$. Secondly, we note that it can be shown that sequences of phase states do converge *weakly* to the eigenstates $\sum_n \exp(in\theta)|n\rangle_H$ [1,5,13] of the Susskind–Glogower phase operator $\widehat{\exp}(i\phi)$ [4]. (Here $|n\rangle_H$ is a photon number state in H and the eigenstates are represented in a rigged Hilbert space[1].) However, while the Pegg–Barnett phase states in Ψ_s are orthogonal these weak limits are not [5]. Also, in order to produce operators of infinite rank, one may be tempted to simply take the infinite-s limits of the Pegg–Barnett operators on H. It has been shown recently [14], however, that the limits of the Pegg–Barnett operators themselves exist only as *weak* limits on H in general, and, for example, the weak limits of the unitary phase operators are the Susskind–Glogower operators [4] which are not unitary. Thus simply taking the limits of the vectors in or operators on H does *not* lead to a faithful infinite-dimensional representation of the Pegg–Barnett formalism. In this Letter we present an alternate approach to the infinite-dimensional extension which does preserve the essential properties of the Pegg–Barnett formalism. This approach sheds new light on the nature of the Pegg–Barnett formalism and the relationship it has with the conventional formalism based on H.

2. Sequences of vectors and operators

The limiting procedure of the Pegg–Barnett formalism involves finding the limit point $\mathcal{F}$ of an infinite sequence of expectation values $\mathcal{F}_0, \mathcal{F}_1, \mathcal{F}_2, \ldots$ where $\mathcal{F}_s$ is an expectation value calculated on the $(s+1)$-dimensional space Ψ_s. Strictly speaking, Pegg and Barnett begin the terms of their sequences at the term $\mathcal{F}_s$ for an arbitrarily large but finite value of s. However, since we are interested ultimately in the limit point as $s \to \infty$ it is not important at what value of s the sequence begins. We choose here to begin the sequences at $s = 0$; equivalent results are obtained by beginning the sequences at any given value of s. In all cases the expectation value $\mathcal{F}$ in the limiting procedure is associated with an infinite sequence of vectors and an infinite sequence of operators. Infinite sequences (as opposed to any specific term) are, therefore, the basic elements of the Ψ-space formalism. We now examine the properties of these sequences in detail.

Ψ_s is spanned by $(s+1)$ orthonormal number-state vectors [11] which we label here as $|0\rangle_s, |1\rangle_s, |2\rangle_s, \ldots, |s\rangle_s$. The subscript s will be used to indicate that a vector or operator belongs to or operates on Ψ_s. Thus the general vector and operator on Ψ_s will be written as

$$|f\rangle_s \equiv \sum_{n=0}^{s} f_n |n\rangle_s \,, \tag{1}$$

$$\hat{A}_s = \sum_{n,m=0}^{s} A_{n,m} |n\rangle_{s\,s}\langle m| \,. \tag{2}$$

For example the phase eigenstates, coherent states, photon number operator, annihilation operator and Hermitian and exponential phase operators are given by [11]

$$|\theta_m\rangle_s \equiv (s+1)^{-1/2} \sum_{n=0}^{s} e^{in\theta_m} |n\rangle_s \,, \tag{3}$$

$$|\alpha\rangle_s \equiv e^{-|\alpha|^2/2} \sum_{n=0}^{s} \alpha^n |n\rangle_s / \sqrt{n!} \,, \tag{4}$$

[1] A concise introduction to the rigged Hilbert space is given by Ballentine [15].

$$\hat{N}_s \equiv \sum_{n=0}^{s} n|n\rangle_s\, {}_s\langle n| \;, \tag{5}$$

$$\hat{a}_s \equiv \sum_{n=0}^{s-1} \sqrt{n+1}|n\rangle_s\, {}_s\langle n+1| \;, \tag{6}$$

$$\hat{\phi}_{\theta,s} \equiv \sum_{m=0}^{s} \theta_m|\theta_m\rangle_s\, {}_s\langle\theta_m| \;, \tag{7}$$

$$\exp(\mathrm{i}\hat{\phi}_\theta)_s \equiv \sum_{m=0}^{s} \mathrm{e}^{\mathrm{i}\theta_m}|\theta_m\rangle_s\, {}_s\langle\theta_m| \;, \tag{8}$$

where $\theta_m = 2\pi m/(s+1)$. For brevity we have assumed $\theta_0 = 0$; the extension to the more general case $\theta_0 \neq 0$ is straightforward. We note for the special case where $s = 0$ that $\exp(\mathrm{i}\hat{\phi}_\theta)_0 = |\theta_0\rangle_0\, {}_0\langle\theta_0|$ which is also the vacuum state projector $|0\rangle_0\, {}_0\langle 0|$, whereas for $s > 0$ the number state matrix elements of $\exp(\mathrm{i}\hat{\phi}_\theta)_s$ are $\sum_{n=0}^{s-1} |n\rangle_s\, {}_s\langle n+1| + |s\rangle_s\, {}_s\langle 0|$.

We define *infinite sequences* of vectors $|f\rangle$ and operators $\hat{A}$ whose terms depend on the value of s by

$$|f\rangle \equiv (|f\rangle_0, |f\rangle_1, |f\rangle_2, \ldots) \;, \tag{9}$$

$$\hat{A} \equiv (\hat{A}_0, \hat{A}_1, \hat{A}_2, \ldots) \;. \tag{10}$$

For example, the sequences representing a coherent state and the number operator are $|\alpha\rangle \equiv (|\alpha\rangle_0, |\alpha\rangle_1, |\alpha\rangle_2, \ldots)$ and $\hat{N} \equiv (\hat{N}_0, \hat{N}_1, \hat{N}_2, \ldots)$. The use of the ket and operator notation in these expressions is justified below where it is shown that $|f\rangle$ and $\hat{A}$ do indeed represent vectors and operators.

We follow the definition of expectation values in the Pegg–Barnett formalism [11] and define the expectation value of the operator sequence $\hat{A}$ for the state sequence $|f\rangle$ as

$$\langle f|\hat{A}|f\rangle \equiv \lim_{s\to\infty} \frac{{}_s\langle f|\hat{A}_s|f\rangle_s}{{}_s\langle f|f\rangle_s} \tag{11}$$

provided the limit exists. (More formally, the expectation value is defined for the pair $|f\rangle$ and $\hat{A}$ only if $\limsup_{s\to\infty} \lambda_s = \liminf_{s\to\infty} \lambda_s$ where $\lambda_s = {}_s\langle f|\hat{A}_s|f\rangle_s / {}_s\langle f|f\rangle_s$.) Thus, for example, the mean and variance of the photon number for the coherent state $|\alpha\rangle$ is $\langle\alpha|\hat{N}|\alpha\rangle = \lim_{s\to\infty} {}_s\langle\alpha|\hat{N}_s|\alpha\rangle_s = |\alpha|^2$ and $\langle\alpha|\Delta\hat{N}^2|\alpha\rangle = \langle\alpha|(\hat{N} - \langle\alpha|\hat{N}|\alpha\rangle)^2|\alpha\rangle = |\alpha|^2$, respectively, which are the same values found in the conventional approach based on H.

3. Vector space E and operator algebra

We define the addition and multiplication of sequences by the corresponding operation on their terms. For example,

$$a|f\rangle + b|g\rangle \equiv (a|f\rangle_0 + b|g\rangle_0,\;\; a|f\rangle_1 + b|g\rangle_1,\;\; a|f\rangle_2 + b|g\rangle_2,\; \ldots) \tag{12}$$

$$= |h\rangle \;, \tag{13}$$

where a, b are complex numbers. Clearly the sequences $|f\rangle$, $|g\rangle$ and $|h\rangle$ belong to a linear vector space which we label as E. They are the *vectors* of this space and represent states of the field. Accordingly, in the remainder of this Letter we refer to them as vectors or states. Similarly

$$a\hat{A} + b\hat{B} \equiv (a\hat{A}_0 + b\hat{B}_0,\ \ a\hat{A}_1 + b\hat{B}_1,\ \ a\hat{A}_2 + b\hat{B}_2,\ \ldots) \tag{14}$$

$$= \hat{C}\ , \tag{15}$$

$$\hat{A} \cdot \hat{B} \equiv (\hat{A}_0 \cdot \hat{B}_0,\ \hat{A}_1 \cdot \hat{B}_1,\ \ \hat{A}_2 \cdot \hat{B}_2,\ \ \ldots) \tag{16}$$

$$= \hat{D}\ , \tag{17}$$

where $\hat{A}$, $\hat{B}$, $\hat{C}$ and $\hat{D}$ are operators on E. Thus the set of operators on E form an algebra over the complex numbers.

We wish to determine the dimensionality of the space E. For this we define the number states $|n\rangle$ in E as follows. The first n terms of $|n\rangle$ are zero and the remaining terms are the vectors $|n\rangle_s \in \Psi_s$,

$$|n\rangle \equiv (0, 0, 0, \ldots, |n\rangle_n, |n\rangle_{n+1}, |n\rangle_{n+2}, \ldots)\ . \tag{18}$$

This gives $\hat{N}|n\rangle = n|n\rangle$ which shows that $|n\rangle$ are indeed photon number eigenstates. The states $|n\rangle$ for $n = 0, 1, 2, \ldots$ form an infinite set of linearly independent vectors. Thus E *is an infinite-dimensional vector space and the operators on* E *can therefore be of infinite rank.*

It is now possible to write results of the Pegg–Barnett formalism found for finite-dimensional spaces in terms of their infinite-dimensional counterparts. For example the polar decomposition of the annihilation operator $\hat{a}$ is given by

$$\hat{a} = \exp(\mathrm{i}\hat{\phi}_\theta)\sqrt{\hat{N}}\ , \tag{19}$$

where $\hat{a} \equiv (\hat{a}_0, \hat{a}_1, \hat{a}_2, \ldots)$ and $\exp(\mathrm{i}\hat{\phi}_\theta) \equiv (\exp(\mathrm{i}\hat{\phi}_\theta)_0, \exp(\mathrm{i}\hat{\phi}_\theta)_1, \exp(\mathrm{i}\hat{\phi}_\theta)_2, \ldots)$. Also the unitarity of the exponential phase operators on Ψ_s, $\exp(\mathrm{i}\hat{\phi}_\theta)_s \cdot \exp(-\mathrm{i}\hat{\phi}_\theta)_s = \exp(-\mathrm{i}\hat{\phi}_\theta)_s \cdot \exp(\mathrm{i}\hat{\phi}_\theta)_s = \hat{1}_s$, ensures that

$$\exp(\mathrm{i}\hat{\phi}_\theta) \cdot \exp(-\mathrm{i}\hat{\phi}_\theta) = \exp(-\mathrm{i}\hat{\phi}_\theta) \cdot \exp(\mathrm{i}\hat{\phi}_\theta) \tag{20}$$

$$= (\hat{1}_0, \hat{1}_1, \hat{1}_2, \ldots) \tag{21}$$

$$\equiv \hat{1}\ , \tag{22}$$

where $\hat{1}$ is the unit operator for E, i.e. $\hat{1}|f\rangle = |f\rangle$, and $\exp(-\mathrm{i}\hat{\phi}_\theta) \equiv (\exp(-\mathrm{i}\hat{\phi}_\theta)_0, \exp(-\mathrm{i}\hat{\phi}_\theta)_1, \exp(-\mathrm{i}\hat{\phi}_\theta)_2, \ldots)$. The expressions in Eqs. (19), (20), (22) involve operators of infinite rank.

The phase eigenvalue equation [11] extends naturally to the infinite-dimensional case. The phase state $|\theta_0\rangle \equiv (|\theta_0\rangle_0, |\theta_0\rangle_1, |\theta_0\rangle_2, \ldots)$ is clearly an eigenstate of the phase operator $\hat{\phi}_\theta \equiv (\hat{\phi}_{\theta,0}, \hat{\phi}_{\theta,1}, \hat{\phi}_{\theta,2}, \ldots)$ since $\hat{\phi}_\theta|\theta_0\rangle = \theta_0|\theta_0\rangle = 0$. We define other phase eigenstates in E for angles $\theta_R = 2\pi R$ where R is a positive rational number less than 1. Let $R = p/q$ where p and q are positive integers sharing no common factors. The phase eigenstate $|\theta_R\rangle$ is the sequence whose $(s+1)$th term is zero unless $(s+1)$ is a multiple of q, say e.g. $(s+1) = nq$, in which case the $(s+1)$th term is the state $|\theta_R\rangle_s$ given by Eq. (3) on replacing θ_m with θ_R (i.e. in this case $|\theta_R\rangle_s = |\theta_m\rangle_s$ with $m = np$ and $\theta_m = 2\pi m/(s+1)$). This yields a *countable infinity* of eigenvalue equations of the form $\hat{\phi}_\theta|\theta_R\rangle = \theta_R|\theta_R\rangle$.

There is a wider class of states which have a zero phase dispersion. We define the phase state $|\theta\rangle$ for any real phase angle θ as the sequence of phase states $|\theta\rangle \equiv (|\Theta_0\rangle_0, |\Theta_1\rangle_1, |\Theta_2\rangle_2, \ldots)$ for which the phase angles are given by $\Theta_s = 2\pi n/(s+1)$ where n is the unique integer satisfying $\theta \leqslant 2\pi n/(s+1) < \theta + 2\pi/(s+1)$ and where $|\Theta_s\rangle_s$ is a phase eigenstate in Ψ_s given by Eq. (3) on replacing θ_m with Θ_s. We note that $\Theta_s \to \theta$ as $s \to \infty$. For each real value of θ in the interval $[0, 2\pi)$ there is a unique phase state $|\theta\rangle$ for which $\langle\hat{\phi}_\theta\rangle = \theta$, $\langle\hat{\phi}_\theta{}^2\rangle = \theta^2$ and so $\langle\Delta\hat{\phi}_\theta{}^2\rangle = 0$. Thus there are an *uncountable number of phase states* $|\theta\rangle$ which have zero phase dispersion.

The phase state $|\theta\rangle$ becomes an eigenstate of $\hat{\phi}_\theta$ when we define an additional operation on E: multiplication by a sequence of numbers. Let $\bar{a}$ be a sequence of numbers $(a_0, a_1, a_2, \ldots)$ and define $\bar{a}|f\rangle \equiv$

$(a_0|f\rangle_0, a_1|f\rangle_1, a_2|f\rangle_2, \ldots)$. We now have $\hat{\phi}_\theta|\theta\rangle = \bar{\theta}|\theta\rangle$ where $\bar{\theta} \equiv (\Theta_0, \Theta_1, \Theta_2, \ldots)$ is a sequence of numbers which converges to θ. Thus there are an *uncountable* number of phase eigenvectors of this more general type.

It is well known [3–7,11] that it is not possible to have unitary phase operators on the Hilbert space H because the operator representing $\exp(\mathrm{i}\phi)$ destroys the vacuum state. In the Pegg–Barnett formalism this problem is avoided by using finite-dimensional spaces where $\exp(\mathrm{i}\hat{\phi}_\theta)_s$ maps the vacuum state to the uppermost number state $|s\rangle_s$. It is interesting to see how this operation is represented on E. The vacuum state in E is given by $|0\rangle \equiv (|0\rangle_0, |0\rangle_1, |0\rangle_2, \ldots)$ and the uppermost number state $|s\rangle_s$ in Ψ_s is represented in E by $|u\rangle \equiv (|0\rangle_0, |1\rangle_1, |2\rangle_2, \ldots)$. We find that $\exp(\mathrm{i}\hat{\phi}_\theta)|0\rangle = |u\rangle$. In fact

$$\hat{N}\exp(\mathrm{i}\hat{\phi}_\theta)|n\rangle = (n-1)\exp(\mathrm{i}\hat{\phi}_\theta)|n\rangle \tag{23}$$

for $n > 0$ and

$$\hat{N}\exp(\mathrm{i}\hat{\phi}_\theta)|0\rangle = \hat{N}|u\rangle = \bar{u}|u\rangle \ , \tag{24}$$

where $\bar{u}$ is the sequence of numbers $(0, 1, 2, \ldots)$. Hence the state $|u\rangle$ is an eigenstate of $\hat{N}$ of the more general type. Eqs. (23), (24) can be compared with the corresponding expressions on Ψ_s: $\exp(\mathrm{i}\hat{\phi}_\theta)_s|n\rangle_s = |n-1\rangle_s$ for $n > 0$ and $\exp(\mathrm{i}\hat{\phi}_\theta)_s|0\rangle_s = |s\rangle_s$.

One of the characteristic features that sets the Pegg–Barnett formalism (and other physically equivalent formalisms [6–10]) apart from earlier attempts such as that of Susskind and Glogower [4] is that it attributes the vacuum state with random phase. This feature is preserved for the Pegg–Barnett phase operators on E: for example, the variance of $\hat{\phi}_\theta$ for the vacuum state $|0\rangle$ is $\pi^2/3$. Indeed, *all the physical results of the Pegg–Barnett formalism carry over to the corresponding calculations on* E by virtue of the definitions in Eqs. (1), (2), (9)–(11). The important new result here is that $\hat{\phi}_\theta$ is an operator of infinite rank and $|0\rangle$ belongs to an infinite-dimensional space E.

4. Comparison with the Hilbert space H

We now come to the question of how the infinite-dimensional space E compares with the infinite-dimensional Hilbert space H. Each vector $|h\rangle_{\mathrm{H}}$ in H can be represented in a natural way in E as a sequence $|h\rangle$ of restricted (or "truncated") vectors,

$$|h\rangle \equiv (|h\rangle_0, |h\rangle_1, |h\rangle_2, \ldots) \ , \tag{25}$$

where

$$|h\rangle_s \equiv \sum_{n=0}^{s} d_n|n\rangle_s \tag{26}$$

and d_n is the number-state coefficient of $|h\rangle_{\mathrm{H}}$. If two vectors $|h\rangle_{\mathrm{H}}, |h'\rangle_{\mathrm{H}}$ in H are different, i.e. if $\||h\rangle_{\mathrm{H}} - |h'\rangle_{\mathrm{H}}\|^2 = \sum |d_n - d'_n|^2 \neq 0$, it then follows that their corresponding representations $|h\rangle, |h'\rangle$ in E are also different because $\||h\rangle_s - |h'\rangle_s\|^2$ cannot vanish as $s \to \infty$ and so $|h\rangle_s - |h'\rangle_s \neq (0, 0, 0, \ldots)$. Thus each vector in H is represented by a unique vector in E. Hence Eq. (25) defines a natural one-to-one mapping, which we shall call M, from H into E.

There are also vectors in E which are not represented in H. For example consider the "flipped" vector $|\mu\rangle_s$ in Ψ_s formed by "flipping" the number-state coefficients of $|h\rangle_s$,

$$|\mu\rangle_s \equiv \sum_{n=0}^{s} d_{s-n}|n\rangle_s \ . \tag{27}$$

The sequence

$$|\mu\rangle = (|\mu\rangle_0, |\mu\rangle_1, |\mu\rangle_2, \ldots) \tag{28}$$

also belongs to E. While each term of this sequence belongs to H the sequence itself does not converge to a limit point in H. Thus there is no element in H which, under the mapping M, corresponds to the infinite sequence $|\mu\rangle$. Similarly the phase state $|\theta\rangle$ and the number state $|\bar{u}\rangle$ belong to E but there are no corresponding vectors in H. Clearly *the Hilbert space* H *is represented as a subspace of* E under the natural mapping M.

5. Probability interpretation

We defined the expectation value $\langle f|\hat{A}|f\rangle$ in Eq. (11) for operator $\hat{A}$ on E and vector $|f\rangle$ in E in accordance with the Pegg–Barnett formalism as the limit $s \to \infty$ of the corresponding calculations on Ψ_s. Let us now compare the probability interpretation underlying this definition with that of the conventional approach.

In the conventional approach the probability interpretation is facilitated by the representation of each physical variable as an observable, i.e. as a Hermitian operator whose eigenvectors form a basis for the state space. The probabilities associated with an ideal measurement of the observable are proportional to the square modulus of the inner products of the vector representing the state of the field mode with the eigenvectors of the observable. We note, however, that the vector space E does not have an inner-product compatible with the usual inner product on the spaces Ψ_s. For example, consider the inner product ${}_s\langle f|g\rangle_s$ on Ψ_s where $|f\rangle_s, |g\rangle_s \in \Psi_s$. This calculation is represented on E by the sequence of complex numbers ${}_0\langle f|g\rangle_0$, ${}_1\langle f|g\rangle_1$, ${}_2\langle f|g\rangle_2, \ldots$ which does not converge, in general, and so there is no straightforward extension of the inner product from Ψ_s to the whole of E.

On the other hand it is possible to define inner products between physically relevant vectors in E, namely the eigenvectors of the operators representing physical variables and the vectors in E representing physical states of the field. We note that physical states [2] of the field mode are represented by vectors in H and, through the mapping M, by vectors $|h_{\text{ph}}\rangle$ in E. Let E_{ph} be the subspace of E containing all such vectors $|h_{\text{ph}}\rangle$. The sequence of inner products on Ψ_s for any pair of vectors in E_{ph} is *convergent*. This leads naturally to the definition of an inner product $\langle h_{\text{ph}}|h'_{\text{ph}}\rangle$ on E_{ph} as the limit point $\lim_{s\to\infty} {}_s\langle h_{\text{ph}}|h'_{\text{ph}}\rangle_s$ where $|h_{\text{ph}}\rangle = (|h_{\text{ph}}\rangle_0, |h_{\text{ph}}\rangle_1, |h_{\text{ph}}\rangle_2, \ldots)$, $|h'_{\text{ph}}\rangle = (|h'_{\text{ph}}\rangle_0, |h'_{\text{ph}}\rangle_1, |h'_{\text{ph}}\rangle_2, \ldots)$ and $|h_{\text{ph}}\rangle_s, |h'_{\text{ph}}\rangle_s \in \Psi_s$. Since, for example, the photon number eigenstates $|n\rangle$ belong to E_{ph} the inner product $\langle h_{\text{ph}}|n\rangle$ is well defined; the probability of finding a field mode in the state $|n\rangle$ after it has been prepared initially in $|h_{\text{ph}}\rangle$ is proportional to the square modulus of this inner product. The definition of the inner product can also be extended to the renormalized phase states $|\theta\rangle' = (|\theta\rangle'_0, |\theta\rangle'_1, |\theta\rangle'_2, \ldots)$, where $|\theta\rangle'_s = \sum_{n=0}^{s} e^{in\theta}|n\rangle_s/\sqrt{2\pi} \in \Psi_s$, as the limit point $\langle h_{\text{ph}}|\theta\rangle' = \lim_{s\to\infty} {}_s\langle h_{\text{ph}}|\theta\rangle'_s$. The phase probability distribution is proportional to the square modulus of this inner product. The same probability interpretations are inferred from the respective probability distributions constructed from the set of all moments $\langle h_{\text{ph}}|\hat{\phi}_\theta{}^m|h_{\text{ph}}\rangle$ and $\langle h_{\text{ph}}|\hat{N}^m|h_{\text{ph}}\rangle$ calculated using Eq. (11) for positive integer m. Thus we find that *the conventional probability interpretation associated with inner products also applies to the extended formalism.*

6. Removing redundancies

The physical results of the Pegg–Barnett formalism are obtained only after the limiting procedure has been carried out. If one changes the first few terms of an operator or vector sequence in a particular calculation the final result will not be altered. This means that many vector sequences represent the same physical state. For example, $|n\rangle$ and $\exp(i\hat{\phi}_\theta)|n+1\rangle$ are both eigenstates of $\hat{N}$ with eigenvalue n but they are not equal. These

[2] Physical states have been defined in Refs. [11,16] as states for which all moments of the photon number are finite.

redundancies can be removed by defining the physical equivalence of two sequences as follows: two sequences are physically equivalent if they differ by a finite number of terms. Thus physically-equivalent operators give the same expectation values. The same holds for physically-equivalent vectors. The vector space E and the set of operators on E are partitioned by this equivalence relation into equivalence classes which contain physically equivalent sequences. In formal terms, the collection of equivalence classes of vectors is the quotient space E/O where O represents the set of vector sequences which have a finite number of non-zero terms. The equivalence classes of the vectors form a vector space and the equivalence classes of the operators form an algebra. The essential features of the above analysis remain unchanged.

7. Discussion and conclusion

We have constructed the infinite-dimensional versions of the vectors and operators of the Pegg–Barnett formalism. We found that the basic elements of the Pegg–Barnett formalism are infinite sequences which arise from the special limiting procedure in the formalism. The infinite sequences of vectors used for calculating expectation values belong to a linear vector space E; that is, the sequences themselves are vectors in E. We have shown that a natural one-to-one mapping M maps the infinite-dimensional Hilbert space H into E so that H is essentially a subspace of E. Since E is larger than H the Pegg–Barnett formalism is clearly more general than the conventional formalism which is based on H. Also, field variables are represented as infinite sequences of finite rank operators; these infinite sequences are themselves linear operators which operate on E. We found that the Pegg–Barnett number and phase operators on E are of infinite rank. Most importantly, all the physical results of the Pegg–Barnett formalism carry over to the corresponding calculations on E, e.g., the vacuum state is attributed with a random phase. *These physical results arise here from calculations involving operators of infinite rank and states belonging to an infinite-dimensional space.*

Our extension of the Pegg–Barnett formalism gives new insight into its relationship with the conventional Hilbert space formalism. We recall that the sequences of Pegg–Barnett phase operators have, in general, only weak limits on H [14]. For example $\exp(i\hat{\phi}_\theta)_s$ and $\exp(-i\hat{\phi}_\theta)_s$ converge weakly to the corresponding Susskind–Glogower operators [4] which are not unitary on H. This situation can now be seen as taking operators $\exp(i\hat{\phi}_\theta)$ and $\exp(-i\hat{\phi}_\theta)$ defined on the more-general space E and representing them on the subspace H; crucial properties of the operators are lost in the process.

We note that there are other approaches that use a vector space larger than the conventional Hilbert space H as a basis for describing the phase of a single-mode light field. For example, Shapiro and Shepard [9] have shown that an extended two-mode model gives the same physical description of the phase of the field as their probability operator measure (SG-POM) approach and, thus, the same physical description as the Pegg–Barnett formalism. Their two-mode model, which can be compared with the signal and image bands in optical heterodyne detection [9], consists of the field mode whose phase is being investigated and an additional apparatus mode. Phase operators are defined on the tensor product of the two vector spaces representing the states of each mode. Ban [10] also introduced an approach which gives identical physical results and is based on a two-mode system. However the advantage of the Pegg–Barnett formalism (and its extension presented here) is that it places phase on an equal footing with all other physical variables of the field mode, such as the photon number, which require only a *single mode* for their theoretical description. This is also true of the single-mode phase formalism introduced some time ago by Newton [6] (see also Ref. [7]) in which phase operators are defined on an extended Hilbert space. However, Newton's extended Hilbert space is spanned by photon number eigenvectors corresponding to both positive and *negative* photon numbers in contrast to the Pegg–Barnett formalism (and its extension here) for which the photon number operator has only *non-negative* integer eigenvalues. Thus the vector space E is special in the sense that it supports a phase operator formalism which gives appropriate physical results and is associated with the conventional unbounded non-negative photon number description of a single-mode field. The next step is to construct the smallest subspace of E which is an infinite-dimensional Hilbert space and which supports an identical phase description. This has already been achieved and is reported elsewhere[3].

Finally we note that the extension presented here can also be applied to other systems described by the Pegg–Barnett Ψ-space formalism such as those involving position–momentum and angle–angular momentum variables [18].

Acknowledgements

J.A.V. thanks Professor D.T. Pegg and Dr. S.M. Barnett for many discussions on the subject of phase. This work was supported by the Max-Planck-Gesellschaft.

References

[1] F. London, Z. Phys. 37 (1926) 915; 40 (1927) 193.
[2] P.A.M. Dirac, Proc. R. Soc. A 114 (1927) 243.
[3] W.H. Louisell, Phys. Lett. 7 (1963) 60.
[4] L. Susskind and J. Glogower, Physics 1 (1964) 49.
[5] P. Carruthers and M.M. Nieto, Rev. Mod. Phys. 40 (1968) 411.
[6] R.G. Newton, Ann. Phys. (NY) 124 (1980) 327.
[7] S.M. Barnett and D.T. Pegg, J. Phys. A 19 (1986) 3849.
[8] C.W. Helstrom, Int. J. Theor. Phys. 11 (1974) 357.
[9] J.H. Shapiro and S.R. Shepard, Phys. Rev. A 43 (1991) 3795.
[10] M. Ban, Phys. Lett. A 152 (1991) 223; 155 (1991) 397; Opt. Commun. 94 (1992) 231.
[11] D.T. Pegg and S.M. Barnett, Europhys. Lett. 6 (1988) 483; Phys. Rev. A 39 (1989) 1665; S.M. Barnett and D.T. Pegg, J. Mod. Opt. 36 (1989) 7.
[12] S.M. Barnett and B.J. Dalton, Phys. Scr. T48 (1993) 13.
[13] R. Loudon, The quantum theory of light, 1st Ed. (Oxford Univ. Press, Oxford, 1973).
[14] J.A. Vaccaro and D.T. Pegg, Phys. Scr. T48 (1993) 22.
[15] L.E. Ballentine, Quantum mechanics (Prentice-Hall, Englewood Cliffs, NJ, 1990); I. M. Gel'fand and G.E. Shilov, Generalized functions, Vol. 4 (Academic Press, New York, 1964); A. Böhm, The rigged Hilbert space and quantum mechanics (Springer, Berlin, 1978).
[16] J.A. Vaccaro and D.T. Pegg, J. Mod. Opt. 37 (1990) 17.
[17] J.A. Vaccaro, Phase operators on Hilbert space, Phys. Rev. A (1995), in press.
[18] D.T. Pegg, J.A. Vaccaro and S.M. Barnett, J. Mod. Opt. 37 (1990) 1703.

[3] We can show that there is a natural one-to-one mapping based on restrictions similar to Eq. (26) which maps the infinite-dimensional Hilbert space H_{sym} described in Ref. [17] into E.

Phase operators on Hilbert space

John A. Vaccaro
Arbeitsgruppe "Nichtklassische Strahlung" der Max-Planck-Gesellschaft an der Humboldt-Universität zu Berlin, Rudower Chaussee 5, 12484 Berlin, Germany
(Received 15 September 1994; revised manuscript received 7 December 1994)

A simple formalism is introduced for describing the quantum phase of a single-mode field (or, equivalently, a harmonic oscillator) using a particular infinite-dimensional Hilbert space H_{sym}. This space is spanned by vectors which represent states of the single field mode containing non-negative numbers of photons. It has a symmetry in the sense that it contains vectors in the neighborhood of both the vacuum state and the infinite photon number state. A striking property of H_{sym} is that it supports Hermitian and unitary phase operators. Indeed the Pegg-Barnett unitary phase operators are shown to converge *strongly* on the Hilbert space H_{sym} and so the corresponding limit points are ordinary operators on H_{sym}. This can be compared with the situation for the infinite-dimensional Hilbert space H conventionally used in quantum optics on which the Pegg-Barnett unitary phase operators converge weakly, in general.

PACS number(s): 42.50.−p, 03.65.Ca

I. INTRODUCTION

The quantum phase of a single-mode field has received increasing attention during recent years [1]. In this paper we consider only the basic problem of defining Hermitian and unitary phase operators. This problem was first investigated by Dirac over 60 years ago [2]. The problem itself can be defined in different ways depending on the level of generality. For example, previous authors have required the phase operators to operate on a vector space whose elements either represent the states of the field mode alone [2–13] or the product of field mode states with the states of some apparatus or other device [14,15] where the vector space is either an infinite-dimensional Hilbert space [2–12,14,15], or a more-general space [13,16], etc. Before we begin, therefore, we must frame the problem we are considering in precise terms. We define it as follows. Find the Hermitian phase operator that is canonically conjugate to the photon number operator and is defined on an infinite-dimensional Hilbert space which is spanned by the (non-negative) photon number states of the single mode field. In this paper we give the solution to this problem. We acknowledge, however, that by defining the quantum phase problem in different terms, it is possible to arrive at different solutions (see, e.g., [3–19]).

The work of many in this field (e.g., Refs. [4,6,10–13]) has as its focus the construction of raising and lowering operators for the photon number states. These operators are required to be unitary in order to represent exponential phase operators consistently. However, it is well known that the infinite-dimensional Hilbert space H, which is conventionally used in quantum optics, presents an impasse for this task. The difficulty is that the lowering operation destroys the vacuum component of every vector. Thus such operators cannot be unitary on H. Two important methods have been introduced recently for overcoming this problem and obtaining unitary phase operators [20]. The first is due to Newton [11] and independently Barnett and Pegg [12] (see also [21,22]), who extended the photon number spectrum to negative infinity so that the vacuum state is mapped to the -1 state when acted on by the lowering operator. We shall refer to this method as the NBP formalism in this paper. The second method was introduced by Pegg and Barnett [13] and involves a special procedure for taking the infinite-dimensional limit of the vector space. We shall refer to this latter method as the Pegg-Barnett formalism. Their limiting procedure allows the use of vectors that belong to a space that is larger than H [16]. In this case the vacuum state vector is mapped to a vector which lies outside H and which has a divergent mean photon number. In fact, the Pegg-Barnett formalism solves the most general statement of the problem of phase in which there is no restriction on the state space other than that it is spanned by photon number states of non-negative number. However, we are interested here in a solution to a more restricted problem as stated above and neither of these two approaches provides the solution.

There is, fortunately, yet another approach that has not been explored previously [23]. Consider for the moment adjoining to H the vector representing the state of "infinite photon number" [24] as the vector to which the vacuum state vector will be mapped by the lowering operation. This would allow the definition of operators of the form $\exp(i\phi)$ and $\exp(-i\phi)$, which do not destroy the vacuum state vector. However, the problem then is that the infinite photon number state vector is destroyed by the lowering operation. To make the raising and lowering operators unitary we need to adjoin a collection of vectors in the neighborhood of the infinite photon number vector. The resulting Hilbert space then contains a symmetry in the sense that vectors representing states in the neighborhood of both zero and infinite photon number are included and this allows it to support the unitary operators required. We label the larger space as H_{sym}.

In this paper we introduce a simple formalism based on

H_{sym} for describing the phase of the single-mode field. It gives the solution to the problem stated above. We give the details of the construction of H_{sym} in Sec. II. In Sec. III we introduce various operators on H_{sym} as the strong limits of Pegg-Barnett operator sequences. Following this is a comparison in Sec. IV between the formalism introduced here and the NBP and Pegg-Barnett formalisms. A discussion is given in Sec. V.

II. THE INFINITE-DIMENSIONAL HILBERT SPACE H_{sym}

We begin by following earlier approaches [8,13] in that we consider a sequence of finite-dimensional spaces but quickly depart from previous treatments in two important ways. The first is in the relationship between spaces of different dimensions: we specify a method of stepping from one space to a space of higher dimension that differs from the previous treatments. The second is that we construct an infinite-dimensional Hilbert space from the finite-dimensional spaces and then consider infinite-rank operators on the infinite-dimensional space: this differs from the Pegg-Barnett formalism where the infinite-dimensional limit is only taken of expectation values and not directly of the operators or the spaces.

Let Λ be an infinite set of orthonormal vectors $|n\rangle_\Lambda$ labeled by the positive integers, i.e., ${}_\Lambda\langle n|m\rangle_\Lambda = \delta_{n,m}$ for $n, m = 1, 2, \ldots$. We wish to stress that the vector $|n\rangle_\Lambda$ does not necessarily represent the state of n photons. The set Λ is any countable infinite orthonormal set labeled in any order. The identification of any vector with a particular photon number state depends on the definition of the number operator which we give later. For example, defining $\hat{N} \equiv 0|\gamma\rangle_{\Lambda\Lambda}\langle\gamma| + 1|\epsilon\rangle_{\Lambda\Lambda}\langle\epsilon| + 2|\chi\rangle_{\Lambda\Lambda}\langle\chi| + \cdots$, where $\gamma, \epsilon, \chi, \ldots$ are unique positive integers, implies that the vectors $|\gamma\rangle_\Lambda, |\epsilon\rangle_\Lambda, |\chi\rangle_\Lambda, \ldots$ represent the 0,1,2,... photon number states, respectively.

We construct finite-dimensional spaces from Λ as follows. First we introduce a new set of symbols which are related to the old by

$$|n\rangle_s \equiv |\xi(n,s)\rangle_\Lambda. \tag{1}$$

Here $\xi(n,s)$ is a positive integer function of n and s, where $n = 0, 1, 2, \ldots, s$ and $s = 0, 1, 2, \ldots$, which we use to label the elements of Λ. We will give specific examples of $\xi(n,s)$ later. For the moment we note that different choices of $\xi(n,s)$ lead to different infinite-dimensional Hilbert spaces. Now let

$$\{|n\rangle_s\}_{n=0,1,2,\ldots,s} \tag{2}$$

be a set of $(s+1)$ linearly independent vectors. This constrains the function $\xi(n,s)$ to the extent that for each given value of $s \geq 0$ it takes on $(s+1)$ *different* values for $n = 0, 1, 2, \ldots, s$. The set in (2) spans a $(s+1)$-dimensional Hilbert space Ψ_s, which is equivalent to that used by Pegg and Barnett. Accordingly we can define on Ψ_s all the Pegg-Barnett operators, including the phase operators. For example, consider the photon number operator $\hat{N}_s$ [13], which is given here by

$$\hat{N}_s = \sum_{n=0}^{s} n|n\rangle_{ss}\langle n| \,. \tag{3}$$

Clearly $\hat{N}_s|n\rangle_s = n|n\rangle_s$ and so in the Pegg-Barnett formalism $|n\rangle_s$ is the eigenvector of $\hat{N}_s$ with eigenvalue n.

So far we have not specified the relationship between Ψ_s and $\Psi_{s'}$ for $s' > s$. If we were to strictly follow the relationship used by Popov and Yarunin [8] and Pegg and Barnett [13] we would define

$$\xi(n,s) = \zeta(n), \tag{4}$$

where $\zeta(n)$ is a positive integer function of n only [e.g., the function $\xi(n,s) = n+1$ satisfies this]. We would then find that the completion of the infinite union $\Theta \equiv \bigcup_{s=0}^{\infty} \Psi_s$, for which the inner product is given by

$$\langle g, s|f, s'\rangle = \sum_{n=0}^{k} g_n^* f_n,$$

where

$$|f, s\rangle = \sum_{n=0}^{s} f_n|n\rangle_s \in \Theta, \tag{5}$$

$$|g, s\rangle = \sum_{n=0}^{s} g_n|n\rangle_s \in \Theta, \tag{6}$$

and k is the smaller of s and s', yields the Hilbert space H'. The general vector $|h\rangle = \sum_{n=0}^{\infty} h_n|\zeta(n)\rangle_\Lambda$ belonging to H' has the property that $\sum_{n=0}^{\infty} |h_n|^2 < \infty$. This restriction [25] implies, in particular, that H' does not contain a vector corresponding to the vector $|s\rangle_s$ in the Pegg-Barnett formalism in the infinite-s limit and so H' cannot support unitary phase operators corresponding to those in the Pegg-Barnett formalism. Indeed, defining operators on H' as the (weak) limit points of the Pegg-Barnett phase operators yields the Susskind-Glogower and the Popov-Yarunin operators [4,8], which attribute the vacuum state with nonrandom phase properties. The reason this situation arises here is that an asymmetry is produced by adding new vectors as s increases only at the end of the sequence of vectors $|0\rangle_s, |1\rangle_s, |2\rangle_s, \ldots, |s\rangle_s$ as given by Eq. (4). This corresponds essentially to adding new vectors at the upper end of the photon number spectrum, which restricts the resulting Hilbert space so that it contains only elements whose number state coefficients h_n vanish as $n \to \infty$. Alternatively, in order to produce a symmetry between both ends of the photon number spectrum, we add new states, in the following, to the *middle* of the sequence of vectors. To simplify the analysis we consider only odd s from now on; the extension to all values of s is straightforward.

We define the relationship between $\xi(n,s)$ and $\xi(m, s+2)$ for odd s by

$$\xi(n, s+2) = \begin{cases} \xi(n,s) & \text{for } 0 \le n \le \frac{1}{2}(s-1) \\ \xi(n-2,s) & \text{for } \frac{1}{2}(s+5) \le n \le s+2 \end{cases} \tag{7}$$

and where the new values $\xi(n, s+2)$ for $n = \frac{1}{2}(s+1), \frac{1}{2}(s+3)$ are different from the values of $\xi(n,s)$ for $n = 0, 1, 2, \ldots, s$. This relationship is depicted graphically in Fig. 1. There are many functions $\xi(n,s)$ that satisfy this; a specific example is given by

$$\xi(n,s) = \begin{cases} 2n+1 & \text{for } 0 \le n \le \frac{1}{2}(s-1) \\ 2(s-n)+2 & \text{for } \frac{1}{2}(s+1) \le n \le s. \end{cases}$$

From the relationship between $\xi(n,s)$ for different s given by Eq. (7) we find that the inner product of two vectors in the infinite union $\Theta \equiv \bigcup_{s=0}^{\infty} \Psi_s$ is now given by

$$\langle g, s | f, s' \rangle = \sum_{n=0}^{(k-1)/2} g_n^* f_n + \sum_{n=(k+1)/2}^{k} g_{s-k+n}^* f_{s'-k+n},$$

where, as before, k is the smaller of s and s' and $|g,s\rangle$ and $|f,s\rangle$ are defined in Eqs. (5) and (6). The completion of Θ with the new function $\xi(n,s)$ defined by Eq. (7) is the infinite-dimensional Hilbert space H_{sym} discussed above. We highlight some of the properties of H_{sym} in the remainder of this section.

It is helpful to introduce a new notation so that we can eliminate reference to the parameter s. Consider the vector given by Eq. (5) with $\xi(n,s)$ now given by Eq. (7) and set

$$\bar{f}_n = \begin{cases} f_n & \text{for } n \ge 0 \\ f_{s+1+n} & \text{for } n < 0 \end{cases}$$

so that

$$|f, s\rangle = \sum_{n=0}^{(s-1)/2} \bar{f}_n |n\rangle_s + \sum_{n=-1}^{-(s+1)/2} \bar{f}_n |s+1+n\rangle_s . \tag{8}$$

We note that in the Pegg-Barnett formalism the vector $|s\rangle_s$ is an eigenstate of $\hat{N}_s$ with eigenvalue s. Since we are ultimately interested only in the infinite-s limit, we see that $|s\rangle_s$ will eventually correspond to a state of *infinite photon number*. Also we find from Eqs. (1) and (7) that $|s-n\rangle_s$ and $|n\rangle_s$ are independent of s for $s > 2n \ge 0$ and so we define

$$|n\rangle \equiv |n\rangle_s, \tag{9}$$

$$|\widetilde{\infty} - n\rangle \equiv |s-n\rangle_s \tag{10}$$

for $0 \le n \le (s-1)/2$. We use the symbol $\widetilde{\infty}$ in expression (10) as a reminder of the divergent nature of the photon number eigenvalues; the properties of the vector $|\widetilde{\infty} - n\rangle$ can be inferred from that of the right-hand side of Eq. (10). Figure 2 illustrates the use of the new notation. From Eqs. (8)–(10) we see that the general vector $|h\rangle$ belonging to H_{sym} can now be written in the form

$$|h\rangle = \sum_{n=0}^{\infty} h_n |n\rangle + \sum_{n=-1}^{-\infty} h_n |\widetilde{\infty} + 1 + n\rangle, \tag{11}$$

where h_n are complex numbers for $n = 0, \pm 1, \pm 2, \ldots$. From the orthogonality of the vectors in Eqs. (9) and (10) we find that the square of the norm of $|h\rangle$ is

$$\langle h | h \rangle = \sum_{n=-\infty}^{\infty} |h_n|^2 < \infty \tag{12}$$

and the innerproduct between two vectors $|f\rangle, |g\rangle \in H_{sym}$ of the form given by Eq. (11) is

$$\langle f | g \rangle = \sum_{n=-\infty}^{\infty} f_n^* g_n .$$

The subspace of H_{sym} containing vectors in the form of Eq. (11) but whose coefficients h_n are zero for $n < 0$ is clearly the Hilbert space H that is conventionally used in quantum optics. Thus, by simply choosing the relationship *between* each of the spaces Ψ_s according to Eq. (7) we are able to construct the Hilbert space H_{sym} which contains H as a subspace. In particular, H_{sym} also contains the vector $|\widetilde{\infty}\rangle$, which does not belong to H. The importance of this vector is that it corresponds to $|s\rangle_s$ in the Pegg-Barnett formalism in the infinite-s limit. Put more precisely, the sequence $|0\rangle_0, |1\rangle_1, |2\rangle_2, \ldots, |s\rangle_s, \ldots$ *converges* to $|\widetilde{\infty}\rangle$ in H_{sym} because, in fact, $|s\rangle_s = |\widetilde{\infty}\rangle$ for all s. The presence of the vectors $|\widetilde{\infty} - n\rangle$ for $n = 0, 1, 2, \ldots$ allows H_{sym} to support the unitary phase operators treated in the next section.

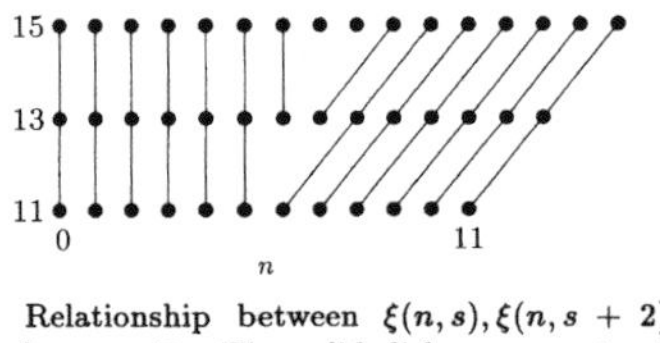

FIG. 1. Relationship between $\xi(n,s), \xi(n, s+2)$, and $\xi(n, s+4)$ for $s = 11$. The solid disks represent values of ξ and the straight lines connect equal values. In particular, $\xi(n, s+2)$ has $s+3$ unique values for $n = 0, 1, 2, \ldots, s+2$ of which $s+1$ values are inherited from $\xi(n,s)$.

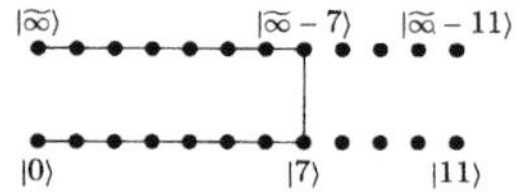

FIG. 2. Diagram representing a portion of the number state basis of H_{sym} using the new notation of Eqs. (9) and (10). Each disk represents a vector in the orthonormal set $\{|n\rangle, |\widetilde{\infty} - n\rangle\}_{n=0,1,2,\ldots}$. The line connects the vectors spanning Ψ_s for $s = 15$. The whole basis of H_{sym} can be represented by extending the diagram indefinitely to the right. The bottom row of disks in the extended diagram represents the basis of the subspace H.

III. OPERATORS ON H_{sym}

We now introduce infinite-rank operators on H_{sym}. In particular we wish to define the photon number and phase operators as a canonically conjugate pair. We adopt the Pegg-Barnett approach [13] for defining these operators via sequences. But whereas in the Pegg-Barnett formalism it is only the limit points of expectation values that are considered, here we find the limit points on H_{sym} of the operator sequences themselves.

The Pegg-Barnett number operator $\hat{N}_s$ given by Eq. (3) takes the form

$$\hat{N}_s = \sum_{n=0}^{(s-1)/2} n|n\rangle\langle n| + \sum_{n=-1}^{-(s+1)/2} (s+1+n)|\widetilde{\infty}+1+n\rangle\langle\widetilde{\infty}+1+n|$$

in the new notation. It is not difficult to show that the sequence $\hat{N}_1, \hat{N}_3, \hat{N}_5, \ldots$ has the strong limit

$$\hat{N} \equiv \sum_{n=0}^{\infty} n|n\rangle\langle n| \tag{13}$$

on the domain $D(\hat{N})$, which contains vectors of the form Eq. (11), but whose coefficients satisfy

$$\sum_{n=0}^{\infty} |nh_n|^2 < \infty$$

and $h_n = 0$ for $n < 0$. Thus we identify the vector $|n\rangle$ with the state of n photons. However, one may ask whether we are justified in identifying $|\widetilde{\infty}\rangle$ as a state of infinite photon number when this vector lies outside the domain of $\hat{N}$. To see that the answer is "yes" consider the sequence of inverse number operators $(1+\hat{N}_s)^{-1}$ for $s = 1, 3, 5, \ldots$, which is easily found to converge strongly on the whole space H_{sym} to

$$\sum_{n=0}^{\infty}(1+n)^{-1}|n\rangle\langle n| + 0\left(\sum_{n=0}^{-\infty}|\widetilde{\infty}+n\rangle\langle\widetilde{\infty}+n|\right) .$$

The vectors $|\widetilde{\infty}-n\rangle$ for $n = 0, 1, 2, \ldots$ are degenerate eigenvectors of this operator with eigenvalue zero. In this respect all of these vectors do, in fact, correspond to states of infinite photon number. It is therefore quite natural that such vectors lie outside the domain of $\hat{N}$. We note also from Eq. (10) that these vectors are orthonormal: $\langle\widetilde{\infty}-n|\widetilde{\infty}-m\rangle = \delta_{n,m}$ for $n, m = 0, 1, 2, \ldots$, which gives further support for our notation. Care should be taken, however, not to confuse the label $\widetilde{\infty}-n$ as representing "infinity minus n." Rather the vectors $|\widetilde{\infty}-n\rangle$ are a collection of orthonormal vectors representing states of infinite photon number.

The Pegg-Barnett unitary phase operators [13] are given here by

$$\begin{aligned}\exp(i\hat{\phi}_{\theta,s}) &= \sum_{n=0}^{s-1}|n\rangle_{ss}\langle n+1| + |s\rangle_{ss}\langle 0| \\ &= \sum_{n=0}^{(s-1)/2}|n\rangle\langle n+1| \\ &\quad + \sum_{n=-1}^{-(s-1)/2}|\widetilde{\infty}+n\rangle\langle\widetilde{\infty}+n+1| + |\widetilde{\infty}\rangle\langle 0| \\ &= [\exp(-i\hat{\phi}_{\theta,s})]^\dagger\end{aligned}$$

for $s = 1, 3, 5, \ldots$. We have set the "phase window" [13] to $[0, 2\pi)$ and used Eqs. (9) and (10) in the second line. The same expression results from other choices of the phase window provided an appropriate sequence of values of s is chosen accordingly [26]. The sequence of operators $\exp(i\hat{\phi}_{\theta,s})$ and $\exp(-i\hat{\phi}_{\theta,s})$ are found to converge strongly on H_{sym} to $\exp(i\hat{\phi})$ and $\exp(-i\hat{\phi})$, respectively, where

$$\begin{aligned}\exp(i\hat{\phi}) &= \sum_{n=0}^{\infty}|n\rangle\langle n+1| \\ &\quad + \sum_{n=-1}^{-\infty}|\widetilde{\infty}+n\rangle\langle\widetilde{\infty}+n+1| + |\widetilde{\infty}\rangle\langle 0| \quad (14)\\ &= [\exp(-i\hat{\phi})]^\dagger . \quad (15)\end{aligned}$$

From the orthogonality of the vectors $|n\rangle$ and $|\widetilde{\infty}-n\rangle$ for $n = 0, 1, 2, \ldots$ we find that

$$\exp(i\hat{\phi})\exp(-i\hat{\phi}) = \exp(-i\hat{\phi})\exp(i\hat{\phi}) = \hat{1},$$

where

$$\hat{1} \equiv \sum_{n=0}^{\infty}|n\rangle\langle n| + \sum_{n=0}^{-\infty}|\widetilde{\infty}+n\rangle\langle\widetilde{\infty}+n|$$

is the unit operator for H_{sym}. Thus the *limit points are unitary on* H_{sym}. This contrasts markedly with the limits of the same sequences on H where the limit points are the Susskind-Glogower operators which are not unitary [27]. The unitarity of $\exp(i\hat{\phi})$ guarantees that there is a corresponding Hermitian phase angle operator on H_{sym}. We introduce this operator via the phase eigenstate basis.

The eigenstates of $\exp(i\hat{\phi})$ do not belong to H_{sym}, but can be accommodated in a rigged Hilbert space H^R_{sym} [28-30]. We describe H^R_{sym} briefly here. Let Ξ be the space of all linear combinations of the vectors $|n\rangle$ and $|\widetilde{\infty}-n\rangle$, for $n = 0, 1, 2, \ldots$, and let Ω_{sym} be the nuclear space consisting of all vectors $|h\rangle$ [given by Eq. (11)] in H_{sym} which satisfy

$$\sum_{n=-\infty}^{\infty} |h_n|^2|n|^m < \infty$$

for $m = 0, 1, 2, \ldots$. The space Ω^x_{sym}, which is conjugate to Ω_{sym}, contains all those vectors $|f\rangle \in \Xi$ satisfying $\langle f|\omega\rangle < \infty$ for all $|\omega\rangle \in \Omega_{sym}$ and for which $\langle f|\omega\rangle$ is a continuous linear functional of $|\omega\rangle$ on Ω_{sym}. The triplet $\Omega_{sym} \subset H_{sym} \subset \Omega^x_{sym}$ is our rigged Hilbert space H^R_{sym}. It is not difficult to show that the vector

$$|\theta\rangle = (2\pi)^{-\frac{1}{2}}\left[\sum_{n=0}^{\infty}\exp(in\theta)|n\rangle + \sum_{n=-1}^{-\infty}\exp(in\theta)|\widetilde{\infty}+1+n\rangle\right],$$

where θ is real, belongs to Ω^x_{sym} and also that it has the property

$$\exp(\mathrm{i}\hat{\phi})|\theta\rangle = \exp(\mathrm{i}\theta)|\theta\rangle .$$

Thus $|\theta\rangle$, which we call a phase state, is clearly an eigenvector of $\exp(\mathrm{i}\hat{\phi})$. The phase states $|\theta\rangle$ for θ in a 2π interval form a complete orthonormal set, e.g.,

$$\hat{1} = \int_{2\pi}|\theta\rangle\langle\theta|\mathrm{d}\theta,$$

$$\langle\theta'|\theta\rangle = \frac{1}{2\pi}\sum_{n=-\infty}^{\infty}\exp[in(\theta-\theta')] \equiv \delta(\theta-\theta')$$

where $\delta(\theta-\theta')$ is the 2π-periodic Dirac delta distribution. The completeness property allows an alternate, equivalent definition of the unitary phase operator:

$$\exp(\mathrm{i}\hat{\phi}) = \int_{2\pi}\exp(\mathrm{i}\theta)|\theta\rangle\langle\theta|\mathrm{d}\theta .$$

It follows from the orthonormal property of the phase states that $\exp(\mathrm{i}\hat{\phi})$ is the exponential of the phase angle operator given by

$$\hat{\phi}_\theta = \int_{\theta_0}^{\theta_0+2\pi}\theta|\theta\rangle\langle\theta|\mathrm{d}\theta$$

for arbitrary θ_0. The value of θ_0 defines a 2π phase window in the usual way [13].

Strictly speaking, the commutator of $\hat{N}$ with $\hat{\phi}_\theta$ does not exist because the range of $\hat{\phi}_\theta$ lies outside the domain $D(\hat{N})$ of $\hat{N}$ and so the product $\hat{N}\hat{\phi}_\theta$ is not an operator on H_{sym}. However, it is still possible to evaluate the expectation value $\langle f|\hat{N}\hat{\phi}_\theta|f\rangle$ for $|f\rangle \in D(\hat{N})$ by operating $\hat{N}$ first on $\langle f|$. Thus we find

$$\langle f|[\hat{N},\hat{\phi}_\theta]|f\rangle = \mathrm{i}(1-2\pi|\langle f|\theta_0\rangle|^2)$$

for all $|f\rangle \in D(\hat{N})$. It can then be shown by a trivial extension of the usual proof [31] that all $|g\rangle \in D(\hat{N}^2)$ satisfy the Heisenberg uncertainty relation

$$\langle\Delta\hat{N}^2\rangle\langle\Delta\hat{\phi}_\theta{}^2\rangle \geq \frac{1}{4}(1-2\pi|\langle g|\theta_0\rangle|^2)^2 .$$

These last two expressions agree with the corresponding expressions found in the Pegg-Barnett formalism.

It remains for us to show that $\hat{\phi}_\theta$ is canonically conjugate to the photon number operator $\hat{N}$. We note that $|\theta\rangle$ contains the broadest spread of number states possible. Moreover, $\hat{N}$ is the generator of phase shifts as given by

$$\langle\theta|\exp(\mathrm{i}\hat{N}\delta)|h\rangle = \langle\theta-\delta|h\rangle$$

for all $|h\rangle \in H \cap \Omega_{sym}$, where δ is real, and $\hat{\phi}_\theta$ is the generator of number shifts as can be seen from the matrix elements of $\exp(\mathrm{i}\hat{\phi})$ in Eq. (14). These results show that $\hat{\phi}_\theta$ is canonically conjugate to $\hat{N}$ in the sense defined by Pegg *et al.* [13,32]. We also note that

$$\langle h|\exp(\mathrm{i}\hat{N})\exp(\mathrm{i}\hat{\phi}_\theta)|h\rangle = \exp(-\mathrm{i})\langle h|\exp(\mathrm{i}\hat{\phi}_\theta)\exp(\mathrm{i}\hat{N})|h\rangle ,$$

where $|h\rangle \in H$, and so $\hat{\phi}_\theta$ is also canonically conjugate to $\hat{N}$ in the Weyl sense [33].

Hence, to summarize, we have constructed a Hilbert space H_{sym} which is spanned by non-negative number states and found on it the Hermitian phase operator canonically conjugate to the photon number operator. Thus *we have solved the phase problem defined in the Introduction.*

We now pursue the nature of the operators on H_{sym} a little further and, in particular, examine the polar decomposition of the annihilation operator sought by Dirac [2]. The annihilation and creation operators are given in the Pegg-Barnett formalism [13] by

$$\begin{aligned}\hat{a}_s &= \sum_{n=0}^{s-1}\sqrt{n+1}|n\rangle_{ss}\langle n+1| \\ &= \sum_{n=0}^{(s-1)/2}\sqrt{n+1}|n\rangle\langle n+1| \\ &\quad + \sum_{n=-1}^{-(s-1)/2}\sqrt{s+n+1}|\widetilde{\infty}+n\rangle\langle\widetilde{\infty}+n+1| \\ &= (\hat{a}_s^\dagger)^\dagger .\end{aligned}$$

The strong limits of the sequences of $\hat{a}_s$ and $\hat{a}_s^\dagger$ are found to be $\hat{a}$ and $\hat{a}^\dagger$, respectively, where

$$\hat{a} = \sum_{n=0}^{\infty}\sqrt{n+1}|n\rangle\langle n+1| = (\hat{a}^\dagger)^\dagger$$

on the domain given by vectors $|h\rangle$ of the form Eq. (11) whose coefficients satisfy

$$\sum_{n=0}^{\infty}|h_n|^2 n < \infty$$

and $h_n = 0$ for $n < 0$. Clearly $\hat{a}^\dagger\hat{a} = \hat{N}$ and

$$[\hat{a},\hat{a}^\dagger] = \sum_{n=0}^{\infty}|n\rangle\langle n| . \qquad (16)$$

Indeed, $\hat{N}$, $\hat{a}$, and $\hat{a}^\dagger$ are, in fact, *identical to the corresponding operators in the conventional approach which uses H as the state space.* The right-hand side of Eq. (16) is the unit operator for H. We note that H contains the domain of $\hat{a}$ and $\hat{a}^\dagger$ and so the above commutator cannot operate on a space larger than H. However, since all physical states belong to H, the commutator is also physically indistinguishable from the unit operator $\hat{1}$ on H_{sym}.

The operator $\exp(i\hat{\phi})$ generates a state of infinite photon number from the vacuum state $\exp(i\hat{\phi})|0\rangle = |\widetilde{\infty}\rangle$. Can such an operation be produced through a physical interaction? The answer has been pointed out previously by Barnett and Pegg [12,13]. All physical interactions between the field mode and other systems involve interaction terms which are functions of the annihilation and creation operators. Since the annihilation operator destroys the vacuum state we see that it is not possible for this operation to occur in physical situations. It follows that a field mode that is initially in a state with a finite mean photon number and that interacts with a system that has a finite energy will always have a finite mean photon number. Thus the state of a field mode in physical situations is always represented by a vector that belongs to H.

Finally, we note that the annihilation operator $\hat{a}$, which operates on a dense subset of H, can be decomposed into unitary phase and photon number operator parts

$$\hat{a} = \exp(i\hat{\phi})\sqrt{\hat{N}} \ . \tag{17}$$

This is *the polar decomposition of the very operator $\hat{a}$ sought by Dirac* [35]. An interesting feature of this decomposition is that $\hat{a}$ is a product of two operators, one of which ($\exp i\hat{\phi}$) operates on a larger space than H while the other ($\sqrt{\hat{N}}$) operates on a dense subset of H. We now see the reason why the previous attempts to find the decomposition of $\hat{a}$ using only H were unsuccessful: the unitary phase operator could not be constructed because it operates on a larger space.

IV. COMPARISON WITH RELATED FORMALISMS

Unitary phase operators are possible on H_{sym} because the vacuum state can be mapped to the state $|\widetilde{\infty}\rangle$, which is orthogonal to all the states $|n\rangle$ for $n = 0, 1, 2, \ldots$. In the NBP formalism [11,12] the vacuum component is also mapped to a state that is orthogonal to the same set of photon number states. This leads us to examine the relationship between the two formalisms.

Newton [11] introduced a fictitious spin-$\frac{1}{2}$ property to the field mode to enable its mathematical description to include a photon number operator $\hat{L}$ that has both positive *and negative* integer eigenvalues. Barnett and Pegg also considered extending the photon number spectrum to negative infinity, but without the spin property [12]. We now construct an operator on H_{sym} with a spectrum equivalent to that of $\hat{L}$ for comparison. Consider the sequence of operators

$$\begin{aligned}\hat{\mathcal{L}}_s &\equiv \sum_{n=0}^{(s-1)/2} n|n\rangle_{ss}\langle n| + \sum_{n=(s+1)/2}^{s} (n-s-1)|n\rangle_{ss}\langle n| \\ &= \hat{N}_s - (s+1)\sum_{n=-1}^{-(s+1)/2} |\widetilde{\infty}+n+1\rangle\langle\widetilde{\infty}+n+1|\end{aligned}$$

for $s = 1, 3, 5, \ldots$. Thus $\hat{\mathcal{L}}_s$ is diagonal in the photon number basis, but with different eigenvalues to $\hat{N}_s$. The sequence converges strongly to

$$\hat{\mathcal{L}} \equiv \sum_{n=0}^{\infty} n|n\rangle\langle n| + \sum_{n=-1}^{-\infty} n|\widetilde{\infty}+n+1\rangle\langle\widetilde{\infty}+n+1| \tag{18}$$

on a dense subspace of H_{sym}. The eigenvalue equation for $\hat{\mathcal{L}}$ is

$$\begin{aligned}\hat{\mathcal{L}}|n\rangle &= n|n\rangle, \\ \hat{\mathcal{L}}|\widetilde{\infty}-n\rangle &= -(n+1)|\widetilde{\infty}-n\rangle\end{aligned}$$

for $n = 0, 1, 2, \ldots$. If we relabel the vectors by their $\hat{\mathcal{L}}$ eigenvalues

$$\begin{aligned}|n\rangle_{\mathcal{L}} &\equiv |n\rangle, \\ |-(n+1)\rangle_{\mathcal{L}} &\equiv |\widetilde{\infty}-n\rangle\end{aligned}$$

for $n = 0, 1, 2, \ldots$, we then have

$$\hat{\mathcal{L}} = \sum_{n=-\infty}^{\infty} n|n\rangle_{\mathcal{L}\mathcal{L}}\langle n| \ .$$

This shows that $\hat{\mathcal{L}}$ and H_{sym} are mathematically equivalent to $\hat{L}$ and the corresponding space used by Newton and also Barnett and Pegg. Does this mean that here we are simply reinterpreting the negative photon number states in the NBP formalism as states of infinite photon number? The answer is quite clearly "no" for the following reasons. Imagine that we map the Hilbert space used in the NBP formalism onto H_{sym} by identifying a state of n photons in the NBP formalism with $|n\rangle$ for $n \geq 0$ and $|\widetilde{\infty}+n+1\rangle$ for $n < 0$. The NBP photon number operator would then be given by $\hat{\mathcal{L}}$ in Eq. (18) which operates on states of infinite photon number whereas our operator $\hat{N}$, given by Eq. (13), is the conventional photon number operator which operates on a dense subset of H only. The time evolution of a state in the NBP formalism [36] would be given by a Hamiltonian that is proportional to $\hat{\mathcal{L}}$ in contrast to that here, where the Hamiltonian is proportional to $\hat{N}$. The NBP annihilation operator would be given by

$$\sum_{n=0}^{\infty}\sqrt{n+1}|n\rangle\langle n+1| + \sum_{n=-1}^{-\infty}\sqrt{|n|}|\widetilde{\infty}+n\rangle\langle\widetilde{\infty}+n+1|, \tag{19}$$

which acts on states of infinite photon number whereas our operator $\hat{a}$ is the conventional annihilation operator acting on a dense subset of H. Clearly the two formalisms differ by more than a simple reinterpretation of the photon number states. Indeed, it is only when the corresponding interpretation of negative or infinite photon numbers is used in each formalism that the respective photon number and annihilation operators have the forms expected of them. For example, vectors that are interpreted as negative photon number states must belong to the domain of the photon number operator, whereas vectors that are interpreted as infinite photon number states cannot belong to the domain of this operator. Thus the interpretations of negative and infinite photon number states are necessarily associated with the NBP and the new formalisms, respectively.

Even so, one may ask how the concept of a state of a negative number of photons differs from that of a state of an infinite number of photons when neither can be occupied physically. We note that both negative and infinite photon number states are orthogonal to states in H and so the distance [based on the norm in Eq. (12)] between any state in H and a negative or infinite photon number state is $\sqrt{2}$. Thus neither negative nor infinite photon number states are approached closely by states in H. Despite this we can describe a procedure, which is based on the limiting procedure of the Pegg-Barnett formalism, that provides a physical interpretation of the states of infinite photon number. Imagine that a field is prepared in the photon number state $|n\rangle$, where n is some given positive integer, and that the physical properties of the field are calculated accordingly. Imagine further that this is repeated for an infinite sequence of increasing values of n. We note that this procedure involves a sequence of states that does not converge strongly to any vector in H; in fact, it can be shown [16] that the sequence is represented by a vector in a space that is *larger* (i.e., more general) than H. The results of some of the calculations of the physical properties will converge as $n \to \infty$, e.g., $\langle(\hat{N}+1)^{-1}\rangle \to 0$ and $\langle\Delta\hat{\phi}_\theta{}^2\rangle = \pi^2/3$ independent of n. The *limit points* of these results represent the properties of a *state of infinite photon number*. This procedure therefore provides a concrete interpretation of states of infinite photon number. In contrast, there is no analogous interpretation of negative photon number states.

Moreover, unlike states of negative photon number, states that have a divergent mean photon number already appear in the conventional approach, which uses the Hilbert space H to represent the state of the field mode. For example, the state $|B\rangle = \sum(n+1)^{-1}|n\rangle$ which belongs to H has a divergent mean photon number. Thus, in this respect, our use of the state $|\widetilde{\infty}\rangle$ is not exceptional [34].

This brings us to an important point regarding the interpretation of quantum phase. We note that if the initial state of the field mode is represented by a vector in H, then in both the NBP and the present formalisms it remains in H for all times under physical interactions. Despite the differences mentioned above, the calculations in both formalisms will yield the same physical results because the corresponding operators are identical for vectors in H. The approach taken in this paper, however, gives a new interpretation of these physical results in terms of Hilbert-space operators which are defined on non-negative photon number states only. Thus it gives *a more conventional interpretation of the representation of quantum phase on an infinite-dimensional Hilbert space.*

We have used the operator sequences from the Pegg-Barnett formalism [13] as a basis for building phase operators on the space H_{sym}. This gives a close relationship between the formalism presented here and the Pegg-Barnett formalism. In particular, the calculations in both give the same physical results. The feature that distinguishes the two is that we have taken *the limit of operator sequences on an infinite-dimensional space*, whereas the corresponding limit in the Pegg-Barnett formalism is taken only of *expectation values calculated on finite-dimensional spaces.* Let us examine this point more closely. Our use of an infinite-dimensional Hilbert space places a restriction, which is absent from the Pegg-Barnett formalism, on the vectors we can treat, as shown by Eq. (12). For example, the vector $|(s+1)/2\rangle_s$, which is easily handled in the Pegg-Barnett formalism in the infinite-s limit, does not have a corresponding representation in H_{sym}. This indicates that the Pegg-Barnett formalism is a more *general* formalism in that it allows more-general vectors [16]. On the other hand, by relaxing the generality a little we are able to retain the physically important features of the Pegg-Barnett formalism and omit fine mathematical features of no physical value in the problem at hand.

More importantly, the existence of the strong limits of the Pegg-Barnett phase operators on H_{sym} means that infinite-rank operators on a Hilbert space can be used directly to represent observables of the field. Since the $s \to \infty$ limit is taken of the operators and vectors before calculating expectation values, the usual mathematical tools associated with infinite-dimensional Hilbert space can be used for the analysis of a system involving a field mode. For example, the equation of motion can be expressed in terms of the standard quasiprobability distributions when the field is in a state represented in H. There is no need for truncating the number state expansion at a finite value of s. This gives the present formalism an important advantage—*it allows the use of conventional mathematical tools.*

We conclude this section with the observation that while the formalism presented here differs from both the NBP and Pegg-Barnett formalisms, nevertheless it can also be viewed as a hybrid containing various features of these related formalisms: an extended infinite-dimensional Hilbert space like that used in the former and the non-negative photon number spectrum and the recipe for canonical conjugate observables of the latter.

V. DISCUSSION AND CONCLUSION

We have introduced here a simple formalism for the quantum phase based on an infinite-dimensional Hilbert space H_{sym} and the strong limits of the Pegg-Barnett operator sequences. An interesting property of H_{sym} is that it is symmetrical in the sense that it contains vectors in the neighborhood of both the vacuum state and the infinite photon number state. This allows it to support Hermitian and unitary phase operators which are canonically conjugate to the photon number operator. Moreover, the space H_{sym} is spanned by non-negative photon number states only. Thus the Hermitian phase operator is the solution to the phase problem defined in the Introduction. The formalism also yields the polar decomposition of the annihilation operator as sought originally by Dirac. We noted how the unitary phase operator in the decomposition operates on a larger space than the annihilation operator itself. This is the reason why the search for unitary phase operators on the infinite-dimensional Hilbert space H conventionally used in quantum optics has been unsuccessful: a larger space is necessary.

In fact, essentially the only different piece here that is not in the conventional treatment of the single mode field is the particular method for dealing with states of infinite photon number. However, states with a divergent mean photon number are not new to quantum optics. We have already noted that the state $\sum(n+1)^{-1}|n\rangle$ belongs to the conventionally used Hilbert space H and has a divergent mean photon number. Also the eigenstates of the field quadrature operators have a divergent mean photon number. Thus our use of *extra* states with a divergent mean photon number is not exceptional [34].

Let us elaborate this point a little further since it is crucial for a complete understanding of the present formalism. Any given physical realization of a single-mode field in a cavity entails basic physical constraints that arise from considerations such as the physical structure of the cavity mirrors, etc., and which are important in the limit of extremely intense fields. These constraints limit the number of photons physically possible in a field mode because, e.g., exceeding the limit would damage the mirrors, etc. For the case of a single mode of the field propagating in free space (where periodic boundary conditions replace the mirrors) one may argue that there are physical constraints associated with, for example, the physical nature of the (remote) source of the energy of the field [13]. However, it is clear that these constraints, whatever their origin, are not included in the conventional theoretical model of the ideal single field mode because, from the commutator $[\hat{a},\hat{a}^\dagger]=1$, we find that the spectrum of the number operator $\hat{N}=\hat{a}^\dagger\hat{a}$ is unbounded above and the sequence of photon number states is infinite. This leaves us with two options: either (a) ignore the physical constraints and retain the conventional commutator $[\hat{a},\hat{a}^\dagger]=1$ or (b) include the physical constraints and adopt a modified commutator $[\hat{a},\hat{a}^\dagger]\neq 1$.

In option (b) the modified commutator gives a modified photon number operator. The phase operator is found as the operator which is canonically conjugate to the modified photon number operator. The Pegg-Barnett formalism, without the infinite-dimensional limit, provides operators of this type [13].

Alternatively, for option (a), which is the one implicitly chosen here, we must use an infinite-dimensional space that supports the commutator $[\hat{a},\hat{a}^\dagger]=1$, that is, a space that contains the domain of $\hat{a}^\dagger\hat{a}$. Many such spaces can be found with H and H_{sym} being the particular examples considered here. Which of these spaces is the more appropriate physically for the model of the field mode? The obvious difference between H and H_{sym} is that H_{sym} contains extra states, such as $|\widetilde{\infty}\rangle$, of infinite photon number. However, we note that the commutator itself implies that the spectrum of $\hat{N}$ is necessarily unbounded and also that both H and H_{sym} contain states, such as $\sum(1+n)^{-1}|n\rangle$, that have a divergent mean photon number. Thus the unboundedness of the photon number does not provide any grounds for regarding either H or H_{sym} as being the more appropriate space. On the other hand, there is an important physical distinction between these spaces on the basis of phase since H_{sym} supports phase operators that attribute the vacuum state to a random phase, whereas H does not support such operators. Indeed all evidence suggests that the space H supports only operators that attribute the vacuum with a nonrandom phase. Thus H_{sym} is clearly the more appropriate Hilbert space for the model of a single-mode field.

There is one last point we wish to make. The *lower bound* of the photon number spectrum has been held responsible for the difficulties previously encountered when attempting to define a well-behaved phase observable on H. Our analysis here (see also [27]) shows, however, that it is the *large photon number behavior* of H that forbids the definition of appropriate phase operators. The lower bound of the photon number spectrum is simply a physical feature that distinguishes the energy of the harmonic oscillator from, say, the angular momentum of a bead on a circular wire. In short, the inability to define appropriate phase operators on H results from the mathematical structure of H and *not* the physical nature of the energy spectrum or the field mode itself. The present formalism overcomes these difficulties by adopting the larger Hilbert space H_{sym} as the state space.

In conclusion, we have presented a simple formalism for describing the quantum phase that is more conventional than the NBP approach and less general than the Pegg-Barnett formalism. It supports the limit points of the Pegg-Barnett operators on an infinite-dimensional Hilbert space and thus allows the use of a Hermitian phase operator as well as the conventional mathematical tools for the analysis of the single-mode field.

ACKNOWLEDGMENTS

I wish to thank Professor H. Paul, Dr. U. Leonhardt, Dr. A. Orlowski, and Dr. A. Wünsche for helpful comments. I am also grateful to Professor D.T. Pegg and Dr. S.M. Barnett for previous discussions concerning the general nature of quantum phase and to Dr. R.F. Bonner for his help with various mathematical constructions associated with this subject. This work was supported by the Max-Planck-Gesellschaft.

[1] See, e.g., papers in Phys. Scr. **T48** (1993), special issue on quantum phase and phase dependent measurements, edited by W.P. Schleich and S.M. Barnett.
[2] P.A.M. Dirac, Proc. R. Soc. London Ser. A **114**, 243 (1927); see also F. London, Z. Phys. **37**, 915 (1926); **40**, 193 (1927).
[3] W.H. Louisell, Phys. Lett. **7**, 60 (1963).
[4] L. Susskind and J. Glogower, Physics **1**, 49 (1964).
[5] P. Carruthers and M.M. Nieto, Rev. Mod. Phys. **40**, 411 (1968).
[6] E.C. Lerner, Nuovo Cimento B **56**, 183 (1968); **57**, 251 (1969).
[7] J.C. Garrison and J. Wong, J. Math. Phys. **11**, 2242 (1970).
[8] V.N. Popov and V.S. Yarunin, Leningrad Univ. J.: Phys. **22**, 7 (1973); J. Mod. Opt. **39**, 1525 (1992).
[9] H. Paul, Fortschr. Phys. **22**, 657 (1974).
[10] J.-M. Lévy Leblond, Ann. Phys. (N.Y.) **101**, 319 (1976).
[11] R.G. Newton, Ann. Phys. (N.Y.) **124**, 327 (1980).
[12] S.M. Barnett and D.T. Pegg, J. Phys. A **19**, 3849 (1986).
[13] D.T. Pegg and S.M. Barnett, Europhys. Lett. **6**, 483 (1988); Phys. Rev. A **39**, 1665 (1989); S.M. Barnett and D.T. Pegg, J. Mod. Opt. **36**, 7 (1989).
[14] J.H. Shapiro and S.R. Shepard, Phys. Rev. **43**, 3795 (1991).
[15] M. Ban, J. Opt. Soc. Am. B **7**, 1189 (1992); Opt. Commun. **94**, 231 (1992).
[16] J.A. Vaccaro and R.F. Bonner, Phys. Lett. **198A**, 167 (1995).
[17] C.W. Helstrom, Int. J. Theor. Phys. **11**, 357 (1974).
[18] U. Leonhardt, J.A. Vaccaro, B. Böhmer, and H. Paul, Phys. Rev. A **51**, 84 (1995).
[19] J.W. Noh, A. Fougères, and L. Mandel, Phys. Scr. **T48**, 29 (1993).
[20] We note that there are also methods that use a tensor product of the state space of the field mode with that of an apparatus or other device. However, we are restricting our attention here to descriptions of phase operators on the state space of the field mode alone.
[21] S. Stenholm, Phys. Scr. **T48**, 77 (1993).
[22] A. Luckš and V. Peřinová, Phys. Scr. **T48**, 94 (1993).
[23] The basic idea underlying this approach was suggested by S.M. Barnett (private communication).
[24] One can compare this vector with the point at the "north pole" of a sphere which maps to the point at infinity in the stereographic projection of the sphere onto the tangent plane at the "south pole."
[25] We note that the Pegg-Barnett formalism [13] itself does not suffer from this restriction since no infinite-dimensional Hilbert space is constructed nor is necessary in their formalism.
[26] For example, we can set the phase window to be $[\theta_0, \theta_0 + 2\pi)$ with $\theta_0 = 2\pi p/q$, where p and q are fixed positive integers, and then let $(s+1)$ be a multiple of $2q$. This allows θ_0 to be 2π times an *arbitrary* rational number. The expression for $\exp(i\hat{\phi})$ and the results which follow remain unchanged.
[27] J.A. Vaccaro and D.T. Pegg, Phys. Scr. **T48**, 22 (1993).
[28] I. M. Gel'fand and G.E. Shilov, *Generalized Functions* (Academic, New York, 1964), Vol. 4.
[29] A. Böhm, *The Rigged Hilbert Space and Quantum Mechanics* (Springer-Verlag, Berlin, 1978).
[30] A concise introduction to the rigged Hilbert space is given in L.E. Ballentine, *Quantum Mechanics* (Prentice-Hall, Englewood Cliffs, NJ, 1990).
[31] See, e.g., W.H. Louisell, *Quantum Statistical Properties of Radiation* (Wiley, New York, 1973).
[32] D.T. Pegg, J.A. Vaccaro, and S.M. Barnett, J. Mod. Opt. **37**, 1703 (1990).
[33] H. Weyl, *The Theory of Groups and Quantum Mechanics* (Dover, New York, 1950), Chap. IV, Sec. 14.
[34] We should also mention that the state $|\tilde{\infty}\rangle$ is, nevertheless, quite different from states in H with a divergent mean photon number such as $|B\rangle = \sum(n+1)^{-1}|n\rangle$. For example, in the Copenhagen interpretation each measurement of the photon number of the state $|B\rangle$ yields a finite number of photons, say n, with probability proportional to $(n+1)^{-2}$; the mean photon number of a finite sample of such measurements diverges as the sample size increases. In contrast, each corresponding measurement of $|\tilde{\infty}\rangle$ is associated with an infinity of photons.
[35] We note that in Refs. [11,12] the polar decomposition is found for the operator equivalent to that given in Eq. (19), which operates on a dense subset of H_{sym}. This operator is very different from the annihilation operator $\hat{a}$, which operates on a dense subset of H.
[36] Newton also considered the case where the Hamiltonian has degenerate eigenvalues.

4

Phase Dynamics and Uncertainties

As soon as it was introduced, the Pegg-Barnett phase formalism was applied to the analysis of quantum phase in a variety of situations. We review the seminal studies in this chapter. We begin with the analyses of special classes of states whose phase properties are optimised in some way, for example, by having the minimum variance of the phase angle subject to various constraints. We then look at the connection between the phase observable and amplitude quadrature observables of an optical field. We end with the dynamics of the phase of a single-mode optical field under diffusive and non-diffusive evolution. These studies paved the way for the application of the new formalism to the theoretical analysis of quantum phase in general.

The non-commutability of observables, in essence, sets quantum physics apart from classical physics. The difference is quantified by the Robertson-Heisenberg uncertainty relation, i.e.

$$\Delta A \Delta B \geq |\langle \hat{C} \rangle|/2 \tag{4.1}$$

for arbitrary Hermitian operators $\hat{A}$, $\hat{B}$ and $\hat{C}$ satisfying $[\hat{A}, \hat{B}] = i\hat{C}$, in contrast to the expression $\Delta A \Delta B \geq 0$ for the corresponding classical quantities. Here the uncertainty ΔQ in $\hat{Q}$ is the standard deviation

$$\Delta Q = \sqrt{\langle \hat{Q}^2 \rangle - \langle \hat{Q} \rangle^2}\,.$$

The uncertainty relation (4.1) distinguishes special classes of states. The states that give the minimum of the uncertainty product $\Delta A \Delta B$ are called *minimum-uncertainty states* and are important because they represent the closest to the classical case in this sense. The minimum here is taken over a specified set of states, which is usually the whole Hilbert space of states. There is another group of states, called *intelligent states* by Aragone *et al.* (1974, 1976), which gives the equality in Eq. (4.1). Intense coherent states were shown to be approximate intelligent states for the pair of observables $(\hat{N}, \hat{\phi}_\theta)$ in Barnett and Pegg (1989 **Paper 2.3**). The first detailed study of the intelligent and minimum uncertainty states associated with the number and phase observable pairs $(\hat{N}, \hat{\phi}_\theta)$, $(\hat{N}, \sin\hat{\phi}_\theta)$ and $(\hat{N}, \cos\hat{\phi}_\theta)$ was undertaken a short time later by Vaccaro and Pegg (1990b **Paper 4.4**).[1] The photon number states were found to be the only *exact* physical intelligent and minimum uncertainty states associated with all three pairs of observables. Approximate physical intelligent states were also found. A state which has Gaussian number-state coefficients and satisfies $\langle \hat{N} \rangle^2 \gg (\Delta N)^2 \gg 1$ is found to be an approximate intelligent state for all three observable pairs. These states can have super-Poisson $(\Delta N)^2 > \langle \hat{N} \rangle$ as well as sub-Poisson photon $(\Delta N)^2 < \langle \hat{N} \rangle$ statistics.

[1] The definition of the minimum uncertainty states in Vaccaro and Pegg (1990b **Paper 4.4**) was formally restricted to the set of intelligent states, that is, a minimum uncertainty state in this context is the intelligent state that yields the minimum uncertainty product. However, this restriction does not change the class of minimum uncertainty states as the minimum uncertainty states for the larger set of all physical states are, in fact, intelligent states.

Jackiw (1968) had previously found the intelligent states associated with the Susskind-Glogower operators; these states are shown to be approximate intelligent states for the pairs $(\hat{N}, \cos\hat{\phi}_\theta)$ and $(\hat{N}, \sin\hat{\phi}_\theta)$ to order $1/\langle\hat{N}\rangle^2$, that is, in the intense field limit.

A quite different class of special states is found by minimising the phase uncertainty subject to various constraints. If the constraints are chosen in terms of restrictions on the energy, i.e. the photon number, then the resulting states represent the optimum phase resolution for specified energy resources. Summy and Pegg (1990 **Paper 4.5**) examined this problem for two such constraints where either the number-state expansion of the state has an upper cutoff at some fixed photon number, or the mean photon number $\langle\hat{N}\rangle$ is fixed at some value. They call the resultant states *phase optimised states*. In the first case the optimum states for $\langle\hat{N}\rangle$ greater than 10 or so have number-state coefficients that are approximately sinusoidal and the phase resolution is

$$\Delta\phi \approx \frac{\pi}{2\langle\hat{N}\rangle}. \tag{4.2}$$

For the second case the number-state coefficients are given approximately by the Airy function for a similar range of $\langle\hat{N}\rangle$ values and have the phase resolution

$$\Delta\phi \approx \frac{1.3}{\langle\hat{N}\rangle}. \tag{4.3}$$

These results are a vast improvement over those for an intense coherent state for which

$$\Delta\phi \approx \frac{1}{2\sqrt{\langle\hat{N}\rangle}}. \tag{4.4}$$

We now turn to studies of the connection between the phase observable and amplitude quadrature observables of an optical field. In the Heisenberg picture, the dimensionless operator $\hat{X} = (\hat{a} + \hat{a}^\dagger)/2$ represents the field amplitude of an optical field mode at time $t = 0$. In a quarter period $\hat{X}$ evolves, under free evolution, to $\hat{Y} = (\hat{a} - \hat{a}^\dagger)/(2i)$, which represents the amplitude in *quadrature* with $\hat{X}$. These two operators had been associated with measured phase operators $\widehat{\cos\phi_{\mathrm{M}}} = \hat{X}/\sqrt{\overline{N}+1}$ and $\widehat{\sin\phi_{\mathrm{M}}} = \hat{Y}/\sqrt{\overline{N}+1}$ where $\overline{N} = \langle\hat{N}\rangle$ (Barnett and Pegg 1986 **Paper 1.7**). Indeed, the measured phase operators do approximate to the corresponding cosine and sine phase operators for a class of physical states with a broad photon number distribution (Vaccaro, Barnett and Pegg 1992 **Paper 4.6**). The minimum uncertainty states associated with the quadrature amplitude uncertainty relation $\Delta X \Delta Y \geq 1/4$ are called *ideal squeezed states, squeezed states* or, less frequently, two-photon coherent states (Yuen 1976, Caves 1981, Loudon and Knight 1987, Barnett and Radmore 1997). For these states the uncertainty in one amplitude operator is "squeezed" at the expense of increased uncertainty in the amplitude operator in quadrature, e.g. $\Delta X < 1/2$ and $\Delta Y > 1/2$. The phase properties of these states are therefore important for understanding the connection between the quadrature amplitudes and the phase. While a number of authors (Sanders, Barnett and Knight 1986, Yao 1987, Fan and Zaidi 1988) had examined squeezed states with respect to the Susskind and Glogower phase operators, Vaccaro and Pegg (1989 **Paper 4.1**) were the first to analyse the phase properties of these states using the new phase formalism. The difference between Vaccaro and Pegg's analysis and the earlier ones were largest for states of weak intensity, e.g. for the weakly-squeezed vacuum state. The reason for this is that the differences between the corresponding Susskind-Glogower and Pegg-Barnett exponential phase operators involve the vacuum state as illustrated, for example, by Eqs. (1.15) and (2.7). The phase probability density for the squeezed vacuum state was found to be bimodal with two peaks separated by an angle of π. In the limit of infinite squeezing the peaks become infinitely sharp. In this limit the phase properties of the squeezed vacuum are identical to the superposition of two phase states whose phase is separated by an angle of π. This later result clearly shows a strong connection between the squeezed states and phase states.

The connection between squeezing and phase fluctuations was further explored in Vaccaro, Barnett and Pegg (1992 **Paper 4.6**). This work first examines the phase fluctuations of a squeezed vacuum state. As noted above, the phase phase properties of this state approach those of a superposition of two phase states which are separated in phase angle by π as the squeezing increases. In particular, the uncertainties $\Delta\cos^{2n}(\phi_\theta)$ and $\Delta\sin^{2n}(\phi_\theta)$ of the even powers of the trigonometric phase operators $\cos(\hat{\phi}_\theta)$ and $\sin(\hat{\phi}_\theta)$ vanish for both the squeezed and phase superposition states in the limit of infinite squeezing. However, the quadrature uncertainties of the phase superposition state are $\Delta X > 1/2$ and $\Delta Y > 1/2$ and do not approach those of a squeezed state. This highlights an *asymmetry* between reducing phase fluctuations and squeezing the field amplitude: states with minimal amplitude fluctuations have minimal fluctuations in the phase operators $\cos^{2n}(\hat{\phi}_\theta)$ and $\sin^{2n}(\hat{\phi}_\theta)$, but states with minimal fluctuations in $\cos^{2n}(\hat{\phi}_\theta)$ and $\sin^{2n}(\hat{\phi}_\theta)$ do not necessarily have minimal fluctuations in the electric field. The paper then examines the converse situation by calculating the fluctuations of the electric field for states with well-defined phase, that is, a state that approaches a phase state to any desired degree of accuracy. The field amplitude for these states is found to change sign at each half period. As the phase resolution increases and the state approaches a phase state, the expectation value of the field amplitude diverges and the times of the sign change become more precise. However, the quadrature amplitudes are found to exhibit no squeezing, i.e. $\Delta X \geq 1/2$ and $\Delta Y \geq 1/2$. These results are consistent with Loudon's analysis (Loudon 1973 **Paper 1.6**) of his phase states as discussed in Chapter 1, and illustrated in Eq. (1.24) and Fig. 2. The precisely defined time of the field amplitude's sign change confirms the phase-like nature of phase states. The use of states with well-defined phase in a quantum clock is also discussed.

The difference between the phase of two harmonic oscillators was first discussed within the Pegg-Barnett formalism in Pegg and Barnett (1989 **Paper 2.4**). A more extensive study of both the phase difference as well as the phase sum was later developed in Barnett and Pegg (1990b **Paper 4.3**). In the latter it is shown that in addition to the simple combinations $\hat{\phi}_{\theta a} - \hat{\phi}_{\theta b}$ and $\hat{\phi}_{\theta a} + \hat{\phi}_{\theta b}$, more-natural definitions of the phase sum and difference operators between two modes are where the result is modulo-(2π), e.g. in terms of the phase states of each mode, the operators are given by

$$(\hat{\phi}_{\theta a} \pm \hat{\phi}_{\theta b})_{2\pi} = \sum_{m_a=0}^{s} \sum_{m_b=0}^{s} [\theta_{m_a} \pm \theta_{m_b}]_{2\pi} |\theta_{m_a}\rangle\langle\theta_{m_a}| \otimes |\theta_{m_b}\rangle\langle\theta_{m_b}| \tag{4.5}$$

where a and b label the two modes and $[\cdots]_{2\pi} \equiv [\cdots] + n2\pi$ for integer n satisfying $\theta_0 < [\cdots]_{2\pi} \leq \theta_0 + 2\pi$. Indeed, this follows naturally from the phase angle operator itself which is, in essence, defined modulo-(2π) as manifest in Eq. (2.38). Moreover, unless a phase reference provided by a third mode is available for comparison, one cannot distinguish between a phase difference of α or $\alpha + 2\pi$ between the two modes. The phase probability distributions associated with the modulo-(2π) operators in Eq. (4.5) are also introduced. As an example application, the properties of the phase sum and phase differences of the two-mode squeezed vacuum are then examined. These states are two-mode extensions of the single mode squeezed states introduced by Caves and Schumaker (1985). The two modes become highly correlated in photon number and quadrature amplitude as the squeezing is increased (Milburn 1984, Barnett and Knight 1987). In Barnett and Pegg (1990b **Paper 4.3**) the two modes are shown to also become highly correlated in phase sum in the same regime.

The first application of the new formalism to the study of a dynamical system was by Barnett, Stenholm and Pegg (1989 **Paper 4.2**). The system was a single-mode optical field undergoing *phase-insensitive* linear amplification or attenuation. The phase probability density $P_{\mathcal{P}}(\theta)$ was shown to exhibit uniform diffusion of the phase, where the rate of diffusion is inversely proportional to the mean photon number of the field, under the condition that the intensity fluctuations of the field are small compared to the mean intensity. An intense coherent field is an example that satisfies this condition. This behaviour was compared to that for quasiprobability representations of the problem. By adopting a polar coordinate representation, quasiprobability distributions give some semblance to a classical amplitude-phase representation, and hence a crude representation of phase. The diffusion of the angular coordinate of the Wigner function was found to be equal to the rate of diffusion of $P_{\mathcal{P}}(\theta)$. On the one hand this shows that the dynamics of $P_{\mathcal{P}}(\theta)$ agrees with the phase-space picture provided by the Wigner function. However, on the other hand, care must be taken not to take this agreement too far and *identify* the angular coordinate of the Wigner function as the phase observable. Indeed, for some states the radially-integrated Wigner function is negative which means that the radial coordinate cannot represent a physical observable at all (Garraway and Knight 1992).

The uniform phase diffusion in this system is not unexpected: the linearity of the dynamics ensures that the amplifier adds noise to the field, and as the amplifier is phase-insensitive, the noise is distributed uniformly over all phase angles. A quite different situation arises for *phase-sensitive* linear amplifiers (Caves 1982). In this case the noise added by the amplifier can be quadrature squeezed. In particular, a phase-sensitive linear attenuator can actually produce a quadrature squeezed output from a vacuum input (Vaccaro and Pegg 1986, Dupertuis, Barnett and Stenholm 1987, Milburn, Steyn-Ross and Walls 1987). That is, the device can produce a field with a non-uniform phase distribution from one with a uniform phase distribution. This implies that the effect on the phase need not be diffusive. Indeed, Vaccaro and Pegg (1994a) showed this non-diffusive behaviour occurs for both phase-sensitive linear amplifiers and attenuators in the weak fields regime. Here, weak field implies the overlap of the field state with the vacuum state and the one photon state are of order unity and ϵ, respectively, where $\epsilon \ll 1$, and the overlap with any other state is of order ϵ^2 or smaller. The analysis was carried out to first order in ϵ.

The analysis was further extended in Vaccaro and Pegg (1994b **Paper 4.7**) to a broader regime for which the mean photon number is at least of order 10. Both phase-insensitive and phase-sensitive amplifiers are found to give time-dependent phase diffusion. The time dependence here is of the same kind found in Barnett, Stenholm and Pegg (1989 **Paper 4.2**) as the mean photon number grows exponentially in time. However, for the phase-sensitive amplifiers, the diffusion is *non-uniform* in the sense that the rate of diffusion depends on the phase angle. That is, compared to the phase-insensitive case, the added amplifier phase noise is reduced for some phase angles at the expense of increased noise at others. This means that the phase sensitivity of the amplifier can be exploited to give higher power gains for the same loss in phase resolution and has application in phase-modulated optical communication.

PHASE PROPERTIES OF SQUEEZED STATES OF LIGHT

J.A. VACCARO and D.T. PEGG
School of Science, Griffith University, Nathan, Brisbane 4111, Australia

Received 28 November 1988

Recently introduced unitary and hermitian phase operators are used to examine the phase properties of squeezed states of light with particular reference to the squeezed vacuum. The results differ markedly from previous calculations involving the Susskind and Glogower operators. The new formalism allows the construction of a phase probability density which, on a polar diagram, is a circle for the vacuum state, becomes elliptical with gentle squeezing and collapses to a line through the origin for full squeezing. This probability density together with the calculation of expectation values of various trigonometrical functions of phase show how squeezing impresses phase information onto the vacuum.

1. Introduction

A considerable amount of interest has been focused on the theoretical and experimental study of squeezed states of light [1]. Squeezed light has reduced quantum fluctuations in the amplitude of one phase component of the field at the expense of increased fluctuations in the amplitude of the phase component in quadrature. Thus, squeezed states have phase sensitive noise and it is important therefore to examine the phase properties of these states. Sanders et al. [2] and more recently, Yao [3] and Fan et al. [4], began a study of this problem by determining particular phase properties of the ideal squeezed state. However, these authors have used the Susskind and Glogower phase formalism [5], the only formalism available at the time, which involves non-unitary phase operators and which leads to peculiar properties especially for fields in states of low excitation. Recently Pegg and Barnett [6–8] have introduced a new formalism based on a unitary phase operator which has properties coincident with those normally associated with phase.

In this paper,we re-examine the problem of the phase properties of squeezed light using the Pegg-Barnett formalism. We re-evaluate various calculations of previous authors and make comparisons with their work. Significant differences are found for fields of low excitation. The Pegg-Barnett formalism supports a phase probability density and a hermitian phase operator, neither of which has a consistent description in the Susskind-Glogower approach. We are able, therefore, to present for the first time the phase probability density for a weakly squeezed vacuum and the associated expectation values and variance of the hermitian phase operator. We also show that, to first order, the phase probability density is an ellipse whose axes are orientated in exactly the same way as those of the elliptical contours of the Q-function for the same state.

2. The unitary phase operator

The new formalism describing the phase properties of a single mode field, recently given by Pegg and Barnett [6–8], overcomes the difficulties associated with previous approaches, including that of Susskind and Glogower [5]. In this section we briefly review the new formalism which is based on a unitary phase operator, making comparisons where appropriate with the Susskind and Glogower approach which is based on non-unitary phase operators.

The unitary phase operator operates on a $(s+1)$ dimensional subspace Ψ spanned by the number states $|0\rangle$, $|1\rangle$, ... $|s\rangle$. The value of s can be made arbitrarily large. Expectation values are first calculated in Ψ before s is allowed to tend to infinity. This approach yields physical results which are indistinguishable from those obtained by using the infinite Hilbert space harmonic oscillator as a mathematical model of the single mode field [6,8]. A complete orthonormal basis of $(s+1)$ phase states is defined on Ψ as

$$|\theta_m\rangle \equiv (s+1)^{-1/2} \sum_{n=0}^{s} \exp(in\theta_m)\,|n\rangle\,, \tag{1}$$

with

$$\theta_m \equiv \theta_0 + 2m\pi/(s+1)\,, \tag{2}$$

and where $m=0, 1, ..., s$. The value of θ_0 is arbitrary; it defines a particular basis set of $(s+1)$ mutually orthogonal phase states. These states are eigenstates of the hermitian phase operator [7,8]

$$\hat{\phi}_\theta \equiv \sum_{m=0}^{s} \theta_m |\theta_m\rangle\langle\theta_m|\,. \tag{3}$$

We note that the eigenvalues of $\hat{\phi}_\theta$ are restricted to lie within a phase window between θ_0 and $\theta_0+2\pi$. This corresponds to choosing a 2π range in which to express the principal values of inverse trigonometric functions which are only defined modulo 2π. The unitary phase operator of which $|\theta_m\rangle$ is an eigenstate with eigenvalue $\exp(\mathrm{i}\theta_m)$ can be written as [6–8]

$$\exp(\mathrm{i}\hat{\phi}) \equiv |0\rangle\langle 1| + |1\rangle\langle 2| + ... + |s-1\rangle\langle s| + \exp[\mathrm{i}(s+1)\theta_0]\,|s\rangle\langle 0|\,, \tag{4}$$

and its hermitian conjugate is

$$[\exp(\mathrm{i}\hat{\phi})]^\dagger = \exp(-\mathrm{i}\hat{\phi})\,, \tag{5}$$

with the same set of eigenstates $|\theta_m\rangle$ but with eigenvalues $\exp(-\mathrm{i}\theta_m)$. Note that the unitary phase operator is the exponential of the hermitian phase operator $\hat{\phi}_\theta$ but we have dropped the subscript θ for convenience.

It can be shown from (4) that

$$\langle \exp(\mathrm{i}m\hat{\phi})\rangle = \langle [\exp(\mathrm{i}\hat{\phi})]^m\rangle$$

$$= \lim_{s\to\infty} \left\langle \left\{ \sum_{n=0}^{s-m} |n\rangle\langle n+m| + \exp[\mathrm{i}(s+1)\theta_0] \sum_{n=0}^{m-1} |s-n\rangle\langle m-1-n| \right\} \right\rangle$$

$$= \langle \widehat{\exp}(\mathrm{i}m\phi_{\mathrm{SG}})\rangle + \lim_{s\to\infty} \left\langle \left\{ \exp[\mathrm{i}(s+1)\theta_0] \times \sum_{n=0}^{m-1} |s-n\rangle\langle m-1-n| \right\} \right\rangle, \tag{6}$$

where the Susskind-Glogower phase operator is

$$\widehat{\exp}(\mathrm{i}m\phi_{\mathrm{SG}}) \equiv \sum_{n=0}^{\infty} |n\rangle\langle n+m|\,. \tag{7}$$

Expression (6) shows an essential difference between the properties of the unitary phase operator and the Susskind and Glogower phase operator. Another important difference which concerns us in this paper occurs when taking the expectation value of products of the phase operators with their hermitian conjugate. From unitarity we have

$$\langle \exp(\mathrm{i}\hat{\phi}) \exp(-\mathrm{i}\hat{\phi})\rangle = \langle \exp(-\mathrm{i}\hat{\phi}) \exp(\mathrm{i}\hat{\phi})\rangle\,,$$

while from the definitions (7) and

$$\widehat{\exp}(-\mathrm{i}m\phi_{\mathrm{SG}}) \equiv [\exp(\mathrm{i}m\phi_{\mathrm{SG}})]^\dagger\,,$$

it is well known that for $m=1$

$$\widehat{\exp}(\mathrm{i}\phi_{\mathrm{SG}})\,\widehat{\exp}(-\mathrm{i}\phi_{\mathrm{SG}})\rangle = 1\,,$$

$$\widehat{\exp}(-\mathrm{i}\phi_{\mathrm{SG}})\,\widehat{\exp}(\mathrm{i}\phi_{\mathrm{SG}})\rangle = 1 - \langle(|0\rangle\langle 0|)\rangle\,.$$

This last equation underlines the non-unitary behaviour of the Susskind-Glogower phase operator.

We will be mainly concerned in this paper with "physical states" $|\mathrm{f}\rangle$ [6]:

$$|\mathrm{f}\rangle = \sum_{n=0}^{\infty} c_n |n\rangle\,,$$

for which the mth moment of the number operator

$$\langle \mathrm{f}|\hat{N}^m|\mathrm{f}\rangle = \sum_{n=0}^{\infty} n^m |c_n|^2\,,$$

(and therefore of the energy) is finite for any given value of m. This imposes the necessary condition

$$\lim_{n\to\infty} |c_n| = 0\,. \tag{8}$$

Such a state can be prepared, in principle, from the vacuum state by interaction for a finite duration with a source whose energy is bounded above. Because of the restriction (8) the last term on the right hand side of equation (6) vanishes for physical states and the expectation value of the unitary phase operator then takes the following simplified form:

$$\langle \exp(im\hat{\phi}) \rangle_p = \left\langle \left[\sum_{n=0}^{\infty} |n\rangle\langle n+m| \right] \right\rangle_p \tag{9}$$

$$= \langle \widehat{\exp}(im\phi_{SG}) \rangle_p . \tag{10}$$

The subscript p refers to a physical state expectation value. For the cosine and sine phase operators defined by

$$\cos\hat{\phi} \equiv \tfrac{1}{2}[\exp(i\hat{\phi}) + \exp(-i\hat{\phi})] ,$$

$$\sin\hat{\phi} \equiv (2i)^{-1}[\exp(i\hat{\phi}) - \exp(-i\hat{\phi})] ,$$

we find

$$\langle \cos\hat{\phi} \rangle_p = \tfrac{1}{2}\langle \exp(i\hat{\phi}) + \exp(-i\hat{\phi}) \rangle_p \tag{11}$$

$$= \langle \widehat{\cos}\,\phi_{SG} \rangle_p , \tag{12}$$

with a similar correspondence for $\langle \sin\hat{\phi} \rangle_p$. Also we find

$$\langle \cos^2\hat{\phi} \rangle_p = \tfrac{1}{4}\langle \exp(i2\hat{\phi}) + \exp(-i2\hat{\phi}) + 2 \rangle_p \tag{13}$$

$$= \langle \widehat{\cos}^2\phi_{SG} \rangle_p + \tfrac{1}{4}\langle (|0\rangle\langle 0|) \rangle_p , \tag{14}$$

$$\langle \sin^2\hat{\phi} \rangle_p = -\tfrac{1}{4}\langle \exp(i2\hat{\phi}) + \exp(-i2\hat{\phi}) - 2 \rangle_p \tag{15}$$

$$= \langle \widehat{\sin}^2\phi_{SG} \rangle_p + \tfrac{1}{4}\langle (|0\rangle\langle 0|) \rangle_p . \tag{16}$$

3. Phase properties of squeezed light

In this paper we consider only the ideal squeezed state, which is also referred to as a two-photon coherent state [1,9]. The single mode ideal squeezed state $|\alpha, \xi\rangle$ can be generated mathematically from the vacuum state $|0\rangle$ by application of the squeeze operator $S(\xi)$ followed by the Glauber displacement operator $D(\alpha)$. That is,

$$|\alpha, \xi\rangle = D(\alpha)\, S(\xi)\, |0\rangle ,$$

where

$$S(\xi) = \exp[\tfrac{1}{2}(\xi^* a^2 - \xi a^{\dagger 2})] ,$$

$$D(\alpha) = \exp(\alpha a^\dagger - \alpha^* a) .$$

These states are physical states and so their phase properties in terms of the unitary phase operator can be found directly by using eqs. (9), (11), (13) and (15) while eqs. (10), (12), (14) and (16) allow a direct comparison with the recently published calculations involving the Susskind and Glogower phase operator [2–4].

We find from eq. (9)

$$\langle \alpha, \xi | \exp(im\hat{\phi}) | \alpha, \xi \rangle = \sum_{n=0}^{\infty} \langle \alpha, \xi | n \rangle \langle n+m | \alpha, \xi \rangle . \tag{17}$$

From Yuen [9]

$$\langle n | \alpha, \xi \rangle = (n!\mu)^{-1/2} (\nu/2\mu)^{n/2} H_n[\beta(2\mu\nu)^{-1/2}] \times \exp[-\tfrac{1}{2}|\beta|^2 + (\nu^*/2\mu)\beta^2] , \tag{18}$$

where $\mu = \cosh r$, $\nu = \exp(2i\eta)\sinh r$, $\beta = \alpha\cos r + \alpha^* \exp(2i\eta)\sinh r$, $\xi = |\xi|\exp(2i\eta)$, $r = |\xi|$, and where $H_n(x)$ is the nth order Hermite polynomial. Using this result in eq. (17) gives

$$\langle \alpha, \xi | \exp(im\hat{\phi}) | \alpha, \xi \rangle = \mu^{-1} \sum_{n=0}^{\infty} [n!(n+m)!]^{-1/2} \times (\nu^*/2\mu)^{n/2} (\nu/2\mu)^{(n+m)/2} \times H_n[\beta^*(2\mu\nu^*)^{-1/2}]\, H_{n+m}[\beta(2\mu\nu)^{-1/2}] \times \exp[-|\beta|^2 + (\nu^*/2\mu)\beta^2 + (\nu/2\mu)\beta^{*2}] . \tag{19}$$

On expanding the Hermite polynomials in (19) using

$$H_n(x) = \sum_{m=0}^{k} \frac{(-1)^m n! (2x)^{n-2m}}{m!(n-2m)!} ,$$

where k is the largest integer not exceeding $n/2$, we find exact agreement with the calculation of $\langle \alpha, \xi | \exp(im\phi_{SG}) | \alpha, \xi \rangle$ by Fan and Zaidi [4] in accordance with eq. (10).

We now consider the squeeze vacuum. This state is of particular interest in the exploration of the effect of squeezing on the phase because any non-random phase character of the squeezed vacuum must be attributable to the squeezing, and not the vacuum which, in the Pegg-Barnett formalism, has random phase. Using the following properties of $H_n(0)$,

$$H_{2n+1}(0) = 0 ,$$

$$H_{2n}(0) = \frac{(-1)^n (2n)!}{n!} ,$$

we find that for the squeezed vacuum $|0, \xi\rangle$

$$\langle 0, \xi| \exp[\mathrm{i}(2m+1)\hat{\phi}]\,|0, \xi\rangle = 0 \,, \tag{20}$$

$$\langle 0, \xi| \exp(\mathrm{i}2m\hat{\phi})\,|0, \xi\rangle = (-1)^m \exp(\mathrm{i}2m\eta)\, \mathrm{sech}\, r$$
$$\times \sum_{n=0}^{\infty} \left[\frac{(2n-1)!!\,(2n+2m-1)!!}{(2n)!!\,(2n+2m)!!}\right]^{1/2} (\tanh r)^{2n+m} \,. \tag{21}$$

In particular, for $m=1$, eq. (21) reduces to the result derived by Fan and Zaidi for the corresponding Susskind-Glogower phase operator, in accordance with eq. (10),

$$\langle 0, \xi| \exp(\mathrm{i}2\hat{\phi})\,|0, \xi\rangle = -\exp(2\mathrm{i}\eta)\, \mathrm{sech}\, r$$
$$\times \sum_{n=0}^{\infty} \frac{(2n-1)!!}{(2n)!!} \left(\frac{2n+1}{2n+2}\right)^{1/2} (\tanh r)^{2n+1} \,. \tag{22}$$

Using eqs. (11), (13), (15), (20) and (22) we find

$$\langle \cos\hat{\phi}\rangle = \langle \sin\hat{\phi}\rangle = 0 \,,$$

$$\langle \cos^2\hat{\phi}\rangle = \tfrac{1}{2}\Bigg[1 - \mathrm{sech}\, r \cos(2\eta)$$
$$\times \sum_{n=0}^{\infty} \frac{(2n-1)!!}{(2n)!!} \left(\frac{2n+1}{2n+2}\right)^{1/2} (\tanh r)^{2n+1} \Bigg] \,, \tag{23}$$

$$\langle \sin^2\hat{\phi}\rangle = \tfrac{1}{2}\Bigg[1 + \mathrm{sech}\, r \cos(2\eta)$$
$$\times \sum_{n=0}^{\infty} \frac{(2n-1)!!}{(2n)!!} \left(\frac{2n+1}{2n+2}\right)^{1/2} (\tanh r)^{2n+1} \Bigg] . \tag{24}$$

The difference between our result for $\langle\cos^2\hat{\phi}\rangle$ (or $\langle\sin^2\hat{\phi}\rangle$) and $\langle\cos^2\phi_{\mathrm{SG}}\rangle$ (or $\langle\sin^2\phi_{\mathrm{SG}}\rangle$) as calculated by Fan and Zaidi is

$$\tfrac{1}{4}\langle 0, \xi|0\rangle\langle 0|0, \xi\rangle = \tfrac{1}{4}\mathrm{sech}\, r \,,$$

as anticipated from eqs. (14) and (16). This difference is proportional to the probability of finding the squeezed vacuum in the vacuum state; it will be greatest for a weakly squeezed vacuum and least for a strongly squeezed vacuum, because the latter has only a small vacuum component.

For weak squeezing r is small and we find from (23) and (24) that

$$\langle 0, \xi| \cos^2\hat{\phi}\,|0, \xi\rangle = \tfrac{1}{2}[1 - r2^{-1/2}\cos(2\eta) + \Delta] \,, \tag{25}$$

$$\langle 0, \xi| \sin^2\hat{\phi}\,|0, \xi\rangle = \tfrac{1}{2}[1 + r2^{-1/2}\cos(2\eta) - \Delta] \,, \tag{26}$$

where Δ is a small term of order r^3. It is clear that as r tends to zero both of these values approach $\frac{1}{2}$ which is precisely in accord with the phase of the vacuum being random. This is not true for previous results [2–4]. On adding eqs. (25) and (26) we get

$$\langle 0, \xi| \cos^2\hat{\phi}\,|0, \xi\rangle + \langle 0, \xi| \sin^2\hat{\phi}\,|0, \xi\rangle = 1 \,, \tag{27}$$

which is in marked contrast to the analogous calculation [2–4] using the Susskind and Glogower phase operators,

$$\langle 0, \xi| \widehat{\cos}^2\phi_{\mathrm{SG}}\,|0, \xi\rangle + \langle 0, \xi| \widehat{\sin}^2\phi_{\mathrm{SG}}\,|0, \xi\rangle$$
$$= \tfrac{1}{2} + \tfrac{1}{4}r^2 + \mathrm{O}(r^3) \,.$$

The fact that the right hand side of (27) is unity is a direct consequence of using a formalism based on a unitary phase operator [6]. Eq. (27) also coincides with what would be expected from a well-defined phase probability density. We shall derive the phase probability density for the weakly squeezed vacuum in the next section.

For a strongly squeezed vacuum (r large), eqs. (23) and (24) used with the approximation technique of Fan and Zaidi [4] yield

$$\langle 0, \xi| \cos^2\hat{\phi}\,|0, \xi\rangle$$
$$\simeq \tfrac{1}{2}\left[1 - \cos(2\eta) \tanh r \left(1 - \frac{\cosh r - 1}{2\sinh^2 r} \right)\right] ,$$

$$\langle 0, \xi| \sin^2\hat{\phi}\,|0, \xi\rangle$$
$$\simeq \tfrac{1}{2}\left[1 + \cos(2\eta) \tanh r \left(1 - \frac{\cosh r - 1}{2\sinh^2 r} \right)\right] . \tag{28}$$

These results differ from that obtained [2,4] with the corresponding Susskind and Glogower operators by $\frac{1}{4}\mathrm{sech}\, r$ in accordance with eqs. (14) and (16). As r is made larger, this difference becomes smaller because of the diminished probability of finding the strongly squeezed field in the vacuum state diminishes. In the limit of very large r the difference between the phase properties of the ideal squeezed vacuum and the superposition of phase states $2^{-1/2}(|\eta+\frac{1}{2}\pi\rangle+|\eta-\frac{1}{2}\kappa\rangle)$ becomes negligible, although these states are not identical. This will be discussed in detail elsewhere but here we can see that for this superposition $\langle\cos\hat{\phi}\rangle$ vanishes and $\langle\cos^2\hat{\phi}\rangle=\cos^2(\eta+\frac{1}{2}\pi)$, to which expression (28) tends as r becomes large.

4. The phase probability density for the weakly squeezed vacuum

Because the vacuum state has random phase as discussed earlier, the phase probability density $P(\theta)$ will be uniform, that is, a circle in a polar representation. We shown in this section that $P(\theta)$ for a weakly squeezed vacuum is elliptical to first order in the squeezing parameter r, illustrating how squeezing impresses phase information on the vacuum. We also calculate the expectation value and variance of the hermitian phase operator for the weakly squeezed vacuum.

Using (18) we can expand the weakly squeezed vacuum as

$$|0,\xi\rangle=\mathrm{sech}^{1/2}r\,[\,|0\rangle-2^{-1/2}\exp(2\mathrm{i}\eta)\tanh r|2\rangle$$
$$+(3/8)^{1/2}\exp(4\mathrm{i}\eta)\tanh^2 r|4\rangle-...]\,,$$

which gives a probability of being found in the phase state $|\theta_m\rangle$ as

$$|\langle\theta_m|0,\xi\rangle|^2=(s+1)^{-1}[1-r2^{1/2}\cos 2(\theta_m-\eta)$$
$$+r^2(3/2)^{1/2}\cos 4(\theta_m-\eta)+\mathrm{O}(r^3)]\,.$$

The density of phase states is $(s+1)/(2\pi)$, so in the continuum limit as s tends to infinity the phase probability density becomes

$$P(\theta)=(2\pi)^{-1}[1-r2^{-1/2}\cos 2(\theta-\eta)$$
$$+r^2(3/2)^{1/2}\cos 4(\theta-\eta)+\mathrm{O}(r^3)]\,. \tag{29}$$

We note that $P(\theta)$ is normalised,

$$\int_{\theta_0}^{\theta_0+2\pi} P(\theta)\,\mathrm{d}\theta=1\,,$$

to order r^3, and that $P(\theta)$ has a minimum at $\theta=\eta+n\pi$ and a maximum at $\theta=\eta+(n+\frac{1}{2})\pi$.

When $P(\theta)$ is represented on a polar diagram with θ as the polar angle and P the radial distance, we find that, to first order in r, $P(\theta)$ traces out an ellipse with a major axis of length $(1+2^{1/2}r)/(2\pi)$ oriented as an angle of $\eta+\frac{1}{2}\pi$ to the x-axis and a minor axis of length $(1-2^{1/2}r)/(2\pi)$. We can compare this with the Q-function which has elliptical contours with axes of exactly the same orientation for the same state [9]. Although the polar diagram is not precisely elliptical to second order in r, it still exhibits a deviation from the circle which manifests the imposed phase information. When r becomes very large, our previous discussion of the equivalence between the squeezed vacuum and an equal superposition of two phase states differing by π predicts that the polar diagram collapses to a straight line through the origin at an angle of $\eta\pm\frac{1}{2}\pi$ to the x-axis.

It is now a simple matter to calculate the expectation value of any function of the phase operator $f(\hat{\phi}_\theta)$ for a weakly squeezed vacuum by

$$\langle f(\hat{\phi}_\theta)\rangle=\int_{\theta_0}^{\theta_0+2\pi} P(\theta)f(\theta)\,\mathrm{d}\theta\,,$$

where $P(\theta)$ is given by eq. (29). Of particular interest are the mean and variance of the hermitian phase operator itself for which there is no corresponding operator in the Susskind and Glogower formalism. We find

$$\langle\hat{\phi}_\theta\rangle=\theta_0+\pi-r2^{-1/2}\sin 2(\theta_0-\eta)$$
$$+r^2(3/32)^{1/2}\sin 4(\theta_0-\eta)+\mathrm{O}(r^3)\,,$$
$$\langle\hat{\phi}_\theta^2\rangle=\theta_0+2\pi\theta_0+(4/3)\pi^2$$
$$-r2^{-1/2}[2(\theta_0+\pi)\sin 2(\theta_0-\eta)+\cos 2(\theta_0-\eta)]$$
$$+r^2(3/128)^{1/2}[4(\theta_0+\pi)\sin 4(\theta_0-\eta)$$
$$+\cos 4(\theta_0-\eta)]+\mathrm{O}(r^3)\,,$$

giving

$$\langle\Delta\hat{\phi}_\theta^2\rangle=\langle\hat{\phi}_\theta^2\rangle-\langle\hat{\phi}_\theta\rangle^2$$
$$=(1/3)\pi^2-r2^{-1/2}\cos 2(\theta_0-\eta)$$
$$+\tfrac{1}{4}r^2\{[(3/8)^{1/2}+1]\cos 4(\theta_0-\eta)-1\}+\mathrm{O}(r^3)\,.$$

We note that these results are dependent on θ_0, that is the choice of the particular 2π phase window, and provide an illustration of a general effect described in refs. [7,8] and also by Judge [10] for periodic probability densities. Judge has suggested that the choice of window which minimizes the variance can be used to specify a unique mean and variance. However, in our case it turns out that this approach would lead to a uniquely specified variance but two possible values of the mean differing by π because the periodicity of the plasma probability density is π. Thus a choice of window is still required to determine the mean.

5. Conclusion

Using the unitary and hermitian phase operator [6–8] in place of the Susskind and Glogower phase operators permits an investigation of the phase properties of ideal squeezed states in terms of a phase probability density, which makes possible a calculation of the mean and variance not only of periodic functions of phase, but also of the phase itself. In this paper we have used these new operators to calculate various expectation values for ideal squeezed states, paying particular attention to the weakly and strongly squeezed vacuum. We have found that, where corresponding results are obtainable from the Susskind-Glogower operators, there are significant differences which are most marked for the weakly squeezed vacuum, which contains a large vacuum component. Use of the new phase operators clearly shows that as the vacuum is squeezed it changes from a state of completely random phase to one which contains an increasing amount of phase information.

Acknowledgement

D.T.P. thanks Dr. S.M. Barnett for earlier discussions on the properties of phase operators. J.A.V. acknowledges the support of a Commonwealth Postgraduate Award.

References

[1] For a review of squeezed light see e.g.: D.F. Walls, Nature 306 (1983) 141;
R. Loudon and P.L. Knight, J. Mod. Opt. 34 (1987) 709.
[2] B.C. Sanders, S.M. Barnett and P.L. Knight, Optics Comm. 58 (1986) 290.
[3] D. Yao, Phys. Lett. A 122 (1987) 77.
[4] Fan Hong-Yi and H.R. Zaidi, Optics Comm. 68 (1988) 143.
[5] L. Susskind and J. Glogower, Physics 1 (1964) 49.
[6] D.T. Pegg and S.M. Barnett, Europhysics Letters 6 (1988) 483.
[7] S.M. Barnett and D.T. Pegg, J. Mod. Opt., (1988) in press.
[8] D.T. Pegg and S.M. Barnett, Phys. Rev. A, (1989) in press.
[9] H.P. Yuen, Phys. Rev. A 13 (1976) 2226.
[10] D. Judge, Nuovo Cimento XXXI (1964) 332.

A NEW APPROACH TO OPTICAL PHASE DIFFUSION

S.M. BARNETT
Department of Engineering Science University of Oxford Parks Road Oxford OX1 3PJ UK

S. STENHOLM
Research Institute for Theoretical Physics University of Helsinki Siltavuorenpenger 20C SF-00170 Helsinki 17 Finland

and

D.T. PEGG
School of Science Griffith University Nathan Brisbane 4111 Australia

Received 16 May 1989

We use the phase states to analyze phase evolution in optical amplifiers and attenuators. The resulting diffusion of the phase probability distribution is compared with that obtained in conventional Fokker-Planck treatments.

1. Introduction

The operation of any amplifier or attenuator must add fluctuations to the output. The minimum noise level associated with these fluctuations is constrained by quantum mechanics [1–3]. In optical devices, this added noise leads to a suppression of any non-classical properties associated with the input field [4–6], although careful preparation of the gain (or loss) medium may reduce the degrading effect of this noise [7–15].

It is often convenient to decompose optical noise into amplitude and phase components. This is particularly true in the theory of the laser where loss, gain and saturation effects constrain the amplitude fluctuations but impose no limitation on the phase [16–19]. The fluctuations associated with the amplification and attenuation mechanisms cause the phase to diffuse away from its original value. This diffusion of the phase is usually quite slow but is the limiting mechanism determining the linewidth in single-mode lasers [17,18].

Previous analyses of phase diffusion have usually been based on either (i) a phenomenological decomposition of the Heisenberg operator equations into polar components or [16,20] (ii) a polar decomposition of a Fokker-Planck equation derived from the master equation for the field density matrix [16–18]. Amplifier phase noise has also been investigated using the Susskind-Glogower phase operators [5]. In this paper we introduce a new approach based on the optical phase states $|\theta\rangle$ defined by [21–24]

$$|\theta\rangle = (s+1)^{-1/2} \sum_{n=0}^{s} \exp(in\theta)\, |n\rangle \tag{1}$$

The $(s+1)$-dimensional linear space Ψ spanned by the first $(s+1)$ number states may also be spanned by the set of orthonormal phase states with phases

$$\theta_m = \theta_0 + \frac{2\pi m}{s+1}, \quad m=0, 1, \ldots, s \tag{2}$$

If we take the limit associated with these states seriously, then it is possible to define unitary exponential [22] and hermitian optical phase operators [23,24].

In the next section we describe our method and illustrate it with an analysis of phase diffusion in a simple linear amplifier. We find that the phase probability distribution (for a field with a low level of intensity fluctuations) will obey a simple diffusion equation. We compare this behaviour with that obtained from a Fokker-Plank analysis of the same model. Our method gives the same phase behaviour as may be derived from a Fokker-Planck analysis of the Wigner function.

2. Diffusion of the phase state probability distribution

The orthonormal phase states exhibit the physical properties expected of a quantum state of well-defined phase. It is natural to associate the diagonal field density matrix elements in a phase state basis with the phase probability distribution [23,24]

$$P(\theta)=\langle\theta|\rho|\theta\rangle=(s+1)^{-1}\sum_{n,n'}\rho_{n'n}\exp[\mathrm{i}(n'-n)\theta], \tag{3}$$

where $\rho_{n,n'}$ is the number state density matrix element $\langle n|\rho|n'\rangle$. The evolution of the density matrix will be determined by the relevant master equation. The dynamics of the phase state probability distribution may be derived from this master equation by selecting the required matrix elements. This procedure is analogous to the common technique of employing number state matrix elements to derive the dynamics of the photon number probability distribution.

The phase state based probability distribution may be used in discussing the phase dynamics of any optical system. Here, we illustrate its use in describing the phase dynamics associated with a simple linear amplifier. The master equation associated with such a device has the simple form [4,6,25]

$$\partial\rho/\partial t=A\,(a^{+}\rho a-\tfrac{1}{2}aa^{+}\rho-\tfrac{1}{2}\rho aa^{+})+C\,(a\rho a^{+}-\tfrac{1}{2}a^{+}a\rho-\tfrac{1}{2}\rho a^{+}a). \tag{4}$$

This master equation takes full account of both stimulated and spontaneous processes associated with the amplifier operation. The detailed form of the gain and loss rates (A and C respectively) depends on the specific amplifier model. In an inverted population amplifier A and C will be proportional to the gain medium ground and excited state populations respectively.

The master equation (4) leads to a simple equation for the number state matrix elements

$$\partial\rho_{nn'}/\partial t=A\,[\sqrt{nn'}\,\rho_{n-1\,n'-1}-\tfrac{1}{2}(n+n'+2)\,\rho_{nn'}]+C\,[\sqrt{(n+1)\,(n'+1)}\,\rho_{n+1\,n'+1}-\tfrac{1}{2}(n+n')\,\rho_{nn'}]. \tag{5}$$

Substitution of this equation into the expression for the phase state probability distribution (3) gives an equation for the evolution of the phase distribution

$$\frac{\partial P(\theta)}{\partial t}=-\frac{A}{2(s+1)}\sum_{n,n'}[\sqrt{(n+1)}-\sqrt{(n'+1)}]^{2}\rho_{nn'}\exp[\mathrm{i}(n-n')\theta]$$
$$-\frac{C}{2(s+1)}\sum_{n,n'}(\sqrt{n}-\sqrt{n'})^{2}\rho_{nn'}\exp[\mathrm{i}(n-n')\theta] \tag{6}$$

It is not easy to relate the right hand side of this equation to the phase probability distribution. This is because the amplitude and phase fluctuations associated with the amplifier do no decouple. A similar problem occurs in other analyses of phase diffusion [16–19]. An often employed solution to this problem is to assume that the intensity fluctuations are small compared to the mean value; that is the photon number probability distribution is sharply peaked (or narrow). This will be the case for a suitably chosen input state such as a moderately intense coherent state (for which $\Delta n/\langle n\rangle=\langle n\rangle^{1/2}$). It will also be true in above threshold devices where gain, loss and the (nonlinear) saturation restrict the intensity to a narrow range [16–19]. We restrict our analysis to states with a narrow photon number distribution and rewrite our phase diffusion equation in the form

$$\frac{\partial P(\theta)}{\partial t} = -\frac{A}{2(s+1)}\sum_{n,n'}\frac{(n-n')^2}{[\sqrt{(n+1)}+\sqrt{(n'+1)}]^2}\rho_{nn'}\exp[i(n-n')\theta]$$

$$-\frac{C}{2(s+1)}\sum_{n,n'}\frac{(n-n')^2}{(\sqrt{n}+\sqrt{n'})^2}\rho_{nn'}\exp[i(n-n')\theta]\,. \tag{7}$$

The presumed narrow photon number distribution allows us to replace the denominators in the above sums by $(4\langle n\rangle)$, where $\langle n\rangle$ is the mean photon number. The resulting equation for the evolution of the phase state probability distribution is

$$\frac{\partial P(\theta)}{\partial t} = -\frac{A+C}{8\langle n\rangle(s+1)}\sum_{n,n'}(n-n')^2\rho_{nn'}\exp[i(n-n')\theta]\,. \tag{8}$$

Direct comparison with the expression for the phase state probability distribution (3) shows that this equation is a diffusion equation for $P(\theta)$

$$\partial P(\theta)/\partial t = [(A+C)/8\langle n\rangle]\partial^2 P(\theta)/\partial\theta^2 \tag{9}$$

The diffusion coefficient associated with this phase diffusion is

$$D_\theta = (A+C)/8\langle n\rangle \tag{10}$$

If the phase probability distribution is initially sharply peaked then this is the rate at which the variance in the hermitian phase operator will grow (if we make a sensible choice of phase operator [23]). The approximation made in deriving this result from the exact expression (6) involves neglecting terms that are smaller than the above diffusion term by a factor of $\Delta n/\langle n\rangle$, where Δn is the width of the photon number probability distribution

3. Comparison with Fokker-Planck methods

A common and popular technique for dealing with master equations, such as (4), is to recast them as a partial differential equation for a c-number quasiprobability distribution (for the complex field amplitude α). The form of the so called Fokker-Planck equation will depend on the choice of quasiprobability distribution which in turn is determined by the choice of operator orderings. The P, Q and Wigner (W) functions will provide normally antinormally and symmetrically ordered expectation values respectively. This difference in operator orderings is reflected in the different forms of the associated Fokker-Planck equations and the different physical interpretations that are associated with them [26]. For the amplifier master equation (4) we find the Fokker-Planck equations

$$\frac{\partial P(\alpha)}{\partial t} = \frac{C-A}{2}\left(\frac{\partial}{\partial\alpha}\alpha + \frac{\partial}{\partial\alpha^*}\alpha^*\right)P(\alpha) + A\frac{\partial^2 P(\alpha)}{\partial\alpha\partial\alpha^*}\,, \tag{11a}$$

$$\frac{\partial W(\alpha)}{\partial t} = \frac{C-A}{2}\left(\frac{\partial}{\partial\alpha}\alpha + \frac{\partial}{\partial\alpha^*}\alpha^*\right)W(\alpha) + \frac{A+C}{2}\frac{\partial^2 W(\alpha)}{\partial\alpha\partial\alpha^*}\,, \tag{11b}$$

$$\frac{\partial Q(\alpha)}{\partial t} = \frac{C-A}{2}\left(\frac{\partial}{\partial\alpha}\alpha + \frac{\partial}{\partial\alpha^*}\alpha^*\right)Q(\alpha) + C\frac{\partial^2 Q(\alpha)}{\partial\alpha\partial\alpha^*}\,. \tag{11c}$$

Clearly the diffusion terms (the second order derivatives) depend on our choice of quasiprobability distribution

We can cast these Fokker-Planck equations into a suitable form for discussing phase diffusion by writing the complex amplitude (α) in polar form

$$\alpha = r\exp(i\theta) \tag{12}$$

With this change of variables the Fokker-Planck equations for the joint amplitude and phase quasiprobability distribution become

$$\frac{\partial P(r,\theta)}{\partial t} = \frac{C-A}{2r}\frac{\partial}{\partial r}r^2 P(r,\theta) + \frac{A}{4r^2}\left(r\frac{\partial}{\partial r}r\frac{\partial}{\partial r} + \frac{\partial^2}{\partial\theta^2}\right)P(r,\theta), \tag{13a}$$

$$\frac{\partial W(r,\theta)}{\partial t} = \frac{C-A}{2r}\frac{\partial}{\partial r}r^2 W(r,\theta) + \frac{A+C}{8r^2}\left(r\frac{\partial}{\partial r}r\frac{\partial}{\partial r} + \frac{\partial^2}{\partial\theta^2}\right)W(r,\theta), \tag{13b}$$

$$\frac{\partial Q(r,\theta)}{\partial t} = \frac{C-A}{2r}\frac{\partial}{\partial r}r^2 Q(r,\theta) + \frac{C}{4r^2}\left(r\frac{\partial}{\partial r}r\frac{\partial}{\partial r} + \frac{\partial^2}{\partial\theta^2}\right)Q(r,\theta) \tag{13c}$$

The phase evolution in these equations is clearly diffusive in nature. However, the diffusion rate depends on the amplitude of the field. If the field amplitude fluctuations are small (that is the marginal amplitude probability distribution is narrow) then we can replace $1/r^2$ by $\langle n\rangle^{-1}$. The phase diffusion coefficients associated with these quasiprobabilities may then be approximated by

$$D_\theta^{P} = A/4\langle n\rangle, \qquad D_\theta^{W} = (A+C)/8\langle n\rangle, \qquad D_\theta^{Q} = C/4\langle n\rangle, \tag{14a,b,c}$$

for the P, Wigner and Q functions respectively. The approximation made in deriving these differential equations involves neglecting terms that are smaller than the diffusion coefficient by a factor of $\Delta n/\langle n\rangle$. This is the same order of approximation as that involved in obtaining the phase state based result described in the previous section.

It is clear that our expression for the phase diffusion coefficient (10) is the same as that associated with the Wigner function (14b). This is not particularly surprising as the phase state probability distribution is associated with the hermitian optical phase operator. It is the Wigner function, rather than the normally or antinormally ordered P and Q functions, that reproduces the correct position (momentum) probability distribution when integrating over the conjugate momentum (position) variable. In the same way, it is the phase dynamics of the Wigner function that has reproduced the diffusive behaviour associated with the phase state probability distribution.

4. Conclusion

We have demonstrated the use of the diagonal phase state density matrix elements as a phase probability distribution. The action of an amplifier (or attenuator) on this distribution is to broaden it via diffusion. We have shown that the rate associated with this diffusion is the same as that obtained in a Fokker-Planck analysis of the Wigner function. This does not mean, of course, that $P(\theta)$ and the angular variation of the Wigner function are the same but simply that they undergo the same diffusive evolution.

The phase states allow us to construct an hermitian optical phase operator [23,24]. The existence of this operator should make it possible to re-analyze earlier phenomenological Heisenberg-picture treatments rigorously. We hope to return to this point elsewhere.

It is a pleasure to thank Claire Gilson for helpful discussions and suggestions S M B thanks G E C and the U K. Fellowship of Engineering for the award of a Senior Research Fellowship

References

[1] J.P. Gordon, L R. Walker and W H. Louisell, Phys. Rev. 130 (1963) 806
[2] H P. Yuen and J H. Shapiro, IEEE Trans. Inf. Theory IT-24 (1978) 657
[3] C M. Caves, Phys. Rev D26 (1982) 1817
[4] S. Friberg and L. Mandel, Optics Comm. 46 (1983) 141.
[5] R. Loudon and T J Shepherd, Optica Acta 31 (1984) 1243.
[6] S. Stenholm Physica Scripta T12 91986) 56.
[7] D T. Pegg and J.A Vaccaro, Optics Comm 61 (1987) 317
[8] G J Milburn, M L Steyn-Moss and D F. Walls, Phys Rev A 35 (1987) 4443
[9] M A Dupertuis and S Stenholm, J Opt Soc. Am. B 4 (1987) 1094
[10] M.-A. Dupertuis S M Barnett and S Stenholm J. Opt. Soc. Am B 4 (1987) 1102.
[11] M -A Dupertuis S M. Barnett and S. Stenholm. J. Opt. Soc Am. B 4 (1987) 1124
[12] J.A Vaccaro and D.t Pegg, J Mod Opt 34 (1987) 855.
[13] C.R Gilson S M Barnett and S Stenholm, J. Mod Opt. 34 (1987) 949
[14] M -A. Dupertuis and S Stenholm, Phys. Rev. 37 (1988) 1226
[15] S.M. Barnett and C R Gilson. (submitted to Phys. Rev A).
[16] H. Haken Handbuch der Physik (ed S Flugge, Springer-Verlag, Berlin) Vol XXV/2c (1970)
[17] W H. Louisell. Quantum statistical properties of radiation (Wiley, New York, 1973) p 469.
[18] M Sargent III, M O Scully and W E. Lamb Jr., Laser physics (Addision-Wesley London, 1974) p 281
[19] R. Loudon, The quantum theory of light. 2nd Edition (Oxford University Press, Oxford 1983) p. 278
[20] M Lax, Phys Rev 112 (1967) 290.
[21] R Loudon The quantum theory of light 1st Edition (Oxford University Press Oxford 1973) p 143
[22] D T Pegg and S M Barnett Europhys Lett. 6 (1988) 483
[23] S M Barnett and D T Pegg J Mod Opt 36 (1989) 7
[24] D T Pegg and S.M Barnett Phys Rev 39 (1989) 1665
[25] W H Louisell Quantum statistical properties of radiation (Wiley New York 1973) p 331
[26] M O Scully and S Stenholm Physica Scripta T21 (1988) 119

Quantum theory of optical phase correlations

Stephen M. Barnett
Department of Engineering Science, University of Oxford, Oxford OX1 3PJ, England

D. T. Pegg
Division of Science and Technology, Griffith University, Nathan, Brisbane 4111, Australia
(Received 19 April 1990)

We extend the theory of the Hermitian optical phase operator to analyze the quantum phase properties of pairs of electromagnetic field modes. The operators representing the sum and difference of the two single-mode phases are simply the sum and difference of the two single-mode phase operators. The eigenvalue spectra of the sum and difference operators have widths of 4π, but phases differing by 2π are physically indistinguishable. This means that the phase sum and difference probability distributions must be cast into a 2π range. We obtain mod(2π) probability distributions for the phase sum and difference that unambiguously reveal the signatures of randomness, phase correlations, and phase locking. We use our approach to investigate the phase sum and difference properties for uncorrelated modes in random and partial phase states and the phase-locked properties of the two-mode squeezed vacuum states. We reveal the fundamental property of two-mode squeezed states that the phase sum is locked to the argument of the squeezing parameter. The variance of the phase sum depends dilogarithmically on $1+\tanh r$, where r is the magnitude of the squeezing parameter, vanishing in the large squeezing limit.

I. INTRODUCTION

The quantum nature of the phase of the electromagnetic field is an old[1] and much studied problem.[2] Recently, we introduced a new approach to the analysis of the quantum optical phase that utilizes the Hermitian optical phase operator.[3-5] This operator does not exist in the conventional harmonic-oscillator infinite Hilbert space, but rather in a state space Ψ of finite but arbitrarily large dimension $(s+1)$. The procedure involves calculating c numbers such as expectation values, variances, and other properties of the field as functions of s *before* allowing s to tend to infinity. Our approach avoids indeterminacies which are inherent in the conventional approach employing an infinite Hilbert space from the outset. We also have shown that this limiting procedure solves many of the problems associated with a quantum formulation of angle variables for rotating systems.[6]

The Hermitian optical phase operator has been applied to a calculate the phase properties of a number of single-mode field states. These include number states and mixed-thermal states,[4,5] coherent states,[4,5] and squeezed states.[7,8] Number-phase minimum uncertainty states[9] and minimum phase-noise states[10] have also been constructed. In addition, the formalism has been used to study the phenomenon of phase diffusion in optical amplifiers[11] and in an analysis of phase dynamics in some nonlinear optical systems.[12]

In this paper we extend our formalism to pairs of field modes and apply it to investigate both uncorrelated and correlated phase behavior. We demonstrate the correlated, phase-locked property of two-mode squeezed states.

II. HERMITIAN OPTICAL PHASE OPERATOR

We have described the properties of the Hermitian optical phase operator ϕ_θ in detail in Refs. 4 and 5. The operator ϕ_θ exists in an $(s+1)$-dimensional states space Ψ spanned by the $s+1$ number states and the $s+1$ orthornormal phase states:

$$|\theta_m\rangle=(s+1)^{-1/2}\sum_{n=0}^{s}\exp(in\theta_m)|n\rangle\ , \tag{2.1}$$

where the $s+1$ phase values θ_m are equally spaced between θ_0 and $\theta_0+2\pi$:

$$\theta_m=\theta_0+\frac{2\pi m}{s+1}\ , \tag{2.2}$$

and m takes integer values from 0 to s. We are free to choose any value for θ_0, giving an uncountable infinity of orthonormal phase-state bases. The Hermitian optical phase operator is defined as

$$\hat{\phi}_\theta\equiv\sum_{m=0}^{s}\theta_m|\theta_m\rangle\langle\theta_m|\ , \tag{2.3}$$

and consequently has phase states as eigenstates with the eigenvalues θ_m. These eigenvalues are restricted to a 2π range and therefore the phase operator is single valued. Different phase-state bases correspond to distinct phase operators with different 2π ranges of eigenvalues.

The restricted range of the phase eigenvalues means that we must be careful when interpreting the calculated moments of the phase operator. The results obtained will depend on the range of eigenvalues and therefore on θ_0.

For example, the expectation value of the phase in a (randomly phased) number or thermal state is $\theta_0+\pi$.[4] Further complications arise when interpreting the sum and difference of the phases associated with two modes.

It is not just the formulation of the phase operator that is important[13] but the limiting procedure is also a crucial part of our approach. We do not take the limit of the states or the operators as s tends to infinity as these are only defined in the state space Ψ. The limit as s tends to infinity is only taken after c-number expressions, such as the moments of operators, are obtained. These limits are well behaved for physical (that is, experimentally accessible) systems.

III. PHASE SUMS AND DIFFERENCES

It is natural to define the phase sum and difference operators for two modes (a and b) to be the sum and difference of the single-mode operators ($\phi_{\theta a}$ and $\phi_{\theta b}$).[5] However, the 4π eigenvalue ranges of these two-mode operators adds further subtlety to the interpretation of the phase probability distributions. If we wish to describe the sum or difference of two single-mode phases, each of which is determined relative to third (reference) phase, then it is natural that this sum or difference will be expressed in a 4π range. This is because each single-mode phase will be expressed in a 2π range. However, if we are interested in only the sum or difference of the two single-mode phases and not the individual phases, then it is more meaningful to restrict the sum or a difference to a single 2π range. This restriction to a 2π range makes it easier to interpret two-mode phase correlations. We will illustrate the use of these two ranges by some examples.

A. Uncorrelated fields of random phase

We begin by examining the simplest possible case of two uncorrelated modes each in a state of random phase (for example, any number or thermal state[4]). The expectation values of the sum and difference of the phase eigenvalues are simply the sum and difference of the individual mean values. Moreover, the variances of the sum and difference are the equal to the sum of individual variances:

$$\Delta(\phi_{\theta a}\pm\phi_{\theta b})^2=\frac{2\pi^2}{3}=\Delta\phi_{\theta a}^2+\Delta\phi_{\theta b}^2 \; . \tag{3.1}$$

This property follows directly from the uncorrelated nature of the two-mode state in question. The compound probability distribution for the sum of the phases is readily derived from the single-mode distributions as follows.

Although it is possible to work with single-mode state spaces of different dimensionality,[5] it is sufficient (and simpler) to let each of these spaces to have dimension $(s+1)$. Then the single-mode phase operators have eigenvalues

$$\theta_{ma}=\theta_{0a}+\frac{2\pi m_a}{s+1} \; , \tag{3.2a}$$

$$\theta_{mb}=\theta_{0b}+\frac{2\pi m_b}{s+1} \; , \tag{3.2b}$$

where m_a and m_b both range from 0 to s. The possible values of the phase eigenvalue sum are

$$\theta_M=\theta_{0a}+\theta_{0b}+\frac{2\pi M}{s+1} \; , \tag{3.3}$$

where M has integer values between 0 and $2s$ inclusive. The probability for finding the two-mode field with a phase sum θ_M is

$$P(\theta_M)=\begin{cases}\displaystyle\sum_{m_a=0}^{M} P(\theta_{ma})P(\theta_M-\theta_{ma}) & (M\le s)\\ \displaystyle\sum_{m_a=M-s}^{s} P(\theta_{ma})P(\theta_M-\theta_{ma}) & (M\ge s)\; .\end{cases} \tag{3.4}$$

The phase distributions for the individual modes are uniform and therefore all the $P(\theta_{ma})$ and $P(\theta_{mb})$ are equal to $1/(s+1)$. Thus we find that the phase eigenvalue sum probability distribution is

$$P(\theta_M)=\begin{cases}\dfrac{M+1}{(s+1)^2} & (M\le s)\\ \dfrac{2s-M+1}{(s+1)^2} & (M\ge s)\; ,\end{cases} \tag{3.5}$$

which gives a triangular probability distribution as shown in Fig. 1. This is analogous to the probability distribution for the sum of the numbers shown on two dice. The probability for scoring any number between 1 and 6 with a single die is uniformly $\frac{1}{6}$, but the probability for the total score has a triangular distribution: $P(2)=\frac{1}{36}, P(3)=\frac{2}{36},\ldots,P(7)=\frac{6}{36},\ldots P(12)=\frac{1}{36}$. The difference between the two single-mode phase eigenvalues $\phi_{\theta a}-\phi_{\theta b}$ will range from $\theta_{0a}-\theta_{0b}-2\pi$ to $\theta_{0a}-\theta_{0b}+2\pi$ and has a similar triangular probability distribution.

While it is perfectly legitimate to calculate the distributions of the phase sums and differences in these 4π ranges, there is a redundancy implicit in using 4π radians to express the result. For example, a phase difference of α radians should be physically indistinguishable from a

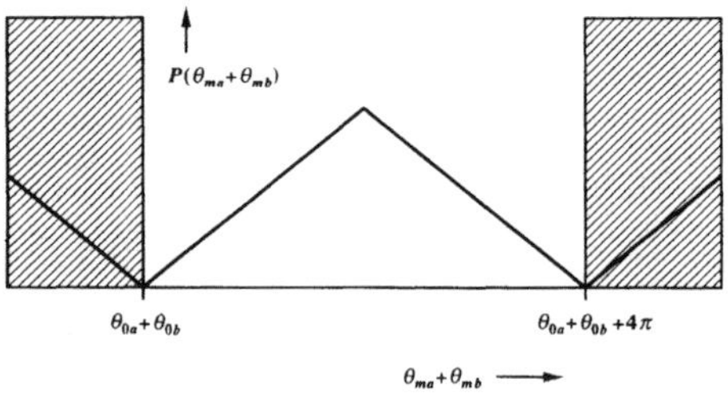

FIG. 1. Probability distribution for sum of the single-mode phase eigenvalues for two uncorrelated modes in states of random phase. The sum has values within a 4π range and the randomness of the phase sum is not immediately evident.

phase difference of $\alpha+2\pi$ radians, but both these values can occur in the same 4π range. It is therefore desirable to reduce the possible values of phase sums and differences to a 2π interval. We can achieve this by selecting a 2π interval within the 4π range and adding or subtracting 2π (as necessary) to or from values outside the selected interval in order to shift these values into the interval. As a result of this procedure we are left with a 2π range for the phase sum and difference probability distributions. Clearly, the moments of the phase sum and differences will, in general, be different when we use a 4π or a 2π range. While we can select any 2π subinterval, we specialize to a simple example by choosing the first half of the range. This reduced-range probability distribution is

$$P_{2\pi}(\theta_M)=P(\theta_M)+P(\theta_M+2\pi) \quad (M \le s)$$
$$=\frac{1}{s+1}, \tag{3.6}$$

which is a uniform distribution characteristic of a phase randomly distributed over a 2π interval. The reduced-range probability distribution is shown in Fig. 2 and can also be obtained from Fig. 1 by adding the second 2π section of Fig. 1 to the first. Naturally, the phase difference distribution will also be uniform when expressed mod(2π). This procedure also has an analogy in the two-dice system. The sum shown on the dice may be cast into the range 1–6 by representing the sum as mod(6).[14] This means that there are two ways of scoring 2, 3, 4, 5, and 6 and that the probability for registering any score between 1 and 6 is uniformly $\frac{1}{6}$.

We now have two methods for expressing the probability distribution of the phase sum or difference and we can calculate phase sum and difference moments for each distribution. Naturally, these will be different and some care is required when interpreting these moments. For the random phases discussed in this section we have already calculated the means and variances of the sum and difference for the 4π distribution. In the mod(2π) distribution for the phase sum, ranging from $\theta_{0a}+\theta_{0b}$ to $\theta_{0a}+\theta_{0b}+2\pi$, the mean and variance are

$$\langle \hat{\phi}_{\theta a}+\hat{\phi}_{\theta b}\rangle_{2\pi}=\theta_{0a}+\theta_{0b}+\pi, \tag{3.7a}$$

$$\Delta_{2\pi}(\phi_{\theta a}+\phi_{\theta b})^2=\frac{\pi^2}{3}, \tag{3.7b}$$

where the 2π subscript denotes the use of the mod(2π) distribution. We note that with this distribution the sum of the means of the phases is *not* the mean of the sum of the phases. This is a consequence of the periodicity of the phase variable and is not of quantum-mechanical origin. Two uncorrelated classical fields with random phases will admit precisely the same choice of 2π or 4π ranges in which to express the sum or difference probability distributions. Moreover the results from an analysis of such a classical system are readily shown to be identical to those obtained above for the quantum phases, when the large-s limit is finally taken.

The 4π and 2π distributions are both valid and useful. The former explicitly reveals the existence of correlations between the single-mode phases; in particular, if the variance in the phase sum or difference is not equal to the sum of the individual variances, then the phases are correlated. The latter is easier to interpret as a phase probability distribution because in it the phase sum or difference is a single-valued variable; for example, the sum and difference distributions for two uncorrelated fields in states of random phase are uniform. This is important because in the 4π distribution there is no unique shape signifying a randomness of the phase sum or difference. There are many distributions in the 4π range that become a uniform 2π distribution.

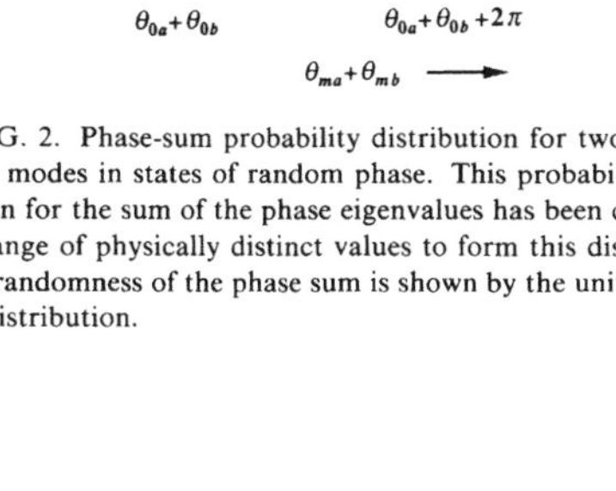

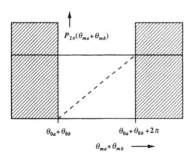

FIG. 2. Phase-sum probability distribution for two uncorrelated modes in states of random phase. This probability distribution for the sum of the phase eigenvalues has been cast into a 2π range of physically distinct values to form this distribution. The randomness of the phase sum is shown by the uniformity of the distribution.

B. Uncorrelated fields in partial phase states

Consider the two modes to be prepared in uncorrelated partial phase states:[5]

$$|a\rangle=\sum_{n=0}^{s} a_n \exp(in\beta_a)|n\rangle, \tag{3.8a}$$

$$|b\rangle=\sum_{n=0}^{s} b_n \exp(in\beta_b)|n\rangle, \tag{3.8b}$$

where a_n and b_n are real and positive. These states are important in the discussion of phase dependence and include coherent states, squeezed states, phase states, and number states as special cases. We have previously shown[5] that for a suitable choice of eigenvalue ranges with

$$\theta_{0a}=\beta_a-\frac{\pi s}{s+1} \tag{3.9}$$

(and a similar choice for mode b), the mean phase values are β_a and β_b. Moreover, the resulting phase probability distributions are symmetric about these mean values. For reasonably well-defined phases, the distributions have a pronounced central peak. When we find the phase eigenvalue sum distribution in the 4π range, we shall obtain another symmetric distribution with a central peak about a mean of $\beta_a+\beta_b$. The variance of the phase eigenvalue sum will be the sum of the individual single-mode variances, showing the lack of phase correlation between the modes. A corresponding result holds for the phase eigenvalue difference where the mean will be $\beta_a - \beta_b$.

The question now arises—what is the best way to cast the phase-sum distribution into a 2π range? Of course, all such choices are valid but the most convenient (and most easily interpreted) is that which reproduces the mean value as $\beta_a + \beta_b$. We realize this distribution by selecting the *central* 2π interval from the full 4π range. If the 4π distribution is sharply peaked (as, for example, when both field modes are prepared in intense coherent states[4]), then inserting the outside parts will have little effect on the moments of the phase sum. If instead we choose to cast the 4π distribution into the *first* 2π interval, then the original single peak is split into two parts, one at each end of the 2π interval. The mean values is shifted by π and the variance is markedly increased. Thus a poor choice of the 2π interval leads to the same interpretational problems encountered for a poor choice of θ_0 in the single-mode case.[4]

A simple, but important, example of a partial phase state is the coherent state. For suitable choices of the phase eigenvalue ranges, the expectation values of the phase sum and difference will simply be the sum and difference of the arguments the two coherent-state amplitudes. For intense coherent states, which have small phase variances,[4] the variances in both the sum and difference will be the sum of the two single-mode variances.

C. Phase locking and two-mode squeezing

Phase sum and differences are most interesting for states in which the two modes exhibit quantum phase correlations. We apply our two-mode phase formalism to analyze the phase sum and difference properties in the two-mode squeezed vacuum state.[15–18] The work of Reid and Drummond[19] suggests that these states might exhibit quantum phase correlations as a result of the correlations between the field quadratures.

The two-mode squeezed state is generalized from the two-mode vacuum by the action of a unitary Bogoliubov transformation and has the form[15–17]

$$|\zeta\rangle = \sum_n \frac{(\tanh r)^n}{\cosh r} \exp(in\xi)|n,n\rangle \ , \tag{3.10}$$

where the squeezing parameter is $\zeta = r\exp(i\xi)$. Expanding the number states in the single-mode phase state bases[4,5] gives

$$|\zeta\rangle = \frac{1}{(s+1)\cosh r} \times \sum_n \sum_{m_a} \sum_{m_b} \{\tanh r \exp[i(\xi - \theta_{ma} - \theta_{mb})]\}^n \times |\theta_{ma}\rangle|\theta_{mb}\rangle \ . \tag{3.11}$$

The compound probability $P(\theta_{ma}, \theta_{mb})$ of finding the modes with phases θ_{ma} and θ_{mb} is obtained from the square modulus of the projection of (3.11) onto $|\theta_{ma}\rangle|\theta_{mb}\rangle$. Evaluating the geometric progression involved, we find for large s that

$$P(\theta_{ma}, \theta_{mb}) = \frac{1}{4\pi^2\cosh^2 r} \left[\frac{1}{1+\tanh^2 r - 2\cos(\xi - \theta_{ma} - \theta_{mb})\tanh r} \right] \delta\theta_a \delta\theta_b \ , \tag{3.12}$$

where $\delta\theta_a$ and $\delta\theta_b$ are the spacings between successive phase eigenvalues for the two modes and have the value $2\pi/(s+1)$. In the limit as s tends to infinity we obtain from this the continuous probability density for the phases θ_a and θ_b:[20]

$$\mathcal{P}(\theta_a, \theta_b) = \frac{1}{4\pi^2\cosh^2 r} \left[\frac{1}{1+\tanh^2 r - 2\cos(\xi - \theta_a - \theta_b)\tanh r} \right] \ . \tag{3.13}$$

This distribution is an explicit function of the phase sum, but not of the phase difference, and suggests that there will be a preferred value for the phase sum corresponding to the maximum of the cosine.

We can now calculate the single-mode, and phase-sum, and difference properties of the two-mode squeezed vacuum state. The phase probability densities for one of the modes is obtained by integrating the joint distribution with respect to the phase of the other mode:

$$\mathcal{P}(\theta_a) = \int_{\theta_{0b}}^{\theta_{0b}+2\pi} \mathcal{P}(\theta_a, \theta_b) d\theta_b \ . \tag{3.14}$$

The integrand is a periodic function of θ_b and therefore the integral will have the same value when evaluated over any 2π range. Reassigning the limits to be $\xi - \theta_a - \pi$ and $\xi - \theta_a + \pi$, we obtain eventually

$$\mathcal{P}(\theta_a) = \frac{1}{2\pi^2\cosh^2 r} \int_0^{\pi} \frac{1}{1+\tanh^2 r - 2\cos\theta_b \tanh r} d\theta_b = \frac{1}{2\pi} \ . \tag{3.15}$$

This uniform probability density shows that the phase of mode a is random, which is consistent with the well-known thermal properties of single modes in a two-mode squeezed state.[16] Clearly, the b mode will also have random phase.

In order to investigate the phase sum and difference properties we introduce the new variables

$$\theta_+ \equiv \theta_a + \theta_b \ , \tag{3.16a}$$

$$\theta_- \equiv \theta_a - \theta_b \ . \tag{3.16b}$$

The Jacobean for this transformation is 2 and therefore the new probability density is

$$\mathcal{P}(\theta_+,\theta_-) = \frac{1}{8\pi^2\cosh^2 r} \times \left[\frac{1}{1+\tanh^2 r - 2\cos(\theta_+ - \xi)\tanh r} \right] . \tag{3.17}$$

The probability densities for the phase sum or difference are obtained by integrating with respect to θ_- or θ_+, respectively. Taking care with the appropriate limits of integration we find that the eigenvalue sum probability density, in the 4π range, is

$$\mathcal{P}(\theta_+) = \begin{cases} \int_{-\theta_+ + 2\theta_{0a}}^{\theta_+ - 2\theta_{0b}} \mathcal{P}(\theta_+,\theta_-)d\theta_- \\ \qquad (\theta_+ \le \theta_{0a} + \theta_{0b} + 2\pi) \\ \int_{\theta_+ - 2\theta_{0b} - 4\pi}^{-\theta_+ + 2\theta_{0a} + 4\pi} \mathcal{P}(\theta_+,\theta_-)d\theta_- \\ \qquad (\theta_+ \ge \theta_{0a} + \theta_{0b} + 2\pi) \ . \end{cases} \tag{3.18}$$

The integrals are straightforward and yield a single expression

$$\mathcal{P}(\theta_+) = \frac{1}{4\pi^2\cosh^2 r} \times \left[\frac{2\pi - |\theta_+ - \theta_{0a} - \theta_{0b} - 2\pi|}{1+\tanh^2 r - 2\cos(\theta_+ - \xi)\tanh r} \right] . \tag{3.19}$$

We are free to choose θ_{0a} and θ_{0b} to have any value and for simplicity we set

$$\theta_{0a} = \theta_{0b} = \frac{\xi}{2} - \pi \ , \tag{3.20}$$

giving a symmetrical distribution for the sum of the phase eigenvalues which is centered on a mean of ξ:

$$\mathcal{P}(\theta_+) = \frac{1}{4\pi^2\cosh^2 r} \times \left[\frac{2\pi - |\theta_+ - \xi|}{1+\tanh^2 r - 2\cos(\theta_+ - \xi)\tanh r} \right] . \tag{3.21}$$

We see that use of the 4π range has led to a distribution with a discontinuity of slope at $\theta_+ = \xi$. There is no physical significance in this discontinuity as we will demonstrate when we turn our attention to the corresponding mod(2π) distribution. This shows the problems that are associated with interpreting the 4π distribution. The phase eigenvalue difference probability density, in the 4π range, is

$$\mathcal{P}(\theta_-) = \begin{cases} \int_{-\theta_- + 2\theta_{0a}}^{\theta_- + 2\theta_{0b} + 4\pi} \mathcal{P}(\theta_+,\theta_0)d\theta_+ \\ \qquad (\theta_- \le \theta_{0a} - \theta_{0b}) \\ \int_{\theta_- + 2\theta_{0b}}^{-\theta_- + 2\theta_{0a} + 4\pi} \mathcal{P}(\theta_+,\theta_-)d\theta_+ \\ \qquad (\theta_- \ge \theta_{0a} - \theta_{0b}) \ . \end{cases} \tag{3.22}$$

The evaluation of these integrals is more complicated, but we find eventually the analytic result

$$\mathcal{P}(\theta_-) = \frac{1}{4\pi^2} \arctan \times \left[\exp(-2r)\tan\left[\frac{\theta_+ - \xi}{2}\right] \right] \Bigg|_{|\theta_-| - 2\pi + \xi}^{-|\theta_-| + 2\pi + \xi} , \tag{3.23}$$

where we have chosen θ_{0a} and θ_{0b} as in (3.20). Again we see the problems associated with using the 4π range. For the case $r = 0$, the two-mode squeezed state becomes the double vacuum and both of the phase eigenvalue sum and difference probability densities become

$$\mathcal{P}(\theta_+) = \frac{1}{4\pi^2}(2\pi - |\theta_+ - \xi|) \quad (\xi - 2\pi \le \theta_+ \le \xi + 2\pi) \tag{3.24}$$

$$\mathcal{P}(\theta_-) = \frac{1}{4\pi^2}(2\pi - |\theta_-|) \quad (-2\pi \le \theta_- \le 2\pi) \ . \tag{3.25}$$

These are precisely the triangular distributions discussed for uncorrelated fields with random phases in Sec. III A. However, as we have previously stated, a triangular distribution does not uniquely signify a random phase sum or difference. We can only determine the true nature of the phase sum and difference from the mod(2π) distribution.

We can obtain the mod(2π) probability densities for the phase sum and difference in two ways. We can either start with the 4π distributions derived above and shift the appropriate parts of this distribution into the chosen 2π interval, or we can return to $\mathcal{P}(\theta_+,\theta_-)$ and shift appropriate parts of this joint distribution into the chosen ranges prior to integration. These methods produce identical distributions, but the former is more complicated and is carried out in the Appendix. From (3.18) and (3.22) the ranges of values that θ_+ and θ_- can take for the joint distribution are

$$|\theta_-| + \xi - 2\pi \le \theta_+ \le -|\theta_-| + \xi + 2\pi \ , \tag{3.26a}$$

$$|\theta_+ - \xi| - 2\pi \le \theta_- \le -|\theta_+ - \xi| + 2\pi \ , \tag{3.26b}$$

where the absolute bounds on θ_+ and θ_- are $\xi - 2\pi$ to $\xi + 2\pi$ and -2π to 2π, respectively. We now cast the joint distribution into the central 2π ranges $\xi - \pi$ to $\xi + \pi$ and $-\pi$ to π for θ_+ and θ_-, respectively by adding or subtracting 2π as necessary to values of θ_+ and θ_- outside these 2π ranges. This gives the joint mod(2π) probability density

$$\mathcal{P}_{2\pi}(\theta_+,\theta_-)=\frac{1}{2\pi}\frac{1}{2\pi\cosh^2 r}\times\left[\frac{1}{1+\tanh^2 r-2\cos(\theta_+-\xi)\tanh r}\right], \quad (3.27)$$

where now

$$\xi-\pi\le\theta_+\le\xi+\pi\ , \quad (3.28a)$$

$$-\pi\le\theta_-\le\pi\ . \quad (3.28b)$$

The first point to notice is that both $\mathcal{P}_{2\pi}(\theta_+,\theta_-)$ and the range of θ_+ are independent of θ_-, so the integral of the distribution over θ_+ will also be independent of θ_-. Indeed, evaluating this integral we find that the mod(2π) probability density for the phase difference is

$$\mathcal{P}_{2\pi}(\theta_-)=\frac{1}{2\pi}\ . \quad (3.29)$$

This is the uniform distribution characteristic of a random phase. Working with the mod(2π) distribution has made it clear that the correlation between the modes is not manifest in the phase difference and that the complicated 4π distribution (3.23) is in fact a random phase distribution. The mod(2π) distribution for the phase sum is obtained by integrating the mod(2π) joint distribution over θ_-:

$$\mathcal{P}_{2\pi}(\theta_+)=\frac{1}{2\pi\cosh^2 r}\times\left[\frac{1}{1+\tanh^2 r-2\cos(\theta_+-\xi)\tanh r}\right]. \quad (3.30)$$

This function is the Airy pattern familiar from the theory of the Fabry-Pérot étalon and has a single central peak, within the defined range $\xi-\pi$ to $\xi+\pi$, at $\theta_+=\xi$. The correlation between the modes is manifest in the preferred value of the phase sum

$$\langle\hat{\phi}_{\theta a}+\hat{\phi}_{\theta b}\rangle_{2\pi}=\xi \quad (3.31)$$

and in the variance of the phase sum

$$\Delta_{2\pi}(\phi_{\theta a}+\phi_{\theta b})^2=\frac{1}{2\pi\cosh^2 r}\times\int_{\xi-\pi}^{\xi+\pi}\frac{(\theta_+-\xi)^2 d\theta_+}{1+\tanh^2 r-2\cos(\theta_+-\xi)\tanh r}\ . \quad (3.32)$$

We can evaluate this integral by first expanding $(\theta_+-\xi)^2$ as a Fourier series in the range of $-\pi$ to π and obtain eventually

$$\Delta_{2\pi}(\phi_{\theta a}+\phi_{\theta b})^2=\frac{\pi^2}{3}+4\sum_{k=1}\frac{(-1)^k}{k^2}\tanh^k r=\frac{\pi^2}{3}+4\,\mathrm{dilog}\,(1+\tanh r)\ , \quad (3.33)$$

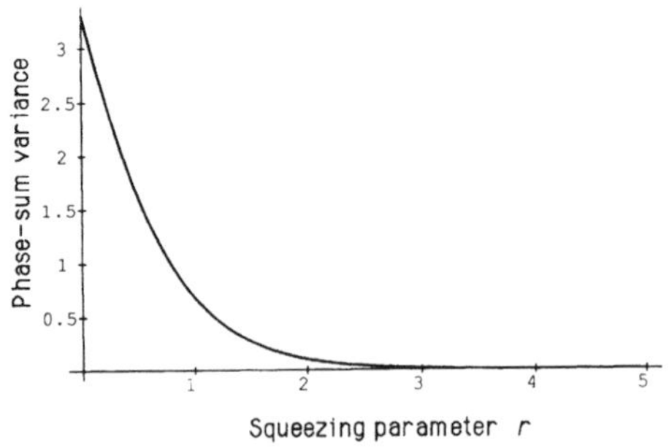

FIG. 3. Phase-sum variance $[\Delta_{2\pi}(\phi_{\theta a}+\phi_{\theta b})^2]$ for a two-mode squeezed vacuum state as a function of r, the modulus of the squeezing parameter. For zero squeezing the phase-sum variance has the value $\pi^2/3$, characteristic of random phase. In the large squeezing limit this variance vanishes.

where dilog( ) is the dilogarithm function.[21] The dilogarithm varies from 0 to $-\pi^2/12$ monotonically as r varies from 0 to ∞. The corresponding variation of the phase-sum variance is from $\pi^2/3$, corresponding to randomness, down to 0. This zero variance denotes the fact that the phase sum (in the 2π range) becomes perfectly locked to the value ξ in the large squeezing limit. This dilogarithmic variation is shown in Fig. 3.

Thus the two-mode squeezed state has simple phase properties that are accessible by exact analytic methods. The progressively stronger locking of the phase sum to the argument of the squeezing parameter with increasing squeezing brings to light another fundamental property of these states.

IV. CONCLUSION

The formulation of the Hermitian optical phase operator has made it possible to examine rigorously the quantum phase properties of light. In this paper we have extended the formalism to discuss two-mode phase properties as revealed in the sum and difference between the single-mode phase operators. This extension involves the added subtlety that the sum and difference operators have a 4π rather than a 2π range of eigenvalues. However, physically distinct phases exist only in a 2π range and therefore the unambiguous interpretation of phase correlations requires us to cast the phase sum and difference probability distributions into a 2π interval. When this is done the physical consequences of phase correlations (or the lack of them) become evident. We have illustrated this procedure with the examples of two uncorrelated modes of random phase, two uncorrelated modes in partial phase states, and the correlated two-mode squeezed state.

It is beyond the scope of this paper to address the experimental implications of our results. Indeed, no precise experimental procedure has yet been suggested for measuring the quantum phase. However, the "measured phase" observables,[22] which are accessible in homodyne detection and other coherent detection techniques, provide a good approximation to the corresponding Hermitian combinations of the unitary phase operators for fields of moderate to high intensity.[9] This strong-field correspondence can also be seen between the phase probability distribution and the angular properties of the Wigner function.[23] The intermode phase correlations exhibited by the two-mode squeezed states increase in strength with the squeezing parameter and thus with the field intensity. Therefore we expect that the effects of these phase correlations will be observable in suitable experiments involving homodyne detection of the individual modes comprising the two-mode squeezed state. In addition, phase-*difference* correlations, similar to the phase-sum correlations studied here, may be present in the output from a correlated emission laser.[24]

Two-mode squeezed states have received a great deal of attention in quantum optics and indeed the two-mode squeezed vacuum was the first squeezed state to be prepared in an experiment.[25] They are of fundamental interest because of the strong quantum correlations between the modes that are responsible for intensity correlations and for the squeezing property itself.[17] The individual modes display random thermal fluctuations and this has led to their application in finite-temperature field theory[26] and in thermodynamic problems in quantum optics.[16] Moreover they are more strongly correlated than any other two-mode state of light.[18] To this impressive list of fundamental properties of this important class of states we can now add their elegant phase properties. They have random single-mode phases and the phase difference is also random, but the phases lock so that their sum has a preferred value. With increasing squeezing this phase locking becomes more rigid until ultimately, in the limit of perfect squeezing, the phase sum can have only one value, which is the argument of the squeezing parameter.

ACKNOWLEDGMENTS

One of us (S.M.B.) thanks the General Electric Company and the United Kingdom Fellowship of Engineering for support.

APPENDIX

Here we derive the mod(2π) distributions $\mathcal{P}_{2\pi}(\theta_+)$ and $\mathcal{P}_{2\pi}(\theta_-)$ for the phase sum and differences from the 4π distributions. In both cases we choose to cast the distribution into the 2π center portion. For the sum distribution this means

$$\mathcal{P}_{2\pi}(\theta_+)=\begin{cases}\mathcal{P}(\theta_+)+\mathcal{P}(\theta_++2\pi) & (\xi-\pi\le\theta_+\le\xi)\\ \mathcal{P}(\theta_+)+\mathcal{P}(\theta_+-2\pi) & (\xi\le\theta_+\le\xi+\pi)\ .\end{cases} \tag{A1}$$

This gives, from (3.21),

$$\mathcal{P}_{2\pi}(\theta_+)=\frac{1}{4\pi^2\cosh^2 r}\left[\frac{2\pi-|\theta_+-\xi|}{1+\tanh^2 r-2\cos(\theta_+-\xi)\tanh r}\right]+\frac{1}{4\pi^2\cosh^2 r}\left[\frac{2\pi-|\theta_+\pm 2\pi-\xi|}{1+\tanh^2 r-2\cos(\theta_+-\xi)\tanh r}\right], \tag{A2}$$

where the $+$ and $-$ signs refer to the two ranges in (A1). The final form of the mod(2π) distribution is

$$\mathcal{P}_{2\pi}(\theta_+)=\frac{1}{2\pi\cosh^2 r}\left[\frac{1}{1+\tanh^2 r-2\cos(\theta_+-\xi)\tanh r}\right], \tag{A3}$$

in agreement with the result derived in III C.

For the mod(2π) phase difference distribution we have

$$\mathcal{P}_{2\pi}(\theta_-)=\begin{cases}\mathcal{P}(\theta_-)+\mathcal{P}(\theta_-+2\pi) & (-\pi\le\theta_-\le 0)\\ \mathcal{P}(\theta_-)+\mathcal{P}(\theta_--2\pi) & (0\le\theta_-\le\pi)\ .\end{cases} \tag{A4}$$

This gives, from (3.23),

$$\mathcal{P}_{2\pi}(\theta_-)=\frac{1}{4\pi^2}\arctan\left[\exp(-2r)\tan\left[\frac{\theta_+-\xi}{2}\right]\right]\Bigg|_{|\theta_-|-2\pi+\xi}^{-|\theta_-|+2\pi+\xi}+\frac{1}{4\pi^2}\arctan\left[\exp(-2r)\tan\left[\frac{\theta_+-\xi}{2}\right]\right]\Bigg|_{|\theta_-\pm 2\pi|-2\pi+\xi}^{-|\theta_-\pm 2\pi|+2\pi+\xi}, \tag{A5}$$

where the $+$ and $-$ signs refer to the two ranges in (A4). It is important to note that the definition of the arctangent function employed here is such that

$$\arctan[\tan(\alpha)]=\alpha \quad \text{for } -\pi\le\alpha\le\pi \tag{A6}$$

but if α is *not* in the range $-\pi$ to π, then it must be shifted by an integer multiple of 2π to bring it into this range. Thus $\arctan[\tan(\alpha+\pi)]$ will be $\alpha+\pi$ if α is between $-\pi$ and 0, but will be $\alpha-\pi$ if α is between 0 and π. Moreover, by writing $\exp(-2r)\tan(\alpha)$ as $\tan(\beta)$ and requiring continuity of the arctangent function with variance or r, we obtain

$$\arctan[\exp(-2r)\tan(\alpha+\pi)]=\arctan[\exp(-2r)\tan(\alpha)]\pm\pi\ . \tag{A7}$$

Clearly, extreme care must be used in evaluating (A5). The result is

$$\mathcal{P}_{2\pi}(\theta_-)=\frac{1}{4\pi^2}\left\{\arctan\left[\exp(-2r)\tan\left(\frac{-|\theta_-|+2\pi}{2}\right)\right]-\arctan\left[\exp(-2r)\tan\left(\frac{|\theta_-|-2\pi}{2}\right)\right]\right\}$$
$$+\frac{1}{4\pi^2}\left\{\arctan\left[\exp(-2r)\tan\left(\frac{-|\theta_-\pm 2\pi|+2\pi}{2}\right)\right]-\arctan\left[\exp(-2r)\tan\left(\frac{|\theta_-\pm 2\pi|-2\pi}{2}\right)\right]\right\}. \quad (A8)$$

The first and fourth terms differ by π as do the second and third terms. Thus this complicated looking expression reduces to the uniform distribution characteristic of random phase:

$$\mathcal{P}_{2\pi}(\theta_-)=\frac{1}{2\pi}\,. \quad (A9)$$

---

[1]P. A. M. Dirac, Proc. R. Soc. London Ser. A **114**, 243 (1927).

[2]W. H. Louisell, Phys. Lett. **7**, 60 (1963); L. Susskind and J. Glogower, Physics **1**, 49 (1964); P. Carruthers and M. M. Nieto, Rev. Mod. Phys. **40**, 411 (1968); R. Loudon, *The Quantum Theory of Light*, 1st ed. (Oxford University Press, Oxford, 1973); J. M. Levy-Leblond, Ann. Phys. (N.Y.) **101**, 319 (1976).

[3]D. T. Pegg and S. M. Barnett, Europhys. Lett. **6**, 483 (1988).

[4]S. M. Barnett and D. T. Pegg, J. Mod. Opt. **36**, 7 (1989).

[5]D. T. Pegg and S. M. Barnett, Phys. Rev. A **39**, 1665 (1989).

[6]S. M. Barnett and D. T. Pegg, Phys. Rev. A **41**, 3427 (1990).

[7]J. A. Vaccaro and D. T. Pegg, Opt. Commun. **70**, 529 (1989).

[8]N. Grombech-Jensen, P. L. Christiansen, and R. S. Ramanujam, J. Opt. Soc. Am. B **6**, 2423 (1989); W. Schleich, R. J. Horowicz, and S. Varro, Phys. Rev. A **40**, 7405 (1989).

[9]J. A. Vaccaro and D. T. Pegg, J. Mod. Opt. **37**, 17 (1990).

[10]G. S. Summy and D. T. Pegg, Opt. Commun. **77**, 75 (1990).

[11]S. M. Barnett, S. Stenholm, and D. T. Pegg, Opt. Commun. **73**, 314 (1989).

[12]S. J. D. Phoenix and P. L. Knight, J. Opt. Soc. Am. B **7**, 116 (1990); C. C. Gerry, Opt. Commun. **75**, 168 (1990).

[13]Similar operators have been obtained before [for example, T. S. Santhanam, Aust. J. Phys. **31**, 233 (1978)], but the limiting procedures used there have only reintroduced the original problems noted by Susskind and Glogower (Ref. 2).

[14]That is, 7 becomes 1, 8 becomes 2, and so on.

[15]G. J. Milburn, J. Phys. A **17**, 737 (1984); C. M. Caves and B. L. Schumaker, Phys. Rev. A **31**, 3068 (1985); B. L. Schumaker and C. M. Caves, *ibid.* **31**, 3093 (1985).

[16]S. M. Barnett and P. L. Knight, J. Opt. Soc. Am. B **2**, 467 (1985).

[17]D. F. Walls and M. D. Reid, Acta Phys. Austriaca **56**, 3 (1984); S. M. Barnett and P. L. Knight, J. Mod. Opt. **34**, 841 (1987).

[18]S. M. Barnett and S. J. D. Phoenix, Phys. Rev. A **40**, 2404 (1989).

[19]M. D. Reid and P. D. Drummond, Phys. Rev. Lett. **60**, 2731 (1988).

[20]As a physical state, such as two-mode squeezed state, has a slowly varying probability distribution (see Ref. 9) we can replace this discrete distribution by a continuous probability density (see Refs. 4 and 5).

[21]*Handbook of Mathematical Functions,* 9th ed., edited by M. Abramowitz and I. A. Stegun (Dover, New York, 1972), p. 1004.

[22]S. M. Barnett and D. T. Pegg, J. Phys. A **19**, 3849 (1986).

[23]S. M. Barnett, S. Stenholm and D. T. Pegg, Opt. Commun. **73**, 314 (1989); W. Schleich, R. J. Horowicz and S. Varo (Ref. 8).

[24]M. O. Scully, Phys. Rev. Lett. **55**, 2802 (1985); S. Swain (private communication).

[25]R. E. Slusher, L. W. Hollberg, B. Yurke, J. C. Mertz, and J. F. Valley, Phys. Rev. Lett. **55**, 2409 (1985).

[26]Y. Takahashi and H. Umezawa, Collect. Phenom. **2**, 55 (1975).

Physical number–phase intelligent and minimum-uncertainty states of light

J. A. VACCARO and D. T. PEGG
School of Science, Griffith University, Nathan, Brisbane 4111, Australia

(*Received 4 May 1989; revision accepted 17 July 1989*)

Abstract. The recently introduced Hermitian phase operator allows a phase-state representation of the single-mode light field. We find the requirement that a light field is in a physical state, that is, has finite energy moments, imposes strict and simple continuity conditions on the phase-amplitude distribution. We exploit these conditions to examine the physical intelligent and minimum-uncertainty states associated with the number-phase uncertainty relations, including those involving the Hermitian phase operator and its sine and cosine forms. The single number-state is found to be the only physical exact intelligent state and also the only physical exact minimum-uncertainty state for all the uncertainty relations considered. We construct states which are both physical states and approximately intelligent states. Under certain conditions coherent states, ideal squeezed states and the number-phase intelligent states associated with the Susskind–Glogower cosine and sine operators are found to be both physical states and approximately number–phase intelligent states for the phase operators considered here.

1. Introduction

The number–phase intelligent and minimum-uncertainty states of a single-mode light field (or, equivalently, the quantum-mechanical harmonic oscillator) have been discussed by Jackiw [1], Carruthers and Nieto [2], Levy-Leblond [3], Sanders *et al.* [4] and Yamamoto *et al.* [5] using the cosine and sine phase operators defined by Susskind and Glogower [6], and by Matthys and Jaynes [7], Shapiro and Wagner [8] and Loudon and Knight [9] using phenomenological descriptions of phase. The Susskind and Glogower formalism for the quantum-mechanical phase, the only formalism available at the time of these discussions, leads to peculiar phase properties, especially for fields of low excitation. Also, there is no unique Hermitian phase operator corresponding to the classical phase angle in this formalism, and so phenomenological arguments have been necessary to describe the uncertainty in the phase angle.

The phase operators recently introduced by Pegg and Barnett [10–12] are based on a *Hermitian* phase operator which has properties normally associated with phase. In this paper we examine the states that give minimum number–phase uncertainty in relation to the Pegg–Barnett phase operators. We also discuss the connection between our results and the corresponding work of previous authors.

To calculate the quantum-mechanical properties of the single-mode field in the Pegg–Barnett formalism, we use a linear space Ψ of a finite but arbitrarily large dimension spanned by the $s+1$ number states $|0\rangle, |1\rangle, \ldots, |s\rangle$. This space is similar to the $(2j+1)$-dimensional space spanned by the eigenstates of the angular momentum operator $\hat{J}^2$ with $j=s/2$. It is beyond the scope of this paper to explore this similarity in detail, but it will be discussed in a future paper. Physical results, such as expectation values, are calculated using the space Ψ. This gives a real number which depends on s. Since a complete description of the single-mode field involves an infinite set of number states, we must take an infinite limit at some stage. The crucial second step of the new formalism is to take this limit only *after* the physical results are calculated, which thus involves finding the limit of a sequence of real numbers. This approach allows us to use normalized orthogonal phase states as the eigenstates of a Hermitian phase operator, which cannot be done if we prematurely replace s by infinity. The number operator $\hat{N}$ is defined in terms of number-state projectors by

$$\hat{N} \equiv \sum_{n=0}^{s} n|n\rangle\langle n|. \quad (1.1)$$

The space Ψ is also spanned by the $s+1$ orthogonal phase states $|\theta_m\rangle$:

$$|\theta_m\rangle \equiv (s+1)^{-1/2} \sum_{n=0}^{s} \exp(\mathrm{i}\theta_m n)|n\rangle, \quad (1.2)$$

where

$$\theta_m \equiv \theta_0 + 2\pi m/(s+1), \quad (1.3)$$

and $m=0, 1, \ldots, s$. The value of θ_0 is arbitrary and defines a particular basis set. The Hermitian phase operator is defined by

$$\hat{\phi}_\theta \equiv \sum_{m=0}^{s} \theta_m|\theta_m\rangle\langle\theta_m|. \quad (1.4)$$

When we calculate the number moments $\langle\hat{N}^q\rangle$ for the phase states we find that as we allow s to tend to inifinity, the moments, except for $q=0$, do not approach a finite limit and we do not obtain, for example, a finite expectation value for the energy. Such a state would therefore be impossible to prepare physically. This leads us to divide the states in Ψ into two categories, the physical states for which $\langle\hat{N}^q\rangle$ approaches a finite limit for any given value of q when s tends to infinity, and the non-physical states for which this is not so. In this paper we will concentrate on results involving physical states and so we begin our analysis by first looking at the properties of physical states in some detail in section 2.

The Heisenberg uncertainty relation for the non-commuting observables $\hat{A}$ and $\hat{B}$ is

$$\langle\Delta\hat{A}^2\rangle\langle\Delta\hat{B}^2\rangle \geqslant \tfrac{1}{4}\langle C\rangle^2, \quad (1.5)$$

where $\hat{C}$ is a Hermitian operator given by $[\hat{A}, \hat{B}] = \mathrm{i}\hat{C}$, and $\langle\Delta\hat{A}^2\rangle \equiv \langle\hat{A}^2\rangle - \langle\hat{A}\rangle^2$ is the variance of $\hat{A}$. The states that give the equality in equation (1.5) are called intelligent states (IS), a term introduced by Aragone *et al.* [13]. The intelligent states that also minimize the product $\langle\Delta\hat{A}^2\rangle\langle\Delta\hat{B}^2\rangle$ are called minimum-uncertainty states (MUS) [14]. In cases where $\hat{C}$ is proportional to the identity operator (as it is for the position and momentum commutator), the left-hand side of equation (1.5) is minimized for all IS and so all IS are also MUS. However this is not the case for the pairs of number and phase operators we consider in this paper. The $\hat{A}$–$\hat{B}$ IS, which we denote as $|f_{AB}\rangle$, can be determined by solving the eigenvalue equation [15]

$$(\hat{A} + \mathrm{i}\gamma\hat{B})|f_{AB}\rangle = \lambda|f_{AB}\rangle, \quad (1.6)$$

where λ and γ are free parameters with γ real. The dependence of $|f_{AB}\rangle$ on these parameters can be seen as follows [1, 15]. Taking the inner product of each side of equation (1.6) with $|f_{AB}\rangle$ gives

$$\lambda = \langle \hat{A} \rangle + i\gamma \langle \hat{B} \rangle, \tag{1.7}$$

and thus λ determines the means in $\hat{A}$ and $\hat{B}$. Taking the inner product of each side of equation (1.6) with itself yields

$$\langle \hat{A}^2 \rangle + \gamma^2 \langle \hat{B}^2 \rangle + i\gamma \langle [\hat{A}, \hat{B}] \rangle = |\lambda|^2.$$

Replacing the right-hand side of this expression with $\langle \hat{A} \rangle^2 + \gamma^2 \langle \hat{B} \rangle^2$ and the commutator with $i\hat{C}$, then on solving the quadratic in γ we obtain $\gamma = \frac{1}{2}\langle \hat{C} \rangle / \langle \Delta \hat{B}^2 \rangle$, or

$$\langle \Delta \hat{B}^2 \rangle = (2\gamma)^{-1} \langle \hat{C} \rangle. \tag{1.8}$$

The state $|f_{AB}\rangle$ satisfies the equality in equation (1.5) and so eliminating $\langle \Delta \hat{B}^2 \rangle$ from equation (1.5) using equation (1.8) yields

$$\langle \Delta \hat{A}^2 \rangle = \tfrac{1}{2}\gamma \langle \hat{C} \rangle, \tag{1.9}$$

and thus

$$|\gamma| = [\langle \Delta \hat{A}^2 \rangle / \langle \Delta \hat{B}^2 \rangle]^{1/2}. \tag{1.10}$$

Hence the sign of γ determines the sign of $\langle \hat{C} \rangle$ while γ^2 gives the ratio of the variances in $\hat{A}$ and $\hat{B}$. In sections 3 and 4 we determine the set of physical states that are also the IS associated with each of the pairs of operators $\hat{N}$–$\hat{\phi}_\theta$, $\hat{N}$–$\cos\hat{\phi}_\theta$ and $\hat{N}$–$\sin\hat{\phi}_\theta$ by considering only physical-state solutions to eigenvalue equations of the form of equation (1.6). We also examine classes of physical states that are approximately the IS associated with each pair of operators.

2. Physical and physically preparable states

Physically preparable states of the single-mode field have been described previously as states which can be prepared from the vacuum state $|0\rangle$ by an interaction for a finite time with a source whose energy spectrum is bounded [10–12]. While in some cases, for example parametric mode–mode coupling, the source is another field mode and in other cases the source is a collection of excited atoms, these sources themselves must at some stage have been excited by a primary source of energy. Assuming that this primary source is bounded, for example, by supposing that there is zero probability that it contains more energy than that of the visible universe, conservation of energy will ensure that the energy of the single-mode field under consideration is also bounded. Hence we are led to define a physically preparable state more precisely as a state whose photon-number probability distribution P_n satisfies

$$P_n = 0, \quad n > t, \tag{2.1}$$

where the upper bound t is some finite, usually very large, integer. The existence of this bound implies, for example, that when we allow s to tend to infinity after finding an expectation value for a field in a physically preparable state, expressions such as n/s, where n refers to any occupied number state, will approach zero.

In addition to states satisfying condition (2.1), there is a wider class of states whose properties can approximate those of physically preparable states to any desired degree of accuracy by adjusting the upper bound t in equation (2.1). We shall call such states 'physical states' and these include physically preparable states as special cases. We define physical states as those states for which the mth moment of the number operator $\langle \hat{N}^m \rangle$ remains finite for any given m when s tends to infinity, that is, those states which have finite energy moments. In the Pegg–Barnett formalism the field annihilation operator $\hat{a}$ is given by

$$\hat{a} \equiv \exp(i\hat{\phi}_\theta)\, \hat{N}^{1/2}.$$

Remembering that we take the limit as s tends to infinity after expectation values are calculated, it is not difficult to show that the requirement of finite energy moments is both sufficient and necessary to ensure that the expectation value $\langle [\hat{a}, \hat{a}^\dagger]^q \rangle$ for any given q is unity, even though the operator $[\hat{a}, \hat{a}^\dagger]^q$ is *not* the unit operator [10, 12]. Indeed, states on which the action of the commutator operator $[\hat{a}, \hat{a}^\dagger]^q$ is equal to the action of the unit operator when s tends to inifinity can be used as an alternative equivalent definition of physical states.

The mth moment of $\hat{N}$ for a physical state $|p\rangle$ with number-state coefficients d_n is given by

$$\langle p|\hat{N}^m|p\rangle = \sum_{n=0}^{\infty} |d_n|^2 n^m,$$

in the limit of infinite s. Since this expectation value is finite for any given m, the tail of the series on the right-hand side,

$$\sum_{n=t}^{\infty} |d_n|^2 n^m,$$

converges to zero as t approaches infinity. Mathematically, for any given $\varepsilon > 0$ we can find a number $M(\varepsilon)$ such that

$$\sum_{n=t}^{\infty} |d_n|^2 n^m < \varepsilon \tag{2.2}$$

for all $t > M(\varepsilon)$. Thus by choosing t sufficiently large, for example $t > M(\varepsilon)$, we can ensure that any moment of $\hat{N}$ for the state $|p\rangle$ approximates to any given degree of accuracy that of the state obtained by truncating the number-state expansion of $|p\rangle$ at $|n\rangle = |t\rangle$. We see therefore that, on the one hand, any physically preparable state is a physical state, because the moments of a state with an upper bound to its component number-states are finite, and on the other hand, any physical state can be approximated to any desired degree of accuracy by a state with such an upper bound.

A state can be shown to be a physical state by evaluating its moment generating function $\langle \exp(x\hat{N}) \rangle$. If this function is finite when s tends to infinity for all values of x in the interval $(-\eta, \eta)$ where η is arbitrarily small but non-zero, then all the moments $\langle \hat{N}^m \rangle$ are finite and can be found from

$$\langle \hat{N}^m \rangle = \frac{d^m}{dx^m} \langle \exp(x\hat{N}) \rangle \bigg|_{x=0}.$$

As examples, we show by using this technique that the coherent state $|\alpha\rangle$ and the squeezed vacuum $|0, \zeta\rangle$ [16] are physical states. Coherent states have number-state coefficients given by

$$\langle n|\alpha\rangle = \exp(-\tfrac{1}{2}|\alpha|^2)\, \alpha^n (n!)^{-1/2},$$

from which we easily obtain, when s tends to infinity,

$$\langle\alpha|\exp(x\hat{N})|\alpha\rangle=\sum_{n=0}^{\infty}\exp(xn)|\langle n|\alpha\rangle|^2$$
$$=\exp\{[\exp(x)-1]|\alpha|^2\}.$$

Since the right-hand side is finite for all finite values of x and $|\alpha|$, it follows that the finite-amplitude coherent state $|\alpha\rangle$ is a physical state.

The ideal squeezed state $|\alpha,\zeta\rangle$ has number-state coefficients of the form [9, 17]

$$\begin{aligned}d_n=\langle n|\alpha,\zeta\rangle&=(n!\cosh r)^{-1/2}[\tfrac{1}{2}\exp(i\theta)\tanh r]^{n/2}\\&\times\exp\{-\tfrac{1}{2}[|\alpha|^2+\alpha^*\exp(i\theta)\tanh r]\}\\&\times H_n\{[\alpha+\alpha^*\exp(i\theta)\tanh r][2\exp(i\theta)\tanh r]^{-1/2}\},\end{aligned} \tag{2.3}$$

where

$$\alpha=|\alpha|\exp(i\phi),\qquad \zeta=r\exp(i\theta),$$

and $H_n(x)$ is the nth-order Hermite polynomial. From the property

$$H_{2n+1}(0)=0,$$
$$H_{2n}(0)=(-1)^n(2n)!/n!,$$

we find for the squeezed vacuum $|0,\zeta\rangle$ (i.e. $\alpha=0$) that $d_{2n+1}=0$ and

$$|d_{2n}|^2=\frac{(\tanh r)^{2n}(2n+1)!!}{\cosh r(2n)!!(2n+1)}.$$

The value of $\langle\exp(x\hat{N})\rangle$ for the squeezed vacuum when s tends to infinity can now be found by evaluating

$$\sum_{n=0}^{\infty}\frac{[\tanh^2 r\exp(2x)]^n(2n+1)!!}{\cosh r(2n)!!(2n+1)},$$

with the aid of

$$(1-y)^{-1/2}=\sum_{n=0}^{\infty}\frac{y^n(2n+1)!!}{(2n)!!(2n+1)},\quad |y|<1,$$

to be

$$\langle 0,\zeta|\exp(x\hat{N})|0,\zeta\rangle=[\cosh^2 r-\exp(2x)\sinh^2 r]^{-1/2}$$

provided $\exp(2x)<\coth^2 r$. This result demonstrates that the squeezed vacuum is a physical state for finite degrees of squeezing (that is, r finite).

The phase-state expansion of a general state $|f\rangle$ is

$$|f\rangle=\sum_{m=0}^{s}c_m|\theta_m\rangle,$$

where

$$\begin{aligned}c_m&=\langle\theta_m|f\rangle\\&=(s+1)^{-1/2}\sum_{n=0}^{s}\langle n|f\rangle\exp(-in\theta_m),\end{aligned} \tag{2.4}$$

by equation (1.2). The dependence of the coefficients c_m on the value of θ_0, in accordance with equations (1.3) and (2.4), may be indicated by writing the coefficients as $c_m(\theta_0)$. Consider the function $\psi(\theta)$ defined by

$$\psi(\theta)\equiv(2\pi)^{-1/2}\sum_{n=0}^{s}\langle n|f\rangle\exp(-in\theta), \tag{2.5}$$

where θ is a *continuous* parameter. It is not difficult to show that the integral of $|\psi(\theta)|^2$ over any 2π interval is unity in the limit of infinite s, and so $\psi(\theta)$ can be compared to a normalized periodic wavefunction. Clearly, from equations (2.4) and (2.5) we find

$$\left[\frac{s+1}{2\pi}\right]^{1/2}c_m(\theta_0)=\psi(\theta_m+2\pi j), \tag{2.6}$$

where j is any integer, so that the scaled phase-state coefficients

$$[(s+1)/(2\pi)]^{1/2}c_m(\theta_0),$$

with θ_0 fixed, represent a countable subset of points on the curve $\psi(\theta)$, for each 2π interval of θ. The remaining points on $\psi(\theta)$, for example for θ between θ_m and θ_{m+1}, can be generated by keeping m constant and varying θ_0 continuously over an interval of $2\pi/(s+1)$. We now consider θ_0 to be fixed and write the phase-state coefficients simply as c_m. As s tends to infinity, the set of points c_m becomes dense. The properties of the curve $\psi(\theta)$ will thus reflect the properties of c_m. We shall return to this point at the end of this section.

For the remainder of this section we shall discuss some general phase properties of physical states, which we denote by replacing f by p in equation (2.4). For a physical state $|p\rangle$ we have, given any positive integer q,

$$\langle p|\hat{N}^{2q}|p\rangle<B^2,$$

where B^2 is some real upper bound. Using the Schwarz inequality [15]

$$|\langle b|d\rangle|^2\leqslant\langle b|b\rangle\langle d|d\rangle$$

for $|d\rangle=\hat{N}^q|p\rangle$ and arbitrary normalized state $|b\rangle$ we find

$$|\langle b|\hat{N}^q|p\rangle|^2\leqslant\langle p|\hat{N}^{2q}|p\rangle,$$

and thus

$$|\langle b|\hat{N}^q|p\rangle|<B. \tag{2.7}$$

Letting $k=q-2$ and replacing $|b\rangle$ with the number state $|n\rangle$ yields

$$|\langle n|\hat{N}^k|p\rangle|<Bn^{-2},$$

and so

$$\sum_{n=0}^{s}|\langle n|\hat{N}^k|p\rangle|<B\sum_{n=1}^{s}n^{-2}<B\pi^2/6 \tag{2.8}$$

for all s. Thus, in the limit of infinite s, the series on the left-hand side is bounded for any given positive integer k. With the aid of equation (1.2) the magnitude of $\langle\theta_m|\hat{N}^k|p\rangle$ is found to satisfy

$$|\langle\theta_m|\hat{N}^k|p\rangle|=\left|(s+1)^{-1/2}\sum_{n=0}^{s}\langle n|\hat{N}^k|p\rangle\exp(-in\theta_m)\right|$$

$$\leqslant(s+1)^{-1/2}\sum_{n=0}^{s}|\langle n|\hat{N}^k|p\rangle|,$$

and thus, using the order symbol O,

$$\langle\theta_m|\hat{N}^k|p\rangle = O[(s+1)^{-1/2}] \tag{2.9}$$

by equation (2.8). Here, the right-hand side is small compared to the right-hand side of equation (2.7) because the magnitudes of all the number-state coefficients of the phase state $|\theta_m\rangle$ are $(s+1)^{-1/2}$, whereas for an arbitrary state they can be of order unity.

We now introduce a new operator $\hat{\vartheta}$:

$$\hat{\vartheta} \equiv \mathrm{i}[\exp(-\mathrm{i}\hat{N}\delta) - 1]/\delta, \tag{2.10}$$

where

$$\delta \equiv 2\pi/(s+1) \tag{2.11}$$

is the step in phase between consecutive phase states, and we examine

$$\langle b|\hat{\vartheta}^q|p\rangle = \langle b|\left[\hat{N} - \frac{\mathrm{i}\hat{N}^2\delta}{2!} + \ldots + \frac{(-\mathrm{i}\delta)^{j-1}\hat{N}^j}{j!} + \ldots\right]^q|p\rangle,$$

where $|b\rangle$ is an arbitrary state. As can be seen from equation (1.1), all number-state matrix elements of $\hat{N}/(s+1)$ are less than unity, and so

$$\delta\langle b|\hat{N}^{q+1}|p\rangle > (2\pi)^{-l}\delta^{l+1}\langle b|\hat{N}^{l+q+1}|p\rangle$$

for any integer $l > 0$, and thus

$$\langle b|\hat{\vartheta}^q|p\rangle = \langle b|\hat{N}^q|p\rangle + O(\delta\langle b|\hat{N}^{q+1}|p\rangle). \tag{2.12}$$

From equation (2.7) $\langle b|\hat{N}^{q+1}|p\rangle$ is bounded, and so the second term on the right-hand side vanishes in the infinite s limit (i.e. for δ zero). Since $|b\rangle$ is an arbitrary state and q is any given positive integer, then it follows that *in this limit $\hat{\vartheta}$ is equivalent to the number operator when operating on physical states*. This result will be important in the next section. Replacing $|b\rangle$ with the phase state $|\theta_m\rangle$ in equation (2.12) and using equation (2.9) yields

$$\langle\theta_m|\hat{\vartheta}^q|p\rangle = \langle\theta_m|\hat{N}^q|p\rangle + O(\delta^{3/2}). \tag{2.13}$$

The exponential operator appearing in equation (2.10) is the displacement operator for phase states:

$$\exp(\mathrm{i}\hat{N}\delta)|\theta_m\rangle = |\theta_{m+1}\rangle, \quad m < s,$$

$$\exp(\mathrm{i}\hat{N}\delta)|\theta_s\rangle = |\theta_0\rangle,$$

which leads to the property

$$\langle\theta_m|\hat{\vartheta} = \mathrm{i}(\langle\theta_{m+1}| - \langle\theta_m|)/\delta, \quad m < s,$$

$$\langle\theta_s|\hat{\vartheta} = \mathrm{i}(\langle\theta_0| - \langle\theta_s|)/\delta.$$

Taking the inner product with an arbitrary state $|f\rangle = \sum c_m|\theta_m\rangle$ yields

$$\langle\theta_m|\hat{\vartheta}|f\rangle = \mathrm{i}\frac{\Delta c_m}{\Delta\theta}, \tag{2.14}$$

where

$$\Delta c_m \equiv c_{m+1} - c_m,$$

$$\Delta\theta \equiv \delta,$$

and where we have let

$$c_{m+s+1} \equiv c_m$$

for simplicity of notation. As s is made very large (i.e. $\Delta\theta$ small) the right-hand side of equation (2.14) can be considered as a derivative. Similarly, we can show that

$$\frac{\Delta^2 c_m}{\Delta\theta^2} = -\langle\theta_m|\hat{\vartheta}^2|f\rangle, \tag{2.15}$$

where

$$\Delta^2 c_m \equiv \Delta c_{m+1} - \Delta c_m,$$

so the left-hand side of equation (2.15) corresponds to the second derivative in the limit of small $\Delta\theta$. In general, for any given q

$$\frac{\Delta^q c_m}{\Delta\theta^q} = (-\mathrm{i})^q \langle\theta_m|\hat{\vartheta}^q|f\rangle, \tag{2.16}$$

where $\Delta^q c_m \equiv \Delta^{q-1} c_{m+1} - \Delta^{q-1} c_m$. If we replace $|f\rangle$ by a physical state $|p\rangle$ and use equations (2.13) and (2.9) we obtain

$$\left|\frac{\Delta^q c_m}{\Delta\theta^q}\right| = O(\delta^{1/2}). \tag{2.17}$$

Thus in the limit as s tends to infinity $\delta^{-1/2}|\Delta^q c_m/(\Delta\theta)^q|$, which approaches the modulus of the qth derivative of the curve $\psi(\theta)$ given by equation (2.5), is bounded. This establishes a simple but important condition on the coefficients c_m for a physical state: in the limit as s tends to infinity, *the modulus of any finite-order derivative of the curve $\psi(\theta)$ is bounded, as is any finite-order derivative of the real and imaginary parts of $\psi(\theta)$*. In particular, for δ small, equation (2.17) allows us to write

$$\delta^{-1/2}|\Delta c_m| = O(\delta), \qquad \delta^{-1/2}\left|\frac{\Delta^2 c_m}{\Delta\theta}\right| = O(\delta),$$

for a physical state. In the infinite-s limit, the first condition eliminates discontinuities in $\psi(\theta)$ and the second eliminates cusps.

3. The physical $\hat{N}$–$\hat{\phi}_\theta$ intelligent and minimum-uncertainty states

The Hermitian phase operator $\hat{\phi}_\theta$ and the number operator $\hat{N}$ given by equations (1.1) and (1.4) do not commute. By Heisenberg's uncertainty principle the variances in $\hat{\phi}$ and $\hat{N}$ must satisfy

$$\langle\Delta\hat{N}^2\rangle\langle\Delta\hat{\phi}_\theta^2\rangle \geqslant \tfrac{1}{4}|\langle[\hat{N}, \hat{\phi}_\theta]\rangle|^2. \tag{3.1}$$

The $\hat{N}$–$\hat{\phi}_\theta$ intelligent states (IS) are those states that give the equality in equation (3.1) and they can be found by solving the eigenvalue equation [15]

$$(\hat{N} + \mathrm{i}\gamma\hat{\phi}_\theta)|f_{N\phi}\rangle = \lambda|f_{N\phi}\rangle. \tag{3.2}$$

The parameters γ and λ determine the means and relative variances in $\hat{N}$ and $\hat{\phi}_\theta$ according to equations (1.7) and (1.10), that is,

$$|\gamma| = [\langle\Delta\hat{N}^2\rangle/\langle\Delta\hat{\phi}_\theta^2\rangle]^{1/2},$$

$$\lambda = \langle\hat{N}\rangle + \mathrm{i}\gamma\langle\hat{\phi}_\theta\rangle.$$

In this paper we concentrate on a restricted class of the $\hat{N}$–$\hat{\phi}_\theta$ IS by examining only those IS that are also physical states. The physical $\hat{N}$–$\hat{\phi}_\theta$ minimum-uncertainty states (MUS) will be found by determining which of the physical IS gives a minimum in $\langle\Delta\hat{N}^2\rangle\langle\Delta\hat{\phi}_\theta^2\rangle$. When operating on physical states the commutator operator on the right-hand side of equation (3.1) can be replaced by [12]

$$[\hat{N}, \hat{\phi}_\theta]_p = \mathrm{i}[1-(s+1)|\theta_0\rangle\langle\theta_0|]. \tag{3.3}$$

The corresponding uncertainty relation satisfied by a physical IS $|p_{N\phi}\rangle$ is then

$$\langle\Delta\hat{N}^2\rangle\langle\Delta\hat{\phi}_\theta^2\rangle = \tfrac{1}{4}|1-(s+1)|c_0|^2|^2, \tag{3.4}$$

where we have used the phase-state expansion,

$$|p_{N\phi}\rangle = \sum_{m=0}^{s} c_m|\theta_m\rangle,$$

which will represent those states satisfying equation (3.2) which are also physical states. In that case, multiplying equation (3.2) on the left-hand side by the phase state $\langle\theta_m|$, which is an eigenstate of $\hat{\phi}_\theta$, and employing equation (2.13) yields

$$\langle\theta_m|\hat{\vartheta}|p_{N\phi}\rangle + \mathrm{i}\gamma\theta_m c_m = \lambda c_m + O(\delta^{3/2}).$$

It is evident that the physical eigenstates of $(\hat{\vartheta} + \mathrm{i}\gamma\hat{\phi}_\theta)$ are also the physical $\hat{N}$–$\hat{\phi}_\theta$ IS in the limit of infinite s (i.e. zero δ). Using equation (2.14) we obtain the difference equation

$$\frac{\Delta c_m}{\Delta\theta} = -(\gamma\theta_m + \mathrm{i}\lambda)c_m + O(\delta^{3/2}). \tag{3.5}$$

When equation (3.5) is considered in conjunction with the normalization condition,

$$\sum_{m=0}^{s} |c_m|^2 = 1, \tag{3.6}$$

we obtain the useful result that there is a lower limit to all phase-state coefficients c_m for all normalizable solutions of equation (3.5). For example, no normalizable solution can have $c_m = O(\delta^{3/2})$ for any m. This can be proved by considering the contrary: if any coefficient, say c_m, *were* of order $\delta^{3/2}$ then by equation (3.5) the difference between it and c_{m+1} would be of the order $\delta^{5/2}$, and thus by equation (3.5) all coefficients would be of order $\delta^{3/2}$ and equation (3.6) cannot be satisfied.

The number states have previously been shown to be $\hat{N}$–$\hat{\phi}_\theta$ MUS with $\langle\Delta\hat{N}^2\rangle = 0$, $\langle\Delta\hat{\phi}_\theta^2\rangle = \pi^2/3$ and $\langle\Delta\hat{N}^2\rangle\langle\Delta\hat{\phi}_\theta^2\rangle = 0$, which is the absolute minimum of this uncertainty product [12]. It is easy to show that here the number states correspond to the solutions with $\gamma = 0$. Similarly the phase states are $\hat{N}$–$\hat{\phi}_\theta$ MUS, but they are not physical. The question immediately arises as to whether there are other physical-state solutions of equation (3.5), that is, other physical intelligent states? We now prove that the number states are the *only* physical $\hat{N}$–$\hat{\phi}_\theta$ IS by showing that there are no physical-state solutions to equation (3.5) when γ is non-zero. The second-order difference can be calculated from equation (3.5) for $m = s$ to be

$$\frac{\Delta^2 c_s}{\Delta\theta} = -[(\gamma\theta_0 + \mathrm{i}\lambda)\,\Delta c_s - \gamma(2\pi - \delta)c_s] + O(\delta^{3/2}). \tag{3.7}$$

In the previous section we found that for physical states both Δc_s and $\Delta^2 c_s/\Delta\theta$ are of the order $\delta^{3/2}$. For non-zero γ these conditions can only be satisfied by equation (3.7) when $c_s = O(\delta^{3/2})$, but we have already observed that no normalizable solution to equation (3.5) exists with any coefficient of order $\delta^{3/2}$. Therefore, there are no physical state solutions to equation (3.5) for non-zero γ and this result proves that the *number states are the only $\hat{N}$–$\hat{\phi}_\theta$ IS which are also physical states.*

This result should perhaps not be unexpected for the following reason: the position–momentum MUS have a Gaussian probability distribution, so one might expect that an $\hat{N}$–$\hat{\phi}_\theta$ IS might also have a Gaussian phase probability distribution, which is indeed suggested by the solution to the differential equation analogous to equation (3.5). However, a periodic phase probability distribution containing Gaussians of finite width cannot be smooth, that is, there will be cusps at 2π intervals, and as section 2 shows, such states would be unphysical. Only when the width of the Gaussians is very small can we closely approximate the periodic Gaussians by a smooth curve representing a physical state.

For the remainder of this section we examine those physical states that are *approximately* $\hat{N}$–$\hat{\phi}_\theta$ IS, that is, states for which the ratio of the left-hand side of the uncertainty relation equation (3.4) to the right-hand side, while not unity, can be made as close to unity as desired. Our method will be to approximate the solution of the difference equation (3.5) with a physical state, and then to show that the physical state is indeed approximately an IS by calculating the terms in Heisenberg's uncertainty relation for physical states given by equation (3.4). We use the solution of the differential equation analogous to equation (3.5) as a guide to construct a trial state $|g\rangle$ with phase-state coefficients given by

$$c_m = A(\gamma)\exp[-\tfrac{1}{2}\gamma(\theta_m - \bar{\phi})^2 - i\bar{n}\theta_m], \tag{3.8}$$

where $\bar{\phi}$ and $\bar{n}$ are real parameters to be determined later, and $A(\gamma)$ is a normalization constant independent of n and determined up to a phase factor from

$$1 = \langle g|g\rangle = \sum_m |c_m|^2.$$

It is not difficult to show that equation (3.8) satisfies the difference equation (3.5) for $0 \leqslant m \leqslant s-1$ with arbitrary $\bar{\phi}$ and $\bar{n}$, and for $m = s$ with particular values of $\bar{\phi}$ and $\bar{n}$. However our main concern here is that $|g\rangle$ is at least an approximate solution of equation (3.5) and therefore we defer imposing restrictions on $\bar{\phi}$ and $\bar{n}$ until a later stage.

The number-state coefficients of $|g\rangle$ can be found from equations (1.2), (3.8) and the number state-expansion

$$|g\rangle = \sum_{n=0}^{s} d_n|n\rangle,$$

to be

$$d_n = A(\gamma)(s+1)^{-1/2}\sum_{m=0}^{s}\exp[-\tfrac{1}{2}\gamma(\theta_m - \bar{\phi})^2 - i(\bar{n}-n)\theta_m].$$

The summation can be approximated by an integral when s is large (i.e. δ is small) according to

$$d_n = A(\gamma)(s+1)^{-1/2}\delta^{-1}\left\{\sum_{m=0}^{s}\exp[-\tfrac{1}{2}\gamma(\theta_m - \bar{\phi})^2 - i(\bar{n}-n)\theta_m]\delta\right\},$$

$$= A(\gamma)(s+1)^{-1/2}\delta^{-1}\left\{\int_{\theta_0}^{\theta_0+2\pi}\exp[-\tfrac{1}{2}\gamma(\theta - \bar{\phi})^2 - i(\bar{n}-n)\theta]\,d\theta + O(\delta)\right\}. \tag{3.9}$$

The regimes where $|g\rangle$ approaches a physical state most closely can be found by considering the function $\psi(\theta)$ as defined by equation (2.5). Using equation (2.6) we generate $\psi(\theta)$, which depends on a continuous variable, from the coefficients c_m given by equation (3.8) in the following way. We recall from equation (1.3) that $\theta_m = \theta_0 + 2\pi m/(s+1)$, and so by allowing θ_0 to vary continuously over a $2\pi/(s+1)$ interval, say $\theta'_0 \leqslant \theta_0 < \theta'_0 + 2\pi/(s+1)$, then θ_m on the right-hand side of equation (3.8) becomes a continuous variable, which we now write as θ and whose value can be any real number in the 2π interval given by $\theta'_0 \leqslant \theta < \theta'_0 + 2\pi$. By equation (2.6), $\psi(\theta)$ on this 2π interval is equal to the right-hand side of equation (3.8) multiplied by $\delta^{-1/2}$ and with θ_m replaced by θ. We use the periodic property of $\psi(\theta)$ given by equation (2.6) to obtain $\psi(\theta)$ outside this interval as:

$$\psi(\theta) = \delta^{-1/2} A(\gamma) \exp\left[-\tfrac{1}{2}\gamma(\theta + 2\pi k - \bar{\phi})^2 - i\bar{n}(\theta + 2\pi k)\right], \tag{3.10}$$

where k is chosen to satisfy $\theta'_0 \leqslant \theta + 2\pi k < \theta'_0 + 2\pi$. $\psi(\theta)$ is a periodic function whose modulus has a smooth Gaussian shape from θ'_0 up to $\theta'_0 + 2\pi$ where it reverts to its value at θ'_0. In general, therefore, there will be discontinuities in $\psi(\theta)$ and its slope at $\theta = \theta'_0$ plus integer multiples of 2π. For $|g\rangle$ to be a physical state, however, $\psi(\theta)$ must be a continuous function with a continuous derivative everywhere. The discontinuities in $\psi(\theta)$ and its slope vanish only in the limit of zero γ or infinite γ, in which case $|\psi(\theta)|$ is a flat line or a delta-function shape respectively. Thus a physical-state approximation to $|g\rangle$ will be most likely to exist for very large or very small values of γ. In the small γ regime the discontinuities only vanish when γ is zero and $\bar{n}$ is an integer, in which case we regain the number-state IS. In the large γ regime, by choosing $\bar{\phi}$ sufficiently different from both θ'_0 and $\theta'_0 + 2\pi$, the Gaussian factor can be made very small near $\theta = \theta'_0$ and $\theta = \theta'_0 + 2\pi$, but the discontinuity in the derivative of $\psi(\theta)$ at these points only vanishes in the limit of infinitely large γ and then $|g\rangle$ is a non-physical phase state. We now examine these two regimes in more detail with the intention of constructing a physical state approximation to $|g\rangle$.

We consider first the behaviour of the number-state coefficients d_n of $|g\rangle$ given by equation (3.9) for small but non-zero γ. The integral in equation (3.9) can be written as a Fourier transform of a product,

$$\int_{-\infty}^{\infty} T(\theta) \exp\left[-\tfrac{1}{2}\gamma(\theta - \bar{\phi})^2 - i(\bar{n} - n)\theta\right] d\theta,$$

where $T(\theta)$ is the 'top hat' function: $T(\theta) = 1$ for $\theta_0 \leqslant \theta < \theta_0 + 2\pi$ and $T(\theta) = 0$ elsewhere. This integral can be written as a convolution yielding

$$d_n = A(\gamma)(\gamma\delta)^{-1/2} \left\{ \int_{-\infty}^{\infty} \frac{\sin[\pi(x + n - \bar{n})]}{\pi(x + n - \bar{n})} \exp[i(x + n - \bar{n})(\theta_0 + \pi)] \right.$$
$$\left. \times \exp(-\tfrac{1}{2}\gamma^{-1}x^2 - i\bar{\phi}x)\, dx + O(\delta) \right\}.$$

For γ small but non-zero, the narrow Gaussian in the integrand of this last expression ensures that the integrand is negligible for $|x| > 1$ and so for $n \gg \bar{n}$ we may approximate the demoninator $\pi(x + n - \bar{n})$ by πn with the accuracy of the approximation increasing with n. We also note that $\sin[\pi(x + n - \bar{n})] \exp(in\pi) = \sin[\pi(x - \bar{n})]$ and thus for very large n the modulus of the integral is approximately

$$(\pi n)^{-1} \left| \int_{-\infty}^{\infty} \sin[\pi(x - \bar{n})] \exp[i(x - \bar{n})(\theta_0 + \pi)] \exp(-\tfrac{1}{2}\gamma^{-1}x^2 - i\bar{\phi}x)\, dx \right|,$$

which shows that $|d_n|$ decays approximately as n^{-1} for large n. This rate of decay is not sufficiently fast for $|g\rangle$ to be a physical state, because $\langle g|\hat{N}|g\rangle$ will not converge, and thus eliminates the possibility of approximating $|g\rangle$ by a physical state for small but non-zero γ. Thus a single number state is the only intelligent *or approximate intelligent* physical state in the regime of small or zero γ.

The above result may seem a little surprising because, given that a number state $|n\rangle$ is an exact intelligent state, one might expect that, for example, $b_n|n\rangle + b_{n+1}|n+1\rangle$ would approximate an intelligent state when b_{n+1} is small but non-zero. Direct substitution into the uncertainty relation (3.4), however, shows that the equality holds only for b_{n+1} or b_n zero and when both of these are non-zero, however small, the ratio of $\langle\Delta\hat{N}^2\rangle\langle\Delta\hat{\phi}_\theta^2\rangle$ to the right-hand side of equation (3.4) is always greater than or equal to $\langle\Delta\hat{\phi}_\theta^2\rangle$ which is greater than 1·2 for all possible values of b_n and b_{n+1}. Thus this ratio does not tend to unity as b_{n+1} approaches zero and so the superposition is not an approximate intelligent state for small non-zero b_{n+1}.

We next consider the physical-state approximation to the trial state $|g\rangle$ in the large γ regime. When γ is sufficiently large and $\bar{\phi}$ is sufficiently different from both θ_0 and $\theta_0 + 2\pi$ so that almost all of the area under the narrow Gaussian factor in the integral in equation (3.9) is between the limits of integration, we can approximate the integral in equation (3.9) reasonably well by extending the limits of integration to positive and negative infinity. We then find

$$d_n \approx d'_n = B(\gamma)\exp[-(2\gamma)^{-1}(n-\bar{n})^2 + \mathrm{i}(n-\bar{n})\bar{\phi}], \tag{3.11}$$

where d'_n are the number-state coefficients of a state $|g'\rangle$ which approximates $|g\rangle$ under these conditions, and $B(\gamma)$ is a normalization constant independent of n. The probability distribution $|d'_n|^2$ has a Gaussian dependence and therefore, provided $\bar{n}$ and γ are finite, $\langle\exp(x\hat{N})\rangle$ will be finite, which indicates that $|g'\rangle$ is a physical state.

For ease of calculation we consider only the situation where $\bar{n}^2 \gg \gamma$, so that

$$|d'_0|^2 = |B(\gamma)|^2 \exp(-\bar{n}^2/\gamma)$$

is negligible and thus the mean and peak of the Gaussian factor in equation (3.11) approximately coincide, that is,

$$\langle\hat{N}\rangle \approx \bar{n}. \tag{3.12}$$

The normalization constant is determined to within an arbitrary phase factor by $\langle g'|g'\rangle = 1$, from which we obtain

$$|B(\gamma)|^{-2} = \sum_{n=0}^{s} \exp[-\gamma^{-1}(n-\bar{n})^2].$$

Comparing the summand with the function $\exp(-\gamma^{-1}x^2)$, whose slope has extreme values of $\pm(2/\gamma)^{1/2}\exp(-\frac{1}{2})$, shows that the magnitude of the difference between successive terms in the summation is less than $\gamma^{-1/2}$, which is small because γ is large. The summation is therefore approximately equal to the integral of

$$\exp[-\gamma^{-1}(n-\bar{n})^2]$$

over n from $n=0$ to $s+1$. Since s is arbitrarily large and $\bar{n}$ $(=\langle\hat{N}\rangle)$ is bounded for a physical state, then $\exp[-\gamma^{-1}(n-\bar{n})^2]$ is arbitarily small for n outside the interval

$[0, s]$ and so we may extend the limits of integration to negative and positive infinity, yielding

$$|B(\gamma)|^{-2} \approx \int_{-\infty}^{\infty} \exp[-\gamma^{-1}(n-\bar{n})^2]\,dn = (\gamma\pi)^{1/2}. \tag{3.13}$$

We now check if the state given by equations (3.11) and (3.13) approximates an intelligent state by finding the appropriate variances. The variance of $\hat{N}$ can be found in a similar manner to the calculation above, with an integral replacing a summation, to yield

$$\langle \Delta\hat{N}^2 \rangle \approx \tfrac{1}{2}\gamma. \tag{3.14}$$

The phase-state coefficients c'_m of $|g'\rangle$ can be found from d'_n by the expression

$$c'_m = \langle \theta_m | g' \rangle = (s+1)^{-1/2} \sum_{n=0}^{s} \exp(-\mathrm{i}\theta_m n)\, d'_n.$$

On substituting for d'_n from equation (3.11) and using the same approximation technique employed in deriving expressions (3.13) and (3.14) we find

$$\begin{aligned} c'_m &\approx (s+1)^{-1/2} B(\gamma) \int_{-\infty}^{\infty} \exp[-(2\gamma)^{-1}(n-\bar{n})^2 + \mathrm{i}n(\bar{\phi}-\theta_m) - \mathrm{i}\bar{n}\bar{\phi}]\,dn \\ &\approx B(\gamma)[2\pi\gamma/(s+1)]^{1/2} \exp[-\tfrac{1}{2}\gamma(\bar{\phi}-\theta_m)^2 - \mathrm{i}\bar{n}\theta_m]. \end{aligned} \tag{3.15}$$

The conditions on γ and $\bar{\phi}$ under which we obtained equation (3.11) imply that $(s+1)|c'_0|^2$ and $(s+1)|c'_s|^2$ are both negligible. The expectation value of functions of phase $f(\hat{\phi}_\theta)$,

$$f(\hat{\phi}_\theta) \equiv \sum_{m=0}^{s} f(\theta_m)|\theta_m\rangle\langle\theta_m|,$$

where $f(\theta)$ represents a power of $\sin\theta$, $\cos\theta$ or θ, can be found from

$$\langle f(\hat{\phi}_\theta) \rangle \equiv \lim_{s\to\infty} \sum_{m=0}^{s} f(\theta_m)|c'_m|^2.$$

Replacing $|c'_m|^2$ using equations (3.15) and (3.13), and approximating the summation by an integral yields

$$\langle f(\hat{\phi}_\theta) \rangle \approx \lim_{s\to\infty} (\gamma/\pi)^{1/2} \left\{ \int_{\theta_0}^{\theta_0+2\pi} f(\theta) \exp[-\gamma(\bar{\phi}-\theta)^2]\,d\theta + O(\delta) \right\}. \tag{3.16}$$

Since both $(s+1)|c'_0|^2$ and $(s+1)|c'_s|^2$ are negligible then the integral in equation (3.16) is approximately equal to the integral in which the limits of integration have been extended to infinity:

$$\begin{aligned} \langle f(\hat{\phi}_\theta) \rangle &\approx \lim_{s\to\infty} (\gamma/\pi)^{1/2} \left\{ \int_{-\infty}^{\infty} f(\theta) \exp[-\gamma(\bar{\phi}-\theta)^2]\,d\theta + O(\delta) \right\} \\ &\approx (\gamma/\pi)^{1/2} \int_{-\infty}^{\infty} f(\theta) \exp[-\gamma(\bar{\phi}-\theta)^2]\,d\theta. \end{aligned} \tag{3.17}$$

Using this result we obtain approximate values for the mean and variance in $\hat{\phi}_\theta$ as

$$\langle \hat{\phi} \rangle \approx \bar{\phi}, \qquad \langle \Delta\hat{\phi}_\theta^2 \rangle \approx (2\gamma)^{-1}. \tag{3.18}$$

From equations (3.14) and (3.18) the product of the uncertainties in $\hat{N}$ and $\hat{\phi}_\theta$ is

$$\langle \Delta \hat{N}^2 \rangle \langle \Delta \hat{\phi}_\theta^2 \rangle \approx \tfrac{1}{4}. \tag{3.19}$$

Comparing equation (3.19) with Heisenberg's relation for physical states, equation (3.4), and recalling that $(s+1)|c_0'|^2$ is negligible, verifies that the physical state $|g'\rangle$ is approximately a $\hat{N}$–$\hat{\phi}_\theta$ IS. Our analysis requires $\bar{n}^2 \gg \gamma \gg 1$ and from equations (3.12) and (3.14) this implies

$$\langle \hat{N} \rangle^2 \gg 2\langle \Delta \hat{N}^2 \rangle \gg 1, \tag{3.20}$$

and so $|g'\rangle$ can have photon statistics ranging from sub-Poissonian ($\langle \Delta \hat{N}^2 \rangle < \langle \hat{N} \rangle$) to super-Poissonian ($\langle \Delta \hat{N}^2 \rangle > \langle \hat{N} \rangle$).

Intense coherent states, which are particular quadrature–amplitude minimum-uncertainty states ($\hat{X}$–$\hat{Y}$ MUS), have recently been shown to be approximately $\hat{N}$–$\hat{\phi}_\theta$ IS [11]; indeed the number-state coefficients of the intense coherent state are given approximately by equation (3.11) with

$$\bar{n} = \tfrac{1}{2}\gamma = \langle \hat{N} \rangle = \langle \Delta \hat{N}^2 \rangle \gg 1.$$

The states $|g'\rangle$ describe a more general class of approximate $\hat{N}$–$\hat{\phi}_\theta$ IS satisfying equation (3.20) and we now compare $|g'\rangle$ with the ideal squeezed state $|\alpha, \zeta\rangle$ which is the most general $\hat{X}$–$\hat{Y}$ MUS and whose number-state coefficients are given by equation (2.3). The state $|g'\rangle$ is a *partial phase state* [12] in that its number-state coefficients (3.11) can be written in the form

$$d_n' = |d_n'| \exp(\mathrm{i}\mu) \exp(\mathrm{i}n\bar{\phi}),$$

where μ and $\bar{\phi}$ are real and independent of n. The ideal squeezed state $|\alpha, \zeta\rangle$ only becomes a partial phase state when the Hermite polynomial in equation (2.3) can be written in the form

$$H_n(z) = |H_n(z)| \exp(\mathrm{i}\beta_1) \exp(\mathrm{i}n\beta_2), \tag{3.21}$$

where β_1 and β_2 are real and independent of n. Now

$$H_n(z) = \sum_{m=0}^{l} \frac{(-1)^m n! (2z)^{n-2m}}{m!(n-2m)!},$$

where l is the largest integer not exceeding $\frac{1}{2}n$. When z is a pure imaginary number $H_n(z)$ can always be factorised in the form of equation (3.21) for all z, which is not necessarily so when z is not imaginary. Thus we concentrate on the case where the argument of the Hermite polynomial in equation (2.3) is imaginary, that is, where $\phi = \frac{1}{2}(\theta \pm \pi)$ and $|\alpha, \zeta\rangle$ is a partial phase state with

$$\langle n|\alpha, \zeta\rangle = |\langle n|\alpha, \zeta\rangle| \exp(\mathrm{i}\nu) \exp(\mathrm{i}n\phi).$$

Here the value of ν turns out to be $\frac{1}{2}|\alpha| \tanh r \sin\phi$. The photon-number probability distribution for the ideal squeezed state with a strong coherent component [9] (which, for convenience, we shall refer to as a strongly coherent squeezed state), that is, the state for which

$$\langle \hat{N} \rangle \approx |\alpha|^2 \gg \exp(2r), \tag{3.22}$$

is given approximately by [9]

$$|\langle n|\alpha, \zeta\rangle|^2 \approx (\Delta\pi)^{-1/2} \exp[-(n - |\alpha|^2)^2/\Delta],$$

where

$$\Delta = 2|\alpha|^2 \exp(2r) \tag{3.23}$$

for $\phi = \frac{1}{2}(\theta \pm \pi)$. The number-state coefficients of the strongly coherent squeezed state in our case are therefore

$$\langle n|\alpha, \zeta\rangle \approx (\Delta\pi)^{-1/4} \exp[-\tfrac{1}{2}(n-|\alpha|^2)^2/\Delta] \exp(\mathrm{i}v) \exp(\mathrm{i}n\phi),$$

which agrees (up to an insignificant phase factor which is independent of n) with the number-state coefficients of $|g'\rangle$ given by equation (3.11) on making the identifications

$$\Delta = \gamma \approx 2\langle \Delta\hat{N}^2\rangle, \tag{3.24}$$

$$|\alpha|^2 = \bar{n} \approx \langle \hat{N}\rangle, \tag{3.25}$$

$$\phi = \bar{\phi} \approx \langle \hat{\phi}_\theta\rangle. \tag{3.26}$$

With $\langle \hat{N}\rangle$ and $\langle \Delta\hat{N}^2\rangle$ given by equations (3.23)–(3.25), the requirement equation (3.20) now becomes

$$|\alpha|^2 \gg 2\exp(2r) \gg |\alpha|^{-2},$$

which is satisfied by equation (3.22) for strongly coherent squeezed states. The strongly coherent squeezed states $|\alpha, \zeta\rangle$ with $\phi = \frac{1}{2}(\theta \pm \pi)$ are therefore approximately $\hat{N}$–$\hat{\phi}_\theta$ IS. From equations (3.14) and (3.18), and with $\gamma = 2|\alpha|^2 \exp(2r)$ according to equations (3.23) and (3.24), we find

$$\langle \Delta\hat{N}^2\rangle \approx |\alpha|^2 \exp(2r), \tag{3.27}$$

$$\langle \Delta\hat{\phi}_\theta^2\rangle \approx \tfrac{1}{4}|\alpha|^{-2} \exp(-2r), \tag{3.28}$$

which shows that the uncertainty in phase reduces as the intensity $|\alpha|^2$ or the squeezing parameter r increases at the expense of increased uncertainty in photon number. We note that here the state $|\alpha, \zeta\rangle$ (with $\phi = \frac{1}{2}[\theta \pm \pi]$) gives either a super-Poisson or a Poisson photon-number probability distribution while $|g'\rangle$ represents a more general physical approximate $\hat{N}$–$\hat{\phi}_\theta$ IS because it can have a sub-Poisson as well as a super-Poisson photon-number probability distribution, according to equation (3.20).

All of our results for the strongly-coherent squeezed states $|\alpha, \zeta\rangle$ are valid in the limit as r tends to zero. In this limit $|\alpha, \zeta\rangle$ are the intense coherent states $|\alpha\rangle$ with $\langle \hat{N}\rangle = |\alpha|^2 \gg 1$, according to equation (3.22), and we regain the result [11] that intense coherent states are approximately $\hat{N}$–$\hat{\phi}_\theta$ IS.

It is interesting to note that Loudon and Knight's calculation [9] of the variance in phase angle for the same state agrees with the right-hand side of equation (3.28). In the absence of a Hermitian phase operator at the time of their work, they took the uncertainty in phase angle to be the angle subtended at the origin by a projection of an 'uncertainty ellipse' which is a particular equal-height contour of the Wigner function $W(x, y)$. The fully quantum-mechanical Hermitian phase operator approach used in this paper shows mathematically that their phenomenologically derived phase uncertainty is correct for high-amplitude field states. The phase operator approach also gives a physical insight into the reason for this close agreement. Under the transformation $\exp(\mathrm{i}\hat{N}\theta)$, the quadrature–amplitude operator $\hat{Y}$ is rotated through an angle θ to become $\hat{Y}(\theta)$. The probability density $P_\theta(0)$ that a measurement of $\hat{Y}(\theta)$ will yield a value of zero is given by integrating $W(x, y)$ along the radial line which is at an angle of θ to the x axis. The angular width of the 'uncertainty ellipse', which is the phenomenological phase uncertainty, is approximately determined by noting the range of θ values for which $P_\theta(0)$ is significant. The corresponding method of determining the quantum-mechanical phase uncertainty would involve shifting the phase of the field by θ while noting the probability density that a measurement of phase yields a given particular value. The range of values of θ for which this probability density is significant gives the quantum-mechanical phase uncertainty. The phase-shift operator is *also* $\exp(i\hat{N}\theta)$. We note that both cases require two steps: a shift and a measurement. In both cases the shift is described by precisely the same operator, however the measurements are different. The former is a measurement of $\hat{Y}(\theta)$ and the latter is a measurement of the phase operator. It is known that $\hat{Y}$ is a good measure of the sine of the phase for large amplitude fields of reasonably well defined phase [18], and so for such fields both of the above cases will yield the same result. On the other hand, for other states, such as very low-intensity fields, it is not difficult to show that the expectation value of the commutator of $\hat{a}$ and $\exp(i\hat{\phi}_\theta)$ is quite significant and thus $\hat{Y}$ and $\hat{\phi}_\theta$ are incompatible. Hence the full quantum-mechanical approach must be used for these states, which are not accessible to the phenomenological approach.

4. The $\hat{N}$–$\cos\hat{\phi}_\theta$ and $\hat{N}$–$\sin\hat{\phi}_\theta$ intelligent and minimum-uncertainty states

In this section we shall examine the IS and MUS associated with the two pairs of operators $\hat{N}$–$\cos\hat{\phi}_\theta$ and $\hat{N}$–$\sin\hat{\phi}_\theta$. The $\cos\hat{\phi}_\theta$ and $\sin\hat{\phi}_\theta$ phase operators can be defined by [10]

$$\cos\hat{\phi}_\theta \equiv \sum_{m=0}^{s} \cos(\theta_m)\,|\theta_m\rangle\langle\theta_m|,$$

$$\sin\hat{\phi}_\theta \equiv \sum_{m=0}^{s} \sin(\theta_m)\,|\theta_m\rangle\langle\theta_m|.$$

The commutators of $\cos\hat{\phi}_\theta$ and $\sin\hat{\phi}_\theta$ with $\hat{N}$ are found from their number-state representations

$$\cos\hat{\phi}_\theta = \frac{1}{2}\left\{\sum_{n=0}^{s-1} |n\rangle\langle n+1| + \exp[i(s+1)\theta_0]|s\rangle\langle 0| + \text{H.c.}\right\},$$

$$\sin\hat{\phi}_\theta = -\frac{1}{2}i\left\{\sum_{n=0}^{s-1} |n\rangle\langle n+1| + \exp[i(s+1)\theta_0]|s\rangle\langle 0| - \text{H.c.}\right\},$$

to be

$$[\cos\hat{\phi}_\theta, \hat{N}] = i\sin\hat{\phi}_\theta - \{\tfrac{1}{2}(s+1)\exp[i\theta_0(s+1)]\,|s\rangle\langle 0| - \text{H.c.}\}, \tag{4.1}$$

$$[\sin\hat{\phi}_\theta, \hat{N}] = -i\cos\hat{\phi}_\theta + \{\tfrac{1}{2}(s+1)\exp[i\theta_0(s+1)]\,|s\rangle\langle 0| + \text{H.c.}\}. \tag{4.2}$$

Accordingly, the Heisenberg uncertainty relations follow:

$$\langle\Delta\hat{N}^2\rangle\langle\Delta\cos\hat{\phi}_\theta^2\rangle \geqslant \tfrac{1}{4}|\langle[\cos\hat{\phi}_\theta, \hat{N}]\rangle|^2, \tag{4.3}$$

$$\langle\Delta\hat{N}^2\rangle\langle\Delta\sin\hat{\phi}_\theta^2\rangle \geqslant \tfrac{1}{4}|\langle[\sin\hat{\phi}_\theta, \hat{N}]\rangle|^2. \tag{4.4}$$

The $\hat{N}$–$\cos\hat{\phi}_\theta$ or $\hat{N}$–$\sin\hat{\phi}_\theta$ IS are those states satisfying the equality in equations (4.3) or (4.4). If the IS are also physical states, $|p_{NC}\rangle$ or $|p_{NS}\rangle$, then they will respectively satisfy

$$\langle p_{NC}|\Delta\hat{N}^2|p_{NC}\rangle\langle p_{NC}|\Delta\cos\hat{\phi}_\theta^2|p_{NC}\rangle=\tfrac{1}{4}|\langle p_{NC}|\sin\hat{\phi}_\theta|p_{NC}\rangle|^2, \tag{4.5}$$

$$\langle p_{NS}|\Delta\hat{N}^2|p_{NS}\rangle\langle p_{NS}|\Delta\sin\hat{\phi}_\theta^2|p_{NS}\rangle=\tfrac{1}{4}|\langle p_{NS}|\cos\hat{\phi}_\theta|p_{NS}\rangle|^2, \tag{4.6}$$

because the expectation value of the terms involving projection operators $|s\rangle\langle 0|$ and $|0\rangle\langle s|$ in equations (4.1) and (4.2) vanish for physical states.

We look first at the $\hat{N}$–$\cos\hat{\phi}_\theta$ IS. These can be determined by solving the eigenvalue equation

$$(\hat{N}+i\gamma\cos\hat{\phi}_\theta)|p_{NC}\rangle=\lambda|p_{NC}\rangle, \tag{4.7}$$

where, by equations (1.7) and (1.10),

$$\lambda=\langle\hat{N}\rangle+i\gamma\langle\cos\hat{\phi}_\theta\rangle, \tag{4.8}$$

$$|\gamma|=[\langle\Delta\hat{N}^2\rangle/\langle\Delta\cos\hat{\phi}_\theta^2\rangle]^{1/2}. \tag{4.9}$$

Taking the inner product of both sides of equation (4.7) with the number state $|n\rangle$ yields

$$(\lambda-n)d_n=i\tfrac{1}{2}\gamma(d_{n+1}+d_{n-1}),\quad 0<n<s, \tag{4.10}$$

$$\lambda d_0=i\tfrac{1}{2}\gamma\{d_1+d_s\exp[-i(s+1)\theta_0]\}, \tag{4.11}$$

$$(\lambda-s)d_s=i\tfrac{1}{2}\gamma\{d_0\exp[i(s+1)\theta_0]+d_{s-1}\}, \tag{4.12}$$

where d_n are the number-state coefficients of $|p_{NC}\rangle$. For physical IS the terms sd_s and d_{s-1} will vanish as s becomes large so, from equation (4.12) γd_0 must also vanish for large s. If $\gamma\neq 0$, then $d_0=0$ and from equation (4.11), $d_1=0$. Substituting into equation (4.10) with $n=1$ then gives $d_2=0$, and subsequent substitution in equation (4.10) with $n=2$ then gives $d_3=0$ and so on, giving $d_n=0$ for all n. Thus a normalizable solution requires $\gamma=0$. In this case the right-hand sides of equations (4.10), (4.11) and (4.12) all vanish and $d_n=0$ unless λ is an integer from 0 to s, that is, the only possible physical IS is a single number state. To show that a single number state is an IS, we note that for such a state [10]

$$\langle\cos\hat{\phi}_\theta\rangle=\langle\sin\hat{\phi}_\theta\rangle=0,$$

$$\langle\Delta\cos\hat{\phi}_\theta^2\rangle=\langle\Delta\sin\hat{\phi}_\theta^2\rangle=\tfrac{1}{2},$$

$$\langle\Delta\hat{N}^2\rangle=0.$$

On substituting these values into equation (4.5) we find that both sides are zero, which demonstrates that the number states are indeed $\hat{N}$–$\cos\hat{\phi}_\theta$ IS. Here the uncertainty product $\langle\Delta\hat{N}^2\rangle\langle\Delta\cos\hat{\phi}_\theta^2\rangle$ is zero, and therefore it is minimized by number states and so number states are also $\hat{N}$–$\cos\hat{\phi}_\theta$ MUS. This analysis shows that the *single number state is the only physical $\hat{N}$–$\cos\hat{\phi}_\theta$ IS and if any other IS exists then it is not a physical state*. For example, it is not difficult to show that a phase state is a $\hat{N}$–$\cos\hat{\phi}_\theta$ IS, but this is not a physical state.

We now develop a set of physical states which are not number states, whose uncertainty ratio

$$\langle\Delta\hat{N}^2\rangle\langle\Delta\cos\hat{\phi}_\theta^2\rangle:\tfrac{1}{4}|\langle\sin\hat{\phi}_\theta\rangle|^2,$$

while not exactly unity, can be made as close to it as desired (i.e. physical states which are approximately IS). We note that for physical states d_s must vanish as s becomes large. If equation (4.12) were to hold exactly for IS other than number states, d_s would need to decrease as s increases at a rate that makes $(\lambda-s)d_s$ finite and non-zero in the limit of large s. Since this rate would produce an infinite mean photon-number, we cannot satisfy equation (4.12) exactly if we are to find approximate IS solutions representing physical states. Consequently our approximation will involve finding the solution of the set of equations which approximate equations (4.10)–(4.12) and which are obtained by setting $d_s=d_{s-1}=0$, as required for physical states, and by replacing the resulting exact expression $d_0=0$ obtained from equation (4.12) with $\gamma\neq 0$ by the approximate expression that d_0 is non-zero but small. That is, we now have

$$(\lambda-n)d_n=\mathrm{i}\tfrac{1}{2}\gamma(d_{n+1}+d_{n-1}),\quad 0<n<s, \tag{4.13}$$

$$\lambda d_0=\mathrm{i}\tfrac{1}{2}\gamma d_1, \tag{4.14}$$

with $\gamma\neq 0$ and d_0 small. These equations are identical to those obtained by using the Susskind–Glogower operator $\cos\hat{\phi}_{\mathrm{SG}}$ in place of $\cos\hat{\phi}_\theta$, and are thus satisfied exactly by the $\hat{N}$–$\cos\hat{\phi}_{\mathrm{SG}}$ IS in the limit of infinite s. This is not surprising when it is remembered that although the operators $\cos\hat{\phi}_\theta$ and $\cos\hat{\phi}_{\mathrm{SG}}$ have very different algebraic properties, they give identical expectation values when taken between states with no vacuum component. Indeed we have shown [19] that for physical states

$$\langle\cos\hat{\phi}_\theta\rangle=\langle\cos\hat{\phi}_{\mathrm{SG}}\rangle, \tag{4.15}$$

$$\langle\sin\hat{\phi}_\theta\rangle=\langle\sin\hat{\phi}_{\mathrm{SG}}\rangle, \tag{4.16}$$

$$\langle\Delta\cos\hat{\phi}_\theta^2\rangle=\langle\Delta\cos\hat{\phi}_{\mathrm{SG}}^2\rangle+\tfrac{1}{4}\langle(|0\rangle\langle 0|)\rangle. \tag{4.17}$$

Therefore physical $\hat{N}$–$\cos\hat{\phi}_{\mathrm{SG}}$ IS will approximate the physical $\hat{N}$–$\cos\hat{\phi}_\theta$ IS provided that the projection on to the vacuum is negligible, that is, when d_0 is very small.

Jackiw [1] has solved the recurrence equations (4.13) and (4.14) in the limit of infinite s for the particular case when λ is real as

$$d_n=A(\gamma)(-\mathrm{i})^n I_{n-\lambda}(\gamma) \tag{4.18}$$

with

$$I_{-\lambda-1}(\gamma)=0, \tag{4.19}$$

and where $I_\nu(x)$ is the modified Bessel function of order ν, and $A(\gamma)$ is a normalization constant. We note that γ is real [15], thus by equation (4.19), we require real zeroes of $I_{-\lambda-1}(\gamma)$ for $\lambda+1$ real and positive which can only be fulfilled if λ satisfies

$$2k-1<\lambda+1<2k,$$

where k is a positive integer. The normalization constant $A(\gamma)$ is determined up to a phase factor by

$$|A(\gamma)|^2\sum_{n=0}^{\infty}I_{n-\lambda}(\gamma)^2=1. \tag{4.20}$$

As equation (4.18) gives exactly the coefficients of the $\hat{N}$–$\cos\hat{\phi}_{SG}$ IS, then by equations (1.8) and (1.9) and noting that $[\hat{N}, \cos\hat{\phi}_{SG}] = -i\sin\hat{\phi}_{SG}$, we find [1]:

$$\left.\begin{aligned} \langle \Delta \cos \hat{\phi}_{SG}^2 \rangle &= -(2\gamma)^{-1} \langle \sin \hat{\phi}_{SG} \rangle, \\ \langle \Delta \hat{N}^2 \rangle &= -\tfrac{1}{2}\gamma \langle \sin \hat{\phi}_{SG} \rangle, \\ \langle \Delta \hat{N}^2 \rangle \langle \Delta \cos \hat{\phi}_{SG}{}^2 \rangle &= \tfrac{1}{4} \langle \sin \hat{\phi}_{SG} \rangle^2, \\ \langle \cos \hat{\phi}_{SG} \rangle = 0, &\qquad \lambda = \langle \hat{N} \rangle. \end{aligned}\right\} \tag{4.21}$$

The sign of γ is the same as $-\langle \sin \hat{\phi}_{SG} \rangle$ and so the right-hand sides of the first two expressions are always non-negative.

A sufficient condition for the states with coefficients given by equation (4.18) to be physical states is that $\langle \exp(x\hat{N}) \rangle$ is finite for some positive x values as s tends to infinity. We find from equation (4.18)

$$\langle \exp(x\hat{N}) \rangle = |A(\gamma)|^2 \sum_{n=0}^{\infty} I_{n-\lambda}(\gamma)^2 \exp(xn). \tag{4.22}$$

For $n > \lambda + \frac{1}{2}$,

$$I_{n-\lambda}(\gamma)^2 \leqslant \frac{4(\frac{1}{2}\gamma)^{2n-2\lambda} \sinh^2 \gamma}{\pi \Gamma(n-\lambda+\frac{1}{2})^2 \gamma^2},$$

where $\Gamma(x)$ is the gamma function [1]. This is sufficient to show that the right-hand side of equation (4.22) converges for finite x, and proves that the states with number-state coefficients given by equation (4.18) are physical states. The relationship between the Susskind–Glogower and Pegg–Barnett phase operators given by equations (4.15)–(4.17) will therefore hold for these states. On comparing equations (4.21) with equations (4.15)–(4.17) we find that the uncertainties in $\cos\hat{\phi}_\theta$ and $\hat{N}$ are

$$\langle \Delta \cos \hat{\phi}_\theta^2 \rangle = -(2\gamma)^{-1} \langle \sin \hat{\phi}_\theta \rangle + \tfrac{1}{4}|d_0|^2,$$

$$\langle \Delta \hat{N}^2 \rangle = -\tfrac{1}{2}\gamma \langle \sin \hat{\phi}_\theta \rangle,$$

$$\langle \Delta \hat{N}^2 \rangle \langle \Delta \cos \hat{\phi}_\theta^2 \rangle = \tfrac{1}{4} \langle \sin \hat{\phi}_\theta \rangle^2 - (\gamma/8)|d_0|^2 \langle \sin \hat{\phi}_\theta \rangle,$$

where

$$\langle \sin \hat{\phi}_\theta \rangle = -|A(\gamma)|^2 \sum_{n=0}^{\infty} I_{n-\lambda+1}(\gamma) I_{n-\lambda}(\gamma),$$

$$\langle \cos \hat{\phi}_\theta \rangle = 0.$$

The value of $|d_0|^2$ is found from equation (4.14) to satisfy

$$|d_0|^2 = (\tfrac{1}{2}\gamma/\lambda)^2 |d_1|^2.$$

The normalization by equation (4.20) ensures that $|d_1|^2 \leqslant 1$, therefore $|d_0|^2 \leqslant (\frac{1}{2}\gamma/\lambda)^2$, and so on, using the fact that $|\langle \sin\hat{\phi}_\theta \rangle| \leqslant 1$ and $\lambda = \langle \hat{N} \rangle$ by equation (4.21), we find

$$(\gamma/8)|d_0|^2 |\langle \sin \hat{\phi}_\theta \rangle| \leqslant \gamma^3/(32\langle \hat{N} \rangle^2),$$

which is negligible when $\langle \hat{N} \rangle^2 \gg \gamma^3$. In that case we may write

$$\langle \Delta \hat{N}^2 \rangle \langle \Delta \cos \hat{\phi}_\theta^2 \rangle = \tfrac{1}{4} \langle \sin \hat{\phi}_\theta \rangle^2 + O(\gamma^3/\langle \hat{N} \rangle^2), \tag{4.23}$$

which when compared with Heisenberg's uncertainty relation (4.5), demonstrates that the IS for the $\hat{N}$–$\cos\hat{\phi}_{SG}$ operators are approximate $\hat{N}$–$\cos\hat{\phi}_\theta$ IS, with the approximation becoming more accurate as the mean photon number increases.

We found in the previous section that the physical states $|g'\rangle$ are approximately $\hat{N}$–$\hat{\phi}_\theta$ IS. We now show that the states $|g'\rangle$ are also approximately $\hat{N}$–$\cos\hat{\phi}_\theta$ IS. The expectation values of the first and second powers of $\cos\hat{\phi}_\theta$ and $\sin\hat{\phi}_\theta$ for $|g'\rangle$ can be calculated approximately from equation (3.17) to be

$$\langle\cos\hat{\phi}_\theta\rangle \approx \exp[-1/(4\gamma)]\cos\bar{\phi}, \tag{4.24}$$

$$\langle\sin\hat{\phi}_\theta\rangle \approx \exp[-1/(4\gamma)]\sin\bar{\phi}, \tag{4.25}$$

$$\langle\cos\hat{\phi}_\theta^2\rangle \approx \tfrac{1}{2}[1+\exp(-1/\gamma)\cos(2\bar{\phi})],$$

$$\langle\sin\hat{\phi}_\theta^2\rangle \approx \tfrac{1}{2}[1-\exp(-1/\gamma)\cos(2\bar{\phi})]. \tag{4.26}$$

Using $\langle\Delta\hat{N}^2\rangle$ for $|g'\rangle$ given by equation (3.14) we find the uncertainty product

$$\langle\Delta\hat{N}^2\rangle\langle\Delta\cos\hat{\phi}_\theta^2\rangle \approx \tfrac{1}{4}\gamma[1-\exp(-\tfrac{1}{2}\gamma^{-1})][1-\cos(2\bar{\phi})\exp(-\tfrac{1}{2}\gamma^{-1})]. \tag{4.27}$$

The uncertainty ratio $\langle\Delta\hat{N}^2\rangle\langle\Delta\cos\hat{\phi}_\theta^2\rangle : \tfrac{1}{4}\langle\sin\hat{\phi}_\theta\rangle^2$ has a minimum value of approximately

$$2\gamma\sinh(\tfrac{1}{2}\gamma^{-1}) = 1+O(\gamma^{-2})$$

at $\cos\bar{\phi}=0$, and thus the physical state $|g'\rangle$ is also an approximate $\hat{N}$–$\cos\hat{\phi}_\theta$ IS here. Expanding equations (4.25) and (4.27) in powers of γ^{-1} for $\cos\bar{\phi}=0$ gives

$$\langle\Delta\hat{N}^2\rangle\langle\Delta\cos\hat{\phi}_\theta^2\rangle \approx \tfrac{1}{4}\langle\sin\hat{\phi}_\theta\rangle^2+O(\gamma^{-2}), \tag{4.28}$$

which can be compared with equation (4.5). We also find

$$\langle\Delta\cos\hat{\phi}_\theta^2\rangle \approx \langle\Delta\hat{\phi}_\theta^2\rangle+O(\gamma^{-2}),$$

indicating that $\cos\hat{\phi}_\theta$ has similar fluctuations to $\hat{\phi}_\theta$ under these conditions. It is interesting to note that $\cos\bar{\phi}=0$ gives the ***maximum*** of both sides of equation (4.28).

In the previous section we found that the strongly coherent squeezed states $|\alpha,\zeta\rangle$ with $\phi=\tfrac{1}{2}(\theta\pm\pi)$ are a subset of $|g'\rangle$ and therefore they are also approximately $\hat{N}$–$\cos\hat{\phi}_\theta$ IS when $\cos\phi=0$.

The properties of the $\hat{N}$–$\sin\hat{\phi}_\theta$ IS follow closely those of the $\hat{N}$–$\cos\hat{\phi}_\theta$ IS and we shall outline their main features only. The number states are again the only exact physical IS and MUS. The $\hat{N}$–$\sin\hat{\phi}_{SG}$ IS [2],

$$|SG_{NS}\rangle = A(\gamma)\sum_{n=0}^{\infty} I_{n-\lambda}(\gamma)|n\rangle,$$

is a physical state that is approximately a $\hat{N}$–$\sin\hat{\phi}_\theta$ IS, as demonstrated by the results

$$\langle\Delta\hat{N}^2\rangle = \tfrac{1}{2}\gamma\langle\cos\hat{\phi}_\theta\rangle,$$

$$\langle\Delta\sin\hat{\phi}_\theta^2\rangle = (2\gamma)^{-1}\langle\cos\hat{\phi}_\theta\rangle+O(\gamma^2/\langle\hat{N}\rangle^2),$$

$$\langle\Delta\hat{N}^2\rangle\langle\Delta\sin\hat{\phi}_\theta^2\rangle = \tfrac{1}{4}\langle\cos\hat{\phi}_\theta\rangle^2+O(\gamma^3/\langle\hat{N}\rangle^2),$$

with $\langle\sin\hat{\phi}_\theta\rangle=0$.

The physical states $|g'\rangle$ are found to give

$$\langle\Delta\hat{N}^2\rangle\langle\Delta\sin\phi_\theta^2\rangle \approx \tfrac{1}{4}\langle\cos\hat{\phi}_\theta\rangle^2+O(\gamma^{-2}),$$

for $\sin\bar{\phi}=0$ and very large γ. The ratio of the left- to the right-hand side of this expression is approximately $1+O(\gamma^{-2})$ which manifests the approximate $\hat{N}$–$\sin\hat{\phi}_\theta$ IS behaviour of the states $|g'\rangle$. We also find $\langle\Delta\sin\hat{\phi}_\theta^2\rangle \approx \langle\Delta\hat{\phi}_\theta^2\rangle+O(\gamma^{-2})$ and so

$\sin\hat{\phi}_\theta$ behaves as $\hat{\phi}_\theta$ when $\sin\bar{\phi}=0$. The strongly coherent squeezed states $|\alpha,\zeta\rangle$ are also approximate $\hat{N}$–$\sin\hat{\phi}_\theta$ IS for $\sin\phi=0$ and $\phi=\frac{1}{2}(\theta\pm\pi)$.

In this section we have shown that there are no exact physical $\hat{N}$–$\sin\hat{\phi}_\theta$ and $\hat{N}$–$\cos\hat{\phi}_\theta$ IS other than number states. Secondly, we have shown that Jackiw's number-phase intelligent-state solutions for the Susskind–Glogower sine and cosine operators are also approximate intelligent states for the corresponding Pegg–Barnett sine and cosine phase operators. This is not unexpected with the underlying physical reason being as follows. The Susskind–Glogower operators and the corresponding Pegg–Barnett exponential phase operators have different properties only when acting on states which contain a significant overlap with the vacuum. Thus Jackiw's intelligent states which have negligible overlap with the vacuum will also be approximate $\hat{N}$–$\sin\hat{\phi}_\theta$ or $\hat{N}$–$\cos\hat{\phi}_\theta$ IS.

5. Discussion and conclusion

The formalism of the Hermitian phase operator permits a phase-state expansion of any field state. The phase-state coefficients c_m in this expansion depend upon the choice of a 2π window, that is, the choice of θ_0 [10–12]. In this paper we have seen that there exists a periodic function $\psi(\theta)$, where θ varies continuously from $-\infty$ to $+\infty$, comprising all possible values of c_m multiplied by a constant scale factor for any choice of θ_0. The requirement that a state of the field be a physical state, that is, that it has finite energy moments, imposes the simple but strict continuity conditions on the curve $\psi(\theta)$ that any finite-order derivative must be bounded. We have exploited these conditions to find states which are physical and minimise number-phase uncertainty. For the three uncertainty relations considered, involving phase and the sine and cosine functions of phase, we find that the single number states take on a particular significance in that they are the only physical intelligent states and also the only physical minimum-uncertainty states.

The physical state $|g'\rangle$, which was shown in particular to be an approximate $\hat{N}$–$\hat{\phi}_\theta$ intelligent state, has a Gaussian dependence in its number-state coefficients and an approximate Gaussian dependence in its phase-state coefficients. The commutator operator $[\hat{N},\hat{\phi}_\theta]$ is equivalent simply to i when acting on these states. We note that the position-momentum commutator is i and so $|g'\rangle$ can be compared with the position–momentum minimum-uncertainty states which have a Gaussian dependence in both their position and momentum representations.

The intelligent states associated with the Susskind–Glogower number-phase uncertainty relations were found to be physical states which are also approximate intelligent states for the corresponding Pegg–Barnett number-phase uncertainty relations when the vacuum-state coefficient was negligible. The reason for this can be traced to the fact that while the algebraic properties of the Pegg–Barnett and the Susskind–Glogower phase operators differ markedly, the expectation values and variances of the respective cosine and sine phase operators are identical for states with no vacuum component, and the respective number–cosine and number–sine commutators are identical for physical states.

The connection between the $\hat{N}$–$\cos\hat{\phi}_\theta$ and $\hat{N}$–$\sin\hat{\phi}_\theta$ IS and the strongly coherent squeezed state can be made clearer in the following way. The operator $\hat{N}$ does not commute with the quadrature–amplitude operator $\hat{Y}$ because

$$[\hat{N},\hat{Y}]=\mathrm{i}\hat{X},$$

where $\hat{Y} = -\frac{1}{2}i(\hat{a} - \hat{a}^\dagger)$ and $\hat{X} = \frac{1}{2}(\hat{a} + \hat{a}^\dagger)$, and therefore the variances in $\hat{N}$ and $\hat{Y}$ must satisfy the Heisenberg uncertainty relation

$$\langle \Delta \hat{N}^2 \rangle \langle \Delta \hat{Y}^2 \rangle \geqslant \tfrac{1}{4} \langle \hat{X} \rangle^2. \tag{5.1}$$

Consider the strongly coherent squeezed state, that is the ideal squeezed state $|\alpha, \zeta\rangle$ with

$$\langle \hat{N} \rangle \approx |\alpha|^2 \gg \exp(2|\zeta|),$$

for $\arg(\alpha) = 0$ and $\arg(\zeta) = \pi$. We have shown that this state is approximately an $\hat{N}$–$\sin\hat{\phi}_\theta$ intelligent state. It can also be shown that for this state [9]

$$\langle \Delta \hat{N}^2 \rangle \approx |\alpha|^2 \exp(2r),$$

$$\langle \hat{Y} \rangle = 0,$$

$$\langle \Delta \hat{Y}^2 \rangle = \tfrac{1}{4} \exp(-2r),$$

$$\langle \hat{X} \rangle^2 = |\alpha|^2,$$

where $r = |\zeta|$. These results give approximately the equality in equation (5.1) and so the state $|\alpha, \zeta\rangle$ is approximately an $\hat{N}$–$\hat{Y}$ intelligent state here. We can show that this state also gives

$$\langle \sin\hat{\phi}_\theta \rangle \approx 0,$$

$$\langle \Delta \sin\hat{\phi}_\theta^2 \rangle \approx \tfrac{1}{4} |\alpha|^{-2} \exp(-2r) + O[|\alpha|^{-4} \exp(-4r)],$$

$$\langle \cos\hat{\phi}_\theta \rangle^2 \approx 1 + O[|\alpha|^{-2} \exp(-2r)],$$

and so $\sin\hat{\phi}_\theta$ behaves approximately as $|\alpha|^{-1}\hat{Y}$ and

$$\langle \cos\hat{\phi}_\theta \rangle^2 \approx |\alpha|^{-2} \langle \hat{X} \rangle^2$$

(although $\langle \Delta \cos\hat{\phi}_\theta^2 \rangle \neq |\alpha|^{-2} \langle \Delta \hat{X}^2 \rangle$) for this state. On making these identifications, equation (5.1) becomes the Heisenberg uncertainty relation describing the variances in $\hat{N}$ and $\sin\hat{\phi}_\theta$ for physical states, which explains why the state $|\alpha, \zeta\rangle$ is *simultaneously* an approximate $\hat{N}$–$\sin\hat{\phi}_\theta$ intelligent state and an approximate $\hat{N}$–$\hat{Y}$ intelligent state. It is interesting to note that $|\alpha|^{-1}\hat{Y}$ is the ***measured sine phase operator*** [18] for the state considered here and this reinforces our approximate identification of $\sin\hat{\phi}_\theta$ with $|\alpha|^{-1}\hat{Y}$. We also found in section 4 for the state considered here that $\sin\hat{\phi}_\theta$ behaves as $\hat{\phi}_\theta$ and this accounts for the fact that this same state is also an approximate $\hat{N}$–$\hat{\phi}_\theta$ intelligent state as well. An analogous relation exists between the approximate $\hat{N}$–$\hat{X}$ intelligent states and the approximate $\hat{N}$–$\cos\hat{\phi}_\theta$ intelligent states.

In conclusion, the Hermitian phase operator formalism, by means of simple and strict phase-state coefficient continuity conditions for physical states, provides a deeper insight into the phase properties of light with particular reference to number-phase uncertainty, and illuminates previous studies involving phenomenological approaches and quadrature–amplitude operators.

Acknowledgment

D.T.P. thanks Dr S. M. Barnett for initial discussions. J.A.V. acknowledges the support of a Commonwealth Postgraduate Research Award.

References

[1] Jackiw, R., 1968, *J. math. Phys.*, **9**, 339.
[2] Carruthers, P., and Nieto, M. M., 1965, *Phys. Rev. Lett.*, **14**, 387; 1968, *Rev. mod. Phys.*, **40**, 411.
[3] Levy-Leblond, J. M., 1976, *Ann. Phys.*, **101**, 319.
[4] Sanders, B. C., Barnett, S. M., and Knight, P. L., 1986, *Optics Commun.*, **58**, 290.
[5] Yamamoto, Y., Machida, S., Imoto, N., Kitagawa, M., and Bjork, G., 1987, *J. opt. Soc. Am.* B, **4**, 1645.
[6] Susskind, L., and Glogower, J., 1964, *Physics*, **1**, 49.
[7] Matthys, D. R., and Jaynes, E. T., 1980, *J. opt. Soc. Am.*, **70**, 263.
[8] Shapiro, J. H., and Wagner, S. S., 1984, *IEEE J. quant. Electron.*, **20**, 803.
[9] Loudon, R., and Knight, P. L., 1987, *J. mod. Optics*, **34**, 709.
[10] Pegg, D. T., and Barnett, S. M., 1988, *Europhys. Lett.*, **6**, 483.
[11] Barnett, S. M., and Pegg, D. T., 1989, *J. mod. Optics*, **36**, 7.
[12] Pegg, D. T., and Barnett, S. M., 1989, *Phys. Rev.* A, **39**, 1665.
[13] Aragone, C., Chalbaud, E., and Salamo, S., 1976, *J. math. Phys.*, **17**, 1963.
[14] Wódkiewicz, K., and Eberly, J. H., 1985, *J. opt. Soc. Am.* B, **2**, 458.
[15] See for example Gottfried, K., 1966, *Quantum Mechanics*, Vol. 1 (New York: W. A. Benjamin), pp. 213–215.
[16] For reviews of squeezed light see for example Walls, D. F., 1983, *Nature*, **306**, 141; and also reference [9].
[17] Yuen, H. P., 1976, *Phys. Rev.* A, **13**, 2226.
[18] Barnett, S. M., and Pegg, D. T., 1986, *J. Phys.* A, **19**, 3849.
[19] Vaccaro, J. A., and Pegg, D. T., 1989, *Optics Commun.*, **70**, 529.

PHASE OPTIMIZED QUANTUM STATES OF LIGHT

G.S. SUMMY and D.T. PEGG
Division of Science and Technology, Griffith University, Nathan, Brisbane 4111, Australia

Received 14 December 1989

We find the states of light which have minimum phase variance both for a given maximum energy state component and for a given mean energy. When these states contain sufficiently many photon number state components, the number state coefficients approximate sinusoidal and Airy functions respectively. The phase of these new states of light is much more sharply defined than is the phase of a coherent state with the same mean energy.

Recently Pegg and Barnett [1] introduced a limiting procedure which allowed an hermitian phase operator to be included in the quantum theory of light. In this procedure, quantum mechanical calculations are carried out using a state space of $s+1$ dimensions, with the limit as s tends to infinity being found only after c-numbers such as means and variances have been derived as a function of s. Properties of the hermitian phase operator were subsequently explored in detail [2,3]. The eigenstates of this operator have precisely defined phase, but their mean energy is infinite [4]. The phase states are therefore not physical in the sense that they can be prepared [1,5]. The question thus arises – given a finite energy, what states of light have the best defined phase, that is, the minimum phase variance? In this paper we examine this problem for two distinct cases. In the first of these there is a fixed upper bound to the possible results of a measurement of the field energy, and in the second case the mean energy of the field is fixed.

The calculations can be simplified by first finding minimum phase variance states for a field with zero mean phase. Then the variance is simply the mean square phase. A simple unitary transformation, discussed later, subsequently yields minimum variance states of arbitrary mean phase. With a mean of zero, the most convenient choice of the arbitrary window in which to express the eigenvalues of phase is the interval $[-\pi, \pi]$, obtained by choosing the lowest phase eigenvalue θ_0 to be $-\pi$ [2,3]. Because we shall be adopting this value throughout this paper we shall write the phase operator, for convenience, simply as $\hat{\phi}$ in place of $\hat{\phi}_\theta$ used previously for arbitrary θ_0 [1–3]. We shall also assume, consistently with the problems we are examining, that $\hat{\phi}$ will be operating only on physical states. These are states which in the limit as s tends to infinity, have significant overlaps only with number states $|n\rangle$ for which n is finite. This being the case, we can write from the corresponding c-number expansion [6],

$$\hat{\phi}^2=\frac{\pi^2}{3}-4\left(\cos\hat{\phi}-\frac{\cos 2\hat{\phi}}{2^2}+\frac{\cos 3\hat{\phi}}{3^2}-...\right)$$

$$=\frac{\pi^2}{3}+2\sum_{n\neq n'}\frac{(-1)^{n-n'}}{(n-n')^2}|n'\rangle\langle n| \,, \qquad (1)$$

where $2\cos m\hat{\phi}=\exp(im\hat{\phi})+\exp(-im\hat{\phi})$.

The general field state can be expanded as

$$|f\rangle=\sum_n b_n|n\rangle \,, \qquad (2)$$

where $b_n=|b_n|\exp(i\zeta_n)$. We firstly seek values of ζ_n which minimize $\langle\phi^2\rangle$, which is equal to the phase variance for zero mean phase, independently of the amplitudes $|b_n|$. The values of ζ_n do not affect the normalization or the mean energy, so this minimization needs no constraints such as those which we shall use later. Differentiation of $\langle\phi^2\rangle$ obtained from (1) and (2) shows that, for the extremum condition to be satisfied independently of the values of $|b_n|$, we require that $\sin(\zeta_n-\zeta_{n'})$ vanishes for all n and n', that is $\zeta_n - \zeta_{n'}$ must be an integer multiple of π. Thus after removing a common factor of modulus unity from all b_n, the coefficients must be positive, zero or negative. This common factor is not detectable, so we can adopt for the extremum condition that the coefficients b_n in (2) are all real. It is not difficult to show from (1) that coefficients b_n with alternating signs in a superpostion of at least two number states produces a maximum in $\langle \phi^2 \rangle$, and hence in the variance, which is greater than $\pi^2/3$. The other extreme, where all the coefficients are real with the same sign, is a partial phase state [3], a limiting case of which is the phase state itself with zero phase variance. In the remainder of this paper we restrict ourselves to examining states with real b_n. It is straightforward to show further that such states have a phase probability distribution which is symmetric about zero, and consequently positive and negative changes in θ_0 will produce the same change in phase variance. Thus these states also have an extremum in the variance as a function of θ_0 at our chosen value of $\theta_0 = -\pi$.

Consider an hermitian operator $\hat{A}$ for which $\langle n|\hat{A}|n'\rangle$ are real, so the corresponding matrix is symmetric. The variation of the expectation value of this operator for a state $|f\rangle$ given by (2) with b_n all real is

$$\begin{aligned} \mathrm{d}\langle f|\hat{A}|f\rangle &= (\mathrm{d}\langle f|)\hat{A}|f\rangle + \langle f|\hat{A}(\mathrm{d}|f\rangle) \\ &= \sum_n \mathrm{d}b_n(\langle n|\hat{A}|f\rangle + \langle f|\hat{A}|n\rangle) \\ &= 2(\mathrm{d}\langle f|)\hat{A}|f\rangle\,, \end{aligned} \tag{3}$$

which follows from the equality of $\langle n|\hat{A}|f\rangle$ and $\langle f|\hat{A}|n\rangle$ for this operator. In order to minimize $\langle f|\hat{A}|f\rangle$ while keeping the expectation values of other hermitian operators $\hat{B}$, $\hat{C}$..., also with real matrix elements, constant, we introduce undetermined multipliers in the usual way and write

$$(\mathrm{d}\langle f|)(\alpha\hat{A} + \beta\hat{B} + ...)|f\rangle = 0\,. \tag{4}$$

For this equation to hold irrespective of particular variations $\mathrm{d}\langle f|$, we require that

$$(\alpha\hat{A} + \beta\hat{B} + ...)|f\rangle = 0\,. \tag{5}$$

Hermitian operators with real matrix elements in the number representation include $\hat{I}$, the unit operator, $\hat{N}$ and $\hat{\phi}^2$. The last follows from (1).

We now study the particular problem of minimizing the variance $\langle \phi^2 \rangle$ for states for which there is an upper bound to the possible results of a measurement of energy. Such states $|f\rangle$ will have number state coefficients b_n which are all zero for values $n > M$, where $M\hbar\omega$ is the maximum energy possible. For this case, the appropriate constraint is the normalization condition that $\langle f|\hat{I}|f\rangle$ is fixed at unity. Applying (5) with $\hat{A} = \hat{I}$, and $\hat{B} = \hat{\phi}^2$ gives

$$\hat{\phi}^2|f\rangle = \lambda|f\rangle\,, \tag{6}$$

where $\lambda = -\alpha/\beta$. Clearly, if we do not impose a limit to the number of non-zero coefficients, a phase eigenstate with zero phase is a solution of this with eigenvalue $\lambda = 0$, that is, a variance of zero. Our problem however, is to find states $|f\rangle$ possessing only $M+1$ non-zero coefficients which satisfy (6). Because $\hat{\phi}^2$ in (6) operates-on a vector with $M+1$ non-zero components, we replace the matrix $\langle n|\hat{\phi}^2|n'\rangle$ by a finite $(M+1)\times(M+1)$ matrix obtained by deleting those matrix elements for which $n, n' > M$. The problem is then reduced to finding the eigenvector of this finite matrix with the minimum eigenvalue λ. Note that it follows easily from (6) that λ is the mean of $\hat{\phi}^2$, that is, the variance. The problem can be solved analytically when $|f\rangle$ is constrained to contain only a small number of photon states. For example, when $M=2$, we easily find that the coefficients of $|0\rangle$, $|1\rangle$ and $|2\rangle$ for minimum variance are 0.48, 0.74, 0.48 with a corresponding value of phase variance of 0.70. This shows immediately that the coefficients are not in general equal, which means that the "rectangular" partial phase state [3] obtained by simply truncating a phase state and renormalizing is not a state of minimum phase variance. For larger values of M, numerical techniques must be used to find the exact eigenvectors and eigenvalues. An example is shown in fig. 1 of a numerical evaluation of the coefficients b_n for the state of minimum phase variance for $M=10$. The phase variance is 0.0713.

For states $|f\rangle$ of zero mean phase and small phase variance, which will occur when M is reasonably large, we can find an approximate analytic solution to (6) by approximating $\hat{\phi}^2$ by $\sin^2\hat{\phi}$. The validity of this approximation can be seen from the expansion of $\sin\hat{\phi}$, and the error involved is related to that arising when the square root of the variance, that is the standard deviation of the phase with a mean of zero, is approximated by its sine. This replacement in (6) gives

$$\lambda b_n = \langle n|\sin^2\hat{\phi}|f\rangle$$

$$= -(\langle n+2|f\rangle + \langle n-2|f\rangle - 2\langle n|f\rangle)/4\,, \quad (7)$$

which follows from expanding $\sin\hat{\phi}$ in terms of the up-shift and down-shift operators $\exp(-i\hat{\phi})$ and $\exp(i\hat{\phi})$ [1]. From (7) we obtain the recurrence relation

$$b_{n+2} = (2-4\lambda)b_n - b_{n-2}\,, \quad (8)$$

which, for b_n real, is satisfied by

$$b_n = K\sin[g(n+\epsilon)]\,, \quad (9)$$

in which case

$$\lambda = \sin^2 g\,. \quad (10)$$

From normalization, we find $K\approx 2^{1/2}(M+1)^{-1/2}$. We need $0<\epsilon<1$ so that $b_n=0$ for $n<0$. An estimate of ϵ can be found by noting from (6) that, for λ very small, $\langle 0|\hat{\phi}^2|f\rangle$ approximates zero. Assuming an approximate linear relation $b_n \propto n+\epsilon$ for small n and substituting into (2), we can obtain a series for $\langle 0|\hat{\phi}^2|f\rangle$ from (1). The vanishing of this series

$$\frac{\pi^2}{3}\epsilon + 2\left(-1-\epsilon+\frac{2}{4}+\frac{\epsilon}{4}-\frac{3}{9}-\frac{\epsilon}{9}+\ldots\right)\approx 0$$

yields $\epsilon\approx 12\ \ln 2/\pi^2 = 0.84$. By selecting $g=\pi/(M+2\epsilon)$ we can ensure that the continuous curve through the points b_n is zero between $n=M$ and $n=M+1$, so $b_n=0$ for $n>M$, and also obtain the symmetry in b_n that we observe in all the numerically calculated results. Thus we have finally

$$b_n = 2^{1/2}(M+1)^{-1/2}\sin[(n+\epsilon)\pi/(M+2\epsilon)]\,, \quad (11)$$

$$\langle\Delta\phi^2\rangle = \sin^2[\pi/(M+2\epsilon)] \approx \pi^2/(4\langle N\rangle^2)\,, \quad (12)$$

where the b_n is symmetric about $M/2$, so $\langle N\rangle = M/2$, and the large M approximation to λ, which is equal to the phase variance, is given. Other solutions are obtainable with integer multiples of π in place of π, but it is not difficult to show that (11) is the solution with the minimum phase variance. In fig. 1 the dotted line shows the function (11) with $M=10$, for comparison with the numerical results. The value for ϵ of 0.84 gives good agreement with numerical results over quite a range of values of M. Indeed as M increases, the precise value of ϵ becomes less important. It is interesting that even with a value of $M=2$ the variance is still sufficiently small to ensure that the approximate results (11) and (12), without the final approximation to $\pi^2/(4\langle N\rangle^2)$, are not grossly in error.

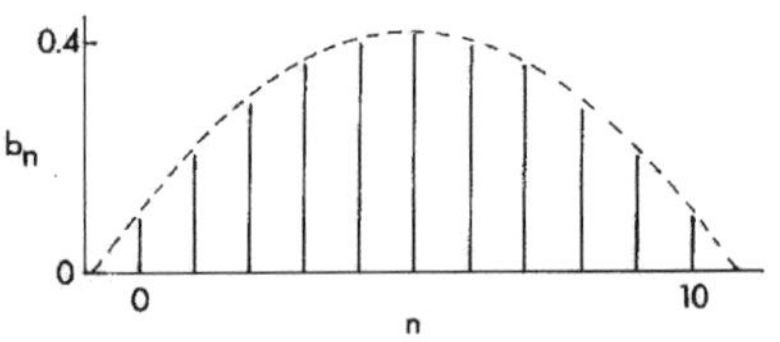

Fig. 1. Number state coefficients b_n of a state with minimum phase variance for a given value of $M=10$, that is, a maximum of 11 non-zero number state components. The full lines show the numerically calculated values of b_n and the dotted line is the sinusoidal function (11) with $\epsilon=0.84$.

The second problem we wish to examine is the case for which the mean photon number of the field, rather than the maximum photon number, is fixed. This is related to the problem studied recently by Shapiro et al. [7], with the major difference being that they were interested in maximizing the reciprocal peak height of the phase distribution, whereas we wish to minimize the phase variance, which leads to very different results. For this problem we must also include the photon number operator to give the eigenvalue equation

$$(\alpha\hat{I} + \beta\hat{\phi}^2 + \gamma\hat{N})|f\rangle = 0\,, \quad (13)$$

or

$$(\hat{N} + a\hat{\phi}^2)|f\rangle = \lambda|f\rangle\,, \quad (14)$$

where $a=\beta/\gamma$ and here $\lambda=-\alpha/\gamma$. Again this equation can be solved numerically, and we can also find analytic approximations.

The first approximation involves allowing a to be small enough for the second term in (14) to be treated as a perturbation on the photon number, or energy, eigenvalue equation. With a single number state $|n\rangle$ as the unperturbed state, the phase variance will be near the unperturbed value of $\pi^2/3$. For $n\neq 0$ we can easily find other states of similar mean number with much smaller phase variances than this, so the perturbation approach will be useful only when the unperturbed state is the vacuum state $|0\rangle$. This places the field very much in the "quantum" regime of small intensities, giving an added importance to this approximation. From standard second-order perturbation theory [8] we obtain

$$|f\rangle = |0\rangle - 2a \sum_{n=1}^{t} (-1)^n n^{-3} |n\rangle , \tag{15}$$

where we have inserted a large number cutoff of t to ensure that all energy moments of $|f\rangle$ are finite, as required for a physical state [3,5]. For t reasonably large, the value of the mean number and mean square number will be insensitive to its precise value:

$$\langle N\rangle = 4a^2 \sum_{n=1} n^{-5} = 4.148a^2 . \tag{16}$$

Similarly we can find $\langle N^2\rangle$ in terms of a and hence in terms of $\langle N\rangle$, giving a number variance eventually of $\langle N\rangle(1.04 - \langle N\rangle)$. The phase variance can be found from (1) and (15) to be, in terms of $\langle N\rangle$ from (16),

$$\langle \Delta\phi^2\rangle \approx \pi^2/3 - 4.07\langle N\rangle^{1/2} . \tag{17}$$

This is slightly less than the phase variance for a coherent state $|\alpha\rangle$ with the same value of $\langle N\rangle$,

$$\langle \Delta\phi^2\rangle \approx \pi^2/3 - 4\langle N\rangle^{1/2} . \tag{18}$$

Numerical calculations and later analytical results show that the relative difference increases markedly at higher values of mean energy.

In the opposite, high intensity, regime where $|f\rangle$ contains sufficient number state components to produce a small phase variance, we can again replace $\hat{\phi}^2$ by $\sin^2\hat{\phi}$, to obtain

$$\langle n|(\hat{N} + a\sin^2\hat{\phi})|f\rangle = \lambda\langle n|f\rangle \tag{19}$$

giving a recurrence relation

$$(b_{n+2} - b_n) - (b_n - b_{n-2}) = 4(n-\lambda)b_n/a . \tag{20}$$

We approach the problem of finding a solution to this unfamiliar recurrence relation by approximating it by a differential equation. Consider a set of points with ordinate $y = b_n$ and abscissa $x = n$. Then, provided the curve through these points varies in height sufficiently slowly with x, its slope at x can be approximated by

$$\mathrm{d}y/\mathrm{d}x \approx (b_{n+1} - b_{n-1})/2 . \tag{21}$$

The relation (20) then becomes

$$\mathrm{d}^2y/\mathrm{d}x^2 = (x-\lambda)y/a . \tag{22}$$

The substitutions

$$X = a^{-1/3}(x-\lambda) , \tag{23}$$

$$Y = a^{1/3}y , \tag{24}$$

reduce (22) to

$$\mathrm{d}^2Y/\mathrm{d}X^2 = XY . \tag{25}$$

The solution of this equation which leads to the state with the best phase definition for the lowest mean intensity is

$$Y = k\mathrm{Ai}(X) \tag{26}$$

where $\mathrm{Ai}(X)$ is an Airy function [9] and k is the normalization constant. As for our previous problem the curve y must have a zero between $n=0$ and $n=-1$, that is, when $x=-\epsilon$ with $0<\epsilon<1$. The expression for $\langle 0|f\rangle$ obtained from (14) is the same as that obtained from (6) with λ/a in place of λ. Thus, provided λ/a is small we again obtain $\epsilon \approx 0.84$. Because the zero in Y occurs at $X=-2.34$ [9], we obtain from (23)

$$\lambda = 2.34a^{1/3} - \epsilon , \tag{27}$$

and thus

$$X = a^{-1/3}(n+\epsilon) - 2.34 , \tag{28}$$

which, together with

$$b_n = ka^{-1/3}\mathrm{Ai}(X) \tag{29}$$

allows us to calculate the coefficients b_n of the minimum phase variance state $|f\rangle$ for any given value of a. Because of the rapid decay of the Airy function with n we do not insert a cutoff t into the summation over b_n.

Fig. 2 shows the values of b_n calculated numerically by diagonalizing (14) for $a=50$ and the corresponding Airy curve with a normalization constant $k=0.742$ and a value of $\epsilon=0.86$, chosen for best fit, which is close to our estimate for ϵ.

From (14) and (27) we obtain

$$\langle N\rangle + a\langle\phi^2\rangle = 2.34a^{1/3} - \epsilon . \tag{30}$$

By relating $\langle N\rangle$ to $\langle X\rangle$ through (28) and eliminating a from this relation and (30), we find eventually

$$\langle\Delta\phi^2\rangle = \langle\phi^2\rangle = C(\langle N\rangle + \epsilon)^{-2} , \tag{31}$$

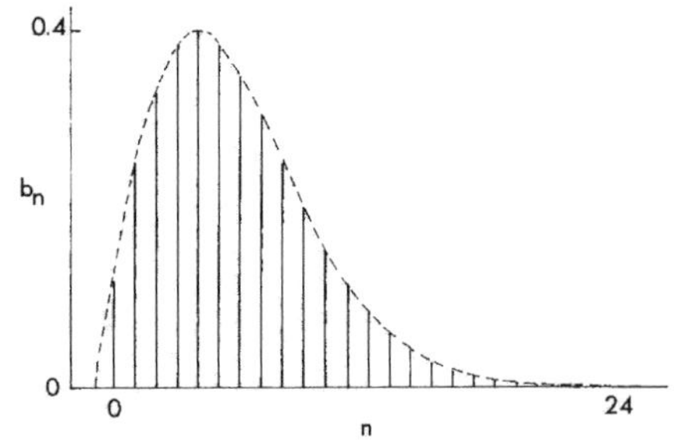

Fig. 2. Number state coefficients of a state with minimum phase variance for a given mean photon number of 4.86. The full lines are numerically calculated values and the dotted line is an Airy function given by eqs. (28) and (29) with $\epsilon=0.86$ and $k=0.742$.

where C is a positive constant equal to $-\langle X\rangle (\langle X\rangle+2.34)^2$. The constant $\langle X\rangle \sim -1$ can be obtained more precisely from a numerical analysis of the Airy function, but it is simpler to find C directly from numerical diagonalizations of (14). We find with $\epsilon=0.86$ that a value $C=1.88$ predicts from (31) phase variances of 0.0575 and 0.0196 for $\langle N\rangle=4.86$, 8.94 respectively, compared with numerically calculated values of 0.574 and 0.0197.

The dependence of phase variance on $\langle N\rangle^{-2}$ given by (31) can be compared with the large amplitude coherent state variance $\langle N\rangle^{-1}/4$ [2], and the truncated phase state variance [3] of $(2\ \ln 2)\langle N\rangle^{-1}$. Thus, although a large amplitude coherent state is approximately a number-phase minimum uncertainty state [2,5], our phase optimized Airy function state clearly has a much better defined phase. Also, comparison of (31) with (12) shows that the Airy function state has a smaller variance than a sinusoidal state with the same mean, although the dependence on $\langle N\rangle$ is similar.

The minimum phase variance states discussed in this paper have a zero mean phase. Operating on these states with the unitary phase shift operator $\exp(i\hat{N}\theta)$ and shifting θ_0 from $-\pi$ to $\theta-\pi$ will give minimum phase variance states with a mean phase of θ, and number state coefficients $b_n \exp(in\theta)$.

In conclusion, we have examined the problem of finding states of light with the best possible defined phase when (a) the maximum energy state component is given and (b) the mean energy of the state is given. Both problems can be solved by analytic approximations for small and large mean energies, and numerically for intermediate regimes. For mean energies which are not too small the phase optimized state solution to (a) has a sinusoidal distribution of number state coefficients and for (b) the distribution is an Airy function. Both of these states have a better defined phase than a coherent state of the same mean energy.

D.T.P. thanks Dr. S.M. Barnett for earlier discussions on minimizing phase variance.

References

[1] D.T. Pegg and S.M. Barnett, Europhysics Letters 6 (1988) 483.
[2] S.M. Barnett and D.T. Pegg, J. Mod. Optics 36 (1989) 7.
[3] D.T. Pegg and S.M. Barnett, Phys. Rev. A 39 (1989) 1665.
[4] R. Loudon, The quantum theory of light, 1st Ed. (Oxford University Press, 1973) p. 143.
[5] J.A. Vaccaro and d.T. Pegg, J. Mod. Optics (1989) in press.
[6] H.B. Dwight, Tables of integrals and other mathematical data, 4th Ed. (Macmillan, New York, 1961) p. 90.
[7] J.H. Shapiro, S.R. Shepard and N.C. Wong, Phys. Rev. Lett. 62 (1989) 2377.
[8] E. Merzbacher, Quantum mechanics, 2nd Ed. (Wiley, New York, 1970) p. 419.
[9] M. Abramowitz and I.A. Stegun, Handbook of mathematical functions (Dover, New York, 1972) p. 446.

Phase fluctuations and squeezing

JOHN A. VACCARO
Division of Commerce and Administration, Griffith University, Nathan, Brisbane 4111, Australia

STEPHEN M. BARNETT† and D. T. PEGG
Division of Science and Technology, Griffith University, Nathan, Brisbane 4111, Australia

(*Received 22 August 1991; accepted 26 September 1991*)

Abstract. We investigate the relationship between squeezing and reduced phase fluctuations for various states of the single-mode electromagnetic field, including the strongly-squeezed vacuum and phase states. We find that, although squeezing the fluctuations of the electric field that arise from the vacuum guarantees a more well-defined phase, reducing phase fluctuations does not guarantee a squeezed electric field. We also investigate the evolution of the electric field and its fluctuations for a phase state. Our results show that even though the electric field fluctuations never vanish for a phase state, the times when the electric field changes sign are precisely defined. We also discuss why it is not always possible to attribute physical properties to certain states, such as simple superpositions of phase states.

1. Introduction

While the concept of phase was incorporated into the earliest theory of quantum electrodynamics [1], it was not until the invention of the laser that interest in the quantum phase properties of light received much attention. It has now become even more important to understand and analyse the phase properties of light following the discovery of squeezed states of light (see, for example, [2] and [3] and references therein). Squeezed light has electric field fluctuations that fall below the vacuum level at particular times during each cycle of the field and this is due to different fluctuations in two quadrature-phase components of the electric field. While the fluctuations in the electric field and its quadrature-phase components are definitely related to the fluctuations in periodic functions of phase, especially for large amplitude fields, nevertheless these two types of fluctuations are certainly *not equivalent*. In this paper we examine the precise nature of the relationship between these two types of fluctuations for the particular situation where the level of both fluctuations is small. We do this in two steps: first, we examine the phase fluctuations of a state with a strongly-squeezed electric field—that is, a state with a well-defined field; and secondly, we look at the converse situation by examining the fluctuations of the electric field for states with well-defined phase.

Our analysis of phase fluctuations requires the use of quantum-mechanical phase operators. However, despite several attempts it has not been possible to give a

† Present address: Department of Physics and Applied Physics, University of Strathclyde, John Anderson Building, 107 Rottenrow, Glasgow G4 0NG, Scotland.

complete set of quantum-mechanical phase operators using the infinite-dimensional Hilbert space [4]. In contrast, the formalism introduced recently by Pegg and Barnett [5], which we shall refer to as the new phase formalism, does give a complete description of quantum-mechanical phase. The crucial features of the new formalism are the use of an $(s+1)$-dimensional linear state space Ψ, and also the particular requirement of only taking the limit $s\to\infty$ of c-number expressions (such as expectation values) after all operator algebra has been performed on Ψ. Delaying the limit $s\to\infty$ until the calculation of expectation values has two important consequences. The first is that more states are able to be treated using the new formalism than those belonging to the infinite-dimensional Hilbert space. For example, it is possible using the new formalism to treat the phase states $|\theta\rangle$

$$|\theta\rangle \equiv (s+1)^{-1/2} \sum_{n=0}^{s} \exp(in\theta)|n\rangle, \qquad (1)$$

as states of precisely-defined phase. Such equally-weighted superpositions of all number states do not belong to the infinite-dimensional Hilbert space. The second consequence, which has not been reported previously, is that the new formalism allows the treatment of particular states for which the calculation of physical properties is indeterminate; that is, state for which certain physical expectation values neither converge nor diverge as $s\to\infty$. We encounter this problem when treating superpositions of phase states and we discuss it more fully in the last section.

2. General phase properties of the strongly-squeezed vacuum

The phase properties of squeezed light have been examined by a number of authors. Sanders *et al.* [6], Yao [7] and Fan Hong-Yi and Zaidi [8] used the Susskind–Glogower phase formalism [4] to investigate particular phase properties of the squeezed vacuum and Loudon and Knight [2] used an approach based on the variances in the amplitudes of quadrature phases of the electric field to calculate approximately the phase fluctuations for ideal squeezed states. Recently we [9] examined the phase properties of the squeezed vacuum using the new phase formalism and found that the strongly-squeezed vacuum has arbitrarily small fluctuations in the $\cos\hat{\phi}_\theta$ operator for a particular choice of quadrature squeezing. Schleich *et al.* [10] have since found that the phase probability density of the same state has relatively sharp peaks at phase angles separated by π. We now extend these recent calculations to show how the phase properties of the strongly-squeezed vacuum approach those of the superposition of phase states given by

$$\begin{aligned}|\psi(\theta)\rangle &\equiv c(|\theta+\pi/2\rangle + |\theta-\pi/2\rangle), \\ &= (1+R)^{-1/2} \sum_{n=0}^{R} (-1)^n \exp(i2n\theta)|2n\rangle,\end{aligned} \qquad (2)$$

where R is the largest integer not greater than $s/2$. Our method is to compare the expectation value of the general unitary operator $\exp(im\hat{\phi}_\theta)$, where m is any integer, for the strongly-squeezed vacuum with that for the state $|\psi(\theta)\rangle$. We choose this method because any function of $\hat{\phi}_\theta$—for example, its square—can be expressed as a series in $\exp(im\hat{\phi}_\theta)$. Also the values of $\langle\cos^n\hat{\phi}_\theta\rangle$ and $\langle\sin^n\hat{\phi}_\theta\rangle$ for arbitrary positive n can be calculated from $\langle\exp(im\hat{\phi}_\theta)\rangle$ and so to compare the phase properties of the two states we need only compare the corresponding expectation values of $\exp(im\hat{\phi}_\theta)$.

The value of $\langle \exp(im\hat{\phi}_\theta)\rangle$ can be calculated using

$$\langle \exp(im\hat{\phi}_\theta)\rangle = \lim_{s\to\infty}\left\{\sum_{n=0}^{s-m} d_n^* d_{n+m} + \exp[i(s+1)\theta_0]\sum_{n=0}^{m-1} d_{s-n}^* d_{m-1-n}\right\},$$

where d_n are the number-state coefficients of an arbitrary state [5]. The squeezed vacuum is represented in the $(s+1)$-dimensional space Ψ by the normalized state

$$|0,\xi\rangle = A_s \sum_{n=0}^{s} c_n|n\rangle,$$

where [2, 11]

$$c_{2n+1}=0,$$

$$c_{2n}=\frac{[-\frac{1}{2}\exp(i\eta)\tanh r]^n((2n)!)^{1/2}}{(\cosh r)^{1/2}n!},$$

$$\xi \equiv r\exp(i\eta),$$

and where A_s is a real normalization constant with the property that $A_s\to 1$ as $s\to\infty$. We find that [9]

$$\langle \exp[i(2m+1)\hat{\phi}_\theta]\rangle = 0, \tag{3}$$

$$\begin{aligned}\langle \exp(i2m\hat{\phi}_\theta)\rangle &= \lim_{s\to\infty} A_s^2 \sum_{n=0}^{R-m} c_{2n}^* c_{2n+2m}\\ &= (-1)^m \exp(im\eta)B,\end{aligned} \tag{4}$$

where

$$B \equiv \lim_{s\to\infty}\frac{1}{\cosh r}\sum_{n=0}^{R-m}(\tanh r)^{2n+m}\frac{[(2n)!(2n+2m)!]^{1/2}}{2^{2n+m}n!(n+m)!}.$$

It is not difficult to show that

$$\frac{(2n+2m)!}{2^{2n+2m}[(n+m)!]^2} < \frac{[(2n)!(2n+2m)!]^{1/2}}{2^{2n+m}n!(n+m)!} < \frac{(2n)!}{2^{2n}(n!)^2}.$$

Multiplying these expressions by $(\tanh r)^{2n+m}/\cosh r$ and then summing over n gives expressions for B and its upper and lower bounds. The expressions for the upper and lower bounds contain series of the form

$$(1-x)^{-1/2} = \sum_{n=0}^{\infty}\frac{x^n(2n)!}{2^{2n}(n!)^2}, \qquad |x|<1;$$

and thus, after some algebra, we find

$$(\tanh r)^{-m}\left(1-\frac{m}{\cosh r}\right) < B < (\tanh r)^m.$$

Expanding the expressions in the last inequality in terms of powers of $\exp(-r)$ shows eventually that $B = 1 + \mathrm{O}\exp(-r)$ where O is the order symbol. Thus substituting into equation (4) yields

$$\langle \exp(i2m\hat{\phi}_\theta)\rangle = (-1)^m \exp(im\eta) + \mathrm{O}\exp(-r). \tag{5}$$

The magnitude of the term $O \exp(-r)$ rapidly approaches zero with increasing squeezing, i.e. increasing r. From this result and equation (3) it is easy to show that the variances in $\cos \hat{\phi}_\theta$ and $\sin \hat{\phi}_\theta$ are

$$\langle \Delta \cos \hat{\phi}_\theta^2 \rangle = \sin^2(\eta/2) + O \exp(-r),$$

$$\langle \Delta \sin \hat{\phi}_\theta^2 \rangle = \cos^2(\eta/2) + O \exp(-r),$$

where $\langle \Delta \hat{A}^2 \rangle = \langle \hat{A}^2 \rangle - \langle \hat{A} \rangle^2$ is the variance in operator A. These expressions are consistent with what we found in our earlier work [9]. Moreover, we also find here that the variances in the operators $\cos^{2n} \hat{\phi}_\theta$ and $\sin^{2n} \hat{\phi}_\theta$ are of the order of $\exp(-r)$ *for all values of the squeezing direction* η and this exemplifies the well-defined phase properties for large degrees of squeezing. These properties are best illustrated graphically by plotting the phase probability density $\mathscr{P}(\theta)$. In figure 1 we represent the infinite-s limit of $\mathscr{P}(\theta)$ for the state $|0, \xi\rangle$ with $r = 2{\cdot}2$ and $\eta = 3\pi/4$ on a polar diagram. $\mathscr{P}(\theta)$ is given by

$$\mathscr{P}(\theta) = |\langle \theta | 0, \xi \rangle|^2 \frac{s+1}{2\pi},$$

where $|\theta\rangle$ is the phase state defined by equation (1). Although it is not shown in the figure, we also find that for larger values of r the infinite-s limit of $\mathscr{P}(\theta)$ is even more sharply peaked at $\theta = \eta/2 \pm \pi/2$. In comparison, the phase probability density for the superposition of phase states $|\psi(\eta/2)\rangle$ (defined by equation (2)) is non-vanishing in the limit of infinite s only for $\theta = \eta/2 \pm \pi/2$. Furthermore, the value of $\langle \exp(im\hat{\phi}_\theta) \rangle$ for the state $|\psi(\eta/2)\rangle$ is found to be

$$\langle \exp[i(2m+1)\hat{\phi}_\theta] \rangle = 0, \tag{6}$$

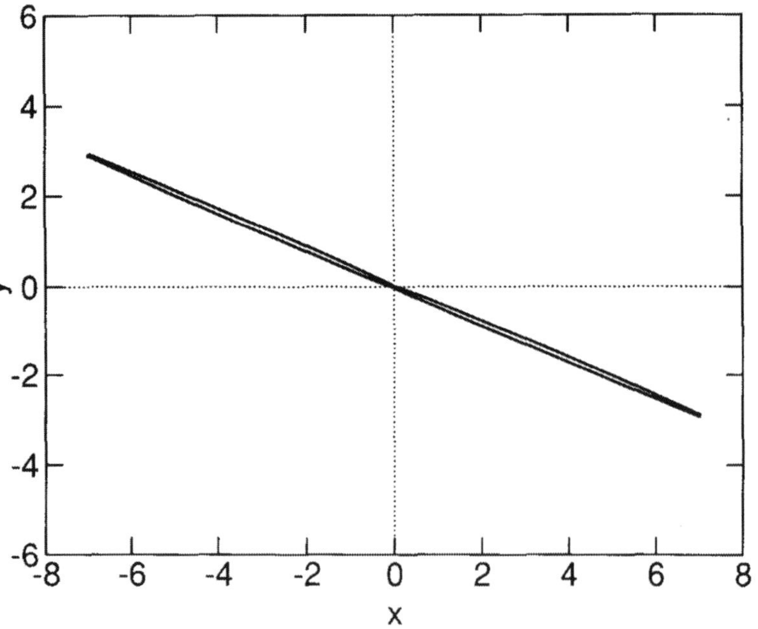

Figure 1. A polar diagram showing the phase probability density $\mathscr{P}(\theta)$ for the squeezed vacuum $|0, \xi\rangle$ with $r = |\xi| = 2{\cdot}2$ and $\eta = \arg(\xi) = 3\pi/4$. The maxima in $\mathscr{P}(\theta)$ lie on a line oriented at an angle of $3\pi/8$ to the y-axis.

$$\langle \exp(i2m\hat{\phi}_\theta)\rangle = (-1)^m \exp(im\eta), \tag{7}$$

which can be compared with equations (3) and (5). Clearly *the phase properties of the squeezed vacuum $|0, \xi\rangle$ become indistinguishable from those of the superposition of phase states $|\psi(\eta/2)\rangle$ as the squeezing parameter r increases.*

3. The electric field for the superposition of phase states $|\psi(\theta)\rangle$

Even though the phase properties of the two states $|0, \xi\rangle$ and $|\psi(\eta/2)\rangle$ become indistinguishable, nevertheless they remain different states. In particular, we show in this section that the two states have different levels of electric field fluctuations. The operator representing the single-mode electric field at the origin is [12]

$$\hat{E} = i\left(\frac{\hbar\omega}{2\varepsilon_0 V}\right)^{1/2} (\hat{a} - \hat{a}^\dagger),$$

where V is the quantization volume, ω is the angular frequency of the mode and $\hat{a}$, $\hat{a}^\dagger$ are the annihilation and creation operators. It is not difficult to show [2] that the minimum variance in $\hat{E}$ for the squeezed vacuum $|0, \xi\rangle$ is $[\hbar\omega/(2\varepsilon_0 V)] \exp(-2r)$ which is well below the vacuum level of $\hbar\omega/(2\varepsilon_0 V)$ for strong squeezing (i.e. large r). We now evaluate the variance of $\hat{E}$ for the superposition of phase states $|\psi(\theta)\rangle$ given by equation (2).

In the new phase formalism the expectation value of $\hat{E}$ and its square, for a state with number-state coefficients d_n, are given by

$$\langle \hat{E}\rangle = -\lim_{s\to\infty} 2\left(\frac{\hbar\omega}{2\varepsilon_0 V}\right)^{1/2} \operatorname{Im} \sum_{n=1}^{s} (n)^{1/2} d_{n-1}^* d_n,$$

$$\langle \hat{E}^2\rangle = \lim_{s\to\infty} \left(\frac{\hbar\omega}{2\varepsilon_0 V}\right)\left\{2\sum_{n=0}^{s} n|d_n|^2 + 1 - (s+1)|d_s|^2 - 2\operatorname{Re}\sum_{n=2}^{s} [n(n-1)]^{1/2} d_{n-2}^* d_n\right\}, \tag{8}$$

where $\operatorname{Re}(z)$ and $\operatorname{Im}(z)$ are the real and imaginary parts of z. The term $(s+1)|d_s|^2$ in the expression $\langle \hat{E}^2\rangle$ is not present in the corresponding expression derived using the infinite-dimensional Hilbert space. (It can be shown that all states belonging to the infinite-dimensional Hilbert space have the property that $n|d_n|^2 \to 0$ as $n\to\infty$ and thus the term $(s+1)|d_s|^2$ appearing in equation (8) vanishes in the infinite-s limit for this space.) Its presence here is a direct consequence of representing the annihilation operator $\hat{a}$ on the $(s+1)$-dimensional space Ψ as [5]

$$\hat{a} = \sum_{n=1}^{s} (n)^{1/2} |n-1\rangle\langle n|,$$

from which the commutator with the hermitian conjugate $\hat{a}^\dagger$ is found to be $[\hat{a}, \hat{a}^\dagger] = 1 - (s+1)|s\rangle\langle s|$. Because the state $|\psi(\theta)\rangle$ contains only even number-state coefficients (see equation (2)) we immediately find that the mean electric field for $|\psi(\theta)\rangle$ is

$$\langle \hat{E}\rangle = 0, \tag{9}$$

and so the variance in $\hat{E}$ is given by

$$\langle \Delta\hat{E}^2\rangle = \langle \hat{E}^2\rangle.$$

On substituting for d_n in equation (8), using equation (2), we find $\langle \hat{E}^2 \rangle$ is the infinite-s limit of

$$\left(\frac{\hbar\omega}{2\varepsilon_0 V}\right)\left\{\frac{2}{1+R}\sum_{n=0}^{R} 2n+1-(s+1)|d_s|^2+\frac{2\cos(2\theta)}{1+R}\sum_{n=1}^{R}[2n(2n-1)]^{1/2}\right\}.$$

This expression diverges as $s\to\infty$ for all θ, except when θ satisfies $\cos(2\theta)=-1$. For these values of θ the expression becomes

$$\left(\frac{\hbar\omega}{2\varepsilon_0 V}\right)\left(\frac{2}{1+R}\left\{\sum_{n=1}^{R} 2n-[2n(2n-1)]^{1/2}\right\}+1-(s+1)|d_s|^2\right). \tag{10}$$

The term containing the summation converges to 1 as $s\to\infty$. However, because the state $|\psi(\theta)\rangle$ contains only even number-state coefficients the last term $(s+1)|d_s|^2$ oscillates between approximately 2 for even s and zero for odd s. Thus, the expression (10) neither converges nor diverges as $s\to\infty$ and hence the limit of this expression as $s\to\infty$ does not exist. The variance in $\hat{E}$ for the state $|\psi(\theta)\rangle$ with $\cos(2\theta)=-1$ is therefore *not able to be determined.*

We discuss the issue of the non-determination of physical properties in more detail in the last section. We note here, however, that the state $|\psi(\theta)\rangle$ cannot actually be prepared physically because to do so would require an infinite amount of energy (i.e. $\langle \hat{N} \rangle$ diverges as $s\to\infty$). Thus the convergence or otherwise of $\langle \Delta\hat{E}^2 \rangle$ for this state has no physical significance. Indeed, it is meaningless to attribute any physical property to states of this type. We must therefore approximate $|\psi(\theta)\rangle$ with a state that can, in principle, be physically prepared.

The class of states called physical states represents fields that can, in principle, be prepared in a laboratory. A state is a physical state [5] if the moment of the energy $(\hbar\omega)^m\langle(\hat{N}+1/2)^m\rangle$ is finite for any given finite integer $m>0$. We find that the expansion of arbitrary moments of the electric field $\hat{E}$, say $\langle \hat{E}^{2k} \rangle$, in terms of number-state coefficients contains terms proportional to $(s+1)^k|d_s|^2$. These terms all vanish for physical states as $s\to\infty$. Thus all moments of the electric field $\hat{E}$ converge for physical states.

For our analysis of the variance in $\hat{E}$ we replace the exact superposition of phase states $|\psi(\theta)\rangle$ with a physical-state approximation $|\psi(\theta)\rangle_\tau$. We obtain $|\psi(\theta)\rangle_\tau$ by truncating the number-state expansion of $|\psi(\theta)\rangle$ given by equation (2) at the number state $|\tau\rangle$:

$$|\psi(\theta)\rangle_\tau \equiv (1+T)^{-1/2}\sum_{n=0}^{T}(-1)^n\exp(i2n\theta)|2n\rangle,$$

where T is the largest integer not greater than $\tau/2$. To verify that the phase properties of the truncated state $|\psi(\theta)\rangle_\tau$ approximate those of the full state $|\psi(\theta)\rangle$ we compare the values of $\langle \exp(im\hat{\phi}_\theta)\rangle$ given by the two states. We find that for the state $|\psi(\theta)\rangle_\tau$:

$$\langle \exp[i(2m+1)\hat{\phi}_\theta]\rangle = 0, \tag{11}$$

$$\langle \exp(i2m\hat{\phi}_\theta)\rangle = (-1)^m\exp(i2m\theta)[1-m/(1+T)], \tag{12}$$

for $m\leqslant T$. These expressions contain the extra factor $[1-m/(1+T)]$ in comparison with the corresponding expressions for $|\psi(\theta)\rangle$ given by equations (6) and (7) on replacing η with 2θ. Clearly the phase properties of the truncated state $|\psi(\theta)\rangle_\tau$ can be made as close as desired to those of the exact state $|\psi(\theta)\rangle$ by choosing a sufficiently large value of the truncation parameter τ compared with any given value of m. There is an appealing physical consequence of this result. The expectation values of $\exp(in\hat{\phi}_\theta)$ provide increasingly finer detail about the phase probability distribution as n increases. Expression (12) shows that this information will be approximately the same as that associated with an exact phase state superposition if $n \ll 2T$ (i.e. $n \ll \tau$). If we only require the expectation values of moments up to $\exp(i2m\hat{\phi}_\theta)$ to approximate the exact phase state values, then we need only construct a physical state of the type given above for which $T \gg m$. Comparing equations (11) and (12) with equations (3) and (5) shows that the phase properties of the truncated state $|\psi(\theta)\rangle_\tau$ and the squeezed vacuum $|0, \xi\rangle$ converge for $2\theta = \eta = \arg(\xi)$ as the values of τ and r increase. In figure 2 we plot the infinite-s limit of the phase probability density for the truncated state $|\psi(\theta)\rangle_\tau$ to illustrate the similarity with the phase probability density of the state $|0, \xi\rangle$ plotted in figure 1.

When evaluating the first and second moments of the electric field for the truncated state $|\psi(\theta)\rangle_\tau$ we find that the calculations are similar to those that resulted in equations (9) and (10) for the exact state $|\psi(\theta)\rangle$. The only differences are that R is replaced with T and the term $(s+1)|d_s|^2$ no longer appears. It follows that for the truncated state $|\psi(\theta)\rangle_\tau$

$$\langle \hat{E} \rangle = 0,$$

and the minimum variance in $\hat{E}$ is

$$\langle \Delta \hat{E}^2 \rangle = \left(\frac{\hbar\omega}{2\varepsilon_0 V} \right) \left(\frac{2}{1+T} \left\{ \sum_{n=1}^{T} 2n - [2n(2n-1)]^{1/2} \right\} + 1 \right). \tag{13}$$

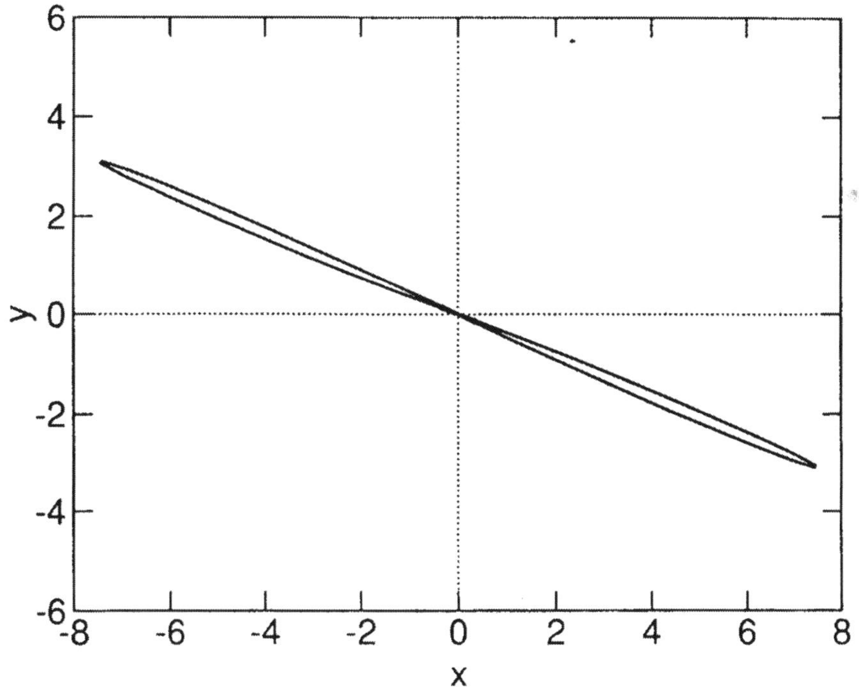

Figure 2. $\mathscr{P}(\theta)$ for the truncated superposition of phase states $|\psi(\eta/2)\rangle_\tau$ with $\eta = 3\pi/4$ and $\tau = 100$. The gross features of this figure are similar to those of figure 1.

The sum is always greater than zero for all values of the truncation parameter and thus

$$\langle \Delta \hat{E}^2 \rangle > \left(\frac{\hbar\omega}{2\varepsilon_0 V} \right).$$

The right-hand side of the last expression is the vacuum level of the electric field fluctuations. We are thus led to the perhaps surprising result that the truncated state $|\psi(\theta)\rangle_\tau$ is ***not squeezed***, even though its phase properties are similar to those of a squeezed vacuum.

Indeed we have shown that the phase properties of the states $|0, \xi\rangle$ and $|\psi(\theta)\rangle_\tau$ converge for $\theta = \eta/2$ as the values of r and τ are increased and thus the fluctuations in the phase operators $\cos^{2n}\hat{\phi}_\theta$ and $\sin^{2n}\hat{\phi}_\theta$ for both the states $|0, \xi\rangle$ and $|\psi(\theta)\rangle_\tau$ approach minimal values. On the other hand, we have also shown that for the same trend in r and τ the fluctuations in the electric field for the state $|\psi(\theta)\rangle_\tau$ are never below the vacuum level for any value of θ. This is in marked contrast to the state $|0, \xi\rangle$ for which the electric field fluctuations become vanishingly small. Hence an asymmetry exists between reducing phase fluctuations and squeezing the electric field: states with minimal electric field fluctuations have minimal fluctuations in the phase operators $\cos^{2n}\hat{\phi}_\theta$ and $\sin^{2n}\hat{\phi}_\theta$, but states with minimal fluctuations in $\cos^{2n}\hat{\phi}_\theta$ and $\sin^{2n}\hat{\phi}_\theta$ do not necessarily have minimal fluctuations in the electric field.

4. The electric field for phase states

In the previous section we considered the electric field for states with minimal phase fluctuations. These states, however, approximate the eigenstates of the ***trigonometric phase*** operators $\cos^{2n}\hat{\phi}_\theta$ and $\sin^{2n}\hat{\phi}_\theta$ rather than the ***phase-angle*** operator $\hat{\phi}_\theta$. The $(s+1)$ eigenstates of $\hat{\phi}_\theta$ are a subset of the set of phase states $|\theta\rangle$ given by equation (1) for all θ in a 2π range. In earlier work [5] properties of the phase states were illustrated by reproducing Loudon's discussion [12] of the mean electric field evolution for phase states. For completeness it remains for us to check here if the phase states themselves, as opposed to the phase-state superposition discussed in the previous section, are squeezed. Rather than follow the last section and analyse the electric field at a fixed time it is more instructive to consider a field freely evolving in time. This also allows us to illustrate how the phase states give rise to changes in the sign of $\hat{E}$ at precisely defined time intervals.

We approximate the unphysical phase state $|\theta\rangle$, whose mean energy is infinite, with the physical state $|\theta\rangle_\tau$ which is obtained by truncating the number-state expansion of $|\theta\rangle$ at $|\tau\rangle$:

$$|\theta\rangle_\tau = (1+\tau)^{-1/2} \sum_{n=0}^{\tau} \exp(in\theta)|n\rangle.$$

The free evolution of the mode causes the state $|\theta\rangle_\tau$ to become the state $|\theta - \omega t\rangle_\tau$ after a time interval t. (The evolved state $|\theta - \omega t\rangle$ may also acquire an additional multiplicative phase factor due to the choice of zero of energy. Naturally this choice has no detectable consequences.) We find that the expectation value of electric field in a truncated phase state is

$${}_\tau\langle \theta - \omega t|\hat{E}|\theta - \omega t\rangle_\tau = -\frac{2}{\tau+1}\left(\frac{\hbar\omega}{2\varepsilon_0 V}\right)^{1/2} \sin(\theta - \omega t) \sum_{n=1}^{\tau} (n)^{1/2}. \qquad (14)$$

The mean electric field diverges with increasing τ at all times, except when $\theta - \omega t$ is an integer multiple of π, at which times it is zero. Thus the expectation value of the electric field oscillates between divergent positive and negative values. It is finite (zero) only at these changeover times. The variance in $\hat{E}$ is found to be

$$\langle \Delta \hat{E}^2 \rangle = \frac{1}{9}\left(\frac{\hbar\omega}{2\varepsilon_0 V}\right)\tau[1 - \cos 2(\theta - \omega t)] + \mathrm{O}(1),$$

which clearly diverges with increasing τ except at the changeover times when the mean field is zero. We note that $\langle \hat{E} \rangle^2 \approx 8\langle \Delta \hat{E}^2 \rangle$ during the periods when $\langle \hat{E} \rangle$ and $\langle \Delta \hat{E}^2 \rangle$ are divergent, and thus the electric field is poorly defined during these times. At the changeover times we find that the variance for large τ is

$$\langle \Delta \hat{E}(\theta - \omega t = k\pi)^2 \rangle = 2\left(\frac{\hbar\omega}{2\varepsilon_0 V}\right) + \mathrm{O}\left(\frac{\ln \tau}{\tau}\right), \tag{15}$$

which is approximately twice the level of fluctuations associated with the vacuum field. Thus, although highly squeezed states can have well-defined phase properties, even the phase states do not exhibit the reduced electric field fluctuations of the squeezed states. In figure 3 we represent the electric field and its uncertainty for the truncated phase state $|\theta\rangle_\tau$ with $\theta = \pi/2$ and $\tau = 300$. Although the electric field is never squeezed, the figure clearly shows that the field becomes relatively more well defined at the changeover times when $\theta - \omega t = k\pi$.

We note that when calculating the variance in $\hat{E}$ for the exact phase state $|\theta\rangle$ (as opposed to the truncated phase state $|\theta\rangle_\tau$) the term $(s+1)|d_s|^2$ in equation (8) does not vanish but instead is equal to unity. Thus, we find the variance of the electric field at

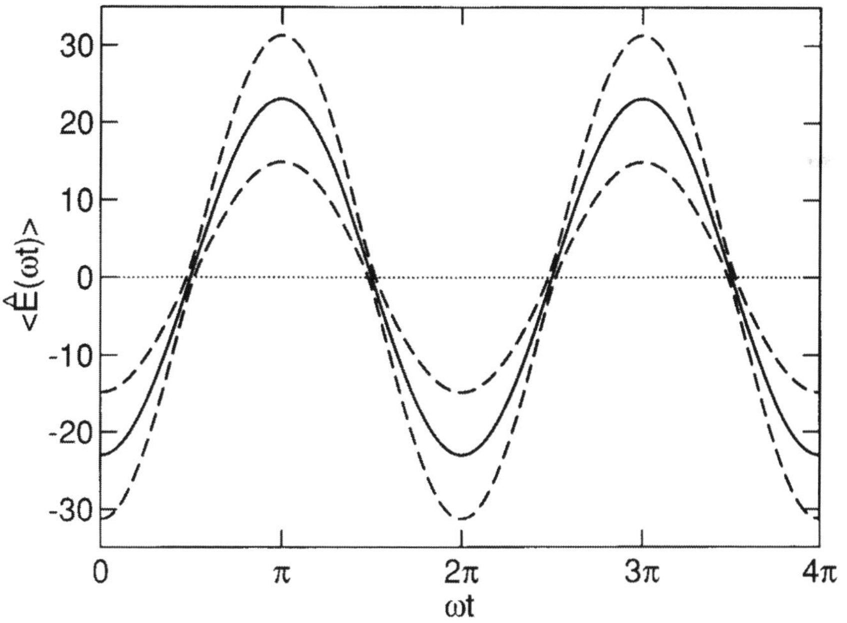

Figure 3. A representation of the electric field and its fluctuations for the truncated phase state $|\theta - \omega t\rangle_\tau$, with $\theta = \pi/2$ and $\tau = 300$. The solid curve represents the mean electric field and the dashed curves represent the mean plus and minus one standard deviation. The units of the vertical axis are $[\hbar\omega/(2\varepsilon_0 V)]^{1/2}$, the root-mean-square vacuum field.

the changeover times for the exact phase state is precisely the vacuum level. Hence, the exact phase states are not squeezed even though they have smaller electric field fluctuations than the truncated phase states.

The well-defined changeover times evident in figure 3 suggest that a single-mode field prepared in a truncated phase state could, in principle, be used as a quantum clock. We take the zero of time to correspond to the instant when the mean electric field passes through zero towards a positive value. A spread in timing obtained from an ensemble of such clocks would arise from the residual vacuum fluctuations in the field at this time. The field of a single member of the ensemble may become zero at any instant between the times when $\langle\hat{E}\rangle-\Delta E$ and $\langle\hat{E}\rangle+\Delta E$ is zero. From equations (14) and (15) we find that the spread in zero times for these clocks is of the order $1/(\tau^{1/2}\omega)$. For increasingly-larger values of τ this tends to zero and *the changeover times become precisely determined.* Thus, even though an electric field prepared in a phase state is never squeezed and so never has a precise value, the sign of the electric field changes at precisely-defined times. This confirms the phase-like behaviour exhibited by a field prepared in a phase state.

5. Discussion

We have shown that squeezing the fluctuations in the electric field arising from the vacuum produces reduced phase fluctuations, whereas reducing phase fluctuations does not necessarily imply a squeezed electric field. The reason for this asymmetry is most readily seen by considering the electric field operator for those physical states with a well-defined phase or electric field. These particular states have very-broad photon-number probability distributions, $P_n=|d_n|^2$, and the property that terms such as $s|d_s|^2$ vanish as $s\rightarrow\infty$. For such states the electric field operator $\hat{E}$ can be approximated in the number-state basis as

$$\hat{E}\approx \mathrm{i}\left(\frac{\hbar\omega}{2\varepsilon_0 V}\right)^{1/2}\sum_{n=1}^{s}\left[(n+1/2)^{1/2}|n-1\rangle\langle n|-(n-1/2)^{1/2}|n\rangle\langle n-1|\right]$$

$$\approx -2\left(\frac{\hbar\omega}{2\varepsilon_0 V}\right)^{1/2}\sin\hat{\phi}_\theta(\hat{N}+1/2)^{1/2}, \qquad (16)$$

where [5] $\hat{N}=\Sigma n|n\rangle\langle n|$ is the number operator and

$$\sin\hat{\phi}_\theta=-\frac{1}{2}\mathrm{i}\left\{\sum_{n=1}^{s}|n-1\rangle\langle n|+\exp\left[\mathrm{i}(s+1)\theta_0\right]|s\rangle\langle 0|-\text{h.c.}\right\}.$$

Thus, squeezing $\hat{E}$ involves reducing the fluctuations in a combination of amplitude $(\hat{N}+1/2)^{1/2}$ as well as phase ($\sin\hat{\phi}_\theta$). Clearly minimizing this combination may well result in minimal phase fluctuations. However, minimizing phase fluctuations alone need not reduce the amplitude fluctuations, and thus need not produce a squeezed electric field.

The physical states considered in this paper have number-state coefficients d_n that are either constant or vary slowly with n over a large range of n values and whose complex argument (states with the property $d_n=|d_n|\exp(in\beta)$, where β is independent of n, are called partial phase states [5]) is a multiple of n (i.e. $d_n=|d_n|\exp(in\beta)$). For these states

$$\sum d_{n-1}^{*}d_n(n+1/2)^{1/2}\approx\left(\sum d_{n-1}^{*}d_n\right)\left(\sum |d_n|^2(n+1/2)^{1/2}\right).$$

Hence, the expectation value of $\hat{E}$ from equation (16) can be further approximated as

$$\langle\hat{E}\rangle \approx -2\left(\frac{\hbar\omega}{2\varepsilon_0 V}\right)^{1/2}\langle\sin\hat{\phi}_\theta\rangle\langle(\hat{N}+1/2)^{1/2}\rangle. \tag{17}$$

We note in passing that the right-hand side of equation (17) corresponds to the exact expression obtained from using the 'measured phase' operator [13]. For a field freely evolving in a truncated phase state $|\theta\rangle_\tau$ we find that $\langle\sin\hat{\phi}_\theta\rangle = \sin(\theta-\omega t)\tau/(1+\tau)$, giving well-defined sign changes, whereas $\langle(\hat{N}+1/2)^{1/2}\rangle = O(\tau^{1/2})$. This confirms that even though the truncated phase state $|\theta\rangle_\tau$ is not squeezed, the mean electric field oscillates between divergent values at precisely defined intervals for large values of the truncation parameter τ.

Finally, when we attempted to evaluate the properties of the electric field for the exact phase state superposition $|\psi(\theta)\rangle$, we found that the expression for the variance neither converged nor diverged. Instead it oscillated between approximately twice the vacuum value and zero as $s\to\infty$. This result suggests that the new formalism cannot describe completely an ideal experiment to measure the electric field for this state because it can predict the mean but not the variance. This leads us to an interesting aspect of the new formalism. The formalism allows the treatment of many more states than contained in the infinite-dimensional Hilbert space H_∞ for the following reason: instead of delaying the $s\to\infty$ limit until after the c-numbers are calculated, we can generate the space H_∞ from the outset. The space H_∞ is the union of the $(s+1)$-dimensional space Ψ for arbitrarily large (but finite) s, with the limit points of all convergent sequences of states belonging to Ψ where, by definition, a sequence of states $|\psi_1\rangle$, $|\psi_2\rangle$, $|\psi_3\rangle\ldots$ converges to a limit point $|\psi\rangle$ if, and only if

$$(\langle\psi| - \langle\psi_n|)(|\psi\rangle - |\psi_n\rangle) \to 0,$$

as n tends to infinity [14]. The limiting process ensures that all the states belonging to H_∞ have the property that the magnitude of the ratio of the number-state coefficients $|\langle n|\psi\rangle/\langle m|\psi\rangle|$ vanishes as $n\to\infty$ for any given m. This excludes from H_∞ many extra states that can be treated in the new formalism. Examples of the extra states include the phase state and simple superpositions of phase states. All the extra states (as well as some others which belong to the infinite-dimensional Hilbert space) are impossible to prepare physically because their mean energy diverges. It is therefore meaningless to attribute physical properties to any of the extra states because they cannot, in principle, have a physical existence. Thus, it is quite admissible for the new formalism *not* to give a complete description of experiments involving the extra states because the experiments are physically impossible to perform. In contrast, we have been able to use the new formalism to describe fully the results of ideal measurements of the electric field for *physical* states. It may appear then that the extra states are superfluous. However, the phase states in particular are useful for mathematical analysis of the phase properties of physical states and also for providing an important conceptual model of the quantum-mechanical phase. In this respect, it should be remembered that position eigenstates, with their divergent momentum fluctuations, are also unphysical but that the position representation is extremely useful in many branches of quantum mechanics.

Acknowledgment

S.M.B. and D.T.P. are grateful to R. Loudon for helpful discussions. S.M.B. also thanks the members of the Division of Science and Technology at Griffith

University for their hospitality during his visit. This work was supported in part by G.E.C. Research Limited, the Fellowship of Engineering and by the Royal Society.

References

[1] DIRAC, P. A. M., 1927, *Proc. R. Soc.* A, **114,** 243.

[2] WALLS, D. F., 1983, *Nature,* **306,** 141; LOUDON, R., and KNIGHT, P. L., 1987, *J. mod. Optics,* **34,** 709.

[3] SLUSHER, R. E., HOLLBERG, L. W., YURKE, B., MERTZ, J. C., and VALLEY, J. F., 1985, *Phys. Rev. Lett.,* **55,** 2409.

[4] SUSSKIND, L., and GLOGOWER, J., 1964, *Physics,* **1,** 49; CARRUTHERS, P., and NIETO, M. M., 1968, *Rev. mod. Phys.,* **40,** 411; LERNER, E. C., HUANG, H. W., and WALTERS, G. E., 1970, *J. Math. Phys.,* **11,** 1679; LEVY-LÉBLOND, J. M., 1976, *Ann. Phys. (N.Y.),* **101,** 319.

[5] PEGG, D. T., and BARNETT, S. M., 1988, *Europhys. Lett.,* **6,** 483; BARNETT, S. M., and PEGG, D. T., 1989, *J. mod. Optics,* **36,** 7; PEGG, D. T., and BARNETT, S. M., 1989, *Phys. Rev.* A, **39,** 1665.

[6] SANDERS, B. C., BARNETT, S. M., and KNIGHT, P. L., 1986, *Optics Commun.,* **58,** 290.

[7] YAO, D., 1987, *Phys. Lett.* A, **122,** 77.

[8] FAN HONG-YI, and ZAIDI, H. R., 1988, *Optics Commun.,* **68,** 143.

[9] VACCARO, J. A., and PEGG, D. T., 1989, *Optics Commun.,* **70,** 529.

[10] SCHLEICH, W., HOROWICZ, R. J., and VARRO, S., 1989, *Phys. Rev.* A, **40,** 7405.

[11] YUEN, H. P., 1976, *Phys. Rev.* A, **13,** 2226.

[12] LOUDON, R., 1973, *The Quantum Theory of Light,* 1st edn (Oxford: Oxford University Press).

[13] BARNETT, S. M., and PEGG, D. T., 1986, *J. Phys.* A, **19,** 3849.

[14] AKHIEZER, N. I., and GLAZMAN, I. M., 1961, *Theory of Linear Operators in Hilbert Space,* Vol. 1, translated from the Russian by M. Nestell (New York: Frederick Ungar).

Phase properties of optical linear amplifiers

John A. Vaccaro*
Faculty of Commerce and Administration, Griffith University, Nathan 4111, Australia

D. T. Pegg
Faculty of Science and Technology, Griffith University, Nathan 4111, Australia
(Received 17 May 1993)

We examine the effects of linear amplification and attenuation on the quantum-mechanical phase properties of light for fields with mean photon numbers of at least the order of 10. The phase probability density is found to satisfy a diffusion equation for both phase-insensitive and phase-sensitive amplifiers and attenuators. The solution is a convolution of the initial phase probability density with an infinite series of expanding Gaussians which clearly illustrates the diffusion of the phase. In particular, we find that for phase-insensitive amplification the phase of the field undergoes *time-dependent* uniform diffusion. In the limit of large amplification the diffusion ceases and the phase variance of the amplified light is given by the input phase variance plus an extra term which is equal to the phase variance of a coherent state of the same intensity as the initial field. We show that the reduced phase variance of phase-optimized states (relative to coherent states of the same intensity) is lost for power gains greater than the photon-cloning value of 2. In contrast, phase-sensitive amplifiers give rise to time-dependent *nonuniform* phase diffusion. The amount of phase diffusion depends on the relative phase angle between the light and the amplifier. If the peak of a relatively narrow phase probability density is near a minimum in the phase diffusion coefficient, then the phase noise added by the amplifier will be less than that found for a phase-insensitive amplifier. Further squeezing of the amplifier reduces the added phase noise proportionally. We find that it is possible, using phase-sensitive amplifiers, to amplify phase-optimized states by power gains considerably larger than 2 and still retain a reduced phase variance.

PACS number(s): 42.50.−p, 42.60.Da

I. INTRODUCTION

The study of the quantum-statistical properties of optical linear amplifiers began over three decades ago. Important contributions have been made by, among others, Haus and Mullen [1], Gordon, Walker, and Louisell [2], and Mollow and Glauber [3] in the 1960s and, more recently, by Caves [4], Friberg and Mandel [5], Loudon and Shepherd [6], and Stenholm [7]. The effect of phase-insensitive amplification on the phase of light has been examined by a number of authors [6,8–11] who have used a variety of descriptions of phase including the Susskind-Glogower phase operators [12], phenomenological decomposition of quasiprobability distributions, and the Ψ-space approach [13–15] introduced by Pegg and Barnett. In particular, Barnett, Stenholm, and Pegg [8] have shown, using the latter phase formalism, that the phase probability distribution undergoes diffusion for small amplifications. Also, Bandilla [9] has shown that the sum of the variances of the phase operators $\sin\hat{\phi}_\theta$ and $\cos\hat{\phi}_\theta$ for an initially coherent field increases by a factor of 2 in the limit of strong amplification. More recently, Thylén *et al.* [10] showed, using a phenomenological description of phase, that the variance in their phenomenological phase angle is increased by an extra term which, under ideal situations, is equal to the phase variance of a coherent state of the same intensity as the initial field. One can conclude from this work that the phase diffusion is time dependent and diminishes as the degree of amplification grows. However, the precise relationship between the phase diffusion and the increased phase variance in the strong amplification limit has not yet been examined.

A few years ago we [16–18] and others [19,20] independently discovered that the noise added to the field by linear amplification can be reduced with phase-sensitive amplifiers. The question that arises from this is whether the phase diffusion and extra phase-variance term can also be reduced with such amplifiers. To answer this question we undertake a study in this paper of the phase properties of phase-sensitive amplifiers. There are now a number of consistent approaches available for describing quantum-mechanical phase properties. We follow Barnett, Stenholm, and Pegg [8] and choose to use the Ψ-space phase formalism of Pegg and Barnett. (We note [21] that the same results would be given by using, for example, the phase probability-operator measure approach [22].) We show that the phase diffusion coefficient is *time dependent* for linear amplification in general, and that it *depends on the relative phase angle* between the light and the amplifier for phase-sensitive amplifiers. We also give the solution to the diffusion equation for the phase probability density. We recover the results of Bandilla and Thylén *et al.* as special cases from the solution for phase-insensitive amplification.

*Present address: Arbeitsgruppe "Nichtklassische Strahlung" der Max-Planck-Gesellschaft an der Humboldt-Universität zu Berlin, Rudower Chaussee 5, 12484 Berlin, Germany.

Thus our solution provides a bridge between their work and previous work on phase diffusion. We also show how the phase noise added by a phase-sensitive amplifier can be reduced to such an extent that, in the limit of large amplification, it can be much less than the phase noise of a coherent state.

II. THE MASTER EQUATION FOR LINEAR AMPLIFIERS

We use the linear amplifier described by Dupertuis, Barnett, and Stenholm [19] as our model of a phase-sensitive amplifier. This amplifier is based on a single-mode field interacting linearly with a reservoir. The reservoir could be constructed from, for example, harmonic oscillators or atomic systems. The condition that sets the treatment by Dupertuis *et al.* apart from usual treatments is that the reservoir is considered to be initially prepared in a special ("rigged") state such as a multimode squeezed vacuum. It is this particular initial state in place of the usual thermal state that produces the phase-sensitive nature of the amplifier. The reduced density operator $\hat{\rho}$ describing the field interacting with such a reservoir satisfies the following master equation in the interaction picture [19]:

$$\begin{aligned}\frac{\partial}{\partial t}\hat{\rho}=&\tfrac{1}{2}A(2\hat{a}^\dagger\hat{\rho}\hat{a}-\hat{a}\hat{a}^\dagger\hat{\rho}-\hat{\rho}\hat{a}\hat{a}^\dagger)\\&+\tfrac{1}{2}C(2\hat{a}\hat{\rho}\hat{a}^\dagger-\hat{a}^\dagger\hat{a}\hat{\rho}-\hat{\rho}\hat{a}^\dagger\hat{a})\\&-B(2\hat{a}\hat{\rho}\hat{a}-\hat{a}\hat{a}\hat{\rho}-\hat{\rho}\hat{a}\hat{a})\\&-B^*(2\hat{a}^\dagger\hat{\rho}\hat{a}^\dagger-\hat{a}^\dagger\hat{a}^\dagger\hat{\rho}-\hat{\rho}\hat{a}^\dagger\hat{a}^\dagger)\ .\end{aligned}\tag{2.1}$$

We should point out that the above master equation was derived using operators defined on the infinite-dimensioned Hilbert space as opposed to operators acting on the $s+1$ dimensional Ψ space of Pegg and Barnett. However, the algebra of the annihilation and creation operators defined on the infinite-dimensioned Hilbert space is equivalent to the algebra of the corresponding Ψ-space operators for physical states [13]. Thus the master equation in both formalisms will be identical for physical states. Hence we can consider Eq. (2.1) as a master equation for physical states in the Ψ-space phase formalism with $\hat{a}$ and $\hat{a}^\dagger$ being the Ψ-space annihilation and creation operators.

The first two terms on the right-hand side of Eq. (2.1) are usual for a linear amplifier of this general type with the constants A and C representing gain and loss rates. The less-usual last two terms are nonzero only for squeezed amplifiers and are absent from the master equation used in earlier studies [8,11] of the phase properties of phase-insensitive amplifiers.

The actual values of the constants A, B, and C for a particular construction of the amplifier do not concern us here. Instead we use only the relationships between A, B, and C that are satisfied by linear amplifiers in general. These relationships can be found from the general expression for the variance of a quadrature component of the electric field as it undergoes linear amplification [4,17,18]:

$$\langle[\Delta\hat{X}(\xi,t)]^2\rangle=G\langle[\Delta\hat{X}(\xi,0)]^2\rangle+\langle[\Delta\hat{A}(\xi,t)]^2\rangle\ ,\tag{2.2}$$

where $\hat{X}(\xi,t)=\frac{1}{2}[\hat{a}(t)e^{i\xi}+\hat{a}^\dagger(t)e^{-i\xi}]$ and $\langle[\Delta\hat{X}(\xi,t)]^2\rangle\equiv\langle[\hat{X}(\xi,t)]^2\rangle-\langle\hat{X}(\xi,t)\rangle^2$. The first term on the right-hand side of Eq. (2.2) is the product of a scaling factor G and the variance before amplification. The second term represents added amplifier noise and must satisfy [17,18]

$$\langle[\Delta\hat{A}(\xi,t)]^2\rangle\langle[\Delta\hat{A}(\xi+\tfrac{1}{2}\pi,t)]^2\rangle\geq\tfrac{1}{16}(1-G)^2\tag{2.3}$$

for all values of ξ in a 2π interval.

It is a straightforward matter to calculate the variance $\langle[\Delta\hat{X}(\xi,t)]^2\rangle$ for the amplifier described by the master equation. For example, solving the differential equations of the form

$$\frac{d}{dt}\langle\hat{Q}\rangle=\mathrm{tr}(\dot{\rho}\hat{Q})\ ,\tag{2.4}$$

with $\dot{\rho}$ replaced with the right-hand side of the master equation and where $\hat{Q}$ is one of $\hat{a}$, $\hat{a}^2$, or $\hat{a}^\dagger\hat{a}$ eventually yields

$$\begin{aligned}\langle\hat{a}(t)\rangle&=\langle\hat{a}(0)\rangle\exp[\tfrac{1}{2}(A-C)t]\ ,\\\langle\hat{a}^2(t)\rangle&=\langle\hat{a}^2(0)\rangle e^{(A-C)t}\\&\quad+[e^{(A-C)t}-1]2B^*/(A-C)\ ,\end{aligned}\tag{2.5}$$

$$\langle\hat{N}(t)\rangle=\langle\hat{N}(0)\rangle e^{(A-C)t}+[e^{(A-C)t}-1]A/(A-C)\ ,\tag{2.6}$$

where $\hat{N}=\hat{a}^\dagger\hat{a}$. The factor $\exp[\frac{1}{2}(A-C)t]$ and its square in these equations indicates that the reservoir acts as an amplifier when $A>C$ and as an attenuator when $A<C$. We find further that

$$\begin{aligned}\langle\hat{X}^2(\xi,t)\rangle&=\tfrac{1}{4}[\langle\hat{a}^2(t)\rangle e^{2i\xi}+\langle\hat{a}^{\dagger2}(t)\rangle e^{-2i\xi}\\&\quad+2\langle\hat{a}^\dagger(t)\hat{a}(t)\rangle+1]\\&=\langle\hat{X}^2(\xi,0)\rangle e^{(A-C)t}+\tfrac{1}{4}[e^{(A-C)t}-1]\\&\quad\times\frac{A+C+2B^*e^{2i\xi}+2Be^{-2i\xi}}{A-C}\end{aligned}$$

and hence

$$\begin{aligned}\langle[\Delta\hat{X}(\xi,t)]^2\rangle&=\langle[\Delta\hat{X}(\xi,0)]^2\rangle e^{(A-C)t}\\&\quad+\tfrac{1}{4}[e^{(A-C)t}-1]\\&\quad\times\frac{A+C+2B^*e^{2i\xi}+2Be^{-2i\xi}}{A-C}\ .\end{aligned}$$

Comparing this result for the Dupertuis amplifier with the equation for a general linear amplifier, Eq. (2.2), shows that the scaling factor in Eq. (2.2) is

$$G=e^{(A-C)t}\tag{2.7}$$

and that the amplifier noise term in Eq. (2.2) is given by

$$\begin{aligned}\langle[\Delta\hat{A}(\xi,t)]^2\rangle&=\tfrac{1}{4}[e^{(A-C)t}-1]\\&\quad\times\frac{A+C+2B^*e^{2i\xi}+2Be^{-2i\xi}}{A-C}\ .\end{aligned}$$

With this specific expression for the amplifier noise inequality (2.3) becomes

$$\frac{A+C+4|B|\cos(\eta-2\xi)}{|A-C|}\,\frac{A+C-4|B|\cos(\eta-2\xi)}{|A-C|}\geq 1\ ,$$

where $B \equiv |B| e^{i\eta}$. This inequality must be satisfied for all values of ξ in a 2π interval and so we find that

$$\frac{A+C+4|B|}{|A-C|}\,\frac{A+C-4|B|}{|A-C|} \geq 1 \ .$$

Our interest in this paper lies with amplifiers that add the minimum possible amount of noise to the field and this requires the equality to be satisfied. For these amplifiers we can write

$$\frac{A+C\pm 4|B|}{|A-C|} = e^{\pm 2r} \ . \tag{2.8}$$

The parameter r is analogous to the squeezing parameter used to describe ideal-squeezed states. Indeed, for the particular cases where the amplifier is constructed from harmonic oscillators r is the squeezing parameter of the multimode squeezed states of the oscillators [19]. We note, however, that the relationship between the values of A, B, and C implied by Eq. (2.8) is satisfied by all linear amplifiers of the general type described by the master equation (2.1) and whose amplifier noise is minimized, regardless of their particular construction.

The special case of phase-insensitive amplification ($B=0$, $A > C$) is given by $r=0$ and $A-C>0$, hence from Eq. (2.8)

$$C=0 \ . \tag{2.9}$$

Alternatively, phase-insensitive attenuation (i.e., $B=0$ and $A<C$) is given by $r=0$ and $A-C<0$ and thus

$$A=0 \ . \tag{2.10}$$

We will use these results in the next section.

For our subsequent analysis, we also need to establish the conditions under which a relatively narrow photon-number probability distribution remains relatively narrow. We determine these conditions by examining the change in the variance of the photon number of the field as it is strongly amplified. Using Eq. (2.4) with $\hat{Q}=\hat{N}^2$ gives

$$\frac{d}{dt}\langle \hat{N}^2\rangle = \tfrac{1}{2}A\langle 4\hat{N}^2+6\hat{N}+2\rangle + \tfrac{1}{2}C\langle 2\hat{N}-4\hat{N}^2\rangle + 2B\langle \hat{a}^2\rangle + 2B^*\langle \hat{a}^{\dagger 2}\rangle \ .$$

On substituting this derivative into

$$\frac{d}{dt}\langle [\Delta\hat{N}]^2\rangle = \frac{d}{dt}\langle \hat{N}^2\rangle - 2\langle \hat{N}\rangle \frac{d}{dt}\langle \hat{N}\rangle \ ,$$

with $\langle \hat{N}(t)\rangle$ given by Eq. (2.6), $\langle \hat{a}^2(t)\rangle$ given by Eq. (2.5), and $d\langle \hat{N}\rangle/dt = A(\langle \hat{N}\rangle+1) - C\langle \hat{N}\rangle$ and then solving for $\langle [\Delta\hat{N}(t)]^2\rangle$ we find

$$\langle [\Delta\hat{N}(t)]^2\rangle = \{\langle (\Delta\hat{N}_0)^2\rangle + u[1-e^{-(A-C)t}] - v\} \times e^{2(A-C)t} + v \ , \tag{2.11}$$

where

$$u = \frac{(A+C)\langle \hat{N}_0\rangle + 2B\langle [\hat{a}(0)]^2\rangle + 2B^*\langle [\hat{a}^\dagger(0)]^2\rangle}{A-C} + \frac{(A+C)A+8|B|^2}{(A-C)^2} \ ,$$

$$v = \frac{AC+4|B|^2}{(A-C)^2} \ ,$$

and where we have written $\hat{N}(0)$ as $\hat{N}_0$. The infinite-t limit of the relative variance $\langle \Delta\hat{N}^2\rangle/\langle \hat{N}\rangle^2$ for the amplification case ($A>C$) is easily found to be

$$\lim_{t\to\infty} \frac{\langle [\Delta\hat{N}(t)]^2\rangle}{\langle \hat{N}(t)\rangle^2} = \frac{\langle (\Delta\hat{N}_0)^2\rangle}{\langle \hat{N}_0\rangle^2} + \frac{u-v}{\langle \hat{N}_0\rangle^2} + O(\langle \hat{N}_0\rangle^{-3}) \ ,$$

where O is the order symbol. We assume that the input field $\sum c_n |n\rangle$ is sufficiently intense for the inequality

$$(A^2+4|B|^2)/(A-C)^2 \ll \langle N_0\rangle^2 \tag{2.12}$$

to hold and has a sufficiently smooth distribution to write

$$|c_{n-2}| \approx |c_n| \ . \tag{2.13}$$

For such a field

$$|\langle \hat{a}^2(0)\rangle| < \sum_n |c_{n-2}|\,|c_n| n \approx \langle \hat{N}_0\rangle \ ,$$

giving

$$\lim_{t\to\infty} \frac{\langle [\Delta\hat{N}(t)]^2\rangle}{\langle \hat{N}(t)\rangle^2} < \frac{\langle (\Delta\hat{N}_0)^2\rangle}{\langle \hat{N}_0\rangle^2} + \frac{A+C+4|B|}{(A-C)\langle \hat{N}_0\rangle} \ .$$

Substituting for the second term from Eq. (2.8), we see that a field which initially has a relatively narrow photon-number probability distribution will retain a relatively narrow distribution provided $e^{2r} \ll \langle \hat{N}_0\rangle$.

An interesting but more specific result is obtained for states of the input field which, in addition to satisfying Eq. (2.12), also give

$$\langle \hat{a}^2(0)\rangle \approx \langle \hat{N}_0\rangle e^{2i\varphi} \ ,$$

where $\varphi = \langle \hat{\phi}_\theta(0)\rangle$ for a suitable choice of θ_0. Examples of such states include intense coherent states and also intense partial-phase states [15] which satisfy Eq. (2.13). For these states

$$\lim_{t\to\infty} \frac{\langle [\Delta\hat{N}(t)]^2\rangle}{\langle \hat{N}(t)\rangle^2} \approx \frac{\langle (\Delta\hat{N}_0)^2\rangle}{\langle \hat{N}_0\rangle^2} + \frac{A+C+4|B|\cos(\eta+2\varphi)}{(A-C)\langle \hat{N}_0\rangle} \ .$$

Using Eq. (2.8) we now find that the second term takes on extreme values when $|\cos(\eta+2\varphi)|=1$ as follows:

$$\lim_{t\to\infty} \frac{\langle [\Delta\hat{N}(t)]^2\rangle}{\langle \hat{N}(t)\rangle^2} \approx \frac{\langle (\Delta\hat{N}_0)^2\rangle}{\langle \hat{N}_0\rangle^2} + \frac{e^{2r}}{\langle \hat{N}_0\rangle} \quad \text{for } \cos(\eta+2\varphi)=1 \ , \tag{2.14}$$

and

$$\lim_{t\to\infty} \frac{\langle [\Delta\hat{N}(t)]^2\rangle}{\langle \hat{N}(t)\rangle^2} \approx \frac{\langle (\Delta\hat{N}_0)^2\rangle}{\langle \hat{N}_0\rangle^2} + \frac{e^{-2r}}{\langle \hat{N}_0\rangle} \quad \text{for } \cos(\eta+2\varphi)=-1 \ . \tag{2.15}$$

These last two expressions show how the amount of photon-number noise added by the amplifier can be *selec-*

tively increased or decreased depending on the relative phase angle between the amplifier and the input field. In the last section we compare this with the corresponding result for the phase noise that is added by the amplifier.

III. THE PHASE PROBABILITY DENSITY UNDER LINEAR AMPLIFICATION

A phase state in the $(s+1)$-dimensional Hilbert space Ψ has the form [13–15]

$$|\theta\rangle = \frac{1}{\sqrt{s+1}} \sum_{n=0}^{s} e^{in\theta} |n\rangle .$$

This state, with phase θ, together with s other phase states of Ψ form a complete orthonormal set. Multiplying the probability $\langle\theta|\hat{\rho}(t)|\theta\rangle$ that a state has a phase of θ by the density of phase eigenvalues $(s+1)/2\pi$ gives the phase probability density $\mathcal{P}_s(t,\theta)$,

$$\mathcal{P}_s(t,\theta) \equiv \langle\theta|\hat{\rho}(t)|\theta\rangle \frac{s+1}{2\pi} = \frac{1}{2\pi} \sum_{k=0}^{s} \sum_{l=0}^{s} \rho_{l,k} e^{i(k-l)\theta} , \tag{3.1}$$

where $\rho_{l,k}$ is an element of the density matrix in the number-state representation.

To find the rate of change in $\mathcal{P}_s(t,\theta)$ we first represent the master equation in the number-state basis as

$$\begin{aligned} \frac{d}{dt}\rho_{l,k} = {} & \tfrac{1}{2} A [2(lk)^{1/2}\rho_{l-1,k-1} - (l+k+2)\rho_{l,k}] + \tfrac{1}{2} C \{2[(l+1)(k+1)]^{1/2}\rho_{l+1,k+1} - (l+k)\rho_{l,k}\} \\ & - B\{2[(l+1)k]^{1/2}\rho_{l+1,k-1} - [(l+1)(l+2)]^{1/2}\rho_{l+2,k} - [k(k-1)]^{1/2}\rho_{l,k-2}\} \\ & - B^*\{2[l(k+1)]^{1/2}\rho_{l-1,k+1} - [l(l-1)]^{1/2}\rho_{l-2,k} - [(k+1)(k+2)]^{1/2}\rho_{l,k+2}\} , \end{aligned}$$

where $\rho_{l,k} = \langle l|\hat{\rho}|k\rangle$, and $|k\rangle$ are the number states. This equation is valid for values of l,k over the entire range $[0,s]$ provided we set $\rho_{n,m}=0$ for n,m equal to -2, -1, $s+1$, or $s+2$. Next, we use this result in the expression

$$\frac{\partial}{\partial t}\mathcal{P}_s(t,\theta) 2\pi = \sum_{k=0}^{s}\sum_{l=0}^{s} \frac{\partial}{\partial t}\rho_{l,k} e^{i(k-l)\theta}$$

to obtain

$$\begin{aligned} \frac{\partial}{\partial t}\mathcal{P}_s(t,\theta) 2\pi = {} & \tfrac{1}{2} A \sum_{k=0}^{s}\sum_{l=0}^{s} \{2(lk)^{1/2}\rho_{l-1,k-1} - (l+k+2)\rho_{l,k}\} e^{i(k-l)\theta} \\ & + \tfrac{1}{2} C \sum_{k=0}^{s}\sum_{l=0}^{s} \{2[(l+1)(k+1)]^{1/2}\rho_{l+1,k+1} - (l+k)\rho_{l,k}\} e^{i(k-l)\theta} \\ & - B \sum_{k=0}^{s}\sum_{l=0}^{s} \{2[(l+1)k]^{1/2}\rho_{l+1,k-1} - [(l+1)(l+2)]^{1/2}\rho_{l+2,k} - [k(k-1)]^{1/2}\rho_{l,k-2}\} e^{i(k-l)\theta} \\ & - B^* \sum_{k=0}^{s}\sum_{l=0}^{s} \{2[l(k+1)]^{1/2}\rho_{l-1,k+1} - [l(l-1)]^{1/2}\rho_{l-2,k} \\ & \qquad - [(k+1)(k+2)]^{1/2}\rho_{l,k+2}\} e^{i(k-l)\theta} , \end{aligned}$$

where, as before, we have set $\rho_{n,m}=0$ for n,m equal to -2, -1, $s+1$, or $s+2$. For very large s the values of $\rho_{l,s-1}$ and $\rho_{l,s}$ and their complex conjugates are negligible for physical states. Thus replacing the dummy indices with more convenient ones and then extending or truncating the summations as appropriate yields

$$\begin{aligned} \frac{\partial}{\partial t}\mathcal{P}_s(t,\theta) 2\pi = {} & \tfrac{1}{2} A \sum_{k=0}^{s}\sum_{l=0}^{s} \{2[(k+1)(l+1)]^{1/2} - (l+k+2)\}\rho_{l,k} e^{i(k-l)\theta} + \tfrac{1}{2} C \sum_{k=0}^{s}\sum_{l=0}^{s} [2(lk)^{1/2} - (l+k)]\rho_{l,k} e^{i(k-l)\theta} \\ & - B \sum_{k=0}^{s}\sum_{l=0}^{s} \{2[l(k+1)]^{1/2} - [(l-1)l]^{1/2} - [(k+2)(k+1)]^{1/2}\}\rho_{l,k} e^{i(k-l+2)\theta} \\ & - B^* \sum_{k=0}^{s}\sum_{l=0}^{s} \{2[(l+1)k]^{1/2} - [(l+2)(l+1)]^{1/2} - [(k-1)k]^{1/2}\}\rho_{l,k} e^{i(k-l-2)\theta} . \end{aligned} \tag{3.2}$$

The right-hand side can be simplified for particular regimes of the state of the field. It is usual to consider the regime where the field is intense and the photon-number distribution is relatively narrow. An example of such a field is the intense coherent state. This regime was treated in previous studies of phase-insensitive amplifiers [6,8–11] and, for comparison, we also restrict our attention to it in the remainder of this paper.

We assume that $\rho_{l,k}$ at time t is non-negligible only when both l and k are close to the large mean photon number $\langle \hat{N}(t)\rangle$. $\langle \hat{N}(t)\rangle$ is related to the initial mean photon number by Eq. (2.6). Thus by first writing $l=\langle \hat{N}(t)\rangle+\mu=\langle \hat{N}(t)\rangle(1+\mu/\langle \hat{N}(t)\rangle)$ and $k=\langle \hat{N}(t)\rangle+\nu=\langle \hat{N}(t)\rangle(1+\nu/\langle \hat{N}(t)\rangle)$, we find factors such as $(lk)^{1/2}$ appearing in the non-negligible terms in the previous equation can be approximated to first order in $\langle \hat{N}(t)\rangle^{-1}$ by

$$(lk)^{1/2}=\frac{l+k}{2}-\frac{(l-k)^2}{8\langle \hat{N}(t)\rangle}+O(\langle \hat{N}(t)\rangle^{-2}) . \qquad (3.3)$$

In the preceding section we found that a field with an initially sharply peaked photon-number distribution retains a relatively small variance $\langle[\Delta\hat{N}(t)]^2\rangle$ under strong amplification provided the "squeezing" of the amplifier was restricted so that $e^{2r}\ll\langle \hat{N}_0\rangle$. We assume this to be the case here. Thus Eq. (3.3) holds even in the limit of strong amplification. We note also that Eq. (3.3) will be valid in the attenuation case ($A<C$) provided the degree of attenuation is restricted to the situation where $\langle[\Delta\hat{N}(t)]^2\rangle$ given by Eq. (2.11) is sufficiently smaller than $\langle \hat{N}(t)\rangle^2$. Thus using Eq. (3.3) in Eq. (3.2) we find the rate of change in $\mathcal{P}_s(t,\theta)$ to order $\langle \hat{N}(t)\rangle^{-2}$ is

$$\begin{aligned}\frac{\partial}{\partial t}\mathcal{P}_s(t,\theta)2\pi=&-\tfrac{1}{8}(A+C)\langle \hat{N}(t)\rangle^{-1}\sum_{k=0}^{s}\sum_{l=0}^{s}(k-l)^2\rho_{l,k}e^{i(k-l)\theta}\\&-\tfrac{1}{4}B\langle \hat{N}(t)\rangle^{-1}e^{i\theta}\sum_{k=0}^{s}\sum_{l=0}^{s}[1-(l-k-1)^2]\rho_{l,k}e^{i(k-l+1)\theta}\\&-\tfrac{1}{4}B^*\langle \hat{N}(t)\rangle^{-1}e^{-i\theta}\sum_{k=0}^{s}\sum_{l=0}^{s}[1-(l-k+1)^2]\rho_{l,k}e^{i(k-l-1)\theta} . \qquad (3.4)\end{aligned}$$

The regime for which this result is valid is illustrated in Fig. 1 where the right-hand sides of Eqs. (3.4) and (3.2) are compared as a function of θ for weak coherent states ($\langle \hat{N}\rangle=5$ and 20). This attaches a quantitative meaning to the term "intense" in our context, that is, a mean photon number of the order of 10. Collecting terms in Eq. (3.4) gives

$$\begin{aligned}\frac{\partial}{\partial t}\mathcal{P}_s(t,\theta)=&\tfrac{1}{8}(A+C)\langle \hat{N}(t)\rangle^{-1}\frac{\partial^2}{\partial\theta^2}\mathcal{P}_s(t,\theta)\\&-\tfrac{1}{4}B\langle \hat{N}(t)\rangle^{-1}e^{i\theta}\left[1+\frac{\partial^2}{\partial\theta^2}\right]e^{i\theta}\mathcal{P}_s(t,\theta)\\&-\tfrac{1}{4}B^*\langle \hat{N}(t)\rangle^{-1}e^{-i\theta}\left[1+\frac{\partial^2}{\partial\theta^2}\right]e^{-i\theta}\mathcal{P}_s(t,\theta)\end{aligned}$$

and further rearranging then yields

$$\frac{\partial}{\partial t}\mathcal{P}_s(t,\theta)=\frac{\partial}{\partial\theta}\left[D(t,\theta)\frac{\partial}{\partial\theta}\mathcal{P}_s(t,\theta)\right] , \qquad (3.5)$$

where we have defined

$$D(t,\theta)\equiv\frac{A+C-4|B|\cos(2\theta+\eta)}{8\langle \hat{N}(t)\rangle} \qquad (3.6)$$

and again used $B=|B|e^{i\eta}$.

Equation (3.5) is in the standard form [23] of a one-dimensional diffusion equation for nonuniform time-dependent diffusion with diffusion coefficient $D(t,\theta)$. It is interesting to compare this result with the diffusion of the Wigner function that describes the quadrature amplitudes of the field. The quadrature-amplitude Wigner function satisfies a Fokker-Planck equation whose diffusion coefficients are *independent* of the mean photon number [19]. Noting that the phase variance of intense fields is given approximately by $\langle\Delta\hat{X}(\xi,t)^2\rangle/\langle \hat{N}(t)\rangle$, where ξ is chosen such that the quadrature amplitude $\langle\hat{X}(\xi+\frac{1}{2}\pi,t)\rangle$ is maximized, suggests that the phase diffusion is *inversely proportional* to the mean photon number, which is consistent with our result here.

The previous analysis of the phase diffusion of linear amplifiers [8] was restricted to levels of amplification which are sufficiently small for the mean photon number not to change appreciably. This gives a time-independent diffusion coefficient. The present analysis, on the other hand, is valid for large amplifications and attenuations. Indeed, the increasing value of $\langle \hat{N}(t)\rangle$ with t for amplification [see Eq. (2.6)] means that the diffusion coefficient reduces approximately exponentially. We also note that the diffusion coefficient is independent of θ when the amplifier or attenuator is phase insensitive (i.e., $|B|=0$). This gives uniform phase diffusion and is the case previously studied by Barnett, Stenholm, and Pegg [8]. In contrast, we also find here that *a phase-sensitive amplifier or attenuator* (i.e., $|B\neq0|$) *gives nonuniform diffusion.* The least diffusion by a phase-sensitive amplifier occurs at the phase angles $\theta=-\frac{1}{2}\eta,\ -\frac{1}{2}\eta+\pi$.

A field with random phase is described by a phase probability density, $\mathcal{P}_s(\theta)$, which is independent of θ. An intense field with random phase will therefore retain its random-phase nature because the right-hand side of the diffusion equation is zero for such a field. This is true regardless of whether the amplifier is operating as an amplifier or attenuator. It may appear at first that this contradicts previous work by us and others showing that a phase-sensitive attenuator can produce squeezed light from arbitrary input fields including those with random phase [16,18,19] and that squeezed light has nonrandom phase [24]. Some reflection will show, however, that this is not so because for a phase-sensitive attenuator to squeeze a noisy input it must attenuate the input light sufficiently to reduce the effect of the noise [16,18] and before the field is attenuated sufficiently for it to be squeezed (and thus have squeezed-light phase properties) the field is no longer in the intense-field regime where our results here are valid. It can be shown that in the weak-

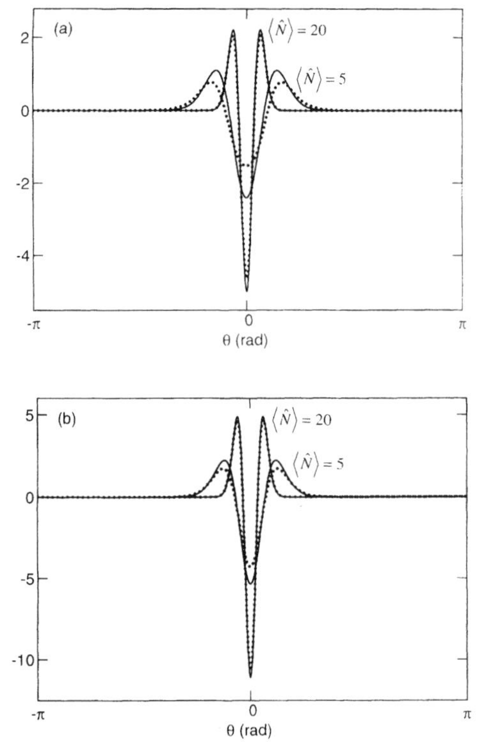

FIG. 1. An illustration of the range of validity of the approximation used in deriving Eq. (3.4) for coherent states with a mean photon number $\langle \hat{N} \rangle$ of 5 and 20 for (a) a phase-sensitive amplifier with $A=1.169$, $C=0.169$, and $B=0.222$ (i.e., $r=0.4$) and (b) a phase-insensitive amplifier with $A=1.0$, $C=0$, and $B=0$. In both cases $A-C=1$. The dotted curves represent the right-hand side of Eq. (3.2) (exact value) and the solid curves represent the right-hand side of Eq. (3.4) (approximate value). Evidently the approximation used in deriving Eq. (3.4) is reasonably accurate even for relatively small $\langle \hat{N} \rangle$.

field regime a nonrandom phase is impressed on a field which has initially random phase [25].

We now proceed to solve the diffusion equation (3.5) for $\mathcal{P}_s(t,\theta)$ in two steps. First we consider the simpler phase-insensitive regime where $B=0$ and the diffusion coefficient is uniform. We then use the phase-insensitive solution to help solve the diffusion equation in the phase-sensitive regime.

A. Phase-insensitive regime

The diffusion coefficient for the phase-insensitive regime ($B=0$) can be written as

$$D(t,\theta)\equiv D(t)=\frac{A+C}{8\langle \hat{N}(t)\rangle}\ , \tag{3.7}$$

where $\langle \hat{N}(t)\rangle$ is given by Eq. (2.6). The diffusion equation (3.5) with the diffusion coefficient given by Eq. (3.7) can be solved using standard techniques [23]. We do not describe the details of these techniques here; instead we give the Green's function solution as follows:

$$f(\theta,t;\phi)=\sum_{n=-\infty}^{\infty}\frac{1}{\sigma\sqrt{2\pi}}\exp\left[\frac{-(\theta-\phi-2n\pi)^2}{2\sigma^2}\right], \tag{3.8}$$

where

$$\sigma^2=\int_0^t 2D(t')dt'=\frac{A+C}{4A}\ln\left[1+\frac{A(1-1/G)}{\langle \hat{N}_0\rangle(A-C)}\right], \tag{3.9}$$

$$G(t)=e^{(A-C)t}\ ,$$

and where ϕ is an arbitrary constant. The fact that $f(\theta,t;\phi)$ is a solution can be checked by substituting into Eq. (3.5). At $t=0$ each term in the series is a Dirac distribution as follows:

$$f(\theta,0;\phi)=\sum_{n=-\infty}^{\infty}\delta(\theta-\phi-2n\pi)\ , \tag{3.10}$$

whereas at later times each term in the series is a Gaussian function with a variance of σ^2 and a mean of $\phi+2n\pi$. The series ensures that the solution is both periodic with a period of 2π and normalized over any 2π interval. In particular, the integral of $f(\theta,t;\phi)$ over the interval $[\alpha,\alpha+2\pi]$ where α is an arbitrary constant is given by

$$\int_\alpha^{\alpha+2\pi} f(\theta,t;\phi)d\theta$$
$$=\sum_{n=-\infty}^{\infty}\int_{\alpha-2n\pi}^{\alpha+2\pi-2n\pi}\frac{1}{\sigma\sqrt{2\pi}}\exp\left[-\frac{(\theta-\phi)^2}{2\sigma^2}\right]d\theta$$
$$=1\ .$$

The contribution to the integral from Gaussians whose peaks lie outside the interval of integration, $[\alpha,\alpha+2\pi]$, can be interpreted as reflections in the boundaries of the interval (i.e., at $\theta=\alpha$ and $\theta=\alpha+2\pi$) of the sole Gaussian whose peak lies within the interval [23].

The general solution of the diffusion equation is found by integrating the product of the Green's function $f(\theta,t;\phi)$ with the initial phase probability density $\mathcal{P}_s(\phi)$ over ϕ,

$$\mathcal{P}_s(t,\theta)=\int_{2\pi} f(\theta,t;\phi)\mathcal{P}_s(\phi)d\phi\ . \tag{3.11}$$

This solution is valid for all initial probability densities belonging to the space of infinitely differentiable functions [26]; we note from Eq. (3.1) that $\mathcal{P}_s(\phi)$ is infinitely differentiable because s is finite and thus Eq. (3.11) represents the most general solution of the diffusion equation as it applies to quantum-mechanical phase. Substituting $f(\theta,t;\phi)$ from Eq. (3.8) shows that the right-hand side is a *convolution* of the initial phase probability density with a series of broadening Gaussians. This clearly shows that the phase probability density is broadened as the field undergoes linear amplification or attenuation and clearly illustrates the diffusion of the phase.

The extent of the broadening for amplification is different from that for attenuation, and so we must treat these cases separately. In the amplification case we find from Eqs. (2.9) and (3.9) that

$$\sigma^2=\tfrac{1}{4}\ln\left[1+\frac{1-e^{-At}}{\langle\hat{N}_0\rangle}\right]. \tag{3.12}$$

In the integral of Eq. (3.11) $f(\theta,t;\phi)$ appears as a function of ϕ; in this respect $f(\theta,t;\phi)$, as given by Eq. (3.8), can be considered as a series of Gaussians each of which is centered on a mean of $\phi=\theta-2n\pi$. Thus increasing the value of θ moves the Gaussians along the positive ϕ axis. For a sufficiently large mean photon number $\langle\hat{N}_0\rangle$ the Gaussians in the series are very narrow and their peaks are separated by many standard deviations σ. Thus only the term in the series of Eq. (3.8) with $n=0$ has a significant contribution to $\mathcal{P}_s(t,\theta)$ in the integral of Eq. (3.11) for $\theta\sim\theta_0+\pi$ when the integral is taken over the interval $[\theta_0,\theta_0+2\pi]$ as illustrated in Fig. 2. We note that in general, because $f(\theta,t;\phi)$ has a period of 2π, the $n=1$ and -1 terms of the series may also have a significant contribution for $\theta\sim\theta_0$ and $\theta\sim\theta_0+2\pi$, respectively. However, if the initial phase probability density $\mathcal{P}_s(\theta)$ of the input field has a sufficiently narrow peak at say, $\theta_0+\pi$, and is negligible elsewhere (see the dotted curve in Fig. 2) then the integral in Eq. (3.11) over the interval $[\theta_0,\theta_0+2\pi]$ will be negligible unless the value of θ is relatively near $\theta_0+\pi$. In this situation all the terms in the series of Eq. (3.8) except the $n=0$ term can be ignored to a good approximation. This simplifies the calculation of the mean and variance of the phase of the output field. We note that moments of the Hermitian phase operator $\hat{\phi}_\theta$ are given by [14,15]

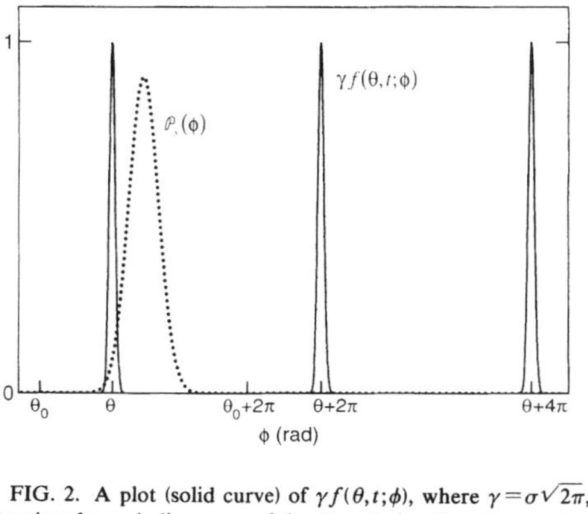

FIG. 2. A plot (solid curve) of $\gamma f(\theta,t;\phi)$, where $\gamma=\sigma\sqrt{2\pi}$, showing the periodic nature of the Green's function. Increasing the value of θ moves the curve along the positive ϕ axis. Also shown (dotted curve) is a plot of an initial phase probability density $\mathcal{P}_s(\phi)$ with a peak centered on $\theta_0+\pi$.

$$\langle\hat{\phi}_\theta^n(t)\rangle=\lim_{s\to\infty}\sum_{m=0}^{s}(\theta_m)^n\mathcal{P}_s(t,\theta_m)2\pi/(s+1),$$

where $\theta_m=\theta_0+2\pi m/(s+1)$. In particular, for states for which the energy is finite this expression becomes [21]

$$\langle\hat{\phi}_\theta^n(t)\rangle=\lim_{s\to\infty}\int_{\theta_0}^{\theta_0+2\pi}\theta^n\mathcal{P}_s(t,\theta)d\theta .$$

Substituting for $\mathcal{P}_s(t,\theta)$ from Eqs. (3.11) and (3.8) and retaining only the term with $n=0$ for $f(\theta,t;\phi)$ yields

$$\langle\hat{\phi}_\theta^n(t)\rangle=\lim_{s\to\infty}\int_{\theta_0}^{\theta_0+2\pi}\theta^n\int_{\theta_0}^{\theta_0+2\pi}\frac{1}{\sigma\sqrt{2\pi}}\exp\left[\frac{-(\theta-\phi)^2}{2\sigma^2}\right]\times\mathcal{P}_s(\phi)d\phi\,d\theta ,$$

where σ is given by Eq. (3.12). Interchanging the order of integration yields, to a good approximation,

$$\langle\hat{\phi}_\theta(t)\rangle\approx\lim_{s\to\infty}\int_{\theta_0}^{\theta_0+2\pi}\phi\mathcal{P}_s(\phi)d\phi=\langle\hat{\phi}_\theta(0)\rangle ,$$

$$\langle\hat{\phi}_\theta^2(t)\rangle\approx\lim_{s\to\infty}\int_{\theta_0}^{\theta_0+2\pi}(\sigma^2+\phi^2)\mathcal{P}_s(\phi)d\phi =\sigma^2+\langle[\hat{\phi}_\theta(0)]^2\rangle$$

and hence

$$\langle[\Delta\hat{\phi}_\theta(t)]^2\rangle\approx\sigma^2+\langle[\Delta\hat{\phi}_\theta(0)]^2\rangle \approx\tfrac{1}{4}\ln\left[1+\frac{1-e^{-At}}{\langle\hat{N}_0\rangle}\right]+\langle[\Delta\hat{\phi}_\theta(0)]^2\rangle , \tag{3.13}$$

where we have used Eq. (3.12).

Further amplification of the output by an identical amplifier gives

$$\langle[\Delta\hat{\phi}_\theta(t')]^2\rangle\approx\tfrac{1}{4}\ln\left[1+\frac{1-e^{-A(t'-t)}}{\langle\hat{N}(t)\rangle}\right]+\langle[\Delta\hat{\phi}_\theta(t)]^2\rangle .$$

Substituting $\langle\hat{N}(t)\rangle=(\langle\hat{N}_0\rangle+1)e^{At}-1$ from Eqs. (2.6) and (2.9) in the first term, using Eq. (3.13) for the last term, and rearranging yields

$$\langle[\Delta\hat{\phi}_\theta(t')]^2\rangle\approx\tfrac{1}{4}\ln\left[1+\frac{1-e^{-At'}}{\langle\hat{N}_0\rangle}\right]+\langle[\Delta\hat{\phi}_\theta(0)]^2\rangle ,$$

which is identical to that obtained from a single amplifier operating for the total period $[0,t']$. An explanation of this result is found by noting that the two-stage amplification is represented by a *convolution* of the two Green's functions representing each stage, and also that the convolution of two Gaussians with variances σ_1^2 and σ_2^2 is another Gaussian with a variance of $\sigma_1^2+\sigma_2^2$. Thus the first term on the right-hand side of the last equation, being the variance of the Green's function representing the two-stage amplification, is the sum of the variances of the Green's functions representing each stage.

The first term on the right-hand side of Eq. (3.13) is added amplifier noise. The value of this term grows with increasing amplification. Its presence has serious implications for the amplification of the phase-optimized states [27]. These states have a phase variance of approximately $c(1/4)\langle\hat{N}\rangle^{-2}$, where c is π^2 or 7.52 depending on whether the optimization constraint is a fixed maximum or mean photon number. Both of these values are much less than the corresponding value of approximately $\frac{1}{4}\langle\hat{N}\rangle^{-1}$ for an intense coherent state. The amplifier noise resulting from strong amplification by a phase-insensitive amplifier will destroy the reduced phase-variance property of a field prepared in a phase-optimized state when, after amplification, the phase variance is equal to that of a coherent state of the same intensity, i.e., when

$$\langle[\Delta\hat{\phi}_\theta(t)]^2\rangle=\tfrac{1}{4}\langle\hat{N}(t)\rangle^{-1}$$
$$\approx\tfrac{1}{4}[(\langle\hat{N}_0\rangle+1)G-1]^{-1},$$

where, here, $G=e^{At}$ is approximately the power gain. Substituting for $\langle[\Delta\hat{\phi}(t)]^2\rangle$ on the left-hand side from Eq. (3.13) and using $\langle[\Delta\hat{\phi}_\theta(0)]^2\rangle\approx c(1/4)\langle\hat{N}_0\rangle^{-2}$ gives

$$\frac{1}{4}\ln\left[1+\frac{1-1/G}{\langle\hat{N}_0\rangle}\right]+\frac{c}{4\langle\hat{N}_0\rangle^2}\approx\frac{1}{4[(\langle\hat{N}_0\rangle+1)G-1]}. \tag{3.14}$$

Expanding the logarithm and the right-hand side to first order in $\langle\hat{N}_0\rangle^{-1}$ yields $G=2+O(\langle\hat{N}_0\rangle^{-1})$. In fact, we have found by numerical analysis that the values of G satisfying Eq. (3.14) approach 2 from below as $\langle\hat{N}_0\rangle$ increases. For example, for the values of $\langle\hat{N}_0\rangle$ of 10, 10^4, and 10^7 the corresponding values of G that satisfy Eq. (3.14) with $c=\pi^2$ are approximately 1.006 251, 1.998 003, and 1.999 998, respectively. Hence we conclude that *the reduced phase fluctuations of phase-optimized states are lost for power gains G greater than 2*. It is interesting to note that all squeezing of the field is also lost for the same value of the power gain for phase-insensitive amplification. We found in previous work [17,18] that the output field can remain squeezed for power gains greater than 2 if a phase-sensitive amplifier is used. Later in this paper when we treat the phase-sensitive amplifier we give a similar result for the retention of a reduced phase variance for power gains greater than 2.

We now examine the large-t limit of the phase variance given by Eq. (3.13) which is, to first order in $\langle\hat{N}_0\rangle^{-1}$,

$$\lim_{t\to\infty}\langle[\Delta\hat{\phi}_\theta(t)]^2\rangle\approx\frac{1}{4\langle\hat{N}_0\rangle}+\langle[\Delta\hat{\phi}_\theta(0)]^2\rangle. \tag{3.15}$$

In this limit the added amplifier noise has a magnitude approximately equal to the phase variance of a coherent state of mean photon number $\langle\hat{N}_0\rangle$ and can be compared with the results of Thylén *et al.* [10]. In particular, if the input field is in an intense coherent state then $\langle[\Delta\hat{\phi}_\theta(0)]^2\rangle\approx\frac{1}{4}\langle\hat{N}_0\rangle^{-1}$ and so

$$\lim_{t\to\infty}\langle[\Delta\hat{\phi}_\theta(t)]^2\rangle\approx2\langle[\Delta\hat{\phi}_\theta(0)]^2\rangle.$$

Thus we regain the specific result found by Bandilla [9] that the phase variance of coherent light increases by a factor of 2 when strongly amplified by a phase-insensitive amplifier. The significance of the analysis given here is that it is more general and shows that the increase in phase variance results from a convolution of a periodic Gaussian series with an *arbitrary* initial phase probability density. In addition our particular result, Eq. (3.15), is also valid in the extreme case where the initial phase probability density is *relatively broad* and zero only near θ_0 and $\theta_0+2\pi$, a case for which the phenomenological description of Thylén *et al.* is not valid.

We turn now to the case of phase-insensitive attenuation where, from Eq. (2.10), $A=0$. The variance σ^2 of the Gaussian functions summed in Eq. (3.8) can be found for the attenuation case from the limit of the right-hand side of Eq. (3.9) as $A\to0$ with $C>0$:

$$\sigma^2=\frac{e^{Ct}-1}{4\langle\hat{N}_0\rangle}.$$

Clearly the Gaussian functions broaden at an exponential rate. For times satisfying $e^{Ct}\ll\langle\hat{N}_0\rangle$ [that is, from Eq. (2.6), $\langle\hat{N}(t)\rangle\gg1$] the Gaussian functions in Eq. (3.8) will still be relatively narrow and so the approximations made when deriving Eq. (3.13) for amplification will also be valid here. Thus we find the phase variance of fields undergoing phase-insensitive attenuation obeys

$$\langle[\Delta\hat{\phi}_\theta(t)]^2\rangle\approx\frac{e^{Ct}-1}{4\langle\hat{N}_0\rangle}+\langle[\hat{\phi}_\theta(0)]^2\rangle.$$

For $\langle\hat{N}_0\rangle\gg e^{Ct}\gg1$ the added noise term on the right-hand side is approximately equal to the phase variance of a coherent state which has the same mean photon number as that of the output field.

B. Phase-sensitive regime

We now use the solution $f(\theta,t;\phi)$ of the diffusion equation for the phase-insensitive regime to construct a solution for the phase-sensitive regime. In the phase-sensitive regime $B\neq0$, thus $r\neq0$ from Eq. (2.8) and the diffusion coefficient $D(t,\theta)$, given by Eq. (3.6), varies with the variable θ as illustrated in Fig. 3. $D(t,\theta)$ has a maximum value at $\theta=\frac{1}{2}(\pi-\eta)$ of $e^{2r}(1/8)|A-C|\langle\hat{N}(t)\rangle^{-1}$ which is e^{2r} times the diffusion coefficient $D(t)$ for the phase-insensitive regime [see Eqs. (3.7), (2.9), and (2.10)]. Consider for the moment the effect of *overestimating* the phase-sensitive diffusion by replacing $D(t)$ in the expression for σ^2 in Eq. (3.9) with this maximum value $e^{2r}D(t)$. The variance σ^2 of the Gaussian functions summed in $f(\theta,t;\phi)$ increases by the factor e^{2r}. However, provided $e^{2r}\ll\langle\hat{N}_0\rangle$ we find that for times satisfying $e^{-(A-C)t}\langle\hat{N}_0\rangle^{-1}\ll1$ the variance σ^2 is relatively small for both the amplification and attenuation cases, and also that the large-t limit of σ^2 is relatively small for the amplification case. Thus for these situations the solution $F(\theta,t;\phi)$ of the phase-sensitive diffusion equation (3.5) analogous to $f(\theta,t;\phi)$ will be a periodic function with a relatively narrow peak at $\theta=\phi$ and a variance no greater than the overestimated value of $e^{2r}(1/4)\langle\hat{N}_0\rangle^{-1}$. Both sides of the diffusion equation (3.5) for this solution are negligible except for a small region near $\theta=\phi$. Moreover, the magnitude of the change in $D(t,\theta)$ over the range $[\phi-\varepsilon,\phi+\varepsilon]$ where $\varepsilon=e^{2r}\langle\hat{N}_0\rangle^{-1}$ can be shown to be approximately $|A-C|\sinh(2r)|\sin(2\phi+\eta)|e^{2r}G^{-1}\langle\hat{N}_0\rangle^{-2}$

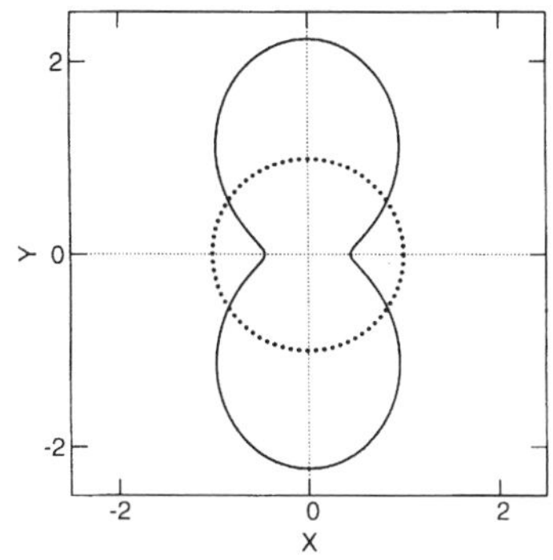

FIG. 3. A polar plot of the scaled phase diffusion coefficient $8\langle\hat{N}(t)\rangle D(t,\theta)$ for a phase-sensitive amplifier (solid curve) with $A=1.169$, $C=0.169$, and $B=0.222$. The value of $8\langle\hat{N}(t)\rangle D(t,\theta)$ is given by the radial distance from the origin at polar angle θ. Also shown for comparison is the scaled coefficient $8\langle\hat{N}(t)\rangle D(t,\theta)$ for a phase-insensitive amplifier (dotted curve) with $A=1.0$, $C=0$, and $B=0$.

where $G=e^{(A-C)t}$. This value is exceedingly small for the situation we are presently considering where $e^{2r}\ll\langle\hat{N}_0\rangle$. Thus to a good approximation we can replace $D(t,\theta)$ in Eq. (3.5) with its value $D(t,\phi)$ at the peak of $F(\theta,t;\phi)$. Hence we now seek the solution of

$$\frac{\partial}{\partial t}F(\theta,t;\phi)=D(t,\phi)\frac{\partial^2}{\partial\theta^2}F(\theta,t;\phi)\ .$$

Following the methods used above we find that the Green's function solution is

$$F(\theta,t;\phi)=\sum_{n=-\infty}^{\infty}\frac{1}{\sigma(\phi)\sqrt{2\pi}}\exp\left[\frac{-(\theta-\phi-2n\pi)^2}{2\sigma^2(\phi)}\right],$$

where

$$\begin{aligned}\sigma^2(\phi)&=\int_0^t 2D(t',\phi)dt'\\&=\frac{A+C-4|B|\cos(2\phi+\eta)}{4A}\\&\quad\times\ln\left\{1+\frac{A[1-\exp(C-A)t]}{(A-C)\langle\hat{N}_0\rangle}\right\}.\end{aligned}\quad(3.16)$$

We note that the variance $\sigma^2(\phi)$ of the Gaussian functions summed in the solution now depends on the value of ϕ and this reflects the nonuniform nature of the diffusion. The function $F(\theta,t;\phi)$ approximates the Green's function solution of Eq. (3.5) and so the general solution of the diffusion equation (3.5) in the phase-sensitive regime is approximated by the integral

$$\mathcal{P}_s(t,\theta)=\int_{2\pi}F(\theta,t;\phi)\mathcal{P}_s(\phi)d\phi\ .$$

Calculating the phase variance for a field whose initial phase variance is very small yields

$$\langle[\Delta\hat{\phi}_\theta(t)]^2\rangle\approx\sigma^2(\langle\hat{\phi}_\theta(0)\rangle)+\langle[\Delta\hat{\phi}_\theta(0)]^2\rangle\ ,$$

where we have used $\langle\cos[2\hat{\phi}_\theta(0)+\eta]\rangle\approx\cos[2\langle\hat{\phi}_\theta(0)\rangle+\eta]$. Two important cases arise from the extreme values of $\sigma(\phi)$ given by Eqs. (3.16) and (2.8): for $\cos[2\langle\hat{\phi}_\theta(0)\rangle+\eta]=1$,

$$\langle[\Delta\hat{\phi}_\theta(t)]^2\rangle\approx\frac{e^{-2r}|A-C|}{4A}\ln\left[1+\frac{A(1-1/G)}{\langle\hat{N}_0\rangle(A-C)}\right]+\langle[\Delta\hat{\phi}_\theta(0)]^2\rangle\ ,\quad(3.17)$$

and for $\cos[2\langle\hat{\phi}_\theta(0)\rangle+\eta]=-1$,

$$\langle[\Delta\hat{\phi}_\theta(t)]^2\rangle\approx\frac{e^{2r}|A-C|}{4A}\ln\left[1+\frac{A(1-1/G)}{\langle\hat{N}_0\rangle(A-C)}\right]+\langle[\Delta\hat{\phi}_\theta(0)]^2\rangle\ .\quad(3.18)$$

The first case corresponds to the peak of the phase probability density lying at a minimum of $D(t,\theta)$ whereas the second case corresponds to the peak lying at a maximum of $D(t,\theta)$.

We now examine in the phase-sensitive regime the effect on the phase variance of amplifying a field which is initially in a phase-optimized state. To find the maximum gain which preserves some reduced phase variance, we determine the gain which increases the phase variance of the output field to that of a coherent state of the same intensity. This maximum occurs in the first case considered above where $\cos[2\langle\hat{\phi}_\theta(0)\rangle+\eta]=1$. Thus substituting $\langle[\Delta\hat{\phi}_\theta(0)]^2\rangle=c(1/4)\langle\hat{N}_0\rangle^{-2}$ for a phase-optimized state into Eq. (3.17), setting $\langle[\Delta\hat{\phi}_\theta(t)]^2\rangle=\frac{1}{4}\langle\hat{N}(t)\rangle^{-1}$, and then replacing $\langle\hat{N}(t)\rangle$ with the right side of Eq. (2.6) yields

$$\frac{1}{4\{G[\langle\hat{N}_0\rangle+A/(A-C)]-A/(A-C)\}}\approx\frac{e^{-2r}|A-C|}{4A}\ln\left[1+\frac{A(1-1/G)}{\langle\hat{N}_0\rangle(A-C)}\right]+\frac{c}{4\langle\hat{N}_0\rangle^2}\ .$$

Expanding both sides to first order in $\langle\hat{N}_0\rangle^{-1}$ and noting that $e^{2r}\ll\langle\hat{N}_0\rangle$ gives an approximate solution for the limiting value of the gain G as

$$G\approx1+e^{2r}+O(\langle\hat{N}_0\rangle^{-1})\ .$$

Clearly the limiting value of G can be made much larger than 2 by increasing the squeezing parameter r of the amplifier reservoir. Hence the field can undergo significantly greater amplification by phase-sensitive amplifiers compared to that by phase-insensitive amplifiers before the field loses the reduced phase-variance property.

The infinite-t limit of $\langle[\Delta\hat{\phi}_\theta(t)]^2\rangle$ given by Eqs. (3.17) and (3.18) for amplification is, to first order in $\langle\hat{N}_0\rangle^{-1}$,

$$\lim_{t\to\infty}\langle[\Delta\hat{\phi}_\theta(t)]^2\rangle\approx\frac{e^{-2r}}{4\langle\hat{N}_0\rangle}+\langle[\Delta\hat{\phi}_\theta(0)]^2\rangle\quad\text{for }\cos[2\langle\hat{\phi}_\theta(0)\rangle+\eta]=1\ ,\quad(3.19)$$

$$\lim_{t\to\infty}\langle[\Delta\hat{\phi}_\theta(t)]^2\rangle \approx \frac{e^{2r}}{4\langle\hat{N}_0\rangle}+\langle[\Delta\hat{\phi}_\theta(0)]^2\rangle$$

$$\text{for } \cos[2\langle\hat{\phi}_\theta(0)\rangle+\eta]=-1\ , \quad (3.20)$$

where we have used $A-C>0$ for amplification and so $G\rightarrow\infty$ as $t\rightarrow\infty$. Comparing Eq. (3.19) with the corresponding expression Eq. (3.15) for phase-insensitive amplification shows that the phase-sensitive amplifier can significantly reduce the phase diffusion arising from linear amplification. Indeed, if the field is initially in an intense coherent state then after phase-sensitive amplification the phase variance, from Eq. (3.19), is

$$\frac{1+e^{-2r}}{4\langle\hat{N}_0\rangle}\ ,$$

which is $(1+e^{-2r})$ times the initial phase variance. Clearly this factor can be reduced significantly below 2, the corresponding value [9,10] for phase-insensitive amplification.

IV. DISCUSSION

We have investigated the diffusion of the phase of light as the light is amplified. We have found that the diffusion coefficient $D(t,\theta)$ decays exponentially with time and thus the light suffers no phase diffusion in the infinite-t limit. This can be interpreted phenomenologically by noting that the variance in phase angle can be approximated for intense fields which have a narrow phase probability distribution by $\langle[\Delta\hat{X}(\xi,t)]^2\rangle/\langle\hat{N}(t)\rangle$. Here, $\hat{X}(\xi,t)$ is the quadrature amplitude which is 90° out of phase with the field and $\langle\hat{N}(t)\rangle$ is the mean photon number of the field. For phase-insensitive amplifiers $\langle[\Delta\hat{X}(\xi,t)]^2\rangle=G\langle[\Delta\hat{X}(\xi,0)]^2\rangle+\frac{1}{4}|G-1|$ and $\langle\hat{N}(t)\rangle \approx G\langle\hat{N}(0)\rangle$ where G is approximately the power gain. Thus both the quadrature-amplitude variance and the photon number increase with the gain. The net result is that as G increases the approximate phase variance, given by

$$\frac{\langle[\Delta\hat{X}(\xi,t)]^2\rangle}{\langle\hat{N}(t)\rangle}\approx\frac{\langle[\Delta\hat{X}(\xi,0)]^2\rangle}{\langle\hat{N}(0)\rangle}+\frac{|1-1/G|}{4\langle\hat{N}(0)\rangle}\ ,$$

approaches a limiting value which is $\frac{1}{4}\langle\hat{N}(0)\rangle^{-1}$ more than its initial value. Thus in the limit the phase diffusion vanishes.

It has been known for some time that phase-insensitive amplification by more than the photon-cloning factor of 2 destroys all squeezing in a squeezed input field. In this paper we have found an interesting parallel result that this is also the value of gain which destroys all phase noise reduction relative to a coherent state for a phase-optimized [27] input state. On the other hand, the phase diffusion coefficient $D(t,\theta)$ for phase-sensitive amplifiers has also been shown to depend on the phase angle. Thus fields with a narrowly peaked phase probability distribution will suffer less phase diffusion if the peak lies at a minimum in $D(t,\theta)$. This permits fields prepared in a phase-optimized state to be amplified by power gains greater than 2 and still retain a phase which is more sharply defined than a coherent state of the same intensity.

Another interesting parallel exists between the work presented in this paper and that of our earlier work [16–18] on the effects on the field from linear amplification. In our earlier work we found that the amplifier noise could be squeezed from one quadrature component of the field to another. Here, on comparing Eqs. (2.14) and (2.15) with (3.19) and (3.20) and identifying φ with $\langle\hat{\phi}_\theta(0)\rangle$ we find that the contribution to the noise from the amplifier can be selectively reduced from the phase and simultaneously increased for the photon number, and vice versa. Thus the phase fluctuations of the amplified light can be reduced only at the expense of increased intensity fluctuations.

In conclusion, we consider briefly how the nonuniform nature of $D(t,\theta)$ illustrated in Fig. 3 might be utilized to maximum advantage in amplifying phase-modulated light used in optical communication. The particular situation we have in mind is where the phase of the light is switched between two values corresponding to the binary digits 0 and 1. For example, consider the case where the phase has a narrow probability distribution and is switched between values of 0 and π. A phase-sensitive amplifier with $\eta=0$ will disturb the phase variance for *both* values of the phase by a minimal amount because the peak in the phase probability density will *always* occur at one of the two minimum values of the diffusion coefficient. The added phase noise from a pretransmission or a predetection amplifier could thereby be reduced considerably.

[1] H. A. Haus and J. A. Mullen, Phys. Rev. **128**, 2407 (1962).
[2] J. P. Gordon, L. R. Walker, and W. H. Louisell, Phys. Rev. **130**, 806 (1963).
[3] B. R. Mollow and R. J. Glauber, Phys. Rev. **160**, 1076 (1967).
[4] C. M. Caves, Phys. Rev. D **26**, 1817 (1982).
[5] S. Friberg and L. Mandel, Opt. Commun. **46**, 141 (1983).
[6] R. Loudon and T. J. Shepherd, Opt. Acta **31**, 1243 (1984).
[7] S. Stenholm, Opt. Commun. **58**, 177 (1986).
[8] S. M. Barnett, S. Stenholm, and D. T. Pegg, Opt. Commun. **73**, 314 (1989).
[9] A. Bandilla, Opt. Commun. **80**, 267 (1991); see also Ann. Phys. (Leipzig) **1**, 117 (1992).
[10] L. Thylén, M. Gustavsson, A. Karlsson, and T. K. Gustafson, J. Opt. Soc. Am. B **9**, 369 (1992).
[11] Ning Lu, Phys. Rev. A **42**, 5641 (1990).
[12] L. Susskind and J. Glogower, Physics (N.Y.) **1**, 49 (1964).
[13] D. T. Pegg and S. M. Barnett, Europhys. Lett. **6**, 483 (1988).
[14] S. M. Barnett and D. T. Pegg, J. Mod. Opt. **36**, 7 (1989).
[15] D. T. Pegg and S. M. Barnett, Phys. Rev. A **39**, 1665 (1989).
[16] J. A. Vaccaro and D. T. Pegg, Opt. Acta **33**, 1141 (1986).
[17] D. T. Pegg and J. A. Vaccaro, Opt. Commun. **61**, 317

(1987).
[18] J. A. Vaccaro and D. T. Pegg, J. Mod. Opt. **34**, 855 (1987).
[19] M.-A. Dupertuis and S. Stenholm, J. Opt. Soc. Am. B **4**, 1094 (1987); M.-A. Dupertuis, S. M. Barnett, and S. Stenholm, *ibid.* **4**, 1102 (1987); **4**, 1124 (1987).
[20] G. J. Milburn, M. L. Steyn-Ross, and D. F. Walls, Phys. Rev. A **35**, 4443 (1987).
[21] J. A. Vaccaro and D. T. Pegg, Phys. Scr. **T48**, 22 (1993).
[22] C. W. Helstrom, *Quantum Detection and Estimation Theory* (Academic, New York, 1976), p. 53; J. H. Shapiro and S. R. Shepard, Phys. Rev. A **43**, 3795 (1991).
[23] J. Crank, *The Mathematics of Diffusion,* 2nd ed. (Clarendon, Oxford, 1975).
[24] J. A. Vaccaro and D. T. Pegg, Opt. Commun. **70**, 529 (1989).
[25] J. A. Vaccaro and D. T. Pegg, J. Mod. Opt. (to be published).
[26] R. D. Richtmyer, *Principles of Advanced Mathematical Physics* (Springer-Verlag, Berlin, 1978), Vol. 1.
[27] G. S. Summy and D. T. Pegg, Opt. Commun. **77**, 75 (1990).

5

Theory of Phase Measurement

The phase operator was introduced as the representation of the phase observable for a single field mode or for a simple harmonic oscillator. As such, it inevitably leads to the question of how to measure it. In the classical domain it is deceptively easy to measure the phase by means of an interference measurement. Superposing two monochromatic fields of the same frequency can be achieved by means of an optical beam splitter and the intensities of the two output fields then depend on the relative phases of the two fields. If the two fields have amplitudes $E_1e^{i\phi_1}$ and $E_2e^{i\phi_2}$ respectively then the intensities measured at the two outputs, a and b, of a balanced symmetric beam splitter are

$$\begin{aligned} I_a &\propto \frac{1}{2}\left|E_1e^{i\phi_1} + iE_2e^{i\phi_2}\right|^2 \\ &= \frac{1}{2}\left[E_1^2 + E_2^2 + 2E_1E_2\sin(\phi_1 - \phi_2)\right] \\ I_b &\propto \frac{1}{2}\left|iE_1e^{i\phi_1} + E_2e^{i\phi_2}\right|^2 \\ &= \frac{1}{2}\left[E_1^2 + E_2^2 - 2E_1E_2\sin(\phi_1 - \phi_2)\right]. \end{aligned} \tag{5.1}$$

We see that the observed interference depends not only on the phase difference $\phi_1 - \phi_2$ but also on the intensities of the interfering fields. It is clear that, even in the *classical theory*, the phase is a quantity that needs to be derived from the resulting intensities, rather than directly measured.

The above conclusion becomes more problematic in the quantum theory and this is fundamentally because, as we have seen, the phase observable is incompatible with the photon number operator. It has been suggested, however, that the phase should itself be defined in terms of a classical interference experiment (Walker and Carrol 1986, Walker 1987). The idea is to analyse an interferometer classically to determine the phase observable and then to treat the operator analogue of this quantity as a phase operator (Noh *et al.* 1991, 1992). There is, however, a fundamental difficulty with this approach which is that it does not lead to a unique phase observable and, moreover, the interpretation of the measurement results themselves cannot always be consistently interpreted as a phase measurement (Barnett and Pegg 1993 **Paper 5.1**). It has been suggested that for this reason the phase is fundamentally a classical quantity without a quantum analogue (Noh *et al.* 1993). This would place the phase in a rather special and indeed strange position amongst physical properties. To avoid this unnecessary situation, it is preferable to reject the idea that phase should be defined by such a direct reference to classical experiments. The phase operator, described in the preceding chapters, is based on clear physical principles and the remaining papers in this chapter discuss the methods that have been devised to measure it.

The study of squeezed states of light led to the development of balanced homodyne detection as the method of choice for measuring phase-sensitive quantum properties of light (Loudon and Knight 1987, Loudon 2000). In this device the field to be measured is mixed on a 50%:50% beam-splitter with an intense, coherent laser field which acts as a local oscillator or phase reference. The two output beams are measured using a pair of high efficiency photodetectors and the difference between the resulting pair of photocurrents is recorded. A careful analysis shows that the statistics of this difference-photocurrent reflects the quantum statistics of one of the quadrature operators for the measured mode (Barnett and Radmore 1997):[1]

$$\hat{x}_{\theta} = \frac{1}{\sqrt{2}}\left(\hat{a}e^{-i\theta} + \hat{a}^{\dagger}e^{i\theta}\right). \tag{5.2}$$

Here the angle θ, and hence the measured quadrature, is determined by the phase of the local oscillator. This quadrature is reminiscent of the product of the field amplitude and the cosine of the phase, $\cos(\phi - \theta)$, and indeed "measured phase operators" have been defined in terms of these (Barnett and Pegg 1986 **Paper 1.7**). The quadratures are not phase operators, of course, but measuring them can provide a very good approximation to measurement of the phase (Vaccaro and Pegg 1994c **Paper 5.2**). Homodyne detection can also provide something close to a single-shot measurement of the phase by actively adjusting the local oscillator phase as the photocurrent is measured (Wiseman 1995 **Paper 5.3**). The idea is to use the accumulated photocurrent data, up to any instant during the measurement, to determine the new value selected for the local oscillator phase.

It is of fundamental interest to determine whether or not the optical phase probability distribution can be measured precisely, even in principle. The state of a light pulse can be measured if we have many copies available. This was demonstrated by Raymer and coworkers who used the technique of optical homodyne tomography (Vogel and Risken 1989, Smithey *et al.* 1993a). The measured state can then be used to give the phase properties and this procedure will be described in the following chapter (Beck *et al.* 1993 **Paper 6.1**, Smithey *et al.* 1993b **Paper 6.2**,1993c). It is highly desirable, however, to have a more direct method for measuring the phase. One possible way is to design an interaction between the mode to be measured and a second specially prepared mode in such a way that the probability for a measurable experimental outcome is proportional to the quantity we desire to measure—in this case a phase probability. This general method is called "projection synthesis" because we are, in effect, engineering the projector we wish to measure. For a phase measurement we need to interfere on a beam splitter the mode to be measured and a second mode that has been prepared in a specially selected state: the reciprocal-binomial state. The measurement is then completed by counting the number of photons emerging in each of the two output arms. If one arm is found to contain precisely N photons and the other to have none, then this outcome is equivalent to projecting the input state onto the truncated phase state (Barnett and Pegg 1996 **Paper 5.4**):[2]

$$|\theta, N\rangle = \frac{1}{\sqrt{N+1}}\sum_{n=0}^{N} \exp(in\theta)|n\rangle. \tag{5.3}$$

By comparing the probabilities for this outcome for values of θ over a full 2π range we can construct the complete phase probability distribution. The reciprocal-binomial states are not readily available, but for measurements of weak fields only small values of N are required and suitable states can be obtained from the familiar coherent and squeezed states (Pegg *et al.* 1997b **Paper 5.5**).

[1] We should note that these quadratures are also commonly used with the prefactor $1/\sqrt{2}$ replaced by $1/2$ or 1, with the former being employed in the introduction to Chapter 4. This choice is, of course, purely a matter of convention.

[2] Note that the upper limit in the summation in Equation (3) of this paper is incorrect as is the sign in the exponent.

The method of projection synthesis is quite powerful and has been used for problems other than phase measurement including the preparation of controlled superpositions of the vacuum and one-photon states (Pegg *et al.* 1998, Babichev *et al.* 2003). A closely related device has been suggested for the purpose of measuring the variance of the phase for weak fields without first obtaining the full probability distribution (Pregnell and Pegg 2001 **Paper 5.6**). It has also been suggested that the method can be adapted to realise measurements of moments of the cosine and sine functions of the phase operator (Pregnell and Pegg 2002a **Paper 5.7**). Remarkably, this proposal requires only the preparation of familiar and easily made coherent states of light. All of these phase measuring devices require the performance of a number of observations in order to build up the phase information and it is highly desirable to have a "single-shot" measurement of the phase itself. This possibility has been demonstrated using a multiport device constructed from beam splitters and phase shifters (Pregnell and Pegg 2002b **Paper 5.8**). The device is based on the observation that such a multiport can be used to construct a desired unitary transformation in the space of the input modes (Reck *et al.* 1994) and in particular a finite unitary transformation that can map a truncated phase state onto a set of photon number states. It follows that a known subset of the possible experimental outcomes can be designed to occur with probabilities proportional to phase probabilities. The measurement scheme is expected to work well for weak fields in the quantum regime, even when realistic experimental effects are taken into consideration (Pregnell and Pegg 2003 **Paper 5.9**).

Phase measurements

Stephen M. Barnett
Department of Physics and Applied Physics, John Anderson Building, University of Strathclyde, Glasgow G4 0NG, Scotland

D. T. Pegg
Faculty of Science and Technology, Griffith University, Nathan, Brisbane 4111, Australia
(Received 20 July 1992)

We address the problem of identifying an operational prescription for quantum phase measurements. As is known, different experiments can lead to different measured phase operators. However, we show that ambiguities of interpretation can arise even if a single experiment, such as that of Noh, Fougères, and Mandel [Phys. Rev. Lett. **67**, 1426 (1991)], is chosen as defining a phase measurement. We show by reference to a simple but fundamental example that it is not possible to deduce a unique phase difference from the measurements.

PACS number(s): 42.50.Wm, 03.65.Bz

I. INTRODUCTION

Recent renewed interest in the theoretical formulation of the quantum optical phase has focused attention on possible phase measurements [1,2]. In designing a suitable experiment, we can follow one of two possible approaches. On the one hand, we can introduce phase as a fundamental mathematical component of the quantum theory of light [3], and only then seek an experiment from which we can infer information about the phase statistics. Alternatively, we could examine classical phase-measuring experiments and define these as measurements of phase in the quantum regime [1,2,4,5]. Unfortunately, in the latter operational approach, the classical description is not sufficient to define a unique quantum phase [2]. Different classical measurement schemes lead to different quantum measured phase operators. While this is disappointing, it might still be possible to choose a particular measurement scheme as the definitive phase measurement. This choice would necessarily represent a *convention* for the definition of phase.

The adoption of an operational definition of quantum phase has been explored by us [4], and more recently advocated by Noh, Fougères, and Mandel [1,2]. The purpose of this paper is to point out that even if one particular measurement scheme is agreed upon, there can still remain an ambiguity in the definition of the phase. In the experiment of Noh, Fougères, and Mandel, this ambiguity leads to inconsistent interpretations of the experiment *as a quantum phase measurement.* Independent methods for extracting the phase from the data lead to equivalent results in the classical regime and therefore to a unique classical phase. However, application of the same methods in the quantum regime can give inconsistent results, precluding the definition of a unique measured phase operator for the chosen experiment.

II. THE EXPERIMENT OF NOH, FOUGÈRES, AND MANDEL

We wish to point out form the outset that the experimental results of Noh, Fougères, and Mandel are in excellent agreement with the predictions of their theoretical analysis, and we do not question the accuracy of either of these. Indeed, the agreement is so good that there is little point in attempting to explain their experimental results in terms of any other theory. We shall, however, question the interpretation of these results as phase measurements. Their approach is to use a classical treatment of the experiment to provide the interpretation of the quantum-mechanical quantities that they measure. That is, their choice of quantum-mechanical operators is based on a direct comparison with the corresponding classical expressions.

A. Classical analysis

Figure 1 shows the experimental scheme of Noh, Fougères, and Mandel, which consists of a similar arrangement to the eight-port homodyne detection scheme of Walker and Carrol [5]. There are four outputs with a photodetector (D_j) at each. There are also four input ports, two of which are used for the fields to be measured. Within the domain of classical optics, the instantaneous amplitudes of the two input fields are

$$V_1 = I_1^{1/2}\exp(i\phi_1) , \tag{2.1a}$$

$$V_2 = I_2^{1/2}\exp(i\phi_2) . \tag{2.1b}$$

The phase difference $\phi_2-\phi_1$ is assumed to remain constant during the measurement time T. The integrated light intensities at the four outputs during the measurement time are [1,2]

$$W_3 = \tfrac{1}{4}[W_1 + W_2 - 2W_{12}\cos(\phi_2-\phi_1)] , \tag{2.2a}$$

$$W_4 = \tfrac{1}{4}[W_1 + W_2 + 2W_{12}\cos(\phi_2-\phi_1)] , \tag{2.2b}$$

$$W_5 = \tfrac{1}{4}[W_1 + W_2 - 2W_{12}\sin(\phi_2-\phi_1)] , \tag{2.2c}$$

$$W_6 = \tfrac{1}{4}[W_1 + W_2 + 2W_{12}\sin(\phi_2-\phi_1)] , \tag{2.2d}$$

where

$$W_l = \alpha\int_t^{t+T} I_l(t')dt' , \tag{2.3a}$$

$$W_{12}=\alpha\int_t^{t+T}[I_1(t')I_2(t')]^{1/2}dt' , \tag{2.3b}$$

and α is the quantum efficiency of the detectors. The integrated intensities are clearly dependent on the phase difference $\phi_2-\phi_1$, and we can use the measured values of these intensities to obtain this phase difference. In particular, we find

$$W_4-W_3=W_{12}\cos(\phi_2-\phi_1) , \tag{2.4a}$$

$$W_6-W_5=W_{12}\sin(\phi_2-\phi_1) , \tag{2.4b}$$

$$W_6-W_4=\frac{1}{\sqrt{2}}W_{12}\sin(\phi_2-\phi_1-\pi/4) , \tag{2.4c}$$

$$W_6-W_3=\frac{1}{\sqrt{2}}W_{12}\cos(\phi_2-\phi_1-\pi/4) , \tag{2.4d}$$

$$W_3-W_5=\frac{1}{\sqrt{2}}W_{12}\sin(\phi_2-\phi_1-\pi/4) , \tag{2.4e}$$

$$W_4-W_5=\frac{1}{\sqrt{2}}W_{12}\cos(\phi_2-\phi_1-\pi/4) . \tag{2.4f}$$

These six quantities, together with the trigonometric identity $\cos^2\theta+\sin^2\theta=1$, are more than enough to determine the value of the classical phase difference from the experimental results. For example, we find that

$$\cos(\phi_2-\phi_1)=\frac{W_4-W_3}{[(W_4-W_3)^2+(W_6-W_5)^2]^{1/2}} , \tag{2.5a}$$

$$\sin(\phi_2-\phi_1)=\frac{W_6-W_5}{[(W_4-W_3)^2+(W_6-W_5)^2]^{1/2}} , \tag{2.5b}$$

$$\cos(\phi_2-\phi_1-\pi/4)=\frac{W_6-W_3}{[(W_6-W_3)^2+(W_6-W_4)^2]^{1/2}} \tag{2.5c}$$

$$=\frac{W_4-W_5}{[(W_4-W_5)^2+(W_3-W_5)^2]^{1/2}} \tag{2.5d}$$

$$=\frac{W_6-W_3}{[(W_6-W_3)^2+(W_3-W_5)^2]^{1/2}} \tag{2.5e}$$

$$=\frac{W_4-W_5}{[(W_4-W_5)^2+(W_6-W_4)^2]^{1/2}} , \tag{2.5f}$$

$$\sin(\phi_2-\phi_1-\pi/4)=\frac{W_6-W_4}{[(W_6-W_4)^2+(W_6-W_3)^2]^{1/2}} \tag{2.5g}$$

$$=\frac{W_3-W_5}{[(W_4-W_5)^2+(W_3-W_5)^2]^{1/2}} \tag{2.5h}$$

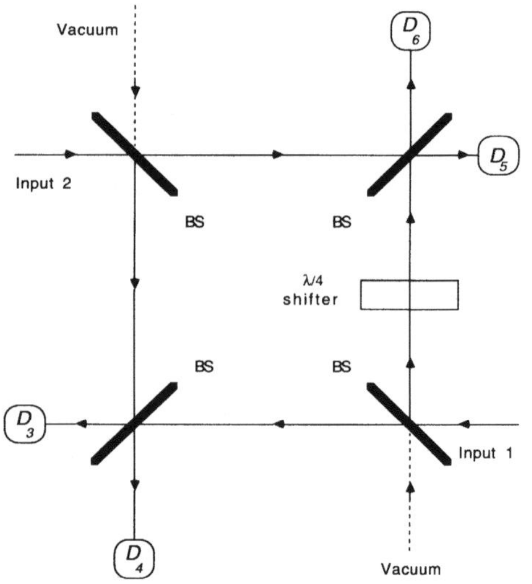

FIG. 1. Outline of the experimental setup used by Noh, Fougères, and Mandel. The thick lines denote 50-50 beam splitters (BS).

$$=\frac{W_3-W_5}{[(W_6-W_3)^2+(W_3-W_5)^2]^{1/2}} \tag{2.5i}$$

$$=\frac{W_6-W_4}{[(W_4-W_5)^2+(W_6-W_4)^2]^{1/2}} . \tag{2.5j}$$

The classical physics of the device ensures the equality between (2.5c), (2.5d), (2.5e), and (2.5f), and between (2.5g), (2.5h), (2.5i), and (2.5j). Any two of the above four trigonometric expressions can be used to determine the phase difference, and the result will be independent of which two are chosen.

B. Quantum analysis

In quantum-mechanical intensity measurements, a discrete number of photoelectron counts are recorded in the counting time. The integrated intensity is proportional to this number and is represented by an operator $\hat{n}_j$, which corresponds to the classical integrated intensity W_j to within a constant of proportionality. From this correspondence, we can define operator functions of $\hat{n}_j$ that relate to the classical trigonometrical expressions (2.5):

$$\hat{C}_M(\phi_2-\phi_1)=\frac{\hat{n}_4-\hat{n}_3}{[(\hat{n}_4-\hat{n}_3)^2+(\hat{n}_6-\hat{n}_5)^2]^{1/2}} , \tag{2.6a}$$

$$\hat{S}_M(\phi_2-\phi_1)=\frac{\hat{n}_6-\hat{n}_5}{[(\hat{n}_4-\hat{n}_3)^2+(\hat{n}_6-\hat{n}_5)^2]^{1/2}} , \tag{2.6b}$$

$$\hat{C}_{M1}(\phi_2-\phi_1-\pi/4)=\frac{\hat{n}_6-\hat{n}_3}{[(\hat{n}_6-\hat{n}_3)^2+(\hat{n}_6-\hat{n}_4)^2]^{1/2}}, \tag{2.6c}$$

$$\hat{C}_{M2}(\phi_2-\phi_1-\pi/4)=\frac{\hat{n}_4-\hat{n}_5}{[(\hat{n}_4-\hat{n}_5)^2+(\hat{n}_3-\hat{n}_5)^2]^{1/2}}, \tag{2.6d}$$

$$\hat{C}_{M3}(\phi_2-\phi_1-\pi/4)=\frac{\hat{n}_6-\hat{n}_3}{[(\hat{n}_6-\hat{n}_3)^2+(\hat{n}_3-\hat{n}_5)^2]^{1/2}}, \tag{2.6e}$$

$$\hat{C}_{M4}(\phi_2-\phi_1-\pi/4)=\frac{\hat{n}_4-\hat{n}_5}{[(\hat{n}_4-\hat{n}_5)^2+(\hat{n}_6-\hat{n}_4)^2]^{1/2}}, \tag{2.6f}$$

$$\hat{S}_{M1}(\phi_2-\phi_1-\pi/4)=\frac{\hat{n}_6-\hat{n}_4}{[(\hat{n}_6-\hat{n}_4)^2+(\hat{n}_6-\hat{n}_3)^2]^{1/2}}, \tag{2.6g}$$

$$\hat{S}_{M2}(\phi_2-\phi_1-\pi/4)=\frac{\hat{n}_3-\hat{n}_5}{[(\hat{n}_4-\hat{n}_5)^2+(\hat{n}_3-\hat{n}_5)^2]^{1/2}}, \tag{2.6h}$$

$$\hat{S}_{M3}(\phi_2-\phi_1-\pi/4)=\frac{\hat{n}_3-\hat{n}_5}{[(\hat{n}_6-\hat{n}_3)^2+(\hat{n}_3-\hat{n}_5)^2]^{1/2}}, \tag{2.6i}$$

$$\hat{S}_{M4}(\phi_2-\phi_1-\pi/4)=\frac{\hat{n}_6-\hat{n}_4}{[(\hat{n}_4-\hat{n}_5)^2+(\hat{n}_6-\hat{n}_4)^2]^{1/2}}. \tag{2.6j}$$

We note that all the $\hat{n}_j$ operators commute with each other, and therefore no ambiguity of ordering results from the division. Moreover, all of these S_M and C_M operators commute with each other, and therefore represent compatible observables. The normalization ensures that these operators obey the identities

$$\hat{C}_M^2(\phi_2-\phi_1)+\hat{S}_M^2(\phi_2-\phi_1)=\hat{1}, \tag{2.7a}$$

$$\hat{C}_{Ml}^2(\phi_2-\phi_1-\pi/4)+\hat{S}_{Ml}^2(\phi_2-\phi_1-\pi/4)=\hat{1}, \quad \forall l, \tag{2.7b}$$

where $\hat{1}$ is the unit operator. Given these features, it seems eminently reasonable to interpret these operators as representing phase observables. Noh, Fougères, and Mandel chose to use their experimental data to calculate the statistics of $S_M(\phi_2-\phi_1)$ and $C_M(\phi_2-\phi_1)$, which they refer to as their measured sine and cosine of the phase difference. However, there is no *a priori* justification for choosing this particular pair to determine the phase difference. Of course, if all the possible choices yield the same result, as they do classically, then all choices would be equivalent. Unfortunately, this is not the case in the quantum regime, and the different pairs of observables can give results that are inconsistent with a single phase probability distribution. We can illustrate this inconsistency with a simple example.

Consider the case where input field 1 contains precisely one photon and input field 2 is in its vacuum state. A quantum-mechanical analysis shows that only one of the four detectors will register a photocount, and that each detector is equally likely to do so. Following Noh, Fougères, and Mandel, we use the knowledge of which detector was triggered to construct the values of the operators (2.6). If for a particular operator the denominator becomes zero, then that result is not used and the moments for that operator are renormalized accordingly [1,2]. In Table I, we show the four possible outcomes of a single measurement. We emphasize that the appearance of indeterminacies (denoted by question marks in the table) for some operators but not for others is not a fundamental distinction, but rather is dependent on the particular state of the input fields. If we interpret the operators as representing the cosines or sines of their argu-

TABLE I. The four possible outcomes for measurement with only one photon in input field 1. The numbers refer to the measured values of the operators defined in Eq. (2.6). The figures in parentheses are the phase differences in the range $(-\pi,\pi]$, which follow if we interpret these operators as measuring the cosines and sines of their argument. Indeterminate results are denoted by a question mark (?).

Operator measured	Photodetector registering photocount			
	D_3	D_4	D_5	D_6
$\hat{C}_M(\phi_2-\phi_1)$	-1 (π)	1 (0)	0 $(\pm\pi/2)$	0 $(\pm\pi/2)$
$\hat{S}_M(\phi_2-\phi_1)$	0 $(0,\pi)$	0 $(0,\pi)$	-1 $(-\pi/2)$	1 $(\pi/2)$
$\hat{C}_{M1}(\phi_2-\phi_1-\pi/4)$	-1 $(-3\pi/4)$	0 $(-\pi/4, 3\pi/4)$	? (?)	$1/\sqrt{2}$ $(0, \pi/2)$
$\hat{S}_{M1}(\phi_2-\phi_1-\pi/4)$	0 $(\pi/4, -3\pi/4)$	-1 $(-\pi/4)$	? (?)	$1/\sqrt{2}$ $(\pi, \pi/2)$
$\hat{C}_{M2}(\phi_2-\phi_1-\pi/4)$	0 $(-\pi/4, 3\pi/4)$	1 $(\pi/4)$	$-1/\sqrt{2}$ $(\pi, -\pi/2)$	? (?)
$\hat{S}_{M2}(\phi_2-\phi_1-\pi/4)$	1 $(3\pi/4)$	0 $(\pi/4, -3\pi/4)$	$-1/\sqrt{2}$ $(0, -\pi/2)$	? (?)
$\hat{C}_{M3}(\phi_2-\phi_1-\pi/4)$	$-1/\sqrt{2}$ $(\pi, -\pi/2)$	? (?)	0 $(-\pi/4, 3\pi/4)$	1 $(\pi/4)$
$\hat{S}_{M3}(\phi_2-\phi_1-\pi/4)$	$1/\sqrt{2}$ $(\pi, \pi/2)$	? (?)	-1 $(-\pi/4)$	0 $(\pi/4, -3\pi/4)$
$\hat{C}_{M4}(\phi_2-\phi_1-\pi/4)$	? (?)	$1/\sqrt{2}$ $(0, \pi/2)$	-1 $(-3\pi/4)$	0 $(-\pi/4, 3\pi/4)$
$\hat{S}_{M4}(\phi_2-\phi_1-\pi/4)$	? (?)	$-1/\sqrt{2}$ $(0, -\pi/2)$	0 $(\pi/4, -3\pi/4)$	1 $(3\pi/4)$

ments, then we can deduce the phase difference. This is shown in parenthesis in Table I. We note that, although the expectation of the square of each operator is $\frac{1}{2}$, there is no single phase difference that is consistent with the outcome of any *single measurement*. For example, if only D_4 registers a photocount, then assigning a value of $\pi/4$ to the phase difference $\phi_2-\phi_1$ is consistent with S_{M2} and C_{M2} (and possibly with S_{M3} and C_{M3}), but not with any of the other operators.

Comparison with the classical analysis provides no objective reason for preferring any of the operators purporting to represent phase over the others. Indeed, classically there would be no inconsistencies and no reason for choosing between the options discussed above. There are no ambiguities in the interpretation of the classical experiment as a phase-difference measurement. However, this is certainly not the case for the quantum-mechanical experiment. There is no consistent interpretation of the result of a single measurement as a phase-difference measurement.

III. CONCLUSION

The example in Sec. II may appear to be a pathological choice. However, if the experiment of Noh, Fougères, and Mandel had a consistent interpretation as a quantum phase-difference measurement, then this should apply to any choice of input states. Moreover, field states with small numbers of excitations are those very states for which a quantum description of phase is especially important.

It is not our main point to address the physical grounds for accepting a quantum formulation of optical phase, although it is worth mentioning that there are strong reasons for requiring the phase difference between number-state fields to be completely random [6]. Our point is this: It is not even possible to address questions such as phase randomness in the approach favored by Noh, Fougères, and Mandel, because this approach does not lead to a consistent phase-difference probability distribution.

We conclude that the interpretation of the experiments of Noh, Fougères, and Mandel as a phase-difference measurement in the quantum regime is inappropriate.

ACKNOWLEDGMENTS

We are grateful to P. L. Knight for his advice. S. M. B. thanks the Royal Society of Edinburgh for financial support.

[1] J. W. Noh, A. Fougères, and L. Mandel, Phys. Rev. Lett. **67**, 1426 (1991).

[2] J. W. Noh, A. Fougères, and L. Mandel, Phys. Rev. A **45**, 424 (1992).

[3] P. A. M. Dirac, Proc. R. Soc. London Ser. A **114**, 243 (1927); L. Susskind and J. Glogower, Physics (NY) **1**, 49 (1964); D. T. Pegg and S. M. Barnett, Europhys. Lett. **6**, 483 (1988); S. M. Barnett and D. T. Pegg, J. Mod. Opt. **36**, 7 (1989); D. T. Pegg and S. M. Barnett, Phys. Rev. A **39**, 1665 (1989); J. H. Shapiro, S. R. Shepard, and N. C. Wong, Phys. Rev. Lett. **62**, 2377 (1989); **63**, 2002 (1989); J. H. Shapiro and S. R. Shepard, Phys. Rev. A **43**, 3795 (1991).

[4] S. M. Barnett and D. T. Pegg, J. Phys. A **19**, 3849 (1986).

[5] N. G. Walker and J. E. Carrol, Opt. Quantum Electron. **18**, 355 (1986); N. G. Walker, J. Mod. Opt. **34**, 16 (1987).

[6] S. M. Barnett and D. T. Pegg, Phys. Rev. A **42**, 6713 (1990).

On measuring extremely small phase fluctuations

John A. Vaccaro
Faculty of CAD, Griffith University, Nathan, Brisbane, 4111 Australia

and

D.T. Pegg
Faculty of Science and Technology, Griffith University, Nathan, Brisbane, 4111 Australia

Received 15 July 1993; revised manuscript received 7 October 1993

The applicability is examined of the measured-phase operator describing a simple homodyne measurement scheme to states of the quantum optical field which have the minimum possible phase fluctuations. Such states necessarily have huge intensity fluctuations, so it is expected that this operator should be inapplicable because intensity fluctuations are deliberately ignored in its derivation. However, the surprising result found is that this operator can still provide a reasonably good approximation to the phase properties of fields even in this extreme limit.

The development of the unitary and hermitian phase operator formalism [1–3] has allowed the properties of phase, which is conjugate to photon number, to be successfully calculated for quantum states of light. A question of increasing interest concerns how these properties can be measured [4–6]. The operationally defined measured-phase operators $\widehat{\cos_M}\phi$ and $\widehat{\sin_M}\phi$ were introduced as the description of phase-like quantities measurable by a simple balanced homodyne experiment [4] in which the local oscillator is in a sufficiently intense coherent state to have negligible fluctuations. Other such operators have also been constructed to describe more sophisticated experiments [5,6]. Such operators do not represent a quantity which is conjugate to photon number, that is phase, but do yield quite good approximations to the phase properties of some states of light. The usefulness of measured-phase operators is determined by the range of states for which these approximations are reasonably accurate. Recently, for example, Ritze [6] has suggested that some measured-phase operators [4,5], while giving good approximations for coherent states, are not useful for phase optimized states [7,8]. These states have the minimum possible phase fluctuations possible for a given mean photon number. In this paper we examine this question more closely and find, remarkably, that the simple measured-phase operator of ref. [4] can still yield a quite good approximation to actual phase even in the extreme limit of small phase fluctuations.

Some results derivable from the hermitian phase operator approach [1–3] which are used in this paper include expressions for the phase probability densities $P_\phi(\theta)$, $P_{\sin\phi}(\theta)$ and $P_{\sin\phi}(\sin\theta)$ and for phase shifts $\Delta\theta$ for a physical state of the field. Here $P_\phi(\theta)\,\mathrm{d}\theta$ is the probability for the phase ϕ to have a value between θ and $\theta+\mathrm{d}\theta$; $P_{\sin\phi}(\theta)\,\mathrm{d}\theta$ is the probability for $\sin\phi$ to have a value between $\sin\theta$ and $\sin(\theta+\mathrm{d}\theta)$; and $P_{\sin\phi}(\sin\theta)\,\mathrm{d}(\sin\theta)$ is the probability for $\sin\phi$ to have a value between $\sin\theta$ and $\sin\theta+\mathrm{d}(\sin\theta)$. In terms of number state coefficients $\langle n|f\rangle$ and phase state coefficients $\langle\theta|f\rangle$, these are, for a physical state $|f\rangle$ [1–3,9]

$$P_\phi(\theta)=\lim_{s\to\infty}\frac{s+1}{2\pi}\,|\langle\theta|f\rangle|^2 \tag{1}$$

$$=\frac{1}{2\pi}\sum_{n,n'}^{\infty}\exp[\mathrm{i}(n'-n)\theta]\,\langle f|n'\rangle\,\langle n|f\rangle\,, \tag{2}$$

$$P_{\sin\theta}(\theta)=\lim_{s\to\infty}\frac{s+1}{2\pi}\left(|\langle\theta|f\rangle|^2+|\langle\pi-\theta|f\rangle|^2\right) \quad (3)$$

$$=\frac{1}{2\pi}\sum_{n,n'}^{\infty}\{\exp[i(n'-n)\theta]$$

$$+(-1)^{n'-n}\exp[-i(n'-n)\theta]\}\langle f|n'\rangle\langle n|f\rangle\,, \quad (4)$$

$$P_{\sin\phi}(\sin\theta)=P_{\sin\phi}(\theta)/\cos\theta\,. \quad (5)$$

In eqs. (1) and (3), $s+1$ is the dimensionality of the Hilbert space spanned by the orthogonal phase states $|\theta_m\rangle$, with $m=0, 1, ..., s$, which are a subset of all the phase states $|\theta\rangle=(1+s)^{-1/2}\sum\exp(in\theta)\ |n\rangle$ where the sum is over $n=0, 1, ..., s$ [1]. In eqs. (3)–(5) θ is between $-\pi/2$ and $\pi/2$ and $\sin\theta$ ranges from -1 to $+1$. The extra term in eq. (3) compared with eq. (1) results from the two-fold degeneracy of the operator $\sin\hat{\phi}_\theta$, of which both $|\theta\rangle$ and $|\pi-\theta\rangle$ are eigenstates with eigenvalue $\sin\theta$.

A shift $\Delta\theta$ in phase of a quantum state is produced by the unitary phase-shift operator $\exp(i\hat{N}\Delta\theta)$ where $\hat{N}$ is the number operator. It is easily checked that application of this operator to $|f\rangle$ translates the probability density (1) along the θ axis by $\Delta\theta$. How can we produce a phase shift experimentally? Writing $\hat{N}\Delta\theta=\hat{N}\omega\Delta t$, where $\Delta t=\Delta\theta/\omega$ and $\hat{N}\hbar\omega$ is the field hamiltonian $\mathscr{H}$, and remembering that $\mathscr{H}/\hbar$ is the generator of a time shift, we see that a phase shift can be produced by inserting an effective time delay. This is, in fact, the usual procedure in most experiments, which leads to the important result that the operational phase shift operator is identical to the actual phase shift operator.

The measured-phase operators are hermitian combinations of the creation and annihilation operators

$$\widehat{\cos_M}\phi=k(\hat{a}+\hat{a}^\dagger)\,, \quad (6)$$

$$\widehat{\sin_M}\phi=-ik(\hat{a}-\hat{a}^\dagger)\,, \quad (7)$$

where k is a state-dependent factor which must approach $(2\langle\hat{N}\rangle^{1/2})^{-1}$ for an intense coherent state to give the correct classical limit [4]. A useful compromise value applicable to a wide range of states was found to be $k=2^{-1}(\langle\hat{N}\rangle+1/2)^{-1/2}$. Clearly in eqs. (6) and (7), $2k\hat{a}$ is used as a quasi-unitary phase operator in place of the unitary phase operator $\exp(i\hat{\phi}_\theta)$. From the relation $\hat{a}=\exp(i\hat{\phi}_\theta)\hat{N}^{1/2}$ [1] we see that $\hat{N}^{1/2}$ is being treated as a state-dependent constant which ignores its operator nature. However, because $n^{-1/2}\,\hat{a}\,|n\rangle=\exp(i\hat{\phi}_\theta)\,|n\rangle$ for $n>0$, this approximation will be reasonable for states with an uncertainty in photon number which is small compared with the mean photon number. This is the case for number states or coherent states of reasonable intensity.

We shall use the term Airy function state, or Airy state, of light to refer to the state whose number state coefficients fit an Airy function curve as described in ref. [7] as the state with minimum phase uncertainty for a given mean photon number $\langle\hat{N}\rangle$ of the order of two and higher. These phase-optimized states have also been studied in ref. [8] and their possible production in ref. [10]. The phase variance is given by $\Delta\phi^2\approx1.88\ (\langle\hat{N}\rangle+0.86)^{-2}$. Thus for a coherent state to have the same phase uncertainty as an Airy state with mean photon number of 50 photons, for example, it must have a mean of over 300 photons. As pointed out by Ritze [6], these extremely small phase fluctuations imply, from the number-phase uncertainty relation [2,3] that the number uncertainty ΔN must be at least of the order $\langle\hat{N}\rangle$. This suggests that the above measured-phase operators, and some other measured-phase operators [5] will not in general represent good measurements of the small phase variance of an Airy state. If an Airy state is to be used successfully to convey a similar amount of phase information as a coherent state of much greater intensity, then a detection experiment capable of resolving the minimum phase uncertainty is essential. In view of the simplicity of the homodyne measurements represented by the measured-phase operators, it is worthwhile examining the problem more closely.

The balanced homodyne experiment incorporates a 50%:50% beam splitter with the signal being the electronically obtained difference between the outputs from two photodetectors [5,6]. The phase of the input field is the phase difference between it and a local oscillator field whose phase we define to be $-\pi/2$. In the classical case the minimum signal magnitude of zero occurs when the input field is in quadrature with the local field, which will be the case when the phase of the input is zero. With this phase the contribution to the signal from intensity fluctuations in the input field is also minimized. A fluctuation from zero of θ in the input field phase, however, produces a signal proportional to $\sin\theta$, which for small fluctuations is approximately θ. Thus if we adjust the phase of the input field until the mean signal is zero then, for small fluctuations, the variance in phase will be $\langle\theta^2\rangle \approx \langle\sin^2\theta\rangle$ which, providing we can ignore the effect of intensity fluctuations, is obtainable from the fluctuations in the signal. In the corresponding quantum mechanical case, assuming a suitable photon counting time interval [5,6], the operator representing the signal is $B\,\widehat{\sin_M\phi}$, where B is a proportionality constant, which allows us to find $\langle(\widehat{\sin_M\phi})^2\rangle$ as the measured-phase variance. (We note that in ref. [4] the signal is proportional to $\widehat{\cos_M\phi}$. This is because the local oscillator phase [4] was defined as being zero instead of $-\pi/2$ as it is here.) In table 1 we give the values of the standard deviation $\langle(\widehat{\sin_M\phi})^2\rangle^{1/2}$ in measured phase calculated with the value of k given above for Airy states with mean photon numbers of 5, 10 and 100. These are compared directly with the values of standard deviation $\langle(\sin\hat{\phi}_\theta)^2\rangle^{1/2}$ obtained from the hermitian phase operator for the same states and the corresponding values of the standard deviation $\langle\hat{\phi}_\theta^2\rangle^{1/2}$ in phase. The agreement is certainly not as good as for coherent states. For example, the corresponding values of $\langle(\widehat{\sin_M\phi})^2\rangle^{1/2}$ and $\langle(\sin\hat{\phi}_\theta)^2\rangle^{1/2}$ for a coherent state with a mean photon number of 10 are 0.154 and 0.161, respectively, and for a mean photon number of 100 are 0.0499 and 0.05006, respectively. (The corresponding values of $\langle\hat{\phi}_\theta^2\rangle^{1/2}$ for means of 10 and 100 photons are 0.163 and 0.05013.) Nevertheless the disagreement as shown in table 1 is not so great as to prevent us from using the measured-phase operator to give a good indication of the actual phase uncertainty. Also, the advantage that the Airy states have in phase resolution as given by the standard deviation is not seriously affected if detected by this means.

In the above calculations we used the compromise value for the factor k given earlier. Although the value of k could be determined by a separate measurement of $\langle\hat{N}\rangle$, a simpler method in practice to measure the standard deviation of phase would be as follows. The possible values S of the signal are divided into a large number of very small bins of size δS. The phase, either of the local oscillator or the input field, is shifter until the probability for the signal to be in the zero-S bin is maximized. Airy states, like other partial phase states [3], have a symmetric phase distribution with the maximum coinciding with the mean. Also the maxima in the measured-phase distribution and in the actual phase distribution coincide when the mean phase is zero. Thus the phase shift needed to maximize the zero-S bin signal is equal to the *actual* original mean phase. This is because the operational phase shift is equal to the actual phase shift. With the phase adjusted so that the mean signal value is zero, the probability $\Pr(S)$ for the signal in each bin in measured. The phase is then shifted through an extra amount $\Delta\theta$ and the shift $\Delta\langle S\rangle$ of the mean of the distribution, with standard deviation σ_s, is found. The standard deviation of the measured phase, that is σ_s/B, is then taken as the measured result $\sigma_s\,\Delta\theta/(\Delta\langle S\rangle)$. We can show that this procedure is equivalent, for small $\Delta\theta$, to using a value for k in eq. (7) of

$$k=[\langle f_0|(\hat{a}+\hat{a}^\dagger)|f_0\rangle]^{-1}, \tag{8}$$

where here $|f_0\rangle$ is the input state after its mean phase has been shifted to zero. Expression (8) follows from equating σ_s/B to $\sigma_s\,\Delta\theta/(\Delta\langle S\rangle)$ for small $\Delta\theta$ which is equivalent to setting $\mathrm{d}\langle\widehat{\sin_M\phi}\rangle/\mathrm{d}\theta$ to unity. This last expression is equal to $-\mathrm{i}\langle[\hat{N}, \widehat{\sin_M\phi}]\rangle$ because $\hat{N}$ is the generator of a phase shift. Substitution from eq. (7) then yields eq. (8). We find that the measured-phase standard deviation based on the alter-

Table 1
Comparison of standard deviations of phase for an Airy state with various values of mean photon number. Columns 2–5 contain the standard deviations of the measured sine operator, the sine of the hermitian phase operator, the hermitian phase operator and the distribution $Q_M(\theta)$, respectively.

$\langle\hat{N}\rangle$	$\langle(\widehat{\sin_M\phi})^2\rangle^{1/2}$	$\langle(\sin\hat{\phi}_\theta)^2\rangle^{1/2}$	$\langle\hat{\phi}_\theta^2\rangle$	σ_Q
5	0.183	0.219	0.234	0.230
10	0.100	0.123	0.126	0.132
100	0.011	0.0136	0.0136	0.0156

native operational value of k given by eq. (8) are not appreciably different from those based on the compromise value of k for the Airy states with the same mean photon numbers given earlier.

The variance represents a combination of only the first and second moments of the probability distribution, and is not the only measure of the phase resolution. Therefore the following question immediately arises. In view of the only approximate agreement in the measured-phase and actual phase variances, how good an agreement is there between measured and actual phase distribution? From the experimental results for the probability density $\Pr(S)/\delta S$ as a function of S, we can obtain a probability density as a function of $\sin\theta$ by using the scaling factor $\Delta\theta/(\Delta S)$ discussed previously. From our comments above, the distribution obtained from this scaling factor is not significantly different from the distribution obtained using the compromise value of k, which we call the measured-phase probability density $P_{\mathrm{M}}(\sin\theta)$. We can calculate $P_{\mathrm{M}}(\sin\theta)$ for a state $|f\rangle$ as follows. Let $|y\rangle$ be the eigenstate of $-\mathrm{i}(\hat{a}-\hat{a}^{\dagger})$ with eigenvalue y, that is, $\widehat{\sin_{\mathrm{M}}\phi}|y\rangle = ky|y\rangle$. This state has number state coefficients

$$\langle n|y\rangle \propto \mathrm{i}^n 2^{-(n/2)}(n!)^{-1/2}\exp(-y^2/4)\,H_n(2^{-1/2}y)\,, \tag{9}$$

where H is a Hermite polynomial (see, e.g. ref. [11]). $P_{\mathrm{M}}(\sin\theta)$ is proportional to the probability of a result $ky=\sin\theta$, that is, to $|\langle f|y\rangle|^2$. Finally the probability density is normalized. A comparison of the predicted measured-phase probability density $P_{\mathrm{M}}(\sin\theta)$ (dashed line) with the corresponding actual phase probability density, which is $P_{\sin\phi}(\sin\theta)$ (solid line), for Airy states with $\langle N\rangle = 5$, 10 and 100 is shown in fig. 1. It can be seen that the measured-phase distributions are not unreasonable representations of the actual phase distributions. We note from eqs. (1) and (3) that for states such as these, for which $\langle f|\pi-\theta\rangle$ is negligible, $P_{\sin\phi}(\sin\theta)$ is proportional to $P_\phi(\theta)/\cos\theta$, which for small phase variances about zero is a approximately $P_\phi(\theta)$.

In order to find the mean phase by the above method, the values of the zero-S bin probability $\Pr(0)$ need to be monitored as the phase of the field is shifted. In view of this, another possible measured-phase distribution immediately suggests itself: simply plot these values of $\Pr(0)$ as a function of the

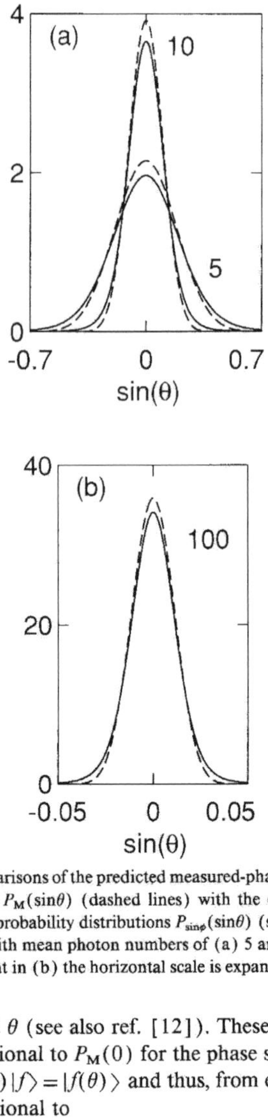

Fig. 1. Comparisons of the predicted measured-phase probability distributions $P_{\mathrm{M}}(\sin\theta)$ (dashed lines) with the corresponding actual phase probability distributions $P_{\sin\phi}(\sin\theta)$ (solid lines) for Airy states with mean photon numbers of (a) 5 and 10 and (b) 100. Note that in (b) the horizontal scale is expanded.

phase shift θ (see also ref. [12]). These values will be proportional to $P_{\mathrm{M}}(0)$ for the phase shifted state $\exp(-\mathrm{i}\hat{N}\pi)|f\rangle = |f(\theta)\rangle$ and thus, from eq. (9), will be proportional to

$$|\langle f(\theta)|y=0\rangle|^2$$

$$\propto \left|\sum_{n=0}^{\infty}\langle f|n\rangle \mathrm{i}^n 2^{-(n/2)}(n!)^{-1/2}H_n(0)\exp(\mathrm{i}n\theta)\right|^2, \tag{10}$$

because the number state coefficients $\langle f(\theta)|n\rangle$ are now $\langle f\exp(i\hat{N}\theta)|n\rangle = \langle f|n\rangle \exp(in\theta)$ and $y=0$ when $S=0$. We then scale the values of Pr(0) so that the area under the curve from $\theta=-\pi/2$ to $\theta=\pi/2$ is unity and call the scaled values $Q_M(\theta)$, which will also be proportional to expression (10). To find the corresponding actual phase probability density $Q(\theta)$, that is for which $\widehat{\sin\phi}$ is used in place of $\widehat{\sin_M\phi}$, we replace $P_M(0)$ by $P_{\sin\phi}(0)$ given by eq. (3) with $\theta=0$ and with the phase-shifted state $|f(\theta)\rangle$ substituted for $|f\rangle$. This gives

$$Q(\theta) = \lim_{s\to\infty} \frac{s+1}{2\pi} (|\langle f|\theta\rangle|^2 + |\langle f|\pi+\theta\rangle|^2), \quad (11)$$

where we have used $\exp(i\hat{N}\theta)$ $|\text{zero phase}\rangle = |\theta\rangle$ and $\exp(i\hat{N}\theta)|\pi\rangle = |\pi+\theta\rangle$. For states $|f\rangle$ with negligible probability of being found in $|\pi+\theta\rangle$, the latter term in eq. (11) vanishes and, from eq. (1), we obtain

$$Q(\theta) = P_\phi(\theta). \quad (12)$$

The operational procedure which yields $Q_M(\theta)$ has the advantage that it is independent of the value used for k and there is no need to use the scaling constant $\Delta\theta/(\Delta S)$. Figure 2 shows a comparison between $Q_M(\theta)$ (dotted line) and $P_\phi(\theta)$ (solid line) for the same states as in fig. 1 over a range near the mean. We might note at this point that the measured-phase operator of Noh et al. [5] also has the advantage of not requiring a scaling constant or k value, but as the results it yields are a function of four input states [13] they will not be identical to those given here.

It is clear that the operational procedure giving the measured-phase probability density $Q_M(\theta)$ yields a better measure of the actual phase probability density than that which gives $P_M(\sin\theta)$. Indeed the agreement in the former case is reasonably good. The standard deviations σ_Q of $Q_M(\theta)$ for the same Airy states considered previously are shown in the last column of table 1. Thus the simple balanced homodyne experiment, if used in the appropriate manner, can give a measured-phase distribution which is a reasonably accurate representation of the actual phase distribution even for Airy states. This is important because, although the phase distribution of any state can be determined by measuring the density matrix [14–16], the above methods are much simpler and far more direct. It is also very surprising

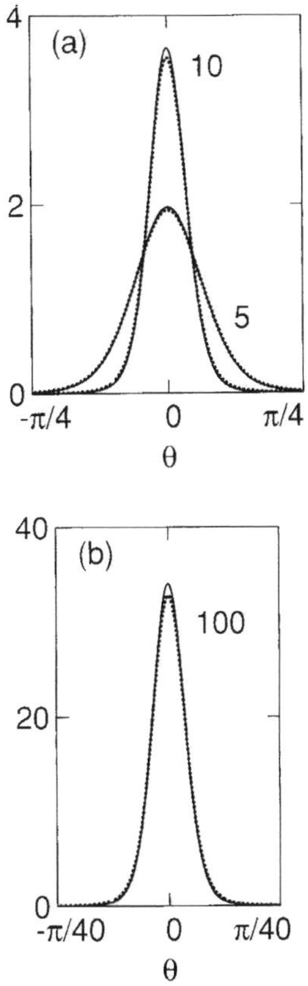

Fig. 2. Comparisons between the distributions $Q_M(\theta)$ (dotted lines) and the actual phase probability distributions $P_\phi(\theta)$ (solid lines) for Airy states with mean photon numbers of (a) 5 and 10 and (b) 100.

since, as pointed out by Ritze [6], the validity of approximating the actual phase operator by the measured-phase operator is expected to break down for such extreme states. Indeed, because Airy states with a mean of more than a few photons are phase-optimized states, it is impossible to find more extreme states in this particular limit of extremely small phase variances. That is, as we move from coherent states to states of decreasing phase variance while keeping the mean photon number fixed, the simple measured-phase operator becomes less accurate, but we reach the limit of such states before we reach the limit of reasonably good applicability for this operator. The work in this paper thus confirms the validity of the simple measured-phase operator over a very useful range of states with a mean from a few photons up to at least hundreds of photons. Applications of this operator by means of the two different experimental procedures described above to states in other regions, such as that of very small photon number, will be examined elsewhere.

References

[1] D.T. Pegg and S.M. Barnett, Europhys, Lett. 6 (1988) 483.
[2] S.M. Barnett and D.T. Pegg, J. Mod. Optics 36 (1989) 7.
[3] D.T. Pegg and S.M. Barnett, Phys. Rev. A 39 (1989) 1665.
[4] S.M. Barnett and D.T. Pegg, J. Phys. A 19 (1986) 3849.
[5] J.W. Noh, A. Fougeres and L. Mandel, Phys. Rev. A 45 (1992) 424.
[6] H.-H. Ritze, Optics Comm. 92 (1992) 127.
[7] G.S. Summy and D.T. Pegg, Optics Comm. 77 (1990) 75.
[8] A. Bandilla, H. Paul and H.-H. Ritze, Quantum Optics 3 (1991) 267.
[9] J.A. Vaccaro and D.T. Pegg, J. Mod. Optics 37 (1990) 17.
[10] A. Bandilla, Optics Comm. 94 (1992) 273.
[11] E. Merzbacher, Quantum Mechanics, 2nd Ed. (Wiley, New York, 1970) p. 362.
[12] W. Vogel and W. Schleich, Phys. Rev. A 44 (1991) 7642.
[13] M. Freyberger, K. Vogel and W. Schleich, Phys. Lett. A 176 (1993) 41.
[14] D.T. Smithey, M. Beck, M.G. Raymer and A. Faridani, Phys. Rev. Lett. 70 (1993) 1244.
[15] K. Vogel and H. Risken, Phys. Rev. A 40 (1989) 2847.
[16] M. Beck, D.T. Smithey, J. Cooper and M.G. Raymer, Optics Lett. 18 (1993) 1259.

Adaptive Phase Measurements of Optical Modes: Going Beyond the Marginal Q Distribution

H. M. Wiseman*

Department of Physics, University of Auckland, Auckland, New Zealand

(Received 20 March 1995; revised manuscript received 31 August 1995)

In standard single-shot measurements of the phase of an optical mode, the phase and amplitude quadratures are jointly measured, and the latter information discarded. These techniques are consequently suboptimal. Here I suggest an adaptive scheme, whereby the phase is estimated from the results so far and fed back to control the phase of the local oscillator so as to measure the (estimated) phase quadrature only. I show that adaptive phase measurements can approach optimal phase measurements for states with both low and high mean photon numbers.

PACS numbers: 42.50.Dv, 03.65.Bz, 42.50.Lc

The phase ϕ of a single mode of the electromagnetic field is the quantity canonically conjugate to the photon number n. It is now generally accepted that there exists a unique *canonical* probability distribution function (PDF) for this variable [1]:

$$P_{\text{can}}(\phi) = \text{Tr}[\rho F_{\text{can}}(\phi)], \qquad (1)$$

where $F_{\text{can}}(\phi)$ is a positive-operator-valued measure (POVM) [1–3] for ϕ defined in terms of the unnormalized phase states $|\phi\rangle$:

$$F_{\text{can}}(\phi) = \frac{1}{2\pi}|\phi\rangle\langle\phi|, \qquad |\phi\rangle = \sum_{n=0}^{\infty} e^{in\phi}|n\rangle. \qquad (2)$$

This PDF is guaranteed to be normalized from the requirement on all POVMs that $\int d\lambda\, F(\lambda) = 1$, where λ is the measurement result. Considerable work has been done showing how this distribution can be inferred from physically realizable homodyne measurements on an arbitrarily large ensemble of identical copies of the system [4]. However, this ability is not at all the same as the ability to make canonical phase measurements. To do the latter, one would have to make a measurement on a *single copy* of the system, the result of which would be a random variable drawn from the canonical PDF (1). There is no known way to achieve this in general, nor is there ever likely to be.

One practical reason for wishing to make canonical phase measurements is for efficient communication [2]. If one encoded information in the phase of a single-mode optical pulse (which is easy to do with an electro-optic modulator), then one would wish the receiver to measure that phase as accurately as possible. In a canonical phase measurement the error in the measured phase would be limited only by the intrinsic quantum uncertainty in the phase [2]. Therefore it is only if a receiver could make a canonical (or near to canonical) phase measurement that schemes for preparing states which have minimum intrinsic phase uncertainty [4] would be able to be fully exploited for efficient communication.

At present, there are a number of practical (noncanonical) schemes for single-shot phase measurements, all of which give equivalent results [1]. One of these schemes (which here stands in place of any of them) is heterodyne detection, which uses a local oscillator highly detuned from the system. The two Fourier components of the photocurrent record yield measurements of both quadratures of the field [5]. These can be converted into results for the intensity and phase, the former of which is discarded. Because half of the measurement information is useless, this phase measurement is far from canonical. The POVM for such standard measurements is

$$F_{\text{std}}(\phi) = \int_0^{\infty} \tfrac{1}{2}\, dn\, F_{\text{het}}(\phi, n). \qquad (3)$$

Here the heterodyne POVM $F_{\text{het}}(\phi, n)$ is defined in terms of coherent states of complex amplitude $A = \sqrt{n}\, e^{i\phi}$:

$$F_{\text{het}}(\phi, n) = \pi^{-1}|A\rangle\langle A|, \qquad (4)$$

where the result A (defined later) encodes both Fourier amplitudes. In other words, the PDF for standard phase measurements is the marginal phase PDF of the Q function $Q(\phi, n) = \text{Tr}[\rho F_{\text{het}}(\phi, n)]$. Such measurements introduce significant extrinsic uncertainty into the measurement result [1]. Thus with standard detection techniques, states with small intrinsic phase uncertainty offer only a modest increase in efficiency over coherent states with the same mean photon number [2].

In this work I am proposing a new technique: adaptive single-shot phase measurements. As I show, such measurements can be much closer to canonical measurements than standard measurements (hence the title of this Letter). The basic idea is to measure the *estimated* phase quadrature of the system by homodyne detection, where the estimate is based on the photocurrent record *so far* from the *single* pulse. That is to say, the local oscillator phase is continuously adjusted by a feedback loop to be in quadrature with the estimated system phase over the course of a single measurement (see Fig. 1). The first part of this Letter explains how this estimate could be made in general. I then present some numerical results for adaptive phase measurements of coherent states. Finally, I present results for a special case which can be solved analytically, in which the adaptive phase measurement is strictly as good as a canonical measurement.

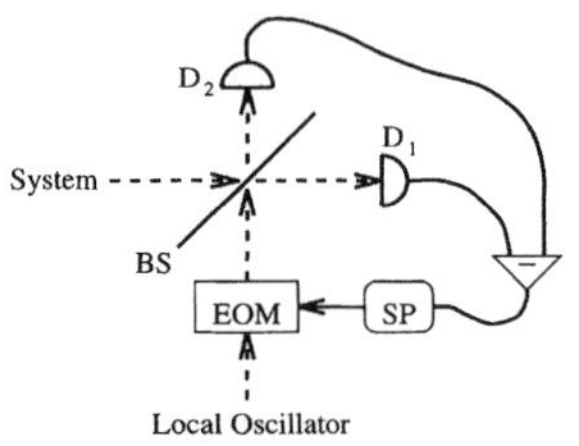

FIG. 1. The adaptive phase measurement scheme. Light beams are indicated by dashed lines and electronics by solid lines. BS denotes a 50/50 beam splitter, D_1 and D_2 photodetectors, and SP a signal processor. The local oscillator phase is controlled by an electro-optic modulator (EOM).

Photodetection with a local oscillator.—An adaptive measurement requires one to estimate the phase of the system based on the measurement record so far. In order to treat this, we require the quantum measurement theory of photodetection with a strong local oscillator, over a finite time interval. This I have derived in generality in Ref. [6], using the recently published theory of linear quantum trajectories of Goetsch and Graham [7]. For simplicity, let the single mode to be measured be prepared initially in a cavity in state ρ. Let the light leak out through an end mirror with decay rate of unity. The emitted light is sent through a 50-50 beam splitter, with a strong local oscillator of complex amplitude $\beta(t) = |\beta|e^{i\Phi(t)}$ entering at the other port. The mean fields at detectors D_2, D_1 are thus proportional to $\beta \pm \langle a\rangle e^{-t/2}$, respectively, where a is the annihilation operator in the cavity mode. The signal photocurrent $I(t)$ in the interval $[t, t+\delta t)$ is defined in terms of the difference between the photocounts $\delta N_2, \delta N_1$ at the two detectors:

$$I(t) = \lim_{|\beta|\to\infty} \frac{\delta N_2(t) - \delta N_1(t)}{|\beta|\delta t}, \tag{5}$$

as in Ref. [5]. It is easy to see that the mean value of this current can be written in terms of $\langle a\rangle = \mathrm{Tr}[a\rho]$ as

$$\langle I(t)\rangle = e^{-t/2}\langle ae^{-i\Phi(t)} + a^\dagger e^{i\Phi(t)}\rangle. \tag{6}$$

It is useful to introduce a new symbol $\mathbf{I}_{[0,t)}$ for the complete photocurrent record $\{I(s) : 0 \le s < t\}$. The quantum measurement theory we require is the POVM for the record $\mathbf{I}_{[0,t)}$ from time 0 (the time of preparation) to time t. This gives the probability for getting the result $\mathbf{I}_{[0,t)}$ given the initial state ρ. Note that $\mathbf{I}_{[0,t)}$ is a continuous infinity of a real number—a very complicated object. Fortunately, it turns out [6] that the POVM depends only on two complex functionals of $\mathbf{I}_{[0,t)}$. These two sufficient statistics are

$$\mathcal{F}_1[\mathbf{I}_{[0,t)}] = \int_0^t e^{i\Phi(s)}e^{-s/2}I(s)ds, \tag{7}$$

$$\mathcal{F}_2[\mathbf{I}_{[0,t)}] = -\int_0^t e^{2i\Phi(s)}e^{-s}ds. \tag{8}$$

It might be thought that the second functional does not even depend on $\mathbf{I}_{[0,t)}$, but it does if the local oscillator phase $\Phi(t)$ depends on $\mathbf{I}_{[0,t)}$ as in adaptive measurements. Denoting the measured values of these functionals R_t and S_t, respectively, the POVM is [6]

$$F_t(R_t, S_t) = P_0(R_t, S_t)G_t(R_t, S_t), \tag{9}$$

where $P_0(R_t, S_t)$ is a positive function defined later, and $G_t(R_t, S_t)$ is a positive operator given by

$$G_t = \exp(\tfrac{1}{2}S_t a^{\dagger 2} + R_t a^\dagger)\exp(-a^\dagger a t) \times \exp(\tfrac{1}{2}S_t^* a^2 + R_t^* a). \tag{10}$$

The POVM (9) is normalized as usual so that

$$\int d^2R_t d^2S_t P(R_t, S_t) = 1, \tag{11}$$

where $P(R_t, S_t) = \mathrm{Tr}[\rho F_t(R_t, S_t)]$ is the *actual* PDF for obtaining the results R_t, S_t given the initial state ρ. By contrast, the function $P_0(R_t, S_t)$ in Eq. (9) can be thought of as the *ostensible* PDF for R_t and S_t [6]. It is the PDF they would have if $dW(t) = I(t)dt$ were a Wiener process [8] satisfying $dW(t)^2 = dt$. Explicitly,

$$P_0(R_t, S_t) = \int d\mathbf{I}_{[0,t)}P_0(\mathbf{I}_{[0,t)})\delta^{(2)}(R_t - \mathcal{F}_1[\mathbf{I}_{[0,t)}]) \times \delta^{(2)}(S_t - \mathcal{F}_2[\mathbf{I}_{[0,t)}]). \tag{12}$$

Here $P_0(\mathbf{I}_{[0,t)})$ equals the continuously infinite product of ostensible distributions for each instantaneous current $I(s)$ over each interval $[s, s+ds)$

$$P_0[I(s)] = \sqrt{ds/2\pi}\exp[-\tfrac{1}{2}dsI(s)^2]. \tag{13}$$

Recovering the standard result.—The theory presented here applies to any sort of detection with a large local oscillator. This includes heterodyne detection for which the local oscillator phase $\Phi(t)$ cycles rapidly in time at rate $\Delta \gg 1$. In this case S_t does not depend on $\mathbf{I}_{[0,t)}$, and from (8) we find that $S_t \to 0$ as $\Delta \to \infty$. The measurement result is thus R_t, which from (7) and (12) has the *ostensible* statistics of the random variable

$$R_t = \int_0^t e^{-s/2}e^{-i\Delta s}dW(s), \tag{14}$$

where $dW(t)$ is a Wiener increment. Being the (continuous) sum of Gaussian random variables, R_t must be a Gaussian random variable. The rapid phase rotation at rate $\Delta \to \infty$ ensures that is has no preferred phase. Writing $A = R_\infty$, it is easy to show from Eq. (14) that the expected value of $|A|^2$ is 1. These three constraints define the ostensible PDF for the final result A at $t = \infty$:

$$P_0^{\text{het}}(A) = \pi^{-1}\exp(-|A|^2). \tag{15}$$

Substituting this into Eq. (9) and using the fact that $\lim_{t\to\infty}\exp(-a^\dagger a t) = |0\rangle\langle 0|$ yields the effect

$$F_\infty^{\rm het}(A) = \pi^{-1}e^{-|A|^2}\exp(Aa^\dagger)\,|0\rangle\langle 0|\exp(A^* a). \quad (16)$$

A little operator algebra confirms that this is identical to the previously stated result (4). The natural phase estimate is thus $\phi = \arg A$. This can be understood as follows. The *actual* mean photocurrent (6) has two counterrotating complex terms. The rotation of the kernel of the integral (7) reinforces that of the second term (which thus averages to zero) but cancels that of the first term, leaving $\langle A\rangle = \int_0^\infty \langle I(t)\rangle e^{i\Phi(t)-t/2}dt = \langle a\rangle$. This result assumes a rapidly varying $\Phi(t)$; in general the second functional (8) is also required to estimate ϕ.

Estimating the phase.—Consider a state ρ_0 with a phase distribution centered around zero. Such a state could be used for communication by encoding a number $\varphi \in [0, 2\pi)$ as a phase shift by the unitary operator $U(\varphi) = \exp(-ia^\dagger a\varphi)$. The PDF for the receiver to get results R_t, S_t for a given φ is thus

$$P(R_t, S_t|\varphi) = {\rm Tr}[U(\varphi)\rho_0 U^\dagger(\varphi)F_t(R_t, S_t)]. \quad (17)$$

The receiver, who wishes to estimate the phase φ, can use Bayesian statistics [3] to find the posterior PDF

$$P(\varphi|R_t, S_t) = \mathcal{N}^{-1}P(R_t, S_t|\varphi)P_{\rm prior}(\varphi), \quad (18)$$

where $P_{\rm prior}(\varphi)$ expresses the *prior* knowledge the receiver has about φ, and $\mathcal{N}$ is a normalization factor. To be unbiased, we assume that the receiver knows ρ_0 but has no idea about φ, so that the prior PDF $P_{\rm prior}(\varphi)$ is flat [3]. Then it follows from Eq. (9) that at time t the maximum likelihood estimate (MLE) $\hat\varphi_t$ for φ is that $\hat\varphi_t(R_t, S_t)$ which maximizes the likelihood function

$$L_t(\hat\varphi_t) = {\rm Tr}[\rho_0 U^\dagger(\hat\varphi_t)G_t(R_t, S_t)U(\hat\varphi_t)]. \quad (19)$$

The MLE is the most convenient estimate of phase because, unlike other estimates (such as the mean), it does not suffer from ambiguity due to the cyclic nature of φ.

Adaptive measurements.—The above result for the MLE $\hat\varphi_t$ of the phase is true no matter how the local oscillator phase $\Phi(t)$ varies. Thus we can use this MLE which emerges from the processing of the signal $\mathbf{I}_{[0,t)}$ in a feedback loop to control $\Phi(t)$ as shown in Fig. 1. The suggestion here is that the local oscillator phase be controlled to be in quadrature with the current estimated phase of the system, so that in the next instant of time the apparatus will make a homodyne measurement of the estimated phase quadrature. Explicitly,

$$\Phi(t) = \hat\varphi_t(R_t, S_t) + \pi/2. \quad (20)$$

Another way of looking at this is that the receiver attempts to make a *null measurement* of phase. At small t, the receiver has very little information. Thus initially $\hat\varphi_t$, and hence $\Phi(t)$, varies wildly in time. This has the same effect as the rapidly cycling $\Phi(t)$ in heterodyne detection: all quadratures are sampled equally. As more information is acquired the phase estimate improves and approaches the value which is finally used as the result of the measurement, $\phi = \hat\varphi_t(R_\infty, S_\infty)$.

In order to understand this process better, it is helpful to look at an example where the function $L_t(\hat\varphi_t)$ has a relatively simple form. If ρ_0 is the coherent state $|r\rangle\langle r|$ with r real, then one finds that one should maximize

$$\ln L_t(\hat\varphi_t) = {\rm Re}[r^2 S_t^* e^{2i\hat\varphi_t} + 2rR_t^* e^{i\hat\varphi_t}] + c, \quad (21)$$

where c is a constant (independent of $\hat\varphi_t$). If $r \gg 1$ then for short times $t \lesssim r^{-2}$ the MLE $\hat\varphi_t$ is approximately equal to $\arg R_t$. This is because for short times $|R_t| \sim \sqrt{t}$, while $S_t \ll t$ due to the rapid variation of $\Phi(t)$. This estimate ($\arg R_t$) is the same as that for heterodyne detection, as expected. For longer times the estimated phase settles down, S_t becomes significant, and hence $\hat\varphi_t$ becomes approximately constant at $\arg\sqrt{S_t}$ (with the ambiguity resolved by the phase of R_t). During this stage the measurement is effectively a homodyne measurement of the phase quadrature.

Even with the relatively simple form (21) of $L_t(\hat\varphi_t)$ for coherent states, it is not possible to solve the scheme analytically. Rather, an ensemble of stochastic numerical simulations is needed. The best way to do this is by using the theory of *nonlinear* quantum trajectories [5,9]. In this particular case the system state need not be simulated; it is simply $|re^{i\varphi-t/2}\rangle$. From this, the photocurrent is generated with the correct *actual* statistics by

$$I(t) = e^{-t/2}2r\cos[\varphi - \Phi(t)] + \xi(t), \quad (22)$$

where $\xi(t)$ is Gaussian white noise [8]. For each simulation, R_t and S_t are calculated and used in Eqs. (20) and (21), and the final MLE $\phi = \hat\varphi_\infty$ stored. An ensemble size of 100 for $r = 50$ gave the mean squared difference between actual and estimated phases to be

$$E[(\phi - \varphi)^2]_{\rm adapt} = (1.0 \pm 0.2)\times 10^{-4}. \quad (23)$$

This is half the variance of the standard result of $(2r^2)^{-1}$ [2]. Within statistical error, it is equal to the error of a canonical phase measurement, $(4r^2)^{-1}$ [2]. This is not unexpected, since for $r \gg 1$ the vast majority of the measurement time is spent in an effective homodyne measurement of the phase quadrature. For states with smaller intrinsic phase uncertainty than coherent states the advantage of adaptive measurements over standard measurements would of course be more dramatic.

An analytic example.—There is one case in which the adaptive measurement may be treated analytically: if the system is known to contain at most one photon. This could occur if the cavity mode were excited by a single atom, in which case the phase of the field is equal to the original phase of the dipole of the atom. Since the harmonic oscillator truncated at one photon is equivalent to a two-level atom, the letters TLA will be used to distinguish this case. First I consider canonical and standard measurements. Projecting (2) into the subspace spanned by $\{|0\rangle, |1\rangle\}$ yields the canonical POVM

$$F_{\rm can}^{\rm TLA} = \frac{1}{2\pi}|\phi\rangle\langle\phi|, \qquad |\phi\rangle = |0\rangle + e^{i\phi}|1\rangle. \quad (24)$$

Similarly, the standard POVM (3) becomes

$$F_{\rm std}^{\rm TLA}(\phi) = \frac{\sqrt{\pi}}{2} F_{\rm can} + \left(1 - \frac{\sqrt{\pi}}{2}\right)\frac{\hat{1}}{2\pi}. \quad (25)$$

That is to say, the standard technique has an efficiency of $\sqrt{\pi}/2 \simeq 88\%$, in the sense that the same POVM would arise from a canonical measurement that worked 88% of the time, and that gave a random answer on the interval $[0, 2\pi)$ the other 22% of the time.

Because of the isomorphism between the at-most-one-photon field and the two-level atom, it is permissible to replace the annihilation operator a with the lowering operator $|0\rangle\langle 1|$ in all operators. One thus finds the very simple expression for the likelihood function (19):

$$L_t(\hat{\varphi}_t) = \mathrm{Re}[2\langle 1|\rho_0|0\rangle R_t^* e^{i\hat{\varphi}_t}] + c', \quad (26)$$

where c' is a constant. This does not depend on S_t, so for any ρ_0 which has a mean phase of zero (i.e., for which $\langle 1|\rho_0|0\rangle$ is real and positive), one has simply $\hat{\varphi}_t = \arg R_t$. Thus using Eq. (20) the local oscillator phase is set to be $\Phi(t) = \arg R_t + \pi/2$. The integral equation $R_t = \mathcal{F}_1[\mathbf{I}_{[0,t)}]$ (7), where $\mathbf{I}_{[0,t)}$ has the ostensible distribution (12), is thus equivalent to the following Itô stochastic differential equation:

$$dR_t = e^{-t/2} i(R_t/|R_t|)dW(t), \quad (27)$$

where as before $dW(t) = I(t)dt$ is ostensibly a Wiener increment. Using the Itô calculus [8] for $|R|^2$ and $\arg R$ quickly yields the solution

$$R_t = \sqrt{1 - e^{-t}} \exp\left[i \int_0^t \frac{e^{-s/2}}{\sqrt{1 - e^{-s}}} dW(s)\right]. \quad (28)$$

For $t \to \infty$ one finds $A = R_\infty$ given by

$$A = \exp(i\phi), \qquad \phi = \int_0^\infty \frac{I(t)dt}{\sqrt{e^t - 1}}. \quad (29)$$

This ϕ is ostensibly a completely random phase (because the integrand diverges as $t \to 0$), so $P_0(\phi) = (2\pi)^{-1}$. Equation (29) shows that the measurement result contains no intensity information, only phase information, as desired. Indeed, substituting these results into Eq. (9) with a replaced by $|0\rangle\langle 1|$ gives the POVM for ϕ

$$F_{\rm adapt}^{\rm TLA}(\phi) = (2\pi)^{-1} \exp(e^{i\phi}|1\rangle\langle 0|)|0\rangle\langle 0| \exp(e^{-i\phi}|0\rangle\langle 1|)$$
$$= F_{\rm can}^{\rm TLA}(\phi). \quad (30)$$

Thus for the two-level atom (or equivalently for a field with at most one photon), a simple adaptive measurement scheme can produce a canonical measurement of the phase, which is significantly better than the best result with no feedback (25).

The results presented here show that adaptive phase measurements can be close to canonical phase measurements for states with both low and high photon numbers. The key idea is to optimize the measurement at each instant of time by using the MLE of the phase to control the local oscillator phase (20). It should be understood that I have not shown that this is the globally optimal algorithm. Also, the present algorithm assumes the receiver has complete knowledge of the initial state except for its phase. Alternative situations will be considered in future work. The important result from this paper is that by using feedback to create an adaptive phase measurement, a great improvement over standard techniques may be found. If there is at most one excitation in the system, the adaptive technique is as good as a canonical one. Since the feedback requires only electronics and an electro-optic modulator, it should be experimentally feasible, and would represent a fundamental methodological advance over standard phase measurements.

I am very grateful to S. M. Tan for many stimulating discussions. This work is funded by the New Zealand Foundation for Research, Science and Technology.

---

*Electronic address: hmw@phy.auckland.ac.nz

[1] U. Leonhardt, J. A. Vaccora, B. Böhmer, and H. Paul, Phys. Rev. A **51**, 84 (1995).

[2] M. J. Hall and I. G. Fuss, Quantum Opt. **3**, 147 (1991).

[3] C. W. Helstrom, *Quantum Detection and Estimation Theory* (Academic, New York, 1976).

[4] *Quantum Phase and Phase Dependent Measurements*, edited by W. P. Schleich and S. M. Barnett [Phys. Scr. **T48**, 13 (1993)].

[5] H. M. Wiseman and G. J. Milburn, Phys. Rev. A **47**, 1652 (1993).

[6] H. M. Wiseman, Quantum Opt. (to be published).

[7] P. Goetsch and R. Graham, Phys. Rev. A **50**, 5242 (1994).

[8] C. W. Gardiner, *Handbook of Stochastic Methods* (Springer, Berlin, 1985).

[9] H. J. Carmichael, *An Open Systems Approach to Quantum Optics* (Springer-Verlag, Berlin, 1993).

Phase Measurement by Projection Synthesis

Stephen M. Barnett and David T. Pegg*
Department of Physics and Applied Physics, University of Strathclyde, Glasgow G4 0NG, Scotland
(Received 12 December 1995)

Experimental determination of the canonical quantum optical phase probability distribution has, until now, required sufficient measurements to determine the complete state of the field. In this Letter we present a more direct means for measuring this distribution which involves synthesizing the projection onto a phase state. Projection synthesis may be applied more generally to measure the probability distribution associated with other observables. [S0031-9007(96)00342-0]

PACS numbers: 42.50.Dv, 03.65.Db

The probability, or probability density, for a quantum mechanical observable to be found with a particular value is the expectation value of the projector formed from the eigenstate of that observable with the corresponding eigenvalue. Thus, for example, the photon number probability distribution for a pure state $|f\rangle = \sum c_n|n\rangle$ of an electromagnetic field mode can be obtained reasonably directly by measuring a quantity proportional to $\langle f|n\rangle\langle n|f\rangle$, where $|n\rangle\langle n|$ is a photon number state projector. For weak fields in the quantum domain a suitable quantity to measure is the probability of the release of n photoelectrons by an ideal photodetector. On the other hand, measuring the canonical phase probability distribution, as defined below, is more difficult. For a weak field, the only method presently available appears to be the reconstruction of the entire state by means of either optical homodyne tomography [1] or other related methods [2]. The phase probability distribution and indeed any other probability distribution can then be calculated. This has raised the question [3] of whether or not it is possible, even in principle, to measure the phase distribution more directly. A direct measurement of the canonical phase distribution would involve measuring a quantity proportional to $\langle f|\theta\rangle\langle\theta|f\rangle$ where $|\theta\rangle$ is a phase state which is complementary to the photon number states [4]. We show in this Letter how this can be done.

The probability density for a field in state $|f\rangle$ to have a phase θ is [5]

$$\langle f|\theta\rangle\langle\theta|f\rangle = \frac{1}{2\pi}\left|\sum_{n=0}^{\infty} c_n \exp(-in\theta)\right|^2 . \tag{1}$$

For any physical state $|f\rangle$, the coefficients c_n must eventually decrease indefinitely with increasing n. It follows that the expectation value (1) can always be approximated to any desired degree of accuracy by setting c_n to zero for $n > N$, if N is suitably large [6]. This allows us to replace (1) with the quantity

$$P_N(\theta) = \frac{1}{2\pi}\left|\sum_{n=0}^{N} c_n \exp(-in\theta)\right|^2 . \tag{2}$$

Of course, the error involved in this replacement will be zero for states which are finite superpositions of number states with $c_n = 0$ for all n greater than N. The probability density (2) is proportional to the expectation value of the projector $|\theta, N\rangle\langle\theta, N|$, where $|\theta, N\rangle$ is the truncated phase state [7]

$$|\theta, N\rangle \equiv \frac{1}{\sqrt{N+1}}\sum_{n=0}^{\infty} \exp(-in\theta)\,|n\rangle . \tag{3}$$

All phase and phase-dependent measurements require the introduction of a reference system to set the zero of phase. The simplest way in which this can be achieved is by coherently mixing the field a in state $|f\rangle_a$ with a reference field b, prepared in a state $|B\rangle_b$, by means of a beam splitter as shown in Fig. 1. If $|B\rangle_b$ is a large amplitude coherent state then this arrangement allows us to perform a measurement of a chosen field quadrature [8]. Our task here is to find a suitable reference state $|B\rangle_b$ such that photocounting in the two outputs of the beam splitter leads to the required probability distribution (2). Using the beam splitter, we can measure the probability for finding n and n' photons in the output modes a and b, respectively, which, for ideal photodetectors, is given by ${}_a\langle f|\hat{\Pi}|f\rangle_a$, where $\hat{\Pi}$ is the projector ${}_b\langle B|\hat{R}^\dagger|n'\rangle_b|n\rangle_{aa}\langle n|_b\langle n'|\hat{R}|B\rangle_b$. Here $\hat{R}$ is the unitary transformation linking the output

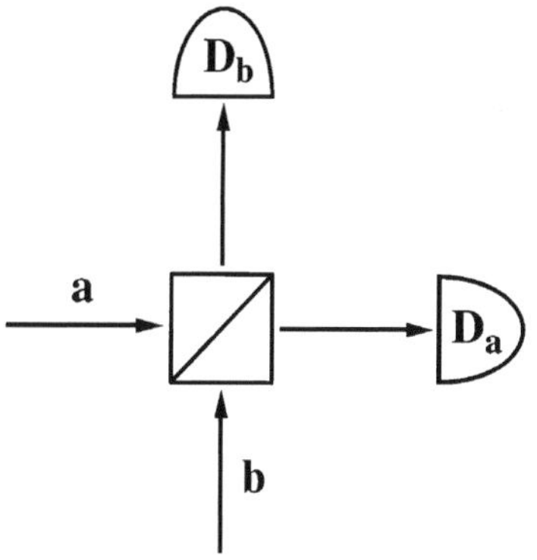

FIG. 1. Schematic representation of a beam splitter with input modes a and b and detectors D_a and D_b.

modes to those for the input [9]. To synthesize a suitable projection which will allow us to find $P_N(\theta)$ by this method, we need to find n, n', $\hat{R}$, and $|B\rangle_b$ such that

$$\hat{\Pi} = K|\theta, N\rangle_{aa}\langle\theta, N|, \quad (4)$$

where K is a positive constant, independent of θ. Remarkably, this problem admits at least one solution.

Consider an ideal 50/50, symmetric beam splitter for which the unitary transformation $\hat{R}$ is [9]

$$\hat{R} = \exp\left[i\frac{\pi}{4}(\hat{b}^\dagger\hat{a} + \hat{a}^\dagger\hat{b})\right]$$
$$= \exp(i\hat{b}^\dagger\hat{a})\exp\left[\frac{\ln 2}{2}(\hat{b}^\dagger\hat{b} - \hat{a}^\dagger\hat{a})\right]\exp(i\hat{a}^\dagger\hat{b}). \quad (5)$$

With this choice, Eq. (4) can be satisfied if we set $n = N$ and $n' = 0$, corresponding to N photocounts in detector D_a and no counts in D_b. The required form of the reference state $|B\rangle_b$ is

$$|B\rangle_b = C\sum_{k=0}^{N}\binom{N}{k}^{-1/2}\exp\left[ik\left(\theta - \frac{\pi}{2}\right)\right]|k\rangle_b, \quad (6)$$

where $\binom{N}{k}$ is the usual binomial coefficient and C is the normalization constant with modulus independent of θ. It is natural to refer to these states as reciprocal-binomial states by analogy with the biminimal states of Stoler, Saleh, and Teich [10]. With this reference state, we find

$$_a\langle N|_b\langle 0|\hat{R}|B\rangle_b = C\exp(iN\theta)2^{-N/2}\sum_{n=0}^{N}{}_a\langle N|\exp(-in\theta), \quad (7)$$

which satisfies Eq. (4) with $K = |C|^2 2^{-N}(N + 1)$. It follows that the probability of registering N counts in D_a and no counts in D_b is proportional to the expectation value of the projector $|\theta, N\rangle\langle\theta, N|$ and hence to the required probability density given in Eq. (2). For a sufficiently large value of N, this yields the phase probability density (1). The full distribution can be obtained by repeating the measurement with reference field states containing different values of θ in Eq. (6). These fields can be prepared by first generating a field in one particular state $|B\rangle_b$ and then changing the value of θ by means of a phase shifter. It is easy to verify that the action of the phase-shift operator $\exp(i\hat{b}^\dagger\hat{b}\Delta\theta)$ on $|B\rangle_b$ is equivalent to adding $\Delta\theta$ to θ.

For simplicity, our analysis has been for a pure state in mode a. It is not difficult to extend this to include any mixed state with density operator $\hat{\rho}_a$. In this case, the probability of registering N counts in D_a and no counts in D_b with reference state $|B\rangle_b$ in Eq. (6) becomes $\mathrm{Tr}(\hat{\rho}_a\hat{\Pi})$, that is, $_a\langle\theta, N|K\hat{\rho}_a|\theta, N\rangle_a$, which is proportional to the probability density of finding the mixed state with phase θ.

In order to perform an experiment it is necessary to select a suitable value for N and then to prepare the required reciprocal-binomial states. The choice of N is determined by the particular state being measured and by the accuracy desired for the measured probability distribution [6]. A practical procedure might be as follows. A preliminary choice of N is made, based either on the mean intensity or on simple knowledge of the source, and the experiment is then performed as described above. The choice of N will lead to an accurate determination of the phase probability distribution if the number of occasions that N counts are registered in D_a and no counts in D_b greatly exceeds those occasions on which the total number of counts registered in the two detectors exceeds $2N$. If this is not the case then N will have to be increased. This procedure ensures that the probability that the field in mode a has more than N photons is sufficiently small for $P_N(\theta)$ to provide an accurate approximation to the phase probability distribution. The probability for registering N counts in D_a with no counts in D_b is then determined as the ratio of the number of these events to the total number of runs. It should be noted that the detection of N counts in D_b and no counts in D_a also provides useful information corresponding to finding the expectation value of the projector $|(\theta - \pi), N\rangle\langle(\theta - \pi), N|$ and hence determining the phase probability density $P_N(\theta - \pi)$. It is only necessary, therefore, to change the phase associated with the reciprocal-binomial states through values in a range of π in order to provide all the information required to reproduce the phase probability distribution. The resulting distribution can be normalized over a 2π range. In our discussion we have, for simplicity, assumed ideal photodetectors. This assumption is not strictly necessary since the ideal detector statistics can be recovered from those measured with sufficiently good detectors [11]. We should note that the experimental procedure described here differs from that used by Noh, Fougères, and Mandel [12]. Their experiments do not measure the canonical phase, that is the complement of the photon number, but rather an operationally defined phase.

Clearly, the most difficult part of the measurement procedure is the preparation of the reciprocal-binomial states. In light of recent work [13], however, it is clear that the problem of generating specific states such as these can and will be solved. Moreover, the realization of measurements based on projector synthesis provides an important motivation for the production of such specially constructed nonclassical states.

Our work answers, in the affirmative, the important question as to whether it is possible in principle to measure the phase distribution without having to obtain sufficient information to reconstruct the complete state. In practice, it would probably be more convenient to measure the photon number probability distribution and then to make a choice of N before applying the projection synthesis technique.

An important application of measuring the phase distribution for a field in a pure state is that, when it is combined with the photon number probability distribution, it provides all the information required to reconstruct the state

[14]. This would provide a possible alternative to existing techniques [1,15] and proposals [2] for experimental state determination. Finally, state projection synthesis provides the means to perform more general measurements than just that of phase. The ability to prepare any chosen reference state for mode b would, in principle, allow the experimental determination of the expectation value of any chosen projector formed from the first $N + 1$ number states.

---

*Permanent address: Faculty of Science and Technology, Griffith University, Brisbane 4111, Australia.

[1] M. Beck, D. T. Smithey, J. Cooper, and M. G. Raymer, Opt. Lett. **18**, 1259 (1993); M. Beck, D. T. Smithey, J. Cooper, M. G. Raymer, and A. Faridani, Phys. Scr. **T48**, 35 (1993), and references therein.

[2] O. Steuernagel and J. A. Vaccaro, Phys. Rev. Lett. **75**, 3201 (1995), and references therein; J. A. Vaccaro, Phys. Rev. A **52**, 3474 (1995).

[3] Z. Bialynicka-Birula and I. Bialynicki-Birula, Appl. Phys. B **60**, 275 (1995).

[4] This phase representation has a long history, see, for example, F. London, Z. Phys. **40**, 193 (1927); R. Loudon, *The Quantum Theory of Light* (Oxford University Press, Oxford, 1973), 1st ed.; I. Bialynicki-Birula and Z. Bialynicka-Birula, Phys. Rev. A **14**, 1101 (1976); L. Susskind and Glogower, Physics (Long Island City, N.Y.) **1**, 49 (1964). Although these phase states are not orthogonal, a genuine probability density can be derived by applying an appropriate limiting procedure to an orthogonal subset, see D. T. Pegg and S. M. Barnett, Europhys. Lett. **6**, 483 (1988); Phys. Rev. A **39**, 1665 (1989); S. M. Barnett and D. T. Pegg, J. Mod. Opt. **36**, 7 (1989).

[5] This is the limiting form of the probability density associated with the Hermitian optical phase operator (see Pegg and Barnett [4]). It is also the probability density associated with phase probability-operator measures, see C. W. Helstrom, *Quantum Detection and Estimation Theory* (Academic Press, New York, 1976), and references therein; J. H. Shapiroand S. R. Shepard, Phys. Rev. A **43**, 3795 (1991).

[6] For example, for a coherent state with mean photon number of unity, replacing the upper summation limit in Eq. (1) by 6 is sufficient to reproduce the exact expression to within about 1% at the peak of the distribution.

[7] It is important to note that N is a specific chosen number. It should not be confused with the parameter s which approaches infinity in the theory of the Hermitian optical phase operator (see Pegg and Barnett [4]).

[8] R. Loudon and P. L. Knight, J. Mod. Opt. **34**, 709 (1987), and references therein.

[9] See, for example, G. Yeoman and S. M. Barnett, J. Mod. Opt. **40**, 1497 (1993).

[10] D. Stoler, B. E. A. Saleh, and M. C. Teich, Opt. Acta **32**, 345 (1985).

[11] C. T. Lee, Phys. Rev. A **48**, 2285 (1993).

[12] J. W. Noh, A. Fougères, and L. Mandel, Phys. Rev. Lett. **71**, 2579 (1993).

[13] K. Vogel, V. M. Akulin, and W. P. Schleich, Phys. Rev. Lett. **71**, 1816 (1993); A. S. Parkins, P. Marte, P. Zoller, and H. J. Kimble, Phys. Rev. Lett. **71**, 3095 (1993); A. S. Parkins, P. Marte, P. Zoller, O. Carnal, and H. J. Kimble, Phys. Rev. A **51**, 1578 (1995).

[14] Z. Bialynicka-Birula and I. Bialynicki-Birula, J. Mod. Opt. **41**, 2203 (1994); J. A. Vaccaro and S. M. Barnett, J. Mod. Opt. **42**, 2165 (1995).

[15] D. T. Smithey, M. Beck, A. Faridani, and M. G. Rayner, Phys. Rev. Lett. **70**, 1244 (1993).

Quantum phase distribution by projection synthesis

DAVID T. PEGG
Faculty of Science and Technology, Griffith University, Nathan, Brisane 4111, Australia

STEPHEN M. BARNETT and LEE S. PHILLIPS
Department of Physics and Applied Physics, University of Strathclyde, Glasgow G4 0NG, Scotland

(*Received 12 November 1996; revision received 23 January 1997*)

Abstract. We show how the recently-introduced method of projection synthesis can be applied to find the phase probability distribution for a single-mode field. The realization of such a measurement scheme requires the production of reciprocal-binomial states of light or suitable alternative states. We describe how such states might be produced in the laboratory.

1. Introduction

State preparation and measurement are closely related ideas. An ideal von Neumann measurement selects the eigenstates of the observable measured corresponding to the eigenvalue found. We can picture the prepared state evolving forward in time and then being measured or the eigenstate corresponding to the result of the measurement propagating backward in time to the point of state preparation. In practice, however, some states are more readily prepared and some measurements more easily performed than others. For example, a single field mode can be prepared in a coherent state by a stabilized laser but no single von Neumann measurement has been devised to show that the mode is in a given coherent state. On the other hand it is relatively straightforward to determine the number of photons in the mode by photon counting but more of a challenge to prepare number states. If we have access to a large number of copies of the original state than we can perform multiple measurements and thereby recreate the probability distribution associated with one or more observables. Given a sufficient quantity of experimental data we can then reconstruct the state for example by converting quadrature measurements into the Wigner function for the state [1] or its density matrix [2]†. Such measurement schemes have recently been realized and applied to coherent, squeezed and vacuum states [4–6]. Homodyne detection to establish the quadrature probability distribution is well suited to these states which all have Gaussian Wigner functions. Optical homodyne tomography and related techniques have a general validity, but the possibility of preparing more exotic states, such as superpositions of coherent states forming so-called Schrödinger cat states

† A number of other methods for determining the state of a field mode or its phase-space distribution have been proposed [3].

[7], raises the questions of whether other measurement strategies might provide the information to reconstruct the state more readily.

In this paper our interest lies mainly in states which are sufficiently weak to lie well within the quantum domain of a small photon number. For such states it is comparatively straightforward to find the photon number probability distribution by photon counting. The complementary information is contained in the quantum phase probability distribution [8, 9] and it has been established that knowledge of these two distributions is sufficient to reconstruct the state completely if the state is pure [10]. When the number of component photon number states in a field is small, the phase probability distribution is also quite simple, so it would appear that reconstructing the state of such fields by measuring the photon number and phase distributions should be reasonably promising. The difficulty, however, lies with measuring the phase distribution. For fields with narrow phase distributions, for example for fields in coherent states with a mean photon number equal to or greater than unity, good approximate methods exist for measuring the phase distribution [11] but fields with a small number of photon number state components do not generally fall within this category. Moreover, there has been recent interest in fields with broad phase distributions such as coherent states with mean photon numbers much less than unity [12]. Until very recently [13], the only way known to determine the phase distribution of such fields was by determining the state of the field first, for example by optical homodyne tomography [5], which defeats the purpose of using the phase distribution to measure the state.

In addition to being used as an intermediate step for quantum state determination, a potential method for the direct measurement of the phase distribution has its own intrinsic interest. For some time it was believed that phase was not a respectable quantum observable in that the measurement of its probability distribution may not be possible, even in principle. While the advent of optical homodyne tomography [1, 4, 5] showed that this could be done, the question still remained as to whether it is possible to measure the phase distribution without first obtaining enough information to determine the complete state. The question then is whether the phase distribution is something which is directly measurable or is it merely a quantity which one calculated from a full knowledge of the state? In this paper we investigate and extend a direct method for measuring the phase distribution based on a technique we have called projection synthesis [13]. The major problem with phase measurement by projection synthesis is that it requires the use of a new state of light, the reciprocal-binomial state, as a reference state. We also address this problem here and investigate mehods by which reciprocal-binomial states might be prepared and used to measure the phase distribution of weak fields.

The methods of projection synthesis and state preparation discussed in this paper have a more general utility than just phase measurement. They can be applied to measure more exotic observables and to prepare yet more unusual states. We will address the more general applications of these ideas elsewhere.

2. The phase probability distribution

We should emphasize here that what we are referring to as phase is the complement of photon number in the spirit of Dirac [14] and London [15] and which can be derived formally by means of a suitable limiting procedure as described in [8].† Although the phase distribution can be derived by means of an appropriate limiting procedure, it can also be obtained more directly from simple physical considerations which better illustrate its essential features. In this section we give a brief outline of this derivation. A fuller discussion is given in [21].

One of the fundamental tenets of quantum mechanics is the concept of a complex probability amplitude which is more fundamental than the real probability which can be obtained from it. For a pure state with a continuous probability distribution for a particular observable, the probability density can be expressed as the square of the modulus of a complex function in the appropriate representation. In the phase representation [21, 22] a pure state of light will be represented by a phase 'wave function' $\psi(\theta)$ such that the phase probability density is given by

$$P(\theta) = |\psi(\theta)|^2. \tag{1}$$

It is not difficult to show that a phase shift of light, for example as caused by inserting an extra optical path length into a travelling beam, either by shifting a mirror or by inserting a retarding plate, is generated mathematically by the action of the unitary operator $\exp(-\mathrm{i}\hat{N}\Delta\theta)$, that is [21, 23]

$$\exp(-\mathrm{i}\hat{N}\Delta\theta)\psi(\theta) = \psi(\theta + \Delta\theta). \tag{2}$$

It is possible to show from (2) that the phase function $\psi_n(\theta)$ representing a photon number state must have the form [21, 23]

$$\psi_n(\theta) = \frac{1}{(2\pi)^{1/2}} \exp \mathrm{i}(\beta_n - n\theta), \tag{3}$$

where β_n is real and can be chosen arbitrarily. This function has a period 2π so wave functions with a phase difference of 2π describe the same state. The modulus of this wave function has been chosen so that the phase probability distribution (1) is normalized over a 2π range. The choice of β_n determines the particular number state basis used. In the phase representation, the operator representing $\exp(-\mathrm{i}\theta)$ will simply be the multiplicative operator $\exp(i\theta)\times$ for the same reason that the operator representing x is a multiplicative operator in the coordinate representation. It follows then that

$$\hat{N}^{1/2} \exp(-\mathrm{i}\theta)\psi_n(\theta) = (n+1)^{1/2}\psi_{n+1}(\theta) \exp \mathrm{i}(\beta_n - \beta_{n+1}). \tag{4}$$

Choosing β_n as a constant allows us to associate $\hat{N}^{1/2}\exp(-\mathrm{i}\theta)$ with the creation operator $\hat{a}^\dagger$ in the usual number state basis used in quantum electrodynamics. There is no physical significance in this constant phase factor which we can associate with the arbitrary phase of the wave function.The phase probability density for a field in a superposition state $\sum c_n|n\rangle$ becomes from (1) and (3)

$$P(\theta) = \frac{1}{2\pi}\left|\sum_{n=0}^{\infty} c_n \exp(-\mathrm{i}n\theta)\right|^2. \tag{5}$$

This probability distribution is the same as that derived by some other methods [8, 9]. In the representation used here, in which the phase operators are simply

† There are other 'operational' definitions of phase defined as the outcomes of particular measurements [16–20]. Each of these 'phases' is different and none of them have been shown to lead to the canonical phase probability distribution [8, 9] associated with the Hermitian optical phase operator.

multiplicative, we have not had to worry about unitarity or Hermiticity. An analysis of the relation of this method to the problems encountered in the more usual representations is given in [21]. We should point out that (5) is not the same as the operationally-defined phase distributions introduced by Mandel and co-workers [12, 17] and by others [16, 18–20]. These operational phase distributions were introduced in order to be able to describe certain optical experiments as phase measurements. The only justification for considering them to be phase measurements is that their classical analogues provide phase information. The problem is that in the quantum domain they do not lead to a unique and hence useful definition of the phase probability distribution. It would, of course, also be possible to take the method for measuring the phase probability distribution, which will be described in this paper, and use it to provide yet another operational definition of this distribution. From a fundamental viewpoint, however, there is nothing to be gained from such an approach. The phase described by (5) is derived as the complement of photon number and is the quantity which is shifted by a phase shifter [21]. Its definition is not dependent on the manner in which it is measured. Another fundamental reason for accepting (5) as the definition of the phase probability distribution is that it is not difficult to show in the phase representation that

$$[\hat{N}\hbar\omega, \exp(-\mathrm{i}\theta)] = \hbar\omega \exp(-\mathrm{i}\theta) \tag{6}$$

which has the precise correspondence with the classical Poisson bracket demanded by the basic postulates of quantum mechanics [16]. Also, the commutator $[\hat{N}\hbar\omega, \theta]$ has the exact correspondence with the appropriate Poisson bracket [21].

The phase probability distribution (5) has been constructed from experimental data by using tomographic data to reconstruct the state [5]. In the following section we show how the phase probability distribution might be determined without first obtaining sufficient data to determine the state. If this method can be realized then it will not only provide a means of measuring the phase distribution but also, when combined with the photon number probability distribution, a means of determining the state.

3. Measurement by projection synthesis

In this section we present the projection synthesis method of phase measurement introduced in [13]. In doing so, we take the opportunity to discuss further some of the subtleties involved and to correct some minor typographical errors that appeared in our earlier letter.

Any pure physical state $|f\rangle$ can be written as a superposition of the number states in the form

$$|f\rangle = \sum_{n=0}^{\infty} c_n |n\rangle, \tag{7}$$

where normalization requires that $\sum |c_n|^2 = 1$ while the requirement that the energy used in making the state is finite leads to the condition that finite moments of the photon number operator are finite so that $\sum n^q |c_n|^2$ is strictly finite for all finite integers q [8]. For such states the phase probability distribution has the form (5) and it is this distribution that we must measure in order to obtain the phase information about the state. The requirement that the moments of $\hat{N}$ be bounded means that the coefficients c_n must eventually become zero or decrease indefinitely with increasing n. It follows that the probability distribution (5) can always be approximated to any desired degree of accuracy by setting c_n to zero for $n > N$, if N is chosen to be suitably large. This allows us to replace (5) with

$$P_N(\theta) = \frac{1}{2\pi k_N}\left|\sum_{n=0}^{N} c_n \exp(-in\theta)\right|^2, \tag{8}$$

where the normalization constant

$$k_N = \sum_{n=0}^{N} |c_n|^2 \tag{9}$$

is inserted to ensure that $P_N(\theta)$ is normalized over a 2π range.† This normalization property allows us to find (8) if we can measure a quantity proportional to it for a sufficient number of different values of θ and then normalize the resulting distribution of results. The quantity $P_N(\theta)$ is proportional to the expectation value of the projector $|\theta, N\rangle\langle\theta, N|$, where $|\theta, N\rangle$ is the truncated phase state‡

$$|\theta, N\rangle = \frac{1}{(N+1)^{1/2}}\sum_{n=0}^{N} \exp(in\theta)|n\rangle. \tag{10}$$

We seek an experimental event for which the probability of occurrence is proportional to the squared modulus of the overlap between the state $|f\rangle$ and $|\theta, N\rangle$. One way of achieving this is to transform the state $|f\rangle$ into another state $\hat{U}|f\rangle$ such that the required probability is equal to $|\langle f|\hat{U}^{-1}\hat{U}|\theta, N\rangle|^2$ where $\hat{U}|\theta, \psi\rangle$ is an eigenstate of some measurement operator. The combination of the transformation together with the measurement then constitutes a synthesis of the projection of $|f\rangle$ onto $|\theta, N\rangle$ using the projection of $\hat{U}|f\rangle$ onto $\hat{U}|\theta, N\rangle$. A standard measurement technique in quantum optics is photon counting for which the measurement state $\hat{U}|\theta, N\rangle$ will be a photon number state. We can then measure the phase probability distribution by photon counting if we can perform the $N+1$ dimensional Fourier transform from the truncated phase states to the number states. Such a transformation could in principle be performed by a quantum computer [24]. An alternative and perhaps more practical approach is outlined below.

Consider a 50%:50% symmetric beam splitter with input modes a and b and output modes c and d as depicted in figure 1. The photon annihilation operators for these four modes are $\hat{a}$, $\hat{b}$, $\hat{c}$ and $\hat{d}$, respectively. By carrying out a large number of photon counting experiments, we can find the probability that N photons are detected in mode c while zero photons are detected in mode d. The state corresponding to this measurement result is

$$|N\rangle_c|0\rangle_d = \frac{(\hat{c}^\dagger)^N}{(N!)^{1/2}}|0\rangle_c|0\rangle_d. \tag{11}$$

† To be an accurate approximation to the true phase probability density, we require N to be large enough so that $k_N \approx 1$. In this limit, (8) tends to the form given in [13].

‡ Note that N is a specific integer. It should not be confused with the parameter s which is the dimension of the field-mode state and tends to infinity in the theory of the Hermitian optical phase operator [8].

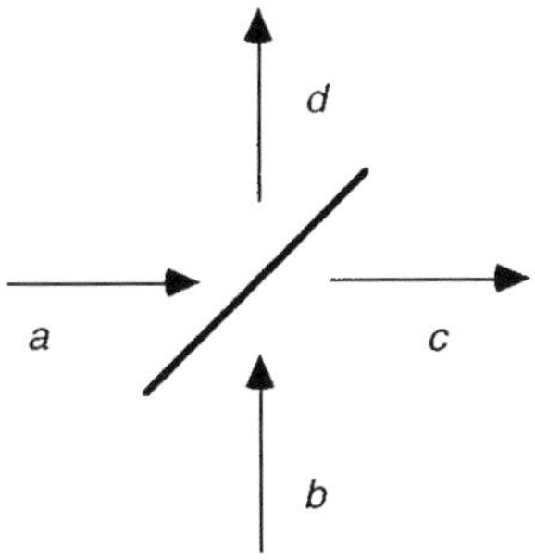

Figure 1. Schematic representation of a beam splitter with input modes a and b and output modes c and d.

In order to see what this outcome can tell us about the states of the input modes, we evolve this state back through the beam splitter and compare it with the input state of modes a and b. We use the beam-splitter relations (see for example [25])

$$\begin{pmatrix} \hat{c} \\ \hat{d} \end{pmatrix} = \begin{pmatrix} t & r \\ r & t \end{pmatrix} \begin{pmatrix} \hat{a} \\ \hat{b} \end{pmatrix}, \tag{12}$$

with $t = 2^{-1/2}$ and $r = \mathrm{i}/2^{1/2}$ to find the state which, under the action of the beam splitter, evolves to (11) in terms of an entangled state of the two input modes

$$\begin{aligned} |M\rangle &= \frac{2^{-N/2}}{(N!)^{1/2}} \sum_{l=0}^{N} \begin{pmatrix} N \\ l \end{pmatrix} (\hat{a}^{\dagger})^{N-l} (-\mathrm{i}\hat{b}^{\dagger})^{l} |0\rangle_a |0\rangle_b \\ &= 2^{-N/2} \sum_{l=0}^{N} \begin{pmatrix} N \\ l \end{pmatrix}^{1/2} (-\mathrm{i})^{l} |N-l\rangle_a |l\rangle_b. \end{aligned} \tag{13}$$

The probability amplitude for N photons to be detected in mode c and none in mode d is just the projection of the input state $|F\rangle = |f\rangle_a |B\rangle_b$ onto the state $|M\rangle$ given by (13), where $|f\rangle_a$ is the state under study and $|B\rangle_b$ is a prepared reference state,

$$|B\rangle_b = \sum_n b_n |n\rangle, \tag{14}$$

for mode b. With the state under investigation given by the general expression (7), the probability for finding N photons in mode c and none in mode d is

$$\begin{aligned} P(N_c, 0_d) = |\langle M|F\rangle|^2 &= \left| 2^{-N/2} \sum_{l=0}^{N} (\mathrm{i})^{l} \begin{pmatrix} N \\ l \end{pmatrix}^{1/2} b_l c_{N-l} \right|^2 \\ &= \left| 2^{-N/2} \sum_{l=0}^{N} (\mathrm{i})^{N-l} \begin{pmatrix} N \\ l \end{pmatrix}^{1/2} b_{N-l} c_l \right|^2. \end{aligned} \tag{15}$$

For this to be proportional to $P_N(\theta)$ we require

$$b_n \propto \begin{pmatrix} N \\ n \end{pmatrix}^{-1/2} \exp[\mathrm{i}n(\theta - \pi/2)] \tag{16}$$

with $0 \leqslant n \leqslant N$ and the constant of proportionality being independent of θ. The values of b_n for $n > N$ can be anything because, apart from the normalization factor which is absorbed into the proportionality constant, they do not affect the projection of $|F\rangle$ onto $|M\rangle$ in (15). Ideally, we require reference states of the form

$$|B;\, \theta, N\rangle = C\sum_{k=0}^{N} \binom{N}{k}^{-1/2} \exp(\mathrm{i}k\theta)|k\rangle \tag{17}$$

where $|k\rangle$ is the k-photon number state and the normalization constant C is independent of θ. We refer to these states as reciprocal-binomial states by analogy with the binomial states of Stoler *et al.* [26]. Similar states have recently been introduced in connection with resolutions of the identity for angular momenta [27]. We will consider the fabrication of reciprocal-binomial states or suitable alternatives in the next section. Here we discuss how we might use a source of reciprocal-binomial states to measure the phase probability distribution.

In order to perform an experiment it is necessary to select a suitable value for N and then to prepare the required reciprocal-binomial state $|B;\, \theta - \pi/2, N\rangle$. The value of N is determined by the form of the state being measured and by the accuracy desired for the measured probability distribution. The choice of N will lead to an accurate determination of the phase probability distribution if the number of occasions on which N counts are found in mode c and none in mode d greatly exceeds those occasions on which the total number of photons found in the output modes exceeds $2N$. If this is not the case then a reciprocal-binomial state with a larger value of N will have to be prepared. This procedure ensures that the probability that the field in mode a has more than N photons is sufficiently small for $P_N(\theta)$ to provide an accurate approximation to the phase probability distribution. The probability for finding N photons in mode c and none in mode d is then determined as the ratio of the number of these events to the total number of runs. The detection of N photons in mode d and none in mode c is also useful as it corresponds to finding the expectation value of the projector $|(\theta - \pi), N\rangle\langle(\theta - \pi), N|$ and hence allows us to determine $P_N(\theta - \pi)$. It is only necessary, therefore, to change the phase associated with the reciprocal-binomial states through values in a range of π in order to provide all the information required to reproduce the phase probability distribution. The resulting distribution can then be normalized over a 2π range.

4. Preparation of reciprocal-binomial states

It is clear that the main problem to be solved in order to use the projection synthesis method is the preparation of reciprocal-binomial states. What is measured by projection synthesis with a reciprocal-binomial state of $(N + 1)$ terms is the exact phase distribution of a state formed by truncating from the signal field state after the component containing $|N\rangle$. Thus for states with a finite number of photon number state components, a reciprocal-binomial state can be found which will yield in principle the exact phase distribution. For other states such as very weak coherent states, reciprocal-binomial states with a small number of terms may give the phase distribution to a very good approximation. As mentioned in the introduction, other methods exist for measuring the phase distribution of these states when the mean photon number is at least unity and we are interested in states much weaker than this. Of specific interest are coherent states with mean photon numbers of 0·047 and 0·076 because operational phase distributions have been determined experimentally for such fields [12]. An interesting question concerns how difficult it will be to measure the actual phase distributions, in the sense described earlier, of these fields. It is not hard to show that a two-component ($N = 1$) reciprocal-binomial state is sufficient to measure the former with a maximum error of 5·8%, which is probably less than the experimental errors which would be involved, and the latter with a maximum error of 9%. A three-component ($N = 2$) reciprocal-binomial state will give maximum errors of 0·7% and 1·5%, respectively. Thus even such simple reciprocal-binomial states will prove useful for fields of this intensity. We discuss in this section how we might use coherent and squeezed states as two- and three-component reciprocal-binomial states and how we might prepare a general reciprocal-binomial state. It is possible that other recent ideas associated with state preparation might also be applied to prepare the required states [28].

4.1. *Coherent and squeezed states*

It is not necessary to construct the reciprocal-binomial states exactly in order to perform a phase measurement by projection synthesis. Any pure state of the form

$$|\psi\rangle = \kappa C \sum_{n=0}^{N} \binom{N}{n}^{-1/2} \exp(in\theta)|n\rangle + \sum_{n=N+1}^{\infty} c_n|n\rangle \tag{18}$$

may be used in place of the reciprocal-binomial state $|B;\ \theta, N\rangle$. The reason for this is that only those situations in which precisely N photons are counted will be used to determine the phase probability distribution. Hence, the components of the reference state corresponding to more than N photons will (at least for an ideal beam splitter and detectors) have no effect on the phase measurement other than to reduce the probability of a useful measurement; the probability for a useful measurement using (18) is less than that based on using the reciprocal-binomial state $|B;\ \theta, N\rangle$ by a factor of $|\kappa|^2$, that is the probability that a suitable measurement on (18) would find it in the reciprocal-binomial state. In this section we show that the $N = 1$ and $N = 2$ reciprocal-binomial states can be replaced by a suitable coherent or squeezed state.

The $N = 1$ reciprocal-binomial state has the form

$$|B;\ \theta, 1\rangle = \frac{1}{2^{1/2}}\{|0\rangle + \exp(i\theta)|1\rangle\}. \tag{19}$$

If we can prepare a state for which the first two number state amplitudes have the ratio $c_1/c_0 = \exp(i\theta)$, then we can use this in place of $|B;\ \theta, 1\rangle$ to perform a phase measurement on any state having no more than one photon. The coherent states $|\alpha\rangle$ have the general form

$$|\alpha\rangle = D(\alpha)|0\rangle = \exp(-|\alpha|^2/2)\sum_{n=0}^{\infty}\frac{\alpha^n}{(n!)^{1/2}}|n\rangle, \tag{20}$$

where α is any complex number. The required ratio of the vacuum and one photon amplitudes is achieved by setting $\alpha = \exp(i\theta)$ corresponding to a mean photon number of unity. We note that we can alter the value of θ by means of a phase shifter. The probability that using a coherent state will lead to a useful phase measurement is less than that found using $|B;\theta,1\rangle$ by a factor of $2\exp(-|\alpha|^2)\approx 0{\cdot}736$.

The $N=2$ reciprocal-binomial state is

$$|B;\theta,2\rangle=\left(\frac{2}{5}\right)^{1/2}\left\{|0\rangle+\frac{\exp(i\theta)}{2^{1/2}}|1\rangle+\exp(i2\theta)|2\rangle\right\}. \tag{21}$$

There is no coherent state for which the ratios of the first three number state amplitudes have the required form. However, there does exist a suitable finite amplitude or bright squeezed state. These states have the general form [29]

$$\begin{aligned}|\alpha,\zeta\rangle &= D(\alpha)S(\zeta)0\rangle\\ &=(\cosh|\zeta|)^{-1/2}\exp\{-\tfrac{1}{2}[|\alpha|^2+(\alpha^*)^2\exp(i\phi)\tanh|\zeta|]\}\\ &\quad\times\sum_{n=0}^{\infty}\frac{[\frac{1}{2}\exp(i\phi)\tanh|\zeta|]^{n/2}}{(n!)^{1/2}}H_n\left\{\frac{\alpha+\alpha^*\exp(i\phi)\tanh|\zeta|}{(2\exp(i\phi)\tanh|\zeta|)^{1/2}}\right\}|n\rangle,\end{aligned} \tag{22}$$

where $\zeta=|\zeta|\exp(i\phi)$ is the squeezing parameter and H_n is the Hermite polynomial of order n [30]. In order to replace the $N=2$ reciprocal-binomial state, we require a squeezed state for which the first three photon number state amplitudes satisfy the conditions $c_1/c_0=\exp(i\theta)/2^{1/2}$ and $c_2/c_0=\exp(i2\theta)$. These conditions correspond to a squeezed state with $\tanh|\zeta|=2^{1/2}-1/2\approx 0{\cdot}914$, $\phi=\pi+2\theta$, $|\alpha|=4+3(2)^{1/2}\approx 8{\cdot}24$ and $\arg(\alpha)=\theta$. This is a strongly squeezed state with a quadrature variance ~22 times lower than the vacuum level corresponding to 95% squeezing. It also has a mean photon number of about 73 which is much greater than two, the maximum number of photons present in the $N=2$ reciprocal-binomial state. The overlap between this squeezed state and the required $N=2$ reciprocal-binomial state is about 0·001 and hence the probability of producing a useful phase measurement is about one thousand times smaller than it would be if a true $N=2$ reciprocal-binomial state were used. This demand for a high level of squeezing can be reduced if say a 60%:40% beam splitter is used in place of the 50%:50% one in the phase measurement.

4.2. *General reciprocal-binomial states*

One possible source of a field in a two-component reciprocal-binomial state is a two level atom excited by a $\pi/2$ pulse to an equal superposition of the excited and ground states. Other pulse areas would produce other two-component states. The relative argument of the two coefficients can be adjusted by use of a phase shifter. In this section we shall not discuss this or other methods, we shall simply assume that we can produce fields in any desired superposition of $|0\rangle$ and $|1\rangle$. We then show how we can coherently mix these in an array of beam splitters to achieve the required multi-component state. Clearly, this will also allow the possibility of preparing states other than reciprocal-binomial states. We begin by examining a single beam splitter with input and output modes as shown in figure 1.

For a symmetric beam splitter the creation operators $\hat{a}^\dagger$ and $\hat{b}^\dagger$ for the input modes are related to $\hat{c}^\dagger$ and $\hat{d}^\dagger$ for the output modes, as shown in figure 1, by

$$\begin{pmatrix}\hat{a}^\dagger\\ \hat{b}^\dagger\end{pmatrix}=\begin{pmatrix}t & -r^*\\ -r^* & t\end{pmatrix}\begin{pmatrix}\hat{c}^\dagger\\ \hat{d}^\dagger\end{pmatrix} \tag{23}$$

where the transmittance t is real and the reflectance r is imaginary. We consider light in a photon number superposition state $\sum_{n=0}^{M} b_n|n\rangle_a$ incident in mode a and light in state $c_0|0\rangle_b + c_1|1\rangle_b$ incident in mode b. The combined input state can then be expressed as

$$|in\rangle = \sum_{n=0}^{M} b_n \frac{\hat{a}^{\dagger n}}{(n!)^{1/2}}(c_0 + c_1\hat{b}^{\dagger})|0\rangle_a|0\rangle_b. \tag{24}$$

The output state can be found by substituting (23) into (24) and replacing the vacuum input state by the vacuum output state $|0\rangle_c|0\rangle_d$. From this we obtain eventually

$$_d\langle 0|out\rangle = b_0c_0\left\{|0\rangle_c + \sum_{n=1}^{M}(y_n + xy_{n-1}n^{1/2})|n\rangle_c + xy_M(M+1)^{1/2}|M+1\rangle_c\right\}, \tag{25}$$

where we have defined new variables

$$x = \frac{-r^*c_1}{c_0}, \qquad y_n = \frac{b_n t^n}{b_0}. \tag{26}$$

We see that, provided no photons are detected in output mode d, this stage adds an $M+1$ photon state to the light travelling to the right, as well as modifying the coefficients of the original component photon states. This condition can be achieved by placing a photodetector in mode d and only accepting the output in mode c if no photon counts are registered. In this way we can prepare any desired superposition of the first $M+2$ number states from a mode prepared in a superposition of the first $M+1$ number states and a mode in a superposition of the vacuum and one-photon states. The normalized form of (25) determines the output state in mode c *conditioned* on no counts being found in mode d. If one or more photons are detected in mode d then a state other than the desired one will be produced so that the final state is dependent on the result of the photon counting measurement [31].

By using a series of beam splitter stages with appropriate two-component state inputs, we can eventually prepare any given reciprocal-binomial state. The above equations allow us to work back from the desired final stage output to determine all the two-component input states which are necessary. We illustrate the procedure by a specific example, preparation of the four-component reciprocal-binomial state. For this we use a two stage device as shown in figure 2. For preparing a five-component reciprocal-binomial state, we would extend this to three beam splitters in series with three photodetectors in the top output modes and so on.

4.3. *Four-component reciprocal-binomial state*

We wish the output from the final stage to be in the state proportional to $|0\rangle + 3^{-1/2}|1\rangle + 3^{-1/2}|2\rangle + |3\rangle$. Thus in (25) we have $M = 2$ and

$$y_1 + x = 1/3^{1/2} \tag{27}$$

$$y_2 + 2^{1/2}xy_1 = 1/3^{1/2} \tag{28}$$

$$3^{1/2}y_2x = 1. \tag{29}$$

By eliminating the other two variables we find that x satisfies the cubic equation

$$6^{1/2}x^3 - 2^{1/2}x^2 + x - 1 = 0. \tag{30}$$

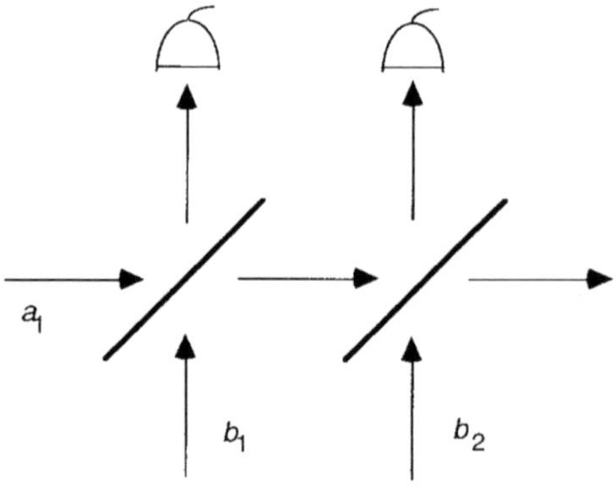

Figure 2. Series of beam splitters used for the preparation of the four-component reciprocal-binomial state. Input fields are in two-component states in the input modes a_1, b_1 and b_2. The field in the output mode to the right is only accepted if both photodetectors in the top output modes register zero photon counts.

For the preparation of the state $|B;\,\theta,N\rangle$ by this method, the parameter x will satisfy an $(N+1)$th order equation. Choosing the real root from the three possibilities yields the values $x=0{\cdot}754, y_1=-0{\cdot}177, y_2=0{\cdot}766$. From (26) we find that the states of light entering from the left and from below are proportional to $|0\rangle - 0{\cdot}177t_2^{-1}|1\rangle + 0{\cdot}766t_2^{-2}|2\rangle$ and $|0\rangle - 0{\cdot}754(r_2^*)^{-1}|1\rangle$, respectively, where the subscript 2 refers to the final or second, stage beam splitter. To produce the three-component state we need one earlier stage to the final stage. To calculate the required input fields for this first stage, we repeat the procedure with the three-component state above as the desired output. For this, $M=1$ and we obtain two equations in x and y_1, giving a quadratic in x. Solving this and then finding y_1 gives

$$x=\frac{1}{t_2}(-0{\cdot}088\pm0{\cdot}73\mathrm{i}) \qquad (31)$$

$$y_1=\frac{1}{t_2}(-0{\cdot}088\mp0{\cdot}73\mathrm{i}) \qquad (32)$$

which allows us to achieve the required output with input fields from the left and below in states proportional to $|0\rangle - (t_1t_2)^{-1}(0{\cdot}088+0{\cdot}73i)|1\rangle$ and $|0\rangle + (r_1^*t_2)^{-1}(0{\cdot}088-0{\cdot}73\mathrm{i})|1\rangle$, where the subscript 1 refers to the first beam splitter.

We see that it is possible to generate a four-component reciprocal-binomial state with two beam splitters and three input fields in selected two-component states. If either of the two detectors situated above the beam splitters, as shown in figure 2, registers one or more photons, then the output will not be in the required state and the subsequent experiment will be aborted. It is important therefore to find the probability that there will be zero photons in these output modes.

If none of the photodetectors registers a photocount then the output will be the desired reciprocal-binomial state. The total number of photons is conserved and it follows, therefore, that the probability for finding no photons in any of the output modes is equal to the probability that there were no photons in the input modes. Hence the probability that the experimental run will successfully produce the required reciprocal-binomial state is the probability that there are no photons in the output divided by the probability that the required reciprocal-binomial state has no protons. If the N input states $v_i|0\rangle + \lambda_i|1\rangle$ are used to prepare the reciprocal-binomial state $|B;\,\theta, N\rangle$ then the probability of a single experimental run successfully producing the state is

$$\frac{\prod_{i=0}^{N-1}|v_i|^2}{|\langle 0|B;\,\theta, N\rangle|^2}. \tag{33}$$

For the two-stage arrangement producing the four-component inverse binomial state, v_0 is the coefficient of the vacuum state for the input field in mode a_1 and v_1 and v_2 are the vacuum state coefficients for the input fields in modes b_1 and b_2. With the input states as calculated (33) reduces to

$$\frac{[1 + 0{\cdot}57/(1 - t_2^2)]^{-1}[1 + 0{\cdot}54/t_1^2t_2^2]^{-1}[1 + 0{\cdot}54/(1 - t_1^2)t_2^2]^{-1}}{[1 + 1/3 + 1/3 + 1]^{-1}} \tag{34}$$

where we have used $|r|^2 = 1 - t^2$. It is not difficult to find the transmittances which maximize this and obtain a maximum value for expression (34) of 0·14 which occurs for $t_1^2 = 0{\cdot}5$ and $t_2^2 = 0{\cdot}66$. In practice, however, it might be simpler to use two 50%:50% beam splitters for which expression (34) has a marginally smaller value of 0·12. With this value, accepting output fields only when both photodetectors register zero photons means that approximately seven in eight experiments will be aborted.

5. Conclusion

In this paper we have considered two related problems; how to perform a general measurement on a field mode and how to prepare an arbitrary superposition of a finite number of number states. We have shown how a beam splitter combined with a specially constructed superposition state, the reciprocal-binomial state, can be used to perform a measurement of the phase probability distribution. This information, when combined with the photon number probability distribution can be used to determine the (pure) state of the field mode [10].

The reciprocal-binomial states are superpositions of the first $N + 1$ number states with photon number probabilities given by the reciprocals of the binomial coefficients. The $N = 1$ and $N = 2$ reciprocal-binomial states can be replaced by a coherent or bright squeezed state. The degree of squeezing required for a $N = 2$ reciprocal-binomial state is demanding. It is by no means necessary, however, to use a 50%:50% beam splitter in the phase measurement. If a different reflectivity is used then a state other than the reciprocal binomial state will be required in order to synthesize the truncated phase state projector. For a 60%:40% beam splitter, the required $N = 2$ state can be obtained as the first three number state components of a bright squeezed state with $\tanh|\zeta| = 0{\cdot}609$ and $|\alpha| = 1{\cdot}48$. The minimum level of squeezing required is for a 62·6%:37·4% beam splitter for which a bright squeezed state with $\tanh|\zeta| = 0{\cdot}545$ and $|\alpha| = 1{\cdot}2$ may be used. Squeezed states with parameters in this range have been realized in recent experiments [6]. If superpositions of the zero and one-photon states can be prepared then it is possible to produce reciprocal-binomial states (or any other superposition of a finite number of photon number states) using a number of beam splitters and photodetectors. The required state will be produced on the condition that there are no photocounts registered in any of the detectors.

The projection synthesis measurement method and the conditional state preparation method described in this paper have a more general applicability than just the measurement of phase and the preparation of reciprocal-binomial states described above. In principle, projection synthesis can be employed to compare the overlap between the state of interest and any two other states constructed from a finite number of photon number states. We will explore this rich potential more fully elsewhere.

Acknowledgments

We are grateful to Gerd Breitenbach for helpful comments. LSP thanks the UK Engineering and Physical Sciences Research Council for the award of a research studentship. SMB thanks the Royal Society for travel support enabling him to visit Brisbane where part of this work was carried out.

References

[1] Vogel, K., and Risken, H., 1989, *Phys. Rev.* A, **40**, R2847.
[2] D'Ariano, G. M., Macchiavello, C., Raymer, M. G., and Paris, M. G. A., 1994, *Phys. Rev.* A, **50,** 4298.
[3] Walker, N. G., 1987, *J. mod. Optics*, **34,** 15; Steuernagel, O., and Vaccaro, J. A., 1995, *Phys. Rev. Lett.*, **75,** 3201; Vaccaro, J. A., 1995, *Phys. Rev.* A, **52,** 3474; Banaszek, K., and Wódkiewicz, K., 1996, *Phys. Rev. Lett.*, **76,** 4344.
[4] Smithey, D. T., Beck, M., Cooper, J., and Raymer, M. G., 1993, *Phys. Rev.* A, **48,** 3159; Smithey, D. T., Beck, M., Raymer, M. G., and Faridani, A., 1993, *Phys. Rev. Lett.*, **70,** 1244.
[5] Beck, M., Smithey, D. T., Cooper, J., and Raymer, M. G., 1993, *Opt. Lett.*, **18,** 1259; Smithey, D. T., Beck, M., Cooper, J., Raymer, M. G., and Faridani, A., 1993, *Phys. Scripta*, **T48,** 35.
[6] Schiller, S., Breitenbach, S., Pereira, S. F., Müller, T., and Mlynek, J., 1996, *Phys. Rev. Lett.*, **77,** 2933.
[7] Brune, M., Hagley, E., Dreyer, J., Maître, X., Maali, A., Wunderlich, C., Raimond, J. M., and Haroche, S., 1996, *Phys. Rev. Lett.*, **77,** 4887.
[8] Pegg, D. T., and Barnett, S. M., 1988, *Europhys. Lett.*, **6,** 483; Barnett, S. M., and Pegg, D. T., 1989, *J. mod. Optics*, **36,** 7; Pegg, D. T., and Barnett, S. M., 1989, *Phys. Rev.* A, **39,** 1665.
[9] Helstrom, C. W., 1976, *Quantum Detection and Estimation Theory* (New York: Academic Press), pp. 53–57; Shapiro, J. H., and Shepard, S. R., 1991, *Phys. Rev.* A, **43,** 3795.
[10] Bialynicka-Birula, Z., and Bialynicki-Birula, I., 1994, *J. mod. Optics*, **41,** 2203; Vaccaro, J. A., and Barnett, S. M., 1995, *J. mod. Optics*, **42,** 2165.
[11] Pegg, D. T., Vaccaro, J. A., and Barnett, S. M., 1994, *Quantum Optics VI* (Berlin: Springer-Verlag), p. 153.
[12] Torgerson, J. R., and Mandel, L., 1996, *Phys. Rev. Lett.*, **76,** 3939.
[13] Barnett, S. M., and Pegg, D. T., 1996, *Phys. Rev. Lett.*, **76,** 4148.
[14] Dirac, P. A. M., 1927, *Proc. R. Soc. London A*, **114,** 243.
[15] London, F., 1926, *Z. Phys.*, **37,** 915.
[16] Barnett, S. M., and Pegg, D. T., 1986, *J. Phys.* A, **19,** 3849.
[17] Noh, J. W., Fougères, A., and Mandel, L., 1991, *Phys. Rev. Lett.*, **67,** 1426; Noh, J. W., Fougères, A., and Mandel, L., 1992, *Phys. Rev.* A, **45,** 424.
[18] Vogel, W., and Schleich, W., 1991, *Phys. Rev.* A, **44,** 7642.
[19] Raymer, M. G., Cooper, J., and Beck, M., 1993, *Phys. Rev.* A, **48,** 4617.
[20] Bužek, V., and Hillery, M., 1996, *J. mod. Optics*, **43,** 1633.
[21] Pegg, D. T., and Barnett, S. M., 1997, *J. mod. Optics*, **44,** 225.
[22] Bialynicki-Birula, I., and Van, C. L., 1980, *Acta Phys. Polonica* A, **57,** 599.

[23] Pegg, D. T., 1997, *Phys. Scripta*, **T70,** 101.
[24] Milburn, G. J., private communication.
[25] Yeoman, G., and Barnett, S. M., 1993, *J. mod. Optics*, **40,** 1497.
[26] Stoler, D., Saleh, B. E. A., and Teich, M. C., 1985, *Opt. Acta*, **32,** 345.
[27] Vourdas, A., 1996, *Phys. Rev.* A, **54,** 4544.
[28] Vogel, K., Akulin, V. M., and Schleich, W. P., 1993, *Phys. Rev. Lett.*, **71,** 1816; Parkins, A. S., Marte, P., Zoller, P., and Kimble, H. J., 1993, *Phys. Rev. Lett.*, **71,** 3095; Parkins, A. S., Marte, P., Zoller, P., Carnal, O., and Kimble, H. J., 1993, *Phys. Rev.* A, **51,** 1578.
[29] Loudon, R., and Knight, P. L., 1987, *J. mod. Optics*, **34,** 709.
[30] Abramowitz, M., and Stegun, I. A., 1970, *Handbook of Mathematical Functions* (New York: Dover).
[31] Ban, M., 1996, *J. mod. Optics*, **43,** 1281 and references therein.

Measuring the phase variance of light

KENNETH L. PREGNELL and DAVID T. PEGG
School of Science, Griffith University, Brisbane 4111, Australia

(*Received 24 November 2000; revision received 3 January 2001*)

Abstract. The results of experiments designed to measure the operational phase cosine and sine variances of weak states of light disagree with the variances predicted by canonical phase formalisms. As these variances are fundamental manifestations of the quantum nature of phase, it is important to be able to measure the canonical variances also. A recent suggestion to do so, based on the use of a two-component probe, involves the difficult preparation of exotic states of light which have not yet been produced. In this paper we show how the variances can be measured with simple coherent state inputs. The retrodictive formalism of quantum mechanics provides useful insight into the physics involved.

1. Introduction

While there are some differences in various theoretical quantum descriptions of the phase of light, a common feature is that there should be some uncertainty relation between photon number and phase. Thus the quantum nature of phase should be manifest as an uncertainty, that is as a non-zero variance in the phase probability distribution. This uncertainty should be most pronounced for states of light with very small photon number variances as must pertain, for example, to states that do not differ very much from the vacuum. By contrast, strong coherent states of light, which approximate classical states, should have sharply defined values of phase. For this reason experimental investigations into the quantum nature of the phase of light [1, 2] have paid particular attention to finding the width of the phase distribution of states of light with low mean photon number. As the variance of the phase angle φ itself depends critically on the 2π window assigned to its range of values, such experiments are usually directed at measuring the phase cosine and sine variances $(\Delta \cos \varphi)^2$ and $(\Delta \sin \varphi)^2$. For small phase variances one might expect from expanding the classical series that $(\Delta \cos \varphi)^2 + (\Delta \sin \varphi)^2 \approx (\Delta\varphi)^2$. Simple balanced homodyne techniques, sometimes referred to as phase measurements, can be used to obtain a distribution of a suitably defined operational, or measured, sine and cosine of phase [3]. For states with a small enough phase variance, this distribution can give a very good approximation to the canonical phase distribution [4] where the canonical phase is defined as the complement of the photon number operator and can be described mathematically by the formalism in [5]. For weak fields in the quantum regime, however, which by necessity have broader phase distributions, significant divergences occur between the canonical and operational phase distributions. This is also true for the operational phase defined and measured by Noh *et al.* [2] whose distribution width for coherent states has a maximum divergence from that of the canonical distribution for mean photon numbers around unity. More recently other techniques have also been suggested that focus on measuring directly the phase properties of weak fields. These include projection synthesis [6] for measuring the canonical phase distribution and a two-component probe technique [7] for measuring the canonical phase cosine or sine variance. These, however, rely on engineering specifically tailored probe states that, although possible in principle, will be very difficult in practice and have so far not been produced. Even producing the two-component probe of [7] by truncating a coherent state with a quantum scissors device [8] is by no means trivial. Thus on one hand there are techniques that use easily prepared states but which do not measure the canonical phase variances and on the other there are techniques that measure canonical phase variances but rely on exotic quantum states.

In this paper we examine the possibility of measuring the canonical phase cosine and sine variances of optical fields, with a particular interest in weak fields, by using input states which are easily produced in the laboratory, that is, coherent states. We find that, even though the two-component probe states needed for the technique of [7] are effectively not available at present, it is not difficult to use a *retrodictive* two-component probe state for our purposes. This is a state which, in the less usual retrodictive formalism of quantum mechanics [9], is assigned on the basis of the output of a measurement and which evolves backwards in time from the measurement event.

2. Mean of the phase cosine

By using the hermitian phase operator formalism of [5], in which the operators $\cos\hat{\varphi}$ and $\sin\hat{\varphi}$ commute, we can show that, for a physical state $|c\rangle = \sum c_n|n\rangle$, where $|n\rangle$ are photon number states,

$$\langle \cos\hat{\varphi}\rangle = \frac{1}{2}\left(\sum_n c_n c^*_{n+1} + c.c.\right), \tag{1}$$

$$\langle \sin\hat{\varphi}\rangle = \frac{\mathrm{i}}{2}\left(\sum_n c_n c^*_{n+1} - c.c.\right), \tag{2}$$

and

$$\langle \cos^2\hat{\varphi}\rangle = \tfrac{1}{2}[1 + \langle \cos(2\hat{\varphi})\rangle], \tag{3}$$

where

$$\langle \cos(2\hat{\varphi})\rangle = \frac{1}{2}\left(\sum_n c_n c^*_{n+2} + \mathrm{c.c.}\right) \tag{4}$$

The formalism of [5] involves first finding the expectation values in a finite dimensional Hilbert space and then finding the limit of these as the dimensionality of the space is allowed to tend to infinity. A physical state is one with finite energy moments. An alternative method is to use the phase operators acting on Vaccaro's space [10]. For a physical mixed state, we can find the mean values from the trace

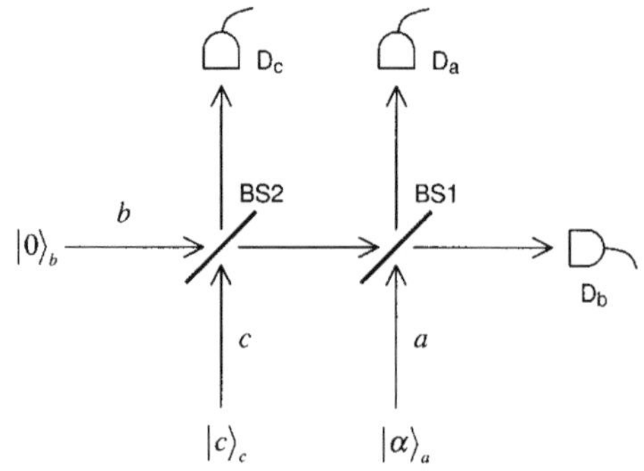

Figure 1. Apparatus for measuring the phase cosine and sine variance. The horizontal mode is b and the vertical modes are c and a. The state $|c\rangle_c$ or $\hat{\rho}^c$ to be measured is in the input mode c of beam splitter BS2 and a vacuum state $|0\rangle_b$ is in input mode b of BS2. A reference field in a coherent state $|\alpha\rangle_a$ is in the input mode a of 50/50 symmetric beam splitter BS1. Photon detectors D_b, D_a and D_c are in the output modes b, a and c.

of the product of the density operator $\hat{\rho}^c$ and the relevant function of the phase operator. We obtain

$$\langle \cos \hat{\varphi} \rangle = \frac{1}{2}\left(\sum_n \rho^c_{n,n+1} + \text{c.c.}\right), \tag{5}$$

$$\langle \sin \hat{\varphi} \rangle = \frac{\mathrm{i}}{2}\left(\sum_n \rho^c_{n,n+1} - \text{c.c.}\right), \tag{6}$$

$$\langle \cos (2\hat{\varphi}) \rangle = \frac{1}{2}\left(\sum_n \rho^c_{n,n+2} + \text{c.c.}\right). \tag{7}$$

The proposed measurement technique uses the beam-splitter arrangement shown in figure 1. A controllable reference field in a coherent state $|\alpha\rangle_a = \sum a_n |n\rangle_a$is in the input mode a of 50/50 symmetric beam splitter BS1. The state $|c\rangle_c$ or $\hat{\rho}^c$ to be measured is in the input mode c of beam splitter BS2 and a vacuum state $|0\rangle_b$ is in input mode b of BS2. For now we do not specify the transmission and reflection coefficients of BS2. Photon detectors D_b, D_a and D_c are in the output mode b and output mode a of BS1 and in the output mode c of BS2. We shall assume for now that these detectors can count photons with perfect efficiency, no dark counts and negligible dead time. We return to this assumption later.

In the usual predictive formalism of quantum mechanics a density operator $\hat{\rho}$ is assigned to describe the (predictive) state of the field between preparation and measurement based on the outcome of the preparation apparatus. This state evolves forward in time until the field interacts with the measurement apparatus. The measurement apparatus is described in general by a probability operator measure (POM) with elements corresponding to possible outcomes of the measurement [11]. The probability of a measurement outcome is given by the projection of the evolved predictive state onto the associated POM element, that is, by the trace of the product of the density operator and the POM element. To avoid unnecessary complications, we assume the predictive states we assign to the input fields of the device in figure 1 describe the fields at their entry to the beam splitters. The free evolution in the intermediate mode b between the beam splitters only changes the phase of the field in this mode so, by choosing the distance between beam splitters to be an integer number of wavelengths, we can ignore this evolution. In practice, even if this is not the case such a phase shift can be compensated by adjusting the phase of $|\alpha\rangle_a$. Finally we can ignore the free evolution in all the output modes, as these do not affect the photocount probabilities. We denote the (forward time) unitary operator for the actions of beam splitters BS2 and BS1 as respectively R_2, which acts on states in modes c and b, and R_1 which acts on states in modes a and b.

The initial combined density operator for the three input fields is

$$\hat{\rho} = \hat{\rho}^c \otimes |0\rangle_{bb}\langle 0| \otimes |\alpha\rangle_{aa}\langle\alpha| \tag{8}$$

and the POM element for the detection of N, n_b and n_a photons in output modes c, b and a respectively is

$$\begin{aligned}\hat{\Pi}(N, n_b, n_a) &= \hat{\Pi}_c \otimes \hat{\Pi}_b \otimes \hat{\Pi}_a \\ &= |N\rangle_{c\,c}\langle N| \otimes |n_b\rangle_{b\,b}\langle n_b| \otimes |n_a\rangle_{a\,a}\langle n_a|.\end{aligned} \tag{9}$$

The probability for the detection of N, n_b and n_a photons in output modes c, b and a respectively is then

$$P(N, n_b, n_a) = \mathrm{Tr}_{abc}\,[R_1 R_2 \hat{\rho} R_2^\dagger R_1^\dagger \hat{\Pi}(N, n_b, n_a)]. \tag{10}$$

Substituting from (8) and (9) and using the cyclic property of the trace we can rewrite this as

$$P(N, n_b, n_a) = \mathrm{Tr}_c\,[\hat{\rho}^c \hat{\Pi}^{\mathrm{retr}}(N, n_b, n_a)]. \tag{11}$$

Here $\hat{\Pi}^{\mathrm{retr}}(N, n_b, n_a)$ is the retrodictive POM element:

$$\begin{aligned}\hat{\Pi}^{\mathrm{retr}}(N, n_b, n_a) &= {}_b\langle 0|R_2^\dagger \hat{\Pi}_c \otimes {}_a\langle\alpha|R_1^\dagger \hat{\Pi}_b \otimes \hat{\Pi}_a R_1 |\alpha\rangle_a R_2 |0\rangle_b \\ &= |r\rangle_{cc}\langle r|\end{aligned}$$

where, from (9),

$$|r\rangle_c = {}_b\langle 0|R_2^\dagger |N\rangle_{ca}\langle\alpha|R_1^\dagger |n_b\rangle_b |n_a\rangle_a. \tag{13}$$

The state $|r\rangle_c$ in (13) can be interpreted as an unnormalized retrodictive state of the field in input mode c associated with the measurement outcome of N, n_b and n_a photons being detected. In the usual predictive formalism the state of a system between preparation and measurement is assigned on the basis of the preparation outcome. In the retrodictive formalism this state is assigned on the basis of the outcome of the measurement, specifically $\hat{\rho}^{\mathrm{retr}} = \hat{\Pi}_j / \mathrm{Tr}\,\hat{\Pi}_j$, where $\hat{\Pi}_j$ is the POM element for the particular outcome j [12], and then evolves backwards in time from the measurement event to the preparation event. In our case the retrodictive fields associated with the measurement outcomes of the photon detectors are simple number state projectors which are already normalized. We can see from (13) that the retrodictive fields $|n_b\rangle_b$ and $|n_a\rangle_a$ associated with the measurements in the output mode b and output mode a evolve backwards in time and are entangled by means of beam splitter BS1. This entangled state is projected onto $|\alpha\rangle_a$ to yield an unnormalized retrodictive probe state

$$|q\rangle_b = {}_a\langle\alpha|R_1^\dagger|n_b\rangle_b|n_a\rangle_a \tag{14}$$

in the intermediate mode b, that is, between the two beam splitters. As we shall see later this is a retrodictive two-component state which performs a similar function to the predictive two-component probe of [7]. The state $|q\rangle_b$ is entangled by beam splitter BS2 with the retrodictive state from the measurement outcome of the detector D_c. This state in turn is projected onto the vacuum in input mode b to give the retrodictive state $|r\rangle_c$ for projection onto the state to be measured. We remark here that if the state $|c\rangle_c$ is a coherent state then in the predictive picture there is no entanglement at all because all input states are coherent. The entanglement mentioned above occurs in retrodiction. In addition to giving new insight, working in terms of the retrodictive probe state (13) has practical calculational advantages in our case in that the fields that evolve backwards originate from single photon number states associated with the measurement outcomes.

The simplest possible retrodictive probe is associated with $n_a = n_b = 0$. In this case we find that the retrodictive probe state $|q\rangle_b$ is just the vacuum and so $|r\rangle_c$ is just proportional to $|N\rangle_c$. Thus only the diagonal matrix elements of $\hat{\rho}^c$ are obtainable from the measured probabilities. The next simplest retrodictive probes are associated with the measurement result $n_a = 0$, $n_b = 1$ and $n_a = 1$, $n_b = 0$. For a symmetric beam splitter with transmission coefficient $\cos\theta$ we have the relations (for a general description of the beam splitter see, for example, [13])

$$R^\dagger a^\dagger R = a^\dagger \cos\theta - \mathrm{i}b^\dagger \sin\theta, \tag{15}$$

$$R^\dagger b^\dagger R = b^\dagger \cos\theta - \mathrm{i}a^\dagger \sin\theta, \tag{16}$$

$$R^\dagger|0\rangle_b|0\rangle_a = |0\rangle_b|0\rangle_a, \tag{17}$$

with $a^\dagger$, $b^\dagger$ and $c^\dagger$ the creation operators for modes a, b and c respectively. For a 50/50 beam splitter $\theta = \pi/4$. Using this value for BS1 we easily find, by writing $|1\rangle_b = b^\dagger|0\rangle_b$, the unnormalized retrodictive probe state for $n_a = 0$, $n_b = 1$ to be

$$|q\rangle_b = (\alpha_o^*|1\rangle_b - \mathrm{i}\alpha_1^*|0\rangle_b)/2^{1/2}. \tag{18}$$

This two-component probe can be used to obtain information about the off-diagonal matrix elements of $\hat{\rho}^c$ as follows. Substituting into (13) allows us to find $|r\rangle_c$ by writing $|N\rangle_c = c^{\dagger N}|0\rangle_c/(N!)^{1/2}$ and using (15) and (16) with $c^\dagger$ in place of $a^\dagger$. We find eventually

$$|r\rangle_c = -\mathrm{i}(\cos^N\theta)[\alpha_1^*|N\rangle_c + (\sin\theta)(N+1)^{1/2}\alpha_0^*|N+1\rangle_c]/2^{1/2}, \tag{19}$$

which allows us to find the retrodictive POM element from (12) and hence, from (11), the measurable probability

$$P(N,1,0) = (\tfrac{1}{2}\cos^{2N}\theta)[|\alpha_1|^2\rho^c_{N,N} + (\sin^2\theta)(N+1)|\alpha_0|^2\rho^c_{N+1,N+1} + (\alpha_0\alpha_1^*(N+1)^{1/2}\rho^c_{N+1,N}\sin\theta + \mathrm{c.c.})]. \tag{20}$$

For now we do not specify the value of θ for BS2.

Similarly we find the probability for $n_a = 1$, $n_b = 0$ and N counts in detector D_c to be

$$P(N, 0, 1) = (\tfrac{1}{2}\cos^{2N}\theta)[|a_1|^2 \rho^c_{N,N} + (\sin^2\theta)(N+1)|a_0|^2 \rho^c_{N+1,N+1} - (a_0 a_1^*(N+1)^{1/2} \rho^c_{N+1,N} \sin\theta + \text{c.c.})]. \tag{21}$$

For the first experiment we choose the phase of the coherent reference state $|\alpha\rangle_a$ so that a_n are all real and positive. From (20), (21) and (5) an expression for the mean of the phase cosine in terms of the measurable probabilities can then be written as

$$\langle\cos\hat{\varphi}\rangle = \sum_N \frac{P_1(N, 1, 0) - P_1(N, 0, 1)}{2(\cos^{2N}\theta)|a_0 a_1|(N+1)^{1/2}\sin\theta}, \tag{22}$$

where the subscript on the probability refers to the first experiment.

3. Variance of the phase cosine and sine

The next simplest possible retrodictive probe originates from the measurement event $n_a = n_b = 1$. Again for the 50/50 beam splitter BS1 we have $\theta = \pi/4$. Writing $|1\rangle_b = \hat{b}^\dagger|0\rangle_b$ and $|1\rangle_a = a^\dagger|0\rangle_a$ we obtain from (15), (16) and (17), the two-component retrodictive probe state

$$|q\rangle_b = -\mathrm{i}(a_o^*|2\rangle_b + a_2^*|0\rangle_b)/2^{1/2} \tag{23}$$

leading to

$$|r\rangle_c = -\mathrm{i}(\cos^N\theta)\left[a_2^*|N\rangle_c - a_0^*(\sin^2\theta)[(N+1)(N+2)/2]^{1/2}|N+2\rangle_c\right]/2^{1/2}. \tag{24}$$

From (11) and (12), this gives the probability

$$P(N, 1, 1) = (\tfrac{1}{2}\cos^{2N}\theta)[|a_2|^2 \rho^c_{N,N} + \tfrac{1}{2}(\sin^4\theta)(N+1)(N+2)|a_0|^2 \rho^c_{N+2,N+2} - (a_0 a_2^*[(N+1)(N+2)/2]^{1/2} \rho^c_{N+2,N} \sin^2\theta + \text{c.c.})]. \tag{25}$$

To allow the experiment with this probe to be conducted simultaneously with the experiment to find $\langle\cos\hat{\varphi}\rangle$, we take a_0 and a_2 to be real and positive and find $P_1(N, 1, 1)$ by simply replacing $a_0 a_2^*$ in (25) by $|a_0 a_2|$.

After measuring the probabilities $P_1(N, 1, 0)$, $P_1(N, 0, 1)$ and $P_1(N, 1, 1)$ the experiment is repeated with a phase shift of $\pi/2$ in the reference state $|\alpha\rangle_a$, which has the effect of changing a_n to $a_n \exp(\mathrm{i}n\pi/2)$. Thus now $a_0 = |a_0|$, $a_1 = i|a_1|$, and $a_2 = -|a_2|$ in (20), (21) and (25), yielding $P_2(N, 1, 0)$, $P_2(N, 0, 1)$ and $P_2(N, 1, 1)$. From (6) we can then obtain the mean phase sine from the measured results as

$$\langle\sin\hat{\varphi}\rangle = \sum_N \frac{P_2(N, 1, 0) - P_2(N, 0, 1)}{2(\cos^{2N}\theta)|a_0 a_1|(N+1)^{1/2}\sin\theta}. \tag{26}$$

We also find from (7) that

$$\langle\cos(2\hat{\varphi})\rangle = \sum_n \frac{P_2(N, 1, 1) - P_1(N, 1, 1)}{2|a_0 a_2|[(N+1)(N+2)/2]^{1/2}\cos^{2N}\theta\sin^2\theta}. \tag{27}$$

After these values are obtained from the measured probabilities, the mean square phase cosine can be found from (3) and finally the phase cosine variance calculated as $\langle\cos^2\hat{\varphi}\rangle - \langle\cos\hat{\varphi}\rangle^2$. Further, we can also write from the phase formalism of [5]

$$\langle\sin^2\hat{\varphi}\rangle = \tfrac{1}{2}[1 - \langle\cos(2\hat{\varphi})\rangle], \tag{28}$$

which allows us also to find the phase sine variance from the measured probabilities.

4. Discussion

We have already assigned a value to the phase of the reference state $|\alpha\rangle_a$ but still have freedom to choose its mean photon number $|\alpha|^2$. To avoid quotients of very small numbers, it is worth maximizing the denominator, and hence the numerator, in (22) and (26). Thus we should choose a reference state to maximize $|a_0 a_1|$ and for (25) we should maximize $|a_0 a_2|$. The former and latter are maximized for mean photon numbers of 0.5 and 1 respectively. The experiment could in principle be run for both these values but in practice it would be simpler to just use a compromise value between 0.5 and 1.

We have yet to choose the reflection to transmission ratio of the beam splitter BS2. Again it is useful to choose a ratio which maximizes the denominators of the terms in (22). The optimum value of $\sin\theta$ for each term is $(1+2N)^{-1/2}$. For (27) the optimum value of $\sin\theta$ for each term is $(1+N)^{-1/2}$. If necessary the experiment could be repeated for different values of N but, given we are mainly interested in weak fields, the spread in values of N should not be huge. Thus a compromise value of around $\sin\theta \approx (1+\langle n\rangle)^{-1/2}$ should be adequate for determining both (22) and (27), where $\langle n\rangle$ is the mean photon number of the field to be measured. Thus for fields with a mean photon number around unity a 50/50 beam splitter would be quite suitable. For stronger fields an increase in the transmission of BS2 would be desirable.

Once we have measured $\langle\cos\hat{\varphi}\rangle$ and $\langle\sin\hat{\varphi}\rangle$ we can easily find $\langle\exp(\mathrm{i}\hat{\varphi})\rangle$ which, for a pure state $|c\rangle = \sum c_n|n\rangle$, is just $c_n^* c_{n+1}$. Then, if the state to be measured is a pure state with no gaps in the photon number distribution, we can use these results to reconstruct the complete state, as shown in [7]. If there are gaps then, provided these only involve the vanishing of a single number state coefficient, the complete state could still be reconstructed using a knowledge of $\langle\exp(\mathrm{i}2\hat{\varphi})\rangle$ to bridge the gaps [7]. This would require an additional measurement of $\langle\sin(2\hat{\varphi})\rangle$ which we could obtain by repeating the measurement of $\langle\cos(2\hat{\varphi})\rangle$ with a phase shift of $\pi/4$ applied to the measured field.

The detection probabilities used above are those that would be obtained by perfect detectors. Practical photodetectors suffer from non-unit efficiencies that reduce the number of counts, the presence of dark counts and a non-zero dead time following a count during which no other counts are registered. Using weak fields, in which we are particularly interested, and sufficiently long gating times reduces the effect of the dead time. Other techniques can also be used for this purpose [14]. For known efficiency and dark count rate the ideal photocount probabilities can be found from the statistics of the experiment. The probability that ideal detectors would have detected N, n_b and n_a photons in output modes c, b and a respectively is

$$P(N, n_b, n_a) = \sum_{M, m_b, m_c} P(M, m_b, m_a)p(N|M)p(n_b|m_b)p(n_a|m_a), \tag{29}$$

where M, m_b and m_a are the actual counts in the detectors and $p(N|M)$, for example, is the probability that an ideal detector would have counted N photons if the actual detector counted M photons. An expression for $p(N|M)$ for non-unit efficiency can be derived using Bernouilli transforms as in, for example, [15, 16,

17]. Allowance for dark counts can also be incorporated using the techniques, for example, in [8, 17].

5. Conclusion

The measurement procedure described here can be used for the complete reconstruction of reproducible pure states of light which have no gaps in their number state distributions. For a mixed state the diagonal, nearest off-diagonal and the next to nearest off-diagonal matrix elements can be measured. These determine the mean sine and cosine of the phase as well as the mean square of the sine and cosine, from which the variance of the phase sine and cosine can be found.

The above quantities can also be obtained by means of a two-component probe field technique suggested in [7]. There are very important differences however. In [7], where states are assigned to the probe fields according to the usual predictive quantum formalism, the required probe states are, as acknowledged in that paper, very difficult to prepare. Indeed, to our knowledge, such two-component states have yet to be produced. The preparation method suggested for [7] was optical truncation using quantum scissors [8], so the measurement would require three beam splitters in all, with separate experiments being run with each different probe. More seriously, the preparation of the probe in a one-photon and vacuum superposition in [7] requires the injection of a single photon state into one input port and the probe in a two-photon and vacuum superposition requires the simultaneous injection of a single photon into two input ports.

In contrast to the technique suggested in [7], the method proposed here has real practical advantages. Only two beam splitters are used and, apart from the state to be measured, the only states injected into other input ports are vacuum and coherent states. These are not only considerably easier to prepare, their coherence lengths can allow longer gating times, reducing the effect of dead times. The retrodictive one-photon states needed to construct the retrodictive probe originate from photon detection events and are thus more readily available than their predictive counterparts which originate from preparation events. Further, because the retrodictive probe states are produced by the detection events, all three probe states, including the retrodictive vacuum state used for measuring the diagonal density matrix elements, are produced in the one experiment. There is no need to run separate experiments for different probe states.

In addition to the improved experimental practicality of the proposal in this paper, we feel that it also has importance as an application of the little used but completely rigorous retrodictive formalism [9]. Until recently this formalism has been treated as having a philosophic interest only but is beginning to find use in quantum communication problems [18] where the receiver and measurer of a quantum state must retrodict what state was prepared and sent. The formalism was also helpful in examining, amongst other problems, the quantum scissors device for truncation optical states [8]. For the present paper we deliberately sought two-component retrodictive probes mirroring the probes of [7]. To emphasize that the retrodictive formalism yields the same measurable probabilities as the predictive formalism we used the standard predictive picture in (10) to begin our mathematical analysis. It is interesting, however, to compare our approach with the retrodictive analysis of the quantum scissors [8]. In the retrodictive picture, the state to be truncated is in one input mode of a beam splitter with detectors in the two output modes. When one of these detects one photon and the other detects zero photons, the retrodictive state in the other input mode is a superposition of the vacuum and one photon states, so the actual cutting out of the higher photon state components is done at this beam splitter. The other beam splitter of the quantum scissors creates a predictive entangled state. Projection of the retrodictive state onto this state effectively converts the retrodictive state into a predictive state with the coefficients of the vacuum and one photon components interchanged. The beam splitter BS1 of the present paper can be regarded as the part of the quantum scissors that creates the retrodictive two-component state. As we wish to use this retrodictive probe directly, there is no need to employ another beam splitter to convert it to a predictive probe. This also dispenses with the necessity to produce and inject single-photon fields.

Finally, the insight provided by the retrodictive formalism of quantum mechanics has enabled us to propose a relatively simple experiment to measure some of the canonical phase properties of light. There is now less need to define separate operational phase properties based on easily performed experiments.

Acknowledgments

DTP thanks the Australian Research Council for financial support.

References

[1] Gerhardt, H., Welling, H., and Frolich, D., 1973, *Appl. Phys.*, **2,** 91; Gerhardt, H., Buchler, U., and Litn, G., 1974, *Phys. Lett. A*, **49,** 119.

[2] Noh, J. W., Fougères, A., and Mandel, L., 1991, *Phys. Rev. Lett.*, **67,** 1426; Noh, J. W., Fougères, A., and Mandel, L., 1992, *Phys. Rev. A*, **45,** 424; Noh, J. W., Fougères, A. and Mandel, L., 1992, *Phys. Rev. A*, **46,** 2840; Noh, J. W., Fougères, A. and Mandel, L., 1993, *Phys. Rev. Lett.*, **71,** 2579.

[3] Barnett, S. M., and Pegg, D. T., 1986, *J. Phys. A*, **19,** 3849.

[4] Vaccaro, J. A., and Pegg, D. T., 1994, *Optics Commun.*, **105,** 335.

[5] Pegg, D. T., and Barnett, S. M., 1988, *Europhys. Lett.*, **6,** 483; Barnett, S. M., and Pegg, D. T., 1989, *J. mod. Optics*, **36,** 7; Pegg, D. T. and Barnett, S. M., 1989, *Phys. Rev. A*, **39,** 1665.

[6] Barnett, S. M., and Pegg, D. T., 1996, *Phys. Rev. Lett.*, **76,** 4148; Pegg, D. T., Barnett, S. M., and Phillips, L. S., 1997, *J. mod. Optics*, **44,** 2135.

[7] Pegg, D. T., and Barnett S. M., 1999, *J. mod. Optics*, **46,** 981.

[8] Pegg, D. T., Phillips, L. S., and Barnett, S. M., 1998, *Phys. Rev. Lett.*, **81,** 1604; Barnett, S. M., and Pegg, D. T., 1999, *Phys. Rev. A*, **60,** 4965.

[9] Aharonov, Y., Bergman, P. G., and Lebowitz, J. L., 1964, *Phys. Rev.*, **134,** B1410; Penfield, R. H., 1966, *Am. J. Phys.*, **34,** 422; Aharonov, Y., and Albert, D. Z., 1984, *Phys. Rev. D*, **29,** 223; Aharonov, Y., and Albert, D. Z., 1984, *Phys. Rev. D*, **29,** 228; Aharonov, Y., and Vaidman, L., 1991, *J. Phys. A: Math. Gen.*, **24,** 2315; Pegg, D. T. and Barnett, S. M., 1999, *Quantum semiclass. Opt.*, **1,** 442; Barnett, S. M., Pegg, D. T., Jeffers, J., and Jedrkiewicz, O., 2000, *J. Phys. B: At. molec. opt. Phys.*, **33,** 3047.

[10] Vaccaro, J. A., 1995, *Phys. Rev. A*, **51,** 3309.

[11] Helstrom, C. W., 1976, *Quantum Detection and Estimation Theory* (New York: Academic Press).

[12] Barnett, S. M., Pegg, D. T., and Jeffers, J., 2000, *J. mod. Optics*, **47,** 1779.

[13] Barnett, S. M., and Radmore, P. M., 1997, *Methods in Theoretical Quantum Optics* (Oxford: Oxford University Press), p. 84.

[14] Paul, H., Torma, P., Kiss, T., and Jex, I., 1997, *Phys. Rev. Lett.*, **76,** 2464.

[15] Lee, C. T., 1993, *Phys. Rev. A*, **48,** 2285.

[16] Pegg, D. T., and Barnett S. M., 1999, *J. mod. Optics,* **46,** 1657.
[17] Barnett, S. M., Phillips, L. S., and Pegg, D. T., 1998 *Optics Commun.,* **158,** 45.
[18] Barnett, S. M., Pegg, D. T., Jeffers, J., Jedrkiewicz, O., and Loudon, R., 2000, *Phys. Rev. A,* **62,** 022313.

Quantum phase distribution by operator synthesis

K. L. PREGNELL and D. T. PEGG
School of Science, Griffith University, Brisbane 4111, Australia

(*Received 1 August 2001*)

Abstract. Methods proposed so far for measuring the canonical phase probability distribution of light are very difficult to implement. In particular the method of projection synthesis, which synthesizes a phase state projector, involves the use of a beam splitter and a specially engineered reference state. Here we extend this method by use of an additional beam splitter to synthesize phase sine and cosine operators. This allows the measurement of all the moments needed to construct the phase distribution. The reference state required for operator synthesis is, remarkably, just an easily prepared coherent state.

1. Introduction

The quantum concept of the phase of light has had a long and interesting history. While it is universally agreed that in the appropriate limit the correct classical behaviour should emerge from the quantum description, this condition alone does not determine a unique description of phase. As a consequence several quantum phase descriptions have been proposed. Some of these involve operational definitions based on phase sensitive measurement techniques [1] and some are purely theoretical concepts. An example of the latter, sometimes referred to as the canonical phase, defines phase as the complement of photon number with the associated phase distribution being shifted without distortion by a phase shifter [2]. We should expect to observe measurable differences between the phase probability distributions predicted from different descriptions of phase. However, as the definition of canonical phase is not based on a measurement procedure, the problem immediately arises as to how to measure its distribution. It turns out that for states with a narrow phase distribution, for example a coherent state with a mean photon number of five or greater, a simple balanced homodyne provides a good approximation to the distribution [3], but this is not applicable to states with a small mean photon number. This is unfortunate because it is in the quantum realm of weak fields, where the phase fluctuations must be large, that we would expect the most significant differences between predictions from different quantum phase descriptions to occur.

Recently a technique, known as projection synthesis, to measure the canonical phase distribution of weak fields was proposed [4]. The practical difficulty with this technique is that it requires the use of an exotic reference field which is difficult to prepare and which to date has not been available. Clausen *et al.* [5] have suggested a method involving, instead of an exotic reference field, a large number of beam splitters with a large number of coherent state input fields and photodetectors in the outputs. In this approach a successful experiment is conditional on one photon being detected at each detector, with zero photons being detected at the final detector. It turns out that the conditional probability associated with this technique is quite small so, in order to obtain accurate sampling for one point on the phase distribution curve, a very large number of experiments must be performed. The problem is exacerbated by the requirement, in common with projection synthesis, for the distribution curve to be generated by a large number of such points obtained by repeating all the experiments as the phase of the field is changed incrementally.

In light of the above remarks, in this paper we address the problem of devising a method for measuring the canonical phase distribution which involves only a small number of beam splitters and which also involves the use of only easily produced coherent reference states. In a preceding paper [6] we found that it is possible to measure the mean and variance of the cosine and sine of the canonical phase with just two beam splitters, one vacuum field input and one coherent state field input. We find here that we can extend this procedure, remarkably with the same apparatus, to measure all the sine and cosine moments needed to construct the canonical phase distribution for weak fields. In the projection synthesis method, the exotic reference field and beam splitter are used to synthesize a phase state projection operator. The trace of the product of this operator and the density operator of the field to be measured gives the appropriate phase probability. In the present paper we show how the double beam-splitter system can be used to synthesize sine and cosine phase operators. The trace of the product of these operators and the density operator of the field to be measured gives the required sine or cosine moments.

2. Phase distribution

Utilizing the properties of the Dirac delta function, the phase probability distribution can be written in general as

$$\begin{aligned} P(\theta) &= \int_{\theta_0}^{\theta_0+2\pi} \delta(\theta-\theta')P(\theta')\mathrm{d}\theta' \\ &= \frac{1}{2\pi}\int_{\theta_0}^{\theta_0+2\pi}\left(1+2\sum_{\lambda=1}^{\infty}\cos\left[\lambda(\theta-\theta')\right]\right)P(\theta')\,\mathrm{d}\theta'. \end{aligned} \tag{1}$$

This can be re-expressed in terms of the sine and cosine phase moments as

$$P(\theta) = \frac{1}{2\pi}+\frac{1}{\pi}\sum_{\lambda=1}^{\infty}[\langle\cos(\lambda\theta)\rangle\cos(\lambda\theta)+\langle\sin(\lambda\theta)\rangle\sin(\lambda\theta)], \tag{2}$$

where, for example,

$$\langle\cos(\lambda\theta)\rangle = \int_{\theta_0}^{\theta_0+2\pi} P(\theta')\cos(\lambda\theta')\,\mathrm{d}\theta'.$$

Different quantum descriptions of phase will yield different values of $\langle\cos(\lambda\theta)\rangle$ and $\langle\sin(\lambda\theta)\rangle$ and therefore will yield different phase distributions. For the canonical phase distribution

$$\langle\cos(\lambda\theta)\rangle = \frac{1}{2}\sum_{p=0}^{\infty}(\hat{\rho}_{p,p+\lambda} + \hat{\rho}_{p+\lambda,p}) \tag{3}$$

and

$$\langle\sin(\lambda\theta)\rangle = \frac{\mathrm{i}}{2}\sum_{p=0}^{\infty}(\hat{\rho}_{p,p+\lambda} - \hat{\rho}_{p+\lambda,p}), \tag{4}$$

where $\rho_{n,m}$ are the elements of the density matrix.

One method for obtaining expressions (3) and (4) is the limiting procedure of [7]. Here

$$\langle\cos(\lambda\theta)\rangle = \lim_{s\to\infty}\langle\cos(\lambda\hat{\phi}_\theta)\rangle, \tag{5}$$

where $\hat{\phi}_\theta$ is the phase operator acting on a $(s+1)$-dimensional Hilbert space. For a mixed state $\hat{\rho}$,

$$\langle\cos(\lambda\hat{\phi}_\theta)\rangle = \mathrm{Tr}\,[\hat{\rho}_s\cos(\lambda\hat{\phi}_\theta)], \tag{6}$$

where $\hat{\rho}_s$ is the truncation of $\hat{\rho}$ onto the $(s+1)$-dimensional subspace. When $\hat{\rho}$ represents a physical state, we obtain expression (3). Other procedures also exist for obtaining (3) [8].

Expressions (3) and (4) are the same results as would be obtained by using

$$\langle\cos(\lambda\theta)\rangle = \mathrm{Tr}\,[\hat{\rho}\hat{C}_\lambda], \tag{7}$$

$$\langle\sin(\lambda\theta)\rangle = \mathrm{Tr}\,[\hat{\rho}\hat{S}_\lambda], \tag{8}$$

where the operators

$$\hat{C}_\lambda = \frac{1}{2}\sum_{p=0}^{\infty}|p\rangle\langle p+\lambda| + |p+\lambda\rangle\langle p|, \tag{9}$$

$$\hat{S}_\lambda = \frac{\mathrm{i}}{2}\sum_{p=0}^{\infty}|p+\lambda\rangle\langle p| - |p\rangle\langle p+\lambda|, \tag{10}$$

act on the usual infinite-dimensional Hilbert space. For any physical state we can make λ sufficiently large for $\langle\cos(\lambda\theta)\rangle$ and $\langle\sin(\lambda\theta)\rangle$ to be zero to any given finite degree of accuracy for this value, and all larger values, of λ. This provides a finite upper limit to the number of moments we need to measure in order to reconstruct the continuous phase probability distribution. For coherent states with a low mean photon number, this value is often quite small. To illustrate, we have calculated the first six moments of the cosine phase operator for three experimentally achievable, low intensity, coherent states each with the argument of the amplitude arbitrarily set to zero. These have mean photon numbers of 0.047, 0.076 and 1. The first two were used by Torgerson and Mandel [9]. From these calculations we find that, to within an accuracy of 5%, the λth cosine moment $\langle\cos\lambda\theta\rangle$ is effectively zero for mean photon numbers of 0.047, 0.076 and 1.000, for all values of $\lambda \geq 2$, 3 and 5 respectively. Thus such low intensity coherent states are particularly amenable to the phase measurement technique outlined in this paper because of this relatively small upper limit to λ.

3. Operator synthesis

To obtain a phase distribution, we need to measure probabilities. For this we require a reproducible state of light which can be subjected to the same measurement procedure a number of times. The probabilities can then be obtained from occurrence frequencies of specific events. From quantum measurement theory [8], the probability of a particular outcome k of some measurement on a system in a state $\hat{\rho}$ is given by

$$P(k) = \mathrm{Tr}\,[\hat{\rho}\hat{\Pi}(k)], \tag{11}$$

where $\hat{\Pi}(k)$ is the element of a probability operator measure (POM). The condition that $P(k) \geq 0$ for any state requires $\hat{\Pi}(k)$ to be a non-negative definite operator [10]. Also, for $P(k)$ to sum to unity for all possible outcomes k, the sum of $\hat{\Pi}(k)$ must be the unit operator.

Comparing equation (7) with equation (11) raises the question as to whether we can construct a measuring device with a POM element $\hat{\Pi}(k)$ equal, or at least proportional, to $\hat{C}_\lambda$. If so, we could then obtain $\langle \cos \lambda\theta \rangle$ immediately from the measured value of $P(k)$. It is clear, however, that because $P(k)$ takes only values from 0 to 1, whereas $\langle \cos \lambda\theta \rangle$ would be expected to range between -1 and 1, this is not possible. On the other hand, if we could realize $\hat{C}_\lambda$ by a linear combination of POM elements, then $\langle \cos \lambda\theta \rangle$ would be the same linear combination of associated probabilities, which could take both positive and negative values. It is important to note that the POM elements used in the linear combination need not belong to the same POM, rather they may represent different outcomes of *different* experiments. We attempt therefore to find measurements such that $\hat{C}_\lambda$ can be written as the linear combination

$$\hat{C}_\lambda = \sum_i C_{\lambda i}\hat{\Pi}_i, \tag{12}$$

where $\hat{\Pi}_i$ represents a particular outcome from a particular experiment. Then,

$$\begin{aligned} \langle \cos \lambda\theta \rangle &= \sum_i C_{\lambda i}\langle \hat{\Pi}_i \rangle \\ &= \sum_i C_{\lambda i} P_i, \end{aligned} \tag{13}$$

where P_i is the associated measurable probability.

Expression (12) effectively synthesizes the operator $\hat{C}_\lambda$ from various elements of different POMs.

4. Measuring the phase distribution

The proposed measurement technique uses the beam-splitter arrangement shown in figure 1. It consists of two symmetric beam splitters, labelled BS1 and BS2 with input modes (b, c) and (b, a) respectively. Positioned at the three outputs are photodetectors D_a, D_b and D_c which detect n_a, n_b and n_c photons respectively. The input fields in modes b and a are respectively in a vacuum state $|0\rangle_b$ and a coherent state $|\alpha\rangle_a$. The optimum value of the amplitude of the coherent state will be discussed later. The field to be measured, in state $\hat{\rho}_c$, is in the input mode c of BS1. Prior to the entry of BS2 is a phase shifter (PS) capable of altering the phase of the coherent state by an arbitrary amount. This has the effect of changing the

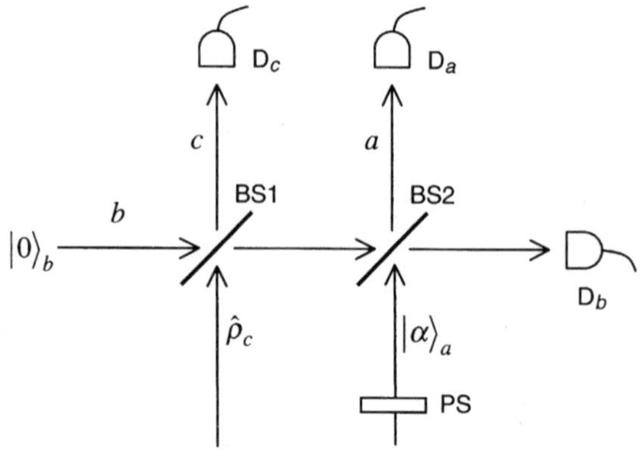

Figure 1. Apparatus for measuring the phase sine and cosine moments. The horizontal mode is b and the vertical modes are c and a. The state to be measured, $\hat{\rho}_c$, is in the input mode c of BS1 and a vacuum state $|0\rangle_b$ is in the input mode b of BS1. A reference field in a coherent state $|\alpha\rangle_a$ is in the input mode a of a 50/50 symmetric beam splitter BS2. Photon detectors D_a, D_b and D_c are in the output modes a, b and c respectively. A phase shifter PS adjusts the phase of $|\alpha\rangle_a$.

argument of the coherent state amplitude. We let the coherent state be $|\alpha\rangle_a$, with amplitude $\alpha = |\alpha| \exp(i\varphi)$, at the entry of BS2, that is the argument φ of α incorporates the phase shift.

In order to simplify the analysis of the experiment, it is convenient to disregard any free evolution of the field caused by the propagation of the fields through the apparatus. This can be justified as follows. First, if we assign the states $|\alpha\rangle_a$, $|0\rangle_b$ and $\hat{\rho}_c$ to the input fields just prior to the entry of the appropriate beam splitters, any evolution prior to this is then contained within the state at entry. In addition, the free evolution between BS1 and BS2 can be disregarded if the experiment is set up such that there is an integer number of wavelengths between the beam splitters. If experimentally this is not possible, then any overall phase shift gained by this evolution can be offset by altering the initial setting of the phase shifter. Finally, the free evolution of the fields upon exit at the beam splitters can be dismissed because we are performing a photon number measurement on the output fields which is unaffected by the phase of the fields. The only remaining unitary evolution to consider is provided by the two beam splitters.

The evolution originating from BS1 can be characterized by the unitary operator $\hat{R}_1$ which is of the form [11]

$$\hat{R}_1 = \exp[i\gamma(\hat{c}^\dagger \hat{b} + \hat{b}^\dagger \hat{c})], \tag{14}$$

where $\hat{b}$ and $\hat{c}$ are the field annihilation operators for modes b and c respectively. A similar expression holds for the BS2 operator $\hat{R}_2$. $\hat{R}_1$ commutes with the annihilation operator $\hat{a}$ for mode a and $\hat{R}_2$ commutes with $\hat{c}$. The effect of $\hat{R}_1$ on the creation operators for modes b and c and on the double vacuum state is given by

$$\hat{R}_1^\dagger \hat{b}^\dagger \hat{R}_1 = t\hat{b}^\dagger - ir\hat{c}^\dagger, \tag{15}$$

$$\hat{R}_1^\dagger \hat{c}^\dagger \hat{R}_1 = t\hat{c}^\dagger - ir\hat{b}^\dagger, \tag{16}$$

$$\hat{R}_1^\dagger |0\rangle_b |0\rangle_c = |0\rangle_b |0\rangle_c, \tag{17}$$

where t and r are the transmission and reflection coefficients of the beam splitter satisfying

$$t = \cos\gamma, \tag{18}$$

$$r = \sin\gamma, \tag{19}$$

where γ is defined in (14). We shall let BS2 be a 50/50 beam splitter, so $\gamma = \pi/4$ but maintain the generality of t and r for BS1.

Our intention is to find a method of obtaining moments from measurements of probabilities of specific detection events. Before identifying these detection events, we introduce the POM element associated with a general photon number measurement. For perfect photodetection, the POM element is the projector formed from the appropriate photon number eigenstate. Then the combined POM element for the detection of n_a, n_b and n_c photons is given by the tensor product

$$\hat{\Pi}(n_a, n_b, n_c) = |n_a\rangle_{a\,a}\langle n_a| \otimes |n_b\rangle_{b\,b}\langle n_b| \otimes |n_c\rangle_{c\,c}\langle n_c|. \tag{20}$$

By writing the POM element in this form we are assuming that the photodetectors have zero dark counts and zero dead times, as well as possessing a quantum efficiency of one. We shall return to this assumption later.

We write the initial three-mode input state as $\hat{\rho}$, where

$$\hat{\rho} = |\alpha\rangle_{a\,a}\langle\alpha| \otimes |0\rangle_{b\,b}\langle 0| \otimes \hat{\rho}_c. \tag{21}$$

This state undergoes evolution given by the action of the unitary operator $\hat{U}$, where

$$\hat{U} = \hat{R}_2\hat{R}_1. \tag{22}$$

From (11), the probability of detecting n_a, n_b and n_c photons with this input state is

$$P_\varphi(n_a, n_b, n_c) = \mathrm{Tr}_{abc}\,[\hat{U}\hat{\rho}\hat{U}^\dagger\hat{\Pi}(n_a, n_b, n_c)], \tag{23}$$

where we have shown explicitly the dependence of the probability on $\varphi = \arg\alpha$.

From the cyclic property of the trace, (23) becomes

$$P_\varphi(n_a, n_b, n_c) = \mathrm{Tr}_c\,[\hat{\rho}_c\hat{\Pi}_\varphi(n_a, n_b, n_c)], \tag{24}$$

and

$$\begin{aligned}\hat{\Pi}_\varphi(n_a, n_b, n_c) &= \mathrm{Tr}_{ab}\,[|\alpha\rangle_{a\,a}\langle\alpha| \otimes |0\rangle_{b\,b}\langle 0|\hat{U}^\dagger\hat{\Pi}(n_a, n_b, n_c)\hat{U}] \\ &= |q\rangle_{c\,c}\langle q|\end{aligned} \tag{25}$$

say, where

$$|q\rangle_c = {}_b\langle 0|\hat{R}_1^\dagger\,{}_a\langle\alpha|\hat{R}_2^\dagger|n_a\rangle_a|n_b\rangle_b|n_c\rangle_c. \tag{26}$$

The reduced operator (25) can be interpreted as a POM element for the extended measuring apparatus consisting of everything in figure 1 except the state to be measured, $\hat{\rho}_c$. We note particularly that the coherent state field is part of the extended measuring device. Thus, altering φ by adjusting the phase shifter changes the extended measuring device and consequently its POM. For this reason we have attached the subscript φ to the POM element in (24).

4.1. *Measuring the λth moment*

We now consider the POM element of the extended measuring apparatus corresponding to photodetectors D_a, D_b and D_c detecting 0, λ and N photons respectively. With BS2 a 50/50 beam splitter we can write [12]

$$\hat{R}_2^\dagger|0\rangle_a|\lambda\rangle_b = [2^{\lambda/2}(\lambda!)^{1/2}]^{-1}(\hat{b}^\dagger - \mathrm{i}\hat{a}^\dagger)^\lambda|0\rangle_a|0\rangle_b \tag{27}$$

and thus

$$_a\langle\alpha|\hat{R}_2^\dagger|0\rangle_a|\lambda\rangle_b|N\rangle_c = \frac{(\hat{b}^\dagger - \mathrm{i}\alpha^*)^\lambda}{2^{\lambda/2}(\lambda!)^{1/2}\exp(|\alpha|^2/2)}|0\rangle_b|N\rangle_c. \tag{28}$$

Then, writing $|N\rangle_c$ as $(\hat{c}^{\dagger N}/(N!)^{1/2})|0\rangle_c$ we obtain, using (15)–(17)

$$\hat{R}_1^\dagger\,{}_a\langle\alpha|\hat{R}_2^\dagger|0\rangle_a|\lambda\rangle_b|N\rangle_c = \frac{[t\hat{b}^\dagger - \mathrm{i}(r\hat{c}^\dagger + \alpha^*)]^\lambda[t\hat{c}^\dagger - \mathrm{i}r\hat{b}^\dagger]^N}{2^{\lambda/2}(\lambda!N!)^{1/2}\exp(|\alpha|^2/2)}|0\rangle_b|0\rangle_c, \tag{29}$$

where we have left the transmission and reflection coefficients of BS1 as t and r. From (26), we obtain

$$|q\rangle_c = \sum_{n=N}^{N+\lambda} q_n|n\rangle_c, \tag{30}$$

where

$$q_n = \frac{(-\mathrm{i})^\lambda t^N r^{n-N}(\alpha^*)^{\lambda+N-n}}{2^{\lambda/2}\exp(|\alpha|^2/2)[(\lambda+N-n)!]^{1/2}}\binom{\lambda}{n-N}^{1/2}\binom{n}{N}^{1/2}, \tag{31}$$

where the last two factors are square roots of binomial coefficients.

The experimental procedure that we propose involves measuring the probabilities $P_\varphi(0,\lambda,N)$ for a discrete number λ of values of φ, that is for $\varphi = \varphi_{\beta j}$, where

$$\varphi_{\beta j} = \frac{\beta\pi}{\lambda} + \frac{2\pi j}{\lambda}, \tag{32}$$

where $\beta\pi/\lambda$ is the initial, that is $j = 0$, value of the argument of α and $j = 0, 1 \ldots \lambda - 1$.

We can form a linear combination of the POM elements $\hat{\Pi}_{\beta j}(0,\lambda,N)$, where we have replaced the subscript φ on the POM element in (24) by βj to denote the separate values, as

$$\hat{L}_\beta(0,\lambda,N) = \frac{1}{\lambda}\sum_{j=0}^{\lambda-1}\hat{\Pi}_{\beta j}(0,\lambda,N) \tag{33}$$

$$= \frac{1}{\lambda}\sum_{j=0}^{\lambda-1}\sum_{n,m=N}^{N+\lambda} q_n q_m^*|n\rangle_{c\,c}\langle m|. \tag{34}$$

We can write

$$q_n q_m^* = |q_n q_m^*|\exp[\mathrm{i}(n-m)(\beta\pi/\lambda + 2\pi j/\lambda)], \tag{35}$$

which gives

$$\hat{L}_\beta(0,\lambda,N) = \sum_{n,m=N}^{N+\lambda}\left(\frac{1}{\lambda}\sum_{j=0}^{\lambda-1}\exp\left[\mathrm{i}(n-m)2\pi j/\lambda\right]\right)|q_n q_m^*|\exp\left[\mathrm{i}(n-m)\beta\pi/\lambda\right]|n\rangle_{c\,c}\langle m|. \tag{36}$$

The term in brackets is a periodic Kronecker delta function which is zero everywhere except at $n-m=0,\pm\lambda,\pm2\lambda\cdots$. Since n and m can at most differ by $\pm\lambda$, (36) reduces to

$$\hat{L}_\beta(0,\lambda,N) = \sum_{n=N}^{N+\lambda}|q_n|^2|n\rangle\langle n| + (|q_N q_{N+\lambda}^*|\exp(-\mathrm{i}\beta\pi)|N\rangle\langle N+\lambda| + h.c.), \tag{37}$$

which resembles the Nth term in the expansion of $\hat{C}_\lambda$ in (9).

This operator depends on the value of β. The unwanted first term in (37) can be removed as follows. After the first set of λ measurements is made with a particular value of β, a second set of measurements is made with the phase shifter altered so that the λ values of φ are now $(\beta+1)\pi/\lambda + 2\pi j/\lambda$ in place of (32). We label the operators corresponding to these two sets of measurements as $\hat{L}_\beta(0,\lambda,N)$ and $\hat{L}_{\beta+1}(0,\lambda,N)$ and form the linear combination

$$\begin{aligned}\sum_N \frac{\hat{L}_\beta(0,\lambda,N)-\hat{L}_{\beta+1}(0,\lambda,N)}{4|q_N q_{N+\lambda}^*|} &= \frac{1}{2}\sum_N[\exp(-\mathrm{i}\beta\pi)|N\rangle\langle N+\lambda| + h.c.]\\ &= \cos\beta\pi\hat{C}_\lambda + \sin\beta\pi\hat{S}_\lambda.\end{aligned} \tag{38}$$

If we choose the initial setting of the phase shifter such that $\beta=0$, we obtain

$$\hat{C}_\lambda = \sum_n \frac{\hat{L}_0(0,\lambda,N)-\hat{L}_1(0,\lambda,N)}{4|q_N q_{N+\lambda}^*|}. \tag{39}$$

Thus we have found a linear combination of experimentally realizable POM elements capable of synthesizing the operator $\hat{C}_\lambda$ for all physical states. However, to provide a complete measure of the phase distribution, the corresponding moments of the operator $\hat{S}_\lambda$ must also be measured. This can be achieved by repeating the above experiments with $\beta=1/2$. This gives from (38)

$$\hat{S}_\lambda = \sum_N \frac{\hat{L}_{1/2}(0,\lambda,N)-\hat{L}_{3/2}(0,\lambda,N)}{4|q_N q_{N+\lambda}^*|}, \tag{40}$$

where the subscripts on $\hat{L}$ denote that the initial values of φ are $\pi/(2\lambda)$ and $3\pi/(2\lambda)$.

The values of the $\langle\cos\lambda\theta\rangle$ and $\langle\sin\lambda\theta\rangle$ required to construct the phase probability distribution can be found from (7) and (8). Specifically, from (24) with the subscript φ replaced by βj and from (39), we find

$$\langle\cos\lambda\theta\rangle = \frac{1}{\lambda}\sum_N\sum_{j=0}^{\lambda-1}\frac{P_{0j}(0,\lambda,N)-P_{1j}(0,\lambda,N)}{4|q_N q_{N+\lambda}^*|}, \tag{41}$$

where P_{0j} and P_{1j} are the measured probabilities of detecting 0, λ and N photons in detectors D_a, D_b and D_c with $\beta=0$ and $\beta=1$, that is with initial arguments of α being 0 and π/λ respectively. A similar expression holds for $\langle\sin\lambda\theta\rangle$ in terms of the measured probabilities of detecting 0, λ and n photons in detectors D_a, D_b and D_c with initial arguments of α being $\pi/(2\lambda)$ and $3\pi/(2\lambda)$ respectively.

5. Some practical considerations

In synthesizing the operators $\hat{C}_\lambda$ and $\hat{S}_\lambda$, we need to perform 4λ different experiments, one for each value $\varphi_{\beta j}$ of the argument α, with $\beta = 0, 1/2, 1, 3/2$ in equation (32). For large values of λ this may become experimentally time consuming. Fortunately, half of these experiments can be made redundant by utilizing the symmetry properties of the 50/50 beam splitter BS2. Up until now we have considered only the detection events $(0, \lambda, N)$. However, we find that detecting the event $(\lambda, 0, N)$ also produces a useful POM element, specifically if λ is odd. This is the event of D_a detecting λ photons and D_b detecting the vacuum, with D_c still detecting N photons. From [12] it is not difficult to show that the effect of detecting the latter event in place of the former is equivalent to changing α to $-\alpha$ in $\hat{\Pi}_\varphi$ in (25). Then from (31) we obtain, in place of (37),

$$\hat{L}_\beta(\lambda, 0, N) = \sum_{n=N}^{N+\lambda} |q_n|^2 |n\rangle\langle n| + (-1)^\lambda [|q_N q^*_{N+\lambda}| \exp(-i\beta\pi)|N\rangle\langle N+\lambda| + h.c.]. \quad (42)$$

Thus

$$\begin{aligned} \hat{L}_\beta(\lambda, 0, N) &= \hat{L}_\beta(0, \lambda, N), && \forall \text{ even } \lambda, \\ &= \hat{L}_{\beta+1}(0, \lambda, N), && \forall \text{ odd } \lambda. \end{aligned} \quad (43)$$

Equation (43) shows that we can obtain the operator $\hat{L}_1(0, \lambda, N)$ by recording the detection events $(\lambda, 0, N)$ in the experiments performed with the phase settings φ_{0j}. We see then that the experiments needed to obtain the expectation value of $\hat{L}_1(0, \lambda, N)$ become redundant for odd values of λ. The same is true for the experiments to obtain the expectation value of $\hat{L}_{3/2}(0, \lambda, N)$. The overall result is a reduction in the total number of experiments needed to be performed, for odd values of λ, by a factor of two.

An equivalent reduction can be obtained for even values of λ by considering the detection event $(\lambda/2, \lambda/2, N)$ in place of $(0, \lambda, N)$. When this detection event is observed, we can show that every second term in the state $|q\rangle_c$ has amplitude zero. This results in the removal of all terms $|n\rangle\langle m|$, for $n - m$ odd, in the POM element $\hat{\Pi}(\lambda/2, \lambda/2, N)$. Since the purpose of performing the phase adjustments $\varphi_{\beta j}$ is to remove all but the $|N\rangle\langle N+\lambda|$ term in $\hat{\Pi}_{\beta j}$, we find that only half the external phase adjustments are required for even values of λ. Thus we find that, by considering a broader range of detection events, we can make redundant half of the experiments needed to be performed when measuring the sine and cosine moments, for all values of λ.

Before any experiments are performed, a value for the $t:r$ ratio of BS1 and the mean photon number of the coherent reference state, $|\alpha|^2$, must be decided upon. In principle, any values for these parameters may be chosen. However, to avoid small quotients in equations (39) and (40) and thus larger error bars, it is desirable to optimize these parameters. For BS1, the optimum value of the $t:r$ ratio can be found by maximizing $|q_N q^*_{N+\lambda}|$, giving

$$\left(\frac{t}{r}\right)^2_{\max} = \frac{2N}{\lambda} \tag{44}$$

with the optimum value of $|\alpha|^2$ found to be

$$|\alpha|^2_{\max} = \frac{\lambda}{2}. \tag{45}$$

If necessary, the experiment could be repeated for different values of N and λ, but given that we are primarily concerned with measuring broad phase distributions, the range of λ needed to be measured is quite small. So in practice, a compromise value of $(t/r)^2 = 2$ and $|\alpha|^2 = \langle n\rangle/2$ would be acceptable, where $\langle n\rangle$ is the mean photon number of $\hat{\rho}_c$.

The detection probabilities used above are those that would be obtained by perfect detectors. As we discussed in [6], practical photodetectors suffer from non-unit efficiencies that reduce the number of counts, the presence of dark counts and a non-zero dead time following a count during which no other counts are registered. We are again particularly interested here in measuring weak fields, so sufficiently long gating times can be used to reduce the effect of the dead time. Other techniques can also be used for this purpose [13]. For known efficiency and dark count rate the ideal photocount probabilities can be found from the statistics of the experiment, as discussed in [6, 14–17].

6. Conclusion

In this paper we have devised a reasonably simple method to measure the canonical phase distribution of a state of light based on a procedure we refer to as operator synthesis. This procedure can be understood in the following general terms. By repeating experiments we can measure the probabilities of various measurement events. Quantum mechanics allows us to calculate these probabilities as the trace of the product of the density operator for the state to be measured and the appropriate element of a probability operator measure (POM). Linear combinations of probabilities can be calculated as the trace of the product of the density operator and the same linear combination of POM elements. From quantum mechanics we also know that the expectation value of an observable is the trace of the product of the density operator and the operator associated with the observable. Thus if we can find a linear combination of POM elements which is equal to this operator we can find the expectation value of the observable from the same linear combination of measured probabilities. We have called this procedure operator synthesis as it is a generalization of projection synthesis [4] which involves synthesizing a projection operator. In quantum optics the usual final detection events are photocounts, but by use of beam splitters and reference fields we can create an extended measurement device which is described by a quite different POM from that of a simple photodetector. We have found in this paper that by this means we can synthesize operators whose expectation values are the mean cosine and sine of multiples of the phase. This provides us with the moments we need to construct the canonical phase distribution. Our new method has the advantage of simplicity over the two methods to measure the phase distribution proposed so far. Only two beam splitters are required and the reference state is an easily prepared coherent state.

References

[1] Noh, J. W., Fougères, A., and Mandel, L., 1991, *Phys. Rev. Lett.*, **67,** 1426; Noh, J. W., Fougères, A., and Mandel, L., 1992, *Phys. Rev.* A, **45,** 424; Noh, J. W., Fougères, A., and Mandel, L., 1992, *Phys. Rev.* A, **46,** 2840; Noh, J. W., Fougères, A., and Mandel, L., 1993, *Phys. Rev. Lett.*, **71,** 2579.

[2] Pegg, D. T., and Barnett, S. M., 1997, *J. Mod. Optics*, **44,** 225.

[3] Vaccaro, J. A., and Pegg, D. T., 1994, *Optics Comun.*, **105,** 335.

[4] Barnett, S. M., and Pegg, D. T., 1996, *Phys. Rev. Lett.*, **76,** 4148; Pegg, D. T., Barnett, S. M., and Phillips, L. S., 1997, *J. Mod. Optics*, **44,** 2135.

[5] Clausen, J., Dakna, M., Knöll, L., and Welsch, D.-G., 2000, *Optics Commun.*, **179,** 189.

[6] Pregnell, K. L., and Pegg, D. T., 2001, *J. Mod. Optics*, **48,** 1293.

[7] Pegg, D. T., and Barnett, S. M., 1988, *Europhys. Lett.*, **6,** 483; Barnett, S. M., and Pegg, D. T., 1989, *J. Mod Optics*, **36,** 7; Pegg, D. T., and Barnett, S. M., 1989, *Phys. Rev.* A, **39,** 1665.

[8] Helstrom, C. W., 1976, *Quantum Detection and Estimation Theory* (New York: Academic Press).

[9] Torgerson, J. R., and Mandel, L., 1996, *Phys. Rev. Lett.*, **76,** 3939.

[10] Messiah, A. M., 1961, *Quantum Mechanics* (Amsterdam: North-Holland), p. 257.

[11] Barnett, S. M., and Radmore, P. M., 1997, *Methods in Theoretical Quantum Optics* (Oxford: Oxford University Press), p. 84.

[12] Phillips, L. S., Barnett, S. M., and Pegg, D. T., 1998, *Phys. Rev.* A, **58,** 3259.

[13] Paul, H., Torma, P., Kiss, T., and Jex, I., 1997, *Phys. Rev. Lett.*, **76,** 2464.

[14] Lee, C. T., 1993, *Phys. Rev.* A, **48,** 2285.

[15] Pegg, D. T., and Barnett, S. M., 1999, *J. Mod. Optics*, **46,** 1657.

[16] Barnett, S. M., Phillips, L. S., and Pegg, D. T., 1998, *Optics Commun.*, **158,** 45.

[17] Pegg, D. T., Phillips, L. S., and Barnett, S. M., 1998, *Phys. Rev. Lett.*, **81,** 1604.

Single-Shot Measurement of Quantum Optical Phase

K. L. Pregnell and D. T. Pegg
School of Science, Griffith University, Nathan, Brisbane, 4111, Australia
(Received 11 April 2002; published 3 October 2002)

Although the canonical phase of light, which is defined as the complement of photon number, has been described theoretically by a variety of distinct approaches, there have been no methods proposed for its measurement. Indeed doubts have been expressed about whether or not it is measurable. Here we show how it is possible, at least in principle, to perform a single-shot measurement of the canonical phase using beam splitters, mirrors, phase shifters, and photodetectors.

DOI: 10.1103/PhysRevLett.89.173601 PACS numbers: 42.50.Dv

Quantum-limited phase measurements of the optical field have important applications in precision measurements of small distances in interferometry and in the emerging field of quantum communication, where there is the possibility of encoding information in the phase of light pulses. Much work has been done in attempting to understand the quantum nature of phase. Some approaches have been motivated by the aim of expressing phase as the complement of the photon number [1]. Examples of these approaches include the probability operator measure approach [2,3], a formalism in which the Hilbert space is doubled [4], a limiting approach based on a finite Hilbert space [5,6], and a more general axiomatic approach [1]. Although these approaches are quite distinct, they all lead to the same phase probability distribution for a field in state $|\psi\rangle$ as a function of the phase angle θ [1]:

$$P(\theta) = \frac{1}{2\pi}\left|\sum_{n=0}^{\infty}\langle\psi|n\rangle\exp(in\theta)\right|^2, \tag{1}$$

where $|n\rangle$ is a photon number state. Leonhardt *et al.* [1] have called this common distribution the "canonical" phase distribution to indicate a quantity that is the canonical conjugate, or complement, of the photon number. This distribution is shifted uniformly when a phase shifter is applied to the field and is not changed by a photon number shift [1]. We adopt this definition here and use the term "canonical phase" to denote the quantity whose distribution is given by (1).

Much less progress has been made on ways to measure the canonical phase. Homodyne techniques can be used to measure phaselike properties of light but are not measurements of the canonical phase. It is possible in principle to measure the canonical phase distribution by a series of experiments on a reproducible state of light [7], but there has been no known way of performing a single-shot measurement. Indeed it is thought that this might be impossible [8]. Even leaving aside the practical issues, the concept that a particular fundamental quantum observable may not be measurable, even in principle, has interesting general conceptual ramifications for quantum mechanics. A different approach to the phase problem, which avoids difficulties in finding a way to measure the canonical phase, is to define "phase" operationally in terms of observables that can be measured [1]. The best known of these operational phase approaches is that of Noh *et al.* [9,10]. Although the experiments to measure this operational phase produce excellent results, they were not designed to measure the canonical phase as defined here and, as shown by the measured phase distribution [10], they do not measure the canonical phase. In this Letter we show how, despite these past difficulties, it is indeed possible, at least in principle, to perform a single-shot measurement of the canonical phase in the same sense that the experiments of Noh *et al.* are single-shot measurements of their operational phase.

A single-shot measurement of a quantum observable must not only yield one of the eigenvalues of the observable, but repeating the measurement many times on systems in identical states should result in a probability distribution appropriate to that state. If the spectrum of eigenvalues is discrete, the probabilities of the results can be easily obtained from the experimental statistics. Where the spectrum is continuous, the probability density is obtainable by dividing the eigenvalue range into a number of small bins and finding the number of results in each bin. As the number of experiments needed to obtain measurable probabilities increases as the reciprocal of the bin size, a practical experiment will require a nonzero bin size and will produce a histogram rather than a smooth curve.

Although the experiments of Noh *et al.* [9,10] were not designed to measure the canonical phase, it is helpful to be guided by their approach. In addition to their results being measured and plotted as a histogram, some of the experimental data are discarded, specifically photon count outcomes that lead to an indeterminacy of the type zero divided by zero in their definitions of the cosine and sine of the phase [9]. The particular experiment that yields such an outcome is ignored and its results are not included in the statistics.

Concerning the discarding of some results, we note in general that the well-known expression for the probability that a von Neumann measurement on a pure state $|\psi\rangle$ yields a result q is $\langle\psi|q\rangle\langle q|\psi\rangle$, where $|q\rangle$ is the eigenstate of the measurement operator corresponding to eigenvalue q. If the state to be measured is a mixed state with a density operator $\hat{\rho}$ the expression becomes $\mathrm{Tr}(\hat{\rho}\hat{\Pi}_q)$, where $\hat{\Pi}_q = |q\rangle\langle q|$. The operator $\hat{\Pi}_q$ is a particular case of an element of a probability operator measure (POM) [2]. The sum of all the elements of a POM is the unit operator and the expression $\mathrm{Tr}(\hat{\rho}\hat{\Pi}_q)$ for the probability is based on the premise that all possible outcomes of the measurement are retained for the statistics. If some of the possible outcomes of an experiment are discarded, the probability of a particular result calculated from the final statistics is given by the normalized expression $\mathrm{Tr}(\hat{\rho}\hat{\Pi}_q)/\sum_p \mathrm{Tr}(\hat{\rho}\hat{\Pi}_p)$, where the sum is over all the elements of the POM corresponding to outcomes of the measurement that are retained.

We seek now to approximate the continuous distribution (1) by a histogram representing the probability distribution for a discrete observable θ_m such that when the separation $\delta\theta$ of consecutive values of θ_m tends to zero the continuous distribution is regained. A way to do this is first to define a state

$$|\theta_m\rangle = \frac{1}{(N+1)^{1/2}}\sum_{n=0}^{N}\exp(in\theta_m)|n\rangle. \tag{2}$$

There are $N+1$ orthogonal states $|\theta_m\rangle$ corresponding to $N+1$ values $\theta_m = m\delta\theta$ with $\delta\theta = 2\pi/(N+1)$ and $m = 0, 1, \ldots, N$. This range for m ensures that θ_m takes values between 0 and 2π. Then, if we can find a measurement technique that yields the result θ_m with a probability of $|\langle\psi|\theta_m\rangle|^2$, the resulting histogram will approximate a continuous distribution with a probability density of $|\langle\psi|\theta_m\rangle|^2/\delta\theta$. It follows that, as we let N tend to infinity, there will exist a value of θ_m as close as we like to any given value of θ with a probability density approaching $P(\theta)$ given by (1). If we keep N finite so that we can perform an experiment with a finite number of outcomes, then the value of N must be sufficiently large to give the resolution $\delta\theta$ of phase angle required and also for $|\psi\rangle$ to be well approximated by $\sum_n \langle n|\psi\rangle|n\rangle$, where the sum is from $n = 0$ to N. The latter condition ensures that the terms with coefficients $\langle n|\psi\rangle$ for $n > N$ have little effect on the probability $|\langle\psi|\theta_m\rangle|^2$. As we shall be interested mainly in weak optical fields in the quantum regime with mean photon numbers of the order of unity, the maximum phase resolution $\delta\theta$ desired will usually be the determining factor in the choice of N.

When N is finite, the states $|\theta_m\rangle$ do not span the whole Hilbert space, so the projectors $|\theta_m\rangle\langle\theta_m|$ will not sum to the unit operator $\hat{1}$. Thus these projectors by themselves do not form the elements of a POM. To complete the POM we need to include an element $\hat{1} - \sum_m |\theta_m\rangle\langle\theta_m|$. If we discard the outcome associated with this element, that is, treat an experiment with this outcome as an unsuccessful attempt at a measurement in a similar way that Noh *et al.* [9] treated experiments with indeterminate outcomes, then the probability that the outcome of a measurement is the phase angle θ_m is given by

$$\mathrm{Prob}(\theta_m) = \frac{\mathrm{Tr}(\hat{\rho}|\theta_m\rangle\langle\theta_m|)}{\sum_p \mathrm{Tr}(\hat{\rho}|\theta_p\rangle\langle\theta_p|)}, \tag{3}$$

where $p = 0, 1, \ldots, N$. We now require a single-shot measuring device that will reproduce this probability in repeated experiments.

As measurements will be performed ultimately by photodetectors, we seek an optical device that transforms input fields in such a way that photon number measurements at the output ports can be converted to phase measurements. We examine the case of a multiport device with $N+1$ input modes and $N+1$ output modes as depicted schematically in Fig. 1. As phase is not an absolute quantity, that is it can be measured only in relation to some reference state, we shall need the field in state $|\psi\rangle_0$ that is to be measured to be in one input and a reference field in state $|B\rangle_1$ to be in another input. We let there be vacuum states $|0\rangle_i$ with $i = 2, 3, \ldots, N$ in the other input modes. We let the device be such that the input states are transformed by a unitary transformation $\hat{R}$. We let $N+1$ photodetectors that can distinguish between zero photons, one photon, and more than one photon be in the output modes.

Consider the case where one photon is detected in each output mode except for mode m, in which zero photons are detected. The amplitude for this detection event is

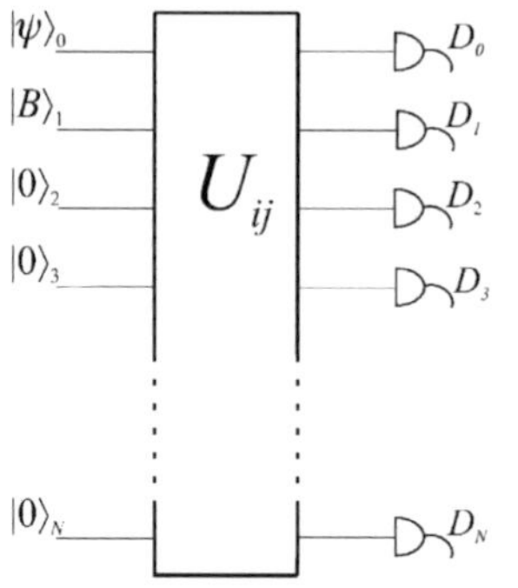

FIG. 1. Multiport device for measuring phase. The input and output modes are labeled $0, 1, \ldots, N$ from the top. In input mode 0 is the field in state $|\psi\rangle_0$ to be measured and in input mode 1 is the reference field in state $|B\rangle_1$. Vacuum states form the other inputs. There is a photodetector in each output mode. If all the photodetectors register one count except the detector D_m in output mode m, which registers no counts, then the detector array acts as a digital pointer mechanism indicating a measured phase angle of θ_m.

$$ {}_m\langle 0| \left(\prod_{j \neq m} {}_j\langle 1| \right) \hat{R} \left(\prod_{i=2}^{N} |0\rangle_i \right) |B\rangle_1 |\psi\rangle_0, \tag{4}$$

where the first product is over values of j from 0 to N excluding the value m. We require the transformation and the reference state to be such that this amplitude is proportional to ${}_0\langle\theta_m|\psi\rangle_0$. We write the photon creation operators acting on the mode i as $\hat{a}_i^\dagger$. Writing $|1\rangle_j$ as $\hat{a}_j^\dagger|0\rangle_j$ we see that we require $|\theta_m\rangle_0$ to be proportional to

$$ {}_1\langle B| \left(\prod_{i=2}^{N} {}_i\langle 0| \right) \left(\prod_{j \neq m} \hat{R}^\dagger \hat{a}_j^\dagger \hat{R} \right) \hat{R}^\dagger \left(\prod_{k=0}^{N} |0\rangle_k \right). \tag{5}$$

We rewrite the unitary transformation in the form

$$ \hat{R}^\dagger \hat{a}_j^\dagger \hat{R} = \sum_{i=0}^{N} U_{ij} \hat{a}_i^\dagger, \tag{6}$$

where U_{ij} are the elements of a unitary matrix. Reck *et al.* [11] have shown how it is possible to construct a multiport device from mirrors, beam splitters, and phase shifters that will transform the input modes into the output modes in accord with any $(N+1) \times (N+1)$ unitary matrix. We choose such a device for which the associated unitary matrix is

$$ U_{ij} = \frac{\omega^{ij}}{\sqrt{N+1}}, \tag{7}$$

where $\omega = \exp[-i2\pi/(N+1)]$ that is, a $(N+1)$th root of unity. Such a device will transform the combined vacuum state to the combined vacuum state so we can delete the $\hat{R}^\dagger$ on the right of expression (5). Substituting (7) into (6) gives eventually

$$ \left(\prod_{i=2}^{N} {}_i\langle 0| \right) \left(\prod_{j \neq m} \hat{R}^\dagger \hat{a}_j^\dagger \hat{R} \right) \left(\prod_{k=0}^{N} |0\rangle_k \right) = \kappa_1 \left[\prod_{j \neq m} (\hat{a}_0^\dagger + \omega^j \hat{a}_1^\dagger) \right] |0\rangle_0 |0\rangle_1, \tag{8}$$

where $\kappa_1 = (N+1)^{-N/2}$.

To evaluate (8) we divide both sides of the identity

$$ X^{N+1} + (-1)^N = (X+1)(X+\omega)(X+\omega^2)\ldots(X+\omega^N) \tag{9}$$

by $X + \omega^m$ to give, after some rearrangement and application of the relation $\omega^{m(N+1)} = 1$,

$$ \prod_{j \neq m} (X + \omega^j) = (-1)^N \omega^{mN} \frac{1 - (-X\omega^{-m})^{N+1}}{1 - (-X\omega^{-m})}. \tag{10}$$

The last factor is the sum of a geometric progression. Expanding this and substituting $X = x/y$ gives eventually the identity

$$ \prod_{j \neq m} (x + \omega^j y) = \sum_{n=0}^{N} x^n (-\omega^m y)^{N-n}. \tag{11}$$

We now expand $|B\rangle_1$ in terms of photon number states as

$$ |B\rangle_1 = \sum_{n=0}^{N} b_n |n\rangle_1 \tag{12}$$

and put $x = a_0^\dagger$ and $y = a_1^\dagger$ in (11). Then from (8) we find that (5) becomes

$$ {}_1\langle B| \left(\prod_{i=2}^{N} {}_i\langle 0| \right) \left(\prod_{j \neq m} \hat{R}^\dagger \hat{a}_j^\dagger \hat{R} \right) \hat{R}^\dagger \left(\prod_{k=0}^{N} |0\rangle_k \right) = \kappa_2 \sum_{n=0}^{N} (-1)^{N-n} \binom{N}{n}^{-1/2} \omega^{-nm} b_{N-n}^* |n\rangle_0, \tag{13}$$

where $\kappa_2 = \kappa_1 \omega^{-m} (N!)^{1/2}$. We see then that, if we let $|B\rangle_1$ be the binomial state

$$ |B\rangle_1 = 2^{-N/2} \sum_{n=0}^{N} (-1)^n \binom{N}{n}^{1/2} |n\rangle_1, \tag{14}$$

then expression (13) is proportional to $\sum_n \omega^{-nm} |n\rangle_0$, that is, to $|\theta_m\rangle_0$. Thus the amplitude for the event that zero photons are detected in output mode m and one photon is detected in all the other output modes will be proportional to ${}_0\langle\theta_m|\psi\rangle_0$. The probability that the outcome of a measurement is this event, given that only outcomes associated with the $(N+1)$ events of this type are recorded in the statistics, will be given by (3), where we note that the proportionality constant κ_2 will cancel from this expression. Thus the measurement event that zero photons are detected in output mode m and one photon is detected in all the other output modes can be taken as the event that the result of the measurement of the phase angle is θ_m. Thus the photodetector with zero photocounts, when all other photodetectors have registered one photocount, can be regarded as a digital pointer to the value of the measured phase angle.

We have shown, therefore, that it is indeed possible in principle to conduct a single-shot measurement of canonical phase to within any given nonzero error, however small. This error is of the order $2\pi/(N+1)$ and will determine the value of N chosen.

While the aim of this Letter is to establish how the canonical phase can be measured in principle, it is worth briefly considering some practical issues. Although we have specified that the photodetectors need only be capable of distinguishing among zero, one, and more than one photons, reflecting the realistic case, there are other imperfections such as inefficiency. These will give rise to errors in the phase measurement, just as they will cause errors in a single-shot photon number measurement. In practice, there is no point in choosing the phase resolution $\delta\theta$ much smaller than the expected error due to

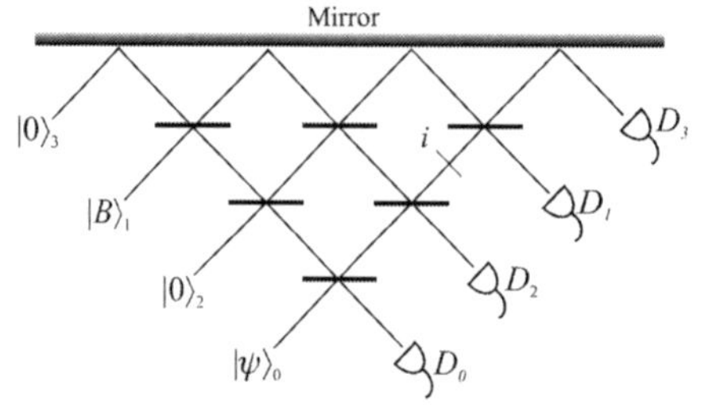

FIG. 2. Triangular array for $N + 1 = 4$. The outside beam splitters in the top row are 50:50, the middle one is fully reflecting. The second row beam splitters are $\frac{2}{3}$ transmitting and the bottom one is $\frac{3}{4}$ transmitting. The phase shifter i produces a $\pi/2$ phase shift.

photodetector inefficiencies; thus there is nothing lost in practice in keeping N finite. A requirement for the measuring procedure is the availability of a binomial state. Such states have been studied for some time [12], but their generation has not yet been achieved. In practice, however, we are usually interested in measuring weak fields in the quantum regime with mean photon numbers around unity [9,10] and even substantially less [13]. Only the first few coefficients of $|n\rangle_0$ in (13) will be important for such weak fields. Also, it is not difficult to show that the reference state need not be truncated at $n = N$, as indicated in (12), as coefficients b_n with $n > N$ will not appear in (13). Thus we need only prepare a reference state with a small number of its photon number state coefficients proportional to the appropriate binomial coefficients. Additionally, of course, in a practical experiment we are forced to tolerate some inaccuracy due to photodetector errors, so it will not be necessary for the reference state coefficients to be exactly proportional to the corresponding binomial state coefficients. These factors give some latitude in the preparation of the reference state. The mutiport device depicted in Fig. 1 can be constructed in a variety of ways. Reck *et al.* [11] provide an algorithm for constructing a triangular array of beam splitters to realize any unitary transformation matrix. Figure 2 shows, for example, such a device that will, with suitable phase shifters in the output modes, realize the transformation U_{ij} in (7) with $N + 1 = 4$. Because we are detecting photons, however, these output phase shifters are not actually necessary and are thus not shown. The number of beam splitters needed for a general triangular array increases quadratically with N. Fortunately, however, our required matrix (7) represents a discrete Fourier transformation and we require only two of the input ports to have input fields that are not in the vacuum state. The device of Törmä and Jex [14] is ideally suited for these specific requirements. This device, which has an even number of input and output ports, is pictured in Ref. [14]. It consists of just $(N + 1)/2$ ordinary 50:50 beam splitters and two plate beam splitters. The latter are available in the form of glass plates with modulated transmittivity along the direction of the incoming beam propagation [14].

As a large fraction of the raw data in this procedure is discarded, an interesting question arises as to whether or not there is a relationship between the method of this Letter and a limiting case of the operational phase measurements of Torgerson and Mandel [15] where it is found that the distribution becomes sharper as more data are discarded. While preliminary analysis indicates that there is not, this will be discussed in more detail elsewhere.

In conclusion, we have shown that it is possible in principle to perform a single-shot measurement of the canonical phase in the same sense that the experiments of Noh *et al.* are single-shot measurements of the operational phase. The technique relies on generating a reference state with some number state coefficients proportional to those of a binomial state.

D. T. P. thanks the Australian Research Council for funding.

[1] U. Leonhardt, J. A. Vaccaro, B. Böhmer, and H. Paul, Phys. Rev. A **51**, 84 (1995).
[2] C.W. Helstrom, *Quantum Detection and Estimation Theory* (Academic Press, New York, 1976).
[3] J. H. Shapiro and S. R. Shepard, Phys. Rev. A **43**, 3795 (1991).
[4] R. G. Newton, Ann. Phys. (N.Y.) **124**, 327 (1980).
[5] D.T. Pegg and S. M. Barnett, Europhys. Lett. **6**, 483 (1988); Phys. Rev. A **39**, 1665 (1989); S. M. Barnett and D.T. Pegg, J. Mod. Opt. **36**, 7 (1989).
[6] For a review, see, for example, D.T. Pegg and S. M. Barnett, J. Mod. Opt. **44**, 225 (1997).
[7] S. M. Barnett and D.T. Pegg, Phys. Rev. Lett. **76**, 4148 (1996); K. L. Pregnell and D.T. Pegg, J. Mod. Opt. **49**, 1135 (2002).
[8] H. M. Wiseman and R. B. Killip, Phys. Rev. A **56**, 944 (1997); **57**, 2169 (1998); M. A. Armen *et al.*, e-print quant-ph/0204005.
[9] J.W. Noh, A. Fougéres, and L. Mandel, Phys. Rev. Lett. **67**, 1426 (1991).
[10] J.W. Noh, A. Fougéres, and L. Mandel, Phys. Rev. Lett. **71**, 2579 (1993).
[11] M. Reck, A. Zeilinger, H. J. Bernstein, and P. Bertani, Phys. Rev. Lett. **73**, 58 (1994).
[12] D. Stoler, B. E. A. Saleh, and M. C. Teich, Opt. Acta **32**, 345 (1985); V.V. Dodonov, J. Opt. B: Quantum Semiclassical Opt. **4**, R1 (2002).
[13] J. R. Torgerson and L. Mandel, Phys. Rev. Lett. **76**, 3939 (1996).
[14] P. Törmä and I. Jex, J. Mod. Opt. **43**, 2403 (1996).
[15] J. R. Torgerson and L. Mandel, Opt. Commun. **133**, 153 (1997).

Binomial states and the phase distribution measurement of weak optical fields

K. L. Pregnell and D. T. Pegg
Center for Quantum Computer Technology, School of Science, Griffith University, Nathan, Brisbane 4111, Australia
(Received 17 December 2002; published 24 June 2003)

We show that the eight-port interferometer used by Noh, Fougères, and Mandel [Phys. Rev. Lett. **71**, 2579 (1993)] to measure their operational phase distribution of light can also be used to measure the canonical phase distribution of weak optical fields, where canonical phase is defined as the complement of photon number. A binomial reference state is required for this purpose, which we show can be obtained to an excellent degree of approximation from a suitably squeezed state. The proposed method requires only photodetectors that can distinguish among zero photons, one photon, and more than one photon and is not particularly sensitive to photodetector imperfections.

DOI: 10.1103/PhysRevA.67.063814 PACS number(s): 42.50.Dv

I. INTRODUCTION

The quantum-mechanical nature of the phase of light has been studied since the beginnings of quantum electrodynamic theory [1], with renewed interest recently. The study of quantum phase is distinguished from the study of many other quantum observables by the difficulties inherent not only in finding a theoretical description but also in finding methods for measuring the phase observable so described [2]. A method of circumventing this latter difficulty is to define phase as the quantity measured by some particular experiment. This is known as an operational approach [3]. A sensible operational phase measurement must, of course, be in accord with a classical phase measurement in the appropriate limit, for example, when the field being measured is in a strong coherent state. This requirement, however, is not sufficient to define a unique operational phase observable and various different operational definitions have been proposed. The best known of these is that of Noh *et al.* [4] who also proposed a means of measuring an operational phase distribution [5], which was further developed in Refs. [6] and [7]. There are also various theoretical approaches for describing the phase observable. What distinguishes these theoretical approaches from one another is their predicted phase distributions for particular states. Some of these approaches are motivated by the aim of expressing phase as the complement of photon number, in the spirit of Dirac's original work [1]. Examples of such approaches include the probability operator measure approach [8], a formalism in which the Hilbert space is doubled [9], a limiting approach based on a finite Hilbert space [10], and a more general axiomatic approach [11]. Although these particular approaches are quite distinct, they all lead to the same phase probability distribution for a field in state $|\psi\rangle$ as a function of phase angle θ [11]:

$$P(\theta)=\frac{1}{2\pi}\left|\sum_{n=0}^{\infty}\langle\psi|n\rangle\exp(in\theta)\right|^{2}, \tag{1}$$

where $|n\rangle$ is a photon number state. Leonhardt *et al.* [11] have called this the "canonical" phase distribution to indicate a quantity that is the canonical conjugate, or complement, of photon number. Irrespective of how it is derived, the canonical phase distribution has properties that one might expect from the complement of photon number: it is shifted uniformly when a phase shifter is applied to the field, it is not changed by a photon number shift [11], and it corresponds to a wave function in the phase representation of which the photon number amplitude is the finite Fourier transform [2,12]. The last property is the natural parallel to momentum-position conjugacy.

Although the canonical phase distribution has interesting theoretical properties, its direct measurement presents difficulties [13]. Good approximate methods exist, based on homodyne techniques, for measuring the canonical phase distribution of states with narrow phase distributions, for example coherent states with mean photon numbers of at least five [14]. Weak fields in the quantum regime, however, must have broad phase distributions because of number-phase complementarity. In principle, the distributions for such fields can be measured by the projection synthesis method proposed in Ref. [15], but this requires the generation of a reciprocal binomial state as a reference state. The generation of this exotic state has still not been achieved. On the other hand, the operational phase measurements of Noh *et al.* are quite practical for the weak fields of interest and have been shown to measure what they are designed to measure very well [5]. Unfortunately, they were not designed to measure, nor do they measure, the canonical phase distribution. Indeed, their results show that their operational phase distribution is significantly different from the canonical phase distribution. Thus, the projection synthesis method measures the canonical phase distribution in principle, while the operational method does not; but the operational method is practical, while the projection synthesis method is not.

In this paper, we show how the apparatus used for measuring the operational phase distribution can be used to measure the canonical phase distribution for weak fields in conjunction with a reference field in a *binomial* state. We show how a binomial state, in contrast to a reciprocal binomial state, can be very well approximated by a squeezed state, making this method much more practical than the original projection synthesis method.

II. PROJECTION SYNTHESIS

A. Beam splitter

In the projection synthesis method of Ref. [15], the aim was to use photodetection in conjunction with a beam splitter, and a special reference state to synthesize the projection of an unknown state onto a truncated phase state

$$|\theta\rangle = \frac{1}{(M+1)^{1/2}} \sum_{n=0}^{M} \exp(in\theta)|n\rangle, \tag{2}$$

where M must be sufficiently large for the density matrix of the reproducible weak field that is to be measured to be well approximated in the number-state basis by a matrix with only the first $(M+1)\times(M+1)$ elements nonzero. This projection event is associated with the detection of M photons in one output mode of the beam splitter and no photons in the other. We label this detection event $(M,0)$. The probability of the event $(M,0)$ is obtained from the occurrence frequency in successive repeated measurements of the field. The procedure is to measure this probability as θ is changed in small steps over the 2π range. The changes in θ can be achieved simply by altering the phase either of the reference field or of the field to be measured. A histogram is then plotted which, when suitably normalized, produces the canonical phase distribution.

The mechanism underlying the projection synthesis can be described as follows. Suppose the measured and reference fields are in the beam splitter input modes 0 and 1, respectively, and the event $(M,0)$ is that for which M photons are detected in output mode 0 and zero photons are detected in output mode 1. The combined output state corresponding to this measurement result is the M-photon state $|M\rangle_0|0\rangle_1$. Following the unitary evolution of this state backwards through the beam splitter transforms this to an M-photon entangled input state of the form

$$|f\rangle = \sum_{n=0}^{M} f_m|n\rangle_0|M-n\rangle_1, \tag{3}$$

which displays photon number conservation. If the measured field is in a pure state $|\psi\rangle_0$ and the reference state is given by

$$|r\rangle_1 = \sum_{n=0}^{\infty} r_n|n\rangle_1, \tag{4}$$

then the amplitude for the detection event $(M,0)$ is ${}_0\langle\psi|\,{}_1\langle r|f\rangle$, that is, the projection of $|\psi\rangle_0$ onto ${}_1\langle r|f\rangle$. By choosing appropriate coefficients r_n for $n=0$ to M, we can make this amplitude proportional to the projection of $|\psi\rangle_0$ onto the truncated phase state $|\theta\rangle$ given by Eq. (2). We note that the values of r_n for $n>M$ are irrelevant, merely affecting the normalization factor for the complete probability distribution. For a 50:50 symmetric, beam splitter, we find that the values of $|r_n|$ need to be proportional to the reciprocal of the square root of the binomial coefficient $\binom{M}{n}$ for $n=0$ to M.

As mentioned earlier, the difficulty with projection synthesis is generating the reciprocal binomial state required for the reference field. As reciprocal binomial states have a finite number of photon number-state coefficients, they can be prepared as traveling fields, in principle, by the generic methods given in Refs. [16,17] by means of beam splitters. Unfortunately, however, in such techniques the state generated is conditioned on measuring particular outputs from the beam splitters and so the method is quite inefficient and difficult and has never been implemented.

B. Multiport device

Projection synthesis relies on transforming projections onto photon number states into at least one projection onto a truncated phase state. As phase is not an absolute quantity, it is necessary to have a transformation device with at least two inputs for phase measurement: one for the field to be measured and one for the reference field. The projection synthesis method of Ref. [15] uses the minimal necessary device, a beam splitter with two inputs and two outputs. More flexibility can be obtained by using a more general multiport device with $N+1$ inputs and $N+1$ outputs with a photodetector in each output. This would require, in addition to the field in state $|\psi\rangle_0$ to be measured being in input mode 0, say, N reference fields being in modes $1,2,\ldots,N$. Rather than exacerbate the problem of preparing special reference states, it is preferable simply to have one reference field in input mode 1 with vacuum fields in the remaining inputs. These input states are transformed by a unitary operator $\hat{R}$ into the output states. If a total of M photons are detected in the output states, then, from photon number conservation, the corresponding M-photon output state evolving backwards through the multiport device will be transformed into an entangled M-photon input state. Projecting the combined vacuum input state $|0\rangle_2|0\rangle_3\cdots|0\rangle_N$ onto this entangled state will result in a two-mode M-photon entangled state in input modes 0 and 1 of the form $|f\rangle$ in Eq. (3). The coefficients f_m will depend on the unitary transformation $\hat{R}$ and on the manner in which the M photons are detected, that is, how they are distributed over the output modes. As these coefficients determine the values of r_n required to synthesize the projection of the state to be measured onto a truncated phase state, use of a multiport device should increase the flexibility in choosing a convenient reference state $|r\rangle_1$.

A natural multiport extension of the 50:50 symmetric beam splitter is the one where $\hat{R}$ is such that a photon entering any input appears with equal probability at any output [18,19] and a photon in any output is equally likely to have come from any input. In this case, it is not difficult to show that, when the M photons are detected in output mode 0 and none are detected in the other modes, the coefficients f_m are similar to those for the beam splitter case for the event $(M,0)$ and we again require the reference field to be in a reciprocal binomial state. Thus, it is worth examining other possible distributions of the detected photons over the output modes.

A particular case of the above device is the one that performs a discrete Fourier transform [19], that is, where the set of mode photon creation operators and the set of transformed operators form a discrete Fourier transform pair:

$$\hat{R}^\dagger \hat{a}_j^\dagger \hat{R} = \sum_{i=0}^{N} U_{ij} \hat{a}_i^\dagger, \tag{5}$$

$$\hat{a}_i^\dagger = \sum_{j=0}^{N} U_{ij}^* \hat{R}^\dagger \hat{a}_j^\dagger \hat{R}, \tag{6}$$

where

$$U_{ij} = \frac{\omega^{ij}}{\sqrt{N+1}}, \tag{7}$$

with $\omega = \exp[-i2\pi/(N+1)]$, that is, a $(N+1)$th root of unity. Recently [20], we found that if the number M of photons detected is equal to N and these photons are distributed such that one is detected in each output of such a multiport device except one, then projection onto a truncated phase state is synthesized, provided the reference field is in a *binomial* state. Specifically, if the detection event is such that it is the mth output detector that records a zero count while all the other detectors record one count each, then the state $|\psi\rangle_0$ to be measured is projected onto a state proportional to

$$|\theta_m\rangle_0 = \sum_{n=0}^{N} U_{mn}^* |n\rangle_0, \tag{8}$$

that is,

$$|\theta_m\rangle_0 = \frac{1}{(N+1)^{1/2}} \sum_{n=0}^{N} \exp(in\theta_m)|n\rangle_0, \tag{9}$$

where

$$\theta_m = m2\pi/(N+1), \tag{10}$$

which is just the required truncated phase state. The amplitude for this detection event is proportional to ${}_0\langle \theta_m|\psi\rangle_0$ [20].

In addition to requiring photodetectors that only need to distinguish among zero photons, one photon, and more than one photon, this multiport device has the advantage that the required binomial reference state is much closer to commonly available states such as coherent states than is a reciprocal binomial state. Binomial states have been studied for some time [21]. They have interesting properties such as interpolating between coherent states and number states and, with the photon number-state coefficients positive, they are partial phase states with a mean phase of zero [22]. It is not difficult to show that they have a smaller phase variance than truncated phase states with the same number of number-state coefficients. As binomial states have a finite number of photon number-state coefficients, they can be prepared as traveling fields by the generic methods given in Refs. [16,17] by means of beam splitters. Unfortunately, however, these inefficient techniques offer no real advance when used for measuring the phase distribution over the use of a reciprocal binomial state in Ref. [15], which can be prepared by the same generic means. For a practical experiment, we require the binomial state to be approximated by a state which is reasonably straightforward to prepare on demand.

As mentioned earlier, the values r_n of the number-state coefficients of the reference state for $n>M$ are irrelevant so here, where $M=N$, only the coefficients r_n with $0 \leq n \leq N$ are important, and thus only these need to be proportional to square roots of binomial coefficients. Further, as we are interested in fields with broad phase distributions, and states with broad number-state distributions tend to have narrow phase distributions, then normally only a small group of number-state coefficients ${}_0\langle n|\psi\rangle_0$ of the state of the measured field will differ significantly from zero. From Eq. (3) with $M=N$ and Eq. (4), if the coefficient ${}_0\langle n|\psi\rangle_0$ is significant then the value of r_{N-n} is important. Thus, for example, if ${}_0\langle n|\psi\rangle_0$ are significant only for a small number of values of n equal to or slightly less than a value n', say, then we would choose $N=n'$ and require the small number of coefficients r_n with n equal to, or near, zero to be proportional to square roots of appropriate binomial coefficients. We show in the Appendix how such a state can be approximated by a squeezed state with squeezing parameter $\tanh^{-1}0.5$. On the other hand, if the significant values of ${}_0\langle n|\psi\rangle_0$ occur for n equal to or near zero, as will be the case for very weak fields, then we require coefficients r_n with n equal to or slightly less than N to be proportional to square roots of the appropriate binomial coefficients. In the Appendix, we find a squeezed state, also with squeezing parameter of $\tanh^{-1}0.5$, that is a very good approximation for the binomial reference state needed for measuring the phase distribution of very weak fields with the eight-port interferometer examined below.

III. EIGHT-PORT INTERFEROMETER

A. Binomial reference state

The eight-port interferometer [23] used by Noh *et al.* [4,5] and Torgerson and Mandel [5,6] is illustrated in Fig. 1. There are four 50:50 symmetric beam splitters at the corners of a square. The phase shifter labeled $-i$ between the two beam splitters on the right shifts the phase by $\pi/2$. The field state $|\psi\rangle_0$ to be measured is in input port 0. The phase shifter in input port 1 allows the phase of the reference field state $|r\rangle_1$ to be changed. The dotted phase shifters in input port 2 and before detectors D_1, D_2, and D_3, which are not present in the original interferometer, are merely inserted here for mathematical convenience. As the field in input port 2 is the vacuum it will not be affected by a phase shift and, as the detectors detect photons, their operation will not be affected by phase shifters in front of them.

A single 50:50 symmetric beam splitter transforms the input creation operators $\hat{b}^\dagger$ and $\hat{c}^\dagger$ in accord with [24]

$$\hat{R}_1 \hat{b}^\dagger \hat{R}_1^\dagger = 2^{-1/2}(\hat{b}^\dagger + i\hat{c}^\dagger), \tag{11}$$

$$\hat{R}_1 \hat{c}^\dagger \hat{R}_1^\dagger = 2^{-1/2}(\widehat{ib}^\dagger + \hat{c}^\dagger), \tag{12}$$

where $\hat{R}_1$ is the unitary operator for the action of the single beam splitter. By using this relation successively, it is not

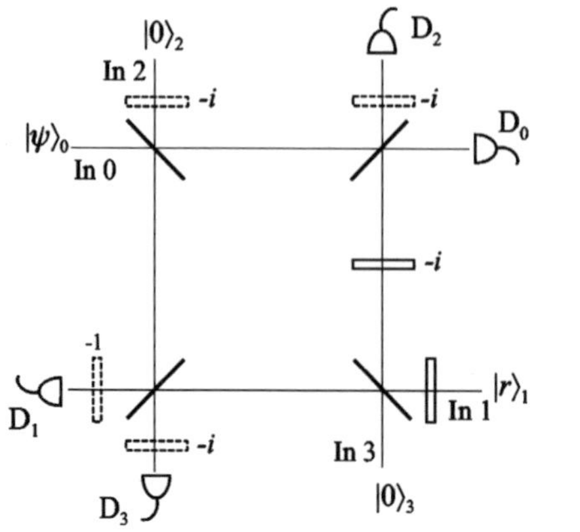

FIG. 1. Eight-port interferometer for measuring the canonical phase distribution of weak fields. The field in state $|\psi\rangle_0$ to be measured is in the input port labeled In 0, the reference field in state $|r\rangle_1$ is in input port In 1, and vacuum state fields are in In 2 and In 3. A photodetector is in each output port. The dotted phase shifters are for mathematical convenience only, and do not affect the results.

difficult to show that the input creation operators for the eight-port interferometer, including the dotted phase shifters, are transformed as

$$\hat{R}\hat{a}_i^\dagger\hat{R}^\dagger=\exp(i\gamma)\sum_{i=0}^{3} U_{ij}^*\hat{a}_j^\dagger, \tag{13}$$

where

$$U_{ij}=\frac{\omega^{ij}}{2}, \tag{14}$$

with ω again being $\exp[-i2\pi/(N+1)]$, provided we set the phase shifter in input 1 to shift the phase by $\pi/2$, that is, to attach a value $-i$ to it. Expressions (13) and (14) are in agreement with Eqs. (6) and (7) for $N=3$ apart from the phase factor $\exp(i\gamma)$, which depends on the difference between the distance between the beam splitters and an integer number of wavelengths. This phase factor does not affect the photocount probabilities and can be ignored.

Thus, we see that the eight-port interferometer, without modification, can be used to synthesize the projection of the state to be measured onto one of four phase states. Specifically, the probability of measurement event (0, 1, 1, 1), that is, the detection of zero photocounts in detector D_0 and one in each of D_1, D_2, and D_3, is proportional to the square of the modulus of the projection of the measured state onto the truncated phase state,

$$|\theta_0\rangle=2^{-1}(|0\rangle+|1\rangle+|2\rangle+|3\rangle), \tag{15}$$

while the probability of the event (1,0,1,1) is proportional the square of the modulus of the projection of the measured state onto the truncated phase state,

$$|\theta_1\rangle=2^{-1}(|0\rangle+i|1\rangle-|2\rangle-i|3\rangle), \tag{16}$$

and so on, in accord with Eq. (9) with $N=3$. Repeating the experiment a number of times with a reproducible state will allow a probability $P_M(\theta_m)$ with $m=0,1,2,3$ to be measured for each of the four events (0,1,1,1), (1,0,1,1), (1,1,0,1), and (1,1,1,0), respectively.

To use these four measured probabilities to construct the phase distribution, we first normalize them to

$$y(\theta_m)=\frac{2P_M(\theta_m)}{\pi\sum_m P_M(\theta_m)}. \tag{17}$$

We note that, as shown, in general, in Ref. [20], $P_M(\theta_m)\propto|\langle\psi|\theta_m\rangle|^2$, where here $|\theta_m\rangle$ is given by Eq. (9) with $N=3$ and we have omitted the subscripts zero for convenience. This allows us to replace $P_M(\theta_m)$ in Eq. (17) by $|\langle\psi|\theta_m\rangle|^2$. Substituting for θ_m from Eq. (10) with $N=3$ yields eventually

$$\sum_{m=0}^{3}|\langle\psi|\theta_m\rangle|^2=|\langle\psi|0\rangle|^2+|\langle\psi|1\rangle|^2+|\langle\psi|2\rangle|^2+|\langle\psi|3\rangle|^2. \tag{18}$$

If the measured field is sufficiently weak for the number-state components $\langle\psi|n\rangle$ to be negligible for $n\geqslant4$, then the right-hand side of Eq. (18) is just unity and so, from Eq. (17),

$$y(\theta_m)=\frac{2}{\pi}|\langle\psi|\theta_m\rangle|^2=P(\theta_m), \tag{19}$$

with the last line being obtained from Eq. (1) in the same weak-field approximation.

Thus, if we normalize the four measured probabilities by dividing each by the sum of the four and then multiplying by $2/\pi$, we obtain four points on the canonical phase probability distribution given by Eq. (1) for a weak field. We note that, with this normalization procedure, if the points $y(\theta_m)$ are used to draw a histogram, the area of the histogram would be unity, because the width of each rectangle is $\pi/2$.

Shifting the phase of the phase shifter in input 1 to change the phase of the reference field by $\Delta\theta$ and repeating the procedure gives four more points of the distribution shifted from the original points by $\Delta\theta$. A 16-point curve, for example, can be constructed by shifting the phase by $\pi/8$ three times and repeating the experiment after each shift.

B. Squeezed reference state

The above analysis and suggested procedure assumes that the reference field is in a perfect binomial state. If, instead, we use the squeezed state approximation to the binomial state as derived in the Appendix, then the vacuum state coefficient differs from the ideal value and the measured state is no longer projected onto the truncated phase state $|\theta_m\rangle$, but is instead projected onto a state proportional to

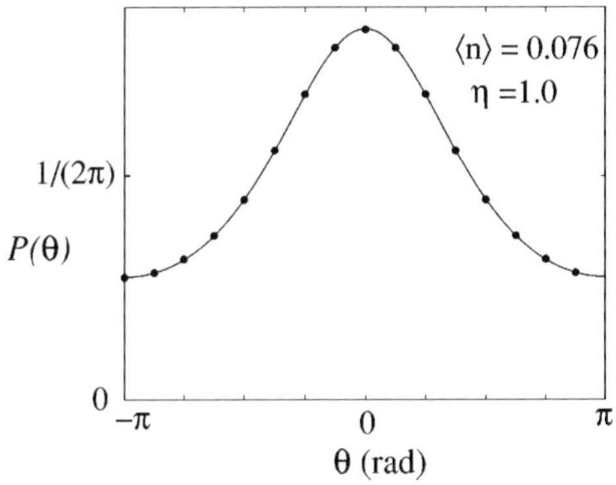

FIG. 2. Canonical phase probability distribution $P(\theta)$ for a coherent-state field with a mean photon number of 0.076. The full line is the theoretical result and the dots are simulated measured results with ideal detectors for four different phase settings of the squeezed reference field.

$$|0\rangle+\exp(i\theta_m)|1\rangle+\exp(2i\theta_m)|2\rangle+1.0146\exp(3i\theta_m)|3\rangle. \tag{20}$$

We would expect that this would lead to some small errors when the procedure suggested above is applied. In practice, if we are measuring a state, such as a coherent or squeezed state, that does not have a truncated photon number distribution, the error caused by the modulus of the three-photon coefficient in expression (20) differing from unity may, in general, be smaller that the error caused by the truncation of expression (20) after the three-photon component. In Fig. 2, we show the points obtained from a simulated experiment for a coherent state with a mean photon number of 0.076, which is is comparable to the field strength of interest in Ref. [6], using a squeezed reference state. The close agreement with the canonical distribution is apparent. For weaker fields, for example, the other field of interest in Ref. [6] with a mean photon number of 0.047, the agreement is even closer. The agreement is still good for stronger coherent-state fields with mean photon numbers of 0.139 and 0.23, as used in Ref. [7], with divergence from the canonical distribution becoming apparent for mean photon numbers of around 0.4. Figure 3 shows simulated results for a coherent state with a mean photon number of 0.5. The error here is almost entirely due to the truncation of the phase state rather than to the nonunit coefficient of the fourth term in Eq. (20). A mean photon number of 0.5 represents the approximate limit to the field strength for a coherent state for which this measurement technique is suitable.

IV. SOME PRACTICAL CONSIDERATIONS

To obtain an idea of the number of experiments needed to measure the required probabilities, we note that, for weak coherent fields of the strengths discussed here, the coefficient of the vacuum component dominates, so the probability of detecting a total of three photons will be approximately the

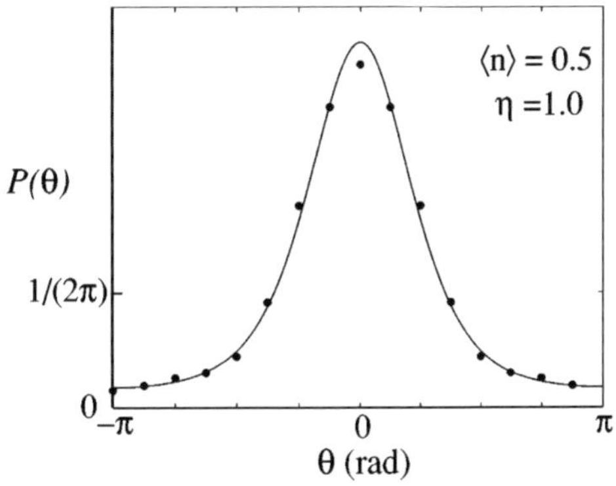

FIG. 3. Simulated measurements (dots) and the theoretical canonical phase distribution (full line) for a coherent state field with a mean photon number of 0.5 with ideal detectors.

probability that the binomial reference state contains three photons, which is 12.5%. These photons can be detected as a variety of events such as (0,1,1,1), (0,0,2,1), or (0,0,0,3). The fraction of these events that are required events, that is three separate counts of one photon and one count of zero, is 3/32. Thus, $\approx$4.7% of the experiments will produce one of the four desired events. After running the experiment sufficient times to obtain the four probabilities with the desired accuracy, the phase of the reference state is then changed and the procedure is repeated to obtain four more probabilities. The question therefore arises: how many probabilities, that is, points on the probability distribution, do we need to determine the distribution with reasonable accuracy? The fact that we are synthesizing a projection onto a truncated phase state with four nonzero photon number state coefficients means that the weak fields being measured must, by necessity, have negligible coefficients for components with photon numbers greater than three. From Eq. (1) this means that the most rapidly oscillating Fourier component of the distribution $P(\theta)$ behaves as $\exp(i3\theta)$. To detect this oscillation we would need a minimum of about 12 equally spaced values of $P(\theta)$. Thus, running the experiment with four different phase settings, giving a total of 16 points on the distribution curve as shown in Fig. 2, should normally be sufficient.

In a practical experimental situation, errors can arise from collection inefficiencies, nonunit quantum efficiencies for one-, two-, and multiple-photon detection; dead times; and accidental counts arising from dark counts and background light. The fact, however, that Noh *et al.* [4] have performed successful experiments involving the measurement of joint detection probabilities with an eight-port interferometer, by means of photon counting, for states with field strengths similar to those of interest in this paper is an encouraging indication that there should be no insurmountable difficulties for the method proposed here arising from such errors. It is worth considering some specific aspects of the sources of error. In the experiments of Noh *et al.* [4], photon count rates were of the order of $10^4\ \mathrm{s}^{-1}$ with a counting interval of about 5 μs, to give the required small mean photon number, and dead-time effects were negligible. In the present proposed experiment, dead times are even less important because it is only necessary to discriminate among zero, one, and many counts rather than among general numbers of counts [25]. Dark counts can be reduced to about 200 s^{-1} [4] or even to 20 s^{-1} [26] by cooling the detectors, and background light can be reduced by appropriate shielding. In the event that the residual dark and background counts are not negligible, the measured joint probabilities of the four photocount events can be corrected by a deconvolution procedure using the data obtained by blocking the input signals [4].

Concerning detector efficiencies, even if collection efficiencies are made to approach unity by, for example, suitable geometry and reflection control, there will still be some detector inefficiency due to nonunit quantum efficiency, so some correction for detector inefficiency may be needed. Conventional single-photon counting module detectors can have an efficiency of around 0.7 [27], while visible light photon counters that distinguish between single-photon and two-photon incidence can have quantum efficiencies of about 0.9 with some sacrifice of the smallness of the dark count rate [27]. We denote the one-photon detection efficiency, that is the probability of recording one photocount if one photon is present, by η. Then, as dead times are not important, the general multiple-detection efficiency is such that the probability of recording n photocounts if N photons are present is $\binom{N}{n}\eta^n(1-\eta)^{N-n}$, where the first factor is the binomial coefficient [28]. If η is the same for all four detectors, the probability for the joint four-count detection event (m,n,p,q) is given by

$$P_c(m,n,p,q)=\sum_{s=m}^{\infty}\sum_{t=n}^{\infty}\sum_{u=p}^{\infty}\sum_{v=q}^{\infty}\binom{s}{m}\binom{t}{n}\binom{u}{p}\binom{v}{q} \times \eta^{m+n+p+q}(1-\eta)^{s+t+u+v-m-n-p-q} \times P_I(s,t,u,v), \tag{21}$$

where $P_I(s,t,u,v)$ is the probability that an ideal detector would have detected the joint four-count event (s,t,u,v). Relation (21) can be inverted by the use of the four-function Bernoulli transform, which is straightforward to derive in a manner similar to that of the two-function transform derived in Ref. [29]. This allows us to calculate the ideal probabilities from the measured probabilities and thus correct for nonunit efficiencies.

Although we can correct for nonunit efficiencies, an analysis shows that the effect of not correcting for them is not as serious as it may first appear. Essentially, this is because the four measured probabilities are always normalized so that their sum is $2/\pi$. The major effect of η not being unity is, as can be seen from Eq. (21), that the probabilities for the four events (0,1,1,1), (1,0,1,1), (1,1,0,1), and (1,1,1,0) to be actually recorded are reduced by a factor η^3. As this affects the event probabilities uniformly, however, the effect vanishes upon normalization. The next-order effect is that some ideal four-count events, such as (1,1,1,1) and (0,2,1,1), are registered, for example, as (0,1,1,1) because of the inefficiency. The effect of this is only partially removed by the normalization. Ideal higher-count events also contribute to the error, but for weak fields the probability of ideal high-count events is not large. A numerical calculation of the total effect of nonunit efficiency, including the effect of normalization, shows that the proposed procedure is not highly sensitive to detector inefficiency, provided the efficiency is reasonable, for the weak states of interest. More precisely, for coherent states with a mean photon number up to 0.5 photons, as discussed above, the error in the final normalized probabilities is less than 2% for $\eta \geqslant 0.9$. For a mean photon number of 0.076, the error is less that 0.5% for such efficiencies. In Fig. 4, we show the effect of a poorer efficiency of $\eta=0.6$ for a mean photon number of 0.076.

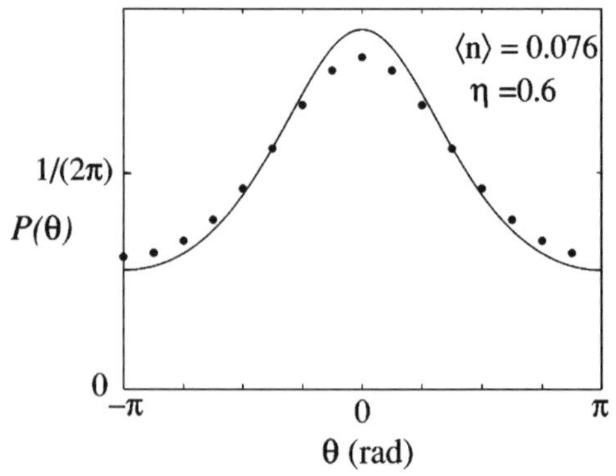

FIG. 4. Uncorrected simulated measurements (dots) and the theoretical canonical phase distribution (full line) for a coherent-state field with a mean photon number of 0.076 where the photodetectors have an efficiency $\eta=0.6$.

To produce the squeezed state required as an approximate binomial state, we note that squeezed vacuum states can be transformed into various types of squeezed states, the squeezing axis can be rotated, coherent amplitude can be added, and the squeezing can be controlled independent of the coherent amplitude. The degree of squeezing needed here is of a magnitude that is a realistic expectation either now or in the near future [30].

Our discussion has focused mainly on measuring the phase probability distribution of pure states of light. It is straightforward, however, to show that our procedure can also be used to measure the distribution of a mixed state. If the mixed state is represented by a density operator that is a weighted sum of pure state-density operators $|\psi\rangle\langle\psi|$ for various states $|\psi\rangle$, then the phase probability density is just the corresponding weighted sum of values of $P(\theta)$ given by expression (1) for the various states $|\psi\rangle$. To measure this, we can use precisely the same procedure as described in this paper with the same binomial reference state.

V. CONCLUSION

We have shown in this paper how the eight-port interferometer used by Noh *et al.* [4,5] to measure their operational phase distribution can used to measure the canonical phase distribution given by Eq. (1), where the canonical phase is defined as the complement of photon number. The procedure is applicable for weak fields in the quantum regime, by which we mean explicitly the states for which number-state components for photon numbers greater than three are negligible. For coherent states, this requirement translates to a mean photon number of a half a photon or less. This is precisely the quantum regime in which large differences between the operational phase and the canonical phase distributions are most apparent. For example, fields of interest in Refs [6,7] are coherent states with mean photon numbers of 0.23, 0.139, 0.076, and 0.047. The success of the experiments in the foregoing references indicates that the procedure proposed in this paper should be viable, given a reliable source of the required reference state.

The procedure in this paper has advantages over the original projection synthesis method proposed for measuring the canonical phase distribution. The most significant of these is that the present procedure requires a binomial reference state rather than a reciprocal binomial state. We have shown that the required binomial state can be approximated by a squeezed state sufficiently well for our purpose. Another advantage is that we require only photodetectors that can distinguish among zero photocounts, one photocount, and more than one photocount. The measurements are not particularly sensitive to photodetector inefficiency and, for reasonably good detector efficiencies, no corrections should be needed. Overall, we feel that the proposal in this paper brings the measurement of the canonical phase distribution for weak optical fields closer to reality.

APPENDIX: BINOMIAL STATES

In this appendix, we show how the required binomial reference state can be approximated by a suitably squeezed state. The particular binomial state of interest to us is given by

$$|B\rangle=\sum_{n=0}^{N}\beta_n|n\rangle=2^{-N/2}\sum_{n=0}^{N}\binom{N}{n}^{1/2}|n\rangle, \tag{A1}$$

where $\binom{N}{n}$ is the binomial coefficient. The binomial state derived in Ref. [20] with alternating signs for the number state coefficients can be obtained by phase shifting this state by π.

The general form for a squeezed state is [31]

$$|\alpha,\zeta\rangle=\sum_{n=0}^{\infty}\alpha_n|n\rangle=(\cosh|\zeta|)^{-1/2}$$
$$\times\exp\{-\tfrac{1}{2}[|\alpha|^2+t(\alpha^*)^2]\}$$
$$\times\sum_{n=0}^{\infty}\frac{(t/2)^{n/2}}{(n!)^{1/2}}H_n\left(\frac{\alpha+t\alpha^*}{(2t)^{1/2}}\right)|n\rangle, \tag{A2}$$

where $\zeta=|\zeta|\exp(i\phi)$ with $|\zeta|$ being the squeezing parameter, $t=\exp(i\phi)\tanh|\zeta|$, and $H_n(x)$ is a Hermite polynomial of order n. α is the complex amplitude of the coherent state obtained in the limit of zero squeezing.

The first case we study is where we are interested in finding a squeezed state whose coefficients α_n are proportional to the coefficients β_n of binomial state for the early terms, that is, for $n\ll N$. In this case, we can approximate the binomial coefficient by

$$\binom{N}{n}^{1/2}=\frac{N^{n/2}}{\sqrt{n!}}\sqrt{\left(1-\frac{1}{N}\right)\left(1-\frac{2}{N}\right)\cdots\left(1-\frac{n-1}{N}\right)}$$
$$\approx\frac{N^{n/2}}{\sqrt{n!}}\left[1-\frac{n(n-1)}{4N}\right]. \tag{A3}$$

We can approximate the Hermite polynomial for large x by its leading terms:

$$H_n(x)\approx(2x)^n-n(n-1)(2x)^{n-2}. \tag{A4}$$

We find remarkably that choosing $t=0.5$ and $\alpha=(2/3)N^{1/2}$ allows us to write

$$\binom{N}{n}^{1/2}\approx\frac{(t/2)^{n/2}}{(n!)^{1/2}}H_n\left[\frac{\alpha+t\alpha^*}{(2t)^{1/2}}\right] \tag{A5}$$

for $n\ll N$. Thus, the first n number-state coefficients of a squeezed state with these values of t and α will be proportional to the required binomial coefficients to a good approximation. With this degree of squeezing, the squeezed quadrature variance is 1/3 that of the vacuum level, that is, 4.77 dB below the standard quantum limit.

The case opposite to the above is where we require a small number of coefficients α_n for $n=N,N-1,N-2,\ldots$ to be proportional to β_n. It is not easy to obtain as general a relationship as the above, so we look at each case individually. In this paper, we are interested in the particular case with four values of β_n, that is, $N=3$. By using the explicit form of the Hermite polynomials in Eq. (A2) and setting $\alpha_2/\alpha_3=\beta_2/\beta_3$ and $\alpha_1/\alpha_3=\beta_1/\beta_3$, we find that the values $t=0.5$ and $\alpha=(2+2^{1/2})/3$ satisfy the two simultaneous equations obtained. We note that the required squeezing parameter $\tanh^{-1}0.5$ is the same as for the first case above, but the value 1.138 of α varies slightly from 1.155, the value of $(2/3)N^{1/2}$ with $N=3$, which is required to make the first few coefficients of $|\alpha,\zeta\rangle$ proportional to binomial coefficients. We also note that with perfect matching of the last three coefficients, the ratio α_0/α_3 becomes 1.0146, a mismatch of only 1.5% with the corresponding binomial coefficient.

ACKNOWLEDGMENT

D.T.P. acknowledges the Australian Research Council for funding.

[1] P.A.M. Dirac, Proc. R. Soc. London, Ser. A **114**, 243 (1927).
[2] For a bibliography see D.T. Pegg, and S.M. Barnett, J. Mod. Opt. **44**, 225 (1997).
[3] For an overview of different approaches to describing phase, see S.M. Barnett and B.J. Dalton, Phys. Scr., **T48**, 13 (1993).
[4] J.W. Noh, A. Fougères and L. Mandel, Phys. Rev. A **45**, 424 (1992); **46**, 2840 (1992).
[5] J.W. Noh, A. Fougères, and L. Mandel, Phys. Rev. Lett. **71**, 2579 (1993).
[6] J.R. Torgerson and L. Mandel, Phys. Rev. Lett. **76**, 3939 (1996).
[7] J.R. Torgerson and L. Mandel, Opt. Commun. **133**, 153 (1997).
[8] C.W. Helstrom, *Quantum Detection and Estimation Theory* (Academic Press, New York, 1976); J.H. Shapiro and S.R. Shepard, Phys. Rev. A **43**, 3795 (1991).
[9] R.G. Newton, Ann. Phys. (N.Y.) **124**, 327 (1980).
[10] D.T. Pegg and S.M. Barnett, Europhys. Lett. **6**, 483 (1988); Phys. Rev. A **39**, 1665 (1989); S.M. Barnett and D.T. Pegg, J. Mod. Opt. **36**, 7 (1989).
[11] U. Leonhardt, J.A. Vaccaro, B. Böhmer, and H. Paul, Phys. Rev. A **51**, 84 (1995).
[12] D.T. Pegg, J.A. Vaccaro, and S.M. Barnett, J. Mod. Opt. **37**, 1703 (1990).
[13] It is possible to obtain the distribution of any optical observable, however defined, by first measuring the complete state and then calculating the defined distribution from this. See, for example, M. Beck, D.T. Smithey, J. Cooper, and M.G. Raymer, Opt. Lett. **18**, 1259 (1993); K.L. Pregnell and D.T. Pegg, J. Mod. Opt. **49**, 1135 (2002); K.L. Pregnell and D.T. Pegg, Phys. Rev. A **66**, 013810 (2002). What is of interest in discussing the measurability of theoretical and operational phase distributions, however, is finding a more direct procedure, that is, one which does not involve obtaining sufficient information to determine the complete state.
[14] J.A. Vaccaro and D.T. Pegg, Opt. Commun. **105**, 335 (1994).
[15] S.M. Barnett and D.T. Pegg, Phys. Rev. Lett. **76**, 4148 (1996).
[16] D.T. Pegg, S.M. Barnett, and L.S. Phillips, J. Mod. Opt. **44**, 2135 (1997); D.T. Pegg, L.S. Phillips, and S.M. Barnett, Phys. Rev. Lett. **81**, 1604 (1998).
[17] M. Dakna, J. Clausen, L. Knöll, and D.-G. Welsch, Phys. Rev. A **59**, 1658 (1999).
[18] K. Mattle, M. Michler, H. Weinfurter, A. Zeilinger, and M. Zukowski, Appl. Phys. B: Photophys. Laser Chem. **60**, S111 (1995).
[19] P. Törmä, S. Stenholm, and I. Jex, Phys. Rev. A **52**, 4853 (1995).
[20] K.L. Pregnell and D.T. Pegg, Phys. Rev. Lett. **89**, 173601 (2002).
[21] D. Stoler, B.E.A. Saleh, and M.C. Teich, Opt. Acta **32**, 345 (1985); Y. Aharonov, E.C. Lerner, H.W. Huang, and J.M. Knight, J. Math. Phys. **14**, 746 (1973).
[22] A. Vidiella-Barranco and J.A. Roversi, Phys. Rev. A **50**, 5233 (1994).
[23] N.G. Walker and J.E. Carroll, Opt. Quantum Electron. **18**, 355 (1986); N.G. Walker, J. Mod. Opt. **34**, 15 (1987).
[24] See, for example, S.M. Barnett and P.M. Radmore, *Methods in Theoretical Quantum Optics* (Oxford University Press, Oxford, 1997).
[25] In cases where dead times are significant their effect can be substantially reduced by use of beam splitters, see H. Paul, P. Törma, T. Kiss, and I. Jex, Phys. Rev. Lett. **76**, 2464 (1996).
[26] A. Trifonov *et al.*, J. Opt. B: Quantum Semiclassical Opt. **2**, 105 (2000).
[27] K. Tsujino, S. Takeuchi, and K. Sasaki, Phys. Rev. A **66**, 042314 (2002); J. Kim, S. Takeuchi, Y. Yamamoto, and H.H. Hogue, Appl. Phys. Lett. **74**, 902 (1999); S. Takeuchi, J. Kim, Y. Yamamoto, and H.H. Hogue, *ibid.* **74**, 1063 (1999).
[28] C.T. Lee, Phys. Rev. A **48**, 2285 (1993); S.M. Barnett, L.S. Phillips, and D.T. Pegg, Opt. Commun. **158**, 45 (1998).
[29] D.T. Pegg and S.M. Barnett, J. Mod. Opt. **46**, 1657 (1999).
[30] H. Bachor, *A Guide to Experiments in Quantum Optics* (Wiley, Brisbane, 1998), p. 288.
[31] R. Loudon and P.L. Knight, J. Mod. Opt. **34**, 709 (1987).

6

Experimental Demonstrations

Measurement of the optical phase has proven to be far from straightforward. Perhaps the simplest reasons for this are that all current optical measurements are based, ultimately, on absorption of the light and so are dependent on the number of photons present. Phase, however, is the conjugate of number and a true phase measurement should be insensitive to the photon number. A number of ingenious schemes for measuring optical phase were presented in the previous chapter, but to date only one of these has been demonstrated (Armen *et al.* 2002 **Paper 6.3**). Optical phase has been studied experimentally, however, through the state-measurement technique of optical homodyne tomography and in particular, the number-phase uncertainty relation has been tested in this way (Beck *et al.* 1993 **Paper 6.1**, Smithey *et al.* 1993b **Paper 6.2**). Azimuthal angle is much more readily observed and the uncertainty relation between orbital angular momentum and this angle has been demonstrated in experiment (Franke-Arnold *et al.* 2004 **Paper 6.4**).

A quantum state has, encoded within it, all of the information it is possible to specify about a quantum system. Although the state is not itself an observable entity, it is possible to experimentally determine the state of the system if we have a sufficient number of identically prepared copies. We can achieve this by performing a sufficiently large number of measurements, one on each system, of a suitable set of observables. For an optical mode, a sufficient set of observables is provided by the quadratures (5.2) measured for a large number of different angles θ. From these it is possible to reconstruct the Wigner function for the mode (Vogel and Risken 1989, Smithey *et al.* 1993a) or, indeed, the state of the mode (Smithey *et al.* 1993a, 1993b **Paper 6.2**, 1993c). The experimentally reconstructed state can be used to obtain the statistics of any desired observable and in particular this has been used to give experimentally derived uncertainties for both the photon number and the phase and also for the expectation value of the phase-photon number commutator for coherent states with a range of mean photon numbers. These were then compared with the uncertainty relation, which for physical states takes the form:

$$\begin{aligned}\Delta\phi_\theta \Delta N &\geq \frac{1}{2}\left|\langle[\hat{\phi}_\theta, \hat{N}]\rangle\right| \\ &= \frac{1}{2}\left|1 - 2\pi P(\theta_0)\right|. \end{aligned} \tag{6.1}$$

For large photon numbers the coherent states saturate this inequality with an uncertainty product of $\frac{1}{2}$ (Barnett and Pegg 1989 **Paper 2.3**). For weaker fields the coherent states have a larger uncertainty product than the allowed minimum. The results derived from the experimentally measured coherent states produced excellent agreement with the theoretically predicted values for the number and phase uncertainties and also for the expectation value of their commutator (Beck *et al.* 1993 **Paper 6.1**, Smithey *et al.* 1993b **Paper 6.2**).

Homodyne measurement is, to a good approximation, a measurement of one of the field quadratues with the quadrature selected depending on the phase of a local oscillator

(Loudon and Knight 1987, Loudon 2000). Heterodyne detection, or alternatively eight-port homodyne detection (Walker and Carrol 1986, Walker 1987), provides a joint measurement of two quadratures differing in phase by $\pi/2$. These are non-commuting observables and hence a joint measurement inevitably introduces additional noise over and above that which is intrinsic to the state (Arthurs and Kelly 1965, Stenholm 1992). The price paid for this joint measurement is a doubling of the uncertainty product derived from the measured outcomes. The resulting enhanced phase uncertainty may be seen in eight-port homodyne measurements (Walker and Carol 1986). Homodyne measurements take a finite time and its was Wiseman (1995 **Paper 5.3**) who first suggested that this could be used to advantage to provide a more accurate phase measurement. The idea is simply that varying the relative phase between the local oscillator and the mode undergoing measurement will cause an imbalance in the photocurrents measured in the two detectors. The resulting difference photocurrent can be used to modify the phase of the local oscillator *during* the measurement. This is essentially an implementation of a phase-lock loop, but one that is operating at the quantum limit. An experimental demonstration of this idea produced a peak in the phase probability distribution with a smaller width than that found using heterodyne detection (Armen *et al.* 2002 **Paper 6.3**) and in good agreement with theoretical predictions (Wiseman and Killip 1998).

The most direct experimental study of phase variables to date has been carried out for angle and angular momentum rather than number and phase variables. The operators representing these are constructed in a similar manner to the number and phase and for physical states they satisfy the commutation relation (2.43) (Barnett and Pegg 1990a **Paper 2.5**). This commutation relation implies the existence of an uncertainty relation, which for physical states takes the form

$$\begin{aligned}\Delta\phi_\theta \Delta L_z &\geq \frac{1}{2}\,|\langle[\phi_\theta, L_z]\rangle| \\ &= \frac{\hbar}{2}\,|1 - 2\pi P(\theta_0)|\,. \end{aligned} \tag{6.2}$$

In the experimental work (Franke-Arnold *et al.* 2004 **Paper 6.4**) two simplifications were made: the first was to introduce a scaled angular momentum $\ell = L_z/\hbar$ and the second was to choose $\theta_0 = -\pi$ so that the range of angles runs from $-\pi$ to π. This leads to a notationally simpler form for the uncertainty relation:

$$\Delta\phi\Delta\ell \geq \frac{1}{2}\,|1 - 2\pi P(\pi)|\,, \tag{6.3}$$

where the subscript $-\pi$ has been omitted from ϕ and, of course, the periodicity of the angle requires that $P(-\pi) = P(\pi)$.

The uncertainty relation (6.3) has a lower bound that itself depends on the state of the system. For this reason it is possible to define at least two distinct types of "minimum uncertainty" states: the intelligent states and the minimum uncertainty product states (Jackiw 1968, Aragone *et al.* 1976, Vaccaro and Pegg 1990b **Paper 4.4**, see also Chapter 4). The intelligent states are those which satisfy the equality in the uncertainty relation. These do not give the smallest possible value of the uncertainty product $\Delta\phi\Delta\ell$, however, as it is possible to reduce this value for a given $\Delta\phi$ (or $\Delta\ell$) and in the process decrease by a greater amount the lower bound in the uncertainty relation. The states which minimise the uncertainty product are the minimum uncertainty product states (or constrained minimum uncertainty product states). The intelligent states (with the phase eigenvalues in the range $-\pi$ to π) have the simple form of a Gaussian, centred at $\phi = 0$ (Galindo and Pascual 1990, Franke-Arnold *et al.*

2004 **Paper 6.4**, Pegg *et al.* 2005 **Paper 6.5**):

$$\Psi(\phi) = \frac{(\lambda/\pi)^{1/4}}{\sqrt{\mathrm{erf}(\pi\sqrt{\lambda})}} \exp(i\bar{\ell}\phi)\exp\left(-\frac{\lambda\phi^2}{2}\right) \qquad (-\pi \le \phi < \pi). \tag{6.4}$$

Here $\bar{\ell}$ is the (integer-valued) mean angular momentum and λ is a real parameter, the value of which determines the uncertainties. This wavefunction is, of course, periodic in the angle and this means that there is a cusp in the wavefunction at $\phi = \pi$, the edge of the range of angles associated with the angle observable. The minimum uncertainty product states have a more complicated form than the intelligent states (6.4) but can be expressed exactly in terms of a confluent hypergeometric function (Pegg *et al.* 2005 **Paper 6.5**). The associated wavefunction has a form that is as close to the sum of a set of overlapping Gaussians, centred on the angles $2n\pi$. These states have the required periodic behaviour but lack the cusps found for the intelligent states. The angular uncertainty can become large and this occurs for states with peaks in the angular probability distribution at the edges of the range of angles (in this case at $-\pi$ and at π). It is also possible to find intelligent states (Galindo and Pascual 1990, Götte *et al.* 2005) and minimum uncertainty product states for such states (Götte *et al.* 2006).

The system of choice for testing the uncertainty relation (6.2) is beams of light carrying orbital angular momentum. In particular the Laguerre-Gaussian modes, with field amplitudes in the form

$$u_{p\ell}^{LG} = \frac{C_{p\ell}}{1+z^2/z_R^2}\left(\frac{\rho\sqrt{2}}{w(z)}\right)^{|\ell|} L_p^{|\ell|}\left(\frac{2\rho^2}{w^2(z)}\right)\exp\left(\frac{-\rho^2}{w^2(z)}\right)\exp(i\ell\phi)$$
$$\times \exp\left(\frac{-ik\rho^2 z}{2(z^2+z_R^2)}\right)\exp[i(2p+|\ell|+1)\tan^{-1}(z/z_R)], \tag{6.5}$$

have been shown to carry $\ell\hbar$ units of orbital angular momentum about the beam axis for each photon (Allen *et al.* 1992, 1999, 2003). Here $C_{p\ell}$ is a normalisation constant z_R is the Rayleigh range, $w(z) = [2(z^2 + z_R^2)/kz_R]^{1/2}$ is the radius of the beam and $L_p^{|\ell|}$ is an associated Laguerre polynomial. We should note that, because of historical accident, the number of units of the z component of angular momentum is denoted by m in quantum theory but by ℓ in optics, so that in optics the eigenvalues of $\hat{L}_z$ (represented as $-i\hbar\partial/\partial\phi$) are usually denoted $\ell\hbar$. The azimuthal-dependent part of the field amplitude plays the role of a wavefunction $\Psi(\phi)$ and the angular uncertainty is its associated $\Delta\phi$ and this means that the Laguerre-Gaussian modes (6.5) have a uniform angle probability distribution, $P(\phi) = (2\pi)^{-1}$, with angular uncertainty $\pi/\sqrt{3}$. In the experiment a mask was used to impose the desired angular (6.4) form on field amplitude of a Gaussian mode (Franke-Arnold *et al.* 2004 **Paper 6.4**). This process imposed the desired angular uncertainty on the light. The angular momentum spectrum of the resulting field was measured by noting that only the zero-angular momentum modes, $\ell = 0$, have an on-axis intensity and so can pass through a pin-hole. Hence it is easy to separate this component from the remaining light. The measurement can be completed by using a hologram with a dislocation at its centre to shift the angular momentum by a controlled amount (Bazhenov *et al.* 1990). Hence it is possible to transform, in turn, each angular momentum component into a zero angular momentum mode and measure how much light is present (Mair *et al.* 2001). The angular momentum uncertainty was calculated from this measured spectrum and multiplied by the angular uncertainty. This product was compared with the theoretical bound with very good agreement.

Experimental determination of number–phase uncertainty relations

M. Beck, D. T. Smithey, J. Cooper,* and M. G. Raymer

Department of Physics and Chemical Physics Institute, University of Oregon, Eugene, Oregon 97403

Received March 18, 1993

An experimental determination of the uncertainty product for the phase and photon number of a mode of the electromagnetic field is performed. The expectation value of the commutator that sets the lower bound for the uncertainty product is also determined experimentally. This is accomplished by using optical homodyne tomography to measure the density matrix of a small-photon-number coherent state. The experimental results agree with the quantum-mechanical predictions.

If a complete measurement of the uncertainty relation for photon number $\hat{n}$ and the field phase $\hat{\phi}$ is to be performed, it is necessary to develop means for experimentally determining the variances of these quantities, $\Delta n^2 = \langle(\hat{n} - \langle\hat{n}\rangle)^2\rangle$ and $\Delta\phi^2 = \langle(\hat{\phi} - \langle\hat{\phi}\rangle)^2\rangle$, where $\langle\ldots\rangle$ denotes an expectation value. The uncertainty relation is

$$\Delta\phi\Delta n \geq (1/2)|\langle\psi|[\hat{\phi}, \hat{n}]|\psi\rangle|, \tag{1}$$

where $|\psi\rangle$ is the state of the field. It can be seen from relation (1) that one also needs to measure the expectation value of the number–phase commutator $[\hat{\phi}, \hat{n}]$, which is an operator and thus is state dependent in general. Previous experiments have measured variances of $\hat{n}$ and $\hat{\phi}$ separately,[1,2] but the commutator expectation value had not been previously measured. We describe an experiment that determines the distributions for the phase and photon number and the commutator expectation for a field containing small numbers of photons. In our approach, the quantities in relation (1) are determined from the experimentally measured quantum state of the field mode. The state is measured with optical homodyne tomography.[3,4] Because the determination of these quantities assumes the validity of quantum mechanics, our method does not provide an independent verification of the uncertainty principle. Our method does provide a measure of how close arbitrary states come to achieving the equality in relation (1). For example, the coherent states are found not to achieve the equality in general. These measurements are in agreement with the quantum-mechanical predictions of the uncertainty principle.

Defining a quantum-mechanical phase of an electromagnetic field has generated much recent interest, and several different phase definitions have been proposed.[1,5,6] In this Letter we use the Pegg–Barnett Hermitian phase operator because of its calculational simplicity.[6] We obtain the phase distribution corresponding to this operator without actually measuring eigenvalues of the operator on individual trials.

The technique of optical homodyne tomography requires that an ensemble of many measurements be performed.[3,4] In this case the relevant interpretation of the uncertainty relation is that the state of the field does not have a well-defined phase and photon number. Quadrature operators for the electric field are defined as $\hat{x} = (\hat{a} + \hat{a}^\dagger)/\sqrt{2}$ and $\hat{p} = (\hat{a} - \hat{a}^\dagger)/i\sqrt{2}$, where $\hat{a}$ is the annihilation operator for a particular spatial–temporal mode of the field. From these operators, rotated quadrature operators can be defined as $\hat{x}_\theta = \hat{x}\cos\theta + \hat{p}\sin\theta$ and $\hat{p}_\theta = -\hat{x}\sin\theta + \hat{p}\cos\theta$. A balanced homodyne detector makes measurements that correspond to the quadrature operator $\hat{x}_\theta$ of the signal field, when the local-oscillator (LO) field is in a large-amplitude coherent state.[7] The spatial–temporal mode of the signal field selected by the homodyne detector is the same as that of the LO,[4] and the measured quadrature is determined by the phase of the LO, θ.

In our experiments an ensemble of measurements of quadrature amplitudes $\hat{x}_\theta$ are made to determine the probability distributions $P_\theta(x_\theta)$ for various values of θ. These distributions can be written in terms of the Wigner function $W(x, p)$ (Ref. 8) of the field mode[3]:

$$P_\theta(x_\theta) = \int_{-\infty}^{\infty} W(x_\theta\cos\theta - p_\theta\sin\theta,\, x_\theta\sin\theta + p_\theta\cos\theta)\mathrm{d}p_\theta\,. \tag{2}$$

For a set of such probability distributions determined on a continuous set of angles θ, Vogel and Risken[3] showed that one could invert the distributions with the inverse Radon transform to obtain the Wigner function.[3] If distributions are available for a finite and discrete set of angles between 0 and π, the inversion of Eq. (2) can still be performed with numerical techniques familiar in tomographic imaging.[9]

From the measured Wigner function, it is possible to construct the density matrix of the measured field mode by numerically performing a one-dimensional Fourier transform[8]:

$$\langle x + x'|\hat{\rho}|x - x'\rangle = \int_{-\infty}^{\infty} W(x, p)\exp(2ipx')\mathrm{d}p\,. \tag{3}$$

Because the density matrix completely determines the quantum-mechanical state of a system, optical homodyne tomography provides a way to determine this state. The density matrix of Eq. (3) can be expressed in the representation of the number states $|n\rangle$

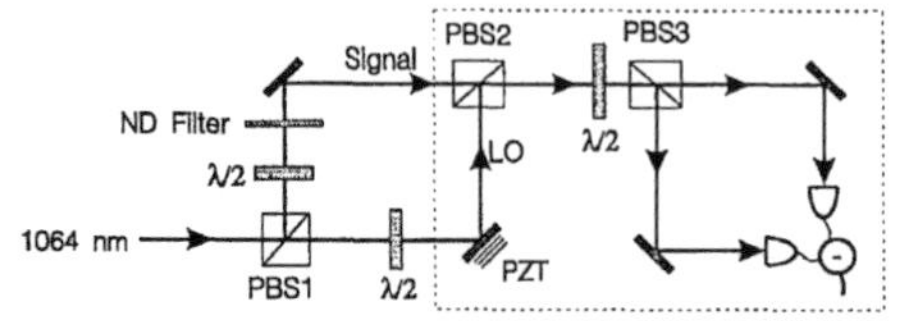

Fig. 1. Experimental apparatus. The dashed lines enclose the balanced homodyne detector.

through a change of basis by computation of overlap integrals of the density matrix in the x representation with Hermite polynomials.

In our experiment, the signal field was prepared in coherent states of differing average photon number. The experimental layout for measuring the field state is shown in Fig. 1. A Nd:YAG laser produces 300-ps, nearly transform-limited pulses at 1064 nm with a repetition rate of 420 Hz and a pulse-energy stability of ±3%. The pulses are split with a polarizing beam splitter (PBS1) to create the signal and the LO beams. Half-wave plates and neutral-density (ND) filters in each arm are used to control the relative intensities of these two beams. The piezoelectric translator (PZT) pushes a mirror that adjusts the phase of the LO. Each LO pulse contains approximately 2×10^6 photons, whereas the number of photons in the signal pulse is varied between 0 and ~10. The signal and LO beams are superimposed with orthogonal polarizations on a polarizing beam splitter (PBS2). Our balanced homodyne detector consists of a half-wave plate that rotates the polarizations of the signal and the LO fields by 45° and a polarizing beam splitter (PBS3) that interferes the signal and the LO to produce two output fields that are detected by high-quantum-efficiency (~85%) InGaAs photodiodes. The photodiode outputs are electronically subtracted and integrated with a low-noise charge-sensitive preamplifier.[10] For each laser pulse this subtracted and integrated signal yields the photoelectron difference number between the outputs of the two detectors. This balancing arrangement removes most of the additive classic noise on the LO and allows us to make measurements at the shot-noise level[11] (SNL). The electronic noise variance is approximately six times lower than the LO SNL, which is defined as the variance of photoelectron counts from a coherent-state LO.

For each coherent state with mean photon number $\bar{n}$, we make 5000 repeated measurements of the photoelectron difference number N_θ at 11 values of the LO phase equally spaced in a $[0, \pi]$ interval and calculate photoelectron distributions $P_{N_\theta}(N_\theta)$. The variance of each of these photoelectron distributions yields the SNL of the LO, denoted by $\bar{n}_{\mathrm{LO},\theta}$. The difference number is scaled to yield the quadrature amplitude with $x_\theta = N_\theta/(2\bar{n}_{\mathrm{LO}})^{1/2}$, where $\bar{n}_{\mathrm{LO}}$ is the average of the 11 values of $\bar{n}_{\mathrm{LO},\theta}$. We can determine the LO SNL, $\bar{n}_{\mathrm{LO}}$, in an independent manner by blocking the signal beam and one detector and directly measuring the number of LO photoelectrons that hit the other detector. Because the detectors are balanced, the total number of LO photoelectrons is twice this measured number. These two methods of obtaining the scaling factor $(\bar{n}_{\mathrm{LO}})^{1/2}$ agree to within 4%.[10] Scaling $P_{N_\theta}(N_\theta)$ then yields the distributions $P_\theta(x_\theta)$ of the quadrature amplitudes. From these distributions we use the inverse Radon transform to calculate the Wigner function, as described above. When performing the numerical inversion to obtain the Wigner function, we use the standard filtered backprojection algorithm for parallel-beam sampling geometry.[9]

We then extract the density matrix by using Eq. (3). A sufficient condition that a density matrix describes a pure state is $\mathrm{Tr}(\hat{\rho}^2) = 1$. When these traces are calculated for our experimentally determined density matrices, they are found to equal 1.00 ± 0.02, and consequently we conclude that our state is pure. From the measured density matrix we can calculate $\Delta\phi$, Δn, and $|\langle[\hat{\phi}, \hat{n}]\rangle|$.

When calculating moments of the signal photon number, we find it best not to use the photon-number distributions obtained from the density matrix. The reason for this is that the distributions do not decay completely to zero, and they must thus be truncated at some point; hence the moments of $\hat{n}$ are sensitive to exactly where the distribution is truncated. To avoid this problem, we write the photon-number operator $\hat{n} = \hat{a}^\dagger\hat{a}$ in terms of the quadrature operators, $\hat{n} = (1/2)(\hat{x}^2 + \hat{p}^2 - 1)$. The quadrature operators are then placed in Weyl order, and the moments are calculated as c-number integrals of the measured Wigner function.[8] Using these moments, we calculate the mean and the variance of the photon number. We estimate the mean to be accurate to better than 8%, limited by drift and noise of the LO pulses and the electronics as well as by the finite number of $P_\theta(x_\theta)$ distributions. An independent, but less accurate, check of the mean photon number yields results a factor of ~1.7 greater than that measured with the Wigner function.

The Pegg–Barnett Hermitian phase operator $\hat{\phi}$ is defined in a finite (but arbitrarily large) dimensional Hilbert space.[6] In an $(s + 1)$-dimensional Hilbert space, the phase states are defined as

$$|\phi\rangle = \frac{1}{\sqrt{s+1}} \sum_{n=0}^{s} \exp(in\phi)|n\rangle . \tag{4}$$

This Hilbert space is spanned by a complete orthonormal set of basis-phase states $|\phi_m\rangle$, given by Eq. (4) with ϕ replaced by $\phi_m = \phi_0 + 2\pi m/(s + 1)$, with $m = 0, 1, \ldots, s$ and ϕ_0 a reference phase. In terms of the states $|\phi_m\rangle$, the Hermitian phase operator is then

$$\hat{\phi} = \sum_{m=0}^{s} \phi_m |\phi_m\rangle\langle\phi_m| . \tag{5}$$

With this definition, the phase states are the eigenstates of the phase operator, i.e., $\hat{\phi}|\phi\rangle = \phi|\phi\rangle$. In our experiments, ϕ actually corresponds to the phase difference between the signal and the LO field. The normalized phase distribution for a state described by $\hat{\rho}$ in the number-state representation is

$$\begin{aligned} P_{pb}(\phi) &= [(s+1)/2\pi]\langle\phi|\hat{\rho}|\phi\rangle \\ &= \frac{1}{2\pi} \sum_{n,m=0}^{s} \exp[i(m-n)\phi]\langle n|\hat{\rho}|m\rangle . \end{aligned} \tag{6}$$

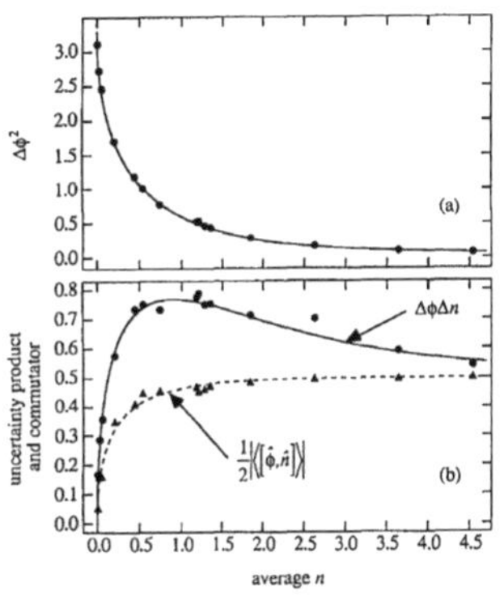

Fig. 2. (a) Variance of the phase plotted against the average number of photons in a coherent state. The points are the experimentally determined values, and the solid curve is calculated from the theory for the Pegg–Barnett phase definition. (b) Points are experimentally determined values for the uncertainty product (circles) and the commutator expectation (triangles). The curves are the theoretical values for the uncertainty product (solid curve) and the commutator expectation (dashed curve).

For all the coherent states we have measured ($\bar{n} \leq 8$), this distribution converges for $s \geq 20$.

Once the phase distributions are computed, moments of the phase can be calculated. The measured variances of the Pegg–Barnett phase as a function of the mean number of photons in the signal field are shown in Fig. 2(a). The theoretical variances for a coherent-state signal field are also plotted in Fig. 2(a).[12] Agreement between the experimental data and the theory is good and indicates that the states that we are measuring in the experiment are well described by pure coherent states.

Both the uncertainty product $\Delta\phi\Delta n$ and the expectation value of the commutator $(1/2)|\langle[\hat{\phi}, \hat{n}]\rangle|$ for our experimentally measured data are plotted in Fig. 2(b). For the experimental data we calculate the commutator expectation value numerically by expressing the matrices for $\hat{\phi}$ and $\hat{n}$ in the number-state representation, evaluate the matrix that corresponds to the commutator, and trace the commutator over the measured-density matrix in the number-state representation. Also plotted are the theoretically predicted values for coherent states. For the theoretical uncertainty product, $\Delta\phi$ is calculated in the same fashion as used to generate Fig. 2(a), whereas the photon-number standard deviation is $\Delta n = \sqrt{\bar{n}}$ for a coherent state. The equation describing the theoretical commutator is found in Ref. 6. Note that the uncertainty relation is satisfied (as it must be with our method of analysis); the uncertainty product is greater than the expectation value of the commutator. This means that, despite their purity, coherent states are not intelligent states with respect to number and phase. Intelligent states are defined as states for which relation (1) is an equality.[13] The only coherent state that is a minimum-uncertainty state is the vacuum, for which $\Delta\phi\Delta n = 0$. It is seen, however, that, in the limit of very small or very large coherent-state amplitude, the coherent states are approximately number–phase intelligent.

In conclusion, we have developed a technique for experimentally determining the standard deviation of the phase $\Delta\phi$ and the photon number Δn, as well as the expectation value of their commutator $|\langle[\hat{\phi}, \hat{n}]\rangle|$, for a mode of an optical field. We have thereby measured, for the first time to our knowledge, the number–phase uncertainty relation, which is in general state dependent. This method of field characterization will work in principle for arbitrary states of the field—the only limitation being available experimental parameters (e.g., detector efficiency, resolution). Agreement between the measured and theoretical points in Fig. 2 is ensured by agreement between the measured and theoretical density matrices. Therefore other phase definitions could be used, also leading to good agreement of experiment and theory. For this reason, our results do not favor one phase definition over another.

We acknowledge several helpful discussions with A. Faridani, H. Carmichael, and W. Schleich. This research is supported by National Science Foundation grant PHY 8921709-01.

*Permanent address, Joint Institute for Laboratory Astrophysics and the Department of Physics, University of Colorado and the National Institute of Standards and Technology, Boulder, Colorado 80309.

References

1. J. W. Noh, A. Fougeres, and L. Mandel, Phys. Rev. Lett. **67,** 1426 (1991); Phys. Rev. A **46,** 2840 (1992).
2. H. Gerhardt, U. Buchler, and G. Litfin, Phys. Lett. A **49,** 119 (1974); M. M. Nieto, Phys. Lett. A **60,** 401 (1977).
3. K. Vogel and H. Risken, Phys. Rev. A **40,** 2847 (1989).
4. D. T. Smithey, M. Beck, A. Faridani, and M. G. Raymer, Phys. Rev. Lett. **70,** 1244 (1993).
5. L. Susskind and J. Glogower, Physics **1,** 49 (1964); J. H. Shapiro and S. R. Shepard, Phys. Rev. A **43,** 3795 (1991).
6. D. T. Pegg and S. M. Barnett, Phys. Rev. A **39,** 1665 (1989); J. Mod. Opt. **36,** 7 (1989).
7. H. P. Yuen and J. H. Shapiro, IEEE Trans. Inform. Theory **IT-26,** 78 (1980).
8. M. Hillery, R. F. O'Connell, M. O. Scully, and E. P. Wigner, Phys. Rep. **106,** 121 (1984).
9. F. Natterer, *The Mathematics of Computerized Tomography* (Wiley, New York, 1986); A. Faridani, in *Inverse Problems and Imaging; Pitman Research Notes in Mathematics,* G. F. Roach, ed. (Wiley, Essex, England, 1991), Vol. 245, p. 68.
10. D. T. Smithey, M. Beck, M. Belsley, and M. G. Raymer, Phys. Rev. Lett. **69,** 2650 (1992).
11. H. P. Yuen and V. W. S. Chan, Opt. Lett. **8,** 177 (1983).
12. R. Lynch, Phys. Rev. A **41,** 2841 (1990).
13. J. A. Vaccaro and D. T. Pegg, J. Mod. Opt. **37,** 17 (1990).

Measurement of number-phase uncertainty relations of optical fields

D. T. Smithey, M. Beck, J. Cooper,* and M. G. Raymer
Department of Physics and Chemical Physics Institute, University of Oregon, Eugene, Oregon 97403
(Received 20 April 1993)

We have experimentally determined all of the quantities involved in the uncertainty relation for the phase and photon number of a mode of the electromagnetic field when the field mode is in a coherent state of small average photon number. This is accomplished by determining the quantum state of the field using optical homodyne tomography, which uses measured distributions of electric-field quadrature amplitude to determine the Wigner function and hence the density matrix. The measured state is then used to calculate the uncertainty product for the number and phase, as well as the expectation value of the commutator of the number and phase operators. The experimental results agree with the quantum-mechanical predictions. We also present measured phase- and photon-number distributions for these weak coherent states, as well as their measured complex wave functions.

PACS number(s): 42.50.Wm, 03.65.Bz

I. INTRODUCTION

The uncertainty principle applied to electromagnetic fields implies that the field phase ϕ and photon number n cannot simultaneously have arbitrarily narrow distributions. Mathematically the uncertainty relation can be written as [1]

$$\Delta\phi\Delta n \geq \tfrac{1}{2}|\langle\Psi|[\hat{\phi},\hat{n}]|\Psi\rangle| , \qquad (1)$$

where the phase variance $\Delta\phi^2=\langle\Delta\hat{\phi}^2\rangle=\langle\hat{\phi}^2\rangle-\langle\hat{\phi}\rangle^2$, Δn^2 is defined similarly, and the expectation values are taken over the state of the system $|\Psi\rangle$. The commutator $[\hat{\phi},\hat{n}]$ is an operator, and its expectation value is in general state dependent. Thus, to set the lower bound of experimental measurements of $\Delta\phi\Delta n$, one needs a way to determine the expectation value on the right-hand side of Eq. (1) for the state of interest.

Here we describe an experiment which determines the variances of phase and photon number for an optical field containing small numbers of photons, as well as the expectation value of the number-phase commutator [2]. This is accomplished by experimentally measuring the quantum state of the field by using optical homodyne tomography (OHT) [3–5]. Since the determination of the quantities in Eq. (1) uses this measured state and assumes the validity of quantum mechanics, the experiment does not provide an independent verification of the uncertainty relation [i.e., a test that is independent of quantum-mechanical principles]. However, the measurements do allow one to determine, for an arbitrary state, how close the inequality of Eq. (1) comes to *equality*. For example, it is found for small-amplitude coherent states that equality in the uncertainty relation [Eq. (1)] is not achieved. States for which Eq. (1) is an equality have been called "intelligent states." States which also minimize the right-hand side of Eq. (1) are called minimum-uncertainty states (MUS). As such, not all intelligent states are MUS [6]. Coherent states (other than the vacuum) are neither intelligent states, nor MUS, with respect to number and phase.

For our measurements, the relevant interpretation of the uncertainty relation is that the state of the field does not have well-defined phase and photon number. Since neither the phase nor photon number of the field are measured on individual trials, the results do not directly address the question of simultaneous measurability or unmeasurability of theses quantities [1,7]. Our experimental method of obtaining the uncertainty relation yields results equivalent to that of measuring photon-number variance directly, and on a separate subensemble, measuring phase variance directly. If an attempt is made at simultaneous (i.e., joint) measurement of two canonically conjugate variables on an individual ensemble member, then the result would be an uncertainty product which is larger than the above method. This result is connected to the requirement that the simultaneous measurement of two canonically conjugate observables necessarily adds additional noise [8,9].

Before discussing in detail number-phase uncertainty relations, one must clarify how the phase variable is to be defined. The problem of defining the phase of an electromagnetic field has generated much recent interest, and several different phase definitions have been proposed. One method of defining the phase is to define operators (Hermitian or operational) which correspond to the phase or to the sine and cosine of the phase [10–15]. Another definition is the Wigner phase distribution, which is a marginal Wigner distribution [16]. In this paper, we measure phase distributions corresponding to the Pegg-Barnett Hermitian phase operator and to the Wigner marginal distribution. In the case of a Hermitian phase operator $\hat{\phi}$, it is easy to formulate the uncertainty relation [Eq. (1)]. In the case of the Wigner phase distribution, there is no corresponding Hermitian phase operator. In fact, for some states the Wigner phase distribution can be negative, which implies that it is not a true phase distribution but a quasidistribution [17]. Thus there is no device which can directly measure the Wigner phase on individual trials. There is no obvious way to define a physically meaningful number-phase uncertainty principle using the Wigner phase definition. It is also possible to formulate number-phase uncertainty relations in terms of a probability operator measure (POM), which is related to a set of complete operators which are not necessarily Hermitian. Shapiro and Shepard have devised a POM for the phase of the field [18]. They show that the results obtained using the POM are equivalent to those obtained using Hermitian operators proposed by Susskind and Glogower [10] and by Pegg and Barnett [11,12].

We will concentrate on the Pegg-Barnett phase operator in this paper. We choose to do this because of the conceptual simplicity of Hermitian operators—it is possible to apply familiar quantum mechanical principles. As stated above, the results obtained using this operator can also be obtained using the POM of Shapiro and Shepard, but the interpretations of the uncertainty relation using the two methods are different.

Previous experiments have measured a quantity called the "dispersion" D [19] of the optical phase for weak coherent-state signals [13–15,20–22]. The dispersion squared is defined as

$$D^2=\langle \Delta \cos^2\phi \rangle + \langle \Delta \sin^2\phi \rangle \ , \tag{2}$$

where $\langle \Delta \cos^2\phi \rangle = \langle \cos^2\phi \rangle - \langle \cos\phi \rangle^2$. In the limit of the well-defined phase, the dispersion squared is approximately equal to the phase variance $\Delta\phi^2=\langle \phi^2 \rangle - \langle \phi \rangle^2$, while for a uniform phase distribution the dispersion is 1 and the variance is $\pi^2/3$ [19]. In these previous experiments, the reason that the dispersion was measured instead of the variance is because the operators being considered were sine and cosine operators for the field. Since our experiments determine the phase distribution, we will work with the phase variance [4,5,23].

We know of no way to directly test the uncertainty relation [Eq. (1)] for the Pegg-Barnett phase operator (or the Shapiro-Shepard POM). This is because we know of no way to directly measure the phase operator (or the POM). Even if a direct measurement scheme for the Pegg-Barnett phase operator could be found, direct measurement of the commutator would be impossible, since in this case it is non-Hermitian. The only known way to determine all of the quantities in the number-phase uncertainty relation is via our state measurement technique.

In the case of the measured coherent states, which are found to be pure, our method yields an experimentally determined complex wave function for the field mode.

In Sec. II we examine the method of state determination via OHT recently developed by us. It is shown that the measurement of the Wigner function can be achieved by measuring many distributions of quadrature amplitude in many different representations and then inverting this data via tomography. Section III describes the experimental measurement technique. Here we present results for the Wigner function, the density matrix (and since the states we measure are pure, the wave function), and the photon-number variance. The Pegg-Barnett and Wigner definitions of phase are described in Sec. IV. The measured probability distributions for both definitions of phase are presented here and compared to theory. In Sec. V we examine the uncertainty relation for number and phase of a single-mode coherent state. The experimental results are shown and compared with the theory based on the Pegg-Barnett formalism. It is seen that the experimental results agree quite well with the theory.

II. STATE DETERMINATION

Our measurements are based on determining the quantum-mechanical state of a field mode. The method we use to do this is called optical homodyne tomography. This technique is discussed in greater detail elsewhere [4,5], and only a brief review will be given here.

Quadrature operators for the electric field are defined as $\hat{x}=(\hat{a}+\hat{a}^\dagger)/\sqrt{2}$ and $\hat{p}=(\hat{a}-\hat{a}^\dagger)/i\sqrt{2}$, where $\hat{a}$ is the annihilation operator for a particular spatial-temporal mode of the field. The operators $\hat{x}$ and $\hat{p}$ correspond to the real and imaginary parts of the electric field, respectively. From these operators, rotated quadrature operators can be defined as $\hat{x}_\theta=\hat{x}\cos\theta+\hat{p}\sin\theta$ and $\hat{p}_\theta=-\hat{x}\sin\theta+\hat{p}\cos\theta$. In terms of these operators, the electric-field operator is

$$\begin{aligned}\hat{E}(t)&=E_0(t)[\hat{x}\cos(\omega t)+\hat{p}\sin(\omega t)]\\ &=E_0(t)[\hat{x}_\theta\cos(\omega t+\theta)+\hat{p}_\theta\sin(\omega t+\theta)] \ ,\end{aligned} \tag{3}$$

where ω is the optical frequency and $E_0(t)$ corresponds roughly to the electric-field strength of a pulse containing a single photon. A balanced homodyne detector makes measurements which correspond to the quadrature operator $\hat{x}_\theta$ of the signal field, when the local-oscillator field is in a large-amplitude coherent state [24,25]. The spatial-temporal mode of the signal field which is selected by the homodyne detector is the same as that of the local oscillator, and the measured quadrature is determined by the phase of the local oscillator θ.

In our experiments, an ensemble of measurements of quadrature amplitudes x_θ are made to determine the probability distributions $P_\theta(x_\theta)$ for various values of θ. These distributions can be written in terms of the Wigner function $W(x,p)$ of the field mode as

$$P_\theta(x_\theta)=\int_{-\infty}^{\infty} W(x_\theta\cos\theta-p_\theta\sin\theta, x_0\sin\theta+p_\theta\cos\theta)dp_\theta \, . \tag{4}$$

For a set of such probability distributions determined on a continuous set of angles θ, it was shown by Vogel and Risken that the distributions could be inverted using the inverse radon transformation to obtain the Wigner function [3]:

$$W(x,p)=\frac{1}{4\pi^2}\int_{-\infty}^{\infty}dx_\theta\int_{-\infty}^{\infty}d\xi|\xi|\int_0^\pi d\theta\, P_\theta(x_\theta)\exp[i\xi(x_\theta-x\cos\theta-p\sin\theta)] \ . \tag{5}$$

If distributions are available for a finite, discrete, set of angles between 0 and π, the inversion of Eq. (5) can still be performed using numerical techniques familiar in tomographic imaging [26,27]. The resulting reconstruction of the Wigner function is unique as long as the measured probability distributions are completely contained within a finite range of x_θ. If $W(x,p)$ does not have structure on a scale finer than the sampling scale, the reconstruction accurately reproduces $W(x,p)$. The dynamic range and resolution of our measurement apparatus ensures that our measured probability distributions satisfy these requirements for the coherent states described here.

From the measured Wigner function, it is possible to construct the density matrix of the measured field mode by numerically performing a one-dimensional Fourier transform [28]

$$\langle x+x'|\hat{\rho}|x-x'\rangle=\int_{-\infty}^{\infty} W(x,p)e^{2ipx'}dp \ . \tag{6}$$

Since the density matrix completely determines the quantum-mechanical state of a system, optical homodyne tomography provides a way to determine this state. The density matrix of Eq. (6) is expressed in the x representation; it will be useful later to have the density matrix in the representation of the number states $|n\rangle$. This change of basis is accomplished by computing overlap integrals of the density matrix in the x representation with Hermite polynomials:

$$\langle n|\hat{\rho}|m\rangle=\left[\frac{1}{\pi 2^n 2^m n!m!}\right]^{1/2} \times\int_{-\infty}^{\infty}dx\int_{-\infty}^{\infty}dx' e^{-(1/2)(x^2+x'^2)}H_n(x) \times H_m(x')\langle x|\hat{\rho}|x'\rangle \ . \tag{7}$$

In our experiment the signal field was prepared in coherent states of differing average photon number. These states were used for two reasons: first, they are easy to produce and measure, and second, the theory for the uncertainty relation is relatively easy to work out and compare to the experiments. The Wigner function of a coherent state $|\alpha\rangle$ is given by

$$W(x,p)=\frac{1}{\pi}e^{-[(x-\bar{x})^2+(p-\bar{p})^2]} , \tag{8}$$

where $\bar{x}$ and $\bar{p}$ are the mean values of x and p [i.e., $\bar{x}=(\alpha+\alpha^*)/\sqrt{2}$ and $\bar{p}=(\alpha-\alpha^*)/i\sqrt{2}$]. Thus the Wigner function is a two-dimensional Gaussian function, centered about the point $(\bar{x},\bar{p})$, with standard deviations of $\sqrt{2}$ in the x and p directions. The mean number of photons in the coherent state described by Eq. (8) is $\bar{n}=\frac{1}{2}(\bar{x}^2+\bar{p}^2)$.

When performing the numerical inversion to obtain the Wigner function, we use the standard filtered back projection algorithm for parallel-beam sampling geometry [26,27]. Using the directly measured distributions $P_\theta(x_\theta)$, the Wigner function reconstruction occurs on a grid in the (x,p) plane that is centered about the origin. This is fine for fields which have zero mean, for which the Wigner function is contained in a region centered about the origin (such as squeezed-vacuum states [4,5]). For fields with nonzero mean, we have found that the numerical inversion is much more efficient and accurate if the reconstruction window is shifted so that its center at (x',p') corresponds approximately to the center of the reconstructed Wigner function. This is accomplished by numerically shifting the measured probability densities according to $P_\theta(x_\theta)\rightarrow P_\theta(x_\theta-x'\cos\theta-p'\sin\theta)$. It can be seen from Eq. (5) that the effect of reconstructing these shifted distributions is to shift the center of the Wigner function according to $W(x,p)\rightarrow W(x-x',p-p')$. The window over which the measured Wigner function is reconstructed is then centered about the point (x',p'). The widths of the reconstruction window in the x and p directions are taken to be approximately eight times the standard deviation of the probability distributions $P_\theta(x_\theta)$, which ensures that all of the structure in the Wigner function is obtained. Outside the reconstruction window, the measured Wigner function is zero.

III. EXPERIMENTAL MEASUREMENTS

The experimental layout is shown in Fig. 1. The mode-locked, Q-switched, cavity-dumped Nd:YAG laser (where YAG denotes yttrium aluminum garnet) produces 300-ps, nearly transform-limited pulses at 1064 nm with a repetition rate of 420 Hz and a pulse-energy stability of $\pm3\%$. The pulses are split with a polarizing beam splitter (PBS1) to create the signal and local oscillator (LO) beams; these two beams form two arms of a Mach-Zhender interferometer. Half-wave plates and neutral density filters in each arm of the interferometer are used to control the relative intensities of these two beams. The piezoelectric translator pushes a mirror which adjusts the phase of the local oscillator. Each LO pulse contains approximately 2×10^6 photons, while the number of photons in the signal pulse is varied between zero and about 10. The signal and LO pulses, which are orthogonally polarized, are superimposed both temporally and spatially on a polarizing beam splitter (PBS2) with nominally

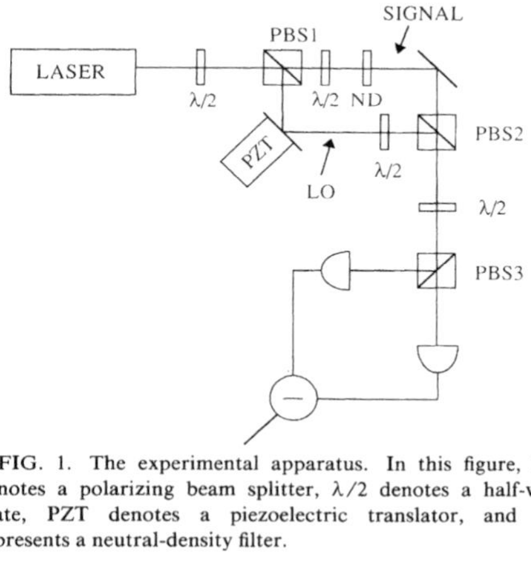

FIG. 1. The experimental apparatus. In this figure, PBS denotes a polarizing beam splitter, $\lambda/2$ denotes a half-wave plate, PZT denotes a piezoelectric translator, and ND represents a neutral-density filter.

zero time delay. Our balanced homodyne detector consists of a $\lambda/2$ plate, a polarizing beam splitter (PBS3) and high-quantum-efficiency ($\eta \sim 85\%$) In-Ga-As photodiodes. The combination of the $\lambda/2$ plate, which rotates the polarizations of the signal and local oscillator fields by 45°, and a polarizing beam splitter (PBS3) serves to interfere the signal and local oscillator fields and to act as a 50/50 beam splitter whose two output fields are detected by the photodiodes. The photodiode outputs are electronically subtracted and integrated using a low-noise charge-sensitive preamplifier [29]. For each laser pulse, this subtracted and integrated signal yields the photoelectron difference number between the outputs of the two detectors. This balancing arrangement removes most of the additive classical noise on the LO and allows us to make measurements at the shot-noise level (SNL) [30]. The electronic noise variance is approximately six times lower than the local-oscillator SNL, which is defined as the variance of photoelectron counts from a coherent-state LO.

A. Wigner function

We have measured Wigner functions for coherent-state signal fields with several different mean numbers of photons, with one of these shown in Fig. 2. For each field strength, we make 5000 repeated measurements of the photoelectron difference number N_θ at 11 values of the local oscillator phase equally spaced in a $[0,\pi]$ interval and calculate photoelectron distributions $P_{N_\theta}(N_\theta)$. Since the signal beam is in a coherent state and the number of photons in the signal beam is much smaller than the number in the local oscillator beam, the variance of each of these photoelectron distributions yields the SNL of the LO, denoted by $\overline{N}_{\mathrm{LO},\theta}$, for a data set with a given value of θ. By averaging the variances of the distributions for different θ, the SNL $\bar{n}_{\mathrm{LO}}$ is determined. The difference number is scaled to yield the quadrature amplitude using $x_\theta = N_\theta/(2\bar{n}_{\mathrm{LO}})^{1/2}$. Thus the $P_{N_\theta}(N_\theta)$ distributions can be scaled to yield distributions $P_\theta(x_\theta)$ of the quadrature

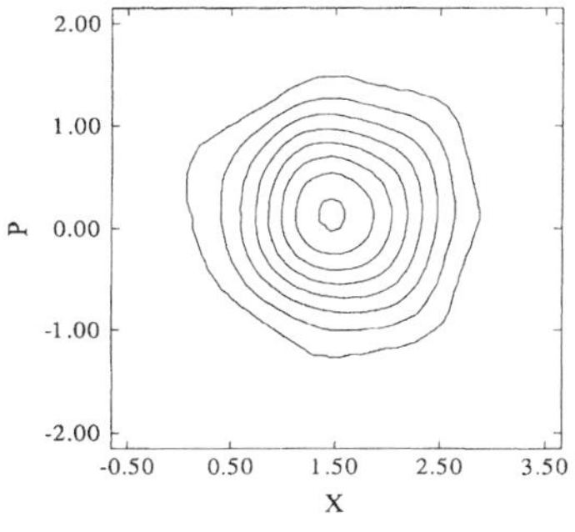

FIG. 2. The measured Wigner function of a coherent state containing an average number of photons of 1.2. The equal-amplitude contours are separated by increments of 0.04.

amplitudes. From these distributions we use the inverse radon transform to calculate the Wigner function, as described above. The LO SNL, $\bar{n}_{\mathrm{LO}}$, can be determined in an independent manner by blocking the signal beam and one detector and directly measuring the number of LO photoelectrons which hit the other detector. Since the detectors are balanced, the total number of LO photoelectrons is twice this measured number. These two methods of obtaining the scaling factor $(\bar{n}_{\mathrm{LO}})^{1/2}$ agree to within 4% [29].

B. Density matrix and wave function

Optical homodyne tomography determines the density matrix for arbitrary states, be they mixed or pure. In our experiments, we anticipate the signal to be in a pure coherent state. A sufficient condition that a density matrix describe a pure state is $\mathrm{Tr}(\hat{\rho}^2)=1$. When these traces are calculated for our experimentally determined density matrices, they are found to equal 1.00 ± 0.02. The measured fields are thus pure states to a high degree. Since there is no reason to expect our measured states to be nonclassical (i.e., having a nonpositive definite Glauber-Sudarshan P distribution), and the only classical states that are pure states are coherent states [31], this indicates that our measured states must be well described by coherent states. This is not surprising, despite the fact that there are losses in our system. The purity of nonclassical states is destroyed by losses, but a coherent state that has suffered loss is still a coherent state, it just has a smaller amplitude [32]. The $\pm3\%$ energy fluctuations from our laser do not destroy the purity of these weak coherent states. It is easily shown that the total noise on a classical signal field is

$$\Delta n^2 = \bar{n} + \Delta E^2 , \tag{9}$$

where $\bar{n}$ is the shot noise and ΔE^2 is the classical noise due to pulse energy fluctuations (normalized to photon number units) [33]. For fields with a fixed $\pm3\%$ classical noise, as the field becomes weaker ΔE^2 becomes smaller more rapidly than $\bar{n}$. For fields with $\bar{n} \lesssim 50$ and classical energy fluctuations of $\pm3\%$, the shot noise dominates and the classical noise has little effect on the field statistics (e.g., with $\bar{n}=50$ and $\pm3\%$ classical fluctuation, $\Delta E^2 \cong 9$).

If the state determined by OHT is found to be pure, it is possible to express the measured density matrix as a wave function. In terms of the density matrix in the x-quadrature representation, the wave function in this representation is

$$\psi(x) = |\psi(x)| e^{i[\beta(x)+\beta_0]} = \frac{\langle x|\hat{\rho}|0\rangle}{\sqrt{\langle 0|\hat{\rho}|0\rangle}} e^{i\beta_0} , \tag{10}$$

where $\beta(x)$ is the complex phase of the wave function and β_0 is an arbitrary constant phase. A similar equation exists for the wave function in the p-quadrature representation $\tilde{\psi}(p)$. Since the coherent states that we have measured have been found to be pure to a high degree, it is thus possible to construct wave functions for these states.

Shown in Fig. 3 is the wave function of a measured

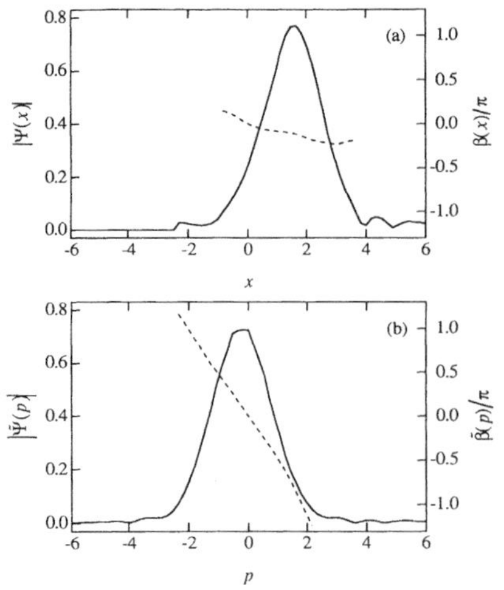

FIG. 3. The measured wave function of a coherent state containing an average of 1.2 photons in (a) the x-quadrature representation and (b) the p-quadrature representation. The solid line is the magnitude $|\Psi|$ of the wave function and is plotted against the left axis, while the dashed line is the phase β of the wave function, plotted against the right axis.

coherent state with a mean number of photons of 1.2. The wave function is plotted in both the x and p representations. The wave function squared $|\psi(x)|^2[|\tilde\psi(p)|^2]$ is the probability density for observing the real [imaginary] part of the electric field with a particular value. It is seen that in the x representation, the wave function is offset from zero and has a phase that is approximately constant. The p-representation wave function is centered at zero and has a phase that is linear in p. The linear dependence of the phase $\beta(x)$ on x [and $\tilde\beta(p)$ on p] implies that the field is in a coherent state. The wave functions in the two representations have these properties because of the Fourier transform relationship that exists between them.

C. Photon number

Information about the photon number in the signal beam can be obtained in several ways. The photon-number distribution $P(n)$ can be obtained by directly integrating the Wigner function with Laguerre polynomial weight functions [34]. Alternatively, the density matrix in the number-state representation [Eq. (6)] can be used to obtain the photon-number distribution $\langle n|\hat\rho|n\rangle$ [5,23]. These two methods are found to yield the same photon-number distributions to within the precision of the procedure. The precision is estimated by examining the $P(n)$ distribution for large values of n. For signal fields with finite numbers of photons, $P(n)$ should approach zero for large n. We find that $P(n)$ decreases to a level of approximately 0.005 at large n and then fluctuates randomly between ± 0.005. We thus estimate our accuracy in determining $P(n)$ to be approximately ± 0.005. Shown in Fig. 4 are measured photon-number distributions for coherent states with average photon numbers of 0.2, 1.2, and 3.6.

When calculating moments of the signal photon number, we find that it is best not to use the photon-number distributions obtained from the above-described methods. The reason for this is that the distributions do not decay completely to zero and they must thus be truncated at some point; hence the moments are sensitive to exactly where the distribution is truncated. The higher the moment, the more sensitive it is to the truncation. To avoid this problem, when calculating moments of the photon number, the photon-number operator $\hat n=\hat a^\dagger \hat a$ is written in terms of the quadrature operators $\hat n=\frac{1}{2}(\hat x^2+\hat p^2-1)$. The quadrature operators are then placed in Weyl order and the moments can be calculated as c-number integrals of the measured Wigner function [28]. For example, the moment $\langle \hat n^2\rangle$ can be calculated using

$$\begin{aligned}\langle \hat n^2\rangle &= \tfrac{1}{4}\langle[\hat x^4+\hat p^4+\hat x^2\hat p^2+\hat p^2\hat x^2-2\hat x^2-2\hat p^2+1]\rangle\\ &= \tfrac{1}{4}\langle[\hat x^4+\hat p^4+2\hat x^2\hat p^2-2\hat x^2-2\hat p^2]_{\mathrm{Weyl}}\rangle\\ &= \tfrac{1}{4}\int_{-\infty}^{\infty}\int_{-\infty}^{\infty} dx\,dp(x^4+p^4+2x^2p^2\\ &\qquad -2x^2-2p^2)W(x,p)\ . \end{aligned} \tag{11}$$

We estimate that the determination of the mean number of photons in the detected mode using this technique is accurate to better than 8%, limited by drift and noise of the LO pulses and of the electronics, as well as the finite number of $P_\theta(x_\theta)$ distributions. The mean number of photons calculated from the measured Wigner distribution can be compared to the directly measured number

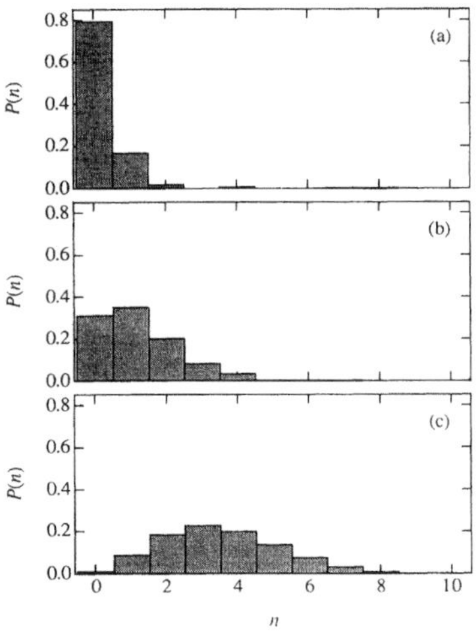

FIG. 4. Photon-number distributions for measured coherent states containing on average (a) 0.2, (b) 1.2, and (c) 3.6 photons.

of photons in the signal beam. We cannot measure the 0.1–10 photons present in the signal beam, but we can directly measure the signal beam when it contains on the order of 10 000 photons, and then attenuate this signal with calibrated neutral density filters. This direct measurement is only accurate to within a factor of 2 because of the accuracy of the filter calibration and the error in measuring the number of photons present. The number of photons calculated from the Wigner distribution is typically smaller than the directly measured number of photons by a factor of 1.7 (e.g., for one particular measurement we obtain a mean of 1.3 photons from the Wigner measurement, while the direct measurement yields two photons).

IV. PHASE

One possible means of describing the phase of a quantum-mechanical field is in terms of the Pegg-Barnett Hermitian phase operator $\hat{\phi}$ [11,12]. This operator is defined in a finite- (but arbitrarily large) dimensional Hilbert space. In an $(s+1)$-dimensional Hilbert space the phase states are defined as

$$|\phi\rangle=\frac{1}{\sqrt{s+1}}\sum_{n=0}^{s} e^{in\phi}|n\rangle\ . \tag{12}$$

This Hilbert space is spanned by a complete orthonormal set of basis phase states $|\phi_m\rangle$, given by Eq. (12) with ϕ replaced by

$$\phi_m=\phi_0+\frac{2\pi m}{s+1},\quad m=0,1,\ldots,s\ , \tag{13}$$

where ϕ_0 is a reference phase. In terms of the states $|\phi_m\rangle$ the Hermitian phase operator is

$$\hat{\phi}=\sum_{m=0}^{s}\phi_m|\phi_m\rangle\langle\phi_m|\ . \tag{14}$$

With this definition, the phase states are the eigenstates of the phase operator, i.e., $\hat{\phi}|\phi\rangle=\phi|\phi\rangle$. In our experiments ϕ actually corresponds to the phase difference between the signal and the LO field.

Using the Pegg-Barnett formalism, one can define a probability distribution for the phase. For a signal mode in a state described by a density operator $\hat{\rho}$, the probability of measuring a particular value of the phase is $P_{\rm PB}(\phi)=[(s+1)/2\pi]\langle\phi|\hat{\rho}|\phi\rangle$, which is normalized so that the integral of $P_{\rm PB}(\phi)$ over a 2π region of ϕ is equal to 1. In terms of the number-state basis, this distribution is

$$P_{\rm PB}(\phi)=\frac{1}{2\pi}\sum_{n,m=0}^{s} e^{i(m-n)\phi}\langle n|\hat{\rho}|m\rangle\ . \tag{15}$$

For all of the coherent states we have measured ($\bar{n}\leq 8$), the phase distribution described by Eq. (15) converges for $s\geq 20$.

Another way to characterize the phase of a field is via the Wigner phase distribution $P_W(\phi)$ [16]. This is defined as the overlap in phase space of the Wigner distribution and a narrow "wedge-shaped" region,

$$P_W(\phi)\Delta\phi=\int_{\phi}^{\phi+\Delta\phi}d\phi'\int_0^{\infty}dr\, rW(r\cos\phi',r\sin\phi')\ . \tag{16}$$

For a coherent state, using Eqs. (8) and (16), we find that the Wigner phase distribution is given by

$$P_{W,{\rm CS}}(\phi)=\frac{e^{-2\bar{n}}}{2\pi}\{1+\sqrt{2\pi\bar{n}}\,e^{2\bar{n}\cos^2\phi}\times\cos\phi[1+{\rm erf}(\sqrt{2\bar{n}}\cos\phi)]\}\ , \tag{17}$$

where erf( ) denotes the error function and the limit $\Delta\phi\rightarrow 0$ has been taken.

Using Eqs. (15) and (16) it is straightforward to calculate the phase distributions for our experimentally measured states. In Fig. 5(a) we compare the measured Wigner and Pegg-Barnett phase distributions for coherent states with different amplitudes. It is seen that as the mean photon number decreases, the phase distribution broadens, as would be expected from the uncertainty principle. It is also seen that the Wigner phase distribution is more peaked than the Pegg-Barnett, as is expected from previous theoretical results [35] and experiments on squeezed states [5,23]. Shown in Fig. 5(b) are the theoretical phase distributions for coherent states with the same mean photon number as in Fig. 5(a). The theoretical results are obtained from Eqs. (15) and (17).

Once the phase distributions are computed, moments of the phase can easily be calculated using these distributions. Shown in Fig. 6 are the measured variances of the Wigner and Pegg-Barnett phase distributions as a function of the mean number of photons in the signal field. As pointed out above, the variance of the Wigner phase distribution is smaller than the corresponding variance for the Pegg-Barnet distribution. This difference is most pronounced when the coherent state has mean number of

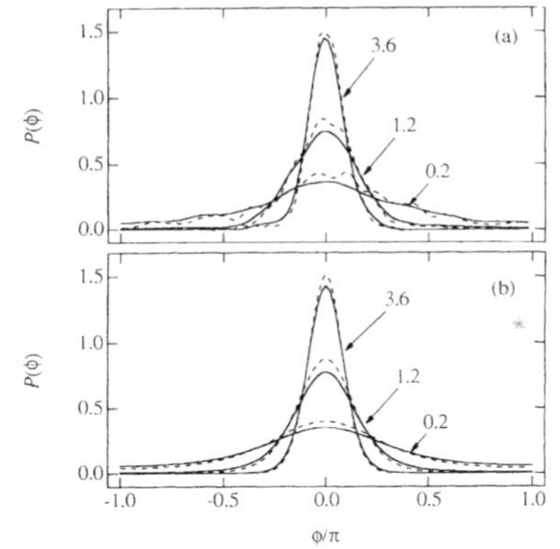

FIG. 5. The solid lines are the Pegg-Barnett and the dashed lines are the Wigner phase distributions for (a) measured and (b) theoretical coherent states. The mean number of photons for the states is indicated.

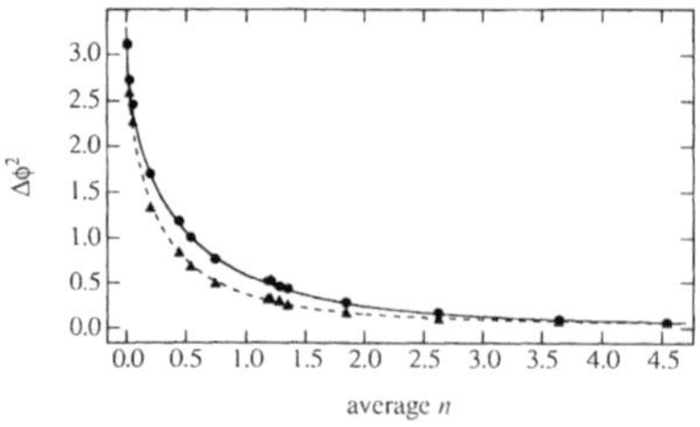

FIG. 6. The variance of the phase is plotted against the average number of photons in a coherent state. The points are experimentally determined values for the Pegg-Barnett (circles) and Wigner (triangles) phase definitions. The curves are the theoretical values for the Pegg-Barnett (solid line) and Wigner (dashed line) phase definitions.

photons approximately equal to 1, while for smaller and larger coherent states the variances are approximately equal for the two distributions. Also plotted in Fig. 6 are the theoretical variances for a coherent-state signal field. Explicit formulas for the phase variance of the Pegg-Barnett phase of a coherent state are given in Ref. [22]. The variance for the Wigner phase is easily calculated numerically using the distribution in Eq. (17). The agreement between the experimental data and the theory is seen to be quite good and again indicates that the states we are measuring in the experiment are well described by coherent states.

V. UNCERTAINTY PRINCIPLE

The uncertainty relation for the phase $\hat{\phi}$ and photon number $\hat{n}$ operators is given by the standard form of Eq. (1) [1]. In terms of number states, the commutator can be written [11]

$$[\hat{\phi},\hat{n}]=\frac{2\pi}{s+1}\times\sum_{n=0}^{s}\sum_{\substack{n'=0\\n'\neq n}}^{s}|n'\rangle\langle n|\frac{(n-n')\exp[i(n'-n)\phi_0]}{\exp[i(n-n')2\pi/(s+1)]-1}. \quad (18)$$

Since this is an operator, its expectation value for the state of interest must be determined to set the lower bound of experimental measurements of $\Delta\phi\Delta n$ [36].

Note that the state dependence of the number-phase uncertainty relation is in contrast to the uncertainty relation for the field quadrature amplitudes (analogous to the position and momentum of a harmonic oscillator, i.e., $\Delta x\,\Delta p\geq 1/2$, which is independent of the state). The uncertainty relation for the quadrature field amplitudes when the field is in a squeezed state was experimentally verified by Wu, Xiao, and Kimble, who compared the measured uncertainty product of the quadrature amplitudes to the theoretical lower bound [37]. For pure squeezed, or coherent, states of the field, the uncertainty relation for the quadrature operators achieves an equality $\Delta x\,\Delta p=\frac{1}{2}$.

Plotted in Fig. 7 are both the uncertainty product $\Delta\phi\Delta n$ and the expectation value of the commutator $\frac{1}{2}|\langle[\hat{\phi},\hat{n}]\rangle|$ for our experimentally measured data. For the experimental data, the commutator expectation value is calculated numerically by expressing the matrices for $\hat{\phi}$ and $\hat{n}$ in the number-state representation, evaluating the matrix which corresponds to the commutator, and tracing the commutator over the density matrix in the number-state representation. Also plotted are the theoretically predicted values for coherent states. For the theoretical uncertainty product, $\Delta\phi$ was calculated in the same fashion as used to generate Fig. 6, while the photon-number standard deviation was calculated using $\Delta n=\sqrt{\bar{n}}$ for a coherent state. Notice that the uncertainty relation is satisfied (as it must be using our method of analysis); the uncertainty product is greater than the expectation value of the commutator. It is interesting, however, that the equality between these two quantities is only achieved for average photon numbers approaching zero or approaching a very large number. This is true for both the theoretical and experimental data, so it is not simply a manifestation of our measurements. This means that, despite their purity, coherent states are not "intelligent" states with respect to number and phase. Intelligent states are defined as states for which Eq. (1) is an equality [6]. Of course, the only coherent state which is a minimum-uncertainty state is the vacuum, for which $\Delta\phi\Delta n=0$. It is seen that in the limit of very small or very large coherent-state amplitude, the coherent states are approximately number-phase intelligent.

It is interesting to speculate as to the origin of the coherent state's nonintelligent property. Coherent states are eigenstates of the annihilation operator, which is closely related to the quadrature operators, and these states are intelligent relative to these variables, i.e., $\Delta x\,\Delta p=\frac{1}{2}|\langle\psi|[\hat{x},\hat{p}]|\psi\rangle|=\frac{1}{2}$. On the other hand, coherent states are not closely related to the phase eigenstates [11,12]. Therefore, there is no reason to expect them to be number-phase intelligent. In contrast, both number states and phase eigenstates are number-phase intelligent [6].

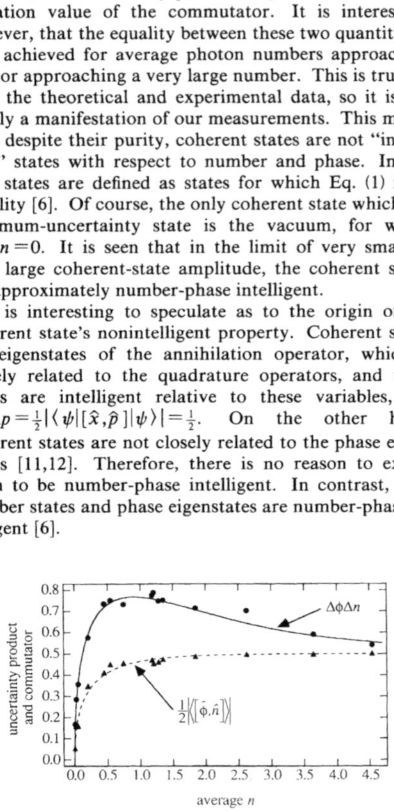

FIG. 7. The number-phase uncertainty product and expectation value of the number-phase commutator are plotted against the average number of photons in a coherent state. The points are experimentally determined values for the uncertainty product (circles) and the commutator expectation (triangles). The curves are the theoretical values for the uncertainty product (solid line) and the commutator expectation (dashed line).

VI. CONCLUSIONS

We have developed a technique for experimentally determining the standard deviation of the phase $\Delta\phi$ and photon number Δn, as well as the expectation value of their commutator $|\langle[\hat{\phi},\hat{n}]\rangle|$, for a mode of an optical field. Although we have concentrated on the results for coherent-state signal fields, nowhere in the measurements was it necessary to assume that the field was in a coherent state [38]. This method of field characterization will in principle work for *arbitrary* states of the field—the only limitation being the available experimental parameters (i.e., detector efficiency, resolution, etc.). It is thus possible to experimentally measure, for any state of the field, how close to uncertainty relation [Eq. (1)] comes to equality (provided, of course, that the field can be identically prepared many times so that an ensemble of measurements can be made to determine its Wigner function).

These results were possible because optical homodyne tomography determines the quantum-mechanical state (as described by a density matrix) of a mode of the field. Since the measured states were found to be relatively pure, it was possible to determine the complex wave functions corresponding to these states.

The agreement between the measured and theoretical points in Figs. 6 and 7 is ensured by the agreement between the measured and theoretical density matrices. Therefore, other phase definitions could be used, also leading to good agreement of experiment and theory. For this reason our results do not favor one phase definition over another.

Our measurements naturally yield the Wigner function of the field mode. Given $W(x,p)$ one can also construct the Q distribution by convolving $W(x,p)$ with a vacuum-state Wigner function [39]. The marginal Q phase distribution can then be constructed in analogy with Eq. (16) [40]. The resulting phase variance is larger than that from the marginal Wigner phase distribution shown in Fig. 6. This Q phase distribution corresponds to that which would be obtained using the eight-port homodyne apparatus of Walker and Carroll [41] and of Noh, Fougéres, and Mandel [13–15] (see also [39,42–44]). The increase in the width of this distribution can be associated with additional noise introduced when measuring both x and p quadratures simultaneously. A rigorous uncertainty relation for the simultaneous (but imprecise) measurement of n and ϕ, which is possible using this apparatus, would have to be developed with this particular experimental arrangement in mind. The result for the uncertainty product $\Delta\phi\Delta n$ for simultaneous number and phase measurements would necessarily be larger than the product $\Delta\phi\Delta n$ that we determined and show in Fig. 7.

ACKNOWLEDGMENTS

We wish to acknowledge several helpful discussions with H. Carmichael, G. Milburn, and W. Schleich. We thank A. Faridani for providing us with the software for performing the inverse Radon transform. This research is supported by the National Science Foundation, Grant No. PHY 8921709-01.

*Permanent address: Joint Institute for Laboratory Astrophysics and the Department of Physics, University of Colorado and the National Institute of Standards and Technology, Boulder, CO 80309.

[1] L. I. Schiff, *Quantum Mechanics,* 3rd ed. (McGraw-Hill, New York, 1968).
[2] M. Beck, D. T. Smithey, J. Cooper, and M. G. Raymer, Opt. Lett. **18,** 1259 (1993).
[3] K. Vogel and H. Risken, Phys. Rev. A **40,** 2847 (1989).
[4] D. T. Smithey, M. Beck, A. Faridani, and M. G. Raymer, Phys. Rev. Lett. **70,** 1244 (1993).
[5] D. T. Smithey, M. Beck, J. Cooper, M. G. Raymer, and A. Faridani, Phys. Scr. (to be published).
[6] J. A. Vaccaro and D. T. Pegg, J. Mod. Opt. **37,** 17 (1990).
[7] V. B. Braginsky and F. Y. Khalili, *Quantum Measurement* (Cambridge University Press, Cambridge, 1992).
[8] E. Aurthurs and J. L. Kelley, Jr., Bell Syst. Tech. J. **44,** 725 (1965).
[9] Y. Yamamoto and H. A. Haus, Rev. Mod. Phys. **58,** 1001 (1986).
[10] L. Susskind and J. Glogower, Physics **1,** 49 (1964).
[11] D. T. Pegg and S. M. Barnett, Phys. Rev. A **39,** 1665 (1989).
[12] S. M. Barnett and D. T. Pegg, J. Mod. Opt. **36,** 7 (1989).
[13] J. W. Noh, A. Fougères, and L. Mandel, Phys. Rev. Lett. **67,** 1426 (1991).
[14] J. W. Noh, A. Fougères, and L. Mandel, Phys. Rev. A **45,** 424 (1992).
[15] J. W. Noh, A. Fougères, and L. Mandel, Phys. Rev. A **46,** 2840 (1992).
[16] W. Schleich, R. J. Horowicz, and S. Varro, Phys. Rev. A **40,** 7405 (1989).
[17] B. M. Garraway and P. L. Knight, Phys. Rev. A **46,** R5346 (1992).
[18] J. H. Shapiro and S. R. Shepard, Phys. Rev. A **43,** 3795 (1991).
[19] Z. Hradil, Quantum Opt. **4,** 93 (1992).
[20] H. Gerhardt, U. Büchler, and G. Litfin, Phys. Lett. **49A,** 119 (1974).
[21] M. M. Nieto, Phys. Lett. **60A,** 401 (1977).
[22] R. Lynch, Phys. Rev. A **41,** 2841 (1990).
[23] M. Beck, D. T. Smithey, and M. G. Raymer, Phys. Rev. A **48,** R890 (1993).
[24] H. P. Yuen and J. H. Shapiro, IEEE Trans. Inf. Theory **IT-26,** 78 (1980).
[25] W. Vogel and W. Schleich, Phys. Rev. A **44,** 7642 (1991).
[26] F. Natterer, *The Mathematics of Computerized Tomography* (Wiley, New York, 1986).
[27] A. Faridani, *Inverse Problems and Imaging; Pitman Research Notes in Mathematics,* edited by G. F. Roach (Wiley, Essex, 1991), Vol. 245, p. 68.
[28] M. Hillery, R. F. O'Connell, M. O. Scully, and E. P. Wigner, Phys. Rep. **106,** 121 (1984).
[29] D. T. Smithey, M. Beck, M. Belsley, and M. G. Raymer, Phys. Rev. Lett. **69,** 2650 (1992).
[30] H. P. Yuen and V. W. S. Chan, Opt. Lett. **8,** 177 (1983).

[31] M. Hillery, Phys. Lett. **111A**, 409 (1985).
[32] Z. Y. Ou, C. K. Hong, and L. Mandel, Opt. Commun. **63**, 118 (1987).
[33] R. Loudon, *The Quantum Theory of Light*, 2nd ed. (Clarendon, Oxford, 1983).
[34] T. Marshall and E. Santos, J. Mod. Opt. **38**, 1463 (1991).
[35] G. S. Agarwal, S. Chaturvedi, K. Tara, and V. Srinivasan, Phys. Rev. A **45**, 4904 (1992).
[36] If the phase is defined in terms of the POM of Ref. [18], then the uncertainty relation is defined in terms of a Fourier transform pair, and it is not necessary to measure the commutator.
[37] L. A. Wu, M. Xiao, and H. J. Kimble, J. Opt. Soc. Am. B **4**, 1465 (1987).
[38] The scaling of the photon-counting distributions to obtain the quadrature distributions as described here relied on the fact that the field was in a coherent state. It is possible to properly scale the distributions without any assumption as to the state of the field, as was done in Refs. [4] and [5].
[39] Y. Lai and H. A. Haus, Quantum Opt. **1**, 99 (1989).
[40] W. Schleich, A. Bandilla, and H. Paul, Phys. Rev. A **45**, 6652 (1992).
[41] N. G. Walker and J. E. Carroll, Opt. Quantum Electron. **18**, 355 (1986).
[42] M. Freyberger and W. Schleich, Phys. Rev. A **47**, R30 (1993).
[43] M. Freyberger, K. Vogel, and W. Schleich (private communication).
[44] The original experiment of Ref. [41] did not make measurements at the shot-noise level, so phase distributions from this experiment would not be true quantum phase distributions. To obtain the Q phase distribution using the technique described in Refs. [13–15], it is necessary for one of the two input beams to be in a large amplitude coherent state [39,42,43].

Adaptive Homodyne Measurement of Optical Phase

Michael A. Armen,* John K. Au, John K. Stockton, Andrew C. Doherty, and Hideo Mabuchi
Norman Bridge Laboratory of Physics 12-33, California Institute of Technology, Pasadena, California 91125
(Received 1 April 2002; published 4 September 2002)

We present an experimental demonstration of the power of feedback in quantum metrology, confirming the predicted [H. M. Wiseman, Phys. Rev. Lett. **75**, 4587 (1995)] superior performance of an *adaptive* homodyne technique for single-shot measurement of optical phase. For measurements performed on weak coherent states with no prior knowledge of the signal phase, adaptive homodyne estimation approaches closer to the intrinsic quantum uncertainty than any previous technique. Our results underscore the importance of real-time feedback for reaching quantum limits in measurement and control.

DOI: 10.1103/PhysRevLett.89.133602 PACS numbers: 42.50.Dv, 03.65.Ta, 03.67.–a, 06.90.+v

Quantum mechanics complicates metrology in two complementary ways. First it forces us to accept the existence of intrinsic uncertainty in the value of observables such as the position, momentum, and phase of an oscillator. Such quantum uncertainty exists even when the *state* of the system being measured has been prepared in a technically flawless way. Even the vacuum state of a single-mode optical field, for example, exhibits "zero-point" fluctuations in the amplitudes of its electric and magnetic components. The same is true of the optical coherent states representative of the output of an ideal laser. These uncertainties limit the sensitivity of ubiquitous measurement techniques such as laser interferometry. Squeezed states of light are challenging to produce but have reduced uncertainties that could be exploited for improved sensitivity in optical metrology [1]. To date, however, the achievable performance gains have not been sufficient to motivate their use in practical applications.

Quantum mechanics presents a second major obstacle to precision measurement by making it generally quite difficult to realize ideal *measurement procedures*, whose inherent inaccuracy is small enough to reveal the intrinsic uncertainty limits associated with quantum states. Let us refer to such ideal measurements as being uncertainty-limited (UL). Clearly, the implementation of UL measurement procedures is essential for any application that seeks to take advantage of exotic quantum states with reduced intrinsic uncertainties. In connection with squeezed states of light, measurements of optical quadrature amplitudes do constitute an important class of UL measurements that actually can be implemented in practice (via homodyne detection). But this is an unusual case as UL measurement schemes have not previously been demonstrated even for closely related observables such as optical phase, despite intense historical interest in their quantum properties [2,3]. This shortcoming is not merely one of achievable signal-to-noise ratio in realistic experiments — there has yet to be any demonstration of a measurement procedure that is capable *even in principle* of achieving UL estimation of true optical phase (as opposed to phase-quadrature amplitude), for coherent or any other pure states of an optical field.

In this Letter we present an experimental demonstration of the surprising efficacy of *real-time feedback* in the development of UL measurement procedures. We do this in the context of measuring the optical phases of weak pulses of light, following a theoretical proposal by Wiseman [4]. This metrological task can be motivated by a coherent optical communication scenario in which information is encoded in the phase of laser pulses that must travel long distances between the sender and receiver. In such a context the receiver is likely to be faced with decoding information carried by optical wave packets whose quantum states correspond to coherent states with low mean photon number (as a result of optical attenuation). If the sender is encoding information in an efficient manner, the variation of phase from one optical wave packet to the next should be uniformly distributed over the entire interval from zero to 2π. Hence, the receiver would ideally like to implement a *single-shot* UL measurement procedure for estimating the phase of each individual pulse [5]. The variance of such optimal phase estimates should be limited only by the intrinsic quantum uncertainty associated with optical coherent states of the given mean photon number.

There is no known experimental procedure to accomplish this goal exactly. Prior to Wiseman's proposal [4] it had been widely believed [6] that the best feasible strategy for single-shot phase decoding should be heterodyne detection, which for coherent signal states can in principle achieve a measurement variance that is only a factor of 2 greater than the intrinsic uncertainty limit [7]. In what follows we will thus consider heterodyne phase estimation as the benchmark we need to surpass. It is important to compare the current measurement scenario with what arises in applications such as the implementation of optical frequency standards (optical clocks), where optical phases must also be measured optimally but where there is a guarantee that the phases vary only over a range much less than $\pi/2$. In such cases, as mentioned above, essentially UL measurement can be achieved by employing fixed-quadrature homodyne detection with a local oscillator whose phase is held at an offset of $\pi/2$ radians from the expected signal phase. Real-time feedback can thus be seen as a key ingredient in formulating an UL scheme for measurement scenarios in which we have no prior knowledge of the signal phase—feedback enables protocols in which the local oscillator phase is *adapted*, in real-time, to the phase of each individual signal pulse [8]. As each optical pulse has some spatiotemporal extent, the measurement signal generated by the leading edge of a given pulse can be used to form a preliminary estimate of its phase, which is used promptly to adjust the local oscillator settings to be optimal for that pulse. (Note that this essentially amounts to the implementation of a quantum-noise limited phase-lock loop; the scheme relies on lock acquisition from the first one or two photons of signal in each pulse.) Detailed theoretical analyses of such schemes by Wiseman and co-workers have led to the striking realization that, despite extremely low signal-to-noise ratio in the feedback loop, adaptive homodyne measurement can be essentially UL for optical pulses with mean photon number of order 10 (or greater), and benefits of real-time adaptation should still be evident for mean photon numbers ~ 1. We turn now to our experimental test of these predictions.

The data plotted in Fig. 1(a) demonstrate the superiority of an adaptive homodyne measurement procedure ("adaptive") to the benchmark heterodyne measurement procedure ("heterodyne") for making single-shot estimates of optical phase. As described in detail below, we perform these measurements on optical pulses of 50 μs duration derived from an intensity-stabilized cw laser. We have also plotted the theoretical prediction for the variance of ideal heterodyne measurement, both with (thin solid line) and without (dotted line) correction for a small amount of excess electronic noise in the balanced photocurrent. The excellent agreement between the heterodyne data and theory indicates that we have no excess phase noise (technically ideal preparation of coherent signal states) and validates electronic calibrations involved in our data analysis [9]. In the range of $\sim$10–300 photons per pulse, the adaptive data lie well below the heterodyne curve that has been corrected for electronic noise (which also has a detrimental effect on the adaptive data), and a few of the data points lie significantly below the absolute limit. Quantitatively, for ≈ 50 photons per pulse the adaptive data point sits 6.5 standard deviations below the absolute heterodyne limit (note logarithmic scale).

For intermediate values of the photon number the adaptive variances should ideally be even lower [10], but the performance of our experiment is limited by finite feedback bandwidth. For signals with large mean photon number, the adaptive scheme is inferior to heterodyne because of excess technical noise in the feedback loop. As the intrinsic phase uncertainty of coherent states becomes large for very low photon numbers, the relative differences among the expected variances for adaptive, heterodyne, and ideal estimation become small. Accordingly, we have been unable to beat the heterodyne limit for phase estimation *variances* with adaptive homodyne measurement for mean photon numbers $N \lesssim 8$. (Note that all theoretical curves in Fig. 1(a) correspond

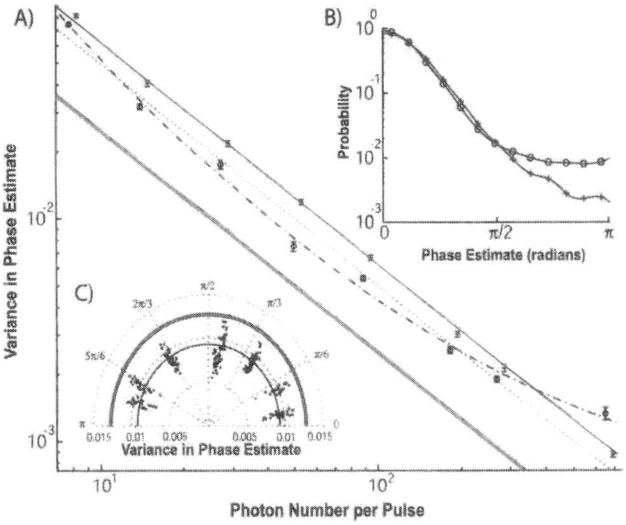

FIG. 1 (color). Experimental results from the adaptive and heterodyne measurements. (a) Adaptive (blue circles) and heterodyne (red crosses) phase estimate variance vs pulse photon number. The blue dash-dotted line is a second order curve through the adaptive data, to guide the eye. The thin lines are the theoretical curves for heterodyne detection with (solid) and without (dotted) corrections for detector electronic noise. The thick solid line denotes the fundamental quantum uncertainty limit, given our overall photodetection efficiency. (b) Phase-estimator distributions for adaptive (blue circles) and heterodyne (red crosses) measurements, for pulses with mean photon number ≈ 2.5. (c) Polar plot showing the variance of adaptive phase estimates (blue dots) for different signal phases (mean photon number ≈ 50). The solid blue line is a linear fit to the data. The double red lines indicate the 1σ scatter for our heterodyne data, averaged over initial phase.

to asymptotic expressions valid for large mean photon numbers; corrections are small for $N \gtrsim 10$.) However, we are able to show that the estimator *distribution* for adaptive homodyne remains narrower than that for heterodyne detection even for pulses with mean photon number down to $N \approx 0.8$. In Fig. 1(b) we display the adaptive and heterodyne phase-estimator distributions for a signal size corresponding to ≈ 2.5 photons. Note that we have plotted the distributions on a logarithmic scale, and that the mean phase has been subtracted off so that the distributions are centered at zero. The horizontal axis can thus be identified with estimation error. The adaptive phase distribution has a smaller "Gaussian width" than heterodyne but exhibits rather high tails, in qualitative agreement with predictions [11] based on quantum estimation theory. Theoretical analysis suggests that the high tails are a consequence of changes in the sign of the photocurrent caused by vacuum fluctuations of the optical field, which become comparable to the coherent optical signal for $N \sim 1$. The feedback algorithm responds to these photocurrent inversions by locking to incorrect phase values with an error that tends towards π.

Accurately assessing the performance of a single-shot measurement requires many repetitions of the measurement under controlled conditions. Figure 2(a) shows a schematic of our apparatus. Light from a single-mode cw Nd:YAG laser is first stripped of excess intensity noise by passage through a high-finesse Fabry-Perot cavity (not pictured) with ringdown time $\approx 16\ \mu$s; the transmitted beam is shot-noise limited above ~50 kHz. This light enters the Mach-Zehnder interferometer (MZI) at beam splitter 1 (BS1), creating two beams with well-defined relative phase. The local oscillator (LO) is generated using an acousto-optic modulator (AOM) driven by an rf synthesizer RF1 at 84.6 MHz, yielding ~230 μW of frequency-shifted light. The signal beam corresponds to a frequency sideband created by an electro-optic modulator (EOM) driven by an rf synthesizer (RF2) that is phase locked to RF1. The power (5 fW to 5 pW) and pulse length (50 μs) of the signal beam are controlled by changing the amplitude of RF2 and by switching it on/off. A pair of photodetectors collect the light emerging from the two output ports of the final 50/50 beam splitter (BS2); the difference of their photocurrents [balanced photocurrent $I(t)$] provides the basic signal used for either heterodyne or adaptive phase estimation [9]. At our typical LO power, the photodetectors supply 6 dB of shot noise (over electronic noise) in the difference photocurrent from ~1 kHz to 10 MHz. We perform adaptive homodyne measurement by feedback to the phase of RF2, which sets the (instantaneous) relative phase between signal and LO [12]. Our feedback bandwidth ~1.5 MHz is limited by the maximum slew rate of RF2. Real-time electronic signal processing for the feedback algorithm is performed by a field programmable gate array (FPGA) that can execute complex computations with very low delay

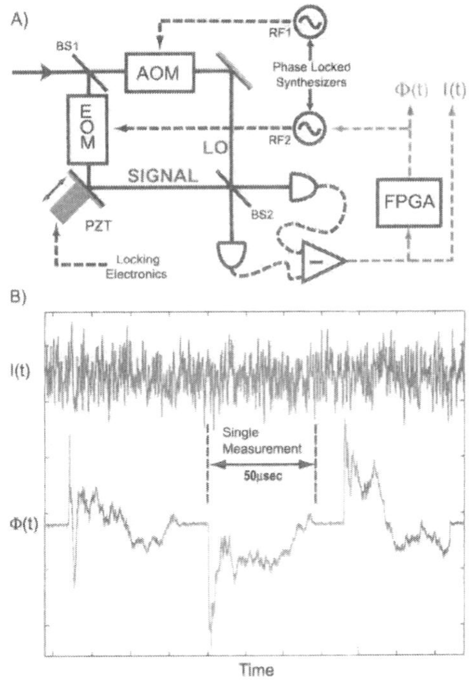

FIG. 2 (color). (a) Apparatus used to perform both adaptive homodyne and heterodyne measurements (see text). Solid lines denote optical paths, and dashed lines denote electrical paths. (b) Photocurrent $I(t)$ (red above), and feedback signal $\Phi(t)$ (green below), for three consecutive adaptive homodyne measurements with $N \approx 8$. The x axis represents time for both signals. The y axis scale indicates the absolute phase shifts made by the feedback signal. The photocurrent is plotted on an arbitrary scale.

[13]. Our feedback and phase-estimation procedures correspond to the "Mark II" scheme of Wiseman and coworkers [11], in which the photocurrent is integrated with time-dependent gain to determine the instantaneous feedback signal: $\Phi(t) - \pi/2 \propto \int_0^t dv\, I(v)/\sqrt{v}$, where t is scaled such that the pulse has duration 1. The final phase estimate depends on integrals of the photocurrent and feedback signal according to the Mark II scheme. For heterodyne measurements we turn off the feedback to RF2 and detune it from RF1 by 1.8 MHz; phase estimates are made by the standard method of I/Q demodulating the photocurrent beat note. For both types of measurement we store the photocurrent $I(t)$ and feedback signal $\Phi(t)$ on a computer for postprocessing.

In Fig. 2(b) we show the photocurrent and feedback signal of three consecutive adaptive phase measurements for $N \approx 8$. At the beginning of each measurement, feedback of the photocurrent shot noise causes the relative phase between the signal and LO to vary randomly. As more of the pulse is detected, the information gained is used to drive the signal-LO phase towards an optimal value. Phase estimation variances are established using ensembles of ≈ 150 consecutive single-shot measurements. Although each optical pulse is the subject of an *independent* single-shot measurement, we fix the signal phase over the length of each ensemble in order to accurately determine the estimation variance. It is important to note that each data point in Fig. 1(a) corresponds to an average of variance estimates from many ensembles, each with a different (random) signal phase.

Ideally, the performance of a single-shot phase measurement procedure should be independent of the signal phase. In our experiment, this implies that the measurement variances should be independent of the initial relative phase between signal and local oscillator. In Fig. 1(c) we display a polar plot of the adaptive variance (radial) versus initial signal phase (azimuth). The data were taken using signals with mean photon number ≈ 50. Each data point corresponds to the variance of an ensemble of phase measurements taken at a fixed phase. The double solid lines indicate the 1σ scatter of our heterodyne data. It is clear from this data that the performance of the adaptive homodyne scheme is essentially independent of initial phase, and is consistently superior to heterodyne measurement.

The photon number per pulse, N, is determined by extracting optical amplitude information from the balanced photocurrent in heterodyne mode. The typical experimental procedure is to fix the signal amplitude, take an ensemble of heterodyne measurements (which yields both heterodyne phase estimates and an estimate of N), and then take an ensemble of adaptive homodyne measurements. The photon number assigned to the subsequent ensemble of adaptive measurements is $0.95N$, where the relative calibration factor arises from the measured response of the EOM.

In conclusion, we have presented an experimental demonstration of adaptive homodyne phase measurement. For pulses with mean photon number $\sim$10–300 our measured variances approach closer to the intrinsic phase-uncertainty limit of coherent states than any previously demonstrated technique. These results establish the feasibility of broadband quantum-noise-limited feedback for adaptive quantum measurement [14], quantum feedback control [15], quantum error correction [16], and studies of conditional quantum dynamics [17,18].

This work was supported by the NSF (PHY-9987541, EIA-0086038) and ONR (N00014-00-1-0479).

*Electronic address: armen@caltech.edu

[1] K. Schneider *et al.*, Opt. Express **2**, 59 (1998).
[2] See special issue of Phys. Scr. **T48** (1993).
[3] U. Leonhardt *et al.*, Phys. Rev. A **51**, 84 (1995).
[4] H. M. Wiseman, Phys. Rev. Lett. **75**, 4587 (1995).
[5] Note that this precludes the use of techniques such as homodyne tomography, which rely on multiple measurements on an ensemble of identically prepared systems.
[6] C. M. Caves and P. D. Drummond, Rev. Mod. Phys. **66**, 481 (1994).
[7] This factor of 2 derives from the fact that heterodyne detection samples both the amplitude and phase-quadrature amplitudes, which are complementary observables.
[8] Y. Yamamoto *et al.*, in *Progress in Optics*, edited by E. Wolf (North-Holland, Amsterdam, 1990), Vol. 28, p. 150.
[9] Our calibration of the horizontal axis has been adjusted for overall photodetection efficiency η, which has no bearing on the comparison of adaptive and heterodyne *procedures* but does affect our placement of curves on the graph; had the measurements been of unit efficiency, the data and theory variances would decrease by a factor of η. We have independently measured $\eta \approx 0.56$ (detector quantum efficiency ≈ 0.85, homodyne efficiency ≈ 0.66). For phase measurements performed on squeezed states η would need to be optimized, but its value and affect on the photon-number calibration are not of direct concern in our current work involving coherent states.
[10] For pulses with mean photon number > 10, an ideal adaptive measurement would lie within 1% of the uncertainty limit. Accounting for technical noise, the variance of the adaptive measurement could at best lie a factor of 2 below the noise-corrected heterodyne curve.
[11] H. M. Wiseman and R. B. Killip, Phys. Rev. A **57**, 2169 (1998).
[12] In principle one can modulate either the signal phase or the LO phase; we do the former to avoid interference between the adaptive homodyne feedback loop and an auxiliary feedback loop used to stabilize the MZI.
[13] J. K. Stockton, M. Armen, and H. Mabuchi, J. Opt. Soc. Am. B (to be published).
[14] D.W. Berry and H. M. Wiseman, Phys. Rev. A **63**, 013813 (2001).
[15] A. C. Doherty *et al.*, Phys. Rev. A **62**, 012105 (2000).
[16] C. Ahn, A. C. Doherty, and A. Landahl, Phys. Rev. A, **65**, 042301 (2002).
[17] W. P. Smith *et al.*, Phys. Rev. Lett. 89, 133601 (2002).
[18] H. Mabuchi and H. M. Wiseman, Phys. Rev. Lett. **81**, 4620 (1998).

Uncertainty principle for angular position and angular momentum

Sonja Franke-Arnold[1], Stephen M Barnett[1], Eric Yao[2], Jonathan Leach[2], Johannes Courtial[2] and Miles Padgett[2]
[1] Department of Physics, University of Strathclyde, Glasgow G4 0NG, UK
[2] Department of Physics and Astronomy, Glasgow University, Glasgow G12 8QQ, UK
E-mail: e.yao@physics.gla.ac.uk

New Journal of Physics **6** (2004) 103
Received 14 May 2004
Published 9 August 2004
Online at http://www.njp.org/
doi:10.1088/1367-2630/6/1/103

Abstract. The uncertainty principle places fundamental limits on the accuracy with which we are able to measure the values of different physical quantities (Heisenberg 1949 *The Physical Principles of the Quantum Theory* (New York: Dover); Robertson 1929 *Phys. Rev.* **34** 127). This has profound effects not only on the microscopic but also on the macroscopic level of physical systems. The most familiar form of the uncertainty principle relates the uncertainties in position and linear momentum. Other manifestations include those relating uncertainty in energy to uncertainty in time duration, phase of an electromagnetic field to photon number and angular position to angular momentum (Vaccaro and Pegg 1990 *J. Mod. Opt.* **37** 17; Barnett and Pegg 1990 *Phys. Rev.* A **41** 3427). In this paper, we report the first observation of the last of these uncertainty relations and derive the associated states that satisfy the equality in the uncertainty relation. We confirm the form of these states by detailed measurement of the angular momentum of a light beam after passage through an appropriate angular aperture. The angular uncertainty principle applies to all physical systems and is particularly important for systems with cylindrical symmetry.

It is a fundamental principle of quantum theory that we cannot establish, with arbitrary precision, all the physical properties of any system. This idea has its most concise quantitative statement in the uncertainty principle, which places a lower bound on the product of the underlying uncertainties associated with a chosen pair of observable quantities. The original uncertainty relation, due to Heisenberg, states that the product of the uncertainties in position and momentum for a particle is bounded by Planck's constant [1]: $\Delta x \Delta p \geqslant \hbar/2$. This inequality has played

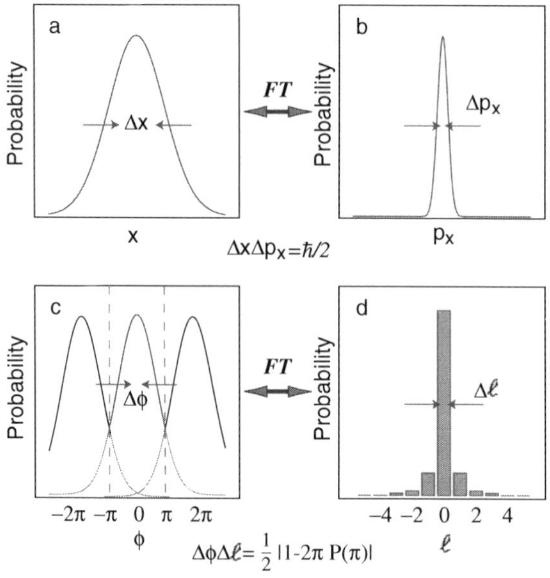

Figure 1. Different forms of the uncertainty principle. The uncertainty principle relates the precision to which various physical quantities of a system can be known. The most familiar form of the principle relates the uncertainty in linear position (a) to the uncertainty in linear momentum (b). Similarly, the uncertainty in angular position (c) is related to the uncertainty in angular momentum (d). For the angular case, the distribution for the intelligent states has a Gaussian form symmetrically truncated within the $-\pi$ to π range. The corresponding distribution of the orbital angular momentum state is the convolution of a Gaussian with a sinc function.

an important role in the study of quantum-limited measurements and in exploring questions concerning the nature of quantum theory [2]. States satisfying the equality in an uncertainty relation are called the intelligent states [3]. If the uncertainty product is bounded by a constant, these intelligent states coincide with the minimum uncertainty-product or critical states [4] that realize the smallest uncertainty product. However, in more complicated relations such as the one between angle and angular momentum, the intelligent states are not necessarily identical with the minimum uncertainty-product states. We will discuss this distinction in more detail elsewhere.

A natural consequence of the continuous Fourier-transform relationship between the linear position and momentum observables is that the intelligent states, which are simultaneously the minimum uncertainty-product states, are Gaussians. Whereas linear position is linked to linear momentum, angular position is linked to angular momentum, and such pairs are called conjugate variables. However, since the angular position is a periodic variable, its relationship with the angular momentum is by a discrete Fourier transform and is more complicated compared with the linear case (see figure 1). For states of precisely defined angular momentum $\Delta L = 0$, there is no restriction on angular position; it can take any value between 0 and 2π with equal probability.

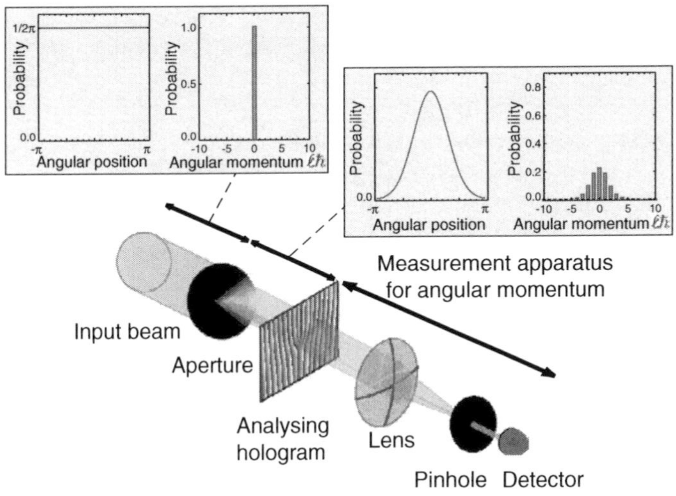

Figure 2. Scheme of the experiment used to observe uncertainties in angular position and angular momentum. Passing a light beam through a 'cake-slice' aperture restricts the angular position, leading to a corresponding broadening of the light's angular momentum states. Experimentally, a spatial light modulator is used to produce both the aperture function and the angular-momentum-analysing hologram. The probability of each angular momentum component is deduced from the fraction of the resulting light transmitted through a pinhole.

Although now unrestricted, the statistical uncertainty in angular position remains finite and readily calculable [5]: $\Delta\phi = \pi/\sqrt{3}$. It is clear, for these states, that the uncertainty product is $\Delta\phi\Delta L = 0$. For other states, where the range of angular position is restricted, it is impossible to know precisely the angular momentum; for small angular uncertainties, the uncertainty principle tends to $\Delta\phi\Delta L = \hbar/2$. Between these extremes, there is a minimum uncertainty-product that varies monotonically from 0 to $\hbar/2$ as the angular uncertainty increases [6]. The general form of the uncertainty principle is a consequence of Robertson's generalization to any pair of observables of Heisenberg's uncertainty principle [1].

In the present paper, by precise measurements on an apertured light beam, we demonstrate for the first time a manifestation of this angular position, angular momentum uncertainty principle. If we place any restriction on the angular position, such as by passing the light beam through an angular aperture (a 'cake-slice'), a distribution of angular momentum states results (see figure 2). We have identified and experimentally verified the form of the aperture that corresponds to the intelligent states for angle and angular momentum.

Light beams carry both linear and angular momenta, which manifest at the macroscopic and single-photon levels. Angular momentum comprises spin and orbital components that are associated with circular polarization and helical-phase fronts respectively. At the photon level, the spin angular momentum can take one of two values, $\pm\hbar$ per photon, corresponding to left- and right-handed circularly polarized light. In contrast, the orbital angular momentum can take one of an unbounded range of values, $\ell\hbar$ per photon [7], where the integer ℓ relates to the azimuthal phase structure of the helically phased beam, $\exp(i\ell\phi)$. At the macroscopic level, both forms of angular momentum can be transferred to particles, causing them to spin about their own axes and orbit about the beam axis, respectively [8]. For single photons, both spin and orbital angular momenta have been shown to be well behaved and to have interesting quantum properties [9, 10].

All physical properties of cylindrical systems are periodic functions of an angular position. For this reason, we must restrict the values of the angle observable to lie within a 2π radian range, and the corresponding angular momentum component L_z can only take on discrete values $\ell\hbar$. The angle operator, $\hat{\phi}_\theta$, will have eigenvalues, ϕ, lying in the range θ to $\theta + 2\pi$, with a common choice being $-\pi \leqslant \phi < \pi$. This dependence on the choice of angular range is denoted by subscript θ in the angle operator. For states with finite uncertainty in angular momentum, the relation between uncertainty in angular position $\Delta\phi_\theta$ and uncertainty in angular momentum ΔL_z has the form [6]

$$\Delta\phi_\theta \Delta L_z \geqslant \tfrac{1}{2}\hbar|1 - 2\pi P(\theta)|, \tag{1}$$

where $\Delta L_z = \Delta\ell\hbar$ and $P(\theta)$ is the angular probability density at the boundary of the chosen angular range. The intelligent states, $|\psi\rangle$, obey the equation

$$[\hat{L}_z - \langle\hat{L}_z\rangle - \mathrm{i}\hbar\lambda(\hat{\phi}_\theta - \langle\hat{\phi}_\theta\rangle)]|\psi\rangle = 0, \tag{2}$$

where $\theta \leqslant \phi < \theta + 2\pi$ and λ is a real constant [11]. By taking its overlap with the state $|\Phi\rangle = (2\pi)^{-1/2}\sum_\ell \mathrm{e}^{-\mathrm{i}\ell\phi}|\ell\rangle$, we can convert equation (2) into a differential equation,

$$\left[\mathrm{i}\frac{\partial}{\partial\phi} + \bar{\ell} + \mathrm{i}\lambda(\phi - \bar{\phi}_\theta)\right]\Psi(\phi) = 0, \tag{3}$$

where $\langle\hat{L}_z\rangle = \bar{\ell}\hbar$ and $\langle\hat{\phi}_\theta\rangle = \bar{\phi}$ denote the mean values of angular momentum and angular position, respectively. The solution of this equation is given by the wavefunction

$$\Psi(\phi) = \frac{(\lambda/\pi)^{1/4}}{\sqrt{\mathrm{erf}(\pi\sqrt{\lambda})}}\mathrm{e}^{\mathrm{i}\bar{\ell}\phi}\exp\left(-\frac{\lambda}{2}(\phi - \bar{\phi})^2\right), \tag{4}$$

where $\theta \leqslant \phi < \theta + 2\pi$. This function is normalized so that $\int_\theta^{\theta+2\pi} \mathrm{d}\phi|\Psi(\phi)|^2 = 1$. Since we are seeking to identify the states giving a finite uncertainty in angular momentum, we require that the wavefunction has no discontinuities, i.e. $\Psi(\theta) = \Psi(\theta + 2\pi)$. Consequently, $\bar{\ell}$ needs to be an integer and the mean value of ϕ must lie at the centre of its allowed range of values, so that $\bar{\phi} = \theta + \pi$. With these restrictions, the truncated Gaussian described by equation (4) is the only realizable intelligent state for angular position (see figure 1(c)).

From equation (4), the corresponding uncertainty in angular position is

$$\Delta\phi = \frac{1}{\sqrt{2\lambda}}\sqrt{1 - \frac{2\sqrt{\pi\lambda}\exp(-\pi^2\lambda)}{\mathrm{erf}(\pi\sqrt{\lambda})}}. \tag{5}$$

Here, we have dropped the subscript in ϕ and selected 0 as the mean value of the angle, so that $-\pi \leqslant \phi < \pi$. We can calculate the angular momentum amplitudes for our intelligent states by taking the Fourier transform of equation (4):

$$\psi(\ell) = \frac{1}{\sqrt{2\pi}} \int_{-\pi}^{+\pi} \mathrm{d}\phi \, \mathrm{e}^{-\mathrm{i}\ell\phi} \psi(\phi)$$

$$= \frac{(\lambda\pi)^{-1/4}}{\operatorname{erf}(\pi\sqrt{\lambda})} \int_{-\infty}^{+\infty} \mathrm{d}k \, \operatorname{sinc}(k\pi) \, \mathrm{e}^{-(\bar{\ell}-\ell-k)^2/(2\lambda)}, \tag{6}$$

which is the convolution of a Gaussian with a sinc function (see figure 1(d)). The corresponding uncertainty in the measured value of the angular momentum quantum number ℓ is

$$\Delta\ell = \lambda\Delta\phi. \tag{7}$$

Taking the product of equations (5) and (7), we confirm that these states do satisfy the equality, i.e.

$$\Delta\phi\Delta\ell = \tfrac{1}{2}|1 - 2\pi P(\pi)|. \tag{8}$$

It is interesting to note that the intelligent states for angular position are truncated Gaussians of equation (4) with a discontinuity in the gradient at $\pm\pi$. The presence of discontinuity is perhaps surprising, but it does not violate the condition for inequality. The minimum uncertainty-product states, in contrast, exhibit no such discontinuity.

The basis of our experiment is the understanding that if a light beam with a single value of orbital angular momentum is passed through an angular aperture, the angular momentum of the transmitted beam will take on a range of values (see figure 2).

Helically phased light beams are readily generated using diffractive optical components, often termed computer-generated holograms. A frequently used design is similar to a diffraction grating, but one that features an ℓ_{holo} pronged fork dislocation at its centre [12]. After illumination with a plane wave, the first-order diffracted beam has an $\exp(\mathrm{i}\ell_{\text{holo}}\phi)$ phase structure and an orbital angular momentum of $\ell_{\text{holo}}\hbar$ per photon. Such holograms also work more generally, such that when the incident beam is already helically phased, e.g. $\exp(\mathrm{i}\ell_{\text{beam}}\phi)$, the diffracted beam has a phase structure $\exp(\mathrm{i}(\ell_{\text{beam}} + \ell_{\text{holo}})\phi)$. When $\ell_{\text{beam}} + \ell_{\text{holo}} = 0$, the diffracted beam has planar phasefronts. Such a system can be used to test for a particular orbital angular momentum state since only a planar diffracted beam can be made to pass through a subsequent pinhole [10].

For accurate measurements, it is essential that both the apex of the aperture and the dislocation within the hologram be centred with respect to the illuminating beam. Consequently, rather than aligning mechanical components, both the aperture and analysing hologram are produced on a spatial light modulator. In principle, separate modulators could be used for the aperture and the analysing holograms; however, it is convenient, in our case, to combine these elements onto a single plane and thereby display them on a single device. The input beam is derived from a 100 mW, 532 nm laser, which is modulated with a mechanical chopper to allow phase-sensitive detection of the light transmitted through the pinhole and, hence, a large dynamic range in the measured angular momentum components. The angular amplitude of the illuminating beam is defined in (4). An angular width for the aperture is picked at random and orientated with a random azimuthal position. The number of dislocations within the analysing hologram is then changed over a wide range of ℓ values to deduce the normalized distribution

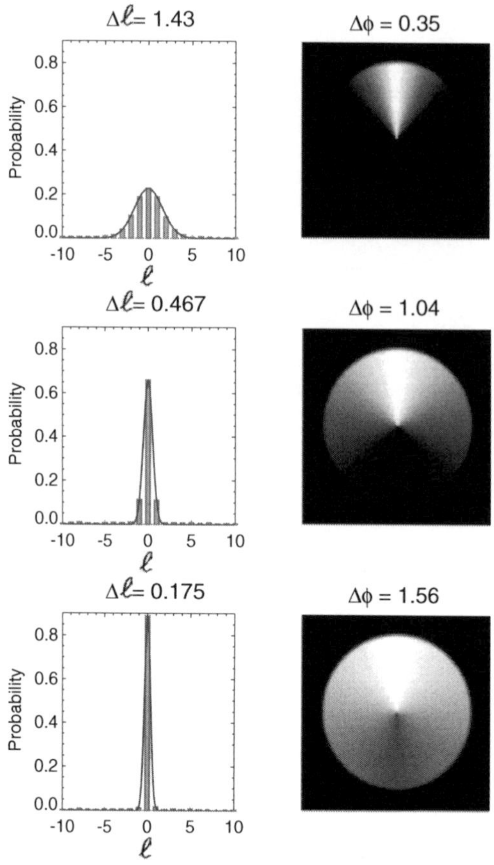

Figure 3. Examples of experimental observation. The observed distribution of the angular momenta (left panel) after transmission through the angular apertures (right panel). The solid curve shows the best fit to the theoretically predicted distribution.

of the orbital angular momentum components of the beam for that particular aperture width. Over several hours, many hundreds of aperture widths can be evaluated and the corresponding distribution of orbital angular momentum states measured.

In our experiment, the incident beam is a Gaussian laser mode with zero angular momentum. Figure 3 shows three angular apertures, the measured distribution of angular momentum components and the statistical best fit to the distribution function $|\psi(\ell)|^2$, obtained from (6). Figure 4 shows the observed product $\Delta\phi\Delta\ell$ plotted against $\Delta\phi$, compared with the theoretical prediction of (8).

Each of the individual measurements is subject to error both in $\Delta\phi$, due to the non-linearity of the spatial light modulator, and in $\Delta\ell$, due to noise in the detection system. For small values of $\Delta\phi$, the total power transmitted through the aperture is low, giving a poor signal-to-noise ratio in the measured ℓ components and resulting in an overestimate of $\Delta\ell$. For large values of $\Delta\phi$, the

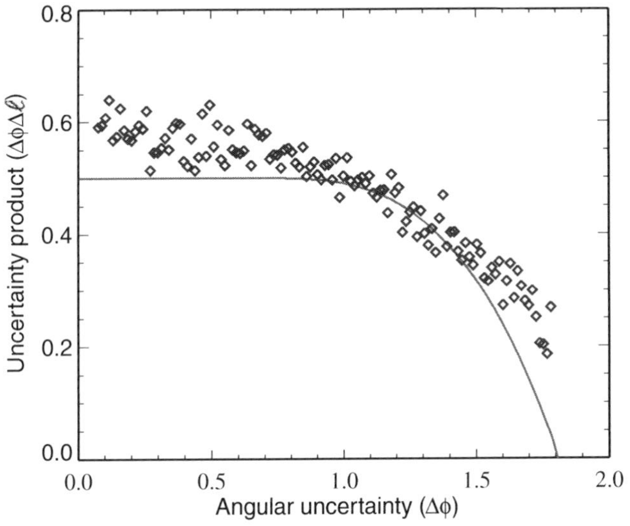

Figure 4. Experimental measurement and theoretical prediction of the angular uncertainty relation. The observed (diamonds) and predicted (line) relationship between the product of the uncertainties in angular position and angular momentum in units of $\hbar$ plotted against the uncertainty in angular position. For small angles, the uncertainty product is constant and equal to 0.5. In the absence of any restricting aperture, there is no uncertainty in angular momentum and the uncertainty product decreases to zero.

angular momentum distribution is dominated by a single value; however, noise for other values again leads to a similar overestimation of $\Delta\ell$.

In addition to supporting the predictions of our theoretical analysis, figure 4 illustrates the detailed behaviour of the angular form of the uncertainty relationship. For no aperture, $\Delta\phi = \pi/\sqrt{3}$ and, as expected, we find that the orbital angular momentum is precisely defined, i.e. $\Delta\ell = 0$. For very small apertures, $P(\pi)$ tends to zero and $\Delta\phi\Delta\ell \approx 1/2$. The most interesting behaviour occurs for intermediate aperture widths when the wings of the Gaussian do not decrease to zero within the 2π range of the distribution. Under these conditions, we see that $\Delta\phi\Delta\ell$ reduces from $1/2$ as $\Delta\phi$ increases.

We note that a similar uncertainty relationship exists between photon number and phase of an electromagnetic field [13], which has been explored experimentally [14]. In this case, however, the fact that the photon number has a minimum value (i.e. zero) means that exact intelligent states, other than the pure photon-number states, do not exist [15]. In this paper, we have observed for the first time the uncertainty relationship governing angular position and angular momentum and have both derived and demonstrated the corresponding intelligent states.

Clearly, experimental results were obtained in the present study from light beams comprising many photons in the same orbital angular momentum state and, hence, are strictly applicable to classical beams only. However, the experiment is fundamentally one in which the measured intensity values arise through interference effects in linear optics. In agreement with other such interference experiments, we can infer that the intensity distribution measured for the many-photon result is proportional to the probability that would be measured if the experiment was repeated with single photons. Consequently, the angular uncertainty relationship between optical orbital angular momentum and the angular aperture applies both in the quantum and classical regimes. This is as anticipated, since the theoretical description relies on linear optics and, thus, is identical for single photons and classical beams.

Acknowledgments

This work was supported by the Engineering and Physical Sciences Research Council, the Royal Society and the Royal Society of Edinburgh.

References

[1] Heisenberg W 1949 *The Physical Principles of the Quantum Theory* (New York: Dover)
Robertson H P 1929 *Phys. Rev.* **34** 127
[2] Wheeler J A and Zurek W H 1983 *Quantum Theory and Measurement* (Princeton, NJ: Princeton University Press)
[3] Aragone C, Guerri G, Salamo S and Tani J L 1974 *J. Phys. A: Math. Gen.* **7** L149
[4] Jackiw R 1968 *J. Math. Phys.* **9** 339
[5] Judge D 1963 *Phys. Lett.* **5** 189
[6] Barnett S M and Pegg D T 1990 *Phys. Rev.* A **41** 3427
[7] Allen L, Beijersbergen M W, Spreeuw R J C and Woerdman J P 1992 *Phys. Rev.* A **45** 8185
Allen L, Barnett S M and Padgett M J 2003 *Optical Angular Momentum* (Bristol: Institute of Physics)
[8] O'Neil A T, MacVicar I, Allen L and Padgett M J 2002 *Phys. Rev. Lett.* **88** 053601
[9] Aspect A, Dalibard J and Roger G 1982 *Phys. Rev. Lett.* **49** 1804
[10] Mair A, Vaziri A, Weihs G and Zeilinger A 2001 *Nature* **412** 313
[11] Merzbacher E 1970 *Quantum Mechanics* 2nd edn (New York: Wiley)
[12] Bazhenov V Yu, Vasnetsov M V and Soskin M S 1990 *JEPT Lett.* **52** 429
[13] Pegg D T and Barnett S M 1989 *Phys. Rev.* A **39** 1665
Barnett S M and Pegg D T 1989 *J. Mod. Opt.* **36** 7
[14] Beck M, Smithey D T, Cooper J and Raymer M G 1993 *Opt. Lett.* **18** 1259
[15] Vaccaro J A and Pegg D T 1990 *J. Mod. Opt.* **37** 17

Minimum uncertainty states of angular momentum and angular position

David T Pegg[1], Stephen M Barnett[2], Roberta Zambrini[2,4], Sonja Franke-Arnold[2] and Miles Padgett[3]

[1] School of Science, Griffith University, Nathan, Brisbane 4111, Australia
[2] Department of Physics, University of Strathclyde, Glasgow G4 0NG, UK
[3] Department of Physics and Astronomy, University of Glasgow, Glasgow G12 8QQ, UK
E-mail: roberta@phys.strath.ac.uk

New Journal of Physics **7** (2005) 62
Received 19 October 2004
Published 17 February 2005
Online at http://www.njp.org/
doi:10.1088/1367-2630/7/1/062

Abstract. The states of linear momentum that satisfy the equality in the Heisenberg uncertainty principle for position and momentum, that is the intelligent states, are also the states that minimize the uncertainty product for position and momentum. The corresponding uncertainty relation for angular momentum and angular position, however, is more complicated and the intelligent states need not be the constrained minimum uncertainty product states. In this paper, we investigate the differences between the intelligent and the constrained minimum uncertainty product states for the angular case by means of instructive approximations, a numerical iterative search and the exact solution. We find that these differences can be quite significant for particular values of angular position uncertainty and indeed may be amenable to experimental measurement with the present technology.

[4] Author to whom any correspondence should be addressed.

1. Introduction

It is well known that the spin angular momentum of a photon associated with circular polarization of a light beam can be found with one of the values $\pm\hbar$. On the other hand, the orbital angular momentum associated with a helical phase front can take a range of values $l\hbar$ per photon, where l is any integer [1]. Recently, it has been shown that the orbital angular momentum of a single photon can be measured [2], and this has led to the experimental confirmation of the uncertainty relation for angular momentum and angular position [3].

The uncertainty relation for angular momentum and angular position is much less well known than the Heisenberg uncertainty principle for linear momentum and position, which is fundamental to quantum mechanics. The latter states that the product of the uncertainties in linear position and momentum has a lower bound such that [4]

$$\Delta x \Delta p \geqslant \hbar/2. \tag{1}$$

The states that satisfy the equality in an uncertainty relation are sometimes referred to as intelligent states [5, 6]. The intelligent states for (1) have Gaussian probability distributions of both position and momentum. Because of the state independence of the right-hand side of (1), for any given Δx, the intelligent states are also the states that minimize the uncertainty product on the left-hand side of (1).

The uncertainty relation between angular momentum and angular position is more complicated than for linear momentum and position (1). If we consider, for example, a bead sliding on a circular wire of known large diameter r, then we might be able to write the angular momentum uncertainty ΔL as $r\Delta p$ and the angular position uncertainty $\Delta\theta$ as $\Delta x/r$. From (1) we might then write $\Delta L\Delta\theta \geqslant \hbar/2$ and find the intelligent states to have Gaussian probability distributions in angle and angular momentum. The difficulty with this simple argument of course is that, unlike the linear position, the angular position takes values only over a finite range of size 2π. Angular positions that differ by 2π represent the same physical state. Thus $\Delta\theta$ must have an upper bound and the relation $\Delta L \Delta\theta \geqslant \hbar/2$ must fail for sufficiently small values of ΔL. For the same reason, no probability distributions of angle can be exactly Gaussian. At best, if this angle probability distribution is sufficiently sharp, that is, if the angular position is sufficiently well defined, the distribution can be approximately Gaussian.

The problems associated with the periodicity of the angular position and probability distributions on a circle were considered by Judge and Lewis [7]. They found that the lower bound on the uncertainty product must be state dependent, of a form given by (4) below. This state dependence leads to the interesting situation in which an intelligent state, even though it has uncertainties that satisfy the equality in (4), need not be a state that minimizes the uncertainty product for a given angular position uncertainty or for a given angular momentum uncertainty [8].

When dealing with an uncertainty relation where the lower bound is state dependent, the question as to what constitutes a minimum uncertainty state can be asked in different ways. Firstly, there is the question about the global minimum. Clearly, here the angular momentum states give this minimum. Beyond this point, one can search for minimum uncertainty states under various additional constraints. One possibility is to consider only states that fulfil the equality in the uncertainty relation, that is, the intelligent states. Another interesting constraint is to consider only states with a given uncertainty in angular position and find which of these minimize the uncertainty product. A third possibility is to consider only states with a given uncertainty in angular momentum. As it turns out, we find that the states that minimize the uncertainty product under the constraint of a given uncertainty in angular position are the same as those that minimize the uncertainty product for a given uncertainty in angular momentum. In the following, we will refer to these states as constrained minimum uncertainty product (CMUP) states.

The experimental confirmation of the uncertainty principle for angular momentum and angular position [3] was carried out for intelligent states, with the uncertainty product plotted against the uncertainty in angle. Because of the state-dependence of the uncertainty bound, a state other than the intelligent states may give an uncertainty product smaller than that measured in [3]. In this paper we examine this question and identify the states that give the minimum uncertainty product. Interestingly, we find that the difference in uncertainty product between intelligent and CMUP states should be measurable with present technology.

2. Intelligent states

In units for which $\hbar = 1$, it is usual to represent the operator for the z-component of angular momentum as [4][5]

$$\hat{L}_z = -\mathrm{i}\frac{\mathrm{d}}{\mathrm{d}\varphi}. \tag{2}$$

We shall refer to the corresponding representation as the angle representation. In this representation, the angle operator $\hat{\varphi}$ is the multiplicative operator $Y(\varphi) = \varphi + 2n\pi$, where n

[5] This representation makes certain assumptions about the differentiability of the angle wavefunction and needs to be used with care. We present a brief analysis of this point in appendix C.

is an integer chosen so that $Y(\varphi)$ has a value within a selected 2π range. $Y(\varphi)$ is a sawtooth function of φ that rises as φ increases but drops sharply by an amount 2π at 2π intervals. This restricts the angle eigenvalues to lie within a particular 2π range as would be expected from the impossibility of distinguishing physically between two states of angle differing by 2π. In this paper we choose the 2π range to be $[-\pi, \pi)$. $Y(\varphi)$ can be expressed in terms of φ plus a series of unit step functions [7], from which follows the commutator

$$[\hat{L}_z, \hat{\varphi}] = -\mathrm{i}\left\{1 - 2\pi \sum_{n=-\infty}^{\infty} \delta[\varphi - (2n+1)\pi]\right\} \tag{3}$$

in the angle representation. From the commutator we can find the uncertainty relation by use of Robertson's general expression [9]. Expressing the expectation value of this operator in terms of the angle wavefunction $\psi(\varphi)$ that is normalized in a 2π interval, allows us to write the corresponding uncertainty relation as

$$\Delta L_z \Delta\varphi \geqslant \tfrac{1}{2}|1 - 2\pi P(\pi)|, \tag{4}$$

where $(\Delta L_z)^2 = \langle \hat{L}_z^2\rangle - \langle \hat{L}_z\rangle^2$ and $P(\pi) = |\psi(\pi)|^2$ is the probability density for finding the system at the angle π. The uncertainty relation (4) has also been derived in [10] for physical states, that is states with finite moments of angular momentum, by use of the angular momentum representation. In that derivation, it arises as a natural consequence of the rigorous commutation relation between the angular momentum and angle operators.

It should be noted that the second term on the right-hand side of (4) depends on our choice of 2π range or window. In general if we choose the range to be $[\theta_0, \theta_0 + 2\pi)$, we would replace $P(\pi)$ by $P(\theta_0)$, which is equal to $P(\theta_0 + 2\pi)$. Thus an intelligent state for a particular choice of θ_0 need not be an intelligent state for another choice. The same will be true for the minimum uncertainty product states discussed later. This may seem somewhat surprising compared with the linear case, where the property of being an intelligent state is independent of the choice of origin of the coordinate system. The reason, however, can be seen as follows. Let us represent the angle probability distribution as a periodic series of peaks at 2π intervals. Then choosing a 2π window that is centred on one peak will give a smaller variance than choosing a window with half of one peak at one end and half of the next peak at the other.

The angular uncertainty relation (4) is of particular interest because it is an example of a case where the intelligent states, which satisfy the equality, do not necessarily minimize the uncertainty product on the left-hand side. This is because the second term on the right-hand side is itself state-dependent. Thus there may be states that, for a given uncertainty $\Delta\varphi$ or ΔL_z, yield a smaller uncertainty product than do the intelligent states but, of course, obey the inequality in (4). This is possible only because $P(\pi)$ for such states is larger than for the intelligent states. This reduces the right-hand side of the uncertainty relation compared with that for the intelligent states, allowing the uncertainty product on the left-hand side to be smaller than that for the intelligent states while still obeying the uncertainty relation.

The intelligent states $|g\rangle$ satisfy the eigenvalue equation [6, 11]

$$(\hat{L}_z - \mathrm{i}\gamma\hat{\varphi})|g\rangle = \mu|g\rangle, \tag{5}$$

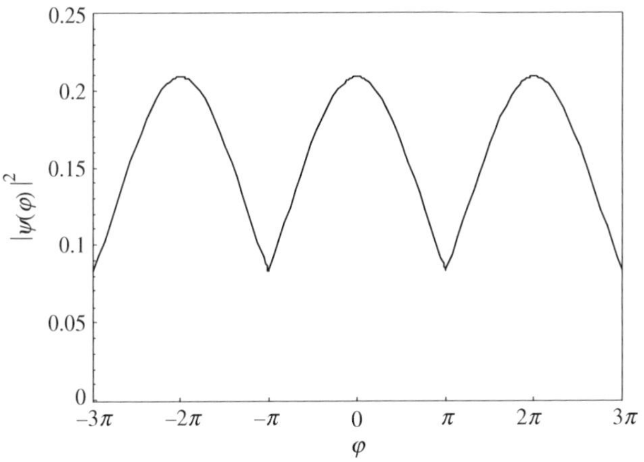

Figure 1. The angular probability distribution for the intelligent state corresponding to an angular uncertainty of $\Delta\varphi = 1.6$. The distribution is Gaussian truncated so as to fit within a chosen 2π range and repeated with a periodicity of 2π.

where γ is real. Without loss of generality, we can restrict ourselves to intelligent states with zero mean angle and angular momentum [3]. Acting on (5) from the left with $\langle g|$ then gives $\mu = 0$, so we have

$$(\hat{L}_z - \mathrm{i}\gamma\hat{\varphi})|g\rangle = 0. \tag{6}$$

In the angle representation $\hat{\varphi} = Y(\varphi)$ so (6) becomes, from (2),

$$\frac{\mathrm{d}\psi(\varphi)}{\mathrm{d}\varphi} = -\gamma Y(\varphi)\psi(\varphi). \tag{7}$$

The solution of this differential equation is

$$\psi(\varphi) = N \exp\left[-\gamma \int Y(\varphi)\,\mathrm{d}\varphi\right] = N \exp\left[-\frac{\gamma}{2} Y^2(\varphi)\right], \tag{8}$$

with N being a normalization constant. Noting that $Y(\varphi + 2n\pi) = Y(\varphi)$, we see that ψ is a Gaussian function between $-\pi$ and π which is duplicated between π and 3π and so on, forming a periodic function with cusps at 2π intervals (see figure 1). It is the same result as was derived by use of the angular momentum representation and obtained experimentally in [3].

We note that the action of the unitary angular momentum shift operator $\exp(\mathrm{i}\hat{\varphi}\,k)$ on the state $|g\rangle$ will not change $|\psi(\varphi)|^2$ and will thus leave $\Delta\varphi$ unaltered. This operation will, however, shift the angular momentum distribution uniformly by an integer amount k [10] but this will not change ΔL_z. Thus the state $\exp(\mathrm{i}\hat{\varphi}k)|g\rangle$ will also be an intelligent state with a non-zero mean angular momentum.

3. Minimum uncertainty product states

In this section, we derive the expression for the states that minimize the uncertainty product $\Delta L_z \Delta\varphi$ either for a given $\Delta\varphi$ or for a given ΔL_z. These are the constrained minimum uncertainty product states. We consider an angular-momentum decomposition

$$|f\rangle = \sum_{m=-\infty}^{\infty} b_m |m\rangle, \tag{9}$$

where $|m\rangle$ is an eigenstate[6] of $\hat{L}_z$ with $m = 0, \pm 1, \pm 2, \ldots$. In appendix A.1 we show that in seeking the CMUP states with a constraint in the angle variance we can assume the b_m to be real and that $\langle \hat{L}_z\rangle = \langle\hat{\varphi}\rangle = 0$. Therefore, the variances simplify to $\langle \hat{L}_z^2\rangle$ and $\langle\hat{\varphi}^2\rangle$. Interestingly, as shown in appendix A.2, exactly the same assumptions can be made in minimizing the uncertainty product for a given angular momentum variance. It follows that the states minimizing the uncertainty product will be of the same form whether we fix $\Delta\varphi$ or ΔL_z.

We require the state $|f\rangle$ that minimizes $\langle \hat{L}_z^2\rangle\,\langle\hat{\varphi}^2\rangle$ subject to the normalization constraint $\langle f|f\rangle = 1$. We approach this by the method of undetermined multipliers [8, 12]. The basic equation is the vanishing of a linear combination of the variations $\delta\langle \hat{L}_z^2\rangle$, $\delta\langle\hat{\varphi}^2\rangle$ and $\delta\langle f|f\rangle$ with the coefficients being the multipliers. This is the equation we obtain whether we minimize $\langle \hat{L}_z^2\rangle$ for a fixed $\langle\hat{\varphi}^2\rangle$ or whether we minimize $\langle\hat{\varphi}^2\rangle$ for a fixed $\langle \hat{L}_z^2\rangle$ and the state $|f\rangle$ that satisfies this equation will minimize $\langle \hat{L}_z^2\rangle\,\langle\hat{\varphi}^2\rangle$ for either a given $\langle\hat{\varphi}^2\rangle$ or a given $\langle \hat{L}_z^2\rangle$. As the b_m are all real we can introduce undetermined multipliers λ and μ and use the arguments of [12] to obtain an eigenvalue equation for the CMUP state $|f\rangle$

$$(\hat{L}_z^2 + \lambda\hat{\varphi}^2)|f\rangle = \mu|f\rangle. \tag{10}$$

Before looking for an analytical solution, it is instructive to examine some of its general properties. To this end we consider the equation in the angular momentum representation, where we find that we can obtain some good analytic approximations.

3.1. General properties of CMUP states

We start by noticing that the CMUP states have a symmetric probability distribution, $|\psi(\varphi)|^2 = |\psi(-\varphi)|^2$ centred at $\varphi = 0$. This property is discussed further in appendix A.

A second property of the CMUP states can be obtained as follows. In the angular momentum representation, we can express $\hat{\varphi}^2$ when operating on the space of physical states as

$$\hat{\varphi}^2 = \frac{\pi^2}{3} + 2 \sum_{\substack{m,m' \\ m\neq m'}} \frac{(-1)^{m-m'}}{(m-m')^2} |m'\rangle\langle m|. \tag{11}$$

This can be obtained either by using the approach of [10] or from the cosine series for φ^2, as used in [12] for the phase of light, together with the cosine angle operators. Using (9) we find

[6] It is an accident of history that the integer associated with the z-component of angular momentum is denoted by m in quantum theory but l in optics.

that (10) becomes, in the angular momentum representation,

$$(m^2 - \mu)b_m + \lambda\langle m|\hat{\varphi}^2|f\rangle = 0. \tag{12}$$

This gives the equation

$$\left(m^2 - \mu + \frac{\pi^2}{3}\lambda\right) b_m + 2\lambda \sum_{m'\neq m} \frac{(-1)^{m-m'}}{(m-m')^2} b_{m'} = 0. \tag{13}$$

Consider the form of this equation for very large $|m|$. Clearly, the first term will tend to $m^2 b_m$. In addition, in order for $\hat{L}_z^2|f\rangle$ to be normalizable we require that b_m must fall off faster than m^{-2}. If we turn to the summation, we see that this will be dominated by values of $|m'| \ll |m|$. It follows that we can approximate $(m-m')^{-2}$ by m^{-2}. We are then left with an equation for the b_m that is valid for very large $|m|$:

$$m^2 b_m = -\frac{(-1)^m}{m^2} 2\lambda \sum_{m'\neq m} (-1)^{m'} b_{m'}. \tag{14}$$

The summation is independent of m and hence for very large $|m|$

$$b_m = \frac{(-1)^m}{m^4} A, \tag{15}$$

where A is some constant. We note that this is precisely the form of the perturbative solutions that we obtain below in (16) and (22). Expression (15) implies that $\hat{L}_z^4|f\rangle$ will not be normalizable and that the eighth moment of the angular momentum will be infinite for a CMUP state. This is associated with discontinuities in the higher derivatives of $\psi(\varphi)$. To be specific, the amplitudes (15) are associated with a discontinuity at $\varphi = \pi$ of $\frac{d^3\psi}{d\varphi^3}$. As discussed in appendix C, these states show the minimum degree of regularity needed for the CMUP states.

3.2. States with small ΔL

It is clear that an eigenstate of (10) for $\lambda = 0$ is just an angular momentum state $|m\rangle$ and thus the CMUP state with zero mean angular momentum is just the $m = 0$ state $|0\rangle$. This is an extreme case of a CMUP state with zero angular momentum variance with an associated uncertainty product of zero. This state is also an intelligent state. Taking $|0\rangle$ as our unperturbed state, we can find the eigenstates of (10) for small λ by perturbation theory. The matrix elements of $\hat{\varphi}^2$ between angular momentum states are easily obtained from (11). Substitution into the standard first-order perturbation expression [4] yields the result

$$|f\rangle = |0\rangle - 2\lambda \sum_{m\neq 0} \frac{(-1)^m}{m^4} |m\rangle. \tag{16}$$

These states will have non-zero but small angular momentum variances. The coefficients b_m of the angular momentum representation and the wavefuction $\psi(\varphi)$ of the angle representation form a finite Fourier transform pair [13]

$$\psi(\varphi) = \frac{1}{\sqrt{2\pi}} \sum_m b_m \exp(\mathrm{i}m\varphi), \tag{17}$$

$$b_m = \frac{1}{\sqrt{2\pi}} \int_{-\pi}^{\pi} \psi(\varphi) \exp(-\mathrm{i}m\varphi)\,\mathrm{d}\varphi. \tag{18}$$

From (17) we find the angle wavefunction for the first-order perturbation solution as

$$\psi(\varphi) = \frac{1}{\sqrt{2\pi}} \left[1 + \frac{\lambda}{6}\left(\frac{\varphi^2}{2} - \pi^2\varphi^2 + \frac{7\pi^4}{30}\right)\right]. \tag{19}$$

The values of ΔL_z and $\Delta\varphi$ are easily obtained from (16) and (19). We find to first order in λ

$$\Delta\varphi = \frac{\pi}{\sqrt{3}}\left(1 - \frac{8\pi^4}{315}\lambda\right), \tag{20}$$

$$\Delta L_z \Delta\varphi = \frac{\lambda\pi^4}{\sqrt{3}}\sqrt{\frac{8}{945}}\left(1 - \frac{8\pi^4}{315}\lambda\right). \tag{21}$$

Remarkably, this first-order approximation, depicted by a dotted line in figure 2, is sufficient to show the difference between the CMUP states and the intelligent states and is quite accurate over more than half the range of values of $\Delta\varphi$, i.e. $1 < \Delta\varphi < \pi/\sqrt{3}$.

Second-order perturbation theory can be used to extend the validity of the small-λ approximation. To second order we obtain the state

$$|f\rangle = \left(1 - 4\lambda^2\frac{\pi^8}{9450}\right)|0\rangle + \sum_{m,m\neq 0}\left[-2\lambda\frac{(-1)^m}{m^4} + 4\lambda^2\frac{(-1)^m}{m^4}\left(\frac{\pi^4}{45} + \frac{4\pi^2}{3m^2} - \frac{15}{m^4}\right)\right]|m\rangle. \tag{22}$$

We note that for large m, coefficients of $|m\rangle$ decrease as m^{-4}. This state is normalized just to second order in λ and, to be accurate, it needs to be renormalized. The resulting uncertainty product is shown by a dashed line in figure 2.

3.3. *States with large ΔL*

The opposite approximation to that above is for states with large $\langle L_z^2\rangle$, with the state of zero angle, for which $\langle L_z^2\rangle$ is infinite, being the extreme case. For this extreme case, $P(\pi) = 0$ and so, from (4), the uncertainty product is $1/2$. In the angular momentum representation, (10) yields, with (9)

$$\lambda\langle m|\hat{\varphi}^2|f\rangle = (\mu - m^2)b_m, \tag{23}$$

where, from (11),

$$\langle m|\hat{\varphi}^2|f\rangle = \frac{\pi^2}{3}b_m + 2\sum_{p>0}\frac{(-1)^p}{p^2}(b_{m+p} + b_{m-p}). \tag{24}$$

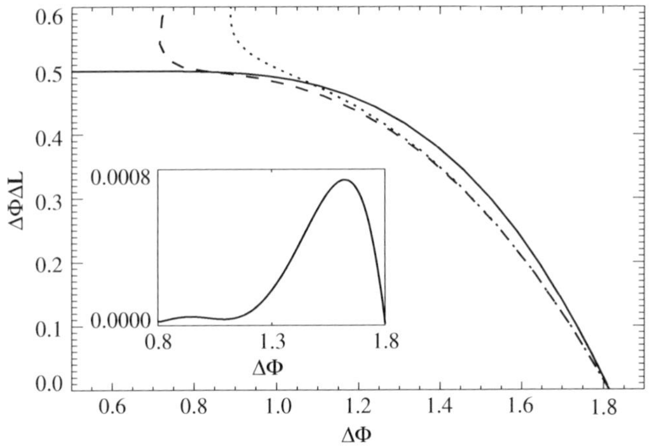

Figure 2. The uncertainty product $\Delta L_z \Delta\varphi$ plotted against $\Delta\varphi$ for the first-order perturbative (······), second-order perturbative (- - - -) and intelligent states (——). At this resolution, the uncertainty product for the exact solution is indistinguishable from the second-order perturbative value for $\Delta\varphi > 0.9$ and from 0.5 for smaller values of $\Delta\varphi$. The inset shows the difference in the product of the uncertainties obtained with the sum of Gaussians and with the exact solution (38), that is $(\Delta\varphi\, \Delta L_z)^G - (\Delta\varphi\, \Delta L_z)^E$, plotted against $\Delta\varphi$.

To solve equation (23), we use an ansatz based on the approximation (B.11) from appendix B, that is

$$\langle m|\hat{\varphi}^2|f\rangle \approx -\frac{\mathrm{d}^2 b(m)}{\mathrm{d}m^2}, \tag{25}$$

where $b(m)$ is a continuous envelope curve for which $b(m) = b_m$ when m is an integer. Our procedure is to use (25) to find a solution of (23) for various values of λ and then test this solution for consistency, by checking that (25) is actually satisfied.

Substituting (25) into (23) gives the differential equation for the envelope as

$$\lambda\frac{\mathrm{d}^2 b(m)}{\mathrm{d}m^2} = (m^2 - \mu)b(m). \tag{26}$$

The solution with the minimum variance of m that satisfies the boundary condition that $b(m) \to 0$ as m becomes infinite is the Gaussian

$$b(m) = N_1 \exp\left(-\frac{m^2}{2\sigma^2}\right), \tag{27}$$

where $\sigma^2 = \mu = \lambda^{1/2}$ and N_1 is a constant to be determined by normalization. Thus, for integer m,

$$b_m = N_1 \exp\left(-\frac{m^2}{2\sigma^2}\right) \tag{28}$$

is our solution of (23), subject to consistency of our ansatz.

To test the ansatz, we substitute (28) into the exact expression (24) and obtain

$$\langle m|\hat{\varphi}^2|f\rangle = \left[\frac{\pi^2}{3} + 4\sum_{p>0}\frac{(-1)^p}{p^2}\cosh\left(\frac{pm}{\sigma^2}\right)\exp\left(-\frac{p^2}{2\sigma^2}\right)\right] b_m. \tag{29}$$

A numerical comparison of (29) and

$$-\frac{\mathrm{d}^2 b(m)}{\mathrm{d}m^2} = \left(\frac{1}{\sigma^2} - \frac{m^2}{\sigma^4}\right) b_m \tag{30}$$

shows that the accuracy of our ansatz (25) is excellent, that is a small fraction of 1%, for $\sigma \geqslant 2$ provided $|m| \leqslant 6\sigma$. Even down to $\sigma^2 = 0.5$, the accuracy is still 10%. The contribution of values of $|m| > 6\sigma$ to $\langle m^2\rangle$ will be negligible, so our solution (27) with $\sigma^2 = \lambda^{1/2}$ will give an accurate value for the uncertainty product for sufficiently broad angular momentum distributions. For $\sigma = 1$ we need $|m| \leqslant 4$, for $\sigma = 0.5$ we need $|m| \leqslant 1$. For $\sigma \leqslant 0.4$, no value of m gives consistency.

To find the angle wavefunction from (17), it is simple to write first

$$b_m = \int_{-\infty}^{\infty} b(x)\,\delta(m - x)\,\mathrm{d}x, \tag{31}$$

with $b(x) \propto \exp\left[-x^2/(2\sigma^2)\right]$. The final result is

$$\psi(\varphi) = N_2 \sum_{n=-\infty}^{\infty} \exp\left[-\frac{(\varphi^2 + 2n\pi)^2}{2/\sigma^2}\right]. \tag{32}$$

This is a sum of Gaussians centred at angles 2π apart. For large σ^2, the values of $\psi(\varphi)$ in the range between $-\pi$ and π reduces to the form of a single Gaussian with $n = 0$. For this case, $P(\pi)$ is effectively zero and the uncertainty product is 1/2.

3.4. Interpolation for all values of ΔL

Our ansatz is quite good for larger values of $\langle \hat{L}_z^2\rangle$, giving the correct value of $\Delta\varphi$ to be well within 1% for values of $\Delta\varphi$ from zero up to about 0.75. We can improve on this accuracy and also interpolate between this value and the values of $\Delta\varphi$ for which perturbation theory is accurate by noting that we can relax the requirement $\sigma^2 = \mu = \lambda^{1/2}$ for the Gaussian representing b_m. We note that as long as

$$\langle m|\hat{\varphi}^2|f\rangle = \alpha m^2 + \xi \tag{33}$$

for all m, where α and ξ are any constants, $|f\rangle$ is a CMUP state. A state for which (33) is approximately true for some value of α and of ξ, for the range of values of m for which b_m are not negligible, will approximate a CMUP state. Thus the Gaussian state (28) will be a good approximation to a CMUP state as long as (33) with $\langle m|\hat{\varphi}^2|f\rangle$ given by (29) is approximately true for some value of α and of ξ for the range of values of m that, for example, make significant contributions to ΔL_z. For large ΔL_z, we have seen that values of α and ξ such that $\sigma^2 = \mu = \lambda^{1/2}$

satisfy (33) for a range of m more than sufficient to ensure that the important values of b_m are quite accurate. As ΔL_z becomes smaller, this range of m decreases but we can find different values of α and ξ that ensure that (33) is reasonably satisfied for the non-negligible values of b_m. In the limit where ΔL_z is very small, only the values of b_m with $m = 0, \pm 1$ will contribute significantly and we can always choose α and ξ such that (33) holds exactly for these values of m. The accuracy of representing b_m by a Gaussian for values of ΔL_z where first-order perturbation theory applies can be seen as follows. Expression (16) shows that the contribution to $\langle \hat{L}_z^2 \rangle$ from the $m = \pm 2$ terms is only $1/64$ of that from the $m = \pm 1$ terms, so a mismatch for these terms will have only a very minor effect. By fitting a Gaussian to have the values b_0, b_1 and b_{-1}, we find from (16) that for such a Gaussian

$$\sigma^2 = -\frac{1}{2\ln(2\lambda)}. \tag{34}$$

From the above discussion, a Gaussian state in the angular momentum representation or, equivalently, a sum of Gaussians in the angle representation should provide a good analytic approximation to a CMUP state across the whole range of values of $\Delta\varphi$. We see below how good such states are in providing an approximation to the uncertainty product.

3.5. Exact CMUP states

We now construct the exact solution of equation (10) by working in the angle representation and making the substitutions

$$x = \sqrt{2}\lambda^{1/4}\varphi, \tag{35}$$

$$a = -\frac{\mu}{2\sqrt{\lambda}}. \tag{36}$$

With $\hat{L}_z^2$ represented by $-\frac{\mathrm{d}^2}{\mathrm{d}\varphi^2}$, (10) becomes, for values of φ between $-\pi$ and π,

$$\frac{\mathrm{d}^2\psi}{\mathrm{d}x^2} - \left(\frac{x^2}{4} + a\right)\psi = 0. \tag{37}$$

The solution for the angle wavefunction ψ will be an even function in this range so that we can write it in the form

$$\begin{aligned}\psi &= \exp(-x^2/4)M\left(\frac{1}{2}a + \frac{1}{4}, \frac{1}{2}, \frac{1}{2}x^2\right)\\ &= \exp(-x^2/4)\left[1 + \left(a + \frac{1}{2}\right)\frac{x^2}{2!} + \left(a + \frac{1}{2}\right)\left(a + \frac{5}{2}\right)\frac{x^4}{4!}\right.\\ &\quad \left. + \left(a + \frac{1}{2}\right)\left(a + \frac{5}{2}\right)\left(a + \frac{9}{2}\right)\frac{x^6}{6!} + \cdots\right],\end{aligned} \tag{38}$$

where M is a confluent hypergeometric function [14]. The evenness ensures that $\psi(-\pi) = \psi(\pi)$, which is required from periodicity and the need for continuity of the function. We can also apply

Table 1. Uncertainty product for four values of the angle variance. The superscripts E, G, I and II refer to the exact solution, the sum of Gaussians, first-order and second-order perturbative solutions, respectively.

$\Delta\varphi$	0.8130954	1.317335	1.633695	1. 809022
$(\Delta\varphi\Delta L_z)^E$	0.498188	0.389542	0.181522	0.0054894678
$(\Delta\varphi\Delta L_z)^G$	0.498210	0.389739	0.182268	0.0055360491
$(\Delta\varphi\Delta L_z)^{\mathrm{I}}$	–	0.392661	0.181683	0.0054894713
$(\Delta\varphi\Delta L_z)^{\mathrm{II}}$	–	0.389559	0.181522	0.0054894678

the boundary condition that the first derivative is continuous at $\varphi = \pi$ (see appendix C). This, together with the fact that the wavefunction is periodic, requires that the first derivative should vanish here. The solution giving the smallest $\Delta\varphi$ will be that whose first minimum occurs when $\varphi = \pi$. The corresponding values of a will be between 0 and -0.5. (When $a = -0.5$, ψ becomes a Gaussian, that is, its first minimum is at infinity; when $a = 0$, the first minimum is at $x = 0$.)

The requirement to have the first minimum of ψ at $\varphi = \pi$ leads to

$$\lambda = \frac{x_0^4}{4\pi^4} \tag{39}$$

and

$$\varphi = \frac{\pi}{x_0}x, \tag{40}$$

so

$$\langle\hat{\varphi}^2\rangle = \frac{\pi^2}{x_0^2}\langle\hat{x}^2\rangle \tag{41}$$

for values of x between $-x_0$ and x_0, where x_0 is the position of the first minimum of $\psi(x)$. The values of ψ can be normalized numerically in this range and the value of $\langle\hat{x}^2\rangle$ and hence $\langle\hat{\varphi}^2\rangle$ computed. The uncertainty product can then be obtained from (10), giving

$$\begin{aligned}\langle\hat{\varphi}^2\rangle\langle\hat{L}_z^2\rangle &= \langle\hat{\varphi}^2\rangle(\mu - \lambda\langle\hat{\varphi}^2\rangle)\\ &= \langle\hat{x}^2\rangle\left(-a - \tfrac{1}{4}\langle\hat{x}^2\rangle\right),\end{aligned} \tag{42}$$

where we have used (41) and (36), and the uncertainty product is the square root of (42). Varying the choice of a allows the calculation of uncertainty products corresponding to a range of $\Delta\varphi$ values.

In table 1 we compare, for a range of $\Delta\varphi$, values of uncertainty products calculated from the exact solution and from the use of sums of Gaussians as CMUP states. We also include the uncertainty products calculated from perturbation theory. For large angle variances, in the last column of table 1, we note the very high precision of the perturbative approximation in relation to the exact solution. As expected, these solutions do not minimize the uncertainty product accurately for small angle variances. On the other hand, the sum of Gaussians is a good approximation to the exact solution for a wide range of angle variances, as shown in figure 2. The maximum deviation between the sum of Gaussians and the exact solution uncertainty product is less than 8×10^{-4}.

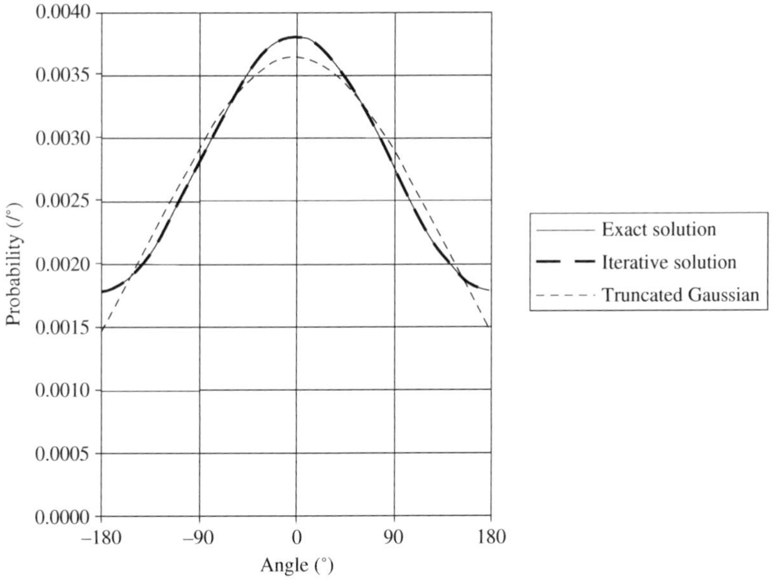

Figure 3. The various forms of the angular wavefunction corresponding to $\Delta\varphi = 1.60186058$. The exact solution (38) and iterative solution agree to 11 significant figures and are clearly distinct from the intelligent state (8).

3.6. Numerical iterative search for CMUP states

To confirm our identification of the correct form of the minimum uncertainty product state, we performed an iterative numerical search for the angle wavefunction that gives the smallest uncertainty product. The overall approach is to set a specific value of $\Delta\varphi$ and then search to find the function giving the smallest possible value of ΔL_z. To reduce the search time required to find the optimum angle wavefunction, we assumed that this function is symmetric and that it increases monotonically from the edge to the centre. The iterative algorithm is seeded with a triangular-shaped function. Upon each iteration, the gradient at a random position within the function is itself randomized and the baseline of the modified function offset to obtain the required standard deviation $\Delta\varphi$. The resulting function is then Fourier-transformed to give the distribution in angular momentum and the corresponding standard deviation ΔL_z from which the uncertainty product is calculated. If the uncertainty product is reduced then the most recent change to the gradient is kept, if not it is discarded. Even using a personal computer, 10 such iterations can be trialled each second with a period of several hours being sufficient to obtain an optimized function. From the inset in figure 2, we see that the largest difference predicted between the uncertainty product corresponding to an overlapping Gaussian wavefunction and the minimum uncertainty product state occurs near a value of $\Delta\varphi \approx 1.6$. The exact solution, equation (38), predicts that the CMUP state with a standard deviation $\Delta\varphi = 1.60186058$ gives an uncertainty product of $\Delta\varphi\Delta L_z = 0.208848427271$. We take this value of $\Delta\varphi$ as the target for the iterative search. Our algorithm obtains an angle probability distribution which agrees with that obtained from the exact solution to 11 significant figures, as shown in figure 3. The angular momentum

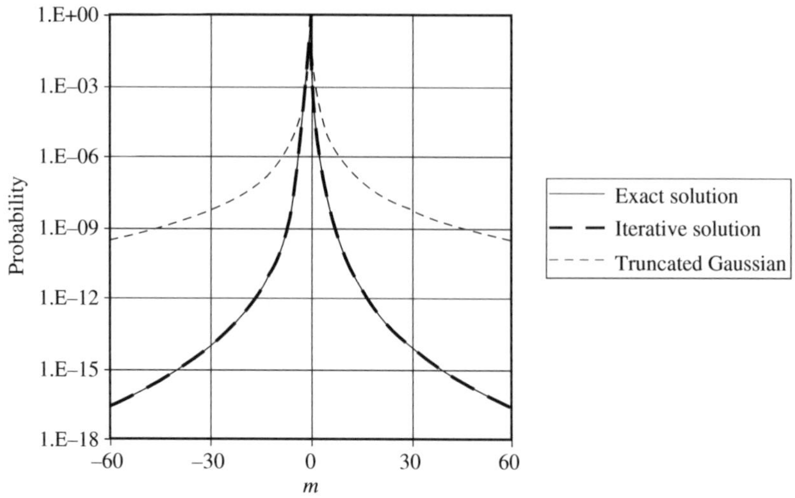

Figure 4. The probability distribution in angular momentum corresponding to the wavefunctions in figure 3.

distributions, as given by the finite Fourier transforms of the iterative and exact solutions, are again numerically indistinguishable over the range $|m| \leqslant 60$, corresponding to a range of probabilities in excess of 16 orders of magnitude. Perhaps the most convincing evidence that the iterative form matches that of the exact solution is that, over the same numerical range, both probability distributions fall with a power dependence of m^{-8}. This is precisely in accord with our asymptotic result (15). The angular momentum probability distributions for our iterative and exact solutions, together with that for the intelligent state, are presented in figure 4.

4. Conclusion

In this paper we have examined the states that minimize the uncertainty product of angular position and angular momentum either for a given variance in angle or for a given variance in angular momentum. We have established that both constraints result in the same CMUP states and that they differ from the intelligent states, that is the states satisfying the equality in the uncertainty relation $\Delta L_z \Delta\varphi \geqslant \frac{1}{2}|1 - 2\pi P(\pi)|$. The constrained minimum uncertainty product (CMUP) states yield a smaller uncertainty product than the intelligent states because they have a larger probability density $P(\pi)$ at the edge of the 2π range than the intelligent states. This allows the uncertainty product to be less than that for the intelligent states while still exceeding $\frac{1}{2}|1 - 2\pi P(\pi)|$. The exact solution for the CMUP states was checked to a very high precision by an iterative minimization algorithm. We also have found that analytic perturbation approximations are useful over a significant part of the range of possible variances $\Delta\varphi$, including the region of the largest deviation from the corresponding intelligent state expression. An analytic approximation that is quite accurate over the whole range of values of $\Delta\varphi$ is a state with a Gaussian distribution in the angular momentum representation. In the angle representation, this corresponds to a wavefunction that is a periodic sum of Gaussians. This behaviour contrasts strongly with the situation for linear momentum and position, for which the intelligent states and CMUP states are identical and have Gaussian distributions in both position and momentum.

In the experimental verification [3] of the uncertainty relation for angular momentum and angular position, the most accurate measurements of the uncertainty product were in the range of $\Delta\varphi$ values between 1.0 and 1.5. It is fortunate that the largest discrepancy between the uncertainty products for the intelligent and the minimum product states occurs around $\Delta\varphi \approx 1.4$, as can be seen from figure 2. This means that the difference between intelligent states and minimum uncertainty product states for angular position and angular momentum should be within the reach of experimental test.

Acknowledgments

This work was supported by the Australian Research Council, the UK Engineering and Physical Sciences Research Council (GR/S03898/01) and the Royal Society of Edinburgh.

Appendix A. Demonstration of the adequacy of considering real b_m

In this appendix, we show that the minimum uncertainty product $\Delta L_z \Delta\varphi$ can be obtained by considering real angular momentum amplitudes b_m, both in the case of a given angle variance and of a given angular momentum variance.

A.1. Minimizing the uncertainty product for a given $\Delta\varphi$

We begin by writing our angle wavefunction in the form

$$\psi(\varphi) = \chi(\varphi)\mathrm{e}^{\mathrm{i}\alpha(\varphi)}, \tag{A.1}$$

where $\chi(\varphi)$ is a real function. It is straightforward to show, using the continuity of $\psi(\varphi)$ and of its derivatives, that

$$\langle \hat{L}_z \rangle = \int_{-\pi}^{\pi} \frac{\mathrm{d}\alpha}{\mathrm{d}\varphi} \chi^2 \mathrm{d}\varphi, \tag{A.2}$$

$$\Delta L_z^2 = -\int_{-\pi}^{\pi} \chi \frac{\mathrm{d}^2\chi}{\mathrm{d}\varphi^2} \mathrm{d}\varphi + \left\langle \left(\frac{\mathrm{d}\alpha}{\mathrm{d}\varphi} - \left\langle \frac{\mathrm{d}\alpha}{\mathrm{d}\varphi} \right\rangle \right)^2 \right\rangle. \tag{A.3}$$

Clearly, the variance in the angular momentum is minimized by choosing $\mathrm{d}\alpha/\mathrm{d}\varphi$ to be a constant so that the second term in (A.3) is zero. The requirement that the wavefunction should be continuous tells us that $\mathrm{d}\alpha/\mathrm{d}\varphi$ must be an integer and (A.2) leads us to identify it as the mean angular momentum. Without loss of generality, but for the sake of definiteness, we choose this mean value to be zero so that α is simply a constant. We have now established that for any given angle probability distribution, and hence for any given angular uncertainty, the corresponding minimum angular momentum variance will be obtained for a wavefunction of the form (A.1) with constant α and we have chosen solutions with $\langle \hat{L}_z \rangle = 0$. We can readily obtain solutions with non-zero mean angular momentum by multiplying our wavefunction by $\exp(\mathrm{i}\bar{m}\varphi)$, where $\bar{m} = \langle \hat{L}_z \rangle$.

We now turn our attention to the expansion (17) which we rewrite as

$$\chi(\varphi) = \frac{1}{\sqrt{2\pi}} \sum_m b_m \mathrm{e}^{-\mathrm{i}\alpha} \exp(\mathrm{i}m\varphi). \tag{A.4}$$

The fact that $\chi(\varphi)$ must be real requires that

$$b_{-m}\mathrm{e}^{-\mathrm{i}\alpha} = b_m^* \mathrm{e}^{\mathrm{i}\alpha}. \tag{A.5}$$

A second relationship between our angular momentum amplitudes follows from the fact that we are seeking the minimum of $\langle \hat{L}_z^2 \rangle$, which is insensitive to the sign of the angular momentum m. It follows that if (A.4) is a state that minimizes this variance, then so too will be the state with each b_m replaced by b_{-m}:

$$\tilde{\chi}(\varphi) = \frac{1}{\sqrt{2\pi}} \sum_m b_{-m} \mathrm{e}^{-\mathrm{i}\alpha} \exp(\mathrm{i}m\varphi). \tag{A.6}$$

If both (A.4) and (A.6) are satisfactory states then, by the linearity of quantum mechanics, so is any superposition of them. It suffices then to consider only symmetric and antisymmetric combinations of these two states, that is $\chi \pm \tilde{\chi}$. These correspond to (A.4) with $b_{-m} = b_m$ for the symmetric case and $b_{-m} = -b_m$ for the antisymmetric case. If we choose the arbitrary phase α in (A.4) to be zero in the symmetric case and $\pi/2$ in the antisymmetric case then, using (A.5), we find in both cases that the b_m are real. It is straightforward to show that both of these states have symmetric angular probability distributions and hence that $\langle \hat{\varphi} \rangle = 0$.

In summary, this short analysis has demonstrated that in seeking to minimize the angular momentum uncertainty for a given angular uncertainty, it suffices to consider states with real angular momentum amplitudes b_m and with a mean angle of zero.

A.2. Minimizing the uncertainty product for given ΔL_z

We start by noting that ΔL_z is independent of a shift of the angular coordinate so $\psi(\varphi)$ and $\psi(\varphi - \Phi)$ will have the same angular momentum variance. The associated angular variance is

$$\Delta\varphi^2 = \int_{-\pi}^{\pi} |\psi(\varphi - \Phi)|^2 \varphi^2 \, \mathrm{d}\varphi - \left(\int_{-\pi}^{\pi} |\psi(\varphi - \Phi)|^2 \varphi \, \mathrm{d}\varphi \right)^2. \tag{A.7}$$

Differentiation with respect to Φ shows that this variance will be minimized by choosing Φ such that $\langle \hat{\varphi} \rangle = 0$. Hence, minimizing $\Delta\varphi^2$ corresponds to minimizing $\langle \hat{\varphi}^2 \rangle$. For the same reasons as given in [12] for minimizing the optical phase variance, we can minimize $\langle \hat{\varphi}^2 \rangle$ by choosing all the b_m to be real and positive or zero. It is interesting to note that we can also obtain the maximum $\Delta\varphi^2$ by setting $\langle \hat{\varphi} \rangle = 0$. It transpires that for an angle distribution peaked at $\varphi = 0$, we find a minimum of $\Delta\varphi^2$ while for an angle distribution peaked at $\varphi = \pm\pi$ we find a maximum.

Appendix B. Derivation of equation (25)

In the angular momentum representation, the angle operator when acting on physical states can be written as [10]

$$\hat{\varphi} = -\mathrm{i} \sum_{\substack{m,m' \\ m \neq m'}} \frac{(-1)^{m'-m}}{m'-m} |m\rangle\langle m'| \tag{B.1}$$

with eigenvalues between $-\pi$ and π.

For the state $|f\rangle = \sum_m b_m |m\rangle$, we have

$$\langle m|\mathrm{i}\hat{\varphi}|f\rangle = \sum_{p\neq 0} (-1)^p \frac{b_{m+p} - b_{m-p}}{2p}. \tag{B.2}$$

Consider a function

$$F_m(x) = \frac{b(m+x) - b(m-x)}{2x}, \tag{B.3}$$

where m and x are continuous variables and $b(m) = b_m$ when m is an integer.

Writing

$$\int_{-\infty}^{\infty} F_m(x)\,\mathrm{d}x = 2\sum_{p\,\mathrm{odd}} \frac{b(m+p) - b(m-p)}{2p} + \varepsilon_1 \tag{B.4}$$

$$= 2\sum_{p\,\mathrm{even}} \frac{b(m+p) - b(m-p)}{2p} + \varepsilon_2. \tag{B.5}$$

(B.5) can also be written as

$$\int_{-\infty}^{\infty} F_m(x)\,\mathrm{d}x = 2F_m(0) + 2\sum_{\substack{p\,\mathrm{even} \\ p\neq 0}} \frac{b(m+p) - b(m-p)}{2p} + \varepsilon_2. \tag{B.6}$$

Hence, from (B.4) and (B.6) we have

$$\sum_{p\,\mathrm{odd}} \frac{b(m+p) - b(m-p)}{2p} - \sum_{\substack{p\,\mathrm{even} \\ p\neq 0}} \frac{b(m+p) - b(m-p)}{2p} = F_m(0) + (\varepsilon_2 - \varepsilon_1)/2 \tag{B.7}$$

so that

$$\langle m|\mathrm{i}\hat{\varphi}|f\rangle = -F_m(0) - (\varepsilon_2 - \varepsilon_1)/2. \tag{B.8}$$

We find $F_m(0)$ from

$$F_m(0) = \lim_{x\to 0} F_m(x) = \frac{\mathrm{d}b(m)}{\mathrm{d}m}, \tag{B.9}$$

which follows from (B.3). Thus

$$\langle m|\mathrm{i}\hat{\varphi}|f\rangle \approx -\frac{\mathrm{d}b(m)}{\mathrm{d}m} \tag{B.10}$$

provided $|F_m(0)| \gg |(\varepsilon_2 - \varepsilon_1)|/2$. The validity of this condition depends on how broad and smooth the function is and also on the value of m.

A parallel derivation starting with expression (11) for $\hat{\varphi}^2$ in place of (B.1) yields

$$\langle m|\hat{\varphi}^2|f\rangle \approx -\frac{\mathrm{d}^2b(m)}{\mathrm{d}m^2}. \tag{B.11}$$

Appendix C. Angle representation of the angular momentum

The replacement (with units in which $\hbar = 1$)

$$\hat{L}_z \to -\mathrm{i}\frac{\mathrm{d}}{\mathrm{d}\varphi} \tag{C.1}$$

can sometimes lead to difficulties. In particular, problems will arise when the action of $\hat{L}_z$ on $|\psi\rangle$ does not lead to a normalizable state.

It is helpful to begin by establishing some asymptotic properties of periodic functions of the angle φ. Consider a function ψ defined by the Fourier series

$$\psi(\varphi) = (2\pi)^{-1/2} \sum_{m=-\infty}^{\infty} c_m \mathrm{e}^{\mathrm{i}m\varphi}. \tag{C.2}$$

We do not demand at present that this should necessarily be a wavefunction. We are interested in the asymptotic properties of the c_m, that is its form as $|m| \to \infty$.

C.1. Divergent functions

Consider first a function, ψ, which contains a delta function at some given angle. The Fourier components of the non-delta-function part will decay away as $|m| \to \infty$, but the delta-function part has constant components. Hence for functions containing a delta function we have

$$c_m \to \alpha \mathrm{e}^{-\mathrm{i}m\varphi_0} \quad \text{as } |m| \to \infty, \tag{C.3}$$

where α is a complex constant and φ_0 is the position of the delta function. Functions containing such delta function components cannot, of course, represent wave functions as they are not square-integrable.

C.2. Discontinuous functions

Consider next a function that contains a discontinuity at some given angle. Such a function might be associated with transmission through a mask having a sharp edge. Such a discontinuity can be described in terms of a Heaviside step function. The derivative, with respect to φ of such a step function is a delta function and hence the Heaviside function is the integral of a delta function. Hence, we can write our discontinuous function as

$$\chi(\varphi) = \int \psi(\varphi)\, \mathrm{d}\varphi, \tag{C.4}$$

where ψ is a function containing a delta function. It follows that for functions containing a discontinuity

$$c_m \to \frac{\beta}{m} \mathrm{e}^{-\mathrm{i}m\varphi_0} \quad \text{as } |m| \to \infty, \tag{C.5}$$

where β is a complex constant and φ_0 is the position of the discontinuity.

Functions containing such discontinuities will represent square-integrable wavefunctions but the associated mean square angular momentum will be divergent:

$$\langle \hat{L}_z^2 \rangle = \infty. \tag{C.6}$$

It is clear that the wavefunctions we seek must contain neither delta functions nor discontinuities.

C.3. Functions having discontinuities of gradient

Consider, finally, a function (such as the truncated Gaussian) which contains a discontinuity of gradient. The derivative, with respect to φ of such a discontinuous gradient is a Heaviside function and it follows that a discontinuity of gradient is the integral of a Heaviside function. It then follows, from the reasoning given above, that for functions containing a discontinuity of gradient

$$c_m \to \frac{\gamma}{m^2} \mathrm{e}^{-\mathrm{i}m\varphi_0} \quad \text{as } |m| \to \infty, \tag{C.7}$$

where γ is a complex constant and φ_0 is the position of the discontinuity of gradient. Such functions will represent square-integrable wavefunctions with finite mean-squared angular momentum.

The argument presented above can be readily extended to higher-order inverse powers of m. This leads us to identify the m^{-3} dependence with discontinuity in the second derivative and the m^{-4} dependence with discontinuity in the third derivative of the wavefunction. The latter occurs for the exact CMUP states (38).

We are now in a position to analyse the success and failing of the differential representation of the angular momentum operator (C.1). In seeking the minimum uncertainty state, we required the solution of the eigenvalue equation:

$$(\hat{L}_z - \bar{l} + \mathrm{i}\lambda\hat{\varphi})|\psi\rangle = 0. \tag{C.8}$$

In order to find this, we used the replacement (C.1) to give a differential equation for $\psi(\varphi)$. This procedure can only work if the solution that is obtained has $-\mathrm{i}\frac{\mathrm{d}\psi}{\mathrm{d}\varphi}$ as a valid wavefunction so that $\hat{L}_z|\psi\rangle$ can be correctly represented by $\psi(\varphi)$ in (C.8). The truncated Gaussian solution has a discontinuity of gradient so that its derivative has a discontinuity. Such a state has a divergent mean-squared momentum, but it is square integrable and so can provide a mathematically sensible probability distribution and hence an acceptable wavefunction. In showing that the solution of the differential equation obtained from (C.8) has nothing worse than a discontinuity of gradient, we are verifying that it is an acceptable solution.

Finding the minimum product state led us to seek the solution of the operator equation

$$(\hat{L}_z^2 + \lambda\hat{\varphi}^2)|\psi\rangle = \mu|\psi\rangle. \tag{C.9}$$

Applying the differential representation (C.1) leads to a number of possible solutions including the truncated Gaussian solution. This solution, however, has discontinuities of gradient, but for this problem, the solution will be acceptable only if $-\mathrm{d}^2\psi/\mathrm{d}\varphi^2$ is a valid wavefunction so that $\hat{L}_z^2|\psi\rangle$ can be correctly represented by $\psi(\varphi)$ in (C.9). The second derivative of a function containing a discontinuity in gradient has a delta-function and so will not be square integrable and cannot be a valid wavefunction. It follows that we need a solution that has a continuous gradient everywhere and this, together with the periodicity of the wavefunction requires that $\mathrm{d}\psi/\mathrm{d}\varphi = 0$ at $\varphi = \pm\pi$. As seen in section 3.1, CMUP states have, not only a continuous gradient, but also a continuous second-order derivative. The latter follows from the evenness and periodicity of these states. Irregularities appear only in the third- and higher-order derivatives.

References

[1] Allen L, Barnett S M and Padgett M J 2003 *Optical Angular Momentum* (Bristol: Institute of Physics Publishing)
[2] Leach J, Padgett M J, Barnett S M, Franke-Arnold S and Courtial J 2002 *Phys. Rev. Lett.* **88** 257901
[3] Franke-Arnold S, Barnett S M, Yao E, Leach J, Courtial J and Padgett M J 2004 *New J. Phys.* **6** 103
[4] Merzbacher E 1998 *Quantum Mechanics* (Brisbane: Wiley)
[5] Aragone C, Chalbaud E and Salam S 1976 *J. Math. Phys.* **17** 1963
[6] Vaccaro J A and Pegg D T 1990 *J. Mod. Opt.* **37** 17
[7] Judge D and Lewis J T 1963 *Phys. Lett.* **5** 190
Judge D 1964 *Nuovo Cimento* **31** 332
[8] Jackiw R 1968 *J. Math. Phys.* **9** 339
[9] Robertson H P 1929 *Phys. Rev.* A **34** 163
[10] Barnett S M and Pegg D T 1990 *Phys. Rev.* A **41** 3427
[11] Gottfried K 1966 *Quantum Mechanics* vol 1 (New York: Benjamin) pp 213–5
[12] Summy G S and Pegg D T 1990 *Opt. Commun.* **77** 75
[13] Pegg DT, Vaccaro J A and Barnett S M 1990 *J. Mod. Opt.* **37** 1703
[14] Abramowitz M and Stegun I A 1972 *Handbook of Mathematical Functions* (New York: Dover) p 686

7

Time

The physical theory of time carries three fundamental problems. David Pegg, the dedicatee of this book, has made important contributions to each of them. We trust that the reader will forgive us if we briefly outline these fascinating issues and his contributions before turning to the link between time and the quantum phase operator, the subject of our book.

The first concerns the asymmetry of the "direction of time". Time is usually associated in physics with a real-valued parameter t. The physical laws of motion, electromagnetism and gravity, largely govern the behaviour of macroscopic objects. These laws are invariant to time inversion, i.e. to t being replaced with $-t$, yet we experience a unidirectional "flow" of time from low t, the past, to larger t, the future. Moreover we remember the past but not the future and so the future appears to be distinct from the past. This suggests an inherent asymmetry in time and, indeed, thermodynamics appears to support this. The problem of how to account for the asymmetry and unidirectional flow is typically addressed in terms of identifying distinct *arrows of time*. The arrows most often discussed are the thermodynamical, psychological, cosmological, electromagnetic and matter-antimatter arrows. The task is to reduce the number of arrows by showing that one is more fundamental in that others can be derived from it. Great care has to be taken, however, not to artificially introduce an asymmetry into an arrow. Price has referred to this fallacy as a temporal double standard (Price 1996). An early example is Boltzmann's use of his H-theorem (viz. the entropy of an isolated ideal gas in a non-equilibrium state will increase in entropy over time) to explain the time asymmetry of the second law of thermodynamics. However, as Loschmidt pointed out, since statistical mechanics is based on time-reversible laws the entropy should increase in either direction of time. The double standard is to associate an arrow of time with the increase in one direction and not the other (Price 1996). Another example is the assertion that the wave function collapse associated with a quantum measurement implies a time asymmetry and thus an arrow of time. Aharonov, Bergmann and Lebowitz (1964) showed that this implication is fallacious for systems with specific preparations. Pegg recently extended the analysis to the general preparation case (Pegg 2006). The assumed direction of time is subtly inserted into the quantum analysis in terms of the type of questions we ask, being either predictive or retrodictive in character.

To avoid the temporal double standard, a meaningful arrow of time should be associated with an asymmetrical physical law or alternatively with asymmetric boundary conditions. The asymmetry of time might then be seen to result from something more fundamental. One example is the matter-antimatter arrow associated with T (time inversion) and CP (charge and parity inversion) violations. Cronin, Fitch and colleagues showed that the decay of neutral kaon was not time symmetric, and that the universe favours the production of matter over anti-matter in the increasing t direction (Christenson *et al.* 1964). However, while CP violation may account for the asymmetry between matter and antimatter we observe in the universe at present, it has no known connection to our experience of the unidirectional flow of time. Another possibility with more direct consequences is the connection between the electromagnetic and cosmological arrows. Using Wheeler and Feynman's time-symmetric absorber theory (Wheeler and Feynman 1945, Pegg 1975), Pegg has linked the thermodynamic and psychological arrows to the asymmetry in the electromagnetic arrow and the latter to the asymmetry in cosmological models (Pegg 1973, 1999). Entropy increase is seen to be due to retarded radiation in an expanding universe and entropy decrease due to advanced radiation in a contracting one. Thus he argues that the cosmological model provides asymmetrical boundary conditions which leads to the asymmetry in the other arrows.

The second fundamental problem is associated with travelling backwards through time. Such travel does not appear to be precluded by current physical theories (Lanczos 1924, van Stockum 1937, Morris *et al.* 1988, Friedman *et al.* 1990). The problem is that time travel can be accompanied with logical inconsistencies. One famous example is the grandfather paradox in which a person supposedly travels back in time and kills a grandparent before the birth of the person's parent; but in that case the person cannot be born to travel back in time and kill the grandparent. The seriousness of this conundrum led Hawking to propose his chronology protection conjecture (1992) which rules out time travel on the basis of the logical inconsistencies it could lead to. However this denial is only necessary for logically inconsistent time travel. It is not relevant to time travel in circumstances where logical inconsistencies are avoided by some natural physical means. Consider, for example, problems involving the advanced potential in time-symmetric absorber theory. Peres and Schulman showed that only self-consistent solutions exist for the associated differential equations, and thus logical inconsistencies are avoided by virtue of the theory itself (Peres and Schulman 1972, Schulman 1974). Moreover, Novikov and collaborators studied the behaviour of classical objects following closed time-like curves. They found that the self-interaction between younger and older versions of the objects leads to multiple solutions for the same initial conditions; it is only the logically consistent solutions that could occur in nature (Friedman *et. al.* 1990, Novikov 1992). Novikov calls this the *principle of self-consistency* (1992). Various quantum scenarios have also been shown to lead to logically consistent solutions (Friedman *et. al.* 1990, Deutsch 1991). More recently Pegg applied *quantum retrodiction* to this problem. Quantum retrodiction was introduced by Aharonov *et al.* (1964) and extensively developed by Barnett, Pegg, Jeffers and coworkers (see e.g. Pegg, Barnett and Jeffers 2002). In quantum retrodiction the preparation of a quantum system is retrodicted from the state implied by a later measurement. This is to be contrasted with the more-conventional formalism where measurement outcomes are predicted from a given preparation state. Pegg used the backwards time evolution of quantum retrodiction and the forwards time evolution of quantum prediction in a quantum optical setting to model a closed time loop (Pegg 2001). He showed that logically inconsistent solutions do not arise because of the destructive interference of quantum amplitudes of the forwards and backwards time evolution. Only logically consistent solutions show constructive interference. With this he suggests that the principle of self-consistency has a basis in quantum physics.

The third fundamental problem of time concerns its lack of a consistent quantum description and leads us to the link between time and the quantum phase operator. Time is represented differently in general relativity and quantum physics. In the former theory space and time are placed on the same footing as external coordinates whereas in the latter the spacial coordinates are special in the sense that every physical system has associated with it spacial operators but not a time operator. The absence of a time operator has precluded the rigourous derivation of the time-energy uncertainty relation. Despite this, the relation is typically found in elementary texts on quantum mechanics due to the conceptual niche it occupies. The time-energy uncertainty relation is often associated with the notion that time is somehow canonically conjugate to energy. But owing to the mathematical difficulty associated with defining a suitable operator, some authors have turned to other possible interpretations of the meaning of time in the time-energy uncertainty relation, as e.g. the lifetime of a state, the duration of an energy measurement etc. These attempts were not without controversy (Aharonov and Bohm 1961). The search for alternatives continues with recent attention given to an uncertainty relation involving the arrival time of a wave packet (Galapon *et al.* 2004). However, given that the phase observable of a harmonic oscillator is indeed canonically conjugate to its energy, there is strong motivation for using the phase observable as a basis for the time observable. Indeed this association can be found in the early work on quantum phase (Susskind and Glogower 1964 **Paper 1.5**). It also appears in the first paper introducing the Pegg-Barnett phase formalism (Pegg and Barnett 1988 **Paper 2.1**) as well as in (Vaccaro, Barnett and Pegg 1992 **Paper 4.7**) where the application of a harmonic oscillator as a clock is discussed.

The question remains, however, as to what observable should represent time for quantum systems other than the harmonic oscillator and, in particular, for systems whose energy spectrum is not evenly spaced. Pegg has defined such an object (Pegg 1998 **Paper 7.2**): the quantity α, which he suggests could be called the "age", has the dimensions of time but does not simply evolve linearly with the parameter time t. For example, α is constant for a system in an eigenstate of energy, that is, in a stationary state. The concept that the system does not "age" in stationary states is very appealing. For finite dimensional systems whose energy eigenvalues E_i have the property that their ratios E_i/E_j are rational, the age is cyclical with a period T, where T is the smallest value to make $TE_i/\hbar$ a multiple of 2π for all energy eigenvalues E_i. This cyclic property is also attractive given the system "rejuvenates" in the sense that it returns to its original state with the same period. Pegg has derived the age-energy uncertainty relation which has some formal similarity with the number-phase uncertainty relation reviewed in Chapter 2. Indeed the Hermitian operator associated with the age property shares similarities with the phase operator, but due to the unevenly spaced energy spectrum its derivation is not a simple extension of the phase formalism. Nevertheless for the cyclical case, one could regard the $\alpha 2\pi/T$ as the phase of the cycle. The details can be found in the original article (Pegg 1998 **Paper 7.2**)

Age is an intrinsic property of a quantum system. The *external* parameter time t associated with the system's evolution (e.g. via Schrödinger's equation) is quite different. It originates in the remainder of the universe which contains devices such as clocks to monitor its passing. This leads to the deeper issue regarding the description of time when the whole universe is treated as a closed quantum system. An external time parameter such as t then has little meaning. Pegg has developed a formalism for treating time in this situation (Pegg 1991 **Paper 7.1**). His starting point is to take the state of the entire universe as the zero energy eigenstate $|E_0\rangle$. This is consistent with the canonical quantisation approach to quantum gravity and the solution of the Wheeler-DeWitt equation $\hat{\mathcal{H}}|E_0\rangle = 0$ where $\hat{\mathcal{H}}$ is the Hamiltonian of the whole universe (Alvarez 1989). $|E_0\rangle$ can be expanded as $(|\phi_0\rangle + |\phi_1\rangle + \ldots + |\phi_s\rangle)/\sqrt{s+1}$ where the set of states $\{|\phi_m\rangle : m = 0 \ldots s\}$ has the property that the Hamiltonian of the universe $\hat{\mathcal{H}}$ is the generator of translations, i.e. $|\phi_m\rangle = \exp(i\hat{\mathcal{H}}m\gamma)|\phi_0\rangle$ for some constant γ. This expansion suggests $|E_0\rangle$ represents the entire history of the universe analogously to how a world line in relativity represents the entire motional history of an object. For this reason it is called the *history state*. Our usual description of a parameter time t external to a quantum system emerges when the universe is divided into two non-interacting parts, a clock device and the system (comprising of the remainder of the universe). These two parts are found to be entangled irrespective of their spatial separation. The entanglement has a pertinent form when the state of the clock is written in terms of phase states as a correlation then emerges between the phase of the clock and the state of the system. This correlation defines the external time t for the system relative to the phase of the clock. Moreover the Schrödinger equation also emerges in a similar natural manner. Hence, while there is no external time parameter for the universe as a whole, an external time parameter emerges for a system within the universe with respect to a clock device, and further the system appears to evolve with respect to the clock. The full analysis is given in Pegg (1991 **Paper 7.1**).

Time in a quantum mechanical world

D T Pegg

Division of Science and Technology, Griffith University, Nathan, Brisbane 4111, Australia

Received 28 February 1991

Abstract. In quantum mechanics it is usual to represent physical reality as a vector in Hilbert space at a particular time, with evolution being governed by Schrödinger's equation which involves an externally imposed time parameter. This leads to difficulties if one wishes to regard the universe as a quantum mechanical system, because there should no time external to such a system. The approach in this paper is to represent the totality of reality as a vector in Hilbert space. We show how time evolution follows, where time now is defined in terms of the states of a quantum mechanical clock which is part of the system. Rather than the correlation between the clock states and the states of the rest of the system arising because both are governed by an imposed law involving an external time parameter, it is seen that this correlation is of the separation-independent Einstein–Podolsky–Rosen type. The total reality vector, which incorporates the whole history of the system, is shown to be a zero-energy eigenstate of the system Hamiltonian. We discuss systems of finite and infinite lifetime, and are able to answer the question: what was the state before the initial state? We conclude that the quantum mechanical system of this paper is a reasonable representation of the observed universe.

1. Introduction

In the canonical quantization approach to gravity (see, for example, Dirac 1958, 1959, Komar 1967, Isham and MacCallum 1975, Alvarez 1989 and references therein) some well known difficulties arise. One of these is that the invariance of the ten classical field equations under arbitrary curvilinear coordinate transformations implies that four of these equations are constraints and the Hamiltonian is effectively zero. The usual quantization procedure yields the quantum mechanical result that the system is in a zero energy eigenstate of the Hamiltonian operator. Application of the Schrödinger time-evolution equation involving the corresponding Hamiltonian operator then leads to a 'frozen' dynamics for which nothing seems to happen. As a possible resolution of this difficulty, it has been considered for some time (Alvarez 1989, Unruh and Wald 1989) that the problem may arise from the imposition on the system of an external time parameter, with the Schrödinger equation being applied as a law of evolution in terms of this external time. While this procedure may be appropriate for a system within the universe, it is doubtful whether it can be applied to the universe itself. If one is dealing with such a system it is reasonable to assume that the Hamiltonian must include all observers and the standard clock which defines time measurement (Page and Wootters 1983, Alvarez 1989, Unruh and Wald 1989). Consequently any reasonable approach should ideally *already incorporate* Schrödinger's equation or its equivalent, that is, there should be no need to postulate the form of any time evolution operator, even that applying to the clock. Another difficulty in matching general relativity to quantum mechanics is that physical reality in usual quantum mechanical theory, that is, that which is represented by a ray in Hilbert space, refers only to a particular instant of time. In general relativity, time is simply a label attached to a spacelike hypersurface and only histories have physical meaning (see, for example, Unruh and Wald 1989).

In this paper we approach the problem entirely from a quantum mechanical point of view. We begin with the proposition that the *totality* of physical reality can be represented by a single vector in Hilbert space. From this we derive a quantum mechanical model which incorporates time evolution and which appears to be applicable to the world we observe.

2. Expansion of the history vector

Without defining precisely what we mean by the totality of physical reality, we denote it by S and let it be described by a vector in some Hilbert space Ψ. We let the dimensionality of this Hilbert space be $s+1$, remembering that for S to pertain to the observed universe, s should be extremely large or infinite. For the present, however, we keep s arbitrary to maintain generality and consider the limit as $s\to\infty$ later.

There are s mutually orthogonal vectors which are orthogonal to the vector describing S. We label this latter vector as $|E_0\rangle$ and the complete set of vectors as $|E_n\rangle$ with $n=0, 1, \ldots, s$. From these vectors we can construct a Hermitian operator

$$\hat{H}=\sum_{n=0}^{s} E_n|E_n\rangle\langle E_n| \tag{2.1}$$

which has eigenvalues E_n, which at present are totally arbitrary. We assign the value $E_0=0$, which gives us

$$\hat{H}|E_0\rangle=0. \tag{2.2}$$

We do not assign particular values to the remaining E_n but restrict them to satisfy

$$E_n=p_n\delta E \tag{2.3}$$

where p_n are arbitrary non-zero integers and δE is a small common factor. From above $p_0=0$. It is clear that E_n with $n\neq 0$ can be made, to within an error of δE, as close to any real value we choose by a suitable choice of p_n.

The vector describing S can be expanded as

$$|E_0\rangle=a_0^{-1}\sum_{n=0}^{r} a_n|E_n\rangle\delta_{n0} \tag{2.4}$$

where $r\leqslant s$. The expansion of the Kronecker delta

$$\delta_{n0}=(s+1)^{-1}\sum_{m=0}^{s}\exp[-imp_n2\pi/(s+1)] \qquad \text{for } p_n\leqslant s \tag{2.5}$$

which can be checked by summing the geometric progression, yields

$$|E_0\rangle=[a_0(s+1)]^{-1}\sum_{m=0}^{s}|\phi_m\rangle \tag{2.6}$$

where

$$|\phi_m\rangle=\sum_{n=0}^{r} a_n\exp[-imp_n2\pi/(s+1)]|E_n\rangle=\exp[-i\hat{H}m\delta E^{-1}2\pi/(s+1)]|\phi_0\rangle \tag{2.7}$$

with

$$|\phi_0\rangle = \sum_{n=0}^{r} a_n |E_n\rangle. \tag{2.8}$$

If we specify that the sum of the squares of the moduli of a_n is unity, then all the $|\phi_m\rangle$ are normalized.

We see, therefore, that the vector $|E_0\rangle$ describing S is proportional to

$$|\phi_0\rangle + |\phi_1\rangle + \ldots + |\phi_m\rangle + \ldots + |\phi_s\rangle \tag{2.9}$$

where each $|\phi_m\rangle$ is a normalized superposition of state vectors $|E_n\rangle$ and which can be obtained from $|\phi_{m-1}\rangle$ by the action of the unitary operator $\exp[-\mathrm{i}H\delta E^{-1}2\pi/(s+1)]$. For reasons which will become apparent later we refer to the superposition (2.9) as a 'history' vector or history state.

3. Time

Let us now suppose that S is such that each of the states $|\phi_M\rangle, |\phi_{M+1}\rangle, \ldots, |\phi_{M+P}\rangle$ in a continuous section of the superposition (2.9) can be factorized into the product of two states which are superpositions of basis states of two subspaces Ψ_B and Ψ_C whose tensor product is the Hilbert space Ψ, where Ψ_B has dimensions $b+1$ and Ψ_C has dimensions $c+1$, with $(s+1)=(b+1)(c+1)$. That is, we can write

$$|\phi_m\rangle = |B_m\rangle|C_m\rangle \qquad \text{for } M \leqslant m \leqslant M+P. \tag{3.1}$$

The fact that $|B_m\rangle$ and $|C_m\rangle$ are completely in Ψ_B and Ψ_C respectively for all the values of m shown in (3.1) allows us to factorize the unitary operator transforming $|\phi_m\rangle$ to another state in this section of the history state into two unitary operators involving only the basis states of Ψ_B and Ψ_C respectively. Explicitly we can define

$$\hat{H}_b = \sum_{i=0}^{b} E_{bi}|E_{bi}\rangle\langle E_{bi}| \tag{3.2}$$

$$\hat{H}_c = \sum_{j=0}^{c} E_{cj}|E_{cj}\rangle\langle E_{cj}| \tag{3.3}$$

where the $(s+1)$ states $|E_{bi}\rangle|E_{cj}\rangle$ form the states $|E_n\rangle$. Then

$$\hat{H}|\phi_m\rangle = (\hat{H}_b + \hat{H}_c)|\phi_m\rangle \qquad \text{for } M \leqslant m \leqslant M+P. \tag{3.4}$$

Consistently with this and with (2.7), we can write

$$|B_{m+1}\rangle = \exp[-\mathrm{i}\hat{H}_b\delta E^{-1}2\pi/(s+1)]|B_m\rangle \tag{3.5}$$

$$|C_{m+1}\rangle = \exp[-\mathrm{i}\hat{H}_c\delta E^{-1}2\pi/(s+1)]|C_m\rangle \tag{3.6}$$

for the range of m in (3.4).

We call the factors in (3.1) the states of subsystems B and C respectively. We now particularize to a specific quantum system C by letting the eigenvalues of $\hat{H}_c$ in (3.3) take equally spaced values

$$E_{cj} = j\Delta E \tag{3.7}$$

where ΔE is constant integer multiple of δE. Further, we let the state $|C_M\rangle$ be an equal superposition of the basis states of Ψ_C:

$$|C_M\rangle = (c+1)^{-1} \sum_{j=0}^{c} |E_{cj}\rangle. \tag{3.8}$$

There are $c+1$ such states which are mutually orthogonal:

$$|C_{M+k}\rangle = (c+1)^{-1} \sum_{j=0}^{c} \exp[-ikj2\pi/(c+1)]|E_{cj}\rangle$$
$$= \exp[-ik\hat{H}_c \Delta E^{-1} 2\pi/(c+1)]|C_M\rangle \tag{3.9}$$

where k takes the values $0, 1, \ldots, c$. The orthogonality of these states, which we call phase states, can be checked directly (compare with the orthonormal set of oscillator phase states discussed by Pegg and Barnett 1988). The next state, that is with $k = c+1$, is identical to $|C_M\rangle$ and so on. We define a special case, which we call an ideal system C, by setting

$$\Delta E = (b+1)\delta E. \tag{3.10}$$

The maximum eigenvalue of $\hat{H}_c$ for this system is $c\Delta E$. The condition $p_n \leqslant s$ in (2.5) restricts the maximum eigenvalue sum E_n to be no greater than $s\delta E$, from which we find that the maximum possible eigenvalue of system B is ΔE. This ideal case is thus quite extreme.

Substituting ΔE from (3.10) into (3.9) and comparing with (3.6), we find that $|C_{M+k}\rangle$ for $k \leqslant P$ is both one of the $c+1$ orthogonal phase states and is also a factor of a term in the history state. Thus the relevant section of the history state (2.9) can be written as

$$|B_M\rangle|C_M\rangle + |B_{M+1}\rangle|C_{M+1}\rangle + \ldots + |B_{M+q}\rangle|C_{M+q}\rangle + \ldots + |B_{M+P}\rangle|C_{M+P}\rangle. \tag{3.11}$$

We note that the state $|C_{M+c+1}\rangle$ is identical with the state $|C_M\rangle$, corresponding to one complete period, so it is convenient to set $P \leqslant c$ to ensure that all phase states of the ideal system C is this history section are orthogonal.

The state (3.11) exhibits an Einstein-Podolsky-Rosen (EPR) type of correlation. In the EPR paradox, two spatially separated systems A and D are in a correlated state $|A_1\rangle|D_1\rangle + |A_2\rangle|D_2\rangle$. Provided $|A_1\rangle$ and $|A_2\rangle$ are orthogonal, then if system A is found in $|A_1\rangle$, the system D must be found in $|D_1\rangle$. In (3.11) the orthogonality of the states $|C_m\rangle$ ensures that there is a distinct phase state for each value of m, that is, there is no overlap between phase states in different terms. We note that the states $|B_m\rangle$ in (3.11) need *not* be orthogonal. Thus, if we have two different states of B, these must be associated with two different phase states, but two different phase states can be associated with the same state of B. For example if $\langle B_m|B_{m+1}\rangle = 1$ then both $|C_m\rangle$ and $|C_{m+1}\rangle$ are associated with $|B_m\rangle$. Thus the system C states, by virtue of their orthogonality, 'set the conditions' for the state of the rest of the system and thus the value of m associated with $|C_m\rangle$ should, in accord with the argument of Unruh and Wald (1989), make a suitable choice for a label to serve as a time value indicator. We therefore define a time interval as being proportional to a difference in m value, that is

$$\Delta t = \delta t \Delta m \tag{3.12}$$

where δt is a constant determined by the units chosen. Then for example, the time difference between the states $|C_M\rangle$ and $|C_{M+q}\rangle$ is $q\delta t$. Because the time differences are defined in terms of phase states of the ideal system C, we call this system an *ideal clock*. Because each state $|B_m\rangle$ in (3.11) is associated with one of the orthogonal clock states, we can also attach to each state the time value of the associated clock state.

Having defined time differences, for states of both systems C and B, it is now reasonable to refer to the generators of the shifts in the time label as Hamiltonians for these systems. From (3.5) and (3.6), it is clear that, at least to within the same constant factor, $\hat{H}_b$ and $\hat{H}_c$ are the Hamiltonians for the system B and the clock respectively. Of course this only holds for the interval, which we can now regard as a time interval, specified by (3.1) during which the clock C and the system B, which we might interpret as the rest of the universe, exist as separately identifiable systems whose states obey (3.1). From (3.4), we see that during this interval these two systems are non-interacting, that is we can write the sum of the two individual Hamiltonians as the total Hamiltonian $\hat{H}$. The correlation between states of B and C at different times arises not through interaction but is an EPR type correlation. A further factorization of the states $|B_m\rangle$ similar to (3.1) for some period enables the introduction of an 'observer' as part of the overall quantum system. The change in state of this observer with m value, that is with time, will also correlate with the change in clock states.

We have mentioned that with a change in the value of m of $c+1$, the state of the clock becomes again identical with its original state. This means that the period of the clock is $(c+1)\delta t$. Equating this to $2\pi/\omega$ where ω is the angular frequency, allows us to write the exponent in (3.9) as

$$-ik\hat{H}_c\Delta E^{-1}2\pi/(c+1) = -i\hbar^{-1}\hat{H}_c k\delta t \tag{3.13}$$

where $\hbar$ is defined as the ratio of ΔE, which we can now call the gap between successive energy states of the clock, to the angular frequency.

Further, from (3.10), (2.7) can then be written as

$$|\phi_m\rangle = \exp[-i\hbar^{-1}\hat{H}m\delta t]|\phi_0\rangle. \tag{3.14}$$

We can, of course, choose our zero of time to correspond to any state and still be consistent with (3.12). For convenience we define $t = m\delta t$, giving

$$|\phi_m\rangle = \exp[-i\hbar^{-1}\hat{H}t]|\phi_0\rangle. \tag{3.15}$$

The total lifetime T_L which can be ascribed to the total system, that is the time period between $|\phi_0\rangle$ and $|\phi_s\rangle$ is equal to

$$s\delta t = s2\pi\hbar/[\Delta E(c+1)]$$

that is

$$T_L\delta E = s2\pi\hbar/(s+1) \tag{3.16}$$

where we have used, from (3.10),

$$(c+1)\Delta E = (s+1)\delta E. \tag{3.17}$$

Having defined time, we now see why we call $|E_0\rangle$, which describes the totality of reality S and which has the form (2.9), a history vector. This vector, which we now know to be an eigenstate of the Hamiltonian $\hat{H}$ with zero eigenvalue, does not represent the state of the universe at a particular time, instead it is the superposition of all such states at all possible distinct times of an ideal clock.

4. Time evolution equation

Writing $|\phi_{m+1}\rangle - |\phi_m\rangle$ as $\delta|\phi_m\rangle$, we obtain from (3.15)

$$\delta|\phi_m\rangle = [\exp(-i\hbar^{-1}\hat{H}\delta t) - 1]|\phi_m\rangle. \quad (4.1)$$

If the energy eigenvalues E_n associated with the states $|E_n\rangle$ with significantly non-zero coefficients in the expansion of $|\phi_m\rangle$ are such that $E_n \ll \hbar/\delta t$, we can obtain from (4.1)

$$\delta|\phi_m\rangle/\delta t = -i\hbar^{-1}\hat{H}|\phi_m\rangle \quad (4.2)$$

with the obvious parallel to the Schrödinger equation. Corresponding expressions would apply for $\delta|C_m\rangle$ and $\delta|B_m\rangle$ in terms of $\hat{H}_C$ and $\hat{H}_B$ under the same restriction. This restriction, however, is quite important. For example, for the ideal clock, the component states of $|C_m\rangle$ are equally weighted and the highest state has, from (3.7), energy E_{cc} such that

$$E_{cc}\delta t = c2\pi\hbar/(c+1), \quad (4.3)$$

where we have used the result, which follows from (3.13), that

$$\Delta E\delta t = 2\pi\hbar/(c+1). \quad (4.4)$$

Thus the Schrödinger equation similar to (4.2) does not apply to the ideal clock itself. The Schrödinger equation may be applicable, however, for isolated subsystems with energies substantially less than $s\delta E$, which from (2.3) and the restriction in (2.5) is the maximum allowed energy $E_{\max}$ of the whole system, and which obeys the relation

$$E_{\max}\delta t = s2\pi\hbar/(s+1) \quad (4.5)$$

which follows from (3.10) and (4.4). The Schrödinger differential equation can only be obtained, of course, if we can let δt tend to zero. Conditions under which this is possible are discussed below.

5. Other clocks

The resolution time δt of the ideal clock, that is the time between successive orthogonal states, is achieved by choosing an impractically high value for ΔE of $(b+1)\delta E$. As discussed earlier this choice restricts the maximum possible energy of system B to be equal to the lowest non-zero energy eigenvalue of the clock. If we wish to use system B to represent the rest of the universe, a more practical clock would have $\Delta E = (b+1)\delta E/K$ where K has a value greater than unity such that $(b+1)/K$ is an integer in accord with (2.3), which means that K is rational. The frequency ω of this clock is correspondingly reduced by a factor K, and the resolution time is $K\delta t$. To study the action of a practical clock, let us choose K to be an integer greater than unity. We then find in the section of the history state corresponding to (3.11) that neighbouring clock states are not orthogonal. However, $|C_M\rangle$ is orthogonal to $|C_{M+K}\rangle$ and $|C_{M+2K}\rangle$ and so on. A clock reading corresponding to a state $|C_{M+K}\rangle$, for example, can be obtained not only if the clock is in this state, but also if it is in any of a number of neighbouring states with a significant overlap with this state. We can show that such states are confined to nearby states $|C_m\rangle$ with m varying in an approximate range of width K centred on M. It follows that the state of B is not precisely determined, but has a high likelihood of being one of the states between $|B_{M+K/2}\rangle$ and $|B_{M+3K/2}\rangle$. Any physical system with equally-spaced energy levels can constitute a clock of the type discussed here. For example as the number of levels increases, the clock states approach the phase states of the harmonic oscillator (Pegg and Barnett 1988), which has long formed the basis for practical clocks. For a finite number of levels, the system also corresponds to a quantum clock similar to that described by Peres (1980).

6. Limiting cases

For $|\phi_m\rangle$ to represent the state of the universe at a particular time, the dimensionality $s+1$ of the Hilbert space Ψ must be extremely large. Whether or not it is infinity is, of course, impossible to determine observationally. Correspondingly we would expect δE and δt to be very small. One other parameter of interest is the lifetime T_L of the universe. All of these are related by the expressions we have derived, and we consider various possibilities below.

We consider first the case where T_L tends to infinity, corresponding to a universe with an infinite lifetime. From (3.16) this implies immediately that $\delta E \to 0$ and thus, because the maximum allowed energy of the system is $s\delta E$, this means that s must approach infinity as well, in order to allow non-zero energy eigenstates. The result $\delta E \to 0$ means that the difference between possible energy configurations of subsystems and that obtainable by having a completely free choice of energy eigenvalues tends to zero. Thus the universe can contain subsystems representing physical systems such as hydrogen atoms for example, with energy levels precisely in accord with present theoretical models. The effect on δt of letting T_L tend to infinity is indeterminate because $\delta t = T_L/s$, so we cannot deduce the size of the minimum time step. However (4.5) shows that it must be very small for any reasonable cosmological model. Thus in a universe with an infinite lifetime, the dimensionality of the Hilbert space is infinite, our condition (2.3) does not restrict possible energy level configurations and the minimum time step of an ideal clock is $2\pi\hbar/E_{max}$.

The other possibility is that T_L is finite. From (3.16) it follows that δE must be non-zero, but there is no restriction on s. This means that there is now a finite difference between the energy eigenvalues of our approach and those obtainable by an entirely free choice. From conventional quantum mechanics, however, we know that an experiment to measure this difference would occupy a time period of at least $2\pi\hbar/\delta E$, which from (3.16) is the lifetime of the universe, so there is no observational reason against the applicability of our approach to a universe with a finite lifetime. There are two possibilities for s with T_L finite: (a) s tends to infinity or (b) s is finite. From $\delta t = T_L/s$, we see that in case (a) δt must approach zero and thus, for example, we can recover Schrödinger's time-dependent differential equation from (4.2) for a finite energy subsystem. In this case there is a countable infinity of component states with different time labels in (2.8) making up the whole history state of the universe. For case (b) δt is non-zero, there are a finite number of component states with different time labels and we have at best a difference, rather than a differential, equation to describe the time evolution.

7. Discussion

In this paper we have examined the properties of a vector $|E_0\rangle$ which satisfies (2.2) where $\hat{H}$ has now been shown to be the Hamiltonian operator. We have found that this vector is a superposition of vectors each of which can represent, to within an error which is in principle unobservable, states of our observed universe at particular times. For subsystems which have an energy much less than the maximum allowed energy of the universe, the variation of these states from time to time can be described by Schrödinger's equation. For the whole universe the time development is still given by the simple unitary operator $\exp(-i\hbar^{-1}\hat{H}\delta t)$, which transforms $|\phi_m\rangle$ to $|\phi_{m+1}\rangle$.

The expansion

$$|E_0\rangle = \ldots + |\phi_{m-1}\rangle + |\phi_m\rangle + |\phi_{m+1}\rangle + \ldots \tag{7.1}$$

has a simple explanation. If (2.2) holds, that is, if $|E_0\rangle$ is a zero energy eigenstate of $\hat{H}$, then

$$\exp(-i\hbar^{-1}\hat{H}\delta t)|E_0\rangle = |E_0\rangle \tag{7.2}$$

and $|E_0\rangle$ is also an eigenstate of the time displacement operator. Applying the time displacement operator in (7.2) to (7.1) means, ignoring the end states for the moment, changing $|\phi_{m-1}\rangle$ to $|\phi_m\rangle$ and $|\phi_m\rangle$ to $|\phi_{m+1}\rangle$ and so on, which clearly leaves (7.1) unaltered. This will be precisely true if the history state (7.1) extends from $-\infty$ to $+\infty$. The interesting question which arises is for a universe with a finite or semi-infinite lifetime. To examine this, we first choose an arrow of time by saying that the state

$$|\phi_{m+1}\rangle = \exp(-i\hbar^{-1}\hat{H}\delta t)|\phi_m\rangle \tag{7.3}$$

is the state 'after' $|\phi_m\rangle$. The arrow thus points in the direction of increasing m in (2.9). To obtain the state before $|\phi_m\rangle$, we apply the inverse operator. There is now meaning to the question: what is the state *before* $|\phi_0\rangle$? It would be tempting, if one were to revert to some external time reference, to say that, because there is no state before $|\phi_0\rangle$ in (2.9), the action of $\exp(-i\hbar^{-1}\hat{H}\delta t)$ on $|\phi_m\rangle$ must be zero, in a way analogous to that in which the boson annihilation operator destroys the vacuum state thus preventing bosons being found in negative energy states. This cannot apply here, however, because from the definition (2.1), $\hat{H}$ is Hermitian and thus the time shift operator and its inverse are definitely unitary. Indeed, the action on $|\phi_0\rangle$ can be calculated directly. From (2.8)

$$\exp(i\hbar^{-1}\hat{H}\delta t)|\phi_0\rangle = \sum_{n=0}^{r} a_n \exp(i\hbar^{-1}E_n\delta t)|E_n\rangle. \tag{7.4}$$

Substituting from (2.3), and using (3.16) with $\delta t = T_U/s$ allows us to write the exponent in (7.4) as

$$i\hbar^{-1}E_n\delta t = ip_n 2\pi/(s+1) = ip_n 2\pi - ip_n 2\pi s/(s+1). \tag{7.5}$$

Thus, because p_n is an integer,

$$\exp(i\hbar^{-1}E_n\delta t) = \exp[-ip_n 2\pi s/(s+1)] = \exp(-i\hbar^{-1}E_n s\delta t). \tag{7.6}$$

Substituting into (7.4) then yields

$$\exp(i\hbar^{-1}\hat{H}\delta t)|\phi_0\rangle = \exp(-i\hbar^{-1}\hat{H}s\delta t)|\phi_0\rangle = |\phi_s\rangle \tag{7.7}$$

which is just the *final* state of the system. Similarly the state just after the final state $|\phi_s\rangle$ is the initial state $|\phi_0\rangle$. This result, though perhaps surprising, is not unreasonable. The expression $\hat{H}|E_0\rangle = 0$ with $\hat{H}$ Hermitian requires that $|E_0\rangle$ must be an eigenstate, with eigenvalue unity, of the time displacement operator. If the series (7.1) is finite, the state $|E_0\rangle$ is indeed invariant under a shift of $|\phi_m\rangle$ to $|\phi_{m+1}\rangle$ for $m \neq s$ and a shift of $|\phi_s\rangle$ to $|\phi_0\rangle$. Indeed this would seem to be an unavoidable consequence. Of course, by letting the lifetime T_L be long enough, we can postpone the return to the initial state indefinitely. We might comment that this mathematical periodicity for a finite lifetime system does not necessarily imply a reversal of the arrow of time in the later part of the life of the system, that is, the next to last states $|\phi_{s-1}\rangle, |\phi_{s-2}\rangle, \ldots$, are not necessarily the same as the first states $|\phi_0\rangle, |\phi_1\rangle, \ldots$. Nor does the apparent jump from the final to the initial state represent a gross physical discontinuity. These states are connected by a simple unitary transformation and the apparent jump is of the type which occurs when the minute hand of a clock 'jumps back' to zero instead of reading 60.

8. Conclusion

This paper is not an attempt to quantize the classical theory of general relativity; instead it is a purely quantum mechanical approach based on representing, by a vector in an abstract Hilbert space, the totality of reality rather than reality at a particular time. While it does make contact with other approaches through (2.2), that is not its main purpose. The universe is essentially quantum mechanical in nature, and classical theories should ideally arise as limiting cases of quantum mechanics. The reasonably simple approach of this paper provides answers to some questions. For example, the correlated behaviour between a clock and another physical system, which occurs irrespective of their spatial separation, is usually considered to arise because both are governed by a law involving some external time parameter. In our approach this correlation is seen to be simply of the EPR type which, as is well known, is also separation independent. This correlation is a natural consequence of the whole history of the universe being expressible as a single vector which is a superposition of a large number of states with different time labels. Further, the way in which these component states vary with the time label is a *consequence* of the theory, rather than an additional postulate. That is, the single history vector already contains the seeds of Schrödinger's equation. Time has been defined in terms of the states of a particular type of quantum clock, of which the harmonic oscillator is a particular example. If a different type of clock structure were used the clock states may not have the same type of spread over history as those of the clock we have chosen, that is, such a clock might not run uniformly relative to our clock and the Schrödinger equation would have a different form.

While we conclude that the quantum mechanical system discussed in this paper can represent the observed universe, there are of course also questions left unanswered. These include the role of the observer and the unidirectionality of the arrow of time, for example, why an observer's amount of memory changes monotonically with a monotonic change in time values. Our approach does not seem to suggest any underlying reason for this. Concerning the question of the collapse of the wavefunction associated with an act of observation, our approach does not accommodate this concept in the literal sense. Any observer is clearly part of the overall quantum system, and the changes of state of the total system comprising the observer, the system observed and the clock have a well defined time evolution which does seem to accommodate discontinuous changes of state of any one component, unless, perhaps, such a change is somehow balanced by a sudden change in another component of the system, or there is a sudden redefinition of what constitutes each component. The simplest postulate consistent with out approach is that the collapse simply does not occur. We do not explore this any further here.

Finally we should remark that, while some of the ideas incorporated in and arising out of this paper have been considered in other contexts and from different viewpoints by a number of other authors, what is presented here is a reasonably straightforward mathematical formalism which unifies such concepts as consequences of the very simple notion that the totality of reality can be represented as a vector in Hilbert space. There is no need to introduce an external time parameter, indeed, time is part of the reality represented by this vector.

Acknowledgments

I thank the Physics Department, University of Queensland, where part of this work was done, for its hospitality. The use of a history state expansion to incorporate time evolution follows a suggestion by Professor G W Series, and the work has benefited from discussions with Dr S M Barnett.

References

Alvarez E 1989 *Rev. Mod. Phys.* **61** 561
Dirac P A M 1958 *Proc. R. Soc.* A **246** 333
—— 1959 *Phys. Rev.* **114** 924
Isham C and MacCallum M A H 1975 *Quantum Gravity* ed C J Isham, R Penrose and D W Sciama (Oxford: Clarendon)
Komar A 1967 *Phys. Rev.* **153** 1385
Page D N and Wootters W K 1983 *Phys. Rev.* D **28** 2960
Pegg D T and Barnett S M 1988 *Europhys. Lett.* **6** 483
Peres A 1980 *Am. J. Phys.* **48** 552
Unruh W G and Wald R M 1989 *Phys. Rev.* D **40** 2598

Complement of the Hamiltonian

D. T. Pegg
Faculty of Science, Griffith University, Nathan, Brisbane 4111, Australia
(Received 26 June 1998)

The much-studied energy-time uncertainty relation has well-known difficulties that are exacerbated for a system with discrete energy levels. The difficulty in representing time in the abstract sense by an operator raises the related question of whether or not there is some other quantity that is complementary to the Hamiltonian of a quantum system. Such a quantity would have dimensions of time but would be a property of the system itself. We examine this question for a system with discrete energy eigenstates for which the ratios of the energy differences are rational. We find that such a quantity does exist and can be represented both by a probability-operator measure and by an Hermitian operator, but in a state space larger than the minimal space needed to include the states of the system. The uncertainty relation with the energy is slightly more complicated than the momentum-position uncertainty relation, but is readily interpretable. To describe such a quantity the name "age" is suggested. [S1050-2947(98)03012-1]

PACS number(s): 03.65.Bz

I. INTRODUCTION

One of the earliest issues in quantum mechanics has been the question of an energy-time uncertainty principle. Fundamental to this problem is that in quantum mechanics the energy is a dynamical variable of the quantum system represented by a Hermitian operator, but time in the normal sense is a parameter, as it is classically. (For a recent overview of this, see Ref. [1] and the references therein.) Another important difficulty is that, whereas the momentum-position uncertainty relation can be derived from the commutation relation

$$[\hat{p}_x,\hat{x}]=-i\hbar, \tag{1}$$

there is apparently no operator $\hat{\tau}$ that is canonically conjugate to the Hamiltonian operator $\hat{E}$ in the sense of Eq. (1) from which a corresponding uncertainty relation can be derived. Indeed, a relation of the form

$$[\hat{E},\hat{\tau}]=-i\hbar \tag{2}$$

simply cannot be correct for quantum systems with discrete energy states. This is because if we find the expectation value of each side of Eq. (2) for an energy eigenstate, we obtain zero on the left-hand side and $-i\hbar$ on the right-hand side. This is exactly the difficulty involved with early attempts to find a phase operator canonically conjugate to the excitation number operator of an harmonic oscillator [2] (for recent reviews and bibliographies of the quantum phase problem, see Refs. [3] and [4]). Commutators of canonically conjugate operators in the form (1) and (2) can be regarded as special cases of a more general form of commutator [5] for conjugate, or complementary, operators. For momentum and position this more general commutator reduces to Eq. (1), but for phase and excitation number it contains an extra term that removes the difficulty described above. Of course this means that the number-phase uncertainty relation is more complicated than the momentum-position uncertainty relation. As Eq. (2) cannot be true, we might expect likewise that for the uncertainty relation for the Hamiltonian with discrete eigenvalues and its conjugate, or complement, a similar complication should apply and Eq. (2) should be generalized. The question then arises as to whether or not there is some operator conjugate or complementary to the Hamiltonian in this more general sense.

In this paper we wish to explore the possibility of the existence of a quantity that can be regarded as the complement of the Hamiltonian for a quantum system with discrete energy levels, recognizing that if this quantity is represented by an operator the appropriate commutator must be a generalized form of Eq. (2). Although this quantity will have dimensions of time, it will not be appropriate to refer to such an operator as a time operator. As an operator it would represent an observable of the quantum system and not time in the abstract or coordinate sense or as a reading on an external clock. Its eigenstates would represent a state of the quantum system and some measurement on the system, at least in principle, should tell us about the quantity involved. Although this quantity is not time, we would hope that for a particular ideal system its expectation value may change linearly with time, for example, so that a measurement of the quantity would also give a measure of time. In this case, we would be using the system as a clock. Before seeking an operator for the complement of the Hamiltonian we first represent this quantity by a nonorthogonal probability-operator measure, which is a more general concept [6].

II. DISCRETE ENERGY AND THE α QUANTITY

We consider a quantum system with $p+1$ energy states $|E_i\rangle$ with $i=0,1,\ldots,p$. To avoid unnecessary complications, for this paper we shall let the states be nondegenerate and we choose our zero of energy so that the lowest-energy eigenvalue $E_0=0$ with the other energies increasing with, but not necessarily proportional to, i. We assume that these include all the accessible, or essential, states of the system. As an example of accessible states, we note that an effective two-level atom can be prepared from a multilevel atom for particular experiments. In this case not all the usual states of the atom are accessible from the initial state by means of the particular interaction applied to the atom. By allowing more interactions that couple more states, we can increase the number of accessible states until eventually we include all physically accessible, or physical, states of the system, that is, states that are accessible from each other by some series of physical interactions.

For now we shall keep p finite, but it is possible to take a limit at a later stage. We let Ψ_p be the $(p+1)$-dimensional state space spanned by these orthogonal energy states. This is the minimal space required for a description of the quantum system. The Hamiltonian operator for the system is

$$\hat{E}=\sum_{i=0}^{p} E_i|E_i\rangle\langle E_i|, \tag{3}$$

where E_i are the energy eigenvalues. A general state of the system can be written as

$$|f\rangle=\sum_{i=0}^{p} f_i|E_i\rangle \tag{4}$$

and will evolve in time according to

$$\exp(-i\hbar^{-1}\hat{E}t)|f\rangle=\sum_{i=0}^{p} f_i\exp(-i\hbar^{-1}E_it)|E_i\rangle. \tag{5}$$

We are interested in a quantity α of the system, which we shall refer to simply as the α quantity, which will have dimensions of time and will be conjugate to the Hamiltonian in the sense that $\hat{E}$ is the generator of shifts in α quantity. Thus we seek a state $|\alpha\rangle$ for which

$$\exp(-i\hbar^{-1}\hat{E}\delta\alpha)|\alpha\rangle=|\alpha+\delta\alpha\rangle. \tag{6}$$

Writing

$$|\alpha\rangle=\sum_{i=0}^{p} c_i(\alpha)|E_i\rangle, \tag{7}$$

we find that we can ensure that Eq. (6) is true by setting

$$c_i(\alpha)\propto\exp(-i\hbar^{-1}E_i\alpha) \tag{8}$$

and thus, with the appropriate normalization factor included, we have

$$|\alpha\rangle=\frac{1}{\sqrt{p+1}}\sum_{i=0}^{p}\exp(-i\hbar^{-1}E_i\alpha)|E_i\rangle. \tag{9}$$

If we replace $\delta\alpha$ on the left-hand side of (6) by δt, we obtain a time translation expression, that is, the state $|\alpha+\delta\alpha\rangle$ is the state to which $|\alpha\rangle$ would evolve in a time $\delta t=\delta\alpha$. Thus, although the quantity represented by α is not the time because it is a property of the system, it bears some relation to time.

The states $|\alpha\rangle$ are not orthogonal and their number exceeds the dimensions of the space Ψ_p spanned by the $p+1$ energy states $|E_i\rangle$, so these cannot be eigenstates of a Hermitian operator on Ψ_p with which we might hope to represent the α quantity. Further, for unequally spaced energy levels, we cannot even pick out a subset of $p+1$ states $|\alpha\rangle$ that are orthogonal. We can, however, make progress for unequally spaced levels if the ratios E_i/E_1 are all rational numbers or can be sufficiently closely approximated by rational numbers. For E_i/E_1 rational

$$\frac{E_i}{E_1}=\frac{C_i}{B_i}, \tag{10}$$

where C_i and B_i are integers with no common factors. We write the lowest common multiple of the values of B_i with $i>1$ as r_1 and define $r_i=r_1C_i/B_i$ for $i>1$ and as zero for $i=0$. Then r_i is an integer for all $i\geqslant 0$ with $r_0=0$. From Eq. (10) we can write

$$E_i=r_i\frac{2\pi\hbar}{T}, \tag{11}$$

where

$$T=2\pi\hbar r_1/E_1. \tag{12}$$

Let us now select $s+1$ states $|\alpha\rangle$ for which the values of α are uniformly spread over the range T. That is, we choose values of α denoted by

$$\alpha_m=\alpha_0+m\frac{T}{s+1}, \tag{13}$$

with $m=0,1,\ldots,s$. We find then that the states $|\alpha_m\rangle$ have the interesting property that

$$\sum_{m=0}^{s}|\alpha_m\rangle\langle\alpha_m|=\frac{1}{p+1}\left\{\sum_{m=0}^{s}\sum_{i}|E_i\rangle\langle E_i| + \sum_{i\neq k}\sum_{m=0}^{s}\exp[i(r_k-r_i)\alpha_m 2\pi/T] \times|E_i\rangle\langle E_k|\right\} \tag{14}$$

and thus

$$\frac{p+1}{s+1}\sum_{m=0}^{s}|\alpha_m\rangle\langle\alpha_m|=\hat{1}, \tag{15}$$

where $\hat{1}$ is the unit operator on the space Ψ_p, provided the second term on the right-hand side of Eq. (14) is zero or at least negligible compared with the first term. For E_i/E_1 rational and thus r_k-r_i an integer, this term will be zero and Eq. (15) true provided r_k-r_i is not a multiple of $s+1$. This follows from substituting Eq. (13) into Eq. (14) and summing the geometric progression involved. We can ensure r_k-r_i is not a multiple of $s+1$ by choosing $s+1>r_p$, which is the largest value of r_i.

If E_i/E_1 is irrational but sufficiently well approximated by a rational number for the second term on the right-hand side of Eq. (14) to be negligible, then this rational number would in general have a large denominator, so r_1 will be very large. This implies from Eq. (12) that T will be very much greater than the natural period $2\pi\hbar/E_i$ associated with the state with energy E_i. When E_i/E_1 are exact rational numbers T has a simple physical interpretation. From Eqs. (5) and (11) T is equal to the smallest time taken for the system to return to its initial state. Thus the state $|\alpha\rangle$ will be the same as the state $|\alpha+T\rangle$. It follows that restricting our selection of states $|\alpha\rangle$ to those for which the values of α are uniformly spread over the range T then prevents us from including the same state twice.

Expression (15) is a resolution of the identity. Thus, although the α quantity is not an observable represented by a Hermitian operator on the space Ψ_p of $p+1$ dimensions, it can be represented by a nonorthogonal probability-operator measure, which is a more general concept [6]. The $s+1$ nonorthogonal elements of this probability-operator measure are $(p+1)(s+1)^{-1}|\alpha_m\rangle\langle\alpha_m|$. Expression (15) can be used to expand the general state given by Eq. (4) as

$$|f\rangle=\frac{p+1}{s+1}\sum_{m=0}^{s}\langle\alpha_m|f\rangle|\alpha_m\rangle \tag{16}$$

and so, using $\langle f|f\rangle=1$, we obtain

$$\sum_{m=0}^{s}\frac{p+1}{s+1}|\langle f|\alpha_m\rangle|^2=1. \tag{17}$$

Thus each term in Eq. (17), all of which are positive, can represent a probability with the total probability correctly normalized. From quantum detection theory [6], this is the probability that the application of the probability-operator measure by means of a suitable measuring instrument yields the result α_m. Of course, this does not indicate immediately how to perform such a measurement, but that is not our concern here, where we are interested merely in identifying the quantity that can be regarded as the complement of the Hamiltonian and in studying some of its properties.

So far the choice of s is arbitrary apart from the requirement that $s\geqslant r_p$. In order for the α quantity to be applicable to all systems of the type considered here, however large the value of r_p, and to be independent of an arbitrary choice of s, we now define the α quantity as that represented by the above probability-operator measure in the limit as $s\rightarrow\infty$. In this limit, the difference between successive values of α_m tend to zero and the probability for a value of α in the small range between α and $\alpha+\delta\alpha$ is $P(\alpha)\delta(\alpha)$, where the probability density is given by

$$P(\alpha)=\frac{1}{T}|\langle f|\bar{\alpha}\rangle|^2, \tag{18}$$

where

$$|\bar{\alpha}\rangle=\sqrt{p+1}|\alpha\rangle=\sum_i \exp(-i\hbar^{-1}E_i\alpha)|E_i\rangle. \tag{19}$$

The resolution of the identity (15) then becomes simply

$$\int_{\alpha_0}^{\alpha_0+T}|\bar{\alpha}\rangle\langle\bar{\alpha}|d\alpha/T=\hat{1} \tag{20}$$

and the α quantity is represented by the probability-operator measure generated by the infinitesimal operators $|\bar{\alpha}\rangle\langle\bar{\alpha}|d\alpha/T$.

The probability density for the system in state $|f\rangle$ to be found with a value α for the α quantity is, from Eqs. (4) and (18),

$$P(\alpha)=\frac{1}{T}\left|\sum_{i=0}^{p}f_i^*\exp(-i\hbar^{-1}E_i\alpha)\right|^2. \tag{21}$$

This expression displays an essential feature of the complementary between energy and the α quantity: If the system is in an energy eigenstate then there is only one term, of modulus unity, in Eq. (21) and $P(\alpha)=T^{-1}$, that is, the α-quantity distribution is constant across the whole period T. Thus if the energy can be determined exactly the α quantity is completely random. The probability density (21) is all that is needed to calculate the statistical properties of the α quantity for any state $|f\rangle$ of our quantum system.

III. RATE OF CHANGE WITH TIME

To find the way in which the properties of the α quantity change with time we first calculate the rate of change of $P(\alpha)$ in (18) from Schrödinger's equation as

$$\frac{dP(\alpha)}{dt}=\frac{1}{T}\left(\frac{d\langle f|}{dt}|\bar{\alpha}\rangle\langle\bar{\alpha}|f\rangle+\langle f|\bar{\alpha}\rangle\langle\bar{\alpha}|\frac{d|f\rangle}{dt}\right)$$
$$=\frac{i}{\hbar T}\langle f|[\hat{E},|\bar{\alpha}\rangle\langle\bar{\alpha}|]|f\rangle. \tag{22}$$

The rate of change of the mean of the α quantity for a state $|f\rangle$ is then

$$\frac{d\langle\alpha\rangle}{dt}=\int_{\alpha_0}^{\alpha_0+T}\alpha\frac{dP(\alpha)}{dt}d\alpha=\frac{i}{\hbar}\langle f|[\hat{E},\hat{A}]|f\rangle, \tag{23}$$

where

$$\hat{A}=\frac{1}{T}\int_{\alpha_0}^{\alpha_0+T}\alpha|\bar{\alpha}\rangle\langle\bar{\alpha}|d\alpha. \tag{24}$$

We can express Eq. (24), from Eqs. (11) and (19), as

$$\hat{A}=\frac{1}{T}\sum_{i,j}\int_{\alpha_0}^{\alpha_0+T}\alpha\exp[i2\pi(r_j-r_i)\alpha/T]d\alpha|E_i\rangle\langle E_j|$$
$$=\alpha_0+T/2+i\hbar\sum_{i\neq j}\frac{\exp[-i\hbar^{-1}(E_i-E_j)\alpha_0]}{E_i-E_j}|E_i\rangle\langle E_j|. \tag{25}$$

Thus

$$[\hat{E},\hat{A}]=i\hbar\sum_{i\neq j}\exp(-i\hbar^{-1}E_i\alpha_0)|E_i\rangle\langle E_j|\exp(iE_j\alpha_0)$$

$$=i\hbar\left(\sum_i \exp(-i\hbar^{-1}E_i\alpha_0)|E_i\rangle\right.$$

$$\left.\times\sum_j \langle E_j|\exp(iE_j\alpha_0)-\hat{1}\right), \tag{26}$$

which can be written from Eq. (19) simply as

$$[\hat{E},\hat{A}]=i\hbar(|\bar{\alpha}_0\rangle\langle\bar{\alpha}_0|-\hat{1}). \tag{27}$$

Thus, from Eq. (23)

$$\frac{d\langle\alpha\rangle}{dt}=1-\langle f|\bar{\alpha}_0\rangle\langle\bar{\alpha}_0|f\rangle. \tag{28}$$

From Eq. (18) the last term of Eq. (28) is just $TP(\alpha_0)$; thus, if the probability distribution $P(\alpha)$ for the state $|f\rangle$ of the system is sufficiently narrow for this last term to be negligible, the expectation value of the α quantity changes exactly as the time parameter. If the system is in an eigenstate of energy, however, it is clear from Eqs. (19) and (28) that the rate of change of $\langle\alpha\rangle$ is zero. This must be true, of course, as an energy eigenstate is a stationary state. Since no observable quantity of a stationary state varies with time its α quantity should not vary. This is a good example displaying the difference between the α quantity and the time parameter. It also shows the need for the second term in Eq. (28) and the inadequacy of a commutator similar to Eq. (2) that would predict $d\langle\alpha\rangle/dt=1$ for all states. A particular example of a system for which $d\langle\alpha\rangle/dt\approx 1$ is a harmonic oscillator in a truncated phase state [7] where the truncation is at a very large excitation number. For such a state $r_i=i$ and p is very large. The mean α quantity of this system changes directly as the time for most of the time and so the harmonic oscillator makes a useful quantum clock. We note that even for this system, this direct variation with time cannot always apply. When the state eventually evolves to have a large overlap with the state $|\alpha_0\rangle$, $d\langle\alpha\rangle/dt$ suddenly becomes very large and negative and $\langle\alpha\rangle$ quickly reverts to the value it had one period of the oscillator earlier. This is just equivalent to the seconds reading on a digital clock jumping from 59 back to 00. On the other hand, $\langle\alpha\rangle$ of a hydrogen atom in a superposition of some of its lower energy states will not be so directly related to the time during one period and the variation of $\langle\alpha\rangle$ with time will be more complicated.

IV. UNCERTAINTY RELATION

As the α quantity is represented by a probability-operator measure and its variance is found from the associated probability distribution, its uncertainty relation with energy cannot be written down immediately as is the case for observables represented by Hermitian operators. Using Eq. (20) we can, however, write the variance of energy as

$$\langle\Delta E^2\rangle=\int_{\alpha_0}^{\alpha_0+T}\langle f|(\hat{E}-\langle E\rangle)|\bar{\alpha}\rangle\langle\bar{\alpha}|(\hat{E}-\langle E\rangle)|f\rangle d\alpha/T. \tag{29}$$

Combining this with the variance of the α quantity obtained from Eq. (18) gives

$$\langle\Delta E^2\rangle\langle\Delta\alpha^2\rangle=\int_{\alpha_0}^{\alpha_0+T}|\langle f|(\hat{E}-\langle E\rangle)|\bar{\alpha}\rangle|^2 d\alpha$$

$$\times\int_{\alpha_0}^{\alpha_0+T}|\langle\bar{\alpha}|f\rangle(\alpha-\langle\alpha\rangle)|^2 d\alpha/T^2$$

$$\geqslant\left|\int_{\alpha_0}^{\alpha_0+T}\langle f|(\hat{E}-\langle E\rangle)|\bar{\alpha}\rangle\right.$$

$$\left.\times\langle\bar{\alpha}|f\rangle(\alpha-\langle\alpha\rangle)d\alpha/T\right|^2 \tag{30}$$

from Schwarz's inequality. Thus the product of the mean square deviations is

$$\Delta E\Delta\alpha\geqslant|\langle f|(\hat{E}-\langle E\rangle)(\hat{A}-\langle\alpha\rangle)|f\rangle|. \tag{31}$$

As α is real, operator $\hat{A}$ defined by Eq. (24) will be Hermitian and we can use the usual inequality [8] to obtain

$$\Delta E\Delta\alpha\geqslant\tfrac{1}{2}|\langle f|[\hat{E},\hat{A}]|f\rangle|=\frac{\hbar}{2}|1-\langle f|\bar{\alpha}_0\rangle\langle\bar{\alpha}_0|f\rangle|. \tag{32}$$

When the system is in an energy eigenstate the uncertainty in energy must be zero, even for a finite $\Delta\alpha$. In this case the right-hand side of Eq. (32) vanishes, ensuring consistency. Thus the energy is exactly measurable in principle and does not suffer from the difficulty discussed in Ref. [1] associated with an uncertainty relation based on Eq. (2). As with Eq. (28), $\langle f|\bar{\alpha}_0\rangle\langle\bar{\alpha}_0|f\rangle$ is just $TP(\alpha_0)$ and so if $|f\rangle$ is orthogonal to $|\alpha_0\rangle$ or at least if the probability distribution $P(\alpha)$ for the state $|f\rangle$ of the system is sufficiently narrow for this last term to be negligible, the uncertainty product takes the more usual form. This is the same condition for the rate of change of $\langle\alpha\rangle$ with time to be unity as discussed earlier. Only under these special circumstances could a commutation relation as given by Eq. (2) be used.

V. THE SPACE Ψ_s

In this section we examine the possible existence of an Hermitian operator to represent the α quantity. In performing the preceding calculations we have found it convenient to define a Hermitian operator $\hat{A}$ by Eq. (24) with a more explicit form given by Eq. (25). It follows immediately from Eqs. (24) and (18) that

$$\langle\alpha\rangle=\langle f|\hat{A}|f\rangle \tag{33}$$

and the rate of change of $\langle f|\hat{A}|f\rangle$ will be the same as that of $\langle\alpha\rangle$. From Eq. (27) the uncertainty relation for the observable represented by $\hat{A}$ and the energy will be the same as Eq. (32). Furthermore, $\hat{A}$ is well defined by Eqs. (24) and (19) even for quantum systems whose energy eigenvalues E_i are not rational multiples of E_1. In view of these properties, why do we not simply take the observable represented by $\hat{A}$ as the complement of the Hamiltonian? The essential reason is that the eigenstates of $\hat{A}$ are not the states $|\alpha\rangle$ and the Hamiltonian does not generate shifts from one eigenstate of $\hat{A}$ to another in a manner similar to Eq. (6). We note that while Eq. (33) is true, in general $\langle f|\hat{A}^2|f\rangle \neq \langle \alpha^2\rangle$ as the states $|\bar{\alpha}\rangle$ are not orthogonal and so, although the variance of $\hat{A}$ satisfies the same uncertainty inequality as $\Delta\alpha^2$, these two variances are not in general equal. Thus, although $\hat{A}$ is interesting in its own right, at best it can only be regarded as conjugate to $\hat{E}$ in a weak sense of satisfying the commutation relation (27). Its expectation value is equal to the expectation value of the α quantity, so it might provide a way of measuring this value, but it can be regarded only as the operator acting on Ψ_p that is nearest to an operator conjugate to $\hat{E}$.

We may, on the other hand, be able to identify the actual Hermitian operator representing the α quantity if we are guided by the work of Naimark [6], which indicates that a general resolution of the identity such as Eq. (15) can be extended to an orthogonal resolution of the identity in a *larger* space Ψ_s of which Ψ_p is a subspace. Returning to Eq. (15), we note that α_m has $s+1$ values, where $s+1>r_p$, the largest of the integers r_i. We thus expect the state space Ψ_s that we seek to have $s+1$ dimensions spanned by $s+1$ eigenstates of the Hermitian operator of this space that represents the α quantity. The energy basis of Ψ_s will therefore also include extra states. As Ψ_p includes all the accessible states of the system, the extra states of Ψ_s will not be accessible, perhaps not even physically accessible. For example, if the matrix elements of any physical interaction Hamiltonian between the states of Ψ_p and the states of Ψ_s orthogonal to Ψ_p are all zero, the system can never evolve from the superposition (4) to include any of the extra states as components. Thus we still use the superposition (4) to describe a general physical state of the system.

We seek orthogonal states $|\theta_m\rangle$ in the space Ψ_s such that

$$\sqrt{\frac{p+1}{s+1}}\,|\alpha_m\rangle = \hat{1}|\theta_m\rangle, \tag{34}$$

where $\hat{1}$ is the unit operator for the space Ψ_p as used in Eq. (15), and such that

$$\sum_{m=0}^{s} |\theta_m\rangle\langle\theta_m| = \hat{I}, \tag{35}$$

where $\hat{I}$ is the unit operator for the space Ψ_s. Then operating on Eq. (35) from the left and from the right with $\hat{1}$ will yield Eq. (15). We can write the states we seek as

$$|\theta_m\rangle = \frac{1}{\sqrt{s+1}} \sum_{n=0}^{s} \exp(-i2\pi n\alpha_m/T)|n\rangle, \tag{36}$$

where $|n\rangle = |E_i\rangle$ for $n=r_i$. It can be shown that these states are orthogonal and satisfy Eq. (35) and, from Eqs. (9) and (11), also Eq. (34). To check Eq. (34) use can be made of the property that all states of Ψ_p are eigenstates of $\hat{1}$ with eigenvalue unity and states orthogonal to Ψ_p are eigenstates of $\hat{1}$ with eigenvalue zero. To check Eq. (35) it is useful to write the states as

$$|\theta_m\rangle = \frac{1}{\sqrt{s+1}} \sum_{n=0}^{s} \exp(-in\theta_m)|n\rangle, \tag{37}$$

where

$$\theta_m = 2\pi\alpha_m/T, \tag{38}$$

which lie in a 2π range between θ_0 and $\theta_0+2\pi$, with $\theta_m = \theta_0 + m2\pi/(s+1)$.

The states $|\theta_m\rangle$ form a complete orthonormal basis for the space Ψ_s. The space Ψ_s is mathematically equivalent to the finite $(s+1)$-dimensional space used to examine the harmonic oscillator, or quantized electromagnetic field mode, in Ref. [7] with angular frequency ω given by $2\pi/T$. Apart from an unimportant change in the sign of i, these states can be seen to be the same as the phase states of Ψ_s [7] and are the eigenstates of the operator

$$\hat{\varphi}_\theta = \sum_{m=0}^{s} \theta_m|\theta_m\rangle\langle\theta_m|, \tag{39}$$

with eigenvalues θ_m, and the eigenstates of the associated Hermitian operator

$$\hat{\tau} = \hat{\varphi}_\theta T/2\pi, \tag{40}$$

with eigenvalues α_m. With the values α_m given by Eq. (13), these are now seen to be the eigenvalues of $\hat{\tau}$, which acts on the space Ψ_s. Each term of Eq. (17), that is, $|\langle f|\alpha_m\rangle|^2(p+1)/(s+1)$, is seen to be the probability of projecting the state $|f\rangle$ onto the state $|\theta_m\rangle$, that is, it is the probability of obtaining the result α_m by a measurement of the Hermitian operator $\hat{\tau}$ given by Eq. (40). This is in accord with the previous interpretation we gave for the terms of Eq. (17), but we can now link the measurement to the Hermitian operator $\hat{\tau}$.

The conjugate of Eq. (39) is the number operator [5]

$$\hat{N} = \sum_{n=0}^{s} n|n\rangle\langle n|, \tag{41}$$

from which we can define an operator

$$\hat{H}_s = \hat{N}2\pi\hbar/T, \tag{42}$$

for which $\hat{1}\hat{H}_s\hat{1} = \hat{E}$. Ψ_s will also be spanned by the $s+1$ eigenstates $|n\rangle$ of $\hat{H}_s$ with $n=0,1,\ldots,s$, with a uniform energy difference of $2\pi\hbar/T$ between successive states. States $|n\rangle$ for which $n=r_i$ will be the same as the states $|E_i\rangle$ of Ψ_p. It is not difficult to show that $\hat{\tau}$ is the conjugate, or complement, of $\hat{H}_s$ in the strict sense of Ref. [5], that is, $\hat{H}_s$ is the generator of shifts in the values of α_m and $\hat{\tau}$ is the generator of energy shifts. The first property follows from Eq. (36),

$$\exp\{-i\hbar^{-1}\hat{H}_s[T/(s+1)]\}|\theta_m\rangle = |\theta_{m+1}\rangle, \tag{43}$$

and the second follows from the relation

$$\exp[i\hbar^{-1}\hat{\tau}(2\pi\hbar/T)] = \exp(i\hat{\varphi}_\theta). \tag{44}$$

Because the right-hand side of Eq. (44) shifts a number state $|n\rangle$ to a neighboring number state [7], the left-hand side shifts an eigenstate of $\hat{H}$ to a neighboring eigenstate. In Eq. (43), $T/(s+1)$ is the difference in α between successive τ states and, from Eq. (42), $2\pi\hbar/T$ in Eq. (44) is the energy step size.

From Eqs. (42), (41), and (40) and with Eq. (36) substituted into Eq. (39), we find eventually

$$[\hat{\tau},\hat{H}_s]=\frac{2\pi\hbar}{s+1}\times\sum_{n\neq n'}\frac{(n-n')\exp[i(n-n')2\pi\alpha_0/T]|n'\rangle\langle n|}{\exp[i(n-n')2\pi/(s+1)]-1}, \tag{45}$$

which, apart from the sign of i, is just $\hbar$ times the expression for $[\hat{\varphi}_\theta,\hat{N}]$ given in [7].

For consistency with our previous definition of the α quantity, we must define $\hat{\tau}$ to represent the α quantity in the limiting sense of $s+1$ being extremely large, that is, $s+1 \gg r_p$. Then the difference between successive values of the α quantity becomes vanishingly small. As with the phase problem [7], we must be careful how we take the infinite-s limit however. We can ensure consistency with the probability distribution derived from Eq. (18) by taking the limit of expectation values of functions of the operator rather than the limit of the operator itself.

The uncertainty relation derived from the commutator (45) is

$$\Delta H_s \Delta\tau \geq \tfrac{1}{2}|\langle f|[\hat{\tau},\hat{H}_s]|f\rangle|, \tag{46}$$

where ΔH_s and $\Delta\tau$ are the uncertainties in the observables represented by $\hat{H}_s$ and $\hat{\tau}$. The rate of change of the expectation value is

$$\frac{d\langle\tau\rangle}{dt}=-\frac{i}{\hbar}\langle f|[\hat{\tau},\hat{H}_s]|f\rangle. \tag{47}$$

The state $|f\rangle$, given by Eq. (4), contains only the states $|E_i\rangle$ of Ψ_p so, when Eq. (45) is substituted, all terms on the right-hand sides of Eqs. (46) and (47) will be zero for $n,n' > r_p$. In our limiting case where $s+1 \gg r_p$, we find

$$\langle f|[\hat{\tau},\hat{H}_s]|f\rangle=-i\hbar\sum_{i\neq k}\exp[i(r_k-r_i)2\pi\alpha_0/T]\times\langle f|E_i\rangle\langle E_k|f\rangle = i\hbar - i\hbar|\langle f|\bar{\alpha}_0\rangle|^2 \tag{48}$$

from Eqs. (9), (13), (11), and (19). It is straightforward to show for the state $|f\rangle$ that $\Delta H_s=\Delta E$ and in the infinite-s limit that $\langle\tau\rangle=\langle\alpha\rangle$. By using the orthogonality of the states $|\theta_m\rangle$ to write from Eqs. (39) and (40)

$$\hat{\tau}^2=\sum_{m=0}^{s}\alpha_m^2|\theta_m\rangle\langle\theta_m|, \tag{49}$$

we can show that in the infinite-s limit

$$\langle f|\hat{\tau}^2|f\rangle=\langle\alpha^2\rangle, \tag{50}$$

where $\langle\alpha^2\rangle$ is the value derived from Eq. (18). It then follows that from Eqs. (46)–(48) we regain the previous expressions (28) and (32) for the rate of change and uncertainty relation for the α quantity. In general, expression (50) is true upon replacing the square with any power so the Hermitian operator $\hat{\tau}$ is indeed a good representation of the α quantity.

We can now understand more clearly the role of the Hermitian operator $\hat{A}$. From Eqs. (25), (40), and (39) we find that $\hat{A}$ is the operator acting on space Ψ_p that has the same matrix elements as $\hat{\tau}$ for the states of Ψ_p in the infinite-s limit. This is analogous to finding the weak limit of the phase operator on the space of physical states [9]. For example, the difference between the mean squares $\langle f|\hat{A}^2|f\rangle$ and $\langle\alpha^2\rangle$ noted earlier, and hence between $\langle f|\hat{A}^2|f\rangle$ and $\langle f|\hat{\tau}^2|f\rangle$, has a parallel in the case of phase arising because the weak limit of the square of the operator is not equal to the square of the weak limit of the operator.

VI. CONCLUSION

The question we have addressed here is basically that if time is not complementary to energy, particularly for a system with discrete energy levels, then what is? Does such a quantity exist? A sensible quantity that is complementary to the energy of a quantum system would have dimensions of time, but would be a property of the system. It would not represent time in the abstract [1] or universal sense [10]. We would expect in some special cases the expectation value of this quantity to vary directly with the time, but not in general. For stationary states the expectation value of this quantity should not vary with time at all.

We have seen that for a quantum system with discrete energy eigenvalues E_i such that E_i/E_1 is rational or can be approximated sufficiently closely by a rational number, the complement of the Hamiltonian is the α quantity that can be represented by the probability-operator measure generated by the operators $|\bar{\alpha}\rangle\langle\bar{\alpha}|d\alpha$. In the $(p+1)$-dimensional minimal state space Ψ_p, which would usually be used to describe the quantum system with $p+1$ accessible energy states, we cannot construct an operator that is the complement of the energy operator $\hat{E}$ in the strict sense. The α quantity can, however, be represented by a Hermitian operator in a larger space Ψ_s with $s+1$ dimensions in the infinite-s limit, provided care is taken in finding this limit in that the limit of expectation values and not the operator itself is found.

We have been dealing with a quantum system with a finite number $p+1$ of energy levels. We can make $p+1$ as large as we please, provided we are careful how we take limits to ensure that $s \gg r_p$ in expressions involving both of these quantities. We have already used the infinite-s limit to derive expression (21) which, of course, is independent of s so we can use Eq. (21) if we wish to examine the properties of the α quantity for, for example, the state of a system involving a continuum of energy values approximated by a very large number of uniformly spaced levels.

In this paper we have limited our considerations to a quantum system for which the ratios E_i/E_1 are rational or at least are represented by rational numbers to a sufficiently good approximation. In this case T is the recurrence period. The general case for a system with irrational energy ratios is not as straightforward. For example, if we apply Eq. (21), we have the problem of identifying what value of T to use. We shall not explore this general case in this paper, but we note that for any nonzero $E_k - E_i$ the left-hand side of Eq. (20) will approach the unit operator for any value of T as long as T is made sufficiently large. This suggests that a limiting approach may be useful. A further modification will be necessary in the event of a degeneracy, but again this is outside the scope of this paper.

Finally, there is the question of a suitable name for the quantity that we have referred to so far as simply the α quantity. This quantity has dimensions of time, but is not time. Rather it is a property of the quantum system and depends on the state of the system. For particular states the rate of change of its mean value can approach unity, but for an energy eigenstate, its mean value does not change with time, precisely as expected for a stationary state. To describe such a quantity the name "age" suggests itself. We could then say, for example, that a system in a stationary state would not age as time goes on.

[1] J. Hilgevoord, Am. J. Phys. **64**, 1451 (1996); A. Peres, *Quantum Theory: Concepts and Methods* (Kluwer Academic, Dordrecht, 1993).
[2] P. A. M. Dirac, Proc. R. Soc. London, Ser. A **114**, 243 (1927).
[3] D. T. Pegg and S. M. Barnett, J. Mod. Opt. **44**, 225 (1997).
[4] R. Tanas, A. Miranowicz, and T. Gantsog, in *Progress in Optics XXXV*, edited by E. Wolf (Elsevier, Amsterdam, 1996), pp. 355–446.
[5] D. T. Pegg, J. A. Vaccaro, and S. M. Barnett, J. Mod. Opt. **37**, 1703 (1990).
[6] C. W. Helstrom, *Quantum Detection and Estimation Theory* (Academic, New York, 1976), pp. 53–80.
[7] D. T. Pegg and S. M. Barnett, Europhys. Lett. **6**, 483 (1988); S. M. Barnett and D. T. Pegg, J. Mod. Opt. **36**, 7 (1989); D. T. Pegg and S. M. Barnett, Phys. Rev. A **39**, 1665 (1989).
[8] E. Merzbacher, *Quantum Mechanics*, 2nd ed. (Wiley, New York, 1970), p. 159.
[9] J. A. Vaccaro and D. T. Pegg, Phys. Scr. **T48**, 22 (1993).
[10] D. T. Pegg, J. Phys. A **24**, 3031 (1991).

References

Aharonov Y, Bergman P G and Lebowitz J L 1964 *Time symmetry in the quantum process of measurement* Phys. Rev. **134** B1410

Aharonov Y and Bohm D 1961 *Time in the quantum theory and the uncertainty relation for time and energy* Phys. Rev. **122** 1649

Alimov A L and Damaskinsky E V 1979 *Self-adjoint phase operator* Theor. Math. Phys. **38** 39, translated from Alimov A L and Damaskinsky E V 1979 Teoreticheskaya i Matematicheskaya Fizika, **38** 58

Allen L, Beijersbergen M W, Spreeuw R J C and Woerdman 1992 *Orbital angular momentum of light and the transformation of Laguerre-Gaussian laser modes* Phys. Rev. A **45** 8185 (Reprinted in Allen *et al.* 2003)

Allen L, Padgett M J and Babiker M 1999 *The orbital angular momentum of light* in *Progress in Optics*, vol. 39, 291

Allen L, Barnett S M and Padgett M J 2003 *Optical Angular Momentum* (Bristol: Institute of Physics)

Alvarez E 1989 *Quantum gravity: an introduction to some recent results* Rev. Mod. Phys. **61** 561

Aragone C, Guerri G, Salamó S, and Tani J L 1974 *Intelligent spin states* **7** L149

Aragone C, Chalbaud E, and Salamó S 1976 *On intelligent spin states* **17** 1963

Armen M A, Au J K, Stockton J K, Doherty A C, and Mabuchi H 2002 *Adaptive homodyne measurement of optical phase* Phys. Rev. Lett. **89** 133602 (**Paper 6.3**)

Arthurs E and Kelly J L Jr 1965 *On the simultaneous measurement of a pair of conjugate observables* Bell Syst. Tech. J. **44** 725

Babichev S A, Ries J, and Lvovsky A I 2003 *Quantum scissors: teleportation of single-mode optical states by means of a nonlocal single photon* Europhys. Lett. **64** 1

Barnett S M and Knight P L 1987 *Squeezing in correlated quantum systems* J. Mod. Optics **34** 841

Barnett S M and Pegg D T 1986 *Phase in quantum optics* J. Phys. A: Math. Gen. **19** 3849 (**Paper 1.7**)

Barnett S M and Pegg D T 1989 *On the Hermitian optical phase operator* J. Mod. Opt. **36** 7 (**Paper 2.3**)

Barnett S M and Pegg D T 1990a *Quantum theory of rotation angles* Phys. Rev. A **41** 3427 (**Paper 2.5**)

Barnett S M and Pegg D T 1990b *Quantum theory of optical phase correlations* Phys. Rev. A **42** 6713 (**Paper 4.3**)

Barnett S M and Pegg D T 1992 *Limiting procedures for the optical phase operator* J. Mod. Optics **39** 2121 (**Paper 3.3**)

Barnett S M and Pegg D T 1993 *Phase measurements* Phys. Rev. A **47** 4537 (**Paper 5.1**)

Barnett S M and Pegg D T 1996 *Phase measurements by projection synthesis* Phys. Rev. Lett. **76** 4148 (**Paper 5.4**)

Barnett S M and Radmore P M 1997 *Methods in Theoretical Quantum Optics* (Oxford: Oxford University Press)

Barnett S M, Stenholm S and Pegg D T 1989 *A new approach to optical phase diffusion* Opt. Commun. **73** 314 (**Paper 4.2**)

Bazhenov V Yu, Vasnetsov M V and Soskin M S 1990 *Laser beams with screw dislocations in their wavefronts* JETP Lett. **25** 429 (Reprinted in Allen *et al.* 2003)

Beck M, Smithey D T, Cooper J, and Raymer M G 1993 *Experimental determination of number-phase uncertainty relations* Opt. Lett. **18** 1259 (**Paper 6.1**)

Böhm A 1978 *The Rigged Hilbert Space and Quantum Mechanics* (Berlin: Springer-Verlag)

Bohr N 1935a *Quantum mechanics and physical reality*, Nature **136** 65 (Reprinted in Wheeler J A and Zurek W H Eds. 1983 *Quantum Theory and Measurement* (Princeton University Press) p 144

Bohr N 1935b *Can quantum-mechanical description of physical reality be considered complete?* Phys. Rev. **48** 696 (Reprinted in Wheeler J A and Zurek W H Eds. 1983 *Quantum Theory and Measurement* (Princeton University Press) p 145

Brif C and Ben-Aryeh Y 1994 *Antinormal ordering of Susskind-Glogower quantum phase operators* Phys. Rev. A **50** 2727

Buot F A 1974 *Method for calculating* $Tr\mathcal{H}^n$ *in solid-state theory* Phys. Rev. B **10** 3700

Busch P, Grabowski M, and Lahti P J 1995 *Operational Quantum Physics* (Berlin: Springer)

Carruthers P and Nieto M M 1968 *Phase and angle variables in quantum mechanics* Rev. Mod. Phys. **40** 411

Caves C M 1981 *Quantum-mechanical noise in an interferometer* Phys. Rev. D **23** 1693

Caves C M 1982 *Quantum limits on noise in linear amplifiers* Phys. Rev. D **26** 1817

Caves C M and Schumaker B L 1985 *New formalism for two-photon quantum optics. I. Quadrature phases and squeezed states* Phys. Rev. A **31** 3068

Christenson J H, Cronin J W, Fitch V L and Turlay R 1964 *Evidence for the 2p decay of the K20 Meson* Phys. Rev. Lett. **13** 138

D'Ariano G M and Paris M G A 1993 *Necessity of sine-cosine joint measurement* Phys. Rev. A **48** R4039

Deutsch D 1991 *Quantum mechanics near closed timelike lines* Phys. Rev. D **44** 3197

Dirac P A M 1925 *The fundamental equations of quantum mechanics* Proc. R. Soc. (Lond.) A **109** 642

Dirac P A M 1927 *The quantum theory of the emission and absorption of radiation* Proc. R. Soc. (Lond.) A **114** 243 (**Paper 1.1**)

Dupertuis M-A, Barnett S M and Stenholm S 1987 *Rigged-reservoir response. II. Effects of a squeezed vacuum* J. Opt. Soc. Am. B **4** 1102

Fain V M 1967 *The quantum mechanical harmonic oscillator in the phase representation and the uncertainty relation between the number of quanta and the phase* Sov. Phys. JETP **25** 1027

Fan H Y and Sun Z H 2000 *Two-mode Wigner operator in number-difference-operational phase representation* Phys. Lett. A **272** 219

Fan H and Zaidi H R 1988 *An exact calculation of the expectation values of phase operators in squeezed states* Opt. Commun. **68** 143

Franke-Arnold S, Barnett S M, Yao E, Leach J, Courtial J and Padgett M 2004 *Uncertainty principle for angular position and angular momentum* New J. Phys. **6** 103 (**Paper 6.4**)

Friedman J, Morris M S, Novikov I D, Echeverria F, Klinkhammer G, Thorne K S, and Yurtsever U 1990 *Cauchy problem in spacetimes with closed timelike curves* Phys. Rev. D **42** 1915

Galapon E A, Caballar R F, and Bahague R T Jr 2004 *Confined quantum time of arrivals* Phys. Rev. Lett. **93** 180406

Galindo A and Pascual P 1990 *Quantum Mechanics I* (Berlin: Springer-Verlag)

Garraway B M and Knight P L 1992 *Quantum phase distributions and quasidistributions* Phys. Rev. A **46** R5346

Garrison J C and Wong J 1970 *Canonically conjugate pairs, uncertainty relations, and phase operators* J. Math. Phys. **11** 2242

Gibbons K S, Hoffman M J, and Wootters W K 2004 *Discrete phase space based on finite fields* Phys. Rev. A **70** 062101

Götte J, Zambrini R, Franke-Arnold S, and Barnett S M 2005 *Large-uncertainty intelligent states for angular momentum and angle* J. Opt. B: Quantum Semiclass. Opt. **7** S563; *ibid* J. Phys. B: At. Mol. Opt. Phys. **39** 2315

Götte J, Radmore P M, Zambrini R, and Barnett S M 2006 *Angular minimum uncertainty states with large uncertainties* J. Phys. B: At. Mol. Opt. Phys. **39** 2791

Hawking S W 1992 *Chronology protection conjecture* Phys. Rev. D **46** 603

Heitler W 1954 *The Quantum Theory of Radiation* (Third edition) (Oxford: Oxford University Press)

Hillery M, O'Connell R F, Scully M O, and Wigner E P 1984 *Distribution functions in physics: fundamentals* Phys. Rep. **106** 121

Jackiw R 1968 *Minimum uncertainty product, number-phase uncertainty product, and coherent states* J. Math. Phys. **9** 339

Jie Q L, Wang S J, and Wei L F 1998 *Partial revivals of wave packets: an action-angle phase-space description* Phys. Rev. A **57** 3262

Joshi A 2001 *Entropic uncertainty relations and the number-phase Wigner function for nonlinear and multiphoton coherent states* J. Opt. B **3** 124

Joshi A, Vaccaro J A, and Hill K E 1998 *Number-phase Wigner representation and entropic uncertainty relations for binomial and negative binomial states* Acta Physica Slovaca **48** 23

Judge D 1963 *On the uncertainty relation for L_z and ϕ* Phys. Lett. A **5** 189 (**Paper 1.3**)

Judge D and Lewis J T 1963 *On the commutator $[L_z, \phi]_-$* Phys. Lett. A **5** 190 (**Paper 1.4**)

Lanczos K 1924 *Über eine stationäre Kosmologie im Sinne der Einsteinschen Gravitationstheorie* Zeitschrift für Physik **21** 73; reprinted as Lanczos K 1997 *On a stationary cosmology in the sense of Einstein's theory of gravitation* Gen. Rel. Grav. **29** 363

Leonhardt U 1995 *Quantum-state tomography and discrete wigner function* Phys. Rev. Lett. **74** 4101 (1995); Leonhardt U 1996 (erratum) Phys. Rev. Lett. 76 4293

Leonhardt U, Vaccaro J A, Böhmer B, and Paul H 1995 *Canonical and measured phase distributions* Phys. Rev. A **51** 84 (**Paper 3.5**)

Lévy-Leblond J M 1976 *Who is afraid of nonhermitian operators? A quantum description of angle and phase* Ann. Phys. (N. Y.) **101** 319

Lighthill M J 1958 *Introduction to Fourier analysis and generalized functions* (Cambridge University Press)

London F 1926 *Winkelvariable und kanonische Transformationen in der Unulationsmechanik* (*Angle variables and canonical transformations in wavemechanics*) Zeits. f. Phys. **40** 193

Loudon R 1973 *The Quantum Theory of Light* (First edition) (Oxford: Oxford University Press) (**Paper 1.6**)

Loudon R 2000 *The Quantum Theory of Light* (Third edition) (Oxford: Oxford University Press)

Loudon R and Knight P L 1987 *Squeezed light* J. Mod. Opt. **34** 709

Louisell W H 1963 *Amplitude and phase uncertainty relations* Phys. Lett. A **7** 60 (**Paper 1.2**)

Lukš A and Peřinová V 1993 *Ordering of "ladder" operators, the Wigner function for number and phase, and the enlarged Hilbert space* Physica Scripta **T48** 94

Mair A, Vaziri A, Weihs G and Zeilinger A 2001 *Entanglement of orbital angular momentum states of photons* Nature **412** 313 (Reprinted in Allen et al. 2003)

Messiah A 1961 *Quantum Mechanics* Volume I, translated by J Potter (North-Holland)

Milburn G J 1984 *Multimode minimum uncertainty states* J. Phys. A: Math. Gen. **17** 737

Milburn G J, Steyn-Ross M L and Walls D F 1987 *Linear amplifiers with phase-sensitive noise* Phys. Rev. A **35** 4443

Morris M S, Thorne K S and Yurtsever U 1988 *Wormholes, time machines, and the weak energy condition* Phys. Rev. Lett. **61** 1446

Newton R G 1980 *Quantum action-angle variables for harmonic oscillators* Ann. Phys. (N. Y.) **124** 327

Noh J W, Fougères A, and Mandel L 1991 *Measurement of the quantum phase by photon counting* Phys. Rev. Lett. **67** 1426

Noh J W, Fougères A, and Mandel L 1992 *Operational approach to the phase of a quantum field* Phys. Rev. A **45** 424

Noh J W, Fougères A, and Mandel L 1993 *Operational approach to phase operators based on classical optics* Physica Scripta **T48** 29

Novikov I D 1992 *Time machine and self-consistent evolution in problems with self-interaction* Phys. Rev. D **45** 1989

Ozawa O 1997 *Phase operator problem and macroscopic extension of quantum mechanics* Ann. Phys. **257** 65

Pegg D T 1973 *Closed time and absorber theory*, Natural Phys. Sci. **243** 143

Pegg D T 1975 *Absorber theory of radiation* Rep. Prog. Phys. **38** 1339

Pegg D T 1991 *Time in a quantum mechanical world* J. Phys. A **24** 3031 (**Paper 7.1**)

Pegg D T 1998 *Complement of the Hamiltonian* Phys. Rev. A **58** 4307 (**Paper 7.2**)

Pegg D T 1999 *Time's arrow* Science Spectra Issue 17

Pegg D T 2001 *Quantum Mechanics and the Time Travel Paradox* in Time's Arrows, *Quantum Measurement and Superluminal Behavior* Mugnai D Ranfagni A and Schulman L S Eds. (Consiglio Nazionale Delle Richerche, Roma) 113; quant-ph/0506141.

Pegg D T 2006 *Causality in quantum mechanics* Phys. Lett. A **349** 411

Pegg D T and Barnett S M 1988a *Unitary phase operator in quantum mechanics* Europhys. Lett. **6** 483 (**Paper 2.1**)

Pegg D T and Barnett S M 1988b *Hermitian phase operator $\hat{\phi}_\theta$ in the quantum theory of light* United Kingdon Atomic Energy Authority Harwell Report No. TP 1290 (**Paper 2.2**)

Pegg D T and Barnett S M 1989 *Phase properties of the quantized single-mode electromagnetic field* Phys. Rev. A **39** 1665 (**Paper 2.4**)

Pegg D T and Barnett S M 1997a *Tutorial review: quantum optical phase* J. Mod. Opt. **44** 225 (**Paper 2.6**)

Pegg D T, Vaccaro J A, and Barnett S M 1990 *Quantum optical phase and canonical conjugation* J. Mod. Opt. **28** 87 (**Paper 3.2**)

Pegg D T, Barnett S M, and Phillips L S 1997b *Quantum phase distribution by projection synthesis* J. Mod. Opt. **44** 2135 (**Paper 5.5**)

Pegg D T, Phillips L S, and Barnett S M 1998 *Optical state truncation by projection synthesis* Phys. Rev. Lett. **81** 1604

Pegg D T, Barnett S M, and Jeffers J 2002 *Quantum theory of preparation and measurement* J. Mod. Optics **49** 913

Pegg D T, Barnett S M, Zambrini R, Franke-Arnold S, and Padgett M 2005 *Minimum uncertainty states of angular momentum and angular position* New J. Phys. **7** 62 (**Paper 6.5**)

Peres A and Schulman L S 1972 *Existence theorem for some differential equations with advanced interactions* Phys. Rev. D **5** 2654

Popov V N and Yarunin V S 1973 Vestnik Leningrad University **22** 7 (in Russian)

Popov V N and Yarunin V S 1992a *Quantum and quasi-classical states of the photon phase operator* J. Mod. Optics **39** 1525

Popov V N and Yarunin V S 1992b *Quantum states of the oscillator phase operator* Sov. J. Nucl. Phys. **55** 1529

Pregnell K L and Pegg D T 2001 *Measuring the phase variance of light* J. Mod. Opt. **48** 1293 (**Paper 5.6**)

Pregnell K L and Pegg D T 2002a *Quantum phase distribution by operator synthesis* J. Mod. Opt. **49** 1135 (**Paper 5.7**)

Pregnell K L and Pegg D T 2002b *Single-shot measurement of quantum optical phase* Phys. Rev. Lett. **89** 173601 (**Paper 5.8**)

Pregnell K L and Pegg D T 2003 *Binomial states and the phase distribution measurement of weak optical fields* Phys. Rev. Lett. **67** 063814 (**Paper 5.9**)

Price H 1996 *Time's Arrow and Archimedes' Point* (New York: Oxford University Press)

Reck M, Zeilinger A, Bernstein H J, and Bertani P 1994 *Experimental realization of any discrete unitary operator* Phys. Rev. Lett. **73** 58

Richtmyer R D 1978 *Principles of Advances Mathematical Physics* Vol. 1 (Berlin: Springer-Verlag)

Sanders B C, Barnett S M, and Knight P L 1986 *Phase variables and squeezed states* Opt. Commun. **58** 290

Schulman L S 1974 *Some differential-difference equations containing both advance and retardation* J. Math. Phys. **15** 295

Shapiro J H and Shepard S R 1991 *Quantum phase measurement: a system-theory perspective* Phys. Rev. A **43** 3795

Shapiro J H, Shepard S R, and Wong N C 1989 *Ultimate quantum limits on phase measurement* Phys. Rev. Lett. **62** 237 *ibid.* **62** 2002

Smithey D T, Beck M, Faridani A, and Raymer M G 1993a *Measurement of the Wigner distribution and density matrix of a light mode using optical homodyne tomography: application to squeezed states and vacuum* Phys. Rev. Lett. **70** 1244

Smithey D T, Beck M, Cooper J, and Raymer M G 1993b *Measurement of number-phase uncertainty relations of optical fields* Phys. Rev. A **48** 3159 (**Paper 6.2**)

Smithey D T, Beck M, Cooper J, Raymer M G, and Faridani A 1993c *Complete characterization of the quantum state of a light mode via the Wigner function and the density matrix: application to quantum phase distributions of vacuum and squeezed-vacuum states* Physica Scripta **T48** 35

Stenholm S 1992 *Simultaneous measurement of conjugate observables* Ann. Phys. N. Y. **218** 233

Stenholm S 1993 *Some formal properties of operator polar decomposition* Physica Scripta **T48** 77

Summy G S and Pegg D T 1990 *Phase optimized quantum states of light* Opt. Commun. **77** 75 (**Paper 4.5**)

Susskind L and Glogower J 1964 *Quantum mechanical phase and time operator* Physics **1** 49 (**Paper 1.5**)

Vaccaro J A 1995a *New Wigner function for number and phase,* Opt. Commun. **113** 421

Vaccaro J A 1995b *Number-phase Wigner function on Fock space* Phys. Rev. A **52** 3474 (**Paper 3.6**)

Vaccaro J A 1995c *Phase operators on Hilbert space* Phys. Rev. A **51** 3309 (**Paper 3.9**)

Vaccaro J A and Ben Aryeh Y 1995 *Antinormal ordering of phase operators and the algebra of weak limits* Opt. Commun. **113** 427 (**Paper 3.7**)

Vaccaro J A and Bonner R F 1995 *Pegg-Barnett phase operators of infinite rank* Phys. Lett. A **198** 167 (**Paper 3.8**)

Vaccaro J A and Joshi A 1998 *Position-momentum and number-phase Wigner functions and their respective displacement operators,* Phys. Lett. A **243** 13

Vaccaro J A and Pegg D T 1986 *Squeezing of light by coherent attenuation* J. Mod. Optics **33** 1141

Vaccaro J A and Pegg D T 1989 *Phase properties of squeezed states of light* Opt. Commun. **70** 529 (**Paper 4.1**)

Vaccaro J A and Pegg D T 1990a *Wigner function for number and phase* Phys. Rev. A **41** 5156 (**Paper 3.1**)

Vaccaro J A and Pegg D T 1990b *Physical number phase intelligent and minimum uncertainty states of light* J. Mod. Opt. **37** 17 (**Paper 4.4**)

Vaccaro J A and Pegg D T 1993 *Consistency of quantum descriptions of phase* Physica Scripta **T48** 22 (**Paper 3.4**)

Vaccaro J A and Pegg D T 1994a *Non-diffusive phase dynamics from linear amplifiers and attenuators in the weak-field regime* J. Mod. Optics **41** 1079

Vaccaro J A and Pegg D T 1994b *Phase properties of optical linear amplifiers* Phys. Rev. A **49** 4985 (**Paper 4.7**)

Vaccaro J A and Pegg D T 1994c *On measuring extremely small phase fluctuations* Opt. Commun. **105** 335 (**Paper 5.2**)

Vaccaro J A Barnett S M, and Pegg D T 1992 *Phase fluctuations and squeezing* J. Mod. Optics **39** 603 (**Paper 4.6**)

van Stockum W J 1937 *The gravitational field of a distribution of particles rotating about an axis of symmetry* Proc. Roy. Soc. Edinburgh **57** 135

Vogel K and Risken H 1989 *Determination of quasiprobability distributions in terms of probability distributions for the rotated quadrature phase* Phys. Rev. A **40** 2847

Walker N G 1987 *Quantum theory of multiport optical homodyning* J. Mod. Opt. **34** 15

Walker N G and Carrol J E 1986 *Multiport homodyne detection near the quantum noise limit* Opt. Quant. Electron. **18** 355

Wheeler J A and Feynman R P 1945 *Interaction with the absorber as the mechanism of radiation* Rev. Mod. Phys. **17** 157

Wigner E 1932 *On the quantum correction For thermodynamic equilibrium* Phys. Rev. **40** 749

Wiseman H M 1995 *Adaptive phase measurements of optical modes: going beyond the marginal Q distribution* Phys. Rev. Lett. **75** 4587 (**Paper 5.3**)

Wiseman H M and Killip R B 1998 *Adaptive single-shot phase measurements: the full quantum theory* Phys. Rev. A **57** 2169

Wootters W K 1987 *A Wigner-function formulation of finite-state quantum mechanics* Ann. Phys. **176** 1

Yao D 1987 *Phase properties of squeezed states of light* Phys. Lett. A **122** 77

Yuen H P 1976 *Two-photon coherent states of the radiation field* Phys. Rev. A **13** 2226

Appendix

Publications by David Pegg

Research papers

1. Pegg D T 1968 *Cosmology and electrodynamics* Nature **220**, 154–155
2. Pegg D T 1969 *Time-symmetric electrodynamics and self-action* Nature **220**, 362–363
3. Pegg D T 1969 *Frequency shifts in double resonance experiments caused by oscillating magnetic fields of intermediate strength* J. Phys. B: Atom. Molec. Phys. **2**, 1097–1103
4. Pegg D T 1969 *Change in effective hyperfine constants of the state* $5^2P_{3/2}$ *of* ^{39}K *induced by an oscillating magnetic field* J. Phys. B: Atom. Molec. Phys. **2**, 1104–1109
5. Pegg D T and Series G W 1970 *Semi-classical theory of Hanle effect with transverse static and oscillating magnetic fields* J. Phys. B: Atom. Molec. Phys. **3**, L33–L35
6. Pegg D T 1971 *Night sky darkness in the Eddington-Lemaître universe* Mon. Not. R. Astr. Soc. **154**, 321–327
7. Pegg D T 1972 *Semi-classical theory of radio-frequency transitions in a dressed atom* J. Phys. B: Atom. Molec. Phys. **5**, L4–L5
8. Pegg D T 1972 *Time direction of information propagation and cosmology* Aust. J. Phys. **25**, 207–214
9. Pegg D T and Series G W 1973 *On the reduction of a problem in magnetic resonance* Proc. Roy. Soc. A **332**, 281–289
10. Pegg D T 1973 *Misalignment effects in magnetic resonance* J. Phys. B: Atom. Molec. Phys. **6**, 241–245
11. Pegg D T 1973 *Semi-classical theory of magnetic resonance in intense radio-frequency fields* J. Phys. B: Atom. Molec. Phys. **6**, 246–253
12. Hannaford P, Pegg D T and Series G W 1973 *Analytical expression for the Bloch-Siegert shift* J. Phys. B: Atom. Molec. Phys. **6**, L222–L225
13. Pegg D T and Series G W 1973 *Comment on quantum-electrodynamic theory of atoms interacting with high-intensity radiation fields* Phys. Rev. A **7**, 371–372
14. Pegg D T 1973 *Closed time and absorber theory* Nature **243**, 143–144
15. Pegg D T 1973 *Semi-classical calculations on multiple quantum transitions* Phys. Rev. A **8**, 2214–2216
16. Pegg D T 1973 *Effect of local absorber in absorber theory* Nature **246**, 40–41
17. Pegg D T 1973 *Misalignment resonance widths* J. Phys. B: Atom. Molec. Phys. **6**, 356–358
18. Heron M L and Pegg D T 1974 *A proposed experiment on absorber theory* J. Phys. A: Math. Nucl. Gen. **7**, 1965–1969

19. Pegg D T 1975 *On a recent experiment to detect advanced radiation* J. Phys. A: Math. Nucl. Gen. **8**, L60
20. Doddrell D M, Pegg D T, Bendall M R and Gregson A K 1976 *Proton spin relaxation in some paramagnetic transition-metal acetylacetonate complexes. Comparison between theory and experiment* Chem. Phys. Lett. **40**, 142–146
21. Pegg D T 1975 *Absorber theory of radiation* Rep. Prog. Phys. **38**, 1339–1383
22. Pegg D T and Doddrell D M 1976 *Theory of spin relaxation in the limit of slow motion. Nuclear and electronic spin relaxation in paramagnetic transition-metal complexes* Aust. J. Chem. **29**, 1869–1884
23. Pegg D T, Doddrell D M, Bendall M R and Gregson A K 1976 *Frequency dependence of nuclear spin relaxation in paramagnetic transition-metal complexes* Aust. J. Chem. **29**, 1885–1897
24. Doddrell D M, Pegg D T, Bendall M R, Gottlieb H P W, Gregson A K and Anker M 1976 13C *spin relaxation in some paramagnetic transition metal acetylacetonate complexes. Importance of ligand-centred relaxation* Chem. Phys. Lett. **39**, 65–68
25. Pegg D T and Doddrell D M 1976 *The asymptotic behaviour of electron spin characteristic decay times induced by strong random magnetic fields* Chem. Phys. Lett. **42**, 607–610
26. Doddrell D M, Bendall M R, Pegg D T, Healy P C and Gregson A K 1977 *Temperature dependence of proton spin-lattice relaxation times in some paramagnetic transition metal acetylacetonate complexes. The possible influences of the Jahn-Teller effect on electron spin relaxation* J. Am. Chem. Soc. **99**, 1281–1282
27. Doddrell D M, Bendall M R, O'Connor A J and Pegg D T 1977 *Measurement, employing* 1H *and* 13C *spin-lattice relaxation times, of the temperature dependence of the correlation time for rotational reorientation of diamagnetic transition-metal bis- and tris-acetylacetonate complexes* Aust. J. Chem. 30, 943–956
28. Doddrell D M, Pegg D T, Bendall M R and Gregson A K 1977 *Electron and nuclear spin lattice relaxation in* $S = 1/2$ *paramagnetic transition-metal complexes* Aust. J. Chem. **30**, 1635–1643
29. Pegg D T 1977 *Magnetic resonance in arbitrarily oriented fields* J. Phys. B: Atom. Molec. Phys. **10**, 1027–1033
30. Pegg D T 1977 *Future variation of Planck's constant* Nature **267**, 408–409
31. Pegg D T and Doddrell D M 1978 *Electronic and nuclear spin relaxation in paramagnetic transition-metal complexes with* $S = 1$*. The effects of a non-zero average zero-field splitting on the spin energy levels* Aust. J. Chem. **31**, 475–481
32. Gregson A K, Doddrell D M and Pegg D T 1978 *Electron spin relaxation in solutions of transition metal ions with* $S \geq 1$*. Conditions where Redfield limit is valid* Aust. J. Chem. **31**, 469–473
33. MacGillivray W R, Pegg D T and Standage M C 1978 *Observation of optical nutational effects in atoms* Optics Commun. **25**, 355–358
34. Doddrell D M, Pegg D T, Bendall M R and Gregson A K 1978 *Intramolecular reorientation, an electron spin relaxation process in solutions of discrete paramagnetic transition-metal complexes* Aust. J. Chem. *31*, 2355–2365
35. Pegg D T 1979 *Absorber theory approach to the dynamic Stark effect* Ann. Phys. N. Y. **118**, 1–17
36. Doddrell D M, Pegg D T and Bendall M R 1979 *Field dependence of chemical shifts in Cobolt(III) n.m.r. spectroscopy. Theoretical predictions* Aust. J. Chem. **32**, 1–10

37. Doddrell D M, Bendall M R, Barron P F and Pegg D T 1979 *Use of* $^{13}CT_{1\rho}$ *measurements to study dynamic processes in solution* J.C.S. Chem. Comm. 77–79

38. Doddrell D M, Pegg D T, Bendall M R and Thomas D M 1979 *Field dependence of* 13C *spin relaxation times. The concept of time-dependent correlation times* Chem. Phys. Lett. **63**, 309–312

39. Pegg D T 1980 *Photoelectron antibunching and absorber theory* J. Phys. A: Math. Gen. **13**, 1389–1394

40. Doddrell D M and Pegg D T 1980 *Assignment of proton-decoupled Carbon-13 spectra of complex molecules by polarization transfer. A superior method to off-resonance decoupling* J. Am. Chem. Soc. 102, 6388–6390

41. Pegg D T 1980 *Zero-point fluctuations in direct-action electrodynamics* Phys. Lett. **76A**, 109–111

42. Pegg D T, Bendall M R and Doddrell D M 1980 *Simplified equation for relaxation of a proton from a rapidly rotating* CHXY *group and application to nuclear Overhauser measurements* Aust. J. Chem. **33**, 1167–1173

43. Pegg D T 1979 *Experiment in cosmological absorber theory* N. Z. J. Sci. **22**, 357–359

44. Doddrell D M, Bergen H, Thomas D M, Pegg D T and Bendall M R 1980 *Enhancement of Carbon-13 resonances in paramagnetic transition-metal complexes using proton polarization transfer spectroscopy* J. Magn. Reson. **40**, 591–594

45. Pegg D T 1980 *Objective reality, causality and the Aspect experiment* Phys. Lett. **78A**, 233–234

46. Thomas D M, Bendall M R, Pegg D T, Doddrell D M and Field J 1981 *Two-dimensional* 13C-1H *polarization transfer J-spectroscopy* J. Magn. Reson. **42**, 298–306

47. Doddrell D M, Pegg D T, Bendall M R, Brooks W M and Thomas D M 1980 *Enhancement of* 14N *nuclear magnetic resonance signals in ammonium nitrate using proton polarization transfer* J. Magn. Reson. **41**, 492–495

48. Pegg D T, Doddrell D M, Brooks W M and Bendall M R 1981 *Proton polarization transfer enhancement for a nucleus with arbitrary spin quantum number from n scalar coupled protons for arbitrary preparation times* J. Magn. Reson. **44**, 32–40

49. Doddrell D M, Pegg D T, Brooks W and Bendall M R 1981 *Enhancement of* ^{29}Si *and* ^{119}Sn *signals in the compound* $M(CH_3)_nCl_{4-n}$(M = Si *or* Sn, $n = 4, 3, 2$) *using proton polarization transfer. Dependence of the enhancement on the number of scalar-coupled protons* J. Amer. Chem. Soc. **103**, 727–728

50. Bendall M R, Pegg D T, Doddrell D M and Field J M 1981 *NMR of protons coupled to* 13C *nuclei only* J. Amer. Chem. Soc. **103**, 934–936

51. Pegg D T 1981 *Pulsed differential excitation of atoms* Optics Commun. **37**, 353–355

52. Pegg D T 1981 *Doppler insensitive differential excitation of isotopic atoms* J. Phys. B: Atom. Molec. Phys. **14**, L343–347

53. Bendall M R, Doddrell D M, and Pegg D T 1981 *Editing of* 13C *n.m.r. spectra. A pulse sequence for the generation of subspectra* J. Amer. Chem. Soc. **103**, 4603–4605

54. Bendall M R, Pegg D T, Doddrell D M and Thomas D M 1982 *Superior pulse sequence for two-dimensional chemical shift correlation spectroscopy* J. Magn. Reson. **46**, 43–53

55. Pegg D T, Bendall M R and Doddrell D M 1981 *Heisenberg vector model for precession via heteronuclear scalar coupling* J. Magn. Reson. **44**, 238–249

56. Bendall M R, Pegg D T, and Doddrell D M 1981 *Polarization transfer pulse sequences for two-dimensional n.m.r. by Heisenberg vector analysis* J. Magn. Reson. **45**, 8–29

57. Pegg D T 1982 *Time-symmetric electrodynamics and the Kocher-Commins experiment* Eur. J. Phys. **3**, 44–49

58. Pegg D T, Bendall M R and Doddrell D M 1982 *Randomization of spins in heteronuclear pulse sequences* J. Magn. Reson. **49**, 32–47

59. Doddrell D M, Pegg D T and Bendall M R 1982 *Quasi-stochastic J cross-polarization in liquids* J. Magn. Reson. **49**, 181–186

60. Pegg D T, Doddrell D M and Bendall M R 1982 *Proton polarization transfer enhancement of a heteronuclear spin multiplet with preservation of phase coherency and relative component intensities* J. Chem. Phys. **77**, 2745–2752

61. Doddrell D M, Pegg D T and Bendall M R 1982 *Distortionless enhancement of NMR signals by polarization transfer* J. Magn. Reson. **48**, 323–327

62. Bendall M R, Pegg D T and Doddrell D M 1982 *Exclusive polarization transfer within methine* (CH) *groups only. Generation of a methine* ^{13}C *n.m.r. subspectrum* J. Chem. Soc. Chem. Commun. 872–874

63. Bendall M R, Pegg D T, Doddrell D M and Williams D M 1982 *Strategy for the generation of* ^{13}C *subspectra. Application to the analysis of the* ^{13}C *spectrum of the antibiotic ristocetin* J. Org. Chem. **47**, 3021–3023

64. Knight P L and Pegg D T 1982 *Double photon transitions induced by antibunched light* J. Phys. B: Atom. Molec. Phys. **15**, 3211–3222

65. Bendall M R, Pegg D T, Doddrell D M, Johns S R and Willing R I 1982 *Pulse sequence for the generation of a* ^{13}C *subspectrum of both aromatic and aliphatic quaternary carbons* J. Chem. Soc. Chem. Commun. 1138–1140

66. Bendall M R, Pegg D T, and Doddrell D M 1983 *Pulse sequences utilizing the correlated motion of coupled heteronuclei in the transverse plane of the doubly-rotating reference frame* J. Magn. Reson. **52**, 81–117

67. Pegg D T, Doddrell D M and Bendall M R 1983 *Correspondence between INEPT and DEPT pulse sequences for coupled spin-half nuclei* J. Magn. Reson. **51**, 264–269

68. Bendall M R, Pegg D T, Doddrell D M and Field J 1983 *Inverse-DEPT sequence. Polarization transfer from a spin-half nucleus to n spin-half heteronuclei via correlated motion in the doubly-rotating reference frame* J. Magn. Reson. **51**, 520–526

69. Bendall M R and Pegg D T 1983 *Re-examination of a sequence designed to cancel* ^{13}C *signals of protonated carbons* J. Magn. Reson. **52**, 136–138

70. Bendall M R, Pegg D T and Doddrell D M 1983 *Comparison of decoupling and spatial randomization methods for use in editing* ^{13}C *spectra* J. Magn. Reson. **52**, 407–423

71. Bendall M R and Pegg D T 1983 *EPT with two variable pulse angles, a universal polorization transfer sequence* J. Magn. Reson. **52**, 164–168

72. Bendall M R and Pegg D T 1983 *Identification of isolated methyl groups by detection of cross-correlation effects using polarization transfer* J. Magn. Reson. **53**, 40–48

73. Bendall M R and Pegg D T 1983 *Complete accurate editing of decoupled* ^{13}C *spectra using DEPT and a quaternary-only sequence* J. Magn. Reson. **53**, 272–296

74. Pegg D T and Bendall M R 1983 *Heisenberg picture approach to polarization transfer by the DEPT sequence* J. Magn. Reson. **53**, 229–234

75. Pegg D T 1983 *Amplitude modulated coherent irradiation of a three-level atom* J. Phys. B: Atom. Molec. Phys. **16**, 2135–2144

76. Bendall M R and Pegg D T 1983 ^{1}H–^{13}C *two-dimensional chemical shift correlation spectroscopy using DEPT* J. Magn. Reson. **53**, 144–148

77. Pegg D T and Bendall M R 1983 *Polarization transfer between two scalar coupled systems of arbitrary numbers of nuclei of arbitrary spins* J. Magn. Reson. **55**, 51–63

78. Pegg D T and Bendall M R 1983 *Two-dimensional DEPT spectroscopy* J. Magn. Reson. **55**, 114–127

79. Bendall M R, Pegg D T, Tyburn G M and Brevard C 1983 *Polarization transfer from and between quadrupolar nuclei by the UPT sequence* J. Magn. Reson. **55**, 322–328

80. Pegg D T and Bendall M R 1984 *Coupled polarization transfer spectra for higher spin nuclei* J. Magn. Reson. **58**, 14–26

81. Bendall M R and Pegg D T 1984 *DEPT at depth. Polarization transfer and sample localization combined using surface coils* J. Magn. Reson. **57**, 337–343

82. Bendall M R and Pegg D T 1984 *Signal to noise considerations and θ-pulse defects when editing ^{13}C spectra. Comparison of DEPT with SEMUT* J. Magn. Reson. **59**, 237–249

83. Pegg D T and Schulz W E 1984 *Decay induced coherence in two-level atoms* J. Phys. B: Atom. Molec. Phys. **17**, 2233–2240

84. Bendall M R, Pegg D T, Wesener J R and Gunther H 1984 *Complete editing of ^{13}C spectra of CD_nHm groups using the UPT sequence plus gated ^{1}H decoupling* J. Magn. Reson. **59**, 223–236

85. Pegg D T and Bendall M R 1984 *Self compensation of pulse error effects in editing ^{13}C spectra with a modified DEPT sequence* J. Magn. Reson. **60**, 347–351

86. Bendall M R and Pegg D T 1985 *Theoretical description of depth pulse sequences, on and off resonance, including improvements and extensions thereof* Mag. Res. Medicine **2**, 91–113

87. Pegg D T 1985 *Two-photon resonance with fields amplitude modulated in quadrature* J. Phys. B: Atom. Molec. Phys. **18**, 415–421

88. Pegg D T and Schulz W E 1985 *Non-linear absorption and dispersion of three-level atoms in a Fabry-Perot etalon* Optics Commun. **53**, 274–278

89. Pegg D T and Bendall M R 1985 *Pulse sequence cycles for editing ^{13}C spectra using an inhomogeneous radio-frequency field* J. Magn. Reson. **63**, 556–572

90. Pegg D T and Bendall M R 1985 *Theory of localized polarization transfer* Mag. Res. Medicine **2**, 453–468

91. Bendall M R and Pegg D T 1985 *Sensitive volume localization for in vivo NMR using heteronuclear spin-echo pulse sequences* Mag. Res. Medicine **2**, 298–306

92. Bendall M R and Pegg D T 1985 *Comparison of depth pulse sequences with composite pulses for spatial selection in in vivo NMR* J. Magn. Reson. **63**, 494–503

93. Garwood M, Schleich T, Bendall M R and Pegg D T 1985 *Improved Fourier series windows for localisation in in vivo NMR spectroscopy* J. Magn. Reson. **65**, 510–515.

94. Krinitzky S P and Pegg D T 1986 *Coherent interaction of multilevel atoms in branched and cyclic configurations* Phys. Rev. A **33**, 403–406

95. Bendall M R and Pegg D T 1986 *Reduction of depth pulse phase cycles for use in in vivo NMR spectroscopy* J. Magn. Reson. **66**, 546–550

96. Pegg D T 1986 *Two-photon resonance with combined amplitude and frequency modulation* Opt. Commun. **57**, 185–189

97. Bendall M R and Pegg D T 1986 *Uniform sample excitation with surface coils for in vivo spectroscopy by adiabatic rapid half passage* J. Magn. Reson. **67**, 376–381

98. Pegg D T 1986 *Interaction of three-level atoms with modulated lasers* Optica Acta **33**, 363–369

99. Bendall M R and Pegg D T 1986 *Further comparisons of simple and composite depth pulses* J. Magn. Reson. **68**, 252–262
100. Pegg D T, Loudon R and Knight P L 1986 *Correlations in light emitted by three-level atoms* Phys. Rev. A. **33**, 4085–4091
101. Pegg D T 1986 *Absorber theory in quantum optics* Physica Scripta **T12** 14–18
102. Pegg D T and MacGillivray W R 1986 *Basis transformations for the sodium D line states* Opt. Commun. **59** 113–118
103. Barnett S M and Pegg D T 1986 *Phase in quantum optics* J. Phys. A: Math. Gen. **19** 3849–3862
104. Knight P L, Loudon R and Pegg D T 1986 *Quantum jumps and atomic cryptograms* Nature **323** 608–609
105. Vaccaro J A and Pegg D T 1986 *Squeezing of light by coherent attenuation* Optica Acta **33** 1141–1147
106. Buckle S J, Barnett S M, Knight P L, Lauder M A and Pegg D T 1986 *Atomic interferometers: phase dependence and multilevel atomic transitions* Optica Acta **33** 1129–1140
107. Pegg D T and Vaccaro J A 1987 *Squeezing in the output of a high gain atomic light amplifier* Opt. Commun. **61** 317–320
108. Vaccaro J A and Pegg D T 1987 *Squeezed atomic light amplifiers* J. Mod. Opt. **34** 855–872
109. Bendall M R, Garwood M, Ugurbil K and Pegg D T 1987 *Adiabatic refocussing pulse which compensates for variable radio frequency power and off-resonance effects* Magn. Reson. Medicine **4** 493–499
110. Pegg D T and Knight P L 1988 *Interrupted fluorescence, quantum jumps and wavefunction collapse* Phys. Rev. A **37** 4303–4308
111. MacGillivray W R and Pegg D T 1988 *Four-state theory of the triple-peaked transmission profile of a sodium-filled Fabry-Perot etalon* Opt. Commun. **66** 299–302
112. Pegg D T and Knight P L 1988 *Interrupted fluorescence, quantum jumps and wavefunction collapse II* J. Phys. D: Appl. Phys. **21** S128–130
113. Pegg D T and Barnett S M 1988 *Unitary phase operator in quantum mechanics* Europhys. Lett. **6** 483–487
114. Barnett S M and Pegg D T 1989 *On the Hermitian optical phase operator* J. Mod. Opt. **36** 7–19
115. Pegg D T and Barnett S M 1989 *Phase properties of the single-mode electromagnetic field* Phys. Rev. A **39** 1665–1675
116. Vaccaro J A and Pegg D T 1989 *Phase properties of squeezed states of light* Opt. Commun. **70** 529–534
117. Pegg D T 1989 *Some new techniques in nuclear magnetic resonance* Contemp. Phys. **30** 101–112
118. Barnett S M, Stenholm S and Pegg 1989 *A new approach to optical phase diffusion* Opt. Commun. **73** 314–318
119. Vaccaro J A and Pegg D T 1990 *Physical number-phase intelligent and minimum-uncertainty states of light* J. Mod. Opt. **37** 17–39
120. Pegg D T, Vaccaro J A and Barnett S M 1990 *Quantum optical phase and canonical conjugation* J. Mod. Opt. **37** 1703–1710
121. MacGillivray W R, Standage M C, Farrell P M and Pegg D T 1990 *Theoretical explanation for the apparent two-state behaviour in Na D_2 line optical signals* J. Mod. Opt. **37** 1741–1746

122. Barnett S M and Pegg D T 1990 *Quantum theory of rotation angles* Phys. Rev. A **41** 3427–3435

123. Vaccaro J A and Pegg D T 1990 *Wigner function for number and phase* Phys. Rev. A **41** 5156–6163

124. Summy G S and Pegg D T 1990 *Phase optimized quantum states of light* Opt. Commun. **77** 75–79

125. Barnett S M and Pegg D T 1990 *Quantum theory of optical phase correlations* Phys. Rev. A **42** 6713–6720

126. Pegg D T 1991 *Wavefunction collapse time* Phys. Lett. A **153** 263–264

127. Pegg D T and Barnett S M 1991 *Reply to comment on phase properties of the quantized single-mode electromagnetic field* Phys. Rev. A **43** 2579–2580

128. Pegg D T 1991 *Time in a quantum mechanical world* J. Phys. A: Math. Gen. **24** 3031–3040

129. Vaccaro J A, Barnett S M and Pegg D T 1992 *Phase fluctuations and squeezing* J. Mod. Opt. **39** 603–614

130. Barnett S M and Pegg D T 1992 *Limiting procedures for the optical phase operator* J. Mod. Opt. **39** 2121–2129

131. Pegg D T 1993 *Wavefunction collapse in atomic physics* Aust. J. Phys. **46** 77–86

132. Barnett S M and Pegg D T 1993 *Phase measurements* Phys. Rev. A **47** 4537–4540

133. Vaccaro J A and Pegg D T 1993 *Consistency of quantum descriptions of phase* Physica Scripta **T48** 22–28

134. Vaccaro J A and Pegg 1994 *Non-diffusive phase dynamics from linear amplifiers and attenators* J. Mod. Opt. **41** 1079–1086

135. Vaccaro J A and Pegg D T 1994 *On measuring extremely small phase fluctuations* Opt. Commun. **105** 335–340

136. Vaccaro J A and Pegg D T 1994 *Phase properties of optical linear amplifiers* Phys. Rev. A **49** 4985–4995

137. Pegg D T 1994 *On Pancharatnam's quantum theory of dispersion* Current Science **67** 272–275

138. Barnett S M, Loudon R, Pegg D T and Phoenix S J D 1994 *Communication using quantum states* J. Mod. Opt. **41** 2351–2374

139. Pegg D T and Vaccaro J A 1995 *Comment on phase-difference operator* Phys. Rev. A **51** 859–860

140. Barnett S M and Pegg D T 1996 *Phase measurement by projection synthesis* Phys. Rev. Lett. **76** 4148–4150

141. Pegg D T and Barnett S M 1997 *Quantum optical phase* J. Mod. Opt. **44** 225–264

142. Pegg D T 1997 *Dressed fields and phase* Physica Scripta **T70** 101–105

143. Pegg D T 1997 *Spontaneous emission and absorber theory* Physica Scripta **T70** 106–111

144. Pegg D T, Barnett S M and Phillips L S 1997 *Quantum phase distribution by projection synthesis* J. Mod. Opt. **44** 2135–2148

145. Phillips L S, Barnett S M and Pegg D T 1998 *Optical measurements as projection synthesis* Phys. Rev. A **58** 3259–3267

146. Pegg D T, Phillips L S and Barnett S M 1998 *Optical state truncation by projection synthesis* Phys. Rev. Lett. **81** 1604–1606

147. Pegg D T 1998 *Complement of the Hamiltonian* Phys. Rev. A **58** 4307–4313

148. Barnett S M, Phillips L S and Pegg D T 1998 *Imperfect photodetection as projection onto mixed states* Opt. Commun. **158** 45–59
149. Pegg D T, Phillips L S and Barnett S M 1999 *Optical state measurement with a two-component probe* J. Mod. Opt. **46** 981–992
150. Pegg D T and Barnett S M 1999 *Retrodiction in quantum optics* J. Opt. B: Quantum and Semiclassical Optics **1** 442–445
151. Pegg D T and Barnett S M 1999 *Optical state measurement by information transfer* J. Mod. Opt. **46** 1657–1667
152. Barnett S M, Pegg D T and Jeffers J 1999 *Equivalence of a lossless beam splitter and a nondegenerate parametric amplifier in conditional measurements* Opt. Commun. **172** 55–57
153. Barnett S M and Pegg D T 1999 *optical state truncation* Phys. Rev. A **60** 4965–4973
154. Vaccaro J A, Pegg D T and Barnett S M 1999 *The problem with "The problem of the Pegg-Barnett phase operator"* Phys. Lett. A **262** 483–495
155. Barnett S M, Pegg D T and Jeffers J 2000 *Bayes' theorem and quantum retrodiction* J. Mod. Opt. **47** 1779–1789
156. Barnett S M, Pegg D T, Jeffers J, Jedrkiewicz O and Loudon R 2000 *Retrodiction for quantum optical communications* Phys. Rev. A **62** 022313
157. Barnett S M, Pegg D T, Jeffers J and Jedrkiewicz O 2000 *Atomic retrodiction* J. Phys. B: At. Mol. Opt. Phys. **33** 3047–3065
158. Pregnell K L and Pegg D T 2001 *Measuring the phase variance of light* J. Mod. Opt. **48** 1293–1302
159. Barnett S M, Pegg D T, Jeffers J and Jedrkiewicz O 2001 *Master equation for retrodiction of quantum communications signals* Phys. Rev. Lett. **86** 2455–2458
160. Jeffers J, Barnett S M and Pegg D T 2002 *Retrodiction with two-level atoms: atomic previvals* J. Mod. Opt. **49** 1175–1184
161. Pregnell K L and Pegg D T 2002 *Quantum phase distribution by operator synthesis* J. Mod. Opt. **49** 1135–1145
162. Pegg D T, Barnett S M and Jeffers J 2002 *Quantum theory of preparation and measurement* J. Mod. Opt. **49** 913–924
163. Jeffers J, Barnett S M and Pegg D T 2002 *Retrodiction as a tool for micromaser field measurements* J. Mod. Opt. **49** 925–938
164. Pregnell K L and Pegg D T 2002 *Measuring the elements of the optical density matrix* Phys. Rev. A **66** 013810
165. Pegg D T, Barnett S M and Jeffers J 2002 *Quantum retrodiction in open systems* Phys. Rev. A **66** 022106
166. Pregnell K L and Pegg D T 2002 *Single-shot measurement of quantum optical phase* Phys. Rev. Lett. **89** 173601
167. Tan E-K, Jeffers J Barnett S M and Pegg D T 2003 *Retrodictive states and two-photon quantum imaging* Eur. Phys. J. D **22** 495–499
168. Pregnell K L and Pegg D T 2003 *Binomial states and phase measurement of weak optical fields* Phys. Rev. A **67** 063814
169. Pregnell K L and Pegg D T 2004 *Retrodictive quantum optical state engineering* J. Mod. Opt. **51** 1613–1626

170. Garretson J L, Wiseman H M, Pope D T and Pegg D T 2004 *The uncertainty relation in 'Which-Way' experiments: how to observe directly the momentum transfer using weak values* J. Opt. B: Quantum and Semiclassical Optics **6** S506–517

171. Pegg D T, Barnett S M, Zambrini R, Franke-Arnold S and Padgett M 2005 *Minimum uncertainty states of angular momentum and angular position* New J. Phys. **7** 62

172. Pegg D T and Jeffers J 2005 *Quantum nature of light* J. Mod. Opt. **52** 1835–1856

173. Pegg D T 2006 *Causality in quantum mechanics* Phys. Lett. A **349** 411

Other publications

Pegg D T 1977 *Mechanics* (Longman Cheshire Ltd.)

Thiel D V, Pegg D T and Nicol J L 1975 *Assessment in first-year physics* Australian Physicist **12** 99

Pegg D T 1980 *Atomic spectroscopy: interaction of atoms with coherent fields* Chap. 2 of Laser Physics eds D F Walls and J D Harvey (Academic Press)

Doddrell D M, Pegg D T and Bendall M R 1980 *Extended use of vectors to describe complex pulse sequences in high resolution NMR* (Bruker Analytische Messtechnik, Karlsruhe, Germany)

Bendall M R, Doddrell D M and Pegg D T 1982 *High resolution multipulse NMR spectrum editing and distortionless enhancement by polarization transfer* (Bruker Analytische Messtechnik, Karlsruhe, Germany)

Pegg D T 1986 *Time-symmetric electrodynamics and quantum measurement* Quantum Optics IV eds J D Harvey and D F Walls (Springer-Verlag, Berlin) p 245–251

Barnett S M and Pegg D T 1989 *Phase in quantum optics* Dynamics of non-linear systems eds L Pesquera and F J Bermejo (World Scientific, Singapore) p 93–100

Pegg D T, Barnett S M and Vaccaro J A 1989 *Phase in quantum electrodynamics* Quantum Optics V eds J D Harvey and D F Walls (Springer-Verlag, Berlin) p 122–132

Barnett S M, Pegg D T and Vaccaro J A 1989 *Applications of the optical phase operator* Coherence and Quantum Optics VI eds J H Eberly, L Mandel and E Wolf (Plenum, New York) p 89–92

Barnett S M, Pegg D T and Phoenix S J D 1992 *Information, quantum correlations and communication* Quantum Measurements in Optics eds P Tombesi and D F Walls (Plenum, New York) p 353–355

Barnett S M, Dalton B J, Pegg D T and Vaccaro J A 1993 *Phase in quantum optics* Optics as a Key to High technology (16th Congress of the International Commission for Optics) eds Gy Ákos, T Lippényi, G Lupkovics and A Podmaniczky (Budapest August 1993). SPIE Vol 1983 pp 63–67

Pegg D T, Vaccaro J A and Barnett S M 1994 *Quantum optical phase and its measurement* Quantum Optics VI eds J D Harvey and D F Walls (Springer-Verlag, Berlin) p 153–161

Pegg D T 1996 *Quantum phase distribution and its determination* Coherence and quantum optics VII eds J H Eberly, L Mandel and E Wolf. (Plenum, New York) p 251–258

Pegg D T 1996 *Tropical quantum optics* Aust. Opt. Soc. News **10** 32–33

Barnett S M and Pegg D T 1997 *The Hermitian optical phase operator* Quantum Fluctuations (Les Houches, Session LXIII, 1995) eds S Reynaud, E Giacobino and J Zinn-Justin (Elsevier Science B.V. 1997) pp 563–575

Barnett S M and Pegg D T 1997 *Quantum optical phase* Quantum Communication, Computing and Measurement eds O Hirota, A S Holevo and C M Caves (Plenum, New York, 1997) pp 415–422

Pegg D T and Barnett S M 1997 *Measurement of quantum phase distribution by projection synthesis* Quantum Communication, Computing and Measurement eds O Hirota, A S Holevo and C M Caves (Plenum, New York, 1997) pp 407–413

Pegg D T 1997 *Can time mend a broken watch?* (Essay Review) Physics World **10** No. 12 (December), 47–48

Pegg D T 1998 *Reflections on time* Contemporary Physics **36** 211–212

Pegg D T 1998 *Black holes, quantum gravity and Schrödinger's cat* Metascience **7** 524–527

Pegg D T 1999 *Time as a river* Contemporary Physics **40** 267–269

Pegg D T 1999 *Time's arrow* Science Spectra **17** 42–47

Barnett S M, Chefles A, Pegg D T and Phillips L S 2000 *Generalised measurements, retrodiction and state manipulation* Quantum Communication, Computing and Measurement 2 eds P Kumar, G M D'Ariano and O Hiroto (Kluwer Academic/Plenum, New York) pp125–136

Pegg D T 2001 *Quantum mechanics and the time travel paradox* Time's Arrows, Quantum Measurement and Superluminal Behaviour eds D Mugnai, A Ranfagni and L S Schulman (Consiglio Nazionale delle Richerche, Roma) pp 113–124 ISBN 88-8080-024-8

Jeffers J, Barnett S M, Pegg D, Jedrkiewicz O and Loudon R 2001 *Quantum retrodiction* Quantum Communication, Computing and Measurement 3, eds P Tombesi and O Hirota (Kluwer Academic/Plenum, New York) pp 143–146 ISBN: 0-306-46609-0

Pregnell K L and Pegg D T, 2003 *Selective measurement of individual elements of the optical density matrix* Quantum Communication, Computing and Measurement eds J H Shapiro and O Hirota (Rinton Press, Princeton, 2003), pp 325 – 328

Barnett S M, Jeffers J, and Pegg D T 2003 *Retrodictive quantum optics*, Coherence and quantum optics VIII, eds N P Bigelow, J H Eberly, C R Stroud and I A Walmsley (Kluwer, New York, 2003) pp 87–94

Jeffers J, Barnett S M, and Pegg D T 2003 *Retrodiction with two-level atoms: atomic previvals and micromaser field measurements* Coherence and Quantum Optics VIII, eds N P Bigelow, J H Eberly, C R Stroud and I A Walmsley (Kluwer, New York, 2003) pp 385–386

Index

S

T

U

V

W